P9-BZZ-626

DISCOVER
BIOLOGY

Using Art as a Learning Tool: How To "Read" the Illustrations

The position of important structures are shown within the organism.

This "zoom" arrow represents an enlargement to a more detailed view.

Consistent color coding is used for elements that repeat. For example, arteries are always red and veins are always blue.

The last part of this diagram relates a schematic functional model to the more realistic drawing just to its left.

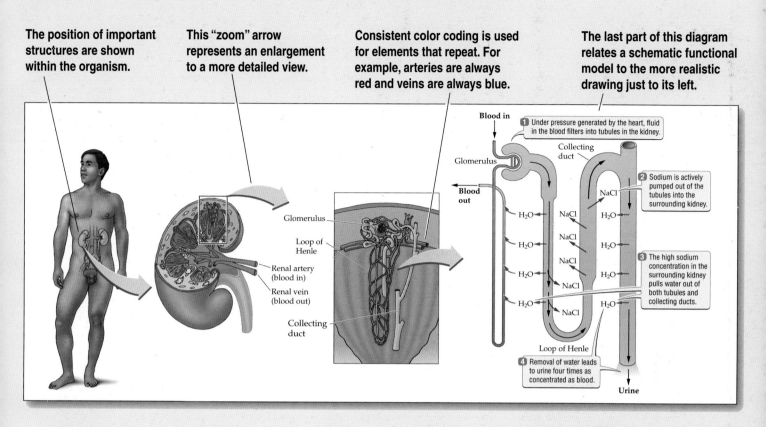

Glomerulus

Loop of Henle

Renal artery (blood in)

Renal vein (blood out)

Collecting duct

Blood in

1 Under pressure generated by the heart, fluid in the blood filters into tubules in the kidney.

Collecting duct

Glomerulus

Blood out

H_2O NaCl H_2O

NaCl

NaCl

NaCl

H_2O NaCl H_2O

2 Sodium is actively pumped out of the tubules into the surrounding kidney.

3 The high sodium concentration in the surrounding kidney pulls water out of both tubules and collecting ducts.

Loop of Henle

4 Removal of water leads to urine four times as concentrated as blood.

Urine

"Balloons" within the art make long captions unnecessary. Often they are numbered so you will read them in a specific order.

This type of arrow is used to show steps in a process.

A "roadmap" icon shows how and where an illustrated concept fits into the big picture. The area or structure of interest is shown in color.

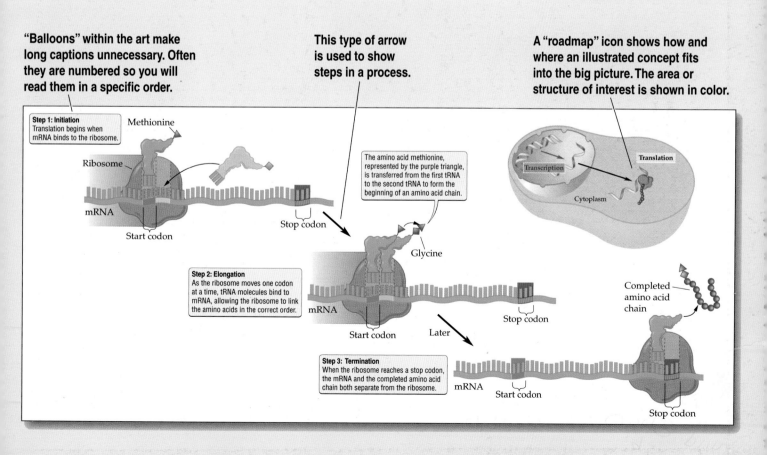

Step 1: Initiation
Translation begins when mRNA binds to the ribosome.

Methionine

Ribosome

mRNA

Start codon

Stop codon

The amino acid methionine, represented by the purple triangle, is transferred from the first tRNA to the second tRNA to form the beginning of an amino acid chain.

Transcription

Translation

Cytoplasm

Glycine

Step 2: Elongation
As the ribosome moves one codon at a time, tRNA molecules bind to mRNA, allowing the ribosome to link the amino acids in the correct order.

mRNA

Start codon

Stop codon

Later

Completed amino acid chain

Step 3: Termination
When the ribosome reaches a stop codon, the mRNA and the completed amino acid chain both separate from the ribosome.

mRNA

Start codon

Stop codon

Biology-100

David Chen

778-8256

Test #3

Ch. 14, 15, 10, 12, 13, 17, 18, 36, 37

Test #4

Ch. 44, 4, 39-45, 2, 32, 24-38.

DISCOVER
BIOLOGY

Michael L. Cain
New Mexico State University

Hans Damman
Calgary, Alberta

Robert A. Lue
Harvard University

Carol Kaesuk Yoon
Bellingham, Washington

Sinauer Associates, Inc.
SUNDERLAND, MASSACHUSETTS

W. W. Norton & Company
NEW YORK • LONDON

ABOUT THE COVER
A mountain goat (*Oreamnos americanus*) in Hidden Lake Glacier
National Park, Alberta, Canada. Photograph by Art Wolfe.

ABOUT THE BOOK
Editor: Andrew D. Sinauer
Project Editors: Nan Sinauer, Kerry L. Falvey
Developmental Editor: James Funston
Copy Editor: Stephanie Hiebert
Permissions Coordinator: Joyce Zymeck
Production Manager: Christopher Small
Book Layout and Production: Jefferson Johnson, Janice Holabird,
    and Joan Gemme
Illustration Program: J/B Woolsey Associates
Pen and Ink Illustrations: Nancy Haver
Design: Susan Brown Schmidler
Book Cover Design: Jefferson Johnson
Photo Research: David McIntyre
Color Separations: Vision Graphics, Inc.
Cover Manufacture: Henry N. Sawyer Co., Inc.
Book Manufacture: Courier Companies, Inc.

Discover Biology

Copyright © 2000 by Sinauer Associates, Inc. All rights reserved.
This book may not be reproduced in whole or in part without permission.

Address editorial correspondence to:
    Sinauer Associates, Inc., P. O. Box 407
    23 Plumtree Road, Sunderland, MA, 01375 U.S.A.
    Fax: 413-549-1118
    Internet: www.sinauer.com; publish@sinauer.com

Address orders to:
    W. W. Norton & Company, 500 Fifth Avenue
    New York, NY, 10110-0017 U.S.A.
    Phone: 1-800-233-4830, Fax: 1-800-458-6515
    Internet: www.wwnorton.com

Library of Congress Cataloging-in-Publication Data

Discover biology / Michael L. Cain ... [et al.].
        p. cm.
    Includes index.
    ISBN 0-393-97377-8 (hbk.)
    1. Biology. I. Cain, Michael L. (Michael Lee), 1956-

QH308.2 .D57 2000
570—dc21

                                        99-048016

10 9 8 7 6 5 4 3 2 1

*To my wife and daughter, Debra and Hannah,
and my parents, Lorraine and Winfield: thanks
to each of you for the love and support that
made this book possible*

M.L.C.

To Rachael, Dulcie, Gilbert, and Sally

H.D.

To Cheri for everything

R.A.L.

*To my mother, June Ginoza Yoon, who made life
and everything else possible*

C.K.Y.

About the Authors

Michael L. Cain is Associate Professor of Biology at New Mexico State University, where he teaches introductory biology and a broad range of other biology courses. He received his Ph.D. in Ecology and Evolutionary Biology from Cornell University in 1989 and did postdoctoral research at Washington University in molecular genetics. Dr. Cain's interest in biology began early, stimulated by walks along the coast of Maine and by the expert guidance of his high school biology teacher, Clayton Farraday. Dr. Cain has published dozens of scientific articles on such topics as genetic variation in plants, insect foraging behavior, long-distance seed dispersal, and factors that promote speciation in crickets. Dr. Cain is the recipient of numerous fellowships, grants, and awards, including the Pew Charitable Trust Teacher-Scholar Fellowship and research grants from the National Science Foundation.

Hans Damman was until 1999 an Associate Professor of Biology at Carleton University in Ottawa and is currently a sessional lecturer at the University of Calgary. He received a Ph.D. in Entomology in 1986 from Cornell University. At Carleton University he twice received the Faculty of Science Award for Excellence in Teaching. His teaching included courses in Animal Form and Function, Plant and Animal Interactions, and Ecology, and he supervised a large number of undergraduate Honors research projects. His research work began with a focus on the interactions between plants and the insects that eat them, and has recently expanded to include the population biology of plants. This research has been funded largely by grants from the Natural Sciences and Engineering Research Council.

Robert A. Lue is Senior Lecturer on Molecular and Cellular Biology and the Director of Undergraduate Studies in the Biological Sciences at Harvard University, where he has received several teaching awards from the Faculty of Arts and Sciences and the Division of Continuing Education. He received his Ph.D. from Harvard and has taught undergraduate courses there since 1988. Dr. Lue has developed a cell biology outreach program for Massachusetts high schools with the support of the Howard Hughes Medical Institute, and has chaired educator conferences on college biology, most recently for the National Science Foundation. Dr. Lue is a cell biologist, and his own research interests include the role of abnormal cell structure in human cancer.

Carol Kaesuk Yoon received her Ph.D. from Cornell University and has been writing about biology for *The New York Times* since 1992. Her articles have also appeared in *The Washington Post*, *The Los Angeles Times*, and *Science* magazine. Dr. Yoon has taught writing as a Visiting Scholar with Cornell University's John S. Knight Writing Program, working with professors to help teach critical thinking in biology classes. She has also served as science consultant to Microsoft for their children's CD-ROM on tropical rainforests, part of the Magic School Bus series.

Preface

Genetically engineered plants and animals. Extinction of species. The search for cures to cancer and AIDS. Global warming. These are just a few of the many ways in which biology touches all of our lives. Biology is a rapidly growing, tremendously exciting scientific discipline, and it has important applications in daily life, which are just as exciting. Biology is at the center of some of the most important issues we face today, such as our impact on the global ecosystem and the implications of new discoveries in genetics for the diagnosis and treatment of human disease.

Biology is such a broad field—ranging as it does from the study of molecules to the study of ecosystem Earth—that it can be a very challenging subject to teach and to learn. When we set out to write this book, we asked ourselves: How can we convey the excitement, breadth, and relevance of biology to introductory students without burying them in definitions and facts? We achieve this goal in *Discover Biology* with clear, streamlined chapters that focus on essential basic concepts but do not overwhelm students with a mass of detail that is hard to sort through and place in context. We also make heavy use of applied topics to drive home important key points. We designed and wrote *Discover Biology* specifically for a non-majors audience, to offer such students a different approach to this dynamic, vital science.

The basic philosophy of this book is "less is more"

We have considerable experience teaching introductory biology, and over the years we have found that students learn best from short, focused chapters. We think introductory students of biology need to understand and retain fundamental biological concepts, not temporarily learn exhaustive lists of terms. Toward that end, we firmly believe that *less is more*; that is, students actually learn more when presented with less material, a view that is well-established in the education literature.

To convey important biological concepts in a manner accessible to introductory students of biology, our chapters begin with a single main message and a small set of key concepts. We then develop the main message and key concepts in a depth that only slightly exceeds what can be covered in one class period.

We place great emphasis on applied topics to illustrate how biology affects our daily lives

As teachers and authors, one of our goals is to improve the biological literacy of our students. Biological topics such as global warming, genetic engineering, new cancer treatments, and the disappearance of rainforests receive increasing coverage in the news. We think it is critically important for students to understand the biology behind the news—only in this way can they make informed choices about biological issues. Throughout *Discover Biology* we introduce such applied topics to drive home basic biological concepts.

These applied topics help students to understand biology as it relates to their day-to-day experience. Biology is relevant to everyone: in understanding ourselves, the food we eat, the medicines we take, our pets, the outdoors, even the movies we watch. Regarding movies, for example, in *Discover Biology*, students delve into what the biology of King Kong would be like if he existed (see pages 393–394). Some other applied topic examples include why cats often survive falls from high buildings (see pages 398 and 409), and how human agricultural systems resemble farming mutualisms found in nature (see page 639).

As part of our effort to demonstrate the applied nature of biology in their daily lives, each chapter in *Discover Biology* also includes a unique feature called "Evaluating the News." Each "Evaluating the News" section features a brief "article," "editorial," or "letter-to-the-editor" based on real news that we have altered to best draw out how applications of biology in real life create issues that confront citizens, both scientists and nonscientists. Each "Evaluating the News" feature includes questions that focus on social and ethical implications of issues in the news, providing an additional format for students to develop critical thinking skills and to apply the knowledge they gain from the text to real-world problems.

The art and text work together to tell a story

We worked hard to create beautiful, instructional art and to select superb photographs that integrate seamlessly with the text to guide students through even the most difficult of material. Like the text, figures in *Discover Biology* are streamlined to convey simple, direct messages that allow students to focus on essential concepts. Most figures in the book include balloons that highlight crucial information or explain the flow of an illustration.

Chapters are organized to improve student understanding

Each chapter begins with a two-page "Hook" designed to capture students' interest and lead them into the material they will learn about in the chapter. "Hook" topics range from mysteries of cat fur color to the puzzling disappearance of the world's frog species to how a biological weapon such as anthrax causes its deadly effects. As a further way of relating to non-majors, each opening spread includes a work of fine art culled from many hundreds of compelling images. Also on the opening spread is the chapter's "Main Message."

Following the opening spread, the third page of each chapter lists key concepts and the text begins with an overview of the contents to come. We use two levels of headings to organize the material in a clear, straightforward fashion. Major headings divide the chapter into a series of topics that flesh out the main message and key concepts. The second level of headings, usually declarative sentences that convey a key point or idea, identify the focus and summarize the message of each subsection within a major heading. Each major heading concludes with a short summary that walks the students through the main points of the material they've just read. About half of the book's chapters include a Box—short essays that illuminate an important biological phenomenon, with the focus either being on human interest ("Biology in our Lives") or on how science works ("The Scientific Process"). The text of each chapter closes with a "Highlight" that either explores the "Hook" in further depth or discusses a separate topic chosen to put the material covered in the chapter in a memorable context.

Chapters close with a summary of important points, a list of key terms, a self-quiz and set of review questions, and an "Evaluating the News" section. Answers to all self-quizzes and review questions are in the back of the text, along with Suggested Readings for further exploration and for enjoyment, a complete Glossary, and an Index.

This book is a work in progress

Like the field of biology itself, any textbook aiming to describe it is a work in progress. We hope you will contact us with your comments, questions, and ideas as you use this book. You can reach any or all of the authors by sending an email message to discoverbiology@sinauer.com or by writing us at Discover Biology, Sinauer Associates, 23 Plumtree Road, Sunderland, MA 01375.

We thank the many people who helped to produce this book

We four were not alone: As every textbook author knows, the process of getting from the kernel of an idea to a finished book involves a veritable army. Here we get to recognize and thank some of the major participants. First, we thank all of the manuscript reviewers who are listed on the next page. Together, their criticism (and encouragement) contributed enormously to the crucial rewriting necessary to achieve a consistent level, clarity, and accuracy. Thanks also to Susan McGlew for arranging and coordinating all of the reviews.

Andy Sinauer at Sinauer Associates and Roby Harrington at W. W. Norton embraced our vision for this book at the outset. They assembled the following team to ensure its realization. The editorial stewardship fell to Nan Sinauer and Kerry Falvey. They both worked intimately with James Funston, the book's developmental editor, and Stephanie Hiebert, the copy editor. James and Stephanie are wordsmiths of the first order; we are grateful for their unsparing focus on finding the clearest path.

Getting the words right is half the challenge in introductory biology. To assure that the visual presentation would be outstanding, John Woolsey and Michael Demaray at J/B Woolsey Associates worked with each author and translated their ideas to the Woolsey team of talented artists. The wonderful photographs in the book were assembled by David McIntyre, who endured many a round of rejects before he and we were satisfied. Christopher Small, in addition to overseeing production, worked with a small team at Steven Diamond to provide the over 50 works of fine art that introduce each unit and chapter. Jeff Johnson and Janice Holabird put all the words and images together, creating enticing page spreads.

Other key roles in the book's gestation were performed by Dean Scudder and Kathaleen Emerson at Sinauer Associates, who helped shape first the plan and then the execution of the extensive array of supplementary material that accompanies the book.

Thanks also to the team of people at W. W. Norton and Company who contributed to the creation of this text and its ancillary package: Vanessa Drake-Johnson, who coordinated the marketing program for the book, consulting editors Richard Mixter and John Byram, DiscoverBiology.com site designer/manager Irene Cheung, and site editor April Lange.

Michael L. Cain
Hans Damman
Robert A. Lue
Carol Kaesuk Yoon

Reviewers

Michael Abruzzo, California State University/Chico
James Agee, University of Washington
Craig Benkman, New Mexico State University
Elizabeth Bennett, Georgia College and State University
Stewart Berlocher, University of Illinois/Urbana
Juan Bouzat, University of Illinois/Urbana
Bryan Brendley, Gannon University
John Burk, Smith College
Kathleen Burt-Utley, University of New Orleans
Naomi Cappuccino, Carleton University
Alan de Queiroz, University of Colorado
Véronique Delesalle, Gettysburg College
Jean de Saix, University of North Carolina/Chapel Hill
Joseph Dickinson, University of Utah
Harold Dowse, University of Maine
John Edwards, University of Washington
Jonathon Evans, University of the South
Marion Fass, Beloit College
Richard Finnell, Texas A & M University
Dennis Gemmell, Kingsborough Community College
Blanche Haning, University of North Carolina/Raleigh
Daniel J. Howard, New Mexico State University
Laura F. Huenneke, New Mexico State University
Laura Katz, Smith College
Katherine C. Larson, University of Central Arkansas

Kenneth Lopez, New Mexico State University
Phillip McClean, North Dakota State University
Amy McCune, Cornell University
Bruce McKee, University of Tennessee
Gretchen Meyer, Williams College
Brook Milligan, New Mexico State University
Kevin Padian, University of California/Berkeley
John Palka, University of Washington
Massimo Pigliucci, University of Tennessee
Barbara Schaal, Washington University
David Secord, University of Washington
Allan Strand, College of Charleston
John Trimble, Saint Francis College
Mary Tyler, University of Maine
Roy Van Driesche, University of Massachusetts/Amherst
Carol Wake, South Dakota State University
Jerry Waldvogel, Clemson University
Elsbeth Walker, University of Massachusetts/Amherst
Stephen Warburton, New Mexico State University
Paul Webb, University of Michigan
Peter Wimberger, University of Puget Sound
Allan Wolfe, Lebanon Valley College
David Woodruff, University of California/San Diego
Robin Wright, University of Washington

Discover These State-of-the-Art Tools for Students and Instructors

Discover Biology Student CD-ROM
By Graham R. Kent, Smith College, and David Demers, University of Hartford

Packaged with each copy of *Discover Biology,* this valuable multimedia tool includes interactive exercises designed to reinforce and link key concepts from the textbook. The CD-ROM also contains quiz modules and a collection of videos and biological animations.

Student Study Guide
By Jerry Waldvogel and Christine Minor (both of Clemson University)

This study aid for students includes chapter-by-chapter essential ideas keyed to the *Discover Biology* textbook plus a wealth of factual and conceptual quiz questions. The guide also includes related activities for key chapter topics.

Art Portfolio
This student aid contains all of the illustrations from the text, suitable for note-taking or reference during lectures.

Student Web Site: DiscoverBiology.com
Contributors: Christine Minor and Jerry Waldvogel (both of Clemson University)

This content-rich online student resource offers Biology-in-the-News postings from Newswise.com with "What's News?" interactivities to help students sharpen their critical thinking skills. The site also provides ample review material, including dynamic matching exercises and interactive self-tests.

Discover Biology Newsletter
Written by the text authors and published once each term, this newsletter features exciting current biology from the news and provides analysis for use in classroom discussions.

Instructor's Manual
By Erica Bergquist, Holyoke Community College

The Instructor's Manual contains ideas for lecture topics, discussion questions, and brief lecture outlines.

Test-Item File/Computerized Test-Item File
By Susan Weinstein, Marshall University

The Test-Item File contains a collection of multiple choice and true/false questions, along with sentence completions. Questions range from factual to conceptual. The computerized Test-Item File is contained on the Instructor's CD-ROM.

Discover Biology Laboratory Manual
By Ralph W. Preszler and Laura Lowell Haas (both of New Mexico State University)

Designed specifically to accompany *Discover Biology,* this lab manual reinforces concepts by having students generate and test predictions derived from hypotheses. In keeping with the text's focus on applied topics, the manual incorporates evaluation of the hypotheses into discussions of societal issues. Also included are unique "Outside the Laboratory" activities, as well as introductions, methods sections, and discussion questions.

Labs on Demand™ Customized Laboratory Manuals
By Alison Morrison-Shetlar and Virginia Bennett (both of Georgia State University), Ralph Preszler and Laura Lowell Haas (both of New Mexico State University), and Robert Shetlar

Available to qualified adopters of *Discover Biology, Labs on Demand™* is an innovative and individualized approach to meeting your lab manual needs. Customizing options include:

- A database of hundreds of proven lab experiments that can be used "as-is" or customized to meet your needs
- Content and design specialists to assist you in creating your lab manual layout
- Access to high-quality illustrations and photographs to include in your lab manual

Discover Biology Instructor's CD-ROM
This resource includes the Computerized Test-Item File, along with all of the illustrations from the text plus a special collection of biological photographs, videos, and animations for lecture display and use on individual, password-protected web sites.

Acetate Transparencies
The *Discover Biology* transparency set includes all of the four-color diagrams and drawings from the text in a relabeled and resized format for optimum projection.

Biology Video Library
This expansive videotape library, available to qualified adopters of *Discover Biology,* includes material on cells and cellular function, development, embryology, predation, and food chains.

Brief Contents

Contents

UNIT 2
Cells: The Basic Units of Life

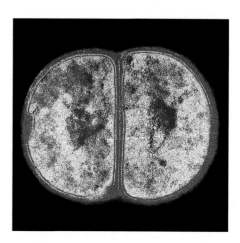

UNIT 3
Genetics

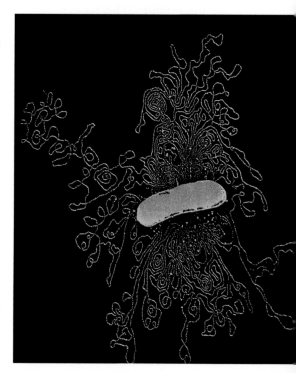

UNIT 4
Evolution

UNIT 5
Form and Function

UNIT 6

Interactions with the Environment

Alexandre-Isidore Leroy de Barde, *Still Life with Exotic Birds*, 1994.

Diversity of Life

chapter 1

The Nature of Science and the Characteristics of Life

Joan Miró, *Carnival of Harlequin*, 1924–25.

What Is Life?

*I*t was January 23, 1998, a cold winter day in Washington, D.C., and tens of thousands of men, women, and children had come, placards in hand, to demonstrate in their nation's capital. Some had come to celebrate and others to protest *Roe v. Wade*, the Supreme Court decision that 25 years earlier had declared abortion legal, one of the most controversial and polarizing rulings in the history of the United States. While pro-life demonstrators decried abortion as murder and pro-choice demonstrators vehemently supported a woman's right to choose, the crucial question still hung in the air: When does life begin?

Like so many controversial ethical issues, this question hinges on an even more basic question, one that has attracted the attention of philosophers, theologians, and scientists alike: What is life and what does it mean to be a living being?

Though deceptively simple, this question is one of the most profound. The search for the definition of life underlies medical controversies ranging from abortion and when life begins, to the right to die and when life ends. The quest reaches deep into the sciences also, as researchers seek to understand when life on our own planet first took hold and whether life has ever existed on other planets.

Any schoolchild can easily distinguish between the inanimate stone and the living being who skips

When Does Life Begin?
Pro-choice and pro-life demonstrators stage competing protests on the twenty-fifth anniversary of *Roe v. Wade*.

it across a pond. Yet scientists and philosophers alike have found that life defies easy, airtight definition. What really does distinguish life from nonlife?

Key Concepts

1. Biology is the scientific study of life. To understand living organisms, biologists develop possible explanations, known as hypotheses, about phenomena observed in nature. These hypotheses are then subjected to rigorous tests, and changed or rejected as appropriate.

2. Because all living organisms, diverse though they are, descended from a single common ancestor, there is a great unity in the characteristics of living organisms.

3. Shared characteristics define life: Living organisms are built of cells, reproduce, grow and develop, gather energy

from their environment, sense stimuli in their environment and respond to them, show a high level of organization, and evolve.

4. A biological hierarchy encompasses all living organisms: from molecules to cells, tissues, organs and organ systems, to individual organisms, populations, communities, ecosystems, and biomes, and finally to the biosphere.

5. More than 3.5 billion years ago, life arose from nonlife. How this happened remains a matter of intense interest and much theorizing.

*B*iology is the scientific study of life. The main goal of biology is to improve our understanding of living organisms, from microscopic bacteria to giant redwood trees to human beings. In this chapter we begin with an exploration of science and how scientists ask and answer questions about living organisms. Then we move on to the most fundamental of biological questions: What is life? We will see that all living things, diverse though they are, share characteristics that unify them, and that all living organisms are part of a greater, biological hierarchy of life.

Asking Questions, Testing Answers: The Work of Science

Science is a method of inquiry, a rational way of discovering truths about the natural world. A powerful way of understanding nature, science holds a central place in modern society. For scientists and nonscientists alike, understanding how nature works can be exciting and fulfilling. In addition, applications of scientific knowledge influence all aspects of modern life. Every time we turn on a light, fly in an airplane, enjoy a vase of hothouse flowers, take medicine, or work at a computer, we are enjoying the benefits of science.

Yet few of us understand how science works, how it generates knowledge, and what its powers and limits are. This lack of understanding is unfortunate, for several reasons. First, an understanding of science can be personally rewarding: It can add to our appreciation of day-to-day events, leading to a sense of awe about how nature works.

A second reason that science is important is the increasingly public nature of how decisions related to science are made. As a society we make decisions on

issues like global warming, the use of new drugs or medical procedures, the control of pollution, the use of genetically engineered organisms, and even whether teachers in our public schools can provide their students with the most current forms of scientific knowledge. To make good decisions on these and many other issues, everyone—not just scientists—benefits by understanding how the scientific process works.

In the sections that follow we first describe, in a general way, the methods scientists use to answer questions and learn about the natural world. Then we give an example of how one scientist answered a question.

Scientists follow well-defined steps to search for answers

The natural world is extremely complex. To deal with this complexity, scientists attempt to explain the natural world by developing a simplified view, or "model," of how nature works. No scientist attempts to study every conceivable thing that could influence the aspects of the natural world that interest him or her. Such a task would be impossible. Instead scientists must master the "art" of simplifying nature in ways that can reveal exciting and important information about the natural world.

To construct a simplified view of nature, scientists begin by observing, describing, and asking questions about the natural world (Figure 1.1). In many ways, this beginning is the most important step of a scientific study. Everything that follows hinges on the quality of these initial observations.

The next step in the scientific process is to come up with a possible explanation—known as a **hypothesis**—of the phenomenon being studied. The hypothesis must have logical consequences whose truth can be tested.

That is, the hypothesis must lead to a prediction that can be tested rigorously. If the results of the test uphold the prediction, the hypothesis is supported but not proven true. Proving a hypothesis true is not possible, because it always might fail if subjected to a different test. If the test does not uphold the prediction, the hypothesis is rejected and must be discarded or changed.

This last point is central to how progress is made in science: Hypotheses are constantly being tested; the good ones are kept, the bad ones rejected. In this way, science can correct itself. Scientists develop a hypothesis, test its predictions by performing experiments, then change or discard the hypothesis if the predictions are not supported by the results of the tests. Together these steps are called the **scientific method**.

An experiment with honeysuckle demonstrates the scientific method

Dr. Katherine Larson, a botanist at the University of Central Arkansas, was interested in two species of vines: a native species called coral honeysuckle, and a species introduced from Asia, the Japanese honeysuckle (Figure 1.2). The Japanese honeysuckle has gorgeous flowers and was originally brought to the southeastern United States as a cultivated plant. But now it is considered a pest species. It has escaped from cultivation and has spread so quickly that scientists are concerned it will drive native plant species to extinction.

Dr. Larson wanted to understand why the Japanese honeysuckle was able to grow so well in the wild. From previous observations, she knew that this plant spread rapidly across the ground, whereas the native honeysuckle tends not to spread nearly as quickly. Since most vines must find supports (for example, growing on trees, shrubs, or fence posts) to survive, she hypothesized that the Japanese honeysuckle was more successful because it was better at locating new supports, especially supports that were far from where the vine was growing.

On the basis of these observations, Dr. Larson designed an **experiment**—a controlled, repeated manipulation of nature—to test her hypothesis. She grew plants from each of the two honeysuckle species in a garden in which supports were both near to and far away from the plants. She predicted (1) that Japanese honeysuckle would grow across regions without supports more rapidly than the native, coral honeysuckle would, and (2) that Japanese

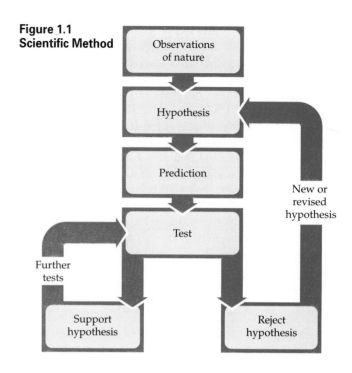

**Figure 1.1
Scientific Method**

Observations of nature → Hypothesis → Prediction → Test → Support hypothesis / Reject hypothesis

New or revised hypothesis

Further tests

honeysuckle would find distant supports more often than coral honeysuckle would.

Though still in progress, this experiment has already shown that Japanese honeysuckle spreads more rapidly across regions without supports than the native species

Figure 1.2 A Scientist and Her Experimental Subjects
Dr. Katherine Larson's experiments with honeysuckles are shedding light on why one pest species of honeysuckle, introduced from Asia, is able to spread more quickly than the honeysuckle species that is native to its environment.

does. Results from this experiment also suggest that Japanese honeysuckle is more effective at locating distant supports than the native species. Thus, the results support both of the predictions from Dr. Larson's hypothesis. In addition, her research has led to some new and unexpected findings.

While conducting the experiment, Dr. Larson noticed differences in the details of how the plants grew. She knew that most plants rotate slowly about a central axis as they grow (Figure 1.3), and that vines show a pronounced version of this behavior. Dr. Larson observed that Japanese honeysuckle rotated differently than the native species.

The way Japanese honeysuckle rotated as it grew along the ground allowed it to form roots more often and to spread more rapidly than the native species. Since plants obtain nutrients and water through their roots, the way it grows probably enables Japanese honeysuckles to survive in areas where the native species cannot.

To summarize, Dr. Larson conducted an experiment to test her hypothesis that the Japanese honeysuckle is outcompeting the native species because it spreads more rapidly and finds distant supports more easily. Thus far, the results of the experiment support her hypothesis. In addition, Dr. Larson made new observations about the details of plant growth that helped explain the results of the experiment. In this way Dr. Larson's work illustrates an important point about how the scientific method works in practice: Chance observations and new discoveries often take the scientist in unexpected directions, and that is part of what makes science so exciting.

Like all other scientific studies, Dr. Larson's work raises as many questions as it answers. What else might be contributing to the success of Japanese honeysuckle? In Asia are there insects that attack the honeysuckle to keep it from growing so heartily and spreading so quickly? Is an absence of enemies in the United States also helping the Japanese honeysuckle thrive in its new environment? Clearly, Dr. Larson's studies are just the beginning of an answer to why these plants have been able to invade so successfully.

> ■ Scientists observe the natural world and then form hypotheses that make predictions about how that world might operate. Scientists then design experiments to test these predictions. When the experiment upholds the predictions, a hypothesis gains strength and support. When predictions are not upheld, hypotheses are discarded or modified.

We have discussed the methods biologists use to study the living world. But what exactly is the living world? How do biologists define life?

The Characteristics that All Living Organisms Share

From the frigid Antarctic to the burning Sahara Desert, from the highest mountaintop to the deepest regions of the sea, life is everywhere. From whales to bacteria, how can all the world's living creatures fit into a single definition of life? In fact, the great diversity of body forms, habits, and sizes of organisms makes a simple, single-sentence definition of life impossible. But despite this diversity, all living organisms are thought to be descendants of a single common ancestor that arose billions of years ago (see the Box on page 10). For this reason, certain characteristics unify all forms of life. Biologists define life by this set of shared characteristics (Table 1.1), which we describe in the sections that follow.

Living organisms are built of cells

The first organisms were single cells that existed billions of years ago. The cell remains the smallest and most basic unit of life, and the simplest of organisms still are made up of just a single cell. Enclosed by a membrane, **cells** are tiny, self-contained units that make up every living organism.

In living organisms cells can be viewed as building blocks. A bac-

> This time-lapse photograph shows the position of a single growing tip of a honeysuckle vine as it rotates on a central axis.

Figure 1.3 Plants Rotate on a Central Axis as They Grow

Bacterium

1.1 *The Shared Characteristics of Life*

All living organisms:
• are built of cells
• reproduce themselves via the hereditary material DNA
• grow and develop
• capture energy from their environment
• sense their environment and respond to it
• show a high level of organization
• evolve

terium is made up of one cell, a single, self-sustaining building block. **Multicellular** organisms like toads or wildflowers are made up of many different kinds of specialized cells. For example, a toad has within it skin cells, muscle cells, brain cells, and so on. Whatever the beast, plant, or bug, it is always at its most basic level a collection of cells (Figure 1.4).

Living organisms reproduce themselves via the hereditary material DNA

One of the key characteristics of living organisms is that they can replicate, or reproduce, themselves. Many single-celled organisms, such as bacteria, reproduce by dividing into two new genetically identical copies of themselves, like a pair of identical twins. (Genetics is discussed in detail in Unit 3.) In contrast, multicellular organisms reproduce in a variety of ways. For example, sunflowers reproduce when their flowers generate seeds that can germinate to become new sunflowers. And humans reproduce by producing infants who eventually grow up to be adult humans, like those that produced them.

Whether organisms produce seeds, lay eggs, or just split in two, they all reproduce by using a molecule that is known as **DNA (deoxyribonucleic acid)**. DNA is the hereditary, or genetic, material that carries information from parents to offspring. We will discuss DNA in detail in Unit 3. Briefly, for our purposes here, the DNA molecule is shaped like a double helix and functions as a blueprint for building an organism (Figure 1.5).

Figure 1.4 The Basic Building Block of Life: The Cell
Like all organisms, this Syke's monkey is composed of cells. The intestinal cells shown here are just one of many different kinds of cells in this monkey.

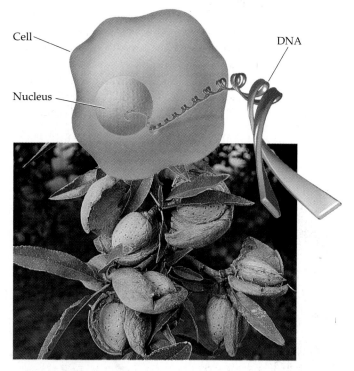

Figure 1.5 The DNA Molecule: A Blueprint for Life
DNA is a hereditary blueprint found in the cells of every living organism. DNA provides a set of instructions by which an individual organism can grow and develop and which it can pass on to its own young so that they can grow and develop. This tree has produced young in the form of almonds, DNA-containing seeds that will eventually develop into new almond trees.

Figure 1.6 Growing and Developing
All organisms develop. A monarch butterfly develops from
egg to caterpillar to chrysalis to flying adult.

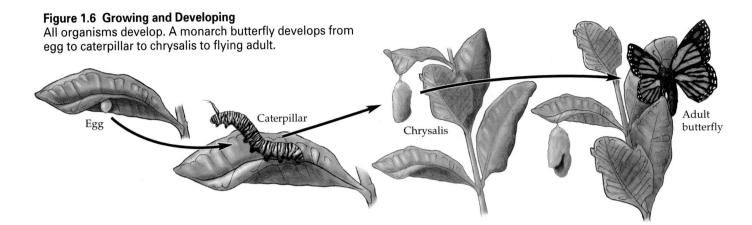

Egg   Caterpillar   Chrysalis   Adult
butterfly

This molecule contains a wealth of information—all the information necessary for an organism to create more cells like itself or to grow from a fertilized egg into a complex, multicellular organism that will eventually produce offspring like itself. DNA is found in every cell in every living thing. Life, no matter how simple or how complex, uses this inherited blueprint. Since all living organisms today reproduce via DNA, we can infer that our original ancestor also reproduced using DNA.

Living organisms grow and develop

Using DNA as the blueprint, organisms grow and build themselves anew every generation. A human begins life as a simple, single cell that eventually grows and develops inside its mother, to emerge after 9 months as a living, breathing baby. This miraculous transformation is part of the process known as **development** (Figure 1.6). All organisms go through a process of growing, whether they complete their growth as a single cell or continue to develop into something as complicated as an octopus.

Living organisms capture energy from their environment

To carry out their growth and development and simply to persist, organisms need energy that they must capture from their environment. Organisms use a great variety of methods to capture energy.

Plants are among the organisms that can capture the energy of sunlight through a chemical process known as photosynthesis, by which they produce sugars and starches. (We discuss photosynthesis in detail in Chapter 8.) Some bacteria can also harness energy from chemical sources such as iron or ammonia through an entirely different chemical reaction. Many organisms, such as animals, can gather energy only by consuming other organisms.

Animals exhibit many different ways of capturing energy efficiently (Figure 1.7). For example, insects have mouthparts that they use to suck out the nutritive juices from plants, and cheetahs can run so quickly that they can chase and capture a fast-moving source of energy like a gazelle.

Living organisms sense their environment and respond to it

In order to survive, different organisms have evolved to sense a wide range of different phenomena in their environment, from danger to mates. Like humans, many organ-

Figure 1.7 All Living Organisms Must Obtain Energy
While plants can photosynthesize, other organisms known as consumers must either eat plants to gather energy or eat other organisms whose energy ultimately derives from producers. Here a boa constrictor captures some energy that it can ingest.

Figure 1.8 Organisms Sense and Respond to Their Environment
All organisms must be able to sense and respond to stimuli in their environment. These South Dakota sunflowers have all detected rays of sunshine and turned toward their light and warmth.

isms can smell, hear, taste, touch, and see their environments. But many organisms can see things humans cannot see, such as ultraviolet or infrared light. Others can hear sounds that humans cannot hear. Still others have senses that are entirely different from any human senses. For example, some bacteria can sense which direction is north and which direction is up or down using magnetic particles within them. All organisms gather information about their environment by sensing it, and then respond appropriately for their continued well-being (Figure 1.8).

Some responses don't need to be learned; for example, a dog automatically pants when it is hot. But learning itself is an excellent example of sensing an environmental stimulus and responding to it. After touching a stinging nettle plant once, many organisms learn never to touch one again. Humans are particularly good at this kind of response because we have large brains relative to our body size. We can even learn very abstract lessons from our environment, such as how to read or sing a song.

Stinging nettle

Living organisms show a high level of organization

Living organisms are complex, functioning beings composed of numerous essential components that are spatially organized in a very specific way. Human bodies, for example, have highly organized internal organs and tissues that allow the body to function properly. Organs or tissues in disarray, whether through disease or accident, can lead to illness or death. In the same way, the structure of a flower or bacterial cell is highly organized. Such organization is required for all organisms to function properly (Figure 1.9).

Living organisms evolve

In the developmental process, individual organisms can change over short spans of time, developing from seeds into mature trees or from eggs into adult fish. Over longer time spans, whole groups of organisms change. A **species** is a group of organisms that can interbreed to produce fertile offspring (that is, offspring that themselves can reproduce) and that do not breed with other organisms. Mountain lions, monarch butterflies, and Douglas fir trees are all distinct species of organisms. Such species can change as a group over time—a process known as **evolution**.

Pronghorn antelope, for example, are the fastest-running

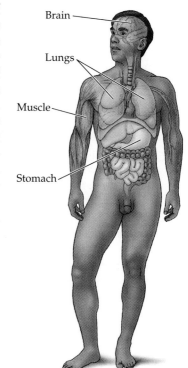

Brain

Lungs

Muscle

Stomach

Figure 1.9 Living Organisms Show a High Level of Structural Organization
Organization—having each part in its proper place—is crucial to an organism's functioning. The internal organs of a human being, like the stomach, lungs, and brain, are highly organized to function properly.

Cooking up a Theory for the Origin of Life: One of Biology's Classic Experiments

At the time this planet formed, 4.6 billion years ago, Earth was lifeless. The first sure signs of cellular life can be found as early as 3.5 billion years ago. But what happened in the meantime—how life arose from nonlife—is one of the most puzzling and hotly debated issues in science.

Scientists have found different degrees of evidence to support various hypotheses. Some researchers have hypothesized that life originated in hot springs deep on the ocean bottom. Others have suggested that life's birthplace lay near underwater volcanoes or perhaps in hot-water geysers on land, like Yellowstone's Old Faithful. Still others have suggested that life, or its building blocks, did not arise on Earth, but that living creatures or key molecules for constructing life reached Earth by traveling through space on an asteroid or meteorite.

All of these competing hypotheses have their strengths and weaknesses. The most widely accepted hypothesis, however, supported by a well-known set of experiments, is what is sometimes referred to as the "soup theory" of life.

In what is now considered a classic experiment in modern biology, in 1953 Stanley L. Miller, then a young graduate student, set up an ingenious test. He attempted to recreate the conditions of early Earth—hot seas and lightning-filled skies—to see if he could recreate the beginnings of life.

In his "primordial soup," Miller began with water, which he kept boiling. The water vapor rose into the simulated atmosphere, which contained the gases methane, ammonia, and hydrogen—some of the gases that Earth's early atmosphere is thought to have harbored and the very sort that

could have belched forth from ancient volcanoes. Miller added an electric spark to simulate lightning. He then cooled the sparking vapors until they turned to liquid, and the resulting liquid, like rain falling back to the seas, was returned to the boiling water. With the apparatus set up to let the water travel continuously from boiling seas to sparking skies and back to the sea again, Miller let the whole thing cook for a week.

Surprisingly, when Miller examined what was in his primordial soup after a week of cooking, he discovered that from nothing but water and simple gases, he had created an array of molecules, including two amino acids, the building blocks of proteins. Since then, other such "soup" experiments recreating the early days on Earth have produced numerous key biological molecules, including all the common amino acids, sugars, and lipids, and the basic building blocks of DNA and RNA. In this exceedingly clever and simple experiment, Miller provided insight into a crucial event in life's history, even though it had occurred billions of years earlier.

Miller's experiment, while groundbreaking, provided only a first step toward understanding how life arose from nonlife. How did this soup of small, floating molecules then become organized into larger molecules and eventually into the first cell able to gather energy and reproduce? How did

the first cell become contained with an outer wall that allowed in important items such as nutrients but kept other things out? Where did the first cell come to be? Did it float freely in a soup or was it organized first on a surface, such as the ocean bottom?

Researchers are trying to answer these questions in various ways. Some study ancient fossils, searching for the first hints of cellular life. Others study the family tree of all living organisms in an attempt to understand better what kind of organism was most likely to have been the ancestor of us all. Still others are studying meteorites and other extraterrestrial objects for hints of what might once have landed on Earth.

Many questions remain, but the soup experiments are still one of the most powerful approaches for scientists studying the crucial leap of Earth from barren stone to cradle of life.

Stanley Miller with the apparatus he used to cook up a primordial soup nearly half a century ago.

creatures in North America. Over time, pronghorn became more fleet of foot because only the ones that could outrace their predators survived to reproduce. The young of these speedy survivors tended to be speedy themselves because they shared much of their DNA with their parents.

In the struggle to survive and the contest to reproduce, characteristics of species—like the average speed at which pronghorn can run—tend to change over time (Figure 1.10). Advantageous features such as the ability of an antelope to run quickly, or the tough, protective covering of an armadillo, are known as **adaptations**.

In evolution, not only can existing species change, but new ones can come into being. For example, one species can split into two different species. Unit 4 focuses on all these aspects of evolution in detail.

> ■ Because all living organisms descended from a common ancestor, they share common characteristics. For example, all organisms are built of cells. They reproduce using DNA, grow and develop, capture energy from their environment, sense and respond to their environment, show a high level of organization, and evolve.

Levels of Biological Organization

Biologists organize the great array of living creatures and the many components that make up any particular creature into a biological hierarchy (Figure 1.11).

At its lowest level, the biological hierarchy begins with molecules—for example, molecules of DNA that carry the blueprint for building an organism. Many such specialized molecules are organized into cells, the basic unit of life. Some organisms, such as bacteria, consist of only a single cell. Other, multicellular organisms can contain within them collections of specialized cells known as **tissues**, such as muscle or nerve tissues, that perform particular, specialized functions in the body. Sometimes these tissues are organized into **organs** such as hearts and brains. Groups of organs can function together in **organ systems**, the way stomach, liver, and intestines are all part of the digestive system. Groups of organ systems work together for the functioning of a single organism—an **individual**.

Each individual organism is a member of a larger group of similar organisms, known as a **population**—for example, a population of field mice in one field or a population of plants on a mountaintop. As we have already learned, the populations of all similar organisms in the world that can successfully breed with one another form a species. For example, all the humans in the

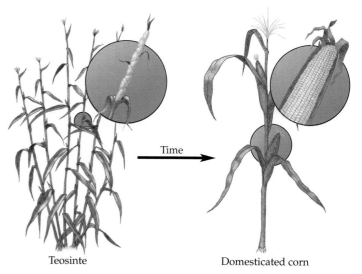

Figure 1.10 Living Organisms Evolve
Over time, species of living organisms evolve. For example, domesticated corn plants evolved from the wild species known as teosinte.

Teosinte   Domesticated corn

world form one species known as *Homo sapiens*. Groups of different species that live and interact in a particular area are known as a **community**—for example, the community of insects living in a particular forest.

Communities, along with the physical environment in which they are situated are known as **ecosystems**; for example, a river ecosystem includes the river itself, as well as all the organisms living in it. At an even larger scale are **biomes**, large regions of the world that on land are defined by vegetation type and in water are defined by the physical characteristics of the environment. The tundra is an example of a biome. Finally, each biome is part of the **biosphere**, which is defined as all the world's living organisms and the places where they live.

> ■ Living organisms are just one part of a biological hierarchy. At the lowest level are molecules, then there are cells, tissues, organs, organ systems, individuals, populations, species, communities, ecosystems, biomes, and finally the biosphere.

Energy Flow through Biological Systems

Plants and other photosynthesizing organisms are called **producers** because they take energy from the light of the sun and convert it to chemical energy in the form of sugars and starches. That energy is then harvested by consumers, such as animals and fungi. These **consumers** eat either plants or other organisms whose energy ultimate-

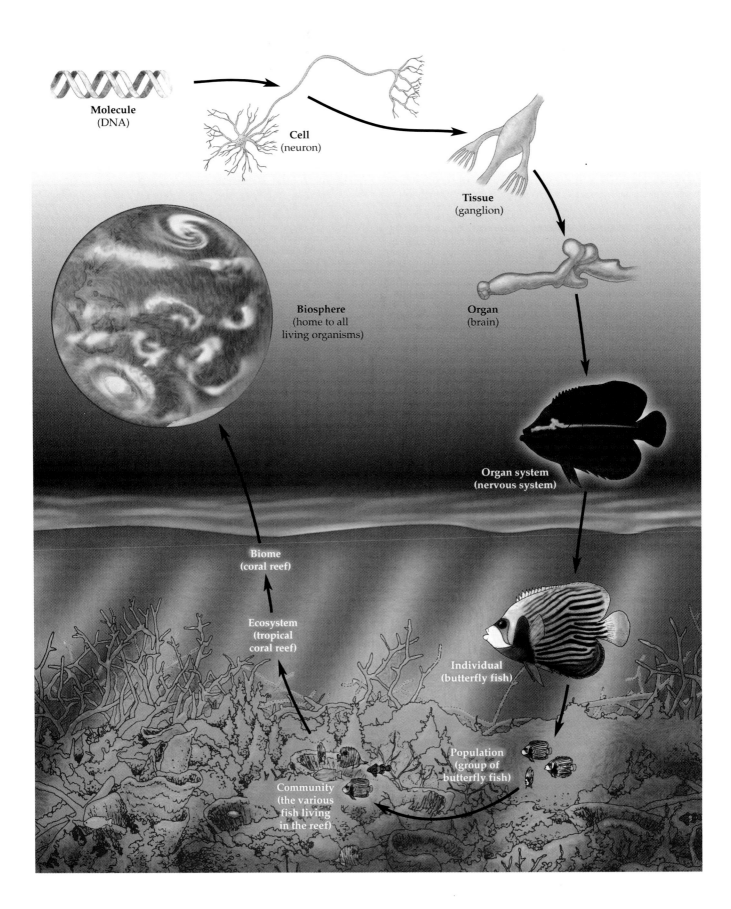

Molecule (DNA)

Cell (neuron)

Tissue (ganglion)

Organ (brain)

Biosphere (home to all living organisms)

Organ system (nervous system)

Biome (coral reef)

Ecosystem (tropical coral reef)

Individual (butterfly fish)

Population (group of butterfly fish)

Community (the various fish living in the reef)

◀ **Figure 1.11 Biological Hierarchy**
Levels of biological organization can be traced from molecules like DNA, to cells like neurons (nerve cells), to tissues like nervous tissue, to organs like the brain, to organ systems like the nervous system, to individuals like an individual butterfly fish, to populations of individuals, to communities of many different kinds of individuals and populations, to ecosystems like tropical coral reefs, to biomes like coral reefs in general, and finally to the biosphere, which includes all living organisms and the places they live.

ly derives from plants. As a result, energy flows almost entirely in one direction through the biosphere: from the sun into producers, which form the basis of the energy flow, and then into consumers that give off energy as heat (Figure 1.12). A depiction of consumers and producers showing which species in an ecosystem are eating which other species is known as a **food web**.

> ■ Energy flows from the sun to producers, such as plants, and then to consumers, such as animals and fungi that consume plants and other organisms. Food webs depict the complex interrelationships of these organisms that eat and are eaten.

Highlight

Life on the Edge: Viruses

Every winter millions of people are laid low by the influenza virus. If you come down with the flu, the symptoms are all too familiar. You are achy. You cough and sneeze. The reason you're suffering is that an influenza virus is infecting your body's cells, reproducing throughout your nose, throat, and lungs.

In response, the specialized cells of your immune system (see Chapter 32) fight back, attacking and destroying cells already infected by the virus and attempting with heat (hence your high fever) to prevent the virus from reproducing. This microscopic flu virus could knock you out for days or weeks, replicating itself again and again at your body's expense while evading your body's defenses. The influenza virus also evolves rapidly over time, changing so quickly that defending against it is extremely difficult. A virus is thus a formidable enemy.

In fact, the flu virus has been a deadly foe throughout human history. In 1918 the Spanish-flu epidemic killed more than half a million people in the United States, 10 times the toll of World War I. Worldwide, the flu epidemic of 1918 killed between 20 million and 40

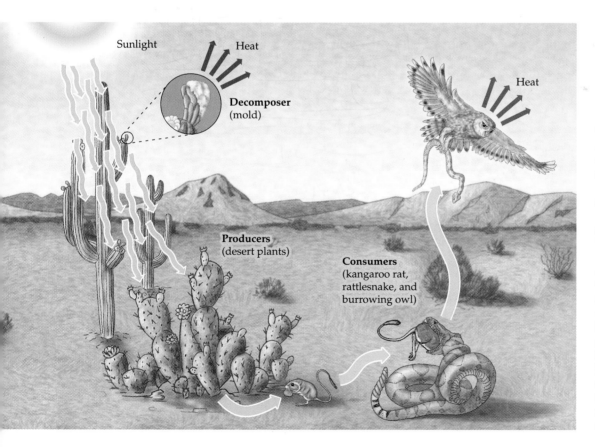

Figure 1.12 Energy Flow
Sunlight is received by producers, in this case desert plants. Energy then flows to consumers, such as a fruit-eating chipmunk. These consumers are then eaten by other consumers, like snakes, which can then be eaten by other consumers. Decomposers, like the mold on fruit, likewise get energy either from plants or from organisms whose energy ultimately derives from plants. Decomposers and consumers give off energy in the form of heat. Energy flows from the sun to producers and then to consumers and decomposers throughout this food web.

million people in just half a year. But there is much more to viruses than influenza viruses that attack humans. There are many different kinds of viruses that infect all the different forms of life.

These powerful foes certainly seem alive, just like the many organisms they attack. Yet all viruses, including the influenza virus, have some very nonlifelike features. A virus is simply a hereditary blueprint, a piece of genetic material wrapped in a coat of protein. Viruses are not made up of cells, so they lack all the structures cells have for gathering and using energy, for reproducing, and for doing all the other things living organisms do. Instead viruses must take over the cellular apparatus of the hosts they invade, using it to carry out their own tasks of living, including replicating their genetic material. Another unusual feature of viruses is that the genetic material they pass from one generation to the next is not always DNA. Some viruses employ another molecule, known as RNA, or ribonucleic acid (we'll learn more about RNA in Unit 3).

What exactly, then, are these viruses—inanimate or alive? Difficult to classify neatly, viruses exist in the netherworld between inanimate objects and living organisms. They share characteristics with living organisms (they reproduce, show a high level of organization, and evolve), but they also lack some of the basic characteristics of life (they are not built from cells, do not capture energy, and do not sense or respond to their environment).

Many scientists think that viruses evolved numerous times from more complex living organisms, but are so highly simplified that they exist on the edge between life and nonlife. As such, they stretch the limits of our definition of living organisms while making it clear just how diverse life on Earth can be.

> ■ Viruses exhibit some characteristics of living organisms but lack others, making it difficult to define whether they are alive or not.

Summary

Asking Questions, Testing Answers: The Work of Science

■ To answer questions about the natural world, scientists begin with observations of nature, from there formulating hypotheses, testing these hypotheses, and then rejecting or modifying them as necessary.

■ Scientists test their hypotheses by performing controlled, repeated manipulations of nature known as experiments.

The Characteristics that All Living Organisms Share

■ The great diversity of life on Earth is unified by a set of shared characteristics.

■ All living organisms are built of cells, reproduce using DNA, grow and develop, capture energy from their environment, sense and respond to their environment, show a high level of organization, and evolve.

Levels of Biological Organization

■ Living organisms are part of a biological hierarchy: from molecules through cells, tissues, organs, organ systems, individuals, populations, species, communities, ecosystems, and biomes, to the biosphere.

Energy Flow through Biological Systems

■ Energy flows through biological systems from producers, such as plants, that create chemical energy in the form of sugars and starches from light, to consumers, such as animals, that eat plants or that eat other organisms whose energy ultimately derives from plants.

Highlight: Life on the Edge: Viruses

■ Viruses occupy a gray zone between living and nonliving things, illustrating how difficult it can be to define life precisely.

Key Terms

adaptation p. 11	food web p. 13
biome p. 11	hypothesis p. 4
biosphere p. 11	individual p. 11
cell p. 6	multicellular p. 7
community p. 11	organ p. 11
consumer p. 11	organ system p. 11
development p. 8	population p. 11
DNA (deoxyribonucleic acid) p. 7	producer p. 11
	science p. 4
ecosystem p. 11	scientific method p. 5
evolution p. 9	species p. 9
experiment p. 5	tissue p. 11

Chapter Review

Self-Quiz

1. Which of the following is *not* an essential element of the scientific method?
   a. observations
   b. conjecture
   c. experiments
   d. hypotheses

2. After reading about Dr. Larson's research, another scientist said, "Japanese honeysuckle might also be doing well in the United States because it isn't being attacked by any of its natural enemies, like insects, from Asia." This statement is an example of
   a. an experiment.
   b. a hypothesis.
   c. a test.
   d. a prediction.

3. Which of the following are both universal characteristics of life?
   a. the ability to grow and the ability to reproduce using DNA
   b. the ability to reproduce using DNA and the ability to capture energy directly from the sun
   c. the ability to move and the ability to sense the environment
   d. the ability to sense the environment and the ability to grow indefinitely

4. Which of the following is the basic unit of life?
   a. virus
   b. DNA
   c. cell
   d. organism

5. Which of the following can reproduce without its own DNA?
   a. human being
   b. virus

   c. single-celled organism
   d. none of the above

6. Which of the following is a multicellular organism?
   a. beetle
   b. brain
   c. bacterium
   d. forest ecosystem

Review Questions

1. What are the observations, hypotheses, and experiments in Dr. Larson's honeysuckle work?

2. What are the elements of the biological hierarchy, and how are they arranged in their proper relationship with respect to one another, from smallest to largest?

3. How does energy flow through biological systems?

1

The Daily Globe

Creationism Fights for a Place alongside Evolution in the Classroom

WEST HARTSVILLE, TENNESSEE. Yesterday some concerned citizens crowded the local elementary school gymnasium to hear arguments about whether "creation science" should be taught alongside evolutionary biology in this small town's classrooms. In contrast to biologists, creation scientists purport that species do not change over time and that all organisms are unchangeable and designed by a divine creator.

"Creation science is not a science at all but a set of religious beliefs," said Dr. Naomi Latte, evolutionary biologist at Tennessee State College, speaking before the school board. "It should not be taught alongside evolutionary biology." Dr. Latte argued that faith "has its place in the home, the church and in private schools, but not in public schools," where the law requires a separation of church and state.

Dr. Tim Garter, creation scientist from the Creation Science Foundation, countered that creation science is a real science. "Though people like Miss Latte will tell you otherwise, creation science is a legitimate field of study, and we should not be censored by university biologists, so many of whom are the worst kind of atheists. Students have the right to know the full range of scientific ideas, however noxious those ideas might be to some people."

Evaluating "The News"

1. The discovery that Japanese honeysuckle can find distant supports more quickly than native honeysuckle is an example of scientific knowledge. The statement that angels are awaiting us in heaven is a religious belief. What distinguishes science from religion?

2. Imagine you are a citizen of West Hartsville and you want the school board to make a fair decision about whether creation science should be taught in schools. Everyone agrees that if creation science is religion it should not be taught in public school, but if it's a science it should. How could you decide whether creation science is a science or not?

3. What kinds of questions can't science answer?

2

Major Groups of Living Organisms

Georgia O'Keeffe, *Jack-in-the-Pulpit No. IV*, 1930.

Here, There, and Everywhere

*I*n 1994 scientists drilling deep into Earth made a shocking finding. Two miles beneath the surface, in what was long thought to be nothing but barren rock, researchers discovered life—abundant life. What they found was a world of microscopic organisms known as bacteria that had been isolated from the rest of the world's organisms for millions of years. Among the species discovered was a bacterium for which scientists immediately proposed the name *Bacillus infernus*, which means literally "bacterium from the inferno" or "bacterium from hell."

Such dramatic finds are becoming more common as biologists probe the last dark corners where life is hiding, as they peer not only into unexplored habitats, like the deep sea, but into internal worlds like a toucan's gut. Everywhere scientists are finding new organisms of all sorts. As the world's species become better known, the lesson biologists continue to learn is that life is hardy and resourceful, thriving almost everywhere, even in the most unlikely places.

In the heat of the Sonoran Desert, cacti point their scorched green limbs to the sky, saved by

Main Message

The world is home to a huge diversity of organisms that exhibit an astounding variety of structures, lifestyles, and behaviors.

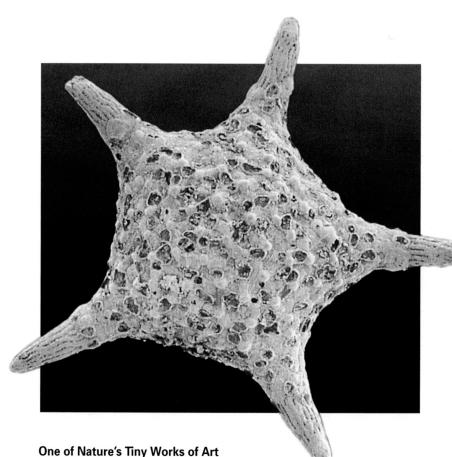

One of Nature's Tiny Works of Art
A microscopic organism called a foraminiferan

the gallons of water that are stored safely inside them. High in the Himalayas, a woolly flying squirrel the size of a woodchuck glides through the skies. Wildflowers in the snow-covered Swiss Alps turn their blossoms to follow the movement of the sun and capture its precious warming rays. In the steaming waters of Yellowstone Park's most famous geyser, Old Faithful, heat-loving bacteria comfortably persist, well suited to such extremes of temperature. And everywhere insects, the world's most abundant animals, fly and crawl, these six-legged creatures making meals out of everything from the nectar of flowers to poison-filled plant leaves to human blood to cow dung.

Life abounds in countless forms on Earth, and biologists are trying to study it all. In this chapter we take a look at the diversity of life, the riot of living organisms that inhabits our planet, and we see how it can be classified.

Key Concepts

1. Biologists use the Linnaean classification hierarchy to group species into larger and larger categories, the largest of which is the kingdom. The five kingdoms are the Monera (bacteria), Protista (amoebas and the like), Plantae (plants), Fungi (mushrooms, molds, and yeasts), and Animalia (animals).

2. The kingdom Monera includes microscopic, single-celled organisms known as bacteria. Bacteria are more diverse in how they obtain their nutrition than are the members of any other kingdom.

3. The cells of all organisms, except bacteria, contain tiny structures with specialized functions known as organelles. Organisms with cells that contain organelles are known as eukaryotes. Bacterial cells do not contain organelles and are known as prokaryotes.

4. The kingdom Protista is the most ancient eukaryotic kingdom. Protists represent early stages in the evolution of the eukaryotic cell and in the evolution of multicellularity.

5. The kingdom Plantae pioneered life on land. Plants are multicellular and they photosynthesize, using the energy of sunlight and carbon dioxide to make sugars. As producers, plants form the base of all food webs on land.

6. The kingdom Fungi includes mushroom-producing species, as well as molds and yeasts. Fungi are among the world's key decomposers, many using dead and dying organisms as their food.

7. The kingdom Animalia includes a wide range of multicellular organisms. Animals have evolved specialized tissues, organs and organ systems, body plans, and behaviors.

*A*s described in Chapter 1, all living organisms share a basic set of characteristics. Life shares this set of common properties because all living organisms are thought to be descended from a single common ancestor. But life first appeared on this planet more than 3.5 billion years ago. Since that time, life has evolved in many directions. From a living world of nothing more than individual cells, a planet full of wildly different organisms has evolved. We are part of a great diversity of life that is still far from completely known, counted, or named.

Most biologists estimate that there are between 3 million and 30 million species on Earth. In this chapter we are not aiming for a comprehensive examination of the world's many species. Instead we will attempt to familiarize you with the diversity of life by introducing the major groups: Monera (bacteria), Protista (amoebas and the like), Plantae (plants), Fungi (mushrooms, molds, and yeasts), and Animalia (animals).

We begin by discussing how biologists deal with the overwhelming diversity of life by using classification systems to name and group organisms. Then we discuss each of the major groups known as kingdoms (a classification category that we will describe shortly). We describe key features that characterize each kingdom. In particular we stress features evolved by the group that allow the members of that kingdom to live and reproduce successfully, features known as **evolutionary innovations**. In addition we describe some of the more important and interesting members of each group.

For each kingdom a photo gallery provides a broad overview, an evolutionary tree, and a description of some of the prominent subgroups within the kingdom. This gallery provides an evolutionary framework for discussion of the evolutionary innovations of each kingdom. It should also serve as a reference as you read later chapters and want to take a second look at groups you are learning about.

Organizing the Diversity of Life: The Linnaean Classification System

When working with a large number of different items, the first thing one must do is put the items in some order, whether that order is an inventory classification system at a store or the Dewey decimal system for organizing books at a library. To study and discuss the world's many species, biologists have organized them into groups in a classification scheme known as the Linnaean hierarchy, the roots of which trace back more than 200 years (Figure 2.1). In the 1700s a Swedish biologist named Carolus Linnaeus developed a system of classifying organisms that we still use today and that is named after him. Working long before biologists understood the process of evolution, Linnaeus created his system of classification to reveal God's plan. Today biologists use the same basic classification scheme, but instead they classify organisms to reflect how closely or distantly related groups are.

In the Linnaean hierarchy, the species is the smallest unit of classification. Closely related species are grouped together to form a genus (plural genera). Closely related genera are grouped together into a family. Closely related families are grouped into an order. Closely related orders are grouped into a class. Closely related classes are grouped into a phylum (plural phyla). Finally, closely related phyla are grouped together into the largest possible category, a **kingdom**. An easy way to remember the hierarchy is to memorize the following sentence, in which the first letter of each word stands for each descending level in the hierarchy: *King Phillip Cleaned Our Filthy Gym Shorts.*

Every species is given a unique, two-part Latin name, also known as a scientific name. For example, humans are called *Homo sapiens*—*Homo* meaning "man" and *sapiens* meaning "wise." *Homo* indicates our genus, and *sapiens* is our particular species name within the genus. We are the only living species in our genus. Other species in the genus are *Homo erectus* ("upright man") and *Homo habilis* ("handy man"), both of which are extinct.

In this chapter we will focus on the broadest level of classification, the kingdom. In the 1700s Linnaeus devised a system with only two kingdoms, plants and animals. In 1969 Robert Whittaker, an ecologist at Cornell University, devised a more complete, five-kingdom system that is still widely used today. Figure 2.2 shows an **evolutionary tree** that depicts the hypothesized evolutionary relationships of some of the major groups of organisms and the five-kingdom system.

Homo habilis

Figure 2.2 also illustrates how these groups can be organized into different kingdom systems. Although we follow the five-kingdom system in this book, some biologists organize life into six or eight kingdoms. In all cases, biologists are attempting to group together what they consider to be groups that are closely related and to separate groups that are not closely related. As we learn more about the relatedness of different groups, scientists will continue to debate which are the most accurate and useful classification schemes, and those schemes will continue to change.

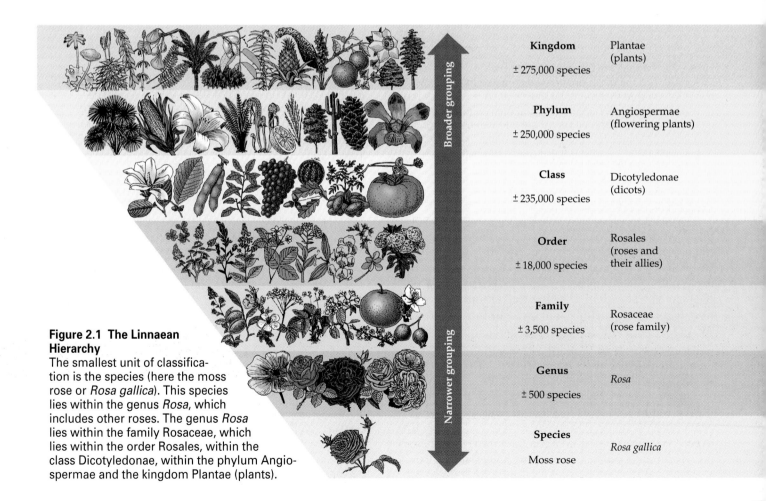

Figure 2.1 The Linnaean Hierarchy
The smallest unit of classification is the species (here the moss rose or *Rosa gallica*). This species lies within the genus *Rosa*, which includes other roses. The genus *Rosa* lies within the family Rosaceae, which lies within the order Rosales, within the class Dicotyledonae, within the phylum Angiospermae and the kingdom Plantae (plants).

Broader grouping

Narrower grouping

Kingdom	Plantae (plants)	± 275,000 species
Phylum	Angiospermae (flowering plants)	± 250,000 species
Class	Dicotyledonae (dicots)	± 235,000 species
Order	Rosales (roses and their allies)	± 18,000 species
Family	Rosaceae (rose family)	± 3,500 species
Genus	*Rosa*	± 500 species
Species	*Rosa gallica*	Moss rose

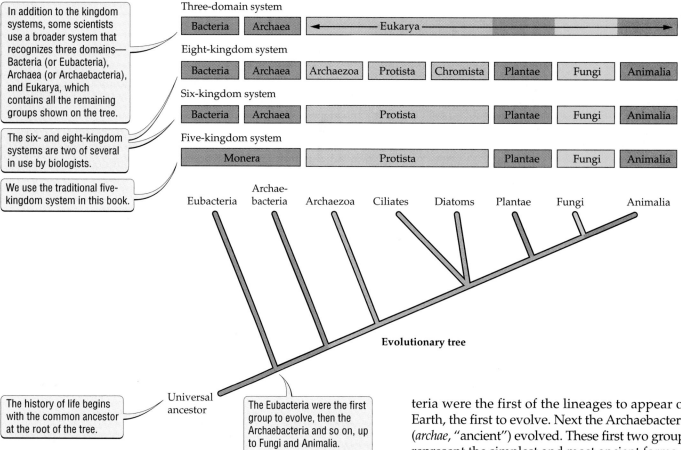

In addition to the kingdom systems, some scientists use a broader system that recognizes three domains—Bacteria (or Eubacteria), Archaea (or Archaebacteria), and Eukarya, which contains all the remaining groups shown on the tree.

The six- and eight-kingdom systems are two of several in use by biologists.

We use the traditional five-kingdom system in this book.

The history of life begins with the common ancestor at the root of the tree.

The Eubacteria were the first group to evolve, then the Archaebacteria and so on, up to Fungi and Animalia.

Figure 2.2 Organizing Life into Kingdoms
The major groups of organisms are depicted here on an evolutionary tree that many biologists believe correctly represents their evolutionary history and their relatedness. Each of the groups branching off the tree can be thought of as a cluster of close relatives—a lineage, just like a lineage in a human family. These major groups, or lineages, can then be placed into kingdoms.

We will put off a detailed discussion of the evolutionary relatedness of the major groups and of evolutionary trees until Chapter 3. Suffice it to say that the tree depicted in Figure 2.2 should be considered a hypothesis. That is, although it is a tree that many biologists believe correctly depicts the evolutionary relationships of the major groups of organisms, like all hypotheses it is subject to change as scientists continue to gather new evidence about these groups.

Notice that the ancestor of all living things is positioned at the base of the tree. From there, one can trace evolutionary history through time by tracing up the tree. Each group of close relatives—for example, the fungi—branching off the tree is considered to be a single **lineage** (like a lineage within a human family). For example, the lineage Eubacteria (or true bacteria; *eu*, "true") is shown on the first branch, indicating that the eubac-

teria were the first of the lineages to appear on Earth, the first to evolve. Next the Archaebacteria (*archae*, "ancient") evolved. These first two groups represent the simplest and most ancient forms of life.

Next a few representative members of the Protista are shown branching off. First the archaezoa evolved. Then ciliates and diatoms are shown branching off the tree together. Biologists use this simultaneous branching off to show that it is unclear which group branched off first. Next plants branch off, and finally the fungi and animals split apart from one another.

■ The Linnaean hierarchy goes from species at the lowest level to genera, families, orders, classes, phyla, and kingdoms. The kingdom system has evolved over time and continues to evolve as we learn more. In this book we use the widely accepted five-kingdom system.

Kingdom Monera: Tiny, Successful, and Abundant

When life arose on Earth more than 3.5 billion years ago, the first living organisms were tiny, simple, single cells. From those first cells evolved the kingdom **Monera**, made up of the microscopic organisms we know as bacteria. The first lineage of the Monera to evolve was the Eubacteria, which was followed by the Archaebacteria (Figure 2.3). The kingdom Monera is the most ancient of

Figure 2.3 The Monera

The Monera are bacteria, microscopic single-celled organisms that are the most ancient forms of life. Eubacteria can be seen branching off first, then the Archaebacteria, from what would become the rest of the living organisms.

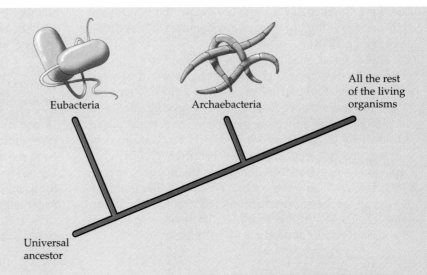

Eubacteria

Archaebacteria

All the rest of the living organisms

Universal ancestor

Number of species discovered to date: ~4,800

Functions within ecosystems: Producers, consumers, decomposers

Economic uses: Many, including producing antibiotics, cleaning oil spills, treating sewage

Funky factoid: The number of bacteria in your digestive tract outnumber all the humans that have lived on Earth since the beginning of time.

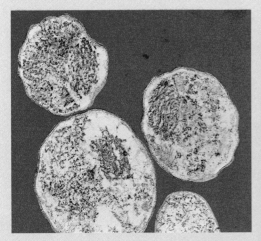

The bacterium *Chlamydia trachomatis* causes the most common sexually transmitted disease, chlamydia.

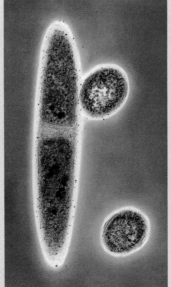

These archaebacteria, known as *Methanospirillum hungatii*, are shown in cross-section (the two circular shapes) and as an elongated cell that is about to fission into two cells.

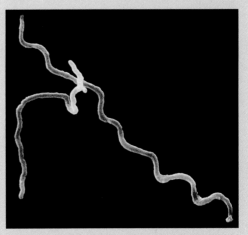

These bacteria, *Borrelia burgdorferi*, known as spirochetes because of their helical cells, cause Lyme disease, which is transmitted to humans through a tick bite.

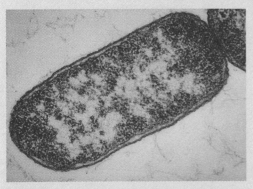

This bacterium, known as *Escherichia coli*, or *E. coli*, is usually a harmless inhabitant of the human gut. However, toxic strains can contaminate and multiply on food, like raw hamburger, and cause illness or death in humans who eat them.

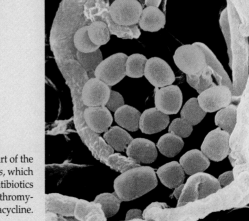

This bacterium is part of the genus *Streptomyces*, which produces the antibiotics streptomycin, erythromycin, and tetracycline.

the kingdoms, made up of what are arguably both the simplest existing organisms and the most successful at colonizing Earth.

All bacterial cells, which can be shaped like spheres, rods, or corkscrews, share a basic, structural plan (Figure 2.4). The picture of efficiency, these stripped-down organisms are nearly always single-celled and small. They typically have much less DNA than cells of organisms in the other kingdoms have. Genomes of organisms in other kingdoms are often full of what appears to be extra DNA that serves no function. In contrast, bacterial genomes contain only DNA that is actively in use for the survival and reproduction of the bacterial cell. Their reproduction is similarly uncomplicated; bacteria typically reproduce by splitting in two.

Simplicity translated into success

Bacteria form the most abundant group of organisms on Earth in terms of numbers of individuals. Scientists estimate that the number of individual bacteria on Earth is about 5,000,000,000,000,000,000,000,000,000,000, or 5×10^{30}. That success is due, in part, to how quickly they reproduce. Overnight, a single *Escherichia coli* (*E. coli*), a bacterium that normally lives harmlessly in the human gut, can divide to produce a population of 10 million bacteria. (Toxic strains of *E. coli* have also become well known as contaminants of food. They can be dangerous, even fatal, to humans who eat them in undercooked meats and other foods.)

Bacteria are also the most widespread of organisms, able to live nearly anywhere. They can persist even in places where most organisms would perish, such as the lightless ocean depths, the insides of boiling hot geysers, and miles below Earth's surface. Because of their small size, bacteria also live on and in other organisms. Scientists estimate that 1 square centimeter of healthy, human skin is home to between 1000 and 10,000 bacteria.

In addition, while many bacteria are aerobes (*aer*, "air") and need the gas oxygen to survive, many others are anaerobes (*an*, "without") and can survive without oxygen. This ability to exist in both oxygen-rich and oxygen-free environments also increases the

number of habitats in which bacteria can persist. But the real key to the success of this group is the great diversity of ways in which they obtain and use nutrients.

Bacteria exhibit unmatched diversity in methods of obtaining nutrition

When humans eat, we consume other species and remove from them the two things that all organisms need to survive: energy and carbon, the chemical building blocks that organisms use to make critical molecules for living, such as proteins and DNA. Consuming other organisms, however, is just one way to get energy and carbon. In the broadest view, the ways in which the world's organisms obtain their nutrition can be divided into four major modes, which are summarized in Table 2.1. Only the kingdom Monera contains species that can do them all.

Notice that in Table 2.1 the first part of the name of each category indicates where an organism gets its energy. *Photo* ("light") means that the organism gets its energy directly from the energy of sunlight. *Chemo* ("chemical") means that the organism gets its energy from chemical compounds.

The second part of each name indicates where the organism gets its carbon compounds. *Hetero* ("other") means that the organism depends on other organisms as sources of compounds that contain carbon. Such carbon-containing compounds are known as organic compounds. *Auto* ("self") means that the organism can produce organic compounds itself using only carbon di-

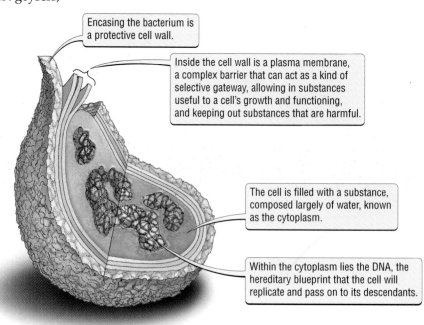

Figure 2.4 The Basics of the Bacterial Cell
Bacterial cells tend to be small, around 10 times smaller than the cells of organisms in other kingdoms with much less DNA.

Encasing the bacterium is a protective cell wall.

Inside the cell wall is a plasma membrane, a complex barrier that can act as a kind of selective gateway, allowing in substances useful to a cell's growth and functioning, and keeping out substances that are harmful.

The cell is filled with a substance, composed largely of water, known as the cytoplasm.

Within the cytoplasm lies the DNA, the hereditary blueprint that the cell will replicate and pass on to its descendants.

2.1 The Major Nutritional Modes			
	Energy source	**Carbon source**	**Kingdoms**
Photoautotrophy	Sunlight	Carbon dioxide	Monera, Protista, Plantae
Photoheterotrophy	Sunlight	Organic compounds	Monera
Chemoautotrophy	Inorganic compounds	Carbon dioxide	Monera
Chemoheterotrophy	Organic compounds	Organic compounds	Monera, Protista, Fungi, Animalia

oxide, a gas in the air. The last part of each name—*trophy*—means "nutrition."

Photoautotrophs use light as an energy source and obtain their carbon from the gas carbon dioxide in the air. Plants are photoautotrophs, as are many bacteria, including the freshwater organisms known as cyanobacteria (Figure 2.5), more familiar to most people as pond scum.

Chemoheterotrophs, despite their foreign-sounding name, are organisms like animals that obtain energy and carbon from organic compounds, typically by consuming or otherwise breaking down the bodies of other organisms. Two chemoheterotrophic bacteria are *Salmonella*, which can cause food poisoning, and

Chlamydia trachomatis, which causes the most common sexually transmitted disease in humans (see Figure 2.3). Most chemoheterotrophs get their organic compounds from other organisms, but some bacteria can live on pesticides and other organic compounds that do not come from organisms.

The most curious nutritional categories are photoheterotrophy and chemoautotrophy, modes of nutrition found only in the kingdom Monera. **Photoheterotrophs** use light as an energy source, as do plants, but they get their carbon from organic compounds rather than from carbon dioxide in the air.

Chemoautotrophs use carbon dioxide to make sugars, just as plants do. But instead of using the energy of light from the sun, these bacteria obtain energy from a chemical reaction that involves inorganic compounds, compounds that do not contain carbon. Chemoautotrophs obtain energy from such unlikely materials as iron and ammonia (Figure 2.6).

Monera can thrive in extreme environments

Bacteria are well known for living in nearly any kind of environment. Within the Archaebacteria, one of two major groups of Monera are extreme thermophiles (*thermo*, "heat"; *phile*, "lover") that thrive in geysers and hot springs. Others are salt lovers, or extreme halophiles (*halo*, "salt"). These bacteria grow where nothing else can live—for example, in the Dead Sea and on fish and meat that have been heavily salted to keep most bacteria away (Figure 2.7).

Not all archaebacteria, however, are so remote from daily experience. Members of one group, the meth-

Figure 2.5 Bacteria That Photosynthesize
These cyanobacteria, sometimes called blue-green algae, can be found growing as mats on freshwater ponds. Early in their evolution, plant cells are thought to have acquired their ability to photosynthesize by engulfing cyanobacterial cells. The descendants of cyanobacteria in plant cells have evolved to become specialized structures known as chloroplasts.

Figure 2.6 An Unusual Eater
The bacterium *Sulfolobus*, a chemoautotroph, needs neither other organisms nor light to make its living. It gets its energy not from sunlight or other organisms, but by chemically processing inorganic materials like iron. Chemoautotrophs get their carbon from carbon dioxide.

anogens (*methano*, "methane"; *gen*, "producer"), inhabit animal guts and produce the methane gas in such things as human flatulence (intestinal gas) and cow burps.

The Monera are important in the biosphere and for human society

Because of their ability to use so many substances as a food source and to live under such a variety of conditions, bacteria play numerous and important roles in ecosystems and in human society. For example, plants require a chemical called nitrate that they cannot make themselves. For this they depend on bacteria that can take nitrogen, a gas in the air, and convert it to nitrate. Without these bacteria there would be no plant life, and without plant life there would be no life on land.

Like plants, some bacteria, such as cyanobacteria, can photosynthesize. (Photosynthesis will be covered in detail in Chapter 8.) This ability makes them producers, the organisms at the base of food webs. Bacteria also are important decomposers, breaking down dead organisms by using them as food. Oil-eating bacteria can be used to clean up ocean oil spills. Bacteria that can live on sewage are used to help decompose human waste so that it can be safely, usefully returned to the environment. Bacteria also live harmlessly in animal guts (including our own), helping digest food.

Of course, not all bacteria are helpful. Many of them cause diseases and are the stuff of nightmares, like the flesh-eating bacteria that can destroy human flesh at frightening rates. With their ability to use almost anything as food, bacteria can also attack crops, stored foods, cattle, and nearly every other living organism.

Figure 2.7 Better than a Bag of Chips
For those who love salt, such as extreme halophiles, nothing beats a salt farm. Here in Thailand seawater is evaporated, making it more and more concentrated in salt and creating an environment that only an archaebacterium could love.

■ The kingdom Monera consists of simple, microscopic, single-celled organisms known commonly as bacteria. There are two lineages: the Eubacteria and the Archaebacteria. The kingdom Monera is the oldest kingdom, and its members are the most numerous (in terms of numbers of individuals) and the most widespread. Bacteria can act as decomposers, producers, and consumers.

The Evolution of More Complex Cells: Eukaryotes Arose from Prokaryotes

Bacteria have played a critical role in the evolution of the cells in all other living organisms. The cells of all other organisms are known as eukaryotic cells and contain tiny structures called organelles. Each organelle performs a different, specialized function. For example, organelles known as chloroplasts in plant cells are the sites at which photosynthesis takes place. The nucleus is an organelle that contains a eukaryotic cell's DNA. While these organelles function as specialized structures within eukaryotic cells today, they are actually descended from what were once free-living bacteria.

Deep in the evolutionary past, some bacterial cells engulfed other bacteria. These captured cells eventually evolved to function as parts of the cells that had engulfed them; that is, they became the specialized tiny structures known as organelles (see Figure 6.14). These organelles became components that were critical to the survival of the eukaryotic cell that housed them, and they were unable to function outside of the cell.

The presence of such organelles is one of the key characteristics that separates the eukaryotes (the kingdoms Protista, Plantae, Fungi, and Animalia) from bacteria. The Monera, or all bacteria, are also known as **prokaryotes** (*pro*, "before"; *karyote*, "kernel," here meaning "nucleus") because they arose before the evolution of the nucleus. All other living organisms, whose cells do contain a nucleus and other organelles, are known as **eukaryotes** (*eu*, "true"), a group in which multicellular, complex organisms evolved.

■ Eukaryotic cells, which include all cells in all organisms other than the Monera, arose when one prokaryotic cell engulfed others. The engulfed prokaryotic cells eventually evolved into organelles. Eukaryotes (Protista, Plantae, Fungi, and Animalia) have organelles in their cells; prokaryotes (Monera) do not.

Kingdom Protista: A Window into the Early Evolution of Eukaryotes

The kingdom **Protista** is the most ancient of the eukaryotic kingdoms and consists largely of single-celled eukaryotes, but it contains some multicellular eukaryotes as well. Included are some familiar groups, like single-celled amoebas and multicellular kelp, and many unfamiliar groups. One of the few generalizations that can be made about this hard-to-define kingdom is that its members are very diverse in size, shape, and lifestyle.

One of the reasons for the great diversity is the fact that the protists are not all members of a single lineage (see Figure 2.2), as plants, animals, and fungi are. We will go into greater detail in Chapter 3 about how biologists decide what lineages should go into what groups in a classification. For now, keep in mind that protists are composed of different groups of organisms that branched off at separate times from the rest of the tree of life (such as the archaezoa, ciliates, and diatoms shown in Figure 2.2), and that variety in the lineages adds to the diversity of the protist kingdom.

There are plantlike protists—for example, algae—that can photosynthesize. There are also animal-like protists—for example, ciliates—that move and hunt for food. Still others—such as the slime molds—are more like fungi. Most protists are single-celled (for example, paramecia), but there are also many multicellular protists (such as the multicellular algae).

Figure 2.8 presents a hypothesis for the evolutionary tree of some of the major groups of protists. Archaezoa are shown branching off first. Then three major groups (one including dinoflagellates, apicomplexans, and ciliates; another including diatoms and water molds; and the last made up of green algae) are shown splitting off at the same time. This simultaneous three-way split indicates that biologists are not sure which of these distinct lineages branched off first.

The evolutionary tree of protists continues to present many challenges to biologists. The protists as a whole fascinate biologists. The reason, as we'll see, is that this diversity of earliest eukaryotes provides snapshots of the evolution of the eukaryotic cell and multicellular organization.

Protists represent early stages in the evolution of the eukaryotic cell

As already described, the eukaryotic cell evolved by a process in which one prokaryotic cell engulfed others. Over time these captured cells evolved and became less like free-living life forms and more like tiny organs, or organelles, within the cells that had captured them. For example, in addition to nuclei, eukaryotic cells typically contain mitochondria (singular mitochondrion),

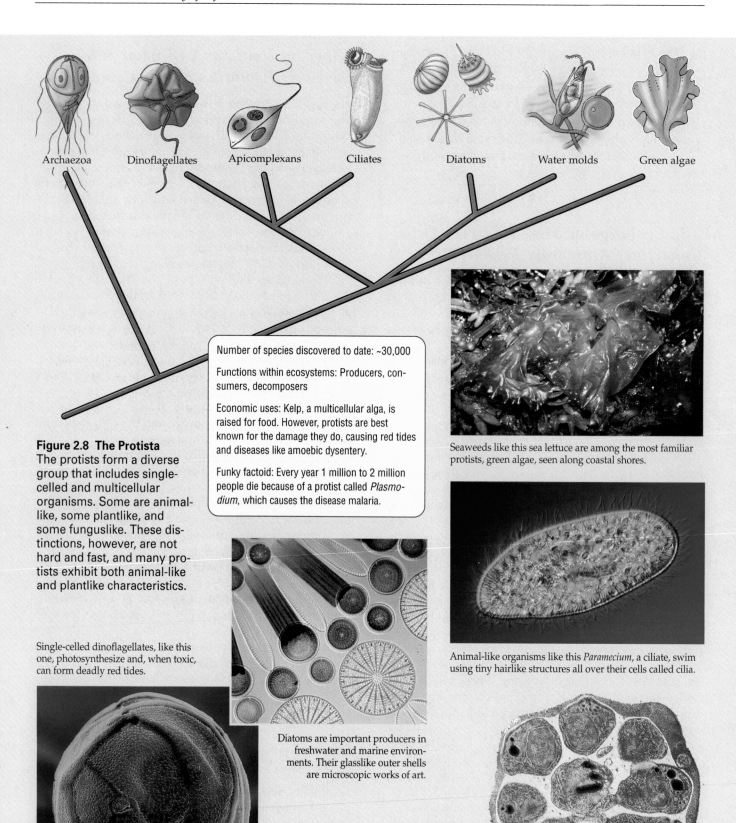

Archaezoa   Dinoflagellates   Apicomplexans   Ciliates   Diatoms   Water molds   Green algae

Number of species discovered to date: ~30,000

Functions within ecosystems: Producers, consumers, decomposers

Economic uses: Kelp, a multicellular alga, is raised for food. However, protists are best known for the damage they do, causing red tides and diseases like amoebic dysentery.

Funky factoid: Every year 1 million to 2 million people die because of a protist called *Plasmodium*, which causes the disease malaria.

Figure 2.8 The Protista
The protists form a diverse group that includes single-celled and multicellular organisms. Some are animal-like, some plantlike, and some funguslike. These distinctions, however, are not hard and fast, and many protists exhibit both animal-like and plantlike characteristics.

Seaweeds like this sea lettuce are among the most familiar protists, green algae, seen along coastal shores.

Single-celled dinoflagellates, like this one, photosynthesize and, when toxic, can form deadly red tides.

Diatoms are important producers in freshwater and marine environments. Their glasslike outer shells are microscopic works of art.

Animal-like organisms like this *Paramecium*, a ciliate, swim using tiny hairlike structures all over their cells called cilia.

A gathering of apicomplexans, a group of protist parasites. Shown here is *Plasmodium*, which causes malaria.

organelles that produce energy. Eukaryotic cells and their organelles now depend entirely on each other (just as a human body and its organs do), and they cannot survive apart from one another.

Scientists now believe that the engulfing of one cell by another was not a onetime event. Increasing evidence shows that bacteria were engulfed by other bacteria many different times in the history of life. Among protists, different species illustrate the variety of experimentation that has gone on over time.

Within the protist group known as archaezoa is the species *Giardia lamblia*. This protist lives in streams and other water sources. It can cause a painful ailment of the digestive tract when it is consumed by humans, causing diarrhea and flatulence that has an offensive rotten-egg odor. Most eukaryotes contain a nucleus and a mitochondrion, and if they photosynthesize, chloroplasts. *Giardia lamblia*, however, appears to be a curious experiment in prokaryotic engulfing: It has two nuclei, but no mitochondria or chloroplasts.

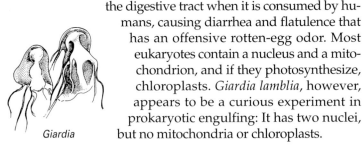

Giardia

Protists provide insight into the early evolution of multicellularity

Protists are of interest to biologists also because many different groups within the kingdom Protista have each evolved from being single-celled creatures, like bacteria, to forming multicellular groupings that function to greater or lesser degrees like more complex multicellular individuals, such as humans. Among the more interesting of these experiments in the evolution of multicellularity are the slime molds, protists that were once thought to be fungi (hence the name, since molds belong to the kingdom Fungi). Commonly found on rotting vegetation, slime molds make their living by digesting bacteria. But these curious organisms can live their lives in two phases: (1) as independent, single-celled creatures and (2) as members of a multicellular body.

Like other protists that do not live strictly as single-celled organisms or as typical multicellular organisms, slime molds are studied by biologists who hope to gain insight into the transition from single-celled to multicellular living. Like some other protist groups, the slime molds are excluded from the evolutionary tree in Figure 2.8 because their relationships to the other protists remain unclear.

Protists had sex first

Like bacteria, many protists reproduce simply by splitting in two, a form of asexual (nonsexual) reproduction. Under some circumstances, bacteria can receive DNA

(genetic material) from other bacteria, a combining of genetic material that some consider to be a form of prokaryotic sexual reproduction. But sex is more often defined as a process in which two individual organisms produce specialized cells known as gametes (for example, human egg and sperm). These gametes fuse together, combining the DNA contributions from both parents into one individual offspring.

One of the most stunning achievements of the protists was the invention of this kind of eukaryotic sex. Although many protists reproduce asexually, it was in the protists that sexual reproduction, with gametes from different individuals that fuse to form a new individual, first appeared.

Protists are best known for their disease-causing abilities

Although most protists are harmless, many of the best-known protists cause diseases. Dinoflagellates (see Figure 2.8) are microscopic plantlike protists that live in the ocean and sometimes have huge population explosions, known as blooms. Occasional blooms of toxic, red dinoflagellates cause dangerous "red tides." During red tides, shellfish eat toxic dinoflagellates, and humans can be poisoned by eating the shellfish. An animal-like protist, *Plasmodium*, is the organism that causes malaria. Finally, the protist kingdom left its mark on human history forever when one of its members, a water mold attacked potato crops in Ireland in the 1800s, causing the disease known as potato blight. The resulting widespread loss of potato crops caused a devastating famine.

> ■ The kingdom Protista is a diverse group of single-celled and multicellular organisms, some animal-like, some plantlike, and some funguslike. The oldest kingdom of eukaryotes, Protista includes species that represent early stages in the evolution of the eukaryotic cell and multicellularity.

Kingdom Plantae: Pioneers of Life on Land

To some people, there is nothing more mundane than a plant, and greenery represents little more than garnishes at meals and decorative houseplants. But plants are among evolution's great pioneers. Life on Earth began in the water, and there it stayed for 3 billion years. It was only when the kingdom **Plantae**, or plants, evolved that life took to land. These first colonists turned barren ground into a green paradise in which a whole new world of land-dwelling organisms evolved.

Figure 2.9 shows the basic structure of a plant. Life on land required special structures to obtain and conserve water. One of the features of plants that allows life on land is a **root system**, a collection of fingerlike growths that absorbs necessary water and nutrients from the soil. Another feature that helps plants live on land is the waxy covering over stem and leaves, known as the **cuticle**. The cuticle prevents the plant tissue from drying out even when exposed to sun and air. Plant cells also have a rigid cell wall that is composed of an organic compound known as cellulose; the cell wall provides support for a stem growing out of the ground.

The key feature of plants, however, is their ability to use light (energy from the sun) and carbon dioxide (a gas in the air) to produce food in the form of sugars. (Recall from Table 2.1 that plants are photoautotrophs.) As described in Chapter 1, plants are producers, and as such these organisms form the basis of essentially all terrestrial (land-based) food webs. In addition, one of the useful by-products of their photosynthesizing is the critical gas oxygen. Most of a plant's photosynthesizing takes place in its leaves, which typically grow in ways that maximize their ability to capture sunlight.

The diversity of plants today ranges from the most ancient lineages, the mosses and their close relatives, to ferns which evolved next, then gymnosperms and angiosperms (Figure 2.10). In the sections that follow we'll examine three major innovations in the evolution of plants: vascular systems, seeds, and flowers. These innovations were critical to what has been a highly successful colonization of land by plants.

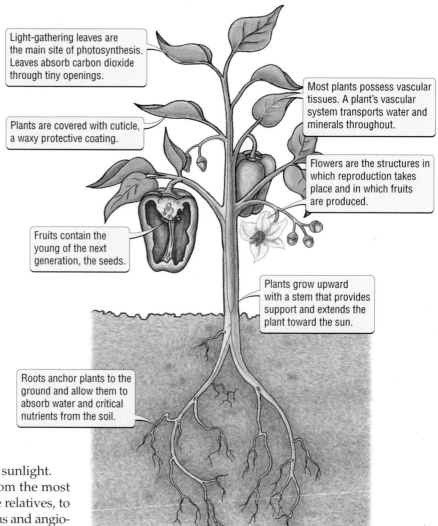

Light-gathering leaves are the main site of photosynthesis. Leaves absorb carbon dioxide through tiny openings.

Plants are covered with cuticle, a waxy protective coating.

Most plants possess vascular tissues. A plant's vascular system transports water and minerals throughout.

Flowers are the structures in which reproduction takes place and in which fruits are produced.

Fruits contain the young of the next generation, the seeds.

Plants grow upward with a stem that provides support and extends the plant toward the sun.

Roots anchor plants to the ground and allow them to absorb water and critical nutrients from the soil.

Figure 2.9 The Basic Form of a Plant

Vascular systems allowed plants to grow to great heights

Early in their evolution, plants grew close to the ground. Mosses and their close relatives represent those early days in the history of plants and are the most ancient of plant lineages. These plants rely on the absorption of water directly by each of their cells. Thus, the innermost cells of their bodies receive water only after it has managed to pass through every cell between them and the outermost layer of the plant. Because such movement of water from cell to cell, like the movement of water through a kitchen sponge, is relatively inefficient, these plants cannot grow to great heights or sizes.

Ferns and their close relatives, which arose later, could grow taller because they had evolved vascular systems. **Vascular systems** are networks of specialized cells, known as tissues, that extend from the roots throughout the body of a plant. These vascular tissues can efficiently transport fluids and nutrients, much as the human circulatory system of veins and arteries transports blood. In addition, this network of water-filled vessels can make a plant firmer. By providing sturdiness and efficient circulation of water, vascular systems allowed plants to grow to new heights and sizes (Figure 2.11). All plants, except mosses and their close relatives, have vascular systems.

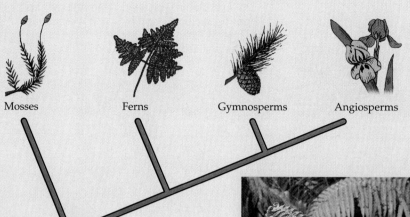

Mosses   Ferns   Gymnosperms   Angiosperms

Number of species discovered to date: ~250,000

Functions within ecosystems: Producers (a few plants, such as the Venus flytrap, also act as consumers)

Economic uses: Flowering plants provide all our crops: corn, tomatoes, rice, and so on. Fir trees and other conifers provide wood and paper. Plants also produce important chemicals, such as morphine, caffeine, and menthol.

Funky factoid: Of the 250,000 species of plants, at least 30,000 have edible parts. In spite of this abundance of potential foods, just three species—corn, wheat, and rice—provide most of the food the world's human populations eat.

Figure 2.10 The Plantae
Plants are multicellular photo-autotrophs, a diverse group that pioneered life on land.

Ferns and their close relatives evolved vascular systems that allowed plants to evolve to greater heights and different shapes. This Ama'uma'u fern grows only in Hawaii.

The most species-rich family of Angiosperms in the world, orchids like these also produce among the world's most beautiful flowers.

On Mt. Sago in Sumatra, the angiosperm *Rafflesia arnoldii* produces the world's largest blossoms, measuring as much as 1 meter across.

The most familiar gymnosperms are the conifers. The most common of the gymnosperms today, conifers like this giant sequoia are important wood and paper producers.

Mosses and their close relatives, the most ancient group of plants, do not have vascular systems and do not grow more than a few inches high.

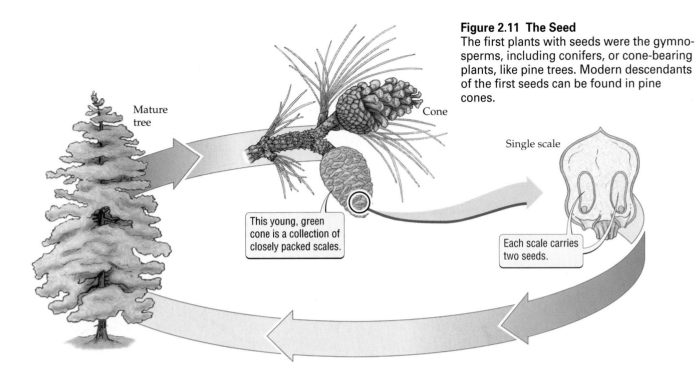

Figure 2.11 The Seed
The first plants with seeds were the gymnosperms, including conifers, or cone-bearing plants, like pine trees. Modern descendants of the first seeds can be found in pine cones.

Mature tree

Cone

Single scale

This young, green cone is a collection of closely packed scales.

Each scale carries two seeds.

Gymnosperms evolved a way to protect their young: Seeds

Following mosses and ferns, the next group of plants to evolve was the **gymnosperms** (*gymno*, "naked"; *sperm*, "seed"), a group that includes pine trees and other cone-bearing plants, known as conifers, as well as cycads and ginkgos (see Figure 2.10).

Gymnosperms were the first plants to evolve **seeds**, structures in which plant young (embryos) are encased in a protective covering and provided with a stored supply of nutrients. Gymnosperms are known as naked seeds because their seeds are relatively unprotected compared with the seeds of angiosperms, which we describe in the next section. Gymnosperms were the dominant plants 250 million years ago, and biologists believe that the evolution of seeds was an important part of their success. Seeds provided nutrients for plant embryos to grow and develop before they were able to produce their own food via photosynthesis. Seeds also afforded protection from drying or rotting, and from attack by predators .

Angiosperms produced the world's first flowers 200 million years ago

Although typically we think of flowers when we think of plants, flowering plants are a relatively recent development in the history of life. Today the flowering plants, known as **angiosperms**, are the most dominant and diverse group of plants on the planet, including such things as orchids, grasses, corn plants, and apple and tulip trees. As already mentioned, angiosperms produce seeds that have more protective tissues than gymnosperm seeds have (*angio* means "vessel," referring to the protective tissues that encase the plant's embryo).

Highly diverse in size and shape angiosperms live in a wide range of habitats—from mountaintops to deserts to salty marshes and fresh water. Almost any plant we can think of that is not a moss, a fern, or a pine tree is an angiosperm. The defining feature of angiosperms is the **flower**, a specialized structure for sexual reproduction, or pollination, in which pollen (the male gamete) and an ovule (the female gamete) meet (Figure 2.12).

Angiosperms display a wide variety of flowers. Flowers often provide food, like the sugary liquid known as nectar, to attract animals that will visit the flowers and in the process transport pollen from one flower to another. Such transported pollen can fertilize a flower's ovules. Thus animals can provide a means of sexual reproduction between even very distant, immobile plants. Flower evolution has reached its extreme in the elaborate blooms of orchids, many of which are designed to deceive animal pollinators. One orchid has evolved flowers that look so much like female bees that male bees repeatedly visit the flowers attempting to mate with them and in the process unintentionally pollinate them (see Figure 2.10).

Angiosperms have also evolved fleshy, tasty fruits that attract animals. When the embryos of angiosperms

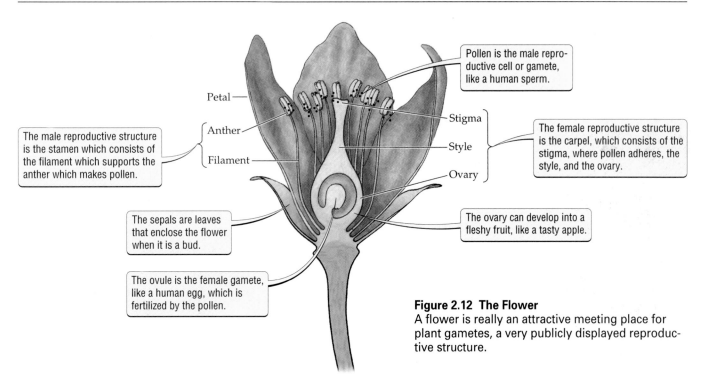

Petal

Anther

Filament

Pollen is the male reproductive cell or gamete, like a human sperm.

Stigma

Style

Ovary

The male reproductive structure is the stamen which consists of the filament which supports the anther which makes pollen.

The female reproductive structure is the carpel, which consists of the stigma, where pollen adheres, the style, and the ovary.

The sepals are leaves that enclose the flower when it is a bud.

The ovary can develop into a fleshy fruit, like a tasty apple.

The ovule is the female gamete, like a human egg, which is fertilized by the pollen.

Figure 2.12 The Flower
A flower is really an attractive meeting place for plant gametes, a very publicly displayed reproductive structure.

are developing, the surrounding ovary (see Figure 2.12) can develop into a ripening fruit. Animals eat the fruits and later excrete the seeds far away in heaps of feces. These nutrient-rich wastes provide a good place for the seeds to sprout and start life, often far from their parent plant so they won't compete with that parent for water, nutrients, or light. But immobile plants don't always need the help of animals to get their young off to a good start; plants have evolved many ways to get their seeds off to distant locations (Figure 2.13).

Plants are the basis of land ecosystems and provide many valuable products

It is difficult to overstate the significance of plants. As photosynthesizing organisms, plants use sunlight and air to make sugars, food that they and the organisms that eat them can use. All other organisms on land ultimately depend on plants, either eating plants or eating other organisms (such as animals) that eat plants, or eating organisms that eat organisms that eat plants, and so on.

(a)

(b)

Figure 2.13 Getting Around
Plants have evolved many ways of spreading to new areas. (*a*) Consider the magnificent fruit known as the coconut. A palm tree seed in a coconut can float for hundreds of miles until it reaches a new beach where it can take root and grow. (*b*) Seeds with wings or other structures that help them to be carried by the wind, such as winged seeds of maple trees or dandelion fluff, can travel great distances.

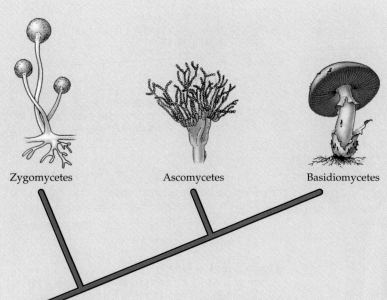

Zygomycetes   Ascomycetes   Basidiomycetes

Number of species discovered to date: ~70,000

Function within ecosystems: Decomposers and consumers

Economic uses: Mushrooms are used for food, yeasts for producing alcoholic beverages and bread. Some fungi also produce antibiotics, drugs that help fight bacterial infections.

Funky factoid: Highly sought-after mushrooms known as truffles can sell for $600 a pound.

Figure 2.14 The Fungi
Fungi are most familiar to us as mushrooms, but the main bodies of such fungi typically are hidden underground. Some fungi are decomposers, attacking dead and dying organisms. Others are parasites, living on or in other organisms and harming them, or mutualists, living with other organisms to their mutual benefit. The three major groups of fungi are shown in this evolutionary tree.

Also known as the stinkhorn mushroom, these foul-smelling basidiomycetes attract flies, which get covered with sticky spores and then scatter the spores as they fly to other locations.

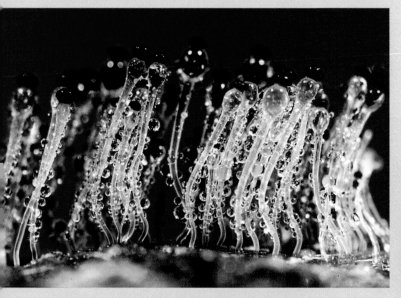

Pilobolus, a zygomycete that lives on dung, can shoot its spores out at an initial speed of 50 kilometers per hour.

This ascomycete *Penicillium* is a relative of the original species that produced the antibiotic penicillin, a drug that fights bacterial infections and has saved the lives of countless people.

Flowering plants provide humans with materials like cotton for clothing and with pharmaceuticals like morphine. Essentially all agricultural crops are flowering plants, and the entire floral industry rests on the work of angiosperms. Gymnosperms like pines, spruce, and fir are the basis of forestry industries, providing wood and paper.

As valuable as plants are when harvested, they are equally valuable when left in nature. By soaking up rainwater in their roots and other tissues, plants prevent runoff and erosion that can contaminate streams and can harm fish populations in areas, like logged hillsides, that are stripped of plants. Plants also produce the crucial gas oxygen.

> ■ The kingdom Plantae was the first kingdom to take to the land. Plants evolved diverse shapes and sizes after they evolved vascular systems. Two other key innovations for plants were seeds and flowers. Plants are essential components of land ecosystems. As producers they provide the food that all other organisms, the consumers, eventually use.

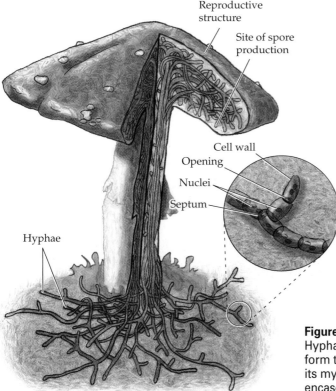

Reproductive structure

Site of spore production

Cell wall

Opening

Nuclei

Septum

Hyphae

Mycelium

Kingdom Fungi: A World of Decomposers

Most people are familiar with fungi as the mushrooms on their pizza or their lawns, but the kingdom **Fungi** includes not only mushroom-producing organisms, but also molds and yeasts. In fact, the familiar mushroom is only the very visible reproductive structure sprouting from what is usually a much larger, main body of a fungus. Most of the tissues of a fungus typically are woven through whatever tissues—whether of a dead or a live organism—it is digesting and making its meal. Because most of the tissues of a fungus are usually hidden from view, fungi are among the most enigmatic and poorly understood of the major groups of organisms.

As Figure 2.14 shows, the fungi can be divided into three distinct groups—zygomycetes, which evolved and branched off first, and the ascomycetes and basidiomycetes. Each differs in and is named for specialized reproductive structures. In the discussion that follows we examine the structure of the main body of a fungus and how that structure makes fungi good decomposers, as well as parasites (harmful organisms that live in or on another organism) and mutualists (organisms that benefit from and provide benefits for the organisms they associate with).

Fungi have evolved a structure that makes them highly efficient decomposers

The main body of a fungus is known as the **mycelium** (plural mycelia), a mat of threadlike projections known as **hyphae** (singular: hypha) that typically grow hidden, either underground or through the tissues of the organism the fungus is digesting (Figure 2.15).

Hyphae are composed of cell-like compartments that are encased in a cell wall. Unlike the cells in other multicellular organisms, however, the cells of fungi are usually only partially separated from one another. In fact, these cell-like compartments typically are separated by a partial divider known as a septum (plural septa), which allows organelles to pass from one compartment to another. Even nuclei can move around from one compartment to another.

Like animals, fungi rely on other organisms for both energy and carbon; thus they are chemoheterotrophs

Figure 2.15 Fungi Grow with Hyphae
Hyphae are threadlike growths that form dense mats. Mats of hyphae form the main feeding body of a fungus and are known collectively as its mycelium. Hyphae are composed of cell-like compartments that are encased in a cell wall. Unlike the plant cell wall, the fungal cell wall is composed of chitin, the same material the makes up the hard outer skeleton of insects.

(see Table 2.1). But unlike animals, which consume food through the mouth and digest it internally, fungi digest their food externally and then absorb the nutrients. In much the same way as our stomachs release digestive juices and proteins, the fungal hyphae growing through the tissues of a plant or animal release special digestive proteins that break down those tissues. The hyphae then absorb the released nutrients for the fungus to use.

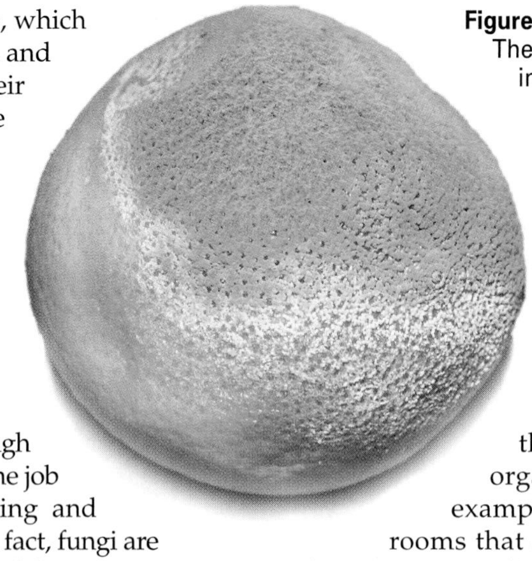

The ability of fungi to grow through things makes them well suited to the job of decomposing—that is, attacking and breaking down dead organisms. In fact, fungi are among the most important groups of decomposers, recycling a large proportion of the world's dead and dying organisms. For example, shelf fungi, organisms that earn their name from their stacked, shelflike formation, are key decomposers of dead and dying trees.

The reproduction of fungi also sets them apart from other organisms. Characterized by complex mating systems, fungi come not in male and female sexes, but in a variety of mating types. Each mating type can mate successfully with only a different mating type.

Another, more familiar aspect of fungal reproduction is **spores**, the reproductive cells of a fungus that typically are encased in a protective coating that shields the cells from drying or rotting. Known to most of us as the powdery dust on molding food (Figure 2.16), spores, like plant seeds, are scattered into the world by wind, water, and animals. Once carried to new locales, spores can begin growing as new, separate individuals.

Figure 2.16 Fungi Spread via Spores
The powdery dust on this orange is a coating of spores.

are trying to use these fungi to kill off insects that are pests of crops (Figure 2.17).

Some fungi live in beneficial associations with other species

Some fungi are **mutualists**; that is, they live in association with other organisms to their mutual benefit. For example, morels are highly prized mushrooms that are the products of fungi living in mutually beneficial associations with plant roots. Such fungi are known as mycorrhizal fungi, and they can be found in all three groups of fungi: zygomycetes, ascomycetes, and basidiomycetes. In these associations, fungi live in or on the roots of plants and form thick, mycelial mats. Fungi receive sugars and amino acids, the building blocks of proteins, from the plants. The roots of the plants, infested with the spongelike hyphae, can absorb more water and nutrients than they could without the fungi.

Morel

Another very familiar fungal association is the lichen. Often lacy, gray-green growths, lichens seen on tree trunks or rocks are actually associations of algae (photosynthetic protists, as we

The same characteristics that make fungi good decomposers make them dangerous parasites

Although most fungi are decomposers, some are dangerous **parasites**, organisms living on or in another organism at that organism's expense. Parasitic fungi grow their hyphae through the tissues of other living organisms, causing disease in animals (including humans) and plants (including crops).

For example, in humans, several different species of fungi can cause athlete's foot, and *Pneumocystis carinii* causes a deadly fungal pneumonia that is a leading killer of people suffering from AIDS. Fungi also attack plants. *Ceratocystis ulmi* causes Dutch elm disease, which has nearly eliminated the elm trees in the United States. Rusts and smuts are fungi that attack crops. Still other fungi are specialized for eating insects, and biologists

Figure 2.17 Fungal Parasites
Some fungi are parasites, making their living by attacking the tissues of other living organisms. This beetle, a weevil in Ecuador, has been killed by a *Cordyceps* fungus, the stalks of which are growing out of its back.

learned earlier) and fungi (Figure 2.18). Both asco-mycetes and basidiomycetes are known to form lichens. The algae and fungi in lichens grow with their tissues intimately entwined in an association in which the fungi receive sugars and other carbon compounds from the algae. In turn, fungi in lichens produce what are known as lichen acids, a mixture of chemicals that scientists believe may function to protect both the fungi and the algae from being eaten by predators.

Fungi act as decomposers in the biosphere, producing important and treasured products

Fungi are crucial components of terrestrial habitats as decomposers of dead and dying organisms. They are also critical to plants; more than 95 percent of ferns and their close relatives, gymnosperms, and angiosperms have mycorrhizal fungi living in association with their roots.

Fungi can be costly to human society. They cause deadly diseases, contaminate crops, rot food, and force us to clean our bathrooms more often than we might like. Fungi also add to the quality of human life. They provide pharmaceuticals, including antibiotics like penicillin. Yeasts (single-celled fungi) can feed on sugars and produce two important products: alcohol and the gas carbon dioxide. What would life be without yeasts like *Saccharomyces cerevisiae*, which produces the carbon dioxide that makes bread rise and the alcohol in beer? Fungi also provide highly sought-after delicacies like the

Figure 2.18 Lichens
A lichen consists of an alga and a fungus living intimately entwined in a mutually beneficial association. These lichens, known as British soldiers, are shown growing on an old log.

mushroom known as the truffle, whose underground growing locations can be found only by specially trained dogs and pigs.

> ■ The main body of a fungus is a mycelial mat composed of hyphae. Hyphae digest food externally and then absorb it. Most fungi are decomposers, though some are parasites and others are mutualists. Fungi are critical components of many ecosystems as decomposers and as mycorrhizal mutualists. Fungi can attack humans, crops, and stored food, among other things. Some fungi are useful to humanity for the pharmaceuticals and foods they provide.

Kingdom Animalia: Complex, Diverse, and Mobile

The animal kingdom, to which we humans belong, is certainly the most familiar of all the kingdoms. All the lineages that make up the kingdom **Animalia** are made up of multicellular creatures, many of them quite complex. The animal kingdom includes flashy creatures like Bengal tigers and your girlfriend or boyfriend. It also includes worms, sea stars, snails, insects, and other creatures that are less obviously animal-like, such as sponges and corals. Figure 2.19 illustrates the major groups of animals.

Figure 2.19 shows sponges, the most ancient of animal lineages branching off first. Next cnidarians (like jellyfish and corals) evolved, and then flatworms. The next group to evolve was the protostomes, which includes molluscs (like snails and clams), annelids (the segmented worms), and arthropods (like crustaceans and insects). The three protostome groups are shown branching off together as a single lineage, indicating that they are a group of close relatives all descended from an ancestor that branched off from the tree after flatworms but before echinoderms. Within the protostomes, however, it is unclear which of the three groups branched off from the others first, so they are shown branching off from one another simultaneously. Next to evolve were the echinoderms (sea stars and the like) and the vertebrates (animals with backbones, like fish, birds, and humans), both deuterostomes.

Like all fungi and some bacteria and protists, animals make their living off the tissues of other organisms, getting carbon and energy from organisms they eat; that is, animals are chemoheterotrophs (see Table 2.1). Animals differ from fungi and plants in that their cells do not have cell walls surrounding their plasma membranes.

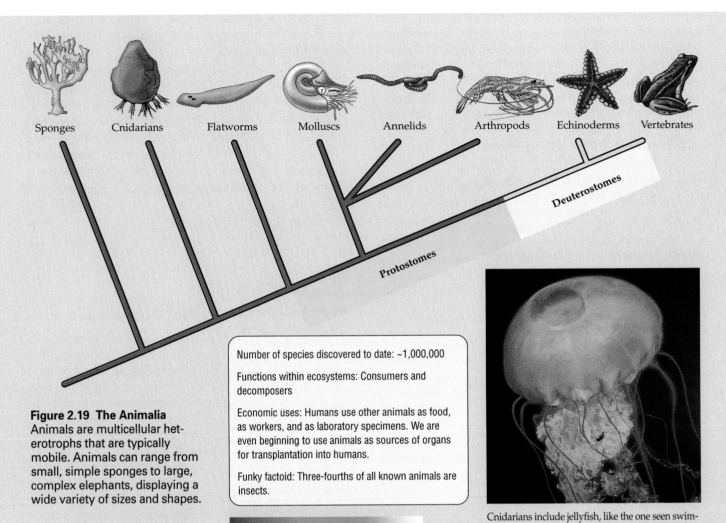

Sponges   Cnidarians   Flatworms   Molluscs   Annelids   Arthropods   Echinoderms   Vertebrates

Deuterostomes

Protostomes

Figure 2.19 The Animalia
Animals are multicellular heterotrophs that are typically
mobile. Animals can range from
small, simple sponges to large,
complex elephants, displaying a
wide variety of sizes and shapes.

Number of species discovered to date: ~1,000,000

Functions within ecosystems: Consumers and
decomposers

Economic uses: Humans use other animals as food,
as workers, and as laboratory specimens. We are
even beginning to use animals as sources of organs
for transplantation into humans.

Funky factoid: Three-fourths of all known animals are
insects.

Cnidarians include jellyfish, like the one seen swimming here, as well as anemones and corals. Members of this group, the first organisms to evolve true
tissues, are named for their stinging cells, which
they use for protection and to disable prey.

Sponges are ancient aquatic animals. They
have evolved some specialized cells but no
true tissues.

Molluscs include snails, slugs, and octopi, as
well as this giant clam from a tropical reef. As
is typical of many molluscs, this clam's tender flesh is protected by a hard, outer shell.

Flatworms, like this oceangoing flatworm from the United
States west coast, were among the earliest animals to evolve
true organs and organ systems.

A key feature of annelids, also known as segmented worms, is segmentation. This body plan of repeating units can be seen as the series of distinct segments in this fire worm. The segmented body plan, which is also seen in arthropods and vertebrates, facilitated the evolution of many different body forms.

Echinoderms include sea stars, like the one from Indonesia shown here, and sea urchins. They are closely related to the vertebrates.

Arthropods include crustaceans, like lobsters and crabs, as well as millipedes, spiders, and the most species-rich of all groups, the insects. This *Morpho* butterfly, an inhabitant of the tropical rainforest, is one of the most spectacularly beautiful insects on Earth.

Amphibians, slimy creatures that include frogs, like these poison arrow frogs from Costa Rica, and salamanders, typically spend part of their lives in water and part on land. Amphibians were the first animals to live on dry land.

Vertebrates are the animals that have backbones, including fish, reptiles (like snakes and lizards), amphibians (like salamanders and frogs), birds, and mammals (a group that includes primates like monkeys and humans). Shown here is a coral reef fish from Thailand. Fish were the earliest vertebrate animals.

Primates include monkeys, apes, and humans. In this group we find our closest relative, the chimpanzee, shown here, and the gorilla.

Mammals are characterized by milk-producing mammary glands in females, as well as young that are born live (rather than being born in an egg that later hatches open). These kangaroos are mammals, as are bears, dogs, lions, and humans.

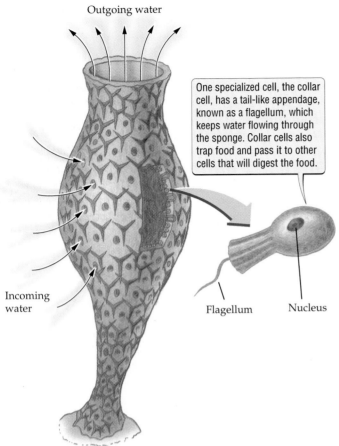

Outgoing water

One specialized cell, the collar cell, has a tail-like appendage, known as a flagellum, which keeps water flowing through the sponge. Collar cells also trap food and pass it to other cells that will digest the food.

Incoming water

Flagellum   Nucleus

Figure 2.20 Sponges Have Specialized Cells but Lack Tissues
Unlike most animals, sponges are loose associations of cells. Although some of these cells are specialized, none are organized as tissues.

Typically mobile and often in search of either food or mates, animals have evolved a huge diversity of ways of getting all they need. As we'll see, through some 500 million years of life on Earth, different lineages of animals have evolved bodies that have collections of specialized and coordinated cells, known as tissues, and collections of specialized tissues organized into organs and organ systems. Like plants, which have specialized tissues like vascular tissues, animals have evolved features that have helped them successfully live in a variety of environments. Animals have also evolved a diversity of sizes and shapes, many of them variations on a few themes. In addition, animals display an astounding array of behaviors that help them survive and reproduce.

Animals evolved tissues

Sponges are among the simplest of animals. They represent a time in the evolution of animals before special-

ized and coordinated collections of cells known as tissues had evolved. Sponges are little more than loose collections of cells. In fact, a sponge can be put through a sieve, completely breaking apart its cells, and it will slowly reassemble itself (Figure 2.20). While widespread and highly successfully, these animals live off the amoebas and other tiny organisms they filter out of the water, filtering a ton of water just to grow an ounce.

One of the earliest animal groups to evolve true tissues was the group that includes jellyfish, corals, and anemones. The name of the group, Cnidaria, comes from the Greek word for "nettle," a stinging plant. Cnidarians are so named because the group is characterized by stinging cells that are used for protection and hunting. Like other cnidarians, jellyfish exhibit specialized nervous tissues (Figure 2.21), musclelike tissues, and digestive tissues that allow them to do things that require coordination of many cells, such as gracefully and rapidly swimming away from predators.

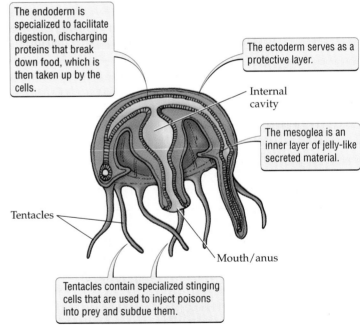

The endoderm is specialized to facilitate digestion, discharging proteins that break down food, which is then taken up by the cells.

The ectoderm serves as a protective layer.

Internal cavity

The mesoglea is an inner layer of jelly-like secreted material.

Tentacles

Mouth/anus

Tentacles contain specialized stinging cells that are used to inject poisons into prey and subdue them.

Figure 2.21 The Tissue Layers of a Jellyfish
Cnidarians were one of the earliest groups to evolve specialized tissues. These animals have an outer layer of tissue known as the ectoderm (*ecto*, "outer"; *derm*, "skin") and an inner layer known as the endoderm (*endo*, "inner"). Sandwiched between the ectoderm and the endoderm is an inner layer of secreted material known as the mesoglea (*meso*, "middle"; *glea*, "jelly"). The endoderm and ectoderm are also organized to contract like muscle tissue. Tentacles bring food into the internal cavity through a single opening, which serves as a mouth and an anus.

Animals evolved organs and organ systems

After tissues the next level of complexity to evolve was organs and organ systems. Organs are body parts composed of different tissues organized to carry out specialized functions. Usually organs have a defined boundary and a characteristic size and shape. An example is the human stomach.

An organ system is a collection of organs functioning together for a specialized task. For example, the human digestive system is an organ system that includes the stomach, as well as other digestive organs, like the pancreas, liver, and intestines. Flatworms, a group of fairly simple wormlike animals, were one of the earliest groups of animals to evolve true organs and organ systems (Figure 2.22).

Animals evolved complete body cavities

Still later in the history of this kingdom, animals evolved the complete body cavity—an interior space with a mouth at one end and an anal opening at the other. The two lineages that exhibit such cavities are known as the protostomes and the deuterostomes (see Figure 2.19). Protostomes include such animals as insects, worms, and snails. Deuterostomes include such animals as sea stars and all the animals with backbones (vertebrates), such as humans, fish, and birds.

Protostomes and deuterostomes are named for the patterns in which they develop from a fertilized egg into an embryo and then into a juvenile animal. Specifically they are named for which of the openings becomes the mouth. In **protostomes** (*proto*, "first"; *stome*, "opening"), the first opening to form in the embryo becomes the mouth, and the anus forms elsewhere later. In **deuterostomes** (*deutero*, "second"), the second opening becomes the mouth and the first is the anus.

This developmental difference has led to very different organizations of tissues in these two groups. Among these two groups, there has been a great flowering of animal sizes and shapes. Interestingly, many of these can be seen as variations on a few basic themes, as described in the next section.

Animal body forms exhibit a variation on themes

Animals exhibit a great variety of shapes and sizes, many of which are variations on a few basic body plans.

Arthropods (*arthro*, "jointed"; *pod*, "foot") are a group that includes millipedes, crustaceans (like lobsters and crabs), insects, and spiders. With their hard outer skeleton (which is made of chitin, the same material that is found in the cell walls of fungi), arthropods are a wonderful illustration of how evolution can take a body plan and run with it (Figure 2.23). Among the arthropods are **insects**, six-legged organisms, such as grasshoppers, beetles, and ants, that live on land. Whereas bacteria dominate Earth in terms of sheer numbers of individuals, insects dominate in terms of the number of species, with many more species than any other group of organisms.

One of the features that has facilitated the evolution of arthropod bodies is the fact that they are segmented. That is, they are divided into segments that have evolved differently over time, resulting in a huge number of different types of animals. Such segmentation can also be seen in the segmented worms, or annelids, such as earthworms whose bodies are made up of a repeated series of segments (see Figure 2.19). **Vertebrates**—animals that have a backbone, such as fish, amphibians (like frogs and salamanders), reptiles (like snakes and lizards), birds, and mammals (like humans and kangaroos)—are also built on a segmented body plan (Figure 2.24).

Vertebrates illustrate other ways in which a variety of very different forms can evolve from one basic plan. The same appendage, the front appendage of vertebrates, has evolved in humans as an arm, in birds as a wing, in whales as a flipper, in snakes as an almost nonexistent nub, and in salamanders as a front leg.

Animals exhibit an astounding variety of behaviors

Another fascinating characteristic of animals is their ability to move and to behave. Animals have evolved many behaviors to capture prey, eat prey, avoid being captured, attract mates, and migrate. Animals are quite useful to immobile organisms, like plants, which have evolved ways to get animals to carry their pollen and seeds. Humans find the mobility of other animals convenient. We travel by horse and camel, and we send messages via carrier pigeon. We will discuss animal behavior in detail in Chapter 35.

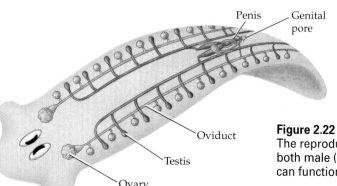

Penis — Genital pore

Oviduct

Testis

Ovary

Figure 2.22 Animals Have Evolved Organs and Organ Systems
The reproductive system, one of the flatworm's organ systems, contains both male (penis) and female (genital pore) structures. Every flatworm can function as both a male and a female.

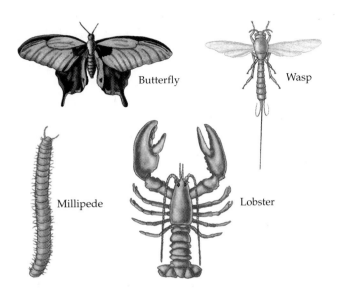

Figure 2.23 Variations on Themes
From simple body plans, arthropods and vertebrates have evolved a huge diversity of body forms and sizes. In arthropods, the widely diverse forms are all variations on a single theme. The millipede can be viewed as the simplest form of these segmented animals, with all segments similar. As segments have evolved and diversified, a variety of organisms have arisen: from lobster to swallow-tailed butterfly, to parasitoid wasp. Looking just at the evolution of the last abdominal segments (the rear ends of these creatures), one can see that the changes in this one set of segments have resulted in a huge variety of shapes and lifestyles. The last abdominal segments have evolved into the delicate abdomen of a butterfly, the abdomen of a wasp, which has a huge structure for laying eggs deep into another animal, and the delicious tail of the lobster.

Figure 2.24 Many Animals Are Segmented
Segmentation, a body plan in which segments repeat and often can evolve independently of one another, is shown here in an annelid (*a*) and a vertebrate (*b*).

(a)

Animals play key roles in ecosystems and provide products for humans

Because they are chemoheterotrophs and because most are mobile, animals play many roles in ecosystems. Most serve as consumers, preying on species of plants and animals. Some animals, like carrion beetles, serve as decomposers of dead animals. Animals also help spread plant seeds and fungal spores. And they can be pests, carrying disease; for example, ticks spread the protist parasite that causes Lyme disease. Animals can also be crop pests, especially insects like the tomato hornworm, a caterpillar that attacks tomatoes. Domesticated animals, like cows, provide food and material for clothing, like leather.

Carrion beetle

Human beings make up the species in the animal kingdom that has the greatest impact on life on Earth. As members of a species with rapidly growing populations and the ability to drastically and rapidly modify Earth with cities, agriculture, and industries, we must take care not to make the planet uninhabitable for ourselves and other species.

■ Animals are multicellular chemoheterotrophs that are typically mobile. Animals include a wide variety of organisms from simple sponges to complex gorillas. Animals have evolved specialized tissues, organs and organ systems, body cavities, and a wide variety of shapes and sizes. Animals also exhibit an astounding variety of behaviors. Insects are the most species-rich group of organisms on Earth. Animals are consumers, but they also act as decomposers. Animals can be pests of crops (like insects) and can supply food and clothing to human society, which is itself composed of animals.

(b)

The Difficulty of Viruses

A virus

Viruses are not classified into any kingdom. Viruses are simply protein wrapped around a fragment of DNA or RNA. They occupy a gray zone between living and nonliving and may have arisen from the genomes (the complete makeup of genetic material) of many different organisms.

As discussed in Chapter 1, viruses neither grow nor reproduce outside of the hosts they infect, and they do not exhibit many of the other characteristics of living organisms. For example, viruses do not gather energy; they neither photosynthesize nor eat other organisms. Difficult to classify and lacking clear evolutionary relationship to any one group, viruses are not easily placed into any of the existing kingdoms.

> ■ Viruses are not classified into any kingdom since they lack clear evolutionary relationships and since it is difficult to define whether they are living or not.

Highlight

Where Kingdoms Meet

The diversity of life can be divided into kingdoms ranging from the single-celled Monera to the creatures of the animal kingdom. But although these divisions are useful for organizing and understanding this explosion of life, organisms in nature often ignore these kingdom boundaries to cohabitate, living together in an association known as symbiosis.

For example, lichens are complex mixtures of organisms from two entirely separate kingdoms (Fungi and Protista) living intimately entwined (see Figure 2.18). Other examples of ways in which kingdoms meet include photosynthetic corals. Corals are animals that have incorporated photosynthesizing algae into their tissues. While most animals have to hunt down their food or seek out plants to eat, corals have set up shop with photosynthesizers living in their bodies. These symbiotic algae can give color, as well as providing the key products of photosynthesis, like sugars, to corals. When corals are stressed—for example, experiencing unusually high water temperatures—these symbionts can abandon their hosts, taking the coloration of the corals with them, leading to an ever more common phenomenon known as coral bleaching.

Even humans can be considered a cross-kingdom grab bag. You're really never alone if you consider that your skin is host to millions of bacteria. Hundreds of microscopic arthropods live on your body. Even your gut houses an amazing diversity of microscopic life, and your body is continually being invaded by viruses. You are a densely populated community even when no one else seems to be around.

> ■ Some apparent individuals are composed of organisms from two entirely separate kingdoms. Examples include lichens and corals.

Summary

Organizing the Diversity of Life: The Linnaean Classification System

- The Linnaean hierarchy is a classification system for organizing the diversity of life into species, genera, families, orders, classes, phyla, and kingdoms.

- There are a number of different systems for dividing the diversity of life into kingdoms. The differences among systems exist, in part, because of the ever-increasing knowledge about and differing interpretations of how groups evolved and are related.

- This text adopts the widely used five-kingdom system, though other systems exist.

Kingdom Monera: Tiny, Successful, and Abundant

- The Monera, also known as bacteria, are microscopic, simple, single-celled organisms and the most ancient kingdom of life.

- Bacteria are the most abundant and widespread form of life on Earth.

- Some bacteria thrive in extreme (for example, very hot or very salty) environments.

- Bacteria exhibit unmatched diversity in their methods of getting and using nutrients.

- Bacteria perform key roles in ecosystems, including providing nitrate to plants and decomposing dead organisms. They are useful to humanity in many ways (for example, in cleaning oil spills), but they also cause deadly diseases.

The Evolution of More Complex Cells: Eukaryotes Arose from Prokaryotes

- In the past, some bacterial cells engulfed others. These engulfed cells eventually evolved into organelles.

- Bacteria contain no organelles and are known as prokaryotes. All other organisms have cells that contain organelles and are known as eukaryotes.

Kingdom Protista: A Window into the Early Evolution of Eukaryotes

- The Protista are highly diverse, in part because they are a collection of separate evolutionary lineages. The kingdom includes plantlike, animal-like, and funguslike organisms.

■ Protists represent early stages in the evolution of the eukaryotic cell.

■ Protists provide examples of the early evolution of multicellularity.

■ The sex that is characteristic of eukaryotes evolved in protists.

■ Although protists include many harmless organisms, like kelp, they also include many disease-causing organisms, like *Plasmodium*, which causes malaria.

Kingdom Plantae: Pioneers of Life on Land

■ The Plantae are multicellular, photosynthesizing organisms.

■ Important evolutionary innovations permitted plants to colonize land successfully, including vascular systems, seeds, and flowers.

■ As producers and the ultimate food source for all other organisms, plants are critical components of land-based food webs.

■ Plants produce food, pharmaceuticals, shelter, oxygen, and many other important products.

Kingdom Fungi: A World of Decomposers

■ Fungi are heterotrophic organisms that grow using hyphae to penetrate, digest, and absorb food.

■ Fungi can be decomposers, parasites, or mutualists.

■ Fungi cause difficult diseases, but they also produce valuable products such as foods and pharmaceuticals.

Kingdom Animalia: Complex, Diverse, and Mobile

■ The Animalia are multicellular heterotrophs that are typically mobile.

■ Important evolutionary innovations in animals were tissues, organs and organ systems, body cavities, and behaviors.

■ Animals exhibit a great variety of forms and sizes, many of them variations on a single theme.

■ Animals play a variety of roles in ecosystems, including consumer and decomposer. Some spread disease; others are pests of crops. Animals also provide food, clothing, and other products to human society.

The Difficulty of Viruses

■ Viruses are hard to categorize and are not placed in any kingdom since they lack clear evolutionary relationships and are difficult to define as living or nonliving.

Highlight: Where Kingdoms Meet

■ Sometimes what appears to be a single organism, like a lichen or coral, is actually a combination of organisms from two separate kingdoms.

Key Terms

angiosperm p. 30	lineage p. 20
Animalia p. 35	Monera p. 20
arthropod p. 39	mutualist p. 34
chemoautotroph p. 23	mycelium p. 33
chemoheterotroph p. 23	parasite p. 34
cuticle p. 28	photoautotroph p. 23
deuterostome p. 39	photoheterotroph p. 23
eukaryote p. 25	Plantae p. 27
evolutionary innovation p. 18	prokaryote p. 25
evolutionary tree p. 19	Protista p. 25
flower p. 30	protostome p. 39
Fungi p. 33	root system p. 28
gymnosperm p. 30	seed p. 30
hypha p. 33	spores p. 34
insect p. 39	vascular system p. 28
kingdom p. 19	vertebrate p. 39

Chapter Review

Self-Quiz

1. Which of the following is *not* a level in the Linnaean hierarchy?
   a. class
   b. group
   c. species
   d. phylum

2. Which of the following is the nutritional mode of animals?
   a. chemoheterotroph
   b. photoheterotroph
   c. photoautotroph
   d. chemoautotroph

3. Which kingdom is the most abundant in terms of numbers of individuals?
   a. Animalia
   b. Plantae
   c. Protista
   d. Monera

4. Eukaryotes differ from prokaryotes in which of the following ways?
   a. They do not have organelles in their cells and prokaryotes do.
   b. They exhibit a much greater diversity of nutritional modes than prokaryotes.
   c. They have organelles in their cells and prokaryotes do not.
   d. They are more widespread than prokaryotes.

5. Which of the following kingdoms contains organisms that represent early stages in the evolution of the eukaryotic cell?
   a. Monera
   b. Protista
   c. Fungi
   d. Animalia

6. Which of the following kingdoms was the first to succeed on land?
   a. Plantae
   b. Animalia
   c. Monera
   d. Archaebacteria

7. Fungi grow using
   a. hyphae.
   b. chloroplasts.
   c. angiosperms.
   d. prokaryotes.

Review Questions

1. Name three factors that likely contributed to the success of the Monera.

2. Why are *Giardia* and slime molds of particular interest to biologists interested in the early evolution of eukaryotes?

3. Describe the evolution of specialized cells, tissues, and organs in the kingdom Animalia, including the name of the animal group that first showed each evolutionary innovation.

4. To what kingdom do viruses belong and why?

2

The Daily Globe

New Method for Treating Malaria?

GLASGOW, MONTANA. Researchers reported today that they may have discovered the key to fighting a huge group of parasites, including the parasite that causes the often-deadly disease malaria.

Deep inside the cells of these parasites, which are quite animal-like, researchers have discovered something that does not seem very animal-like at all. Biologists say that inside these parasites' cells are organelles known as plastids, ancient chloroplasts similar to those found in plants.

Researchers say the finding raises the possibility that malaria and other diseases caused by this group could be treated with herbicides, chemicals normally used to kill plants.

"It was like a bolt from the blue," said Dr. Bob D. Wartheim, a biologist at Central Montana University, who was one of the authors of the new paper and who described the finding as a complete shock. Researchers say they still don't know what the function of the plastid is. So far it appears to be non-functional and is clearly unable to carry out photosynthesis. "We were about to give up studying these organelles because we had been unable to get funding from any source for the past five years. Everyone rejected the research as uninteresting and without important applications or ramifications. You just never know what you're going to find."

Evaluating "The News"

1. Why would herbicides be a potentially powerful drug for treating malaria in humans?

2. Some elected officials have complained bitterly about the use of taxes for basic research—for example, the study of odd little organelles inside of little-understood organisms—proclaiming them a clear waste of hard-earned dollars. Should taxpayers be funding basic research or should all research be aimed at solving a particular societal woe, like human disease or environmental problems?

3. Scientists point to cases like this one as proof that basic research is valuable to society. But the vast majority of basic scientific research does not result in such applied findings. Is it still useful to fund basic research? Why or why not?

3

Relationships among the Major Groups of Organisms

Alexis Rockman, *The Bounty*, 1991.

The Iceman Cometh

On a September day in 1991, hikers high in the Alps near the border between Austria and Italy came across a remarkable sight: the body of an ancient, mummified man in the melting glacial ice. Instantly dubbed the Iceman, the well-preserved body appeared to be a pre-

historic hiker dating to the Stone Age, possibly a shepherd or traveler, who died on the mountainside, his body captured by the ice, which preserved it for some 5,000 years.

Extremely well preserved—right down to his underwear—this one-and-a-half-meter-tall visitor from another time promised a tan-

All living things are related in a tree of life that scientists are continuing to decipher and complete.

talizing peek into the past. Sporting tattoos on his body, he wore a loincloth, a leather belt and pouch, leggings and a jacket made of animal skin, a cape of woven grasses, and calfskin shoes lined with grass. He carried a bow and arrows, an axe, knives, and two pieces of birch fungus, possibly a kind of prehistoric penicillin.

Perfectly suited for scientific examination, this first-ever corpse from the Stone Age found right in the path of hikers seemed too good to be true. Researchers began to fear that the Iceman might be an elaborate hoax. Speculation was rising that the body was a transplanted Egyptian or pre-Columbian American mummy. How could biologists determine the true identity of this long-lost wanderer?

Studying the DNA still preserved in his tissues, researchers were able to show that the Iceman was not a transplant. In fact, he turned out to be a close relative of contemporary northern European peoples from the same area, much more closely related to them than to any other group. The authenticity of the Iceman was established by comparison of his DNA to that of other groups of people. Thus he was placed in the human family tree as a northern Europeaner.

In the same way, every organism can be classified by a determination of its closest relatives. Biologists can ask of any newly discovered organism, Is it most closely related to plants, animals, fungi, or bacteria? Is a particular plant a close relative of a cycad or a

The Iceman
The prehistoric mummy known as the Iceman as he was discovered in melting ice by hikers in the Alps in 1991, some 5,000 years after he died.

bluebell? Is a particular fungus more closely related to morels or to brewer's yeast? Whether scientists are collecting new species from the rainforest or sleuthing out a disease bacterium, the first question always is, Where does this organism fit in the family tree of life?

1. Systematics is the science of naming and classifying groups of organisms, and determining the relationships among them.

2. Because living things have evolved from a common ancestor, their relationships can be depicted on a kind of family tree called an evolutionary tree.

3. Cladistics is now widely accepted as the method of choice for building evolutionary trees. In cladistic analyses, scientists use shared, novel features of organisms to determine evolutionary relationships and assemble evolutionary trees.

4. Biologists use all sorts of characteristics of organisms, including the shape of structures on the body, behaviors, and DNA, to help them decipher the evolutionary relationships among species.

5. Once researchers know the closest relatives of an organism and where that organism belongs on an evolutionary tree, they can begin to make predictions about the biology of the organism, with expectations that much of its biology will be similar to that of its close relatives.

Genealogists show how different people are related by using a family tree that illustrates which people are most closely related to which others and how they are related (see, for example, Figure 3.1a). Biologists called **systematists** also study relationships, but with groups of different organisms rather than people, figuring out which groups are most closely related to which others. Using this information, they create what are known as **evolutionary trees** (Figure 3.1b). Systematists also give groups of organisms their scientific names and organize the groups into classification schemes—for example, the kingdoms, families, genera, and species described in Chapter 2.

In this chapter we examine how researchers build evolutionary trees, looking at what kinds of features of organisms are useful and what kinds are not useful for determining relatedness. We also look at some of the more interesting and important branches on the evolutionary tree of life that scientists are working to decipher.

Let's begin by imagining a systematist with a diverse array of plants or animals. Her goal is to figure out their family tree. Where should she begin?

Looking for Clues to Evolutionary Relationships

As anyone who has been to a family reunion knows, the more closely related two people are the more similar they tend to be to one another, in the way they look and sometimes even in the way they act—showing the same smile or sneezing in just the same way. Close relatives even tend to be more similar in how their bodies work, often exhibiting the same physical strengths or being prone to the same kinds of illnesses. In fact, the similarities extend right down to the level of a person's genetic material or DNA, and for good reason.

Recall that DNA is the molecule passed from one generation to the next, the genetic blueprint for the development of an individual passed down from its parents. We inherited these blueprints for body structures and behaviors from our parents, who inherited theirs from their parents, and so on. As a result, we exhibit many of the characteristics—a particular smile or sneeze or tendency to hay fever, for example—that our relatives exhibit. In the same way, closely related groups of organisms (those that arose from the same ancestor) tend to resemble each other.

Although we don't discuss evolution and the origin of species in detail until Unit 4, for now suffice it to say that over time one species can evolve, splitting into two different species. That is, an ancestor species gives rise to new species that are its descendants. On evolutionary trees, these descendants are depicted as the tips of branches that trace back and meet at a point where they shared their last common ancestor (see Figure 3.1b).

Descendants often share key features because they share a common ancestor. In human families, for example, a father's distinctive nose may be seen on the faces of all his children. Similarly, all the vertebrate animals—including humans, birds, snakes, and fish—can be placed together on an evolutionary tree because they all have the backbone that was a feature of their **most recent common ancestor**, the ancestral animal from which all these descendant animals sprang.

Systematists can find such key, shared features in various aspects of an organism's biology. Traditionally systematists have compared species by looking at inherited structural characteristics of the body: numbers of legs,

(a) Family tree of Britain's royalty

(b) Swallowtail butterfly evolutionary tree

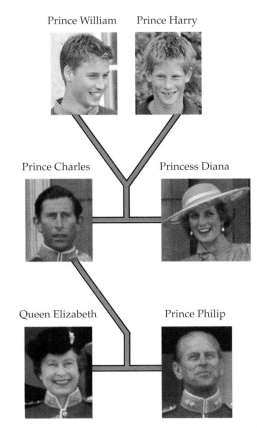

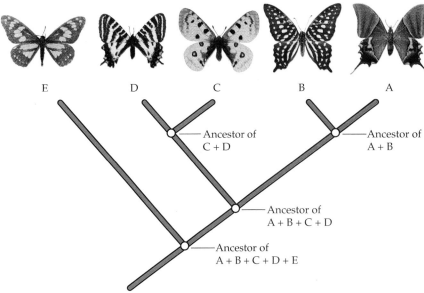

Figure 3.1 Family Trees versus Evolutionary Trees
(*a*) Like other family trees, this tree depicts relationships between ancestors and descendants, in this case some of the familiar members of Britain's royal family, including Queen Elizabeth and Prince Philip, their son Charles, Princess Diana, and Charles and Diana's two sons. (*b*) An evolutionary tree is read quite differently but also depicts relationships, in this case the evolutionary relationships of swallowtail butterflies. Each group is depicted by a photograph at the tip of a branch of the tree. Trace any branch back to where it meets with another group's branch. That meeting point represents the common ancestor from which those groups descended. When we trace backward, we are really tracing back through time, looking for the point at which that lineage hooks up with another lineage. The farther down the tree we go, the farther into history we are delving. The arrangement of groups in this tree indicates that butterfly groups A and B are each other's closest relatives. These two butterfly groups descended from a common ancestor depicted on the tree at the point at which their branches meet. Thus from the common ancestor, groups A and B branched off and evolved to be what the photographs show. Groups A, B, C, and D all descended from the common ancestor depicted at the next branching point down the tree and so on.

structure of a flower, the anatomy of an animal's heart, and so on. In recent years, however, more researchers have begun searching for similarities in other features, including the behaviors of organisms. The most powerful new tool available to systematists is an organism's genetic material, or DNA. Recent advances in techniques for studying DNA have revolutionized all of biology, including the study of evolutionary relationships.

All living organisms use DNA as their hereditary material. As a result, systematists have been able to study the relationships of many different groups that they were unable to study before, by comparing their DNA. For example, by looking at DNA researchers have been able to make stunning progress in understanding the relationships among major groups such as bacteria, plants, and animals—groups whose body parts are so different that they are very difficult to compare to one another. In addition, researchers are learning much more about

Parasitic worm

groups, such as parasitic worms, that were difficult to study previously because they are so simple structurally that they do not exhibit many distinctive features that can be used for comparison.

■ Closely related groups of organisms that descended from a common ancestor tend to share features that they inherited from that common ancestor. In the process of trying to determine which groups of organisms are most closely related, systematists examine all aspects of an organism's biology—its body structure, its behavior, its DNA—to look for such inherited similarities.

Just looking for similarities in different features, however, does not provide the whole answer. What researchers often find is that while plant species A shares certain features with plant species B, it also shares certain other features with plant species C. Which then is A's closest relative? Should A and B sit closest together on the evolutionary tree, or should A and C (Figure 3.2). Because there is only one true evolutionary history, there is only one correct evolutionary tree for any given group of organisms and, for that matter, for all the living things on Earth. Which shared features are most likely to show us the true tree of life?

Assembling Evolutionary Trees

Over the years there have been several different schools of thought on which features of organisms best reveal their evolutionary relationships. In this section we discuss two of them: evolutionary taxonomy (the oldest school of thought on this issue) and cladistics, the school of thought that has come to prominence today.

Evolutionary taxonomy is the oldest school of systematics

In the most traditional school of systematics, known as **evolutionary taxonomy**, scientists used their expert knowledge of a group of organisms to decide whether groups were closely or distantly related. Without using a codified formula or rationale, these scientists determined the relatedness of two groups by attempting to assess their general, overall similarity.

In addition, these researchers sometimes singled out particular characteristics of groups as the most "important" and gave those characteristics greater weight. An evolutionary taxonomist might decide that two species were each other's closest relatives because they shared one special feature that the scientist considered either extreme-

Figure 3.2 Showing Relationships on an Evolutionary Tree
(*a*) In this tree, A and B are shown as more closely related, sharing their most recent common ancestor at the point depicted. The two also share a common ancestor with C, but farther down the tree, meaning deeper in the evolutionary past. Note that switching the positions of A and B would not change how the tree is read. (*b*) This tree indicates that species A and C are more closely related to one another than either one is to B, because they share a common ancestor more recently than either one does with B. Again, switching the positions of A and C here does not change how the tree is read.

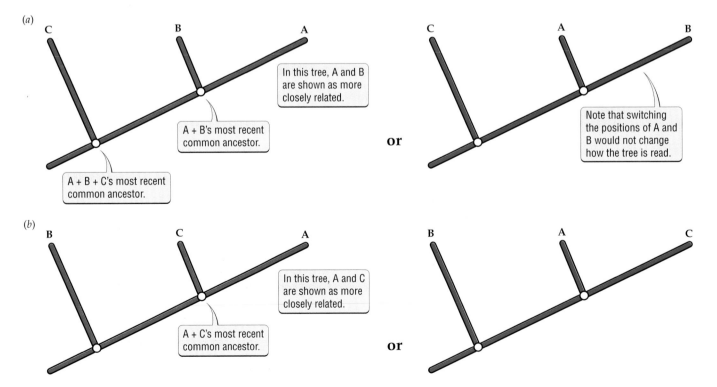

ly important to the biology of the group or so unusual that it was a very reliable indicator of their relatedness. For example, two species might share a certain feather structure, or an unusual arrangement of internal organs.

Choosing the most important features without explicit criteria quickly becomes a subjective endeavor, leading to controversies that are difficult to resolve. More importantly, such decisions, along with the use of overall similarity, do not necessarily lead to the most accurate evolutionary tree. Since the 1950s, evolutionary taxonomy has slowly been replaced by a more explicit, rigorous method, known as cladistics.

Cladistics uses shared, derived features

In **cladistics**, systematists rely on one very particular class of features: those novel, distinctive features that evolved in an ancestor (that is, unique features that evolved in the ancestor organism for the first time) that were then were passed down to all of its descendant species. The presence of such novel traits clearly identifies an ancestor organism's descendants as a closely related group. Systematists recognize that groups of organisms that share many such novel, distinctive features are more closely related to one another than they are to groups that were not descended from that common ancestor. The less closely related groups, not descended from the same common ancestor, do not display the novel traits in those features. Instead they display the original, ancestral traits and other novel features that are unique to themselves and their close relatives.

Figure 3.3 depicts the relationships of some familiar animals, identifying some of the features that define these groups. The evolutionary tree shows some of the shared, distinctive features at each level of grouping. Starting at the upper right-hand corner of the tree, the first shared trait is opposable thumbs (thumbs that are capable of being placed against one or more of the other fingers on the hand), one of the many features that set chimpanzees and humans apart from the other animals on the tree. No other organisms on the tree share this feature. Most likely we share it with chimpanzees because our common ancestor had an opposable thumb,

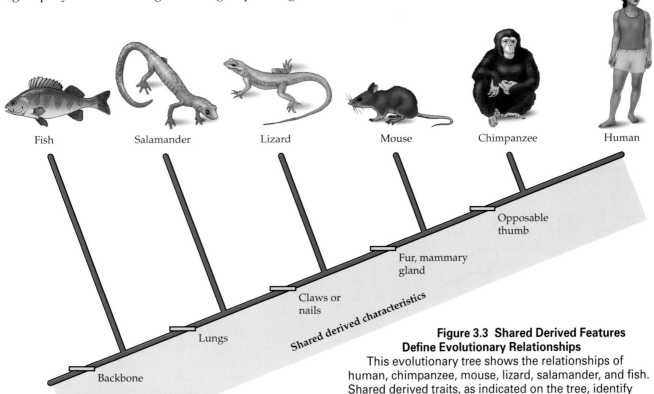

Figure 3.3 Shared Derived Features Define Evolutionary Relationships
This evolutionary tree shows the relationships of human, chimpanzee, mouse, lizard, salamander, and fish. Shared derived traits, as indicated on the tree, identify groupings of animals. For example, chimpanzees and humans share an opposable thumb. Chimpanzees, humans, and mice share fur and mammary glands. Chimpanzees, humans, mice, and lizards share claws or nails. Chimpanzees, humans, mice, lizards, and salamanders all share lungs. All animals depicted share a backbone.

making it one of the pieces of evidence that humans and chimpanzees share a recent common ancestry when compared with the other animals on the tree.

Working down the tree, the next-closest relatives of chimpanzees and humans are mice. Distinctive features shared by chimpanzees, humans, and mice are their hair and the mammary glands that females use to produce milk for their young. These are features shared by the most recent common ancestor of these three groups, the original mammal. Other features define larger groups as we move down the tree. At each point where lineages trace back to a common ancestor, a shared feature derived from that ancestor is indicated. These features—unique to that common ancestor and then passed down to all the descendants, clearly defining the descendants as a group—are called **shared, derived features**.

Shared, ancestral features are not useful

In addition to sharing derived traits, the organisms of any particular group share many traits that are *not* novel traits derived from their most recent common ancestor—**shared, ancestral features**. Shared ancestral features are not useful for identifying close evolutionary relationships. For example, if a systematist were comparing mice, chimpanzees, and humans, she might consider the fact that mice and chimpanzees both lack the ability to use written language. Without distinguishing between ancestral and derived traits, she could easily decide that mice and chimpanzees are more closely related to each other than either one is to humans and that they more recently shared a common ancestor—simply because neither group can write. An evolutionary tree based on this information would be wrong.

The reason is that the lack of written language is not a novel trait that evolved in the most recent, common ancestor of mice and chimpanzees, defining the two as a closely related group. Instead, the lack of an ability to use written language is an ancestral trait that any number of organisms—from bacteria to plants to mice and chimpanzees—share, and thus it does not distinguish any pair of them as being more closely related to one another than they are to humans.

Shared, ancestral features do not help systematists understand the correct relationships between groups of organisms and cannot be used to help build trees. There is one more class of shared features that is not useful for building tress, convergent features.

Convergent features do not indicate relatedness

Sometimes organisms share features not because they share a common ancestor, but because they all evolved the same feature independently. Such features are known as **convergent features**. For example, many desert plants (such as cacti and some spurges) resemble one another because they have all evolved characteristics that help them live under the same parching sun. Desert-dwelling plants typically have very small leaves or no leaves at all, and a shape that reduces water loss. These features evolved independently in each separate lineage of plants. A systematist might be tempted to say that cacti and spurges are each other's closest relatives, because they share so many features. However, they share these traits only because these plants "converged" on a similar strategy for surviving in the desert. Convergent features can mislead systematists rather than guiding them to the correct evolutionary tree.

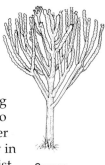

Spurge

While convergent features might seem obvious and easy to avoid, in practice they can be difficult to identify, whether one is looking at leaf shape or at DNA. This difficulty in identifying features as convergent is one of the major obstacles that systematists face in trying to decipher the true tree of life from the data at hand.

> ■ Cladistics is the most prominent school of systematics today. In cladistics, systematists use one particular set of shared features to build evolutionary trees: shared, derived features. Two types of shared features that are not good indicators of relatedness are shared, ancestral features and convergent features.

Using Evolutionary Trees to Predict the Biology of Organisms

Once systematists discover the correct evolutionary relationships among the organisms in a group, the resulting evolutionary tree is useful for more than just describing and organizing knowledge already gained about the organisms. The tree also has predictive power, because researchers can expect that close relatives will share many of the same novel features passed down by their common ancestor.

As surprising as it might seem, there is now overwhelming evidence from living and fossil animals that the closest relatives of birds are the extinct creatures we know as dinosaurs (Figure 3.4). Of the animals shown in Figure 3.4, the next-closest relatives of birds, after dinosaurs, are crocodiles and alligators, a group known as crocodilians. Knowing the relationships among these

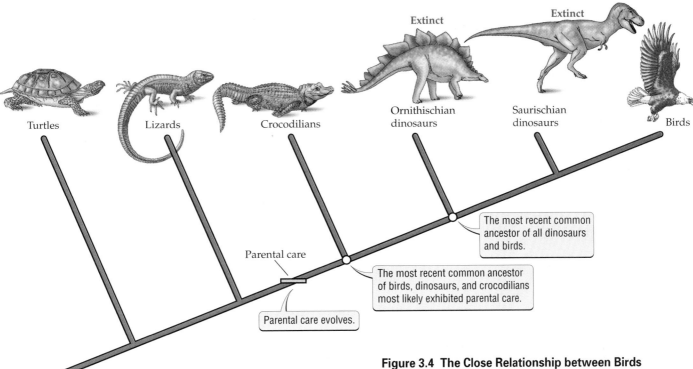

Figure 3.4 The Close Relationship between Birds and Dinosaurs
As this tree shows, birds and the two lineages of dinosaurs—ornithischians (plant eaters characterized by horny beaks and a lack of teeth) and saurischians (characterized by a long, mobile S-shaped neck)—are most closely related. Ornithischian dinosaurs include *Stegosaurus* and *Iguanodon*; dinosaurs such as *Apatosaurus* and *Tyrannosaurus* are saurischians. The next most closely related group depicted on the tree is the crocodilians, including crocodiles and alligators.

groups has made it possible for biologists to make predictions about the behavior of long-extinct dinosaurs.

Crocodilians and birds are known to be dutiful parents. They build nests and defend their eggs and young. Scientists reasoned that if crocodilians and birds both exhibit extensive and complex parental care, then most likely their common ancestor exhibited this behavior as well. Because dinosaurs share this same common ancestor, scientists were able to predict that these long-gone and unobservable creatures likely tended their eggs and hatchlings as well, a shocking notion for creatures with a reputation for being big, vicious, and pea-brained.

In 1994, Mark Norell, a paleontologist at the American Museum of Natural History in New York City, and an international team of colleagues confirmed that dinosaurs exhibited parental care. They discovered a dinosaur that died 80 million years ago—while sitting on a nest of its own eggs. This dinosaur, which had actually been unearthed from the Gobi Desert in 1923, was originally thought to have been eating the eggs and hence was given the name *Oviraptor,* which means "egg seizer." Biologists at the time did not know that birds and dinosaurs were close relatives, so they did not expect the two to show similar behaviors. As a result, the idea that a dinosaur could brood a nest of eggs was inconceivable. In fact, however, *Oviraptor* appears to have died in a

sandstorm protecting its nest, its limbs encircling its unhatched young in a posture as protective as that of any bird (Figure 3.5).

■ Evolutionary trees not only depict the relationships of groups of organisms, but they can also be used to predict behaviors and other attributes of organisms on the tree. Using evolutionary trees, biologists were able to predict that dinosaurs would care for nests of eggs.

Classification versus Evolutionary Relationships

It might seem logical for systematists to name and classify together into groups only those sets of organisms that constitute all the descendants of a common ances-

(a)

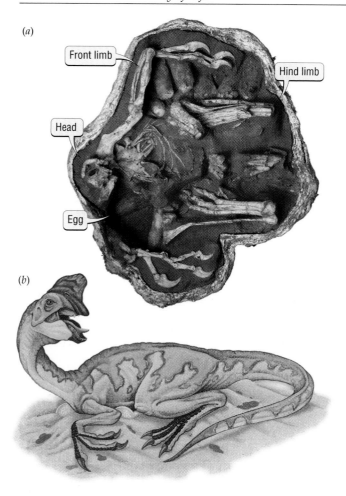

(b)

Figure 3.5 Parental Care by Dinosaurs

(*a*) The fossil shows the remains of an *Oviraptor* dinosaur, which died sitting on its nest of eggs. (*b*) An artist's rendition of the dinosaur brooding its eggs shows how the dinosaur might have looked shortly before the sandstorm began. Compare the dinosaur brooding its eggs (*b*) with an ostrich as it would look on a nest of eggs today (*c*).

(c)

tor. That is, we might expect them never to use partial groupings that are missing some descendants or mixed groupings that include some descendants from one ancestor and some from another. With the advent of cladistic analyses, naming only such complete groups is exactly what some systematists have proposed.

However, the basis of the current classification system, described in Chapter 2 (kingdoms, phyla, classes, and so on), has been in place since the 1700s, long before both the theory of evolution and cladistic analyses came to prominence. In fact, we still use more than 11,000 names and groupings that Linnaeus originally proposed. As a result, many of the named and familiar groups (such as protists and reptiles) are not the complete groups of descendants of a single, common ancestor, the "**real groups**" that many systematists are now calling for. In the controversy over what to do about such groups, a growing number of scientists want them to be abandoned. For many others, abandoning groupings like the reptiles and reorganizing the classification of all living things to include only real groups as we now understand them would be too radical a move.

For example, dinosaurs, a well-known group of creatures that most elementary school students can identify, are not a complete group consisting of all the descendants of their most recent common ancestor. As Figure 3.4 shows, the most recent common ancestor of both lineages of dinosaurs has as its other descendants the birds. Thus dinosaurs are not a "real group" unless they include the living birds.

If groups are to correspond only to complete groups of descendants, either the term "dinosaurs" must be eliminated from our language—a practical impossibility—or we must accept that the sight of a robin in the yard means that dinosaurs have not really gone extinct after all. In fact, many scientists now consider the modern birds, along with such beasts as *Tyrannosaurus rex*, to be types of saurischian dinosaurs.

Systematists are encountering an increasing number of such problems as new information about evolutionary relationships and the push for nothing but real groups clashes with traditional classifications. Doing away with familiar groups remains controversial, and scientists continue to debate the merits and problems of different systems.

> ■ There is a controversy in biology today about the naming and classification of groups. Some biologists believe that only real groups should be named. Others believe that doing away with long-standing and familiar groupings will cause more problems than it will solve.

Branches on the Tree of Life: Interpreting Relatedness

What have scientists learned about the relationships of different groups of organisms? The study of evolutionary relationships is one of the fastest-moving areas of biology, with new evolutionary trees being deciphered all the time. In this section we examine two of the trees of greatest interest to scientists: the tree that includes all living organisms, and the tree that depicts the relationships of humans and their close primate relatives.

The tree of all living things reveals some surprising relationships

The more distantly related two organisms are—for example, a bacterium and a human—the more difficult it is to compare them. What part of a bacterium would show any similarity to a person? One of the few features that can easily be compared across such wide divides is DNA, something all organisms have, using it as their hereditary blueprint. By using DNA to make their comparisons, researchers have revolutionized the study of evolutionary relationships. They can compare even extremely distantly related kingdoms to reconstruct the arrangement of the major branches on the tree of life.

The tree that has emerged (Figure 3.6), like the tree that includes the dinosaurs (see Figure 3.4), is an example of the problems that can arise as we gain new information about the evolutionary relationships of familiar groups. Here the eukaryotes are clearly shown as the descendants of a single ancestor, thus making up a "real group." But the Monera or prokaryotes, a kingdom we discussed in Chapter 2, consists of the Archaebacteria and the Eubacteria. As Figure 3.6 shows, the Monera or prokaryotes are not the complete set of descendants of a single ancestor. Their most recent common ancestor—the universal ancestor of all living things at the base of the tree—includes all other living things among its complete set of descendants. So the prokaryotes are not a "real group" then unless they include the eukaryotes, a useless grouping that would simply be all the living organisms. As a result, a growing number of scientists object to the kingdom Monera and the group "prokaryotes."

The fact that prokaryotes are not a real group should not be surprising. Remember from Chapter 2 that the

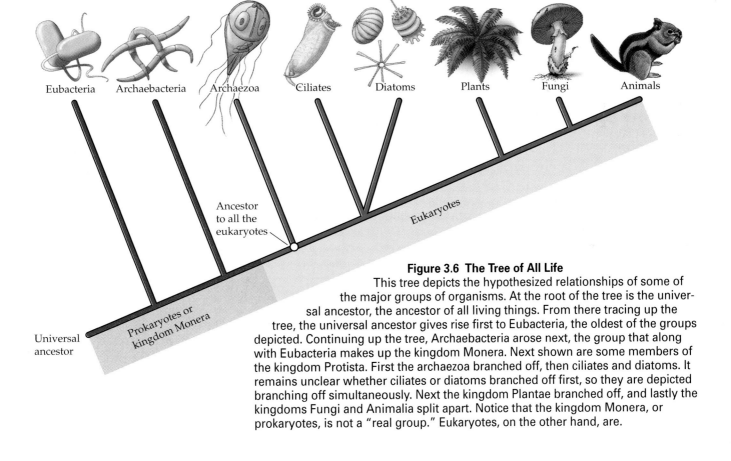

Figure 3.6 The Tree of All Life
This tree depicts the hypothesized relationships of some of the major groups of organisms. At the root of the tree is the universal ancestor, the ancestor of all living things. From there tracing up the tree, the universal ancestor gives rise first to Eubacteria, the oldest of the groups depicted. Continuing up the tree, Archaebacteria arose next, the group that along with Eubacteria makes up the kingdom Monera. Next shown are some members of the kingdom Protista. First the archaezoa branched off, then ciliates and diatoms. It remains unclear whether ciliates or diatoms branched off first, so they are depicted branching off simultaneously. Next the kingdom Plantae branched off, and lastly the kingdoms Fungi and Animalia split apart. Notice that the kingdom Monera, or prokaryotes, is not a "real group." Eukaryotes, on the other hand, are.

prokaryotes were defined as a group because they lack certain features—compartmentalized organelles like a nucleus inside their cells. That is, they have been classified together not because they share novel features but because they lack the novel features—organelles—that define the eukaryotes. What they share are ancestral features—the kind that systematists now try to avoid using to define groups. Nevertheless, many biologists continue to find the designation "prokaryotes" very useful, thus illustrating how difficult it can be to change classification schemes that have been in place for years.

A search for protists on the tree in Figure 3.6 reveals why biologists refer to this group as an evolutionary grab bag. Protists include at least three separate lineages, the archaezoa, ciliates, and diatoms.

In general, though, the relationships among the kingdoms are not surprising. The most distant relatives of plants, animals, and fungi, are the bacteria, single-celled creatures that seem like very distant cousins of ours, at best. But notice in Figure 3.6 the relationships of plants, fungi, and animals; the two most closely related groups among the three are fungi and animals.

For years, fungi have been thought to be closely related to plants. Unable to move and very unlike animals, these faceless organisms, most familiar to us as mushrooms, seem more akin to trees, shrubs, and mosses. As a result, it came as a huge surprise when recent studies showed that fungi are actually more closely related to animals, including people, than they are to plants. That is, fungi share a common ancestor more recently with animals than with plants (see Figure 3.6). Put another way, the mushrooms on your pizza are more closely related to you than they are to the green peppers sitting next to them.

Plants and fungi were lumped together as closest relatives not because they shared unique, derived features, but because biologists mistakenly based their grouping on shared, ancestral features. Plants and fungi are similar simply because they have not evolved some of the unique characteristics that animals have evolved. Animals and fungi, on the other hand, belong together because they share certain novel features.

Could the slime in your bathroom shower really be more closely related to you than it is to a plant? How much do we really have in common with the likes of bread mold or yeast or a mushroom at the grocery store? A lot, it turns out.

The finding that fungi and animals are more closely related to one another than either are to plants solved the long-standing mystery of why doctors often have such a difficult time treating fungal infections, particularly internal infections, in which a fungus has begun living inside the human body. The reason is that because fungi and animals are such close relatives, human cells and fungal cells work similarly. Thus anything a doctor might use to to kill off a fungus could kill or nearly kill the person as well.

So similar are humans and fungi that there is even a disorder in yeasts that is similar to Lou Gehrig's disease, a fatal disease of humans in which the nervous system quickly degenerates. Thus yeast is the perfect, if surprising, model for studying this dangerous human disease.

The primate family tree reveals the closest relative of humans

When we visit the ape house at a zoo, the striking similarities between the beings standing outside the cage looking in and those inside the cage looking out become obvious. But which of our primate relatives is our closest relation? Over the years this question has generated intense interest. Researchers have studied everything from bone structure to behavior to DNA in attempts to determine which primate is humankind's closest kin.

While controversy still remains, something of a consensus has been reached on the basis of DNA evidence. The emerging consensus suggests that we humans have as our closest relatives the chimpanzee, fellow tool user, with whom we share a remarkable degree of similarity in our DNA (Figure 3.7). More distantly related are gorillas, and beyond that orangutans, gibbons, Old and New World monkeys like the spider monkey, and, most distantly, lemurs.

The evolutionary tree of life is a work in progress

Although many of the major branching patterns on the evolutionary tree of life are well established and unlikely to change, scientists view evolutionary trees as working hypotheses, the best approximations given what we know today. Biologists continue to study and reevaluate the relationships of all organisms using new data to understand the tree of life. In fact, the vast majority of relationships among the world's millions of different species of plants, animals, fungi, bacteria, and protists remain to be worked out in detail.

■ Biologists are making dramatic progress in deciphering the evolutionary relationships of the world's many organisms, in part due to the increasing use of DNA by systematists. As biologists study the tree of all living organisms, they are finding surprising relationships, like the unexpectedly close relationship between animals and fungi. Biologists studying the most controversial of evolutionary trees are beginning to come to a consensus that humanity's closest relative is the chimpanzee.

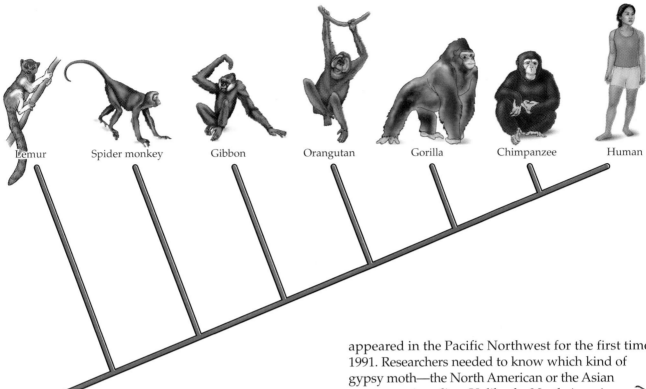

| Lemur | Spider monkey | Gibbon | Orangutan | Gorilla | Chimpanzee | Human |

Figure 3.7 The Primate Family Tree
This tree shows what many biologists now believe are the correct relationships between humans and the rest of the primates. Humans and chimpanzees are shown as most closely related, having shared a common ancestor more recently with each than with any of the other groups. Our most distant relatives depicted are the lemurs, followed by spider monkeys, gibbons, orangutans, and gorillas.

Highlight

Everything in Its Proper Place: The Contribution of Systematics

Systematics has long been considered a rather esoteric endeavor, but in fact, it is one of biology's most vital sciences, crucial for the study of everything from agriculture to research on human diseases. The reason why systematics is so important is that before researchers can begin to ask serious questions about an organism they must know what the organism is, which means identifying its close relatives and determining where it belongs on the tree of life.

Like the Iceman described at the beginning of this chapter, new, unidentified organisms often find their way into our lives. Gypsy moths, which had long been ravaging the trees of the northeastern United States,

appeared in the Pacific Northwest for the first time in 1991. Researchers needed to know which kind of gypsy moth—the North American or the Asian type—was invading. Unlike the North American gypsy moth female, the Asian gypsy moth female can fly, making the species much more mobile and quick to spread. Asian gypsy moths can also feed on and damage a wider variety of tree species.

Gypsy moth

Studying the moth's DNA, Richard Harrison and Steven Bogdanowicz of Cornell University were able to show that the newly invading moths were close relatives of moths from Asia. By determining which pests they were up against, and thus knowing how quickly the moths could spread and which plants they might attack, researchers were able to give forest managers a better chance of controlling the moths effectively.

It was also systematists who solved the mystery of whether a dentist infected with HIV (the human immunodeficiency virus, which causes AIDS) had transmitted the virus to his patients. Against a backdrop of controversy about whether health care workers with HIV are a risk to their patients, 10 patients of one infected Florida dentist tested positive for HIV. (Four, however, had lifestyles or habits that put them at risk for HIV infection by other means.) Was the dentist to blame or not?

Researchers created an evolutionary tree of the viruses taken from each of the patients, from the dentist, and from others by comparing DNA in each person that was derived from the virus. If the dentist had infected his patients, the viruses from his patients should be mostly closely related to his virus and less closely related to viruses sampled from other people.

The results showed that all six patients who were HIV-positive but were not at risk of contracting the HIV infection for other reasons were carrying a virus very closely related to the virus from the dentist. Those whose lifestyles or habits had exposed them to HIV in other ways had viruses that were not closely related to the dentist's virus. Scientists concluded that, at least in the case of six of the patients, the dentist had given them their infection.

The reach of systematics is wide and becoming ever wider, as these examples illustrate. When it comes to solving biological problems in the real world, systematics—knowing where an organism fits in the tree of life—is the first order of business.

> ■ Systematics is an active and vital part of biology which is helping answer a vast array of questions in areas as wide-ranging as evolutionary biology, insect pest management, and the study of the spread of AIDS.

Summary

Looking for Clues to Evolutionary Relationships

■ Species give rise to descendant species with which they share characteristics.

■ Descendant species often share key features, such as distinctive physical structures or behaviors, because they share DNA inherited from a common ancestor.

■ Evolutionary trees depict the evolutionary relationships of groups of organisms.

■ Systematists look for features that groups of organisms share to study their relatedness. DNA has proven to be a particularly useful new feature in such studies.

Assembling Evolutionary Trees

■ Evolutionary taxonomy, the oldest approach to systematics, lacks a well-defined method.

■ Cladistics uses shared, derived features to identify close relatives and to determine evolutionary relationships.

■ Shared, ancestral features are not useful for determining evolutionary relationships.

■ Convergent features can mislead systematists into thinking that distantly related groups are closely related and vice versa.

Using Evolutionary Trees to Predict the Biology of Organisms

■ Evolutionary trees can predict and give insight into the biology of organisms and lead to the discovery of such surprising things as the behavior of long-extinct dinosaurs.

Classification versus Evolutionary Relationships

■ Some systematists want to name only "real groups" and do away with other names, some of which are very familiar. Such a move remains controversial.

Branches on the Tree of Life: Interpreting Relatedness

■ Studies using DNA have allowed scientists to make great strides in deciphering the tree of life, which has provided some interesting surprises.

■ Scientists have greatly improved our understanding of the primate family tree. There is growing agreement that chimpanzees are humans' closest relatives.

■ Evolutionary trees can best be thought of as working hypotheses or works in progress.

Highlight: Everything in Its Proper Place: The Contribution of Systematics

■ Systematics places organisms in their proper place in the tree of life, answering what are sometimes otherwise unanswerable biological questions and providing a starting point for asking other biological questions.

Key Terms

cladistics p. 49	"real group" p. 52
convergent feature p. 50	shared, ancestral feature p. 50
evolutionary taxonomy p. 48	shared, derived feature p. 50
evolutionary tree p. 46	systematist p. 46
most recent common ancestor p. 46	

Chapter Review

Self-Quiz

1. In Figure 3.2, which evolutionary trees depict A and B as most closely related?
   a. *a*
   b. *b*
   c. *a* and *b*
   d. none of the above

2. The most powerful new tool being studied by systematists today is
   a. behavior.
   b. the cell.
   c. DNA.
   d. organs.

3. Evolutionary taxonomy
   a. is the newest school of systematics.
   b. is more rigorous than cladistics.
   c. requires the study of DNA.
   d. uses no codified formula or rationale to define its methodology.

4. In a cladistic analysis the most useful features are
   a. convergent features.
   b. shared, ancestral features.
   c. shared, derived features.
   d. shared features of any kind.

5. In Figure 3.7, the closest relative of humans is the
   a. chimpanzee.
   b. gorilla.
   c. orangutan.
   d. lemur.

6. Dinosaurs and crocodilians most likely exhibit similar parental behaviors because
   a. they are both scaly.
   b. they both lay eggs.
   c. they are both closely related to birds.
   d. they share a common ancestor that exhibited parental behaviors.

Review Questions

1. How do systematists identify an unknown organism and place it on the tree of life?

2. What is a "real group," and why do some systematists think real groups are important?

3. How is an evolutionary tree like a hypothesis?

---

3   𝔗𝔥𝔢 𝔇𝔞𝔦𝔩𝔶 𝔊𝔩𝔬𝔟𝔢

New Species of Deer Discovered in Da Nang, Vietnam

DA NANG, VIETNAM. A team of scientists from the University of Eastern Texas emerged from the Annamite Mountains yesterday saying they had discovered the world's newest species of large mammal.

"It is an incredible thrill to see a huge beast that you had no idea even existed," said Dr. Sean Vanderveld, mammalogist at the University of Eastern Texas. "She was clearly a new kind of deer. She stood just ten feet away and munched grass while we photographed her." Researchers then shot the animal, took blood samples, and car-ried the entire carcass back to the city of Da Nang. Reseachers said they plan to do systematics studies by examining the DNA they get from the deer's blood as well as by studying the carcass itself.

The region in which the animal was found, known as the Annamite Mountains, is remote and rugged, forming a forbidding natural border between Vietnam and Laos. Dr. Vanderveld and colleagues note that while new to western science, the animal is well known to locals, who call it a honinh. Researchers said they went in search of this ap-parently rare mammal whose antlers are displayed in a number of homes in the small mountain villages.

Amidst the excitement, Dr. Morris Berger, a mammalogist at the University of Los Angeles, said that the new deer, whose photographs he has now seen, is no different from a Vietnamese subspecies of deer described in 1945.

"People shouldn't be going and getting all excited just yet," said Dr. Berger, who has already requested a chance to examine the blood samples and the carcass.

Evaluating "The News"

1. How can scientists determine if the newly discovered deer is really a different species from the one known since 1945?

2. What does systematics have to offer society—and the researchers studying the new deer—other than the names of organisms?

3. There is a growing controversy among scientists as to whether newly discovered and apparently rare animals should be killed for study. Some say that without the organism in hand there is no way to definitively identify it. Others ask, What good is a name if you kill the last one, or perhaps the last male or female? Should these researchers have photographed their discovery and then set the deer free?

4

Biodiversity

Martin Johnson Heade, *Study of an Orchid*, 1872.

Where Have All the Frogs Gone?

*I*n 1987, in a tropical forest high in the mountains of Costa Rica, the golden toad, a spectacularly beautiful creature, could be found in abundance. That year hundreds of the brightly colored animals were seen in the forests at Monteverde, the only spot in the world from which the toads were known. The next year just a few were found. Within a few years the golden

toad had disappeared entirely, never to be seen again.

While there is always a concern when a species plummets into extinction, most extinctions are easier to understand than the loss of the golden toad. When a forest-dwelling bird goes extinct because its forest is cut down, there is no lingering mystery. But the golden toads were living in a pris-

The diversity of life has risen and fallen naturally and drastically in the past, but is now declining rapidly as a result of human activity.

tine area far from deforestation or development. These frogs, it seemed, should not have gone extinct.

Since the time when the golden toad was last seen, biologists around the world have documented declines in populations of other amphibians (a group that includes frogs, toads, and salamanders), many in preserved areas. In the United States, for example, in and around Yosemite National Park, where frogs and toads were once abundant, numerous species have declined or disappeared. For frogs living at high altitudes, increased ultraviolet light, perhaps due to deterioration of the ozone layer in the upper atmosphere, may be a problem. In Australia and Central America, a fungal disease is killing huge numbers of frogs. In other cases, researchers believe that pollution is the cause of frogs' decline.

Whatever the specific problem, amphibians are being lost around the world. Many scientists say that these animals probably are more sensitive than other animals to environmental deterioration, and that their deaths are signs of an ever more poisoned environment.

Meanwhile, as amphibians mysteriously disappear, biologists are finding that many other species are rapidly going extinct. Everywhere we hear warnings about the loss of species around the globe. How serious are these species losses, and in the end, do they really matter?

The Golden Toad
Once abundant on a mountaintop in Costa Rica, this amphibian mysteriously went extinct in the 1980s.

Key Concepts

1. Estimating the number of species in the world is difficult but not impossible. Although the total number remains a question, biologists generally agree that the vast majority of species have yet to be discovered.

2. Biologists generally agree that we are on our way toward a mass extinction, one of a few times in the history of Earth during which huge numbers of species have been lost. If species losses continue at their current rate, the result could be the most rapid mass extinction in the history of the planet.

3. Deforestation and other habitat deterioration, invasion of foreign species, climate change, and the resulting decrease in species numbers are problems not just in the tropical rainforest but in our own backyards.

4. Biodiversity matters to the health of forests, grasslands, rivers, oceans, and other ecosystems on which human society depends. Therefore, biodiversity matters to the health of humans.

We hear constantly that the world's species are rapidly becoming extinct, and that the world's **ecosystems**—the Earth's many habitats and the organisms that live in them—are under threat. But how many species are there really, and what exactly is happening to them? In this chapter we examine how species numbers are changing now and how they have changed during Earth's history. We also explore the question of what value, if any, the world's many species—which constitute its **biodiversity**—have to humanity.

How Many Species Are There on Earth?

Despite intense worldwide interest, scientists do not know the exact number of species on Earth. Estimates range widely from 3 million to 100 million species. Most estimates, however, fall in the range of 3 million to 30 million.

Scientists use indirect methods to estimate species numbers

Up to now, the total number of species that have been collected, identified, named, and placed in the taxonomic hierarchy is around 1.5 million. But despite this great cataloging of organisms, which has taken more than two centuries to complete, many researchers believe that biologists have barely scratched the surface. Some estimates suggest that 90 percent or more of all living organisms remain to be identified and named by biologists.

Biologists use indirect methods to estimate how many more species remain unknown. For example, in 1952 a researcher at the U.S. Department of Agriculture estimated, on the basis of the rate at which unknown insects were pouring into museums, that there were 10 million insect species in the world. Scientist Terry Erwin, an insect biologist at the Smithsonian Institution, shocked the world in 1982 with his estimates that the arthropods alone (a grouping that includes insects, spiders, and their allies) numbered more than 30 million species, 20 times the number of all the previously named species of all organisms on Earth. Most of these arthropods, Erwin said, were living in the nearly inaccessible tops of rainforest trees, a region of the forest known as the **canopy** (see Box on page 62).

Erwin based his estimate on actual counts that were obtained by a method known as fogging (Figure 4.1), in which a biodegradable insecticide was blown high into the top of a single rainforest tree. The dead and dying insects that rained down to the ground were collected and the number of different species counted. Erwin found 163 species of beetles alone living in the top of one particular kind of tropical tree. Knowing that, he was able to come up with a minimum estimate of the number of arthropod species in the world.

There are an estimated 50,000 or so tropical tree species. Thus if the tree species Erwin was studying was typical, beetle species in tropical trees should number 8,150,000 (50,000 × 163). Beetle species are thought to make up about 40 percent of all arthropod species. If that is the case, then the total number of arthropod species should be about 20 million. Many scientists believe that the total number of arthropod species in the canopy is double the number found in other parts of tropical forests, suggesting that there are another 10 million arthropod species in noncanopy tropical environments. Assuming these numbers, the total number of arthropod species in the tropics should be about 30 million. That means of course that there should be even more species of all kinds in the tropics and in the world.

Like all such estimates, this one is based on numerous assumptions. Changes in these assumptions could dras-

Figure 4.1 Fogging to Count Species
Tropical biologist Terry Erwin fogs the canopy of a tree to collect its many insects. On the basis of his studies of insects from tropical treetops, Erwin estimated that there could be as many as 30 million species of tropical arthropods alone.

tically alter the final number. While Erwin's is among the most famous of these estimates, such indirect measures are typical of how the numbers of species yet to be discovered are estimated, since it is impossible to count the numbers of unknown species directly. Scientists continue to argue over the exact figures and the assumptions on which they are based, but one thing is certain: The 1.5 million species discovered and named so far are but a fraction of the species out there.

Figure 4.2 Numerical Breakdown of the Species Known on Earth
Animals (particularly insects) and higher plants make up the vast majority of the known species on Earth, but many more remain to be discovered, particularly in other groups.

Some groups are well known and others are poorly studied

About half of the 1.5 million known species (750,000) are insects. All the remaining animals make up a mere 280,000 species or so. The next-largest group is the plants, with 250,000 known species. There are also approximately 69,000 named fungi, 30,000 protozoans, 27,000 algae, and some 3000 prokaryotes, including the Eubacteria and the Archaebacteria (Figure 4.2).

Among these groups, some have been very well studied because they are large or easy to capture or popular with biologists. Others have been studied very poorly, often because they are microscopic or otherwise hard to collect and identify. The birds, for example, total 9000 species and are among the best-studied organisms on Earth with fewer new species remaining to be discovered. Insects, on the other hand, remain poorly known; the majority, possibly the vast majority, are still undiscovered and unidentified. Other groups, including whole kingdoms, such as the Fungi and Monera, are also poorly known. In a single gram of Maine soil, scientists have esti-

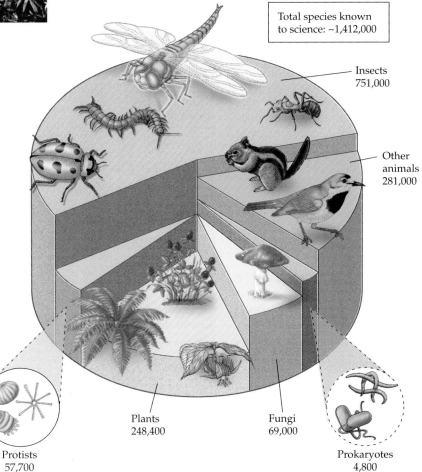

Total species known to science: ~1,412,000

Insects
751,000

Other animals
281,000

Plants
248,400

Fungi
69,000

Protists
57,700

Prokaryotes
4,800

Biological Exploration: *When Getting There Is Half the Fun*

With so many of the world's species still undiscovered, biologists have much work ahead of them. But many of the remaining species have remained elusive, not just because they are among so many still unknown but because they live in inaccessible habitats.

To find most of the remaining species, biologists need to do more than just doggedly collect every organism they come across. They must also become daring explorers, venturing into extreme habitats, a feat that often requires not only a sense of adventure but more than the usual laboratory equipment.

Among the most challenging of biological explorations are expeditions into the forest canopy, an aerial world of leaves and tree branches, a complex habitat that biologists estimate has the same total surface area as the entire planet (see Figure A). Biologists who have managed to make their way into the canopy have found that life abounds in the upper stories of these forest trees.

To reach these heights, some canopy biologists use dirigibles (blimps) floating through the air and work on inflated rafts that are gently lowered onto treetops (see Figure B). Others use hooks attached to their ankles to work their way up a tree like a clawed animal. Still others use huge cranes that move them around their treetop study sites as if they were doing construction work.

The ocean bottom is another entirely new world, most of which is still unexplored. In the crushing depths, beneath miles of water, biologists descend in submersibles (small submarine vehicles) or dredge what they can from boats. One researcher described hauling in net after net full of creatures, with each dredge pulling up an

entirely different group of organisms—lampshells, peanut worms, moss animals, ribbon worms, beard worms, and numerous others without such common names, many of which he had never seen before.

Yet despite the increased effort to reach these remote areas, many creatures still have never been seen alive in their natural habitats. For example, a squid the length of a city bus, known as the giant squid or *Architeuthis*, a monstrous representative of the molluscs, has become known to scientists only because it occasionally washes ashore, dead. Still elusive in its natural surroundings, the living, giant squid is a symbol of the plentiful lingering mysteries of the biosphere.

Peanut worm

A

B

Scientists study canopies like this one (A) using dirigibles (B) in the lowland rainforest in Costa Rica where many species have remained hidden from view.

mated that there may be as many as 10,000 species of bacteria. That's about 5000 more species than have so far been named by biologists. The studies that form the basis of this estimate examined the numbers of different types of DNA contained in the soil, each of which represents a different species (see Figure 4.3).

Biologists are discovering new species even in relatively well-known groups. About 100 new fish species, for instance, are discovered each year. In another case, although scientists had thought that all the large land mammals were accounted for, in 1992 a large deerlike species was found in Vietnam. Just 2 years later another large deer species, known as a barking deer, was discovered there as well.

The barking deer

■ The exact number of species on Earth remains unknown. Scientists use indirect methods to estimate the total number of species, with estimates ranging from 3 million to 100 million; most fall between 3 million and 30 million. Half of the 1.5 million known species are insects. There is great variation in how well documented different groups of organisms are, and many, including fungi and bacteria, remain extremely poorly known.

The Beginnings of a Present–Day Mass Extinction

The history of life on Earth includes a handful of drastic events during which huge numbers of species have gone extinct, events known as **mass extinctions**. Today,

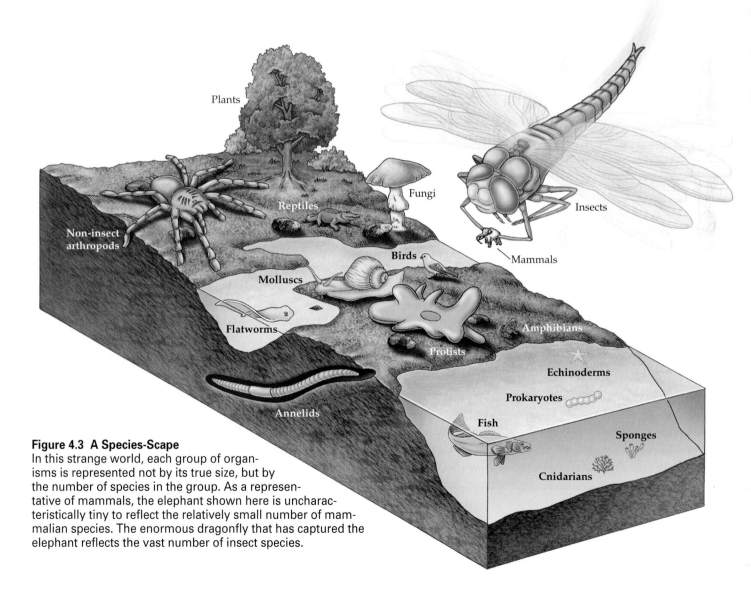

Figure 4.3 A Species-Scape
In this strange world, each group of organisms is represented not by its true size, but by the number of species in the group. As a representative of mammals, the elephant shown here is uncharacteristically tiny to reflect the relatively small number of mammalian species. The enormous dragonfly that has captured the elephant reflects the vast number of insect species.

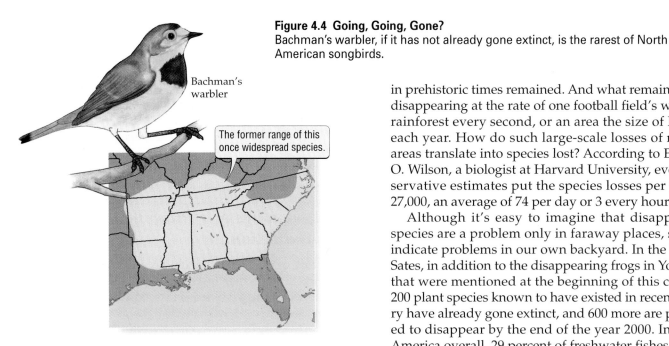

Figure 4.4 Going, Going, Gone?
Bachman's warbler, if it has not already gone extinct, is the rarest of North American songbirds.

Bachman's
warbler

The former range of this once widespread species.

in prehistoric times remained. And what remained was disappearing at the rate of one football field's worth of rainforest every second, or an area the size of Florida each year. How do such large-scale losses of natural areas translate into species lost? According to Edward O. Wilson, a biologist at Harvard University, even conservative estimates put the species losses per year at 27,000, an average of 74 per day or 3 every hour.

Although it's easy to imagine that disappearing species are a problem only in faraway places, studies indicate problems in our own backyard. In the United Sates, in addition to the disappearing frogs in Yosemite that were mentioned at the beginning of this chapter, 200 plant species known to have existed in recent history have already gone extinct, and 600 more are predicted to disappear by the end of the year 2000. In North America overall, 29 percent of freshwater fishes and 20 percent of freshwater mussels are endangered or extinct.

In fact, evidence suggests that humans have been driving species to extinction for a very long time. The fossil record shows that around the time humans arrived in North America, Australia, Madagascar, and New Zealand, large animal species began to disappear. Although some people suggest that species extinctions may also be due to climate changes, the coincidence of species loss with the arrival of humans in three different parts of the world is striking and consistent. Of the

even as biologists struggle to get a total species count, many biologists believe that we are on our way toward a mass extinction of many of those species. In fact, many biologists believe that if extinction continues unabated it will lead to the most rapid mass extinction in the history of Earth. As with the total number of species on Earth, extinction rates are estimates. But even by conservative estimates, species are being lost at a staggering rate. The cause of this mass extinction is abundantly clear: the activity of the ever-increasing numbers of human beings living on Earth.

Numerous studies have documented recent species declines caused by humans. One large-scale study found that 20 percent of the world's species of birds that existed 2000 years ago are no longer alive. Of the remaining bird species, 10 percent are estimated to be endangered—that is, in danger of extinction (Figure 4.4). Twenty percent of the freshwater fish species known to be alive in recent history have either gone extinct or are nearly extinct. And in 1998, scientists announced that one in every eight plant species in the world is under threat of extinction.

The most devastating and obvious losses are occurring in wet, lush forests in the tropical regions, known as tropical **rainforests** (Figure 4.5). Known to be cradles of huge numbers of unique species, found nowhere else in the world, rainforests are quickly being burned or cut. By 1989, less than half of the rainforest that had existed

Figure 4.5 Rainforest
Tropical rainforests, like this one in Hawaii, are typically home to numerous species found nowhere else.

genera of large mammals that roamed Earth 10,000 years ago, 73 percent are now extinct.

Many of these large mammals may have made a hearty meal for prehistoric hunters, creatures like mammoths, ground sloths, camels, horses, and saber-toothed cats. Such creatures also may have suffered from having to compete with humans for prey. A similar number of birds, particularly those that would have been the easiest of prey for humans, disappeared. One species lost, for example, was a flightless duck—literally a sitting duck. Numerous other animals that depended on the larger animals for survival, such as vulture species that fed on the carcasses of the dead beasts, also went extinct.

> ■ Though estimating extinction rates is difficult, there are many studies that indicate that species are currently going extinct at an extremely rapid rate and that we are in the midst of a mass extinction—possibly the most rapid mass extinction in the history of Earth.

Like estimates of total species number, the specific rates of species loss are impossible to determine with certainty. The main point is that biologists have amassed a wealth of data on species already gone or on their way out, and even conservative analyses suggest that huge numbers of species have been and continue to be lost. Why are these species disappearing?

The Many Threats to Biodiversity

The remaining species of the world face continuing challenges to their survival. In this section we examine some of the forces that are threatening and destroying biodiversity around the globe, including in our own backyards.

Habitat loss and deterioration are the biggest threats to biodiversity

Foremost among the direct threats to biodiversity is the destruction or deterioration of places where different species can live—that is, their **habitats**. Habitats for non-human species continue to disappear or become radically altered as human homes, farms, and industries spring up where natural areas once existed. The term "habitat loss" quickly conjures up images of burning rainforest in the Amazon, but the problem is much more widespread.

Every time a suburban development of houses goes up where once there was a forest or field, habitat is destroyed. So widespread is the impact of growing human populations in urban and suburban areas that species are disappearing even from parks and reserves in these areas. For example, Richard Primack and Brian Drayton, ecologists at Boston University, found that in a large preserve in the midst of increasing suburban development, 150 of a park's native plant species had disappeared. The species were most likely lost as a result of trampling and other disturbances as more and more people used the park (Figure 4.6*a*). In addition, pollution, erosion, and other effects of human activity and human population growth are altering natural habitats to the point where many species can no longer inhabit them (Figure 4.6*b*).

Figure 4.6 The Threats of Habitat Loss and Deterioration
(*a*) Researchers discovered that 150 species had disappeared from the Middlesex Fells Reservation in Massachusetts, a 100-year-old preserve, despite the ban on development within the park. Many of the plants (which still exist in other locations) were suspected to have been killed as the result of increasing use by people from the growing suburban area surrounding the park. (*b*) Usually habitat loss or deterioration is more obvious.

(*a*)

(*b*)

Introduced foreign species can wipe out native species

Another devastating problem is the introduction of **non-native**, or **foreign**, **species**—that is, species that do not naturally live in an area but are brought there on purpose or accidentally by humans. Often these invaders are able to sweep through a landscape, eating, overtaking, or otherwise wiping out native species as they go.

In Africa's Lake Victoria, more than 300 species of native cichlid fish evolved in some 10,000 years, making this lake a treasure trove of fish diversity. Fewer than half of those species now remain, and many of the surviving species are close to extinction. The Nile perch, a voracious predator, which was introduced to the lake as a food fish, is largely responsible for the extinction of these species.

In Guam, the brown tree snake, another introduced species, has drastically reduced the numbers of most of Guam's forest birds, leaving an eerie quiet where once the forest was noisy with tropical birdsong. The tree snake is thought to have been introduced accidentally, brought by U.S. military planes, from New Guinea, where it occurs naturally.

In Hawaii, introduced pigs that have escaped into and are living in the wild (Figure 4.7a) are eating away the native plant species. Domesticated cats and mongooses, also introduced species, have killed many of Hawaii's native birds, especially the ground-dwelling species.

Purple loosestrife (Figure 4.7b), eucalyptus trees, and scotch broom are introduced plant species that are choking out native plants in various parts of the United States.

Climate changes also threaten species

Changes in climate, believed by many scientists to be caused by human activity, constitute another threat that seems to be affecting some species. For example, in Austria biologists have found whole communities of plants moving slowly up the Alps at a rate of about 3 feet per decade, only able to survive at ever higher, cooler elevations, apparently in response to global warming. If the climate continues to warm, these plants, which exist nowhere but on these mountaintops, will eventually run out of mountain and become extinct (Figure 4.8).

Some threats are difficult to identify or define

Although biologists now agree that many frogs, salamanders, and other amphibians are disappearing around the world, the reason for these disappearances remains unclear. In some cases pollution appears to be the culprit. In other cases, increased ultraviolet radiation due to a thinning ozone layer in the atmosphere seems to be killing frogs off. In still other instances diseases are wiping out amphibians. But it remains unclear why so many amphibians all around the world are dying off at the same time. Is there something that is weakening them all? Lacking an obvious "smoking gun," biologists continue to confront a variety of potential threats without knowing which ones cause the most problems for the populations they study.

Human population growth underlies many, if not all, of the major threats to biodiversity

Perhaps the biggest threat to nonhuman species is the growth of human populations, since many of the problems that we have mentioned here are the direct result of the growth of the human species. For example, the

(a)

(b)

Figure 4.7 The Threat of Nonnative Species Introduced species threaten native ecosystems everywhere. (*a*) Feral pigs brought by humans from the mainland to Hawaii are eating away the native plants. (*b*) Purple loosestrife brought from Europe to the United States threatens native plants across the country.

Figure 4.8 Plants with Nowhere to Go
The mountain wildflower Alpine androsace has been moving up the mountainsides of the Alps. For example, on Mount Hohe Wilde in the Austrian Alps, the plant has been moving at a rate of about 6 feet per decade. Researchers say that if global warming continues, this species will soon run out of mountaintop to climb and go extinct.

growth of human populations is what spurs continuing habitat deterioration as natural areas are converted to farms, roads, and factories that support human life. As more people drive more cars, burn more oil, and use more paper, plastic, and other products, more pollution finds its way into water and air, and more waste crowds the landfills, all of which hastens the demise of other species. In the search for solutions to these problems, we need to look not only at deforestation itself, for example, but also at the human population growth that drives it.

The factors we have discussed in this section are causing extinctions around the globe. Many more species have dwindled to dangerously low levels because of these same problems. While such species continue to hang on, their low numbers make them much more susceptible to future extinction.

> ■ The world's species are threatened by numerous well-known forces, including habitat loss and deterioration, the introduction of nonnative species, climate change, and other forces still unidentified. The ever increasing number of humans on Earth underlies most, if not all, of these challenges to the persistence of biodiversity.

The current mass extinction is not, however, the first in the history of life on Earth. Next we'll examine the other times in Earth's history when species numbers have been devastated.

Mass Extinctions of the Past

Since the time when life first took hold on this planet, more than 3.5 billion years ago, there have been five previous, well-documented mass extinctions. These mass extinctions took place around 440, 350, 250, 206, and 65 million years ago, respectively.

The most devastating mass extinction occurred 250 million years ago, at the end of the Permian geologic time period. At that time, some 80 to 90 percent of all marine species went extinct. The most famous mass extinction took place 65 million years ago, at the end of the Cretaceous and the beginning of the Tertiary periods. At that time, the last of the dinosaurs (not including the birds!—see Chapter 3) disappeared.

Unlike the current extinction, previous mass extinctions were not caused by humans (which had not yet evolved). Scientists have hypothesized other causes for the different extinctions, including climate change, increased volcanic activity, reductions in sea level, and, in the case of the dinosaurs' demise, the aftereffects of an asteroid's hitting Earth and filling the planet's atmosphere with a thick cloud of dust.

During these mass extinctions, certain groups of organisms disappeared while others survived apparently unscathed. And after each mass extinction, species numbers rebounded as whole new groups of organisms evolved and colonized the planet. For example, when dinosaurs were the dominant creatures on Earth, there were only a few kinds of small mammals around. But when the dinosaurs went extinct, mammals evolved into many new species. These species began living in new parts of their environment, or habitat, and exhibiting new habits. In fact, this great diversification of the mammals eventually resulted in the evolution of our own species.

If Earth has recovered from mass extinctions in the past, why should people be concerned about the current mass extinction? Scientists suggest a variety of reasons for concern. First, it may be much more difficult to recover from the rapid kinds of change we humans are creating, such as large-scale cutting of forests or building of cities, than it was to recover from the natural changes of past mass extinctions. Second, although new species might eventually evolve to increase species numbers again, those that are lost will be gone forever.

In addition, recoveries from past mass extinctions required many millions of years. Ocean reefs, for example, have been destroyed and have recovered from mass extinctions multiple times in the past 500 million years, but the recovery time was on the order of 5 to 10 million years. Are we really willing to wait that long? And are we willing to take the chance that one particular species, *Homo sapiens*, might not survive the ongoing extinction?

> ■ There have been five previous well-documented mass extinctions in the history of life on Earth. Unlike the current mass extinction, the previous ones were not caused by humans. Species numbers recovered from each mass extinction over many millions of years.

Besides our own potential extinction, are there other reasons why conserving biodiversity is important? In the next section we will examine some of the evidence and the arguments that can be made for the value of biodiversity.

The Importance of Biodiversity

Many people wonder whether the loss of one mouse species here or one beetle species there really makes a difference to humanity. One question that has particularly interested biologists is how, if at all, biodiversity affects the forests, fields, wetlands, oceans, rivers, and other ecosystems, whose plants produce the oxygen we breathe, clean the water we drink, produce much of the food we eat, and so on. In fact, scientists have long wrestled with the question of the value of diverse species within particular habitats. During the 1990s in various experimental studies, researchers have shown that biodiversity can indeed contribute to the health and stability of ecosystems.

Biodiversity can make ecosystems more productive, stable, and resilient

From tiny experimental ecosystems in the laboratory in England (Figure 4.9) to experimental prairies in the midwestern United States, researchers have found that the more species that are present in an ecosystem, the healthier the ecosystem appears to be. One measure of the health or vigor of an ecosystem is its **productivity**, the actual mass of plant matter—leaves, stems, fruits—that a plot can produce from the available nutrients and sunlight. Researchers looking at a variety of ecosystems have found that the more biodiversity there is, the more productive the ecosystem tends to be.

The thinking is that different species are good at using the different resources in a habitat. Some plant species grow best in bright sunlight. Others do best in the shade, and still others grow best in dappled light. The more different kinds of species there are, the more productively all the resources, from shade to bright sun, can be used. The same goes for other resources, such as water, the availability of which varies considerably from being abundant in muddy areas, to being largely absent in dry, parched areas, to being variable as in areas that are occasionally very dry and at other times soaked. The more species there are in a habitat, the more likely it is that at least one of those species can use each of the different resources productively.

Researchers have also found evidence that the more species a prairie contains, the better able it appears

Figure 4.9 Testing the Importance of Biodiversity
The ecotron is a tiny experimental ecosystem created in England to test the importance of biodiversity under controlled laboratory conditions. These constructed ecosystems include grasses, wildflowers, snails, flies, and other organisms.

to recover and return to a healthy state following a drought. Researchers are continuing to study the effects of biodiversity, looking at its possible effects on an ecosystem's ability to resist disease and invasion to foreign species.

The rivet hypothesis suggests ecosystems can collapse quickly and unexpectedly

Although the presence of more species has enabled experimental ecosystems to function better and to be more stable, the question remains: How many species are required to keep an ecosystem from collapsing? That is, do we really need them all?

The **rivet hypothesis** provides some insight into this question. First put forward by Paul Ehrlich, an ecologist at Stanford University, this notion suggests that as species drop out of ecosystems, at first there may be no easily noticeable effect, just as a few rivets lost from a plane might not be noticed. After losing enough rivets, though, a plane abruptly becomes a disintegrating bucket of metal crashing down out of the sky. Loss of species from an ecosystem might follow the same pattern as the loss of rivets in an airplane. Many species might be lost without any visible effect, then as a few more are lost the ecosystem could suddenly, unexpectedly collapse.

Researchers already know there are many different interactions between different groups of organisms, creating a broad interdependence among them, as exists among the many parts and pieces of a plane. Put another way, just as rivets and other plane parts perform specific functions in a plane, species perform particular functions in ecosystems. For example, bacteria provide nitrogen in a usable form to plants, and plants form the basis of most terrestrial food webs (see Chapter 1 for a discussion of food webs). Many animals, fungi, and bacteria are important decomposers, returning nutrients back to ecosystems in usable form. Plants provide food for many animals, which in turn harbor numerous forms of bacterial, protozoan, arthropod, and other life.

Like the airplane that loses small part after small part, after losing enough species an ecosystem that has withstood the previous insults of species extinctions with no sign of harm or drastic change could suddenly, seemingly inexplicably, collapse—like lakes that have died, no longer able to support plant or animal life—under the cumulative loss of its members.

Given the many interconnections in the habits of living organisms in the ecosystems that provide so much for humanity—from oxygen to clean water—conserving species clearly will help maintain the health of the world's many, important ecosystems and consequently the continued existence of humanity.

> ■ Biodiversity increases the productivity of ecosystems, and some evidence suggests it makes ecosystems more stable and resilient as well. The rivet hypothesis suggests that while the loss of individual species may not appear to matter, a cumulative loss of apparently insignificant species may lead to unexpected and sudden ecosystem collapse.

Highlight

Harvesting the Fruits of Nature

In addition to oxygen and clean water, the world's non-human species have much to offer humanity, from drugs to foods to spiritual sustenance. Let's look at some of these important contributions to our existence.

The biosphere directly provides many products used by humans. One-fourth of all prescription drugs dispensed by pharmacies are extracted from plants. Nearly as many come from animals, fungi, or microscopic organisms like bacteria. Quinine, used as an antimalaria drug, comes from a plant called yellow cinchona. Taxol, an important drug for treating cancer, comes from the Pacific yew tree. Bromelain, a substance that controls tissue inflammation, comes from pineapples. Wild species also provide chemicals that are useful as glues, fragrances, pesticides, and flavorings.

Wood for constructing homes and furniture is provided by many different tree species. Even the most basic requirements of human life are provided by other organisms. Plants produce the oxygen that we breathe. Every bit of food we eat is provided by other species. A wide variety of crops and livestock, such as tomatoes and cows, have been domesticated from wild species.

In addition, in many societies other wild species provide important foodstuffs. Insects are an important source of protein for many peoples around the world. In Central America, some people dine on the green iguana, a huge lizard that likes to sunbathe in treetops. Known as the chicken of the trees, this lizard has been a food source for 7000 years. In some parts of South America, the guanaco (a relative of the llama) is, like the cow in the United States, an important source of meat and hides. Also in South America, the capybara, the world's largest rodent, is a prized source of meat.

Capybara

Biodiversity also provides the world with aesthetic gifts. Scarlet macaws, parrot fish, sea anemones, tulip trees, and shooting-star wildflowers are among the species that make it clear that if there is a value to beauty, then biodiversity is worth a lot. For many people, the existence of such a rich, living world goes beyond mere beauty, to providing spiritual refreshment and rejuvenation. Consider also the argument that biodiversity does not have to fulfill a human need in order to have value and to have the right to exist unperturbed and not be destroyed by other species, such as our own.

As great as the bounty of nature already is, with so many species yet undiscovered, the vast majority of nature's wealth remains untapped. If the sheer numbers of species yet to be discovered are any indication, much more awaits—beauty, food, shelter, nutriment—if we can just find it before it has disappeared from the planet.

> ■ Biodiversity provides a number of products to human society: food, shelter, clothing, drugs, and, for some, spiritual rejuvenation.

Summary

How Many Species Are There on Earth?

- Despite intense worldwide interest, scientists do not know the exact number of species on Earth.

- Using indirect methods, scientists have generated estimates of the total number of Earth's species ranging from 3 million to 100 million species, with most estimates falling between 3 million and 30 million.

- A large proportion of the world's species remain undiscovered.

- Of the 1.5 million species already discovered, about half are insects.

The Beginnings of a Present-Day Mass Extinction

- Although it is difficult to measure extinction totals and rates precisely, abundant evidence suggests that Earth is in the early stages of a mass extinction, which could end up being the most rapid in the history of life.

- Humans are the cause of the current mass extinction.

The Many Threats to Biodiversity

- Among the threats to biodiversity today are habitat loss and deterioration, invasion of foreign species, and climate change.

- Many of the threats to biodiversity today are a result of the continuing growth of human populations.

Mass Extinctions of the Past

- Each of the five previous mass extinctions on Earth wiped out millions of species, devastating the planet's biodiversity.

- The previous mass extinctions were not caused by humans.

- After each mass extinction of the past, the number of species on Earth has rebounded slowly, over millions of years.

The Importance of Biodiversity

- Biodiversity can make the ecosystems on which humanity depends more productive, stable, and resilient.

- The rivet hypothesis suggests that although the loss of one species here or there may have no clear effects, the cumulative effect of continual species loss could result in the sudden, seemingly inexplicable collapse of ecosystems.

- Humanity depends on healthy ecosystems and the diversity of the world's species for oxygen and clean water, among many other things.

Highlight: Harvesting the Fruits of Nature

- The many species of Earth provide a multitude of products, including foods, medicines, material for shelter and clothing, flowers, and other useful and beautiful items.

Key Terms

biodiversity p. 60	mass extinction p. 63
canopy p. 60	nonnative species p. 66
ecosystem p. 60	productivity p. 68
foreign species p. 66	rainforest p. 64
habitat p. 65	rivet hypothesis p. 69

Chapter Review

Self-Quiz

1. Estimates of the total number of species on Earth range between
   a. 1 billion and 3 billion.
   b. 50 million and 500 million.
   c. 3 million and 100 million.
   d. 1000 and 3000.

2. The number of species that have been collected, identified, and named is
   a. 15,000.
   b. 150,000.
   c. 1,500,000.
   d. 15,000,000.

3. The known species of insects, plants, and fungi each number in the
   a. hundreds.
   b. thousands.
   c. millions.
   d. billions.

4. Which of the following has been hypothesized to be a cause of past mass extinctions?
   a. human population growth
   b. water pollution
   c. an asteroid
   d. disappearance of rainforests

5. Which of the following is *not* considered a threat to species today?
   a. rainforest destruction
   b. human population growth
   c. climate change
   d. scientific research

6. Which of the following have scientists shown to be a benefit that biodiversity offers ecosystems?
   a. increased ability to withstand climate change
   b. increased productivity
   c. increased resistance to pollution
   d. increased resistance to disease

Review Questions

1. How did Terry Erwin estimate the number of insect species in the world, and why are such estimates so difficult to do?

2. How has biodiversity risen and fallen over time—when and in what way?

3. How has human population growth helped spur other threats to biodiversity?

4　　　　　　𝕿𝖍𝖊 𝕯𝖆𝖎𝖑𝖞 𝕲𝖑𝖔𝖇𝖊

Protect the Rights of Humans, Not Animals

To the Editor:

The most useful thing Congress could do this year is throw out the Endangered Species Act. This law, which supposedly protects endangered species from being destroyed by human activities, has everything completely upside down. I live in the Pacific Northwest, where many of us have long and proud family traditions of logging trees. Now all of a sudden we can't cut trees because conservationists are worried that a bird, like the spotted owl or the marbled murrelet, might be killed off in the process. People who are living in houses built of wood that my grandfather risked his life to harvest are telling us we can't go into these forests to make a living anymore!

The same kind of thing is happening everywhere, every day, with insects, worms, and fish getting in the way of progress. What's more important—a person making a living, or a bug that no one's ever heard of? No one would care if these species went extinct, but I care about providing for my family. I am sick of reading in your paper about how the world's species are going extinct and how terrible that is.

When you write about the destruction of tropical rainforests, what you never say is that most of the species in rainforests are insects that we all hate anyway! Biologists don't even know exactly how many species there are on Earth, let alone how many are going extinct. I have no doubt that most species—except humans—are doing just fine. What's really endangered are the rights of people.

Steven S. Gotling

Evaluating "The News"

1. Do nonhuman species have a right to exist? If so, where do the rights of human beings end and the rights of other species begin? Should logging or other industries or development ever be curtailed to protect, for example, a "bug that no one's ever heard of"?

2. In the past when endangered species were threatened by a new building or other development, that development was often entirely forbidden by the courts and the Endangered Species Act. Now, more often, when development conflicts with the survival of an endangered species, developers and conservationists compromise. Some land is used for building, and some is set aside for the endangered species. Why might such compromise be a good idea? Why not?

3. The author of the letter to the editor is correct in stating that biologists don't know either the exact number of species on Earth or the exact rate at which species are going extinct. Given their lack of precise information, should we take seriously biologists' concerns about worldwide extinction of species? Why or why not?

4. Write a short letter to the editor, either arguing for or against the importance of preserving biodiversity on the planet and backing up your statements with evidence wherever possible. This issue is alive and kicking. Every community has its conflicts between biodiversity and development, and every voice counts. Send your letter to the editor of the local paper.

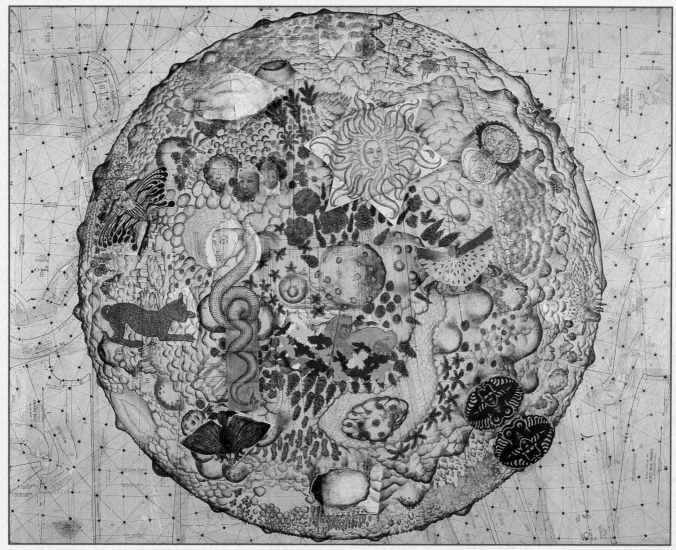

James Barsness, *The World All Around*, 1998.

Cells: The Basic Units of Life

chapter 5

Chemical Building Blocks

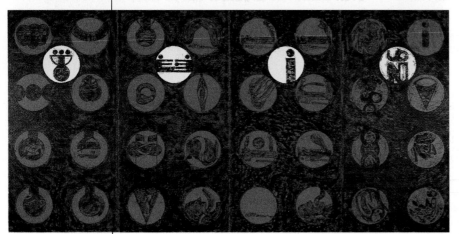

Matt Mullican, *Dallas Project, Cosmology Model*, panels 1–8, 1987.

The Martians Have Landed!

Ever since we discovered the existence of other planets, we have wondered whether Earth is the only planet in the universe that has life. The many stories of alien encounters and abductions found in the popular media reveal our fascination with the idea of life on other planets. In 1996, people who believe that life may exist elsewhere received a tremendous boost from studies of a piece of rock with the undistinguished name ALH84001.

This was not just any rock. It was an asteroid that was ejected from the planet Mars by a powerful volcanic event more than 100 million years ago. After traveling through space, this cosmic traveler landed in Antarctica about 13,000 years ago and was buried in glacial ice until its discovery in the twentieth century. Now, after millions of years as an anonymous rock, first in space and then in the Antarctic ice, ALH84001 has been thrust into the scientific and public spotlight as evidence that life exists on other planets in our solar system.

How can a simple rock provide such evidence? The answer lies inside the rock: the chemical remains of what may have been Martian life. Perhaps the most dramatic discovery thus far has

Main Message

Every living organism is made up of a limited number of chemical components that interact with each other in well-defined ways.

been shapes embedded in the rock that look like microscopic fossils. These shapes look remarkably like bacteria, and one in particular appears almost wormlike. The shock of seeing what looks like the remains of a living thing from another planet jolted the scientific world and injected new life into the public's belief in extraterrestrial life.

The idea that the observed forms were once alive has been met with some skepticism. Some scientists point to the need for other kinds of evidence besides the mere appearance of lifelike shapes. The search for more evidence has focused on the chemical compounds found in ALH84001, which may represent the residue of living organisms. If certain compounds that contain carbon are found, they might be

ancient traces of biological activity on Mars. With this in mind, scientists are applying what they have learned about the chemical compounds that make up life on Earth and are searching for similar compounds in ALH84001. This approach assumes that all life consists of a limited number of chemical compounds that interact with each other in well-established ways—a reasonable assumption, as we'll see in this chapter.

So far, the results of this quest for chemical evidence of Martian life have been inconclusive and even controversial. Simple carbon compounds have been found in ALH84001, together with traces of a few amino acids, which are the building blocks of proteins that are necessary for life. However, some scientists claim that these chemical traces could have come from the Antarctic ice in which the asteroid was buried for thousands of years, and thus might have originated on Earth. But even in those studies showing that the chemical traces are Earthly contaminants, a small percentage of the carbon compounds and amino acid compounds discovered are unknown on Earth and could not be found in Antarctic ice. These compounds may have originated on Mars.

Though the controversy is destined to continue for several years, chemical detective work combined with careful examination will eventually answer the question of whether there are traces of Martian life on asteroids that have fallen to Earth.

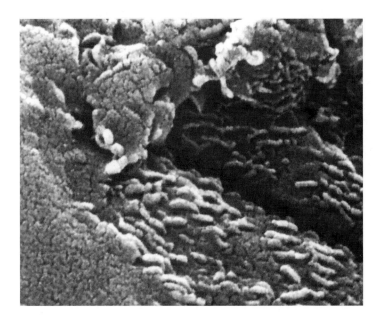

Tiny Lifelike Shapes Can Be Seen in the ALH84001 Meteorite

Key Concepts

1. Living organisms are composed of atoms linked together by chemical bonds. Arrangements of linked atoms (molecules) are essential for the processes of life.

2. Covalent bonds are the strongest chemical linkages that can form between two atoms. Most molecules found in living organisms are arrangements of atoms such as carbon, nitrogen, hydrogen, oxygen, phosphorus, and sulfur that are held together by covalent bonds.

3. Weaker, noncovalent bonds form between two or more separate molecules and between parts of a single molecule. Types of noncovalent bonds include hydrogen bonds and ionic bonds.

4. The chemical characteristics of water are essential to the chemistry of life. Water is both the primary medium for and a key participant in life-supporting chemical reactions.

5. Chemical reactions change the arrangement of atoms in molecules. These changes are responsible for the many different processes observed in living organisms.

6. The four major classes of chemical building blocks found in living organisms are nucleotides, amino acids, sugars, and fatty acids. Each class has several functions in living systems, ranging from information storage to energy transfer to structural support.

For all its remarkable diversity, life as we know it is based on a rather limited number of chemical elements found throughout the universe. The fact that such complexity can come from simple components underscores how humans could have arisen from the soup of a primitive Earth, using the chemical elements that make up the very Earth we stand on. More than anything else, this origin in Earth's chemical elements reminds us of the fundamental unity between our bodies and our surroundings.

In this chapter we begin to explore the tremendous complexity of living organisms, by first identifying their simplest chemical components. Then we examine how these simple components are linked together to form the many levels of organization in both the physical structures and chemical processes of life.

It stands to reason that the properties of an element like oxygen must depend on the properties of oxygen atoms. So what makes the atoms of one element different from those of another? The answer lies in the different combinations of what are called subatomic particles, bits of matter that make up the structure of individual atoms. Two subatomic particles of particular interest are **protons** and **electrons**. Both are electrically charged: Protons have a positive charge (+), and electrons have a negative charge (−). These opposite charges determine how atoms behave in the physical world and interact with each other: Like charges repel; opposite charges attract one another.

A single atom consists of a central core that contains protons and is thus positively charged. The core is sur-

How Atoms Make Up the Physical World

All components of the physical world are made up of 92 natural chemical elements, such as oxygen and calcium. **Elements** are the simplest building blocks of the physical world, and each is identified by a one- or two-letter code. For example, oxygen is identified as O, calcium as Ca. Each element exists as units so small that more than a trillion such units could easily fit on the head of a pin. These tiny units are called **atoms**, and there are 92 different kinds, one for each natural element. An atom is therefore the smallest unit of an element that still has the characteristic chemical properties of that element.

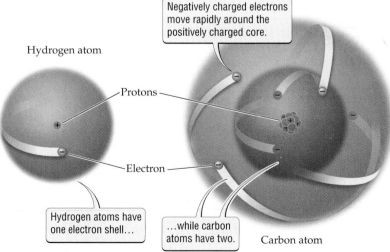

Figure 5.1 Atoms
The electrons, protons, and nucleus of these carbon and hydrogen atoms are shown greatly enlarged in relation to the size of the whole atom.

Hydrogen atom

Negatively charged electrons move rapidly around the positively charged core.

Protons

Electron

Hydrogen atoms have one electron shell...

...while carbon atoms have two.

Carbon atom

rounded by one or more negatively charged electrons (Figure 5.1). The positive core is called the nucleus of the atom, and electrons move around it like moons orbiting a planet. However, the troupe of electrons in a given atom is not fixed; the atom can lose, gain, and even share electrons with another atom. When an atom loses one or more of its negative electrons, it becomes positively charged. Likewise when an atom gains one or more electrons, it becomes negatively charged. Atoms that become charged due to the loss or gain of electrons are called ions.

The number of electrons associated with an atom determines the chemical properties of the element. These properties allow the atoms of one element to form linkages either with each other or with atoms of other elements. Orderly associations of atoms from different elements form what are known as **chemical compounds**, and the smallest unit of a compound with the required arrangement of atoms is called a **molecule**.

Chemists have developed a simple molecular formula to represent compounds made up of molecules ranging in size from two atoms to many thousands of atoms. For example, table salt is a compound that has equal numbers of sodium (Na) and chlorine (Cl) atoms. The molecular formula for salt is therefore NaCl. When the ratio of the different atoms is not one to one, a sub-script number is listed after any atoms that are present more than once. For example, each molecule of water is made up of two hydrogen (H) atoms and one oxygen (O) atom, so the molecular formula of water is H_2O. The same notation holds true for more complex compounds, such as table sugar (sucrose), which has 12 carbons, 22 hydrogens, and 11 oxygens ($C_{12}H_{22}O_{11}$) per molecule.

> ■ All components of the physical world, including living organisms, are made up of atoms that are linked together into specific associations called molecules.

Covalent Bonds: The Strongest Linkages in Nature

All chemical linkages that hold atoms together to form molecules are called bonds. The strongest of these linkages are **covalent bonds**, which link atoms within a single molecule. In covalent bonds, atoms share electrons (Figure 5.2). The electrons around the nucleus of every atom move in defined orbits that lie in layers, or shells, around the nucleus. Each shell can contain a certain maximum number of electrons, and the atom is in its most stable state when all shells are filled to capacity. One way

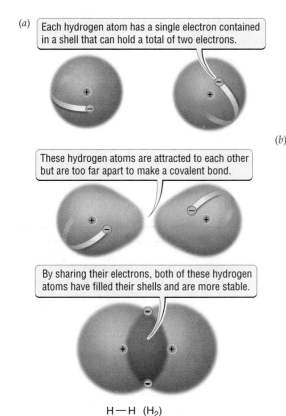

(a) Each hydrogen atom has a single electron contained in a shell that can hold a total of two electrons.

These hydrogen atoms are attracted to each other but are too far apart to make a covalent bond.

By sharing their electrons, both of these hydrogen atoms have filled their shells and are more stable.

H—H (H$_2$)

Figure 5.2 Covalent Bonds
(*a*) If close enough to each other, two hydrogen atoms can share electrons, forming a covalent bond. The length of a covalent bond is such that the electrons can be shared, while the two positively charged nuclei are still kept far enough apart that they don't repel each other. (*b*) The number of covalent bonds that an atom can form depends on the number of extra electrons needed to fill its outermost shell. An oxygen atom requires two more electrons to fill its shell and thus can form two covalent bonds. A carbon atom requires four more electrons and thus can form four covalent bonds.

(*b*)

Atom	Symbol	Number of possible bonds	Examples	
Oxygen	O	2 bonds	H—O—H	Water
Carbon	C	4 bonds	H—C—H (with H above and below)	Methane
Nitrogen	N	3 bonds	H—N—H (with H above)	Ammonia
Sulfur	S	2 bonds	H—S—H	Hydrogen sulfide
Hydrogen	H	1 bonds	H—H	Hydrogen gas

for an atom to fill its outermost shell is to share electrons with a neighboring atom. This sharing of electrons links the two atoms by forming a strong covalent bond.

Water and the natural gas called methane have the molecular formulas H_2O and CH_4, respectively. These formulas reveal the atomic components of each compound, but they say nothing of how the various atoms are bonded together and arranged in space. Another type of notation, known as a structural formula, is used to indicate both the atoms and the bonding arrangement of compounds. As Figure 5.2b shows, individual water molecules are held together by covalent bonds between the single oxygen atom and each of two hydrogen atoms. Likewise, methane has four covalent bonds that link the lone carbon atom to each of four hydrogens.

The bonding arrangement of atoms in a particular molecule is never accidental or random. Thus, there will never be a water molecule in which one hydrogen is bonded to another hydrogen, which is then bonded to oxygen. When two hydrogen atoms share electrons, they form hydrogen gas (H_2) and have no electrons left to share with an oxygen atom. Oxygen, on the other hand, has the room in its outer shell of electrons to form two covalent bonds with hydrogen atoms (H_2O). These examples show that the bonding properties of each atom are determined by its electrons. Therefore, covalent bonds occur in well-defined ways that establish the spatial arrangement of the atoms and hence the three-dimensional shape of the molecule.

Carbon, nitrogen, hydrogen, oxygen, phosphorus, and sulfur are the most common elements found in living organisms, and their atoms all form covalent bonds. Consequently, combinations of all or some of these atoms are found in the molecules that make up living organisms.

> ■ Covalent bonds are the strongest chemical linkages between atoms and are required for the assembly of molecules.

Noncovalent Bonds: Dynamic Linkages between Molecules

Whereas covalent bonds link atoms to form molecules, **noncovalent bonds** are the most common linkages between separate molecules and between different parts of a single large molecule. In fact, by bringing molecules together in specific configurations, noncovalent bonds promote the complex organization that we perceive as the diverse physical structures and activities characteristic of living organisms. Despite this important role, noncovalent bonds are far weaker than covalent bonds and do not involve the direct sharing of electrons between atoms. Instead, the various kinds of noncovalent bonds are based on other chemical properties of two or more atoms.

Because of the relative weakness of noncovalent bonds, many such bonds are often required to hold two molecules together, and even to link two parts of the same large molecule. In this case, however, weakness is a virtue because it also allows molecules to come together and pull apart more easily. The resulting changeability of molecular arrangements is necessary for many living processes. For example, when skin is pinched and then released, its ability to stretch and then spring back depends on the breaking and re-forming of noncovalent bonds between many different classes of molecules.

Hydrogen bonds are extremely important temporary bonds in nature

The **hydrogen bond** is one of the most important kinds of noncovalent bonds in nature. Each hydrogen bond is about 20 times weaker than a covalent bond. During rapidly changing life processes such as muscle contraction, hydrogen bonds are broken and re-formed moment by moment.

The simplest example of a hydrogen bond can be found between water molecules (Figure 5.3). Hydrogen bonding is an important attribute of water and is based on a particular chemical property of water molecules. Most organisms consist of more than 70% of water by weight, and nearly every chemical process associated with life occurs in water.

As discussed earlier, each molecule of water is made up of two hydrogen atoms and one oxygen atom that are held together by covalent bonds. The positive nucleus of the oxygen atom tends to attract the negative electrons of the two hydrogen atoms, causing each water molecule to have an uneven distribution of electrical charge. That is, the oxygen atom carries a slightly negative charge, while the hydrogen atoms in turn carry a slightly positive charge. Molecules with an uneven distribution of charge are described as **polar**.

Figure 5.3 Hydrogen Bonds Determine the Properties of Water and How Other Compounds Interact with Water Each water molecule forms temporary hydrogen bonds with its neighbors, resulting in a network of molecules that account for water being liquid at room temperature. Water molecules are also attracted to electrically charged ions. Thus, compounds that are held together by ionic bonds, such as table salt (NaCl), will dissolve in water as each ion becomes surrounded by water molecules.

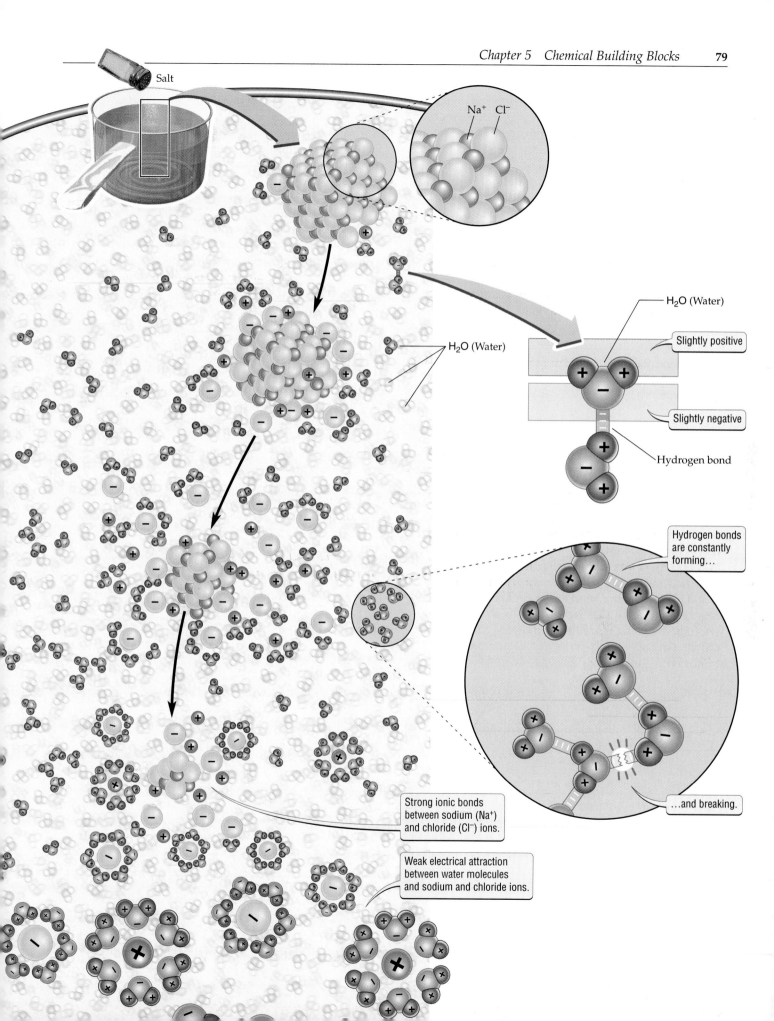

Salt

Na⁺ Cl⁻

H₂O (Water)

H₂O (Water)

Slightly positive

Slightly negative

Hydrogen bond

Hydrogen bonds
are constantly
forming…

…and breaking.

Strong ionic bonds
between sodium (Na⁺)
and chloride (Cl⁻) ions.

Weak electrical attraction
between water molecules
and sodium and chloride ions.

The slightly positive hydrogen atoms of one water molecule are attracted to the slightly negative oxygen of a neighboring water molecule because opposite charges attract. This attraction forms a hydrogen bond linkage between the two water molecules, which in turn can form hydrogen bonds with other neighboring water molecules (see Figure 5.3). The resulting network of hydrogen-bonded water molecules is responsible for the liquid properties of water.

Any polar compound that contains hydrogen atoms can form hydrogen bonds with a neighboring polar compound that contains a partially negative atom. Thus water molecules can form hydrogen bonds with other polar compounds, with the result that these compounds dissolve in the water. Such compounds are said to be **soluble** in water, since each of the compound's molecules is surrounded by water molecules and can move freely throughout the liquid. For example, when sugar is added to water, the solid crystals break apart as each molecule of sugar is surrounded by water molecules and is then scattered uniformly throughout the water.

The hydrogen-bonding properties of water molecules also mean that they will not associate with other molecules that are not charged—that is, that are nonpolar. When nonpolar molecules are added to water, instead of being separated they are pushed into clusters. This is exactly what happens when olive oil is added to a salad dressing. The distribution of electrons between carbon and hydrogen is nearly equal in oil molecules, making them nonpolar. In the salad dressing, water molecules do not mix with nonpolar oil molecules. Instead the oil molecules tend to cluster into tiny floating droplets when the dressing is shaken. Waxes are also nonpolar, and automobile enthusiasts wax their cars not just to make them look shiny, but also to repel water and reduce the risk of rusting.

Molecules such as sugar that freely interact with water are **hydrophilic** (*hydro*, "water"; *philic*, "loving"), while those such as oil that are repelled by water are **hydrophobic** (*phobic*, "fearing").

Ionic bonds form between atoms of opposite charge

Ionic bonds are important noncovalent bonds that, like hydrogen bonds, rely on the fact that opposite electrical charges attract. In this case, the attraction between two atoms of opposite charge has different consequences from those of hydrogen bonding. The electrons in the outer shell of one atom are completely transferred to the outer shell of the second atom. As mentioned earlier, the loss of electrons by one atom and the gain of electrons by another means that both become charged ions.

Ionic bonds between molecules dissolved in water are relatively weak. Like hydrogen bonds, however, they are essential for many temporary associations between molecules. For example, our ability to taste what we eat depends on the ionic bonds that form between food molecules dissolved in water and other specialized molecules in our taste buds. The rapid association and dissociation of multiple food molecules also enable us to discern several different tastes in a short amount of time. If these associations were not brief, every meal would be dominated by the taste of the first bite.

Ionic bonds between molecules under dry conditions can be very strong. For example, dry table salt consists of countless sodium ions (Na^+) linked by ionic bonds to chloride ions (Cl^-). In the absence of water, these ions pack tightly to form the hard, three-dimensional structures we know as salt crystals. When salt is added to water, the polar water molecules are attracted to the charged ions surrounding both types of ions. This interaction with water breaks up and dissolves the salt crystals, scattering both positive and negative ions throughout the liquid (see Figure 5.3).

Like sugar, then, salt is soluble in water. But there is one key difference: In the case of salt, water molecules are attracted to and surround each ion because of its charge, but hydrogen bonds do not form. Hydrogen bonds form only between polar compounds and not between water molecules and ions. Therefore, ions and polar molecules form two different classes of hydrophilic compounds that dissolve in water.

> Noncovalent bonds such as hydrogen bonds and ionic bonds are temporary linkages that are responsible for the changeability of biological molecules that is necessary for living processes.

How Chemical Reactions Rearrange Atoms in Molecules

The arrangements of atoms found in molecules are the simplest examples of physical structures found in living organisms. We have identified the covalent bonds that link atoms into molecules and the noncovalent bonds that exist between molecules. However, molecules are not static arrangements of atoms. Living processes require atoms to break existing connections and form new ones with other atoms. Consider the additional molecules that must be made to provide a growing plant with components for new leaves and stems. The plant does not produce these new molecules from raw atoms; rather it acquires and rearranges pre-existing molecules. The processes that break and form bonds between atoms are known as **chemical reactions**.

The standard notation for chemical reactions describes changes in the arrangement of atoms. Nitrogen and hydrogen can combine to produce ammonia gas (NH_3), which gives laundry bleach its sharp odor. The chemical reaction is written as follows:

$$3\,H_2 + N_2 \rightarrow 2\,NH_3$$

The arrow indicates that hydrogen and nitrogen are converted to ammonia, and the numbers before molecules define how many molecules participate in the reaction. In this case three molecules of hydrogen gas (H_2) combine with one molecule of nitrogen gas (N_2) to produce two molecules of ammonia.

The reaction begins and ends with the same number of each type of atom. Thus the ratio of atoms that make up the molecules (reactants) at the start of the reaction must be the same as the ratio of atoms that make up the molecules (products) at the end of the reaction. That is, the total number of each kind of atom on the left side of the reaction must match the number of that atom on the right side. In the example here, each side of the equation has six hydrogen molecules and two nitrogen molecules. All chemical reactions rearrange atoms and can neither create nor destroy them.

Later in this chapter we'll see how chemical reactions promote more complex molecular organization. Now, however, we'll focus on the chemical reactions that determine how ions and molecules interact with water.

Acids, bases, and pH

All chemical reactions that support life occur in water. Of particular importance are those that involve two classes of compounds: acids and bases. An **acid** is a polar substance that dissolves in water and loses one or more hydrogen ions (H^+). These hydrogen ions prefer to bond with the surrounding water molecules, forming a positively charged hydronium ion, H_3O^+.

Because the formation of hydronium ions is easily reversed, hydrogen ions are constantly being exchanged between water molecules and other molecules dissolved in water.

Bases are also polar and are compounds that accept hydrogen ions from water molecules. Thus acids and bases interact with water molecules in different ways and have opposite effects on the amount of free hydrogen ions in water.

The concentration of free hydrogen ions in water influences the chemical reactions of many other molecules. This hydrogen ion concentration is expressed as a scale of numbers from 0 to 14, where 0 represents the highest concentration of free hydrogen ions and 14 rep-

resents the lowest. This is called the **pH** scale, and each unit represents a 10-fold increase or decrease in the concentration of hydrogen ions in a sample of water.

When water contains no acids or bases, the concentrations of free hydrogen ions and hydroxyl ions are equal, and the pH is said to be neutral, or in the middle of the scale, at pH 7. Below pH 7, the solution is said to be acidic because the concentration of free hydrogen ions is higher as a result of donations from an acid. Above pH 7, the solution is said to be basic because the concentration of free hydrogen ions is lower as a result of the acceptance of hydrogen ions by a base.

We have all experienced acidic and basic substances. The tartness of lemon juice in a good homemade lemonade is due to the acidity of the juice (about pH 3). Our stomach juices are able to break down food because they are very acidic (about pH 2). At this low pH, many bonds between molecules are disrupted by the high concentration of free hydrogen ions that associate with atoms that would otherwise be bonded to other atoms. At the other extreme is a substance such as oven cleaner, which is very basic (about pH 13).

Buffers prevent large changes in pH

Most living systems function at an internal pH that is close to 7. Any change in pH to a value significantly below or above 7 adversely affects many life processes. Because hydrogen ions can move so freely from one molecule to another during normal life processes, organisms need to control the pH levels of their internal environments. This need is met by compounds called **buffers**, which maintain the concentration of hydrogen ions between narrow limits. Most buffers are weak acids or bases or both, and can therefore release or accept free hydrogen ions to maintain a relatively constant concentration in an internal environment.

> ■ Chemical reactions alter the atomic composition and/or arrangement of molecules, thereby generating new properties. The chemical characteristics of water dictate the properties of acids and bases, and the pH of the environment.

The Chemical Building Blocks of Living Systems

If we removed the water in any living organism, we would be left with four major groups of chemical compounds: nucleotides, amino acids, sugars, and fatty acids. These chemical compounds consist of different

(a)

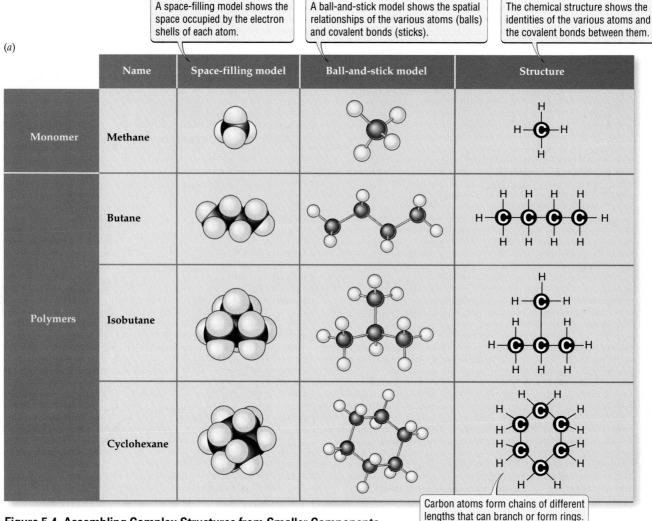

	Name	Space-filling model	Ball-and-stick model	Structure
Monomer	Methane			
Polymers	Butane			
	Isobutane			
	Cyclohexane			

A space-filling model shows the space occupied by the electron shells of each atom.

A ball-and-stick model shows the spatial relationships of the various atoms (balls) and covalent bonds (sticks).

The chemical structure shows the identities of the various atoms and the covalent bonds between them.

Carbon atoms form chains of different lengths that can branch or form rings.

Figure 5.4 Assembling Complex Structures from Smaller Components In living organisms, the important principle of building large and complex structures from smaller components is applied to both atoms and molecules. (*a*) A single carbon atom can form four covalent bonds with other atoms. When carbon atoms form bonds with other carbon atoms, a variety of chains can be formed, with different structures such as branches and rings. (*b*) Small chemical compounds also form covalent bonds, giving rise to larger molecules with specific chemical properties. The small constituent compounds are called monomers, while the resulting large assemblage is called a polymer.

(*b*) Monomers

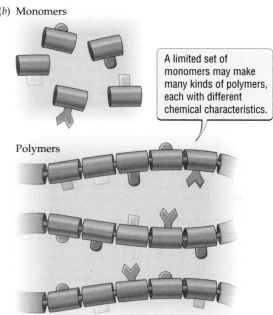

A limited set of monomers may make many kinds of polymers, each with different chemical characteristics.

Polymers

arrangements of carbon atoms, though hydrogen, oxygen, nitrogen, phosphorus, and sulfur atoms may also be present.

Carbon is the predominant atom in living systems partly because it can form large molecules that contain thousands of atoms. A single carbon atom can form strong covalent bonds with up to four different atoms. Even more importantly, carbon can bond to carbon, forming long chains, branched trees, or even rings. Living processes can therefore use the wide variety of large and complex structures that are possible with carbon-containing molecules. All carbon compounds found in living organisms are described as organic compounds or molecules.

Small carbon compounds, containing about 20 atoms, can either remain as individual molecules or bond with other small molecules to form larger structures called macromolecules (*macro*, "large"). Carbon compounds in living organisms often follow this principle of building very large structures from small units. Individual small molecules are called **monomers** (*mono*, "one"; *mer*, "part"). Macromolecules, which can contain hundreds of monomers bonded together (Figure 5.4), are called **polymers** (*poly*, "many"). Polymers account for most of an organism's dry weight and are essential for every structure and chemical process that we associate with life.

In living organisms, fewer than 70 different biological monomers are combined in a nearly endless variety of ways, to produce polymers with many different properties. Polymers are therefore a step up from monomers in organizational complexity, and they have chemical properties that are not possible for a monomer. Furthermore, many carbon polymers acquire chemical properties from attached **functional groups** (Table 5.1).

As the name implies, functional groups are groups of covalently linked atoms that have specific chemical properties wherever they are found. Some functional groups help establish covalent linkages between monomers; others have more general effects on the chemical characteristics of a polymer.

The importance of these properties is illustrated in each of the four groups of organic compounds found in all living systems: nucleotides, amino acids, sugars, and fatty acids. As we consider each of these groups of compounds in the sections that follow, we will see how tremendously complex and changeable structures can be based on such a limited range of organic compounds.

Nucleotides are the building blocks of DNA and RNA

Nucleotides are organic compounds that consist of linked rings of atoms. In each nucleotide (Figure 5.5), one ring structure, known as a **nucleotide base**, is covalently bonded to another ring structure, known as a sugar. The five different bases are named adenine, cytosine, guanine, thymine, and uracil. Two different sugars can be attached to the bases. Deoxyribose sugars can bond to adenine, cytosine, guanine, or thymine bases, forming the building blocks for a stringlike polymer called deoxyribonucleic acid, or **DNA**. Ribose sugars can attach to adenine, guanine, cytosine, and uracil bases, forming the building blocks for a polymer called ribonucleic acid, or **RNA**. (We will learn a lot more about DNA and RNA

5.1 **Some Important Functional Groups Found in Biological Molecules**

Functional group	Formula	Ball-and-stick model
Amino	NH_2	Bond to carbon atom
Carboxyl	COOH	
Hydroxyl	OH	
Phosphate	PO_4	

in Unit 3, on genetics.) A third component of nucleotides is the phosphate group, which consists of a phosphate atom and four oxygen atoms (see Table 5.1).

Nucleotides have two essential functions in the cell—information storage and energy transfer—both of which highlight key characteristics of a living system. Every organism is defined by a coded "blueprint" that dictates what chemical building blocks are produced to make up the organism, and how they are assembled. The order in which different nucleotides are hooked together in the polymers of DNA or RNA determines all the physical attributes of and chemical reactions that occur in living organisms. The structure of DNA and its role as the storehouse of genetic information will be discussed in Chapter 14.

In addition to storing genetic information, nucleotides assist in energy transfer. A key player is adenosine triphosphate, or **ATP**, which is an adenine nucleotide with three attached phosphate groups. Although not directly eaten, ATP is the universal fuel for living organisms, and many chemical reactions require energy from ATP in order to proceed. The energy of ATP is contained in the covalent bonds that link its three phosphate groups (Figure 5.6). The breaking of the bonds between

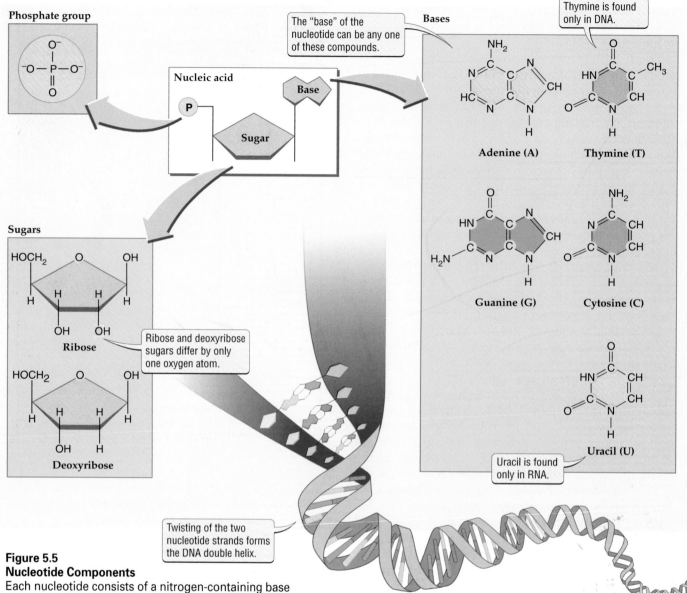

Figure 5.5
Nucleotide Components
Each nucleotide consists of a nitrogen-containing base linked to a sugar and one or more phosphate groups. The bases adenine, guanine, cytosine, and thymine, when linked to deoxyribose sugar, form the building blocks of DNA. The bases adenine, guanine, cytosine, and uracil, when linked to ribose sugar, form the building blocks of RNA.

the second and third phosphates is what releases energy used to drive other chemical reactions. The production of ATP and how it is used as an energy currency will be discussed in detail in Chapter 8.

Amino acids are the building blocks of proteins

Of the many different kinds of chemical compounds found in living organisms, **proteins** may be the most familiar. These polymers make up more than half the dry weight of living things. The steaks we throw on the grill in the summer consist mainly of proteins, and our ability to run to the finish line of a race depends on the coordinated actions of many proteins in our muscles. Our bodies are made up of thousands of different proteins. Some of these proteins form physical structures like hair; others, called enzymes, help regulate the chemical reactions that drive living processes. We will discuss how enzymes work in Chapter 7.

Amino acids are the monomers used to build proteins. There are 20 different amino acids, all of which share some structural characteristics. All amino acids have a carbon atom called the alpha carbon, which forms a central attachment site for several functional groups: an amino group (—NH$_2$), a carboxyl group

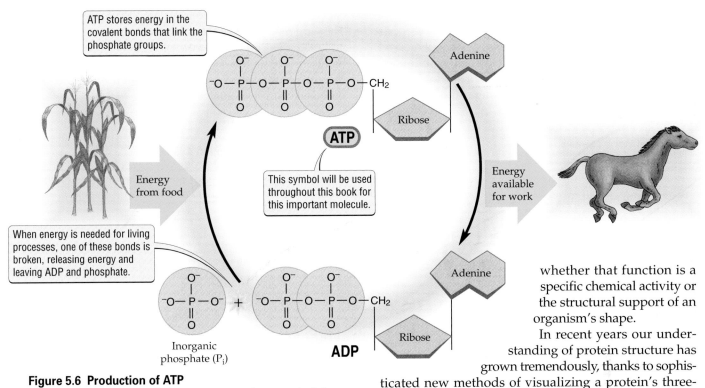

ATP stores energy in the covalent bonds that link the phosphate groups.

ATP

This symbol will be used throughout this book for this important molecule.

When energy is needed for living processes, one of these bonds is broken, releasing energy and leaving ADP and phosphate.

Adenine

Ribose

Energy from food

Energy available for work

Adenine

Ribose

Inorganic phosphate (P$_i$)

ADP

Figure 5.6 Production of ATP
ATP is the major form of short-term stored energy in living organisms. The energy obtained from food is used to form ATP from ADP (adenosine diphosphate) and phosphate.

(—COOH), and a chemical side chain called the reactive or R group (Figure 5.7).

The different R groups give different amino acids different properties. R groups range from a single hydrogen atom to complex arrangements of carbon chains and ring structures (Figure 5.8). Thus organisms have a diverse pool of building blocks from which they can make proteins with many different properties.

To form proteins, linear chains of amino acids are linked together by covalent bonds between the amino group of one amino acid and the carboxyl group of another. When amino acids are linked together in a chain, the chain bends and folds so that the R groups can interact to define the overall structure and properties of the protein. Both noncovalent and covalent bonds between the R groups of amino acids in the same protein help fold the linear chain and maintain its compact three-dimensional structure. For most proteins, this specific folded structure is necessary for normal function,

whether that function is a specific chemical activity or the structural support of an organism's shape.

In recent years our understanding of protein structure has grown tremendously, thanks to sophisticated new methods of visualizing a protein's three-dimensional shape. These breakthroughs have helped unravel the mysterious chemical forces that shape proteins, paving the way for the design of improved, synthetic proteins. We return to this important idea at the end of the chapter.

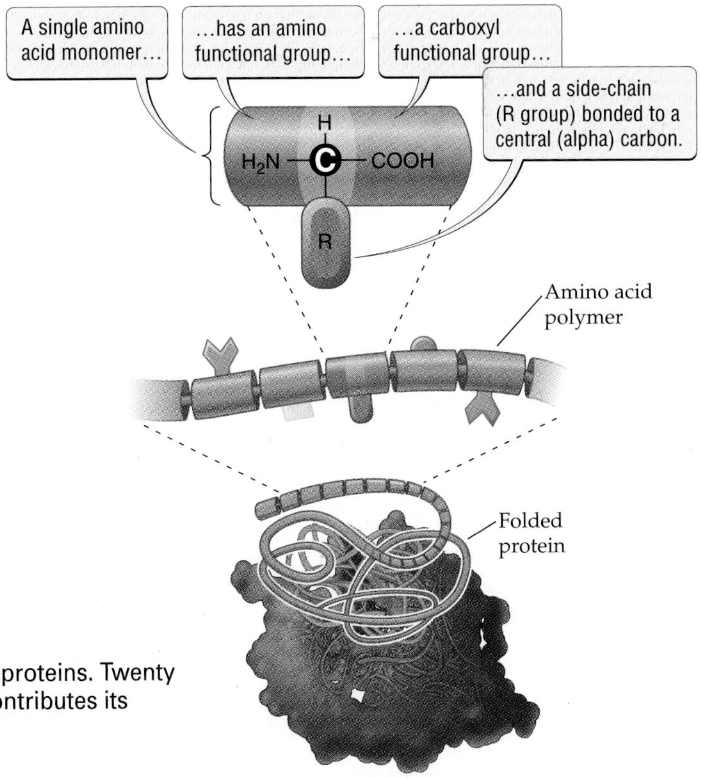

A single amino acid monomer…

…has an amino functional group…

…a carboxyl functional group…

…and a side-chain (R group) bonded to a central (alpha) carbon.

H$_2$N—C—COOH

R

Amino acid polymer

Folded protein

Figure 5.7 Amino Acids Make Up Proteins
Amino acids are the building blocks that form polymers called proteins. Twenty amino acids are known, each with a different side chain that contributes its chemical properties to the protein.

Sugars provide energy for living processes

Sugars are familiar to us as compounds that make foods taste sweet. Although only some sugars are perceived as sweet by human taste buds, most sugars are important food sources and a major means of storing energy in living organisms. The simplest sugar molecules are called **monosaccharides** (*mono*, "one"; *sacchar*, "sugar"). They are made up of units containing carbon, hydrogen, and oxygen atoms in the ratio of 1:2:1. This ratio can also be expressed as the chemical formula $(CH_2O)_n$, with *n* ranging from 3 to 7. The best-known example of a monosaccharide is glucose.

Figure 5.8 The Diversity of Amino Acids
Each of the 20 amino acids has a different R group bonded to the alpha carbon. The various R groups give each amino acid chemical properties that can be classified as either hydrophobic, hydrophilic, or special.

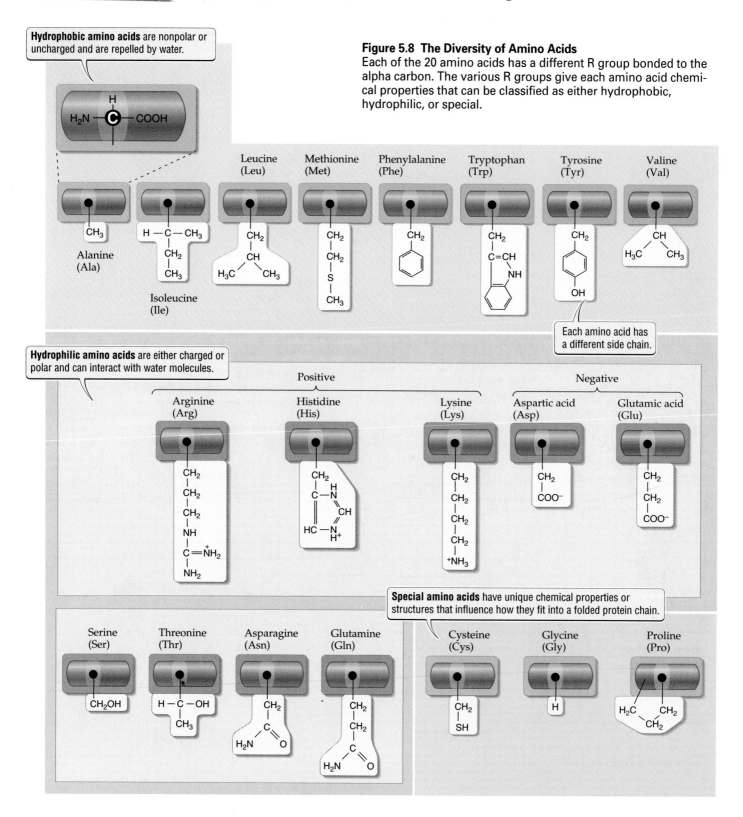

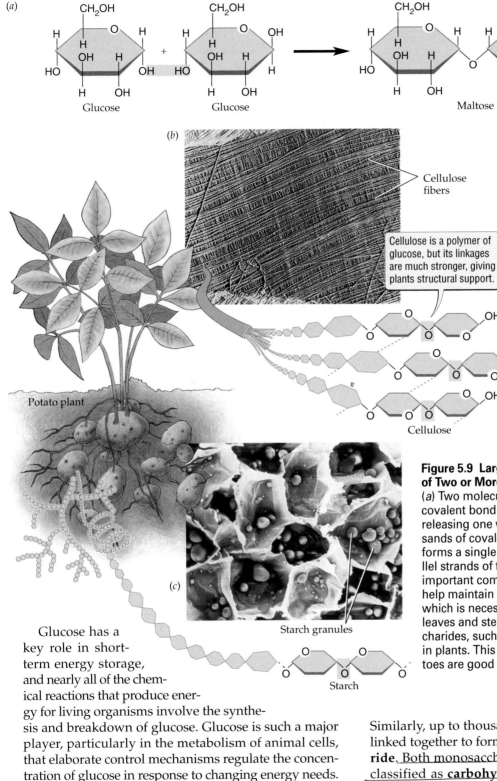

(a) Glucose + Glucose → Maltose + H₂O

(b) Cellulose fibers

Cellulose is a polymer of glucose, but its linkages are much stronger, giving plants structural support.

Cellulose

Potato plant

(c) Starch granules

Starch

Figure 5.9 Larger Carbohydrates Are Made Up of Two or More Sugar Molecules Joined Together
(*a*) Two molecules of glucose can be linked by a covalent bond to form the disaccharide maltose, releasing one water molecule. A chain of thousands of covalently bonded glucose molecules forms a single polysaccharide molecule. (*b*) Parallel strands of the polysaccharide cellulose are important components of plant cell walls. They help maintain the rigid structure of cell walls, which is necessary for the physical soundness of leaves and stems. (*c*) Highly branched polysaccharides, such as starch, are used to store energy in plants. This is why starch-rich foods like potatoes are good sources of energy.

Glucose has a key role in short-term energy storage, and nearly all of the chemical reactions that produce energy for living organisms involve the synthesis and breakdown of glucose. Glucose is such a major player, particularly in the metabolism of animal cells, that elaborate control mechanisms regulate the concentration of glucose in response to changing energy needs. The role of glucose in providing energy for life will be discussed further in Chapter 8.

Monosaccharides can combine to form larger, more complex molecules. For example, two covalently bonded molecules of glucose form maltose (Figure 5.9*a*).

Similarly, up to thousands of monosaccharides can be linked together to form a polymer called a **polysaccharide**. Both monosaccharides and polysaccharides are classified as **carbohydrates**, since each carbon atom (*carbo-*) is linked to the equivalent of a molecule of water (*-hydrate*).

Carbohydrates perform several different functions in living organisms. For example, a polysaccharide called cellulose forms strong parallel fibers that help

(a)

> Stearic acid is a saturated fatty acid. It contains no double bonds between its carbon atoms.

(b)

> Oleic acid is an unsaturated fatty acid. It has one double bond between two of its carbon atoms.

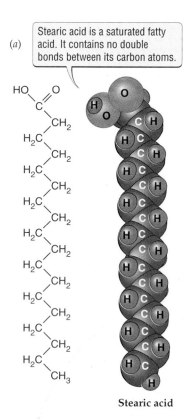

Stearic acid

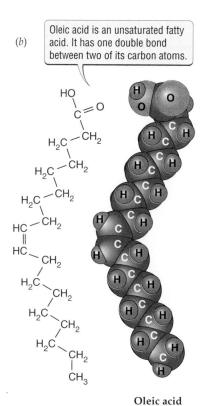

Oleic acid

Figure 5.10 Saturated and Unsaturated Fatty Acids
(*a*) This space-filling model of stearic acid, with the chemical structure shown to the left of it, shows that stearic acid is straight and can pack tightly to form a waxy solid at room temperature. (*b*) The double bond in oleic acid forms a kink, preventing its molecules from packing as tightly as they do in stearic acid. Oleic acid therefore tends to be more liquid at room temperature and is commonly found in the storage fat of humans.

The long hydrocarbon chains found in fatty acids usually contain 16 or 18 carbon atoms that can vary in the way they are covalently bonded together. Fatty acids in which all the carbon atoms are linked together by single covalent bonds are said to be **saturated** because each carbon is also bonded to a full complement of hydrogens (Figure 5.10*a*). When one or more of the carbon atoms are linked together by double covalent bonds, the fatty acid is said to be **unsaturated** because a bond to hydrogen must be sacrificed to create each double covalent bond between two carbon atoms (Figure 5.10*b*).

The significance of the double bonds in unsaturated fatty acids goes beyond simple differences in the number of hydrogens linked to the carbon chain. Single bonds tend to adopt a straight configuration in space, while double bonds tend to introduce kinks in the hydrocarbon chain. Consequently, the straighter saturated fatty acids can pack very tightly together, forming solids such as fats and waxes at room temperature. In contrast, the double-bond kinks in unsaturated fatty acids prevent such tight packing, so these compounds tend to be liquid (oils) at room temperature.

The role of fatty acids in energy storage and membrane structure depends on the covalent bonding of fatty acids to a simple three-carbon compound named glycerol. Glycerol has three OH groups, each of which can form a covalent bond with the carboxyl group (—COOH) at the end of a fatty acid chain. When all three OH groups are bonded to a fatty acid, the resulting compound is the most common storage fat in animals and plants (Figure 5.11). Fats contain significantly more energy than does an equal amount of glucose.

The linkage of two fatty acid chains and a negatively charged phosphate group to glycerol produces anoth-

support the leaves and stems of plants (Figure 5.9*b*). On the other hand, carbohydrates provide fuel (energy), and are the basis for "carbo loading" before a strenuous sporting activity like a marathon. Starch, found in so many foods, is one example of a polysaccharide that provides energy (Figure 5.9*c*). Just as nucleic acids have more than one essential function in the life of an organism, sugars play multiple roles, ranging from providing energy to forming important structural components of the body.

Fatty acids store energy and form membranes

Fatty acids are molecules composed primarily of long carbon-and-hydrogen chains, referred to as **hydrocarbons**. They are also the key components of fats and lipids. **Fats** function in the long-term storage of energy for living organisms and are familiar to some of us as the prime targets of weight loss programs. The role of fats in energy storage will be discussed in Chapter 7. **Lipids**, on the other hand, help form essential physical boundaries both inside organisms and between an organism and the external environment. Lipids establish the surface structure of living cells and control the exchange of molecules between cells and their environment. These lipid-based boundaries, called **membranes**, are discussed shortly.

er class of compounds, called **phospholipids**. Phospholipids are the major component of all membranes in living organisms. The negatively charged phosphate group on one end of the phospholipid means that this region of the molecule can interact with polar water molecules or with positively charged ions and thus is hydrophilic. In contrast, the fatty acid chains are hydrophobic. They are entirely nonpolar and therefore are repelled by water molecules.

The resulting dual character of phospholipids allows them to form double-layered sheets in water that expose hydrophilic heads to the water while keeping the hydrophobic tails isolated in the middle. This double-layered sandwich of molecules is called a phospholipid bilayer, and is the basis of all biological membranes (Figure 5.12). Membranes are a clear demonstration of how the chemical properties of molecules in water define a physical structure that is essential for living organisms. The structure and roles of biological membranes will be discussed in Chapter 6.

> ■ Nucleotides, amino acids, sugars, and fatty acids are the molecular building blocks that make up all living organisms.

Highlight

Proteins That Can Take the Heat

Chemical reactions are as common in the kitchen as they are in the laboratory. For example, a hard-boiled egg is so different in texture from a raw egg because chemical changes during cooking cause the egg to change from the thick, sticky liquid rich in the nutrients needed by a developing chick embryo to the firm white and yellow mass that tastes so good. What compounds are responsible for these changes? The answer is proteins, which simply cannot withstand heat.

Proteins are very particular about the temperatures at which they will remain folded and function normally. Most properly folded proteins tend to have hydrophilic R groups exposed on their surfaces and hydrophobic R groups isolated on the inside. This careful arrangement allows the protein to form hydrogen bonds with surrounding water molecules and remain dissolved.

When heated beyond a certain temperature, weak noncovalent bonds break and proteins unfold, losing their regular three-dimensional structure. However, the covalent bonds that link one amino acid to another remain intact. In the ensuing chaos, hydrophobic R groups from several protein molecules randomly cluster together away from water, forming a featureless

Glycerol

3 molecules of stearic acid
(glyceryl tristearate)

> A single molecule of glycerol can be covalently linked to three fatty acids to form a fat molecule.

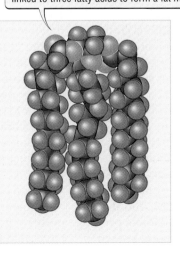

Figure 5.11 Fats
When the fatty acids that bond to glycerol are stearic acid, the resulting fat is glyceryl tristearate, the most common storage form of fat in animals and plants.

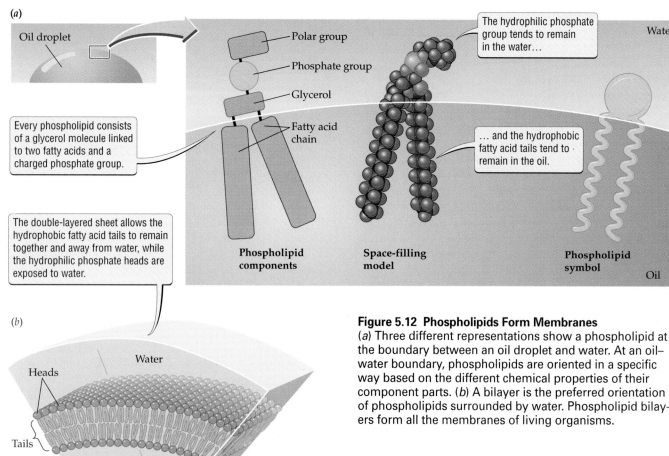

(a)

Oil droplet

Every phospholipid consists of a glycerol molecule linked to two fatty acids and a charged phosphate group.

Polar group

Phosphate group

Glycerol

Fatty acid chain

The hydrophilic phosphate group tends to remain in the water...

... and the hydrophobic fatty acid tails tend to remain in the oil.

Water

Phospholipid components

Space-filling model

Phospholipid symbol

Oil

The double-layered sheet allows the hydrophobic fatty acid tails to remain together and away from water, while the hydrophilic phosphate heads are exposed to water.

(b)

Water

Heads

Tails

Water

Phospholipid bilayer

Figure 5.12 Phospholipids Form Membranes
(*a*) Three different representations show a phospholipid at the boundary between an oil droplet and water. At an oil–water boundary, phospholipids are oriented in a specific way based on the different chemical properties of their component parts. (*b*) A bilayer is the preferred orientation of phospholipids surrounded by water. Phospholipid bilayers form all the membranes of living organisms.

blob. In the case of an egg in boiling water, proteins such as albumin that were previously dissolved in the watery environment of the egg at room temperature are unfolded by the heat and randomly cluster together into the firm white of the boiled egg.

Until the 1970s it was accepted that proteins could not be heated significantly above body temperature and be expected to function normally. Then came the discovery of bacteria that live and thrive in hot springs at temperatures just at the boiling point of water (100°C). To survive, these bacteria must have proteins that remain functional at temperatures as high as 100°C. To improve our understanding of how proteins might be able to withstand such high temperatures, biologists set out to isolate and study the proteins from these heat-loving organisms. They discovered several proteins that promote chemical reactions in these bacteria, and that are fully active at 100°C. One of these heat-stable proteins is even the basis of the forensic technique known

as the polymerase chain reaction that has figured so prominently in recent criminal trials in which matching of DNA samples is important evidence.

Biologists are now learning how these proteins are able to take the heat, and they are trying to alter proteins from other organisms that live at room temperature to be just as active at 100°C. This endeavor is not meant merely to satisfy scientific curiosity. Learning how to engineer a heat-resistant protein means discovering what kinds of chemical bonds form a folded structure that is exceptionally stable. Thus, proteins that are engineered to survive higher temperatures are also far more stable at room temperature and could last longer on a supermarket shelf.

How do biologists make a protein more heat-resistant, or in other words, better folded? The answer lies in understanding how the chemical properties of amino acids form the bonds that fold proteins into their highly stable three-dimensional shapes. One recent test case that proves the possibility of such an approach involves a protein called thermolysin.

Found in bacteria that live at room temperature, thermolysin helps break down other proteins. Biologists have changed 8 out of a total of 319 amino acids to produce a reengineered thermolysin that is fully active at 100°C. In making these amino acid changes, the researchers caused the formation of a new covalent bond that locked the three-dimensional structure of the enzyme in place, making it heat-resistant. In comparison to the normal protein, the reengineered thermolysin was 340 times more stable at 100°C.

Although we may not want to engineer an egg that will not hard-boil, the ability to engineer proteins that remain active at high temperatures opens up exciting new possibilities. Not only could protein-rich food products be engineered to last far longer without chemical preservatives, but even medically relevant proteins like vaccines could be modified so that they no longer required cold storage conditions, an important benefit for developing countries.

> ■ Understanding how the chemical properties of amino acids define protein shapes has allowed biologists to engineer proteins that are more stable.

Summary

How Atoms Make Up the Physical World

- The physical world is made up of atoms. There are 92 different kinds of atoms, one for each chemical element known.

- Individual atoms are made up of subatomic particles, which include positively charged protons in the core and negatively charged electrons surrounding the core. The particular combination of subatomic particles associated with an atom determines its chemical properties.

- When atoms lose or gain electrons, they become either positively or negatively charged ions.

- Chemical compounds are formed by specific arrangements of atoms from different elements. These arrangements depend on the unique chemical characteristics of the atoms involved.

Covalent Bonds: The Strongest Linkages in Nature

- Atoms are linked together by a variety of different chemical bonds.

- Covalent bonds, formed by the sharing of electrons between atoms, connect the atoms within a molecule. They are the strongest bonds in nature.

- The proton core of every atom is surrounded by a specific arrangement of electrons that move in defined orbits or shells. Atoms share electrons with other atoms to fill their outer electron shells to capacity. The bonding properties of an atom are determined by its particular arrangement of electrons.

- Covalent bonds establish the spatial arrangement of atoms in a molecule and hence its three-dimensional shape.

Noncovalent Bonds: Dynamic Linkages between Molecules

- Noncovalent bonds link separate molecules and are secondary links between atoms within a single large molecule. They are weaker than covalent bonds and do not involve the sharing of electrons between atoms. Important types of noncovalent bonds include hydrogen bonds and ionic bonds.

- Hydrogen bonds form between polar compounds that have an unequal distribution of charge. Partially positive hydrogen atoms in one compound form hydrogen bonds with partially negative atoms in another compound. Hydrogen bonding between water molecules accounts for the physical properties of water and is essential for life.

- Compounds that interact freely with water and each other are hydrophilic; nonpolar compounds, which are repelled by water, are hydrophobic. In hydrophobic interactions, nonpolar compounds tend to group together in water.

- Ionic bonds form between two atoms when electrons from one atom are transferred to another atom.

How Chemical Reactions Rearrange Atoms in Molecules

- Chemical reactions break and form bonds between atoms. They neither create nor destroy atoms, but merely alter the arrangement of atoms in molecules.

- All chemical reactions that support life occur in water. Specialized classes of compounds called acids and bases affect the amount of free hydrogen ions in water. Acids donate hydrogen ions to water molecules, forming H_3O^+; bases accept hydrogen ions from water molecules, forming OH^-.

- The concentration of free hydrogen ions in water is expressed as the pH scale of numbers. The pH scale ranges from 0 (highest concentration of hydrogen ions—that is, most acidic) to 14 (lowest concentration—that is, most basic). Most living organisms prefer a pH of about 7.

- Buffers can both donate and accept hydrogen ions to and from water molecules. They help maintain the constant internal pH that is necessary for the chemical reactions of life.

The Chemical Building Blocks of Living Systems

- Carbon compounds form the physical framework for all biological molecules. The ability of carbon atoms to form large and complex polymers has an important role in the generation of diverse biological building blocks.

- Nucleotides are the building blocks of DNA and RNA. DNA polymers made up of four types of nucleotides form the blueprint for life and dictate all the physical features and chemical reactions of a living organism. Specialized nucleotides function in the storage and transfer of energy from one chemical reaction to another.

- Amino acids are the building blocks of proteins. The chemical properties of individual amino acids are determined by their different R groups. The function and three-dimensional shape of a protein are defined by the chemical properties of the amino acids it contains.

- Sugars provide both energy and physical support for living organisms. Carbohydrates consist of simple sugars (monosaccharides) and more complex polymers (polysaccharides).

- Fatty acids are the building blocks of fats and lipids. Fats are an important means of long-term energy storage; lipids are the basic components of biological membranes.

Highlight: Proteins That Can Take the Heat

- Most proteins lose their three dimensional folded structure when heated.

- Changing a few amino acids in a protein may allow it to withstand high temperatures.

- Reengineered heat-resistant proteins will have important uses in food products and medicine.

Key Terms

acid p. 81	lipid p. 88
amino acid p. 84	membrane p. 88
atom p. 76	molecule p. 77
ATP p. 83	monomer p. 83
base p. 81	monosaccharide p. 86
buffer p. 81	noncovalent bond p. 78
carbohydrate p. 87	nonpolar p. 80
chemical compound p. 77	nucleotide p. 83
chemical reaction p. 81	nucleotide base p. 83
covalent bond p. 77	pH p. 81
DNA p. 83	phospholipid p. 88
electron p. 76	polar p. 78
element p. 76	polymer p. 83
fat p. 88	polysaccharide p. 87
fatty acid p. 88	protein p. 84
functional group p. 83	proton p. 76
hydrocarbon p. 88	RNA p. 83
hydrogen bond p. 78	saturated p. 88
hydrophilic p. 80	soluble p. 80
hydrophobic p. 80	sugar p. 86
ion p. 77	unsaturated p. 88
ionic bond p. 80	

Chapter Review

Self-Quiz

1. The atoms of a single element
   a. have the same number of electrons.
   b. can form linkages only with atoms of the same element.
   c. can have different numbers of electrons.
   d. can never be part of a chemical compound.

2. Two atoms can form a covalent bond
   a. by sharing protons.
   b. by swapping nuclei.
   c. by sharing electrons.
   d. by sticking together on the basis of opposite electrical charges.

3. Which of the following statements about molecules is true?
   a. A single molecule can contain atoms from only one element.
   b. The chemical bonds that link atoms into a molecule are arranged randomly.
   c. Molecules are found only in living organisms.
   d. Molecules can contain as few as two atoms.

4. Which of the following statements about ionic bonds is *not* true?
   a. They cannot exist without water molecules.
   b. They are not the same as hydrogen bonds.
   c. They require the loss of electrons.
   d. They are more temporary than covalent bonds.

5. Hydrogen bonds are especially important for living organisms because
   a. they occur only inside of organisms.
   b. they are very strong and maintain the physical stability of molecules.
   c. they allow biological molecules to dissolve in water, which is the universal medium for living processes.
   d. once formed, they never break.

6. Glucose is an important example of a
   a. protein.
   b. carbohydrate.
   c. fatty acid.
   d. nucleotide.

Review Questions

1. All the major chemical building blocks found in living organisms form polymers. Why are polymers especially useful in the organization of living systems?

2. A sample of pure water contains no acids or bases. Predict the pH of the water and explain your reasoning.

3. Describe the chemical properties of carbon atoms that make them especially suitable for forming biological polymers.

4. Polymers of amino acids contain chemical information that is important for life. How is this information used by a living organism?

5   **The Daily Globe**

Hard Candy Could Save Your Life

RESEARCH TRIANGLE PARK, NC. How would you like to preserve yourself forever? Imagine a time when individuals with a fatal disease can be preserved in suspended animation until a cure is found. Today this idea may conjure up images from Saturday morning science fiction movies, but thanks to recent discoveries made in plants, the idea may soon be more than just fantasy.

Corn seeds can survive very adverse conditions by entering a sort of suspended animation. The corn seed can remain dry for decades and then sprout vigorously when it receives water. Biologists have discovered that the delicate proteins necessary for all the seed's living functions are carefully preserved in a protective coating of hard sugar. When the seed comes in contact with water, the sugar dissolves, the proteins become active, and the seed sprouts.

Certain insects also encase themselves in sugars to survive long winters in a frozen state of suspended animation. But will this ever work for humans? The answer from Ivana Livalot, president and CEO of Methuselah Technologies, is a resounding yes.

In a recent press conference, Livalot announced a joint venture with researchers at Yoakum University that would test the use of complex sugars in the suspended animation of small mammals. When asked about the source of funding for this venture, Livalot admitted that most of it came from private investors with a "strong personal interest" in the technology.

Evaluating "The News"

1. If preserving proteins is so important for the success of suspended animation, what does this tell you about the role of proteins in living processes?

2. For Methuselah Technologies to develop a suspended animation method for humans, it must eventually be tested on human subjects. Which individuals should have the highest priority as test subjects for suspended animation?

3. Do you think that the greater complexity of our brains is likely to make waking up from suspended animation more difficult for humans than for insects? Would this affect your willingness to be a human volunteer in a clinical trial?

chapter 6

Cell Structure and Compartments

Terry Winters, *Direction Field*, 1996.

Inner Space Travel

*I*n 1997, planet Earth was treated to a spectacular astronomical event: the comet Hale-Bopp. You did not need to be an astronomer with a telescope to observe this event, since the comet was easily visible to the naked eye. In fact, it was a fixture in the night sky for weeks on end as a bright particle with a trailing luminous tail.

Many of the millions of people who gazed up at this magnificent galactic intruder would have been surprised to know that another invader with a remarkably similar appearance can sometimes be found in the human body. For those unfortunate enough to have been exposed to a bacterium named *Listeria monocytogenes*, comet-shaped invaders move around various organs of the body much as the comet Hale-Bopp wanders from one solar system to another.

Listeria infects humans and can cause severe gastroenteritis and damage to the nervous system. Nearly 2000 people in the United States become seriously ill from *Listeria* each year, and more than 400 die from the infection. The bacterium is found in contaminated foods, and recent outbreaks have been traced to sources as seemingly harmless as chocolate milk. What is most remarkable about *Listeria*, however, is that through a microscope it can be seen trav-

Main Message

Prokaryotes and eukaryotes differ in the way they form internal compartments to concentrate and facilitate chemical reactions.

eling around in organ samples taken from infected individuals.

Since every organ in the body is made up of small, membrane-enclosed units called cells, the

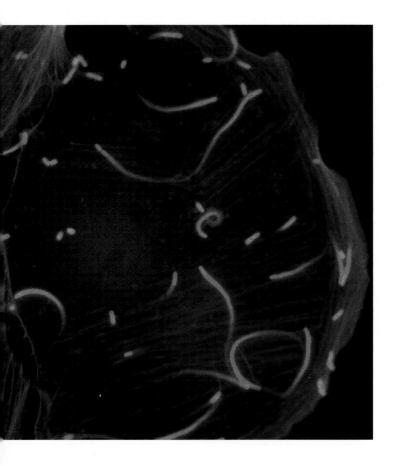

The Bacteria *Listeria monocytogenes* Moving inside of an Infected Cell

bacteria must cross membranes to move from cell to cell. In the microsope, these moving *Listeria* look very much like comets with rod-shaped bodies and trailing tails. Biologists now know that the "comet tail" of *Listeria* is made up of proteins that are captured from the infected cell and made to form fibers that extend from the body of the bacterium. The growth of these fibers provides the propulsive force for the bacterium, literally pushing it through the cell. This force is so significant that when *Listeria* hits the inside of the membrane, it pushes the membrane out to form a spike, allowing it to leap to another cell and spread the infection.

Although the beautiful sight of the comet moving through space bears a coincidental resemblance to the movement of *Listeria* through the inner space of the cell, the relationship of the tail to the body is very different in the comet. The tail of a comet is nothing more than a trail of gases and ice crystals that contributes nothing to its movement. In the case of *Listeria*, the tail drives the movement.

The evolution of *Listeria*'s ability to use the cell's own proteins against it underscores the rich possibilities for function that lie in the components of the cell. In this chapter we will explore some of these components and their normal function in the life of a cell.

Key Concepts

1. All living organisms are made up of one or more basic units called cells.

2. The plasma membrane forms a boundary around every cell. It limits the movement of molecules both into and out of the cell, and it determines how the cell communicates with the external environment.

3. Prokaryotes are single-celled organisms that lack a nucleus and have very little internal organization. Eukaryotes are single-celled or multicellular organisms that have a nucleus and several other specialized internal compartments.

4. The specialized internal compartments of the eukaryotic cell are called organelles. They concentrate and transport the macromolecules necessary for such processes as the production of energy and the breakdown of food material.

5. Organelles like mitochondria and chloroplasts are probably descendants of primitive prokaryotes that were engulfed by the ancestors of the eukaryotic cell.

6. The cytoskeleton is composed of distinct filament systems and their associated proteins. It has an important role in supporting cell shape and movement.

Every complex structure imaginable can be broken down into smaller and simpler parts. Even something as impressive as the Empire State Building in New York City can be reduced to simple components such as concrete blocks and steel girders. Based on this assumption, we can begin to understand a complex system by first identifying and examining its elementary components.

The basic principle of building complex structures out of simple components applies to living systems as much as it applies to skyscraper buildings. Chapter 5 discussed the atoms and other chemical components that make up the building blocks of life. These components must be further arranged into living units before an organism like a human being can exist. In other words, macromolecules like proteins and DNA must be organized into more complex arrangements that promote the chemical reactions required for life. This chapter explores the next level of organization in living systems that ultimately defines the simplest unit of life.

Cells: The Simplest Units of Life

The **cell** is the basic unit of life. In the same way that molecules like proteins and fatty acids are made up of atoms, every living organism known is made up of from one to billions of membrane-enclosed units called cells. Bacteria are single cells, while a complex organism such as a human being contains billions of cells. Cells make up every organ in our bodies, and they determine how we look, move, and function as organisms.

Given the wide variety of different organs and specialized parts in the human body, it is not surprising that more than 200 different kinds of cells are required to make up all these different parts. Let's compare the cells that make up muscles with those that form the clear lens of the eye. Muscle cells have the important task of generating the movements we experience as muscle contractions and relaxations. They have specific protein components that allow them to change shape and generate physical force.

In contrast, the cells that make up the lens of the eye do not need to generate physical force. Instead, their task is to help focus light into the eye, allowing for vision. They therefore contain specific protein components that focus light as it passes through the cell in much the same way that a glass lens focuses light into a camera.

The diversity of cells is far greater than the range of cell types found in the human body. In fact, none of the cell types found in our bodies is found in any plant. Furthermore, with millions of different species on Earth, the variety of cell types is enormous. Yet even with so many different kinds of cells, certain basic components and structures are shared by all cells. Our ability to see these structures in the microscope led to the discovery of cells (see the Box on page 97).

In this chapter we tour the cell and examine its major structural components. We begin at the physical boundary of the cell and work our way inside to discover the structures and compartments that allow cells to support living processes.

Cells are the basic units that make up all living organisms.

Exploring Cells in the Microscope

To see something is to begin to know it. This simple statement is just as true of biology as it is of fine art. Our awareness of cells as the basic units of life is based largely on our ability to see them. The instrument that opened the eyes of the scientific world to the existence of cells is known as the light microscope and was invented in the last quarter of the sixteenth century. The key components of early light microscopes were ground-glass lenses that bent incoming rays of light to produce magnified images of tiny specimens.

The study of magnified images began in the seventeenth century when Robert Hooke examined a piece of cork in a microscope and saw that it was made up of little compartments. Hooke described these structures as small rooms, or cells, originating the term we use today. Ironically, what Hooke saw in the early microscope were not living cells, since cork is dead plant tissue. Instead, the small chambers that he saw were nothing more than empty cell walls. However, the discovery of previously invisible living things proceeded rapidly, opening up a new world to scientific exploration.

While the light microscope has a place in the early history of biology, similar instruments are just as important in ongoing research today. The basic principles that enable light microcopes to magnify the image of a specimen remain the same as shown in the figure, but the quality of current lenses has improved significantly. Thus, the 200- to 300-fold magnification achieved in the seventeenth century has been improved to the well over a 1000-fold magnification achieved by today's standard light microscopes. This degree of magnification allows us to distinguish structures as small as 1/2,000,000 of a meter, or 0.5 micrometer (μm). Light microscopes therefore reveal not just animal and plant cells (5–100 μm), but also organelles such as mitochondria and chloroplasts (2–10 μm), and bacteria (1 μm).

Since the 1930s, an even more dramatic increase in magnification has been achieved by the replacement of visible light with streams of electrons that are focused by powerful magnets instead of glass lenses. Called electron microscopes, these instruments can magnify a specimen by more than 100,000-fold, revealing the internal structure of cells and even individual molecules such as proteins and nucleic acids. Both types of microscopy give us insights into how cells are organized and how one type of cell is physically adapted to a specific function in the body of a multicellular organism.

The ability to distinguish the various parts of a specimen in the microscope depends on the existence of a way to create contrast. This has been a challenge for light microscopy because cells are generally very transparent and different cellular structures do not contrast with one another, tending as a result to be indistinguishable. The earliest solution to this problem was to stain specific parts of cells with dyes that would alter the light passing through the specimen and produce different colors distinguishing one structure from another. Today similar methods are still used to visualize cellular structures and even the distribution of specific proteins in the cell.

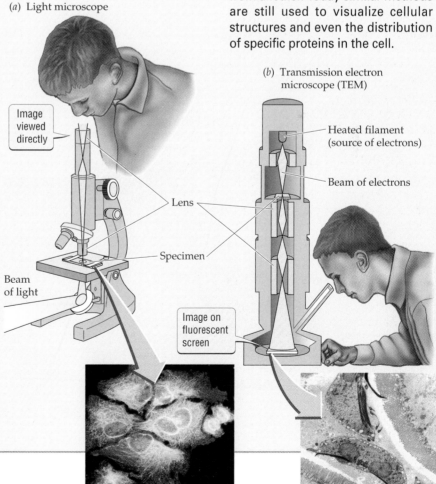

(*a*) Light microscope

Image viewed directly

Lens

Specimen

Beam of light

(*b*) Transmission electron microscope (TEM)

Heated filament (source of electrons)

Beam of electrons

Image on fluorescent screen

The Plasma Membrane: Separating Cells from the Environment

One of the key characteristics of life is the existence of a boundary that separates the organism from its nonliving environment. If all the molecules and proteins that were discussed in Chapter 5 were allowed to diffuse freely in the environment, they would not encounter each other frequently enough for life-sustaining reactions to take place. Thus, an enclosed compartment—the cell—concentrates all the required compounds in a limited space and allows the appropriate chemical reactions of life.

The boundary structure that defines all cells is called the **plasma membrane**. As introduced in Chapter 5, biological membranes are composed mainly of a bilayer of phospholipids oriented such that their hydrophilic (water-loving) heads are exposed to the watery environments both inside and outside of the cell. In turn, the hydrophobic (water-fearing) fatty acid tails are grouped together in the interior of the membrane.

If the plasma membrane had no other function but to define the boundary of the cell and keep all its contents inside, a simple phospholipid bilayer would certainly suffice. However, the plasma membrane must also allow living cells to communicate with the outside environment, selectively capture essential molecules, and release unwanted waste products. Cells therefore need a plasma membrane that can selectively allow some molecules to pass through, while (1) preventing others from entering or leaving the cells and (2) receiving signals from the outside environment.

This selective permeability of plasma membranes depends on various proteins that are inserted either all the way through or partially through the phospholipid bilayer. As Figure 6.1 shows, some proteins form gateways that allow the passage of selected ions and molecules into and out of the cell. Other proteins are used by the cell to recognize changes in the outside environment, including signals from other cells.

Unless they are anchored to structures inside the cell, most proteins that are embedded in the plasma mem-

Figure 6.1 Proteins Embedded in the Plasma Membrane
Proteins can span the phospholipid bilayer in a variety of ways.

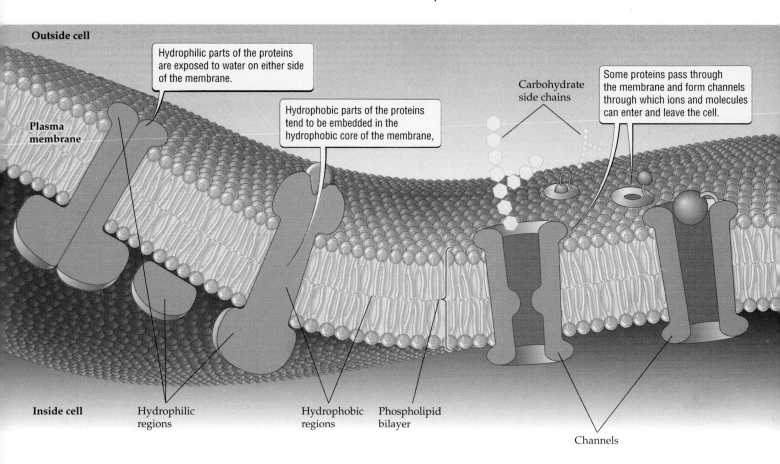

Outside cell

Hydrophilic parts of the proteins are exposed to water on either side of the membrane.

Plasma membrane

Hydrophobic parts of the proteins tend to be embedded in the hydrophobic core of the membrane,

Carbohydrate side chains

Some proteins pass through the membrane and form channels through which ions and molecules can enter and leave the cell.

Inside cell   Hydrophilic regions   Hydrophobic regions   Phospholipid bilayer

Channels

brane are free to move in the plane of the phospholipid bilayer. This freedom of movement supports what is known as the **fluid mosaic model**, which describes the plasma membrane as a highly mobile mixture of phospholipids and proteins. Although the plasma membrane is a common feature of all cells, the set of proteins found in the membrane varies from one cell type to another. The specific combination of proteins determines how cells interact with the external environment and contributes to the unique properties of each cell type.

> ■ Every cell is surrounded by a plasma membrane that separates the chemical reactions of life from the nonliving environment. Proteins in the plasma membrane allow the cell to communicate and respond to the environment.

Comparing Prokaryotes and Eukaryotes

All the living organisms that we know today can be classified into two groups based on whether their cells contain internal membrane-enclosed compartments. Organisms with cells that lack internal membranes are known as **prokaryotes** (*pro*, "before"; *karyote*, "kernel"). Those that have internal membrane-enclosed compartments are known as **eukaryotes** (*eu*, "true").

Prokaryotes have very little internal organization, and they were probably the first cells to arise during evolution. Today, all prokaryotes are single-celled bacteria that are usually spherical or rod-shaped and have a tough cell wall. The cell wall is made up of polysaccharides and protein, and it forms outside of the plasma membrane (Figure 6.2). It helps maintain the shape and structural integrity of the bacterium.

A typical bacterial cell consists of a watery jellylike substance known as the **cytosol**. The cytosol contains a multitude of molecules, such as DNA, RNA, proteins, and enzymes, all suspended in water. These components, together with a host of free ions, support the chemical reactions necessary for life. There are so many small and large molecules crowded into the cytosol that it behaves more like a thick jelly than like a free-flowing liquid. In fact, the cytosol is probably similar to the rich soup that first supported life billions of years ago.

A well-studied bacterium, *Escherichia coli*, is a common resident of the human intestine and is only 2 millionths of a meter (2 micrometers) long. The relatively small size of bacteria may account for their ability to get along without further internal organization. Since the chemical components are contained in such a small vol-

ume of cytosol, they do not need to be further concentrated for their particular activities.

Eukaryotes can exist as single cells (like yeasts) or large multicellular organisms (like humans). All multicellular organisms are collections of eukaryotic cells that are specialized for different functions. The major distinction between prokaryotic and eukaryotic cells is the presence in eukaryotic cells of an internal membrane-enclosed compartment called the **nucleus**. This structure houses most of the cell's DNA (see Chapter 5), effectively separating it from the remainder of the cell's components. The nucleus is therefore a cellular compartment with a specific function.

All eukaryotic cells are even further organized by the presence of several other membrane-enclosed compartments (see Figure 6.2). Like the nucleus, each compartment in the eukaryotic cell is formed from internal membranes and has a specific function. This arrangement is important since most eukaryotic cells are about a hundred times larger than bacteria, and they cannot rely on chemical components being close enough together for the necessary reactions to occur.

> ■ Living organisms can be classified as either prokaryotes or eukaryotes. Prokaryotes are made up of cells that lack internal membrane-enclosed compartments; the cells of eukaryotes have internal compartments such as the nucleus. The larger size and greater diversity of eukaryotic cells require a more complex internal organization than is found in prokaryotes.

The Specialized Internal Compartments of Eukaryotic Cells

The typical eukaryotic cell can be compared to a large corporate office building with many rooms and different departments. Each department must have a specific function and internal organization that contributes to the overall "life" of the corporation. Specific subdivisions represent an effective means of accomplishing particular tasks. For example, if all the members of the shipping department were scattered throughout the building, it would be difficult for the corporation to ship out packages of goods in a timely fashion. Thus, the packers and shippers are all located in a centralized shipping department, effectively concentrating and coordinating their efforts.

The eukaryotic cell faces challenges similar to our hypothetical corporation, since specific processes need to be carried out more efficiently by specialized "depart-

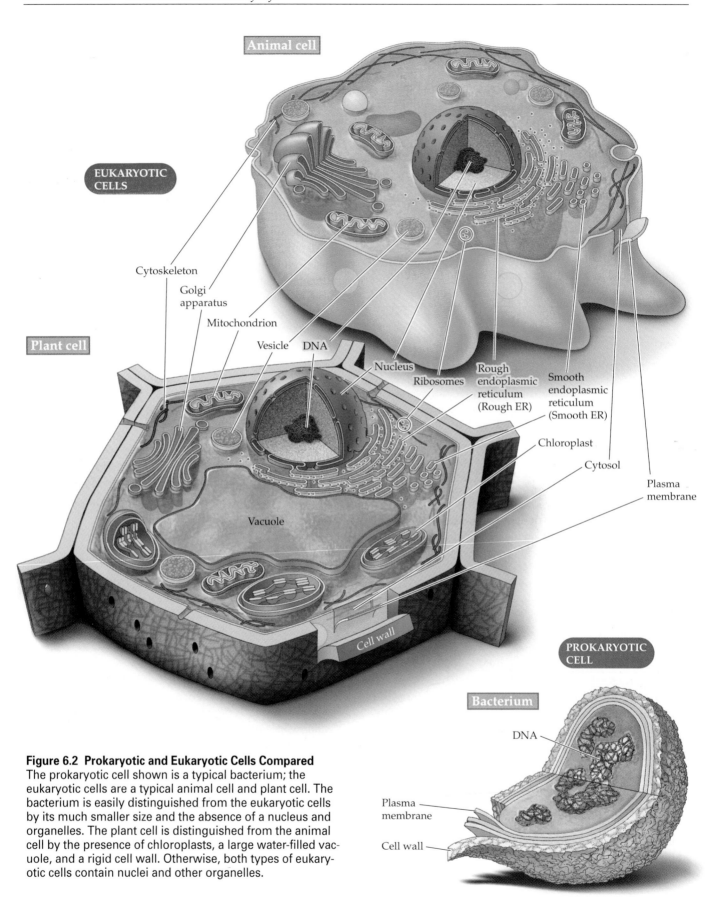

Figure 6.2 Prokaryotic and Eukaryotic Cells Compared
The prokaryotic cell shown is a typical bacterium; the eukaryotic cells are a typical animal cell and plant cell. The bacterium is easily distinguished from the eukaryotic cells by its much smaller size and the absence of a nucleus and organelles. The plant cell is distinguished from the animal cell by the presence of chloroplasts, a large water-filled vacuole, and a rigid cell wall. Otherwise, both types of eukaryotic cells contain nuclei and other organelles.

ments." The goals of enhanced efficiency are similar for both a corporation and the cell. They include the faster manufacturing of products, which for the cell could be energy-rich compounds like ATP. The specialized departments of the cell are the various membrane-enclosed compartments that divide its contents into smaller spaces. These smaller spaces contain, isolate, and concentrate the proteins and smaller molecules necessary for different processes. For example, processes such as ATP production and protein synthesis occur in different compartments.

The cell's membrane-enclosed compartments are called **organelles**, a name that is especially appropriate since they are the "little organs" of the cell. In the same way that organs such as the heart and lungs have different and unique functions in the human body, each organelle has specific duties in the life of the cell. Unlike bacteria, in which all the contents of the cell inside the plasma membrane form the cytosol, the contents of a eukaryotic cell are divided between the cytosol and organelles. In other words, the eukaryotic cytosol consists of all the cell contents inside the plasma membrane, excluding the organelles. **Cytoplasm** is another common term used to describe all the contents of the eukaryotic cell, excluding only the nucleus.

The nucleus is the storehouse for genetic material

The nucleus is the most distinctive organelle in eukaryotic cells. As discussed earlier, this membrane-enclosed compartment distinguishes eukaryotes from prokaryotes, which do not have a nucleus. Returning to the comparison with a corporation, the nucleus of the cell is equivalent to the administrative offices of the corporation. In other words, the nucleus is the specialized compartment that directs the activities and physical appearance of the cell. It fulfills this function by housing the cell's DNA, which carries the information necessary for all the activities and structures of the cell (Figure 6.3).

Inside the nucleus, long polymers of DNA are packaged with proteins into the remarkably small space. Since eukaryotic cells can have more than a thousand times more DNA than bacteria, the careful packing of DNA with protein is necessary for it to fit inside the nucleus.

The arrangement of membranes that surround the nucleus is different from that of the plasma membrane that surrounds the whole cell. The boundary of the nucleus is called the **nuclear envelope** and is a double membrane sandwich that contains small openings called

nuclear pores (see Figure 6.3). These pores are the gateways that allow the movement of molecules into and out of the nucleus, enabling it to communicate with the rest of the cell.

The pores are essential features of the nuclear envelope because the transfer of information encoded by DNA depends on the movement of molecules out of the nucleus. Likewise, how and when this DNA information is used depend on specialized proteins that move into the nucleus. How the DNA in the nucleus dictates the activities of the cell will be discussed in Chapter 15.

The endoplasmic reticulum manufactures proteins and lipids

If the nucleus functions as the administrative offices of the cell, the endoplasmic reticulum is the factory floor where many of the cell's chemical building blocks are manufactured. The **endoplasmic reticulum (ER)** is surrounded by a single membrane that is connected to the outer membrane of the nuclear envelope. Unlike the nucleus, which is usually an irregular spherical structure, the ER is an extensive and complex network of interconnected tubes and flattened sacs. You can visualize it as a series of mem-

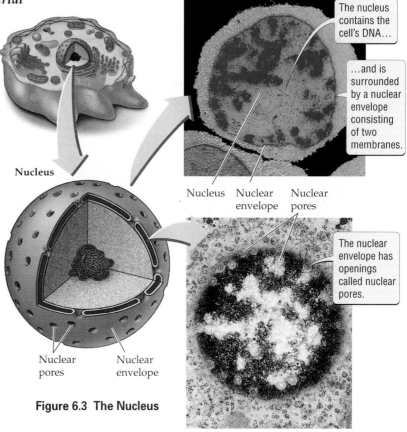

Figure 6.3 The Nucleus

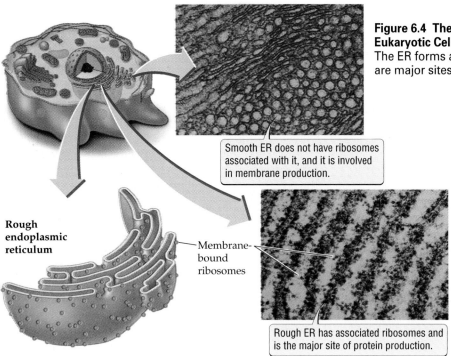

Figure 6.4 The Endoplasmic Reticulum of a Eukaryotic Cell
The ER forms a series of flattened membrane sacs that are major sites for the synthesis of proteins and lipids.

Smooth ER does not have ribosomes associated with it, and it is involved in membrane production.

Rough endoplasmic reticulum

Membrane-bound ribosomes

Rough ER has associated ribosomes and is the major site of protein production.

When viewed in a microscope, ER membranes have two different appearances: rough and smooth (see Figure 6.4). In most cells, the majority of ER membranes appear to have small rounded particles associated with them that are exposed to the cytosol. Such ER is referred to as **rough ER**, and the particles are called **ribosomes**. Each ribosome can manufacture proteins from amino acid building blocks using instructions from the DNA in the nucleus. Ribosomes attached to the rough ER manufacture proteins that are destined for the ER lumen or for insertion into a membrane. Ribosomes that float free in the cytosol manufacture proteins that remain in the cytosol. The process of protein synthesis by ribosomes will be discussed in Chapter 15.

In most cells, a small percentage of ER membrane lacks ribosomes and is called **smooth ER** (see Figure 6.4). The smooth ER is connected to the rough ER and marks sites where portions of the membrane actively bud off to produce small enclosed membranous bags called **vesicles**. Since each vesicle is formed from a patch of ER membrane that encloses a portion of the ER lumen, vesicles are an effective means of moving proteins that either are embedded in the ER membrane or float free in the lumen (Figure 6.5). You could think of vesicles as the carts that are used to move goods between different departments of a company, since they are used to move proteins and lipids from the ER to other organelles.

branous shapes—some like tubes, others like hot water bottles—all stacked and connected to each other.

As Figure 6.4 shows, the ER has a multichambered appearance. These chambers produce the various lipids and proteins destined for other cellular compartments or for export from the cell. The internal space enclosed by the ER membrane is called the **lumen**, and it contains free-floating proteins. In contrast, proteins and lipids that are meant to reside in membranes are inserted into the membrane of the ER.

The Golgi apparatus sorts proteins and lipids to their final destinations in the cell

Another membranous organelle, the **Golgi apparatus**, directs proteins and lipids produced by the ER to their final destinations either

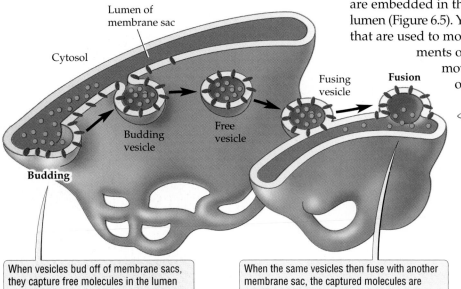

Lumen of membrane sac

Cytosol

Budding vesicle

Free vesicle

Fusing vesicle

Fusion

Budding

When vesicles bud off of membrane sacs, they capture free molecules in the lumen and molecules embedded in the membrane.

When the same vesicles then fuse with another membrane sac, the captured molecules are transferred to the second lumen and membrane.

Figure 6.5 How Vesicles Move Different Kinds of Proteins

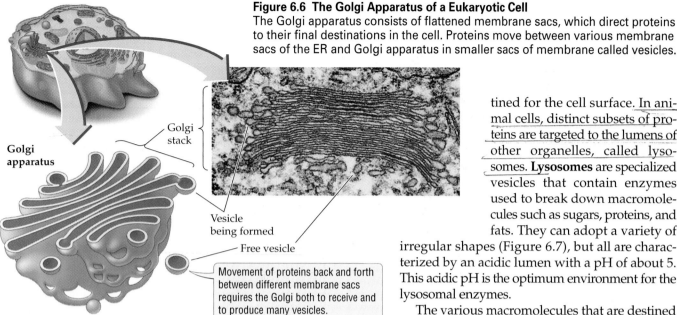

Figure 6.6 The Golgi Apparatus of a Eukaryotic Cell
The Golgi apparatus consists of flattened membrane sacs, which direct proteins to their final destinations in the cell. Proteins move between various membrane sacs of the ER and Golgi apparatus in smaller sacs of membrane called vesicles.

Golgi apparatus

Golgi stack

Vesicle being formed

Free vesicle

Movement of proteins back and forth between different membrane sacs requires the Golgi both to receive and to produce many vesicles.

tined for the cell surface. In animal cells, distinct subsets of proteins are targeted to the lumens of other organelles, called lysosomes. **Lysosomes** are specialized vesicles that contain enzymes used to break down macromolecules such as sugars, proteins, and fats. They can adopt a variety of irregular shapes (Figure 6.7), but all are characterized by an acidic lumen with a pH of about 5. This acidic pH is the optimum environment for the lysosomal enzymes.

The various macromolecules that are destined to be broken down are delivered to lysosomes by vesicles. The breakdown products, which include amino acids and lipids, are then transported across the lysosomal membrane into the cytosol for use by the cell.

Plants and fungi have a different class of organelles, called **vacuoles**, which are related to lysosomes. Vacuoles are significantly larger than lysosomes, usually occupying more than a third of a plant cell's total volume (Figure 6.8). Besides containing enzymes that break down substances, some vacuoles can store nutrients for later use by the plant cell. In the case of seeds, vacuoles in specialized cells store protein nutrients that are later broken down to provide amino acids for the growth of the plant embryo during germination. As shown by the

inside the cell or out to the external environment. The Golgi therefore functions as a sorting station, much as a shipping department in a company does. In a shipping department, goods destined for different locations must have address tags that indicate where they should be sent. Something similar happens in the Golgi, where the addition of specific chemical groups to proteins and lipids helps target them to other destinations in the cell. These cellular address tags include carbohydrates and phosphate groups.

In the electron microscope, the Golgi looks like a series of flattened membrane sacs stacked together and surrounded by many small vesicles (Figure 6.6). These vesicles transport proteins from the ER to the Golgi, and between the various sacs of the Golgi. Vesicles are therefore the primary means by which proteins and lipids move through the Golgi apparatus and to their final destinations.

Lysosomes and vacuoles are specialized compartments

Not all the proteins produced and sorted by the ER and Golgi apparatus are des-

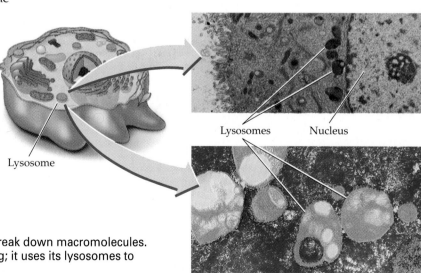

Lysosome

Lysosomes   Nucleus

Figure 6.7 Lysosomes in an Animal Cell
Lysosomes are vesicles full of enzymes that break down macromolecules. The cell shown here is from the stomach lining; it uses its lysosomes to break down food materials.

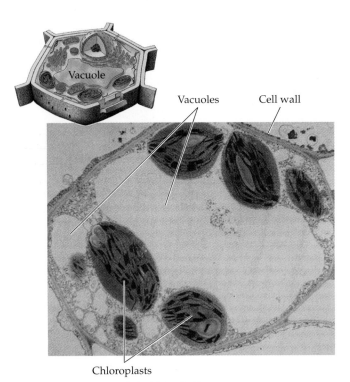

Figure 6.8 Vacuoles in a Plant Cell

Vacuoles are large vesicles in plant cells that are used to store water, nutrients, or enzymes.

fact that seeds from the burial tombs of Egyptian kings have been successfully germinated thousands of years later, storage vacuoles are very good at preserving their contents.

In many cases large vacuoles filled with water also contribute to the overall rigidity of the plant by applying pressure against cell walls. Vacuoles therefore make many different contributions to the life of the plant cell, and a single cell can easily have several vacuoles with different functions.

Mitochondria are the power stations of the cell

The previous discussion shows that various organelles are responsible for the production of cellular components and their transport to the right places in the cell. The nucleus directs what proteins are to be produced, the ER is the site of production for most proteins, and the Golgi apparatus modifies and sorts proteins to their final destinations. Thus, we have explored the administrative offices, factory floor, and shipping departments of the company. However, none of these specialized departments could function without a source of energy to run the machines that produce the goods. The eukaryotic cell is no different: All the cellular processes discussed so far require a source of energy.

The producers of this energy are organelles called **mitochondria** (singular: mitochondrion), which are like the cell's power plants. They use chemical reactions to transform the energy from many different molecules into ATP. As discussed in Chapter 5, ATP is the universal fuel of the cell and is able to release the energy stored in covalent bonds. This energy is used to drive the many chemical reactions in the cell.

Mitochondria are pod-shaped and are surrounded by a double membrane. Unlike the nucleus, which also has a double membrane, the inner mitochondrial membrane forms distinct folds (called cristae; singular crista) that project into the lumen of the organelle (Figure 6.9). The

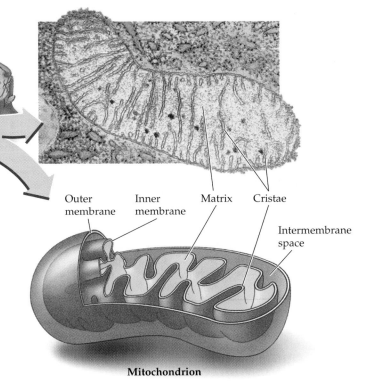

Figure 6.9 Energy-Transforming Organelles: Mitochondria

Mitochondria are the major energy-converting organelles in all eukaryotic cells—that is, in animals, plants, fungi, and protists. Each mitochondrion has a double membrane, and the inner membrane is highly folded. The infoldings (cristae) that result from the twisting of the inner membrane contain enzymes that participate in energy production. The inner lumen of the mitochondrion is called the matrix.

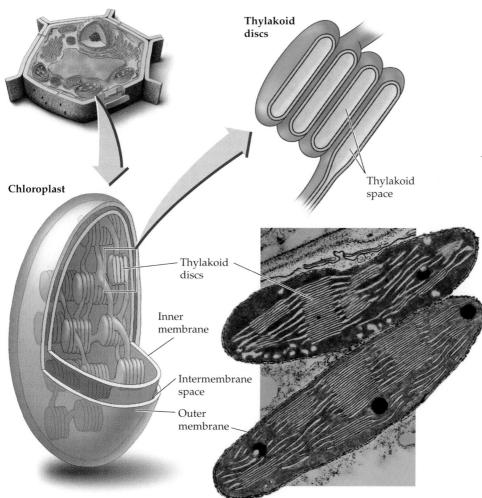

Chloroplasts capture energy from sunlight

Both animal and plant cells have mitochondria to provide them with life-sustaining ATP, but plant cells have additional energy-producing organelles, called **chloroplasts**. Unlike mitochondria, which break down sugars to produce ATP, chloroplasts capture the energy of sunlight to synthesize sugar molecules from CO_2 and H_2O. This process is called **photosynthesis** and results in the release of oxygen as a waste product. This release of oxygen explains why hiking in any dense forest on a sunny day is so refreshing, since all the plant life is flooding the surroundings with oxygen.

Chloroplasts are enclosed by a double membrane, within which lies a separate internal membrane system that is arranged like stacked pancakes (Figure 6.10). These stacked membranes, called thylakoids, contain specialized pigments, of which **chlorophyll** is the most common example. Chlorophylls enable chloroplasts to capture energy from sunlight. The green color of chlorophyll also accounts for the green coloration of most plants. Enzymes present in the space surrounding the thylakoids use energy, water, and carbon dioxide to produce carbohydrates. The mechanism of energy production by photosynthesis in plants will be discussed in Chapter 8.

Figure 6.10 Energy-Transforming Organelles: Chloroplasts
Found only in plant cells, chloroplasts capture energy from sunlight. Each chloroplast has both a double outer membrane and a third inner membrane structure, which consists of stacked discs called thylakoids. Each stack of thylakoid discs contains the proteins and pigments used to harness energy from light.

production of ATP by mitochondria depends on both the activities of proteins embedded in the inner mitochondrial membrane and the separation of the mitochondrial lumen from the space formed between the two membranes (the intermembrane space).

Using these membrane proteins and the physical structure of the organelle, mitochondria are able to harness the energy released by the chemical breakdown of sugar molecules to synthesize energy-rich ATP. In the process oxygen is consumed and carbon dioxide is released, just as in our own breathing or respiration. The details of cellular respiration in mitochondria will be discussed in Chapter 8.

■ Each organelle inside a eukaryotic cell makes a specific contribution to the life of the cell. The nucleus stores the genetic blueprint of the cell. The endoplasmic reticulum and Golgi apparatus produce and direct proteins to various destinations inside and outside of the cell. Lysosomes and vacuoles break down macromolecules, and some vacuoles also provide physical support for plant cells. Finally the production of energy for use by the cell depends on mitochondria in animal cells, and on both mitochondria and chloroplasts in plant cells.

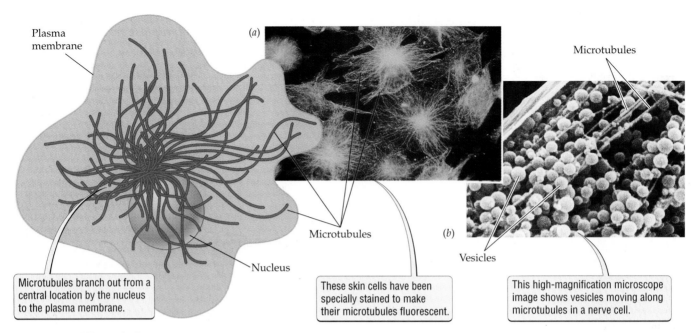

Plasma membrane

Microtubules

Microtubules

Nucleus

(a)

(b)

Vesicles

Microtubules branch out from a central location by the nucleus to the plasma membrane.

These skin cells have been specially stained to make their microtubules fluorescent.

This high-magnification microscope image shows vesicles moving along microtubules in a nerve cell.

Figure 6.11 Microtubules
(*a*) Microtubules form radial patterns in most cells, spanning from the center of the cell to the plasma membrane. (*b*) Microtubules also function as tracks along which vesicles are shuttled around the cell.

The Cytoskeleton: Providing Shape and Movement

If a cell consisted only of the plasma membrane and organelle compartments, it would be a limp bag of cytosol with no sustainable organization. Thanks to a system of structural supports called the **cytoskeleton**, however, such an unfortunate state of affairs is not the case. As the name "skeleton" implies, the cytoskeleton is an internal support system for the cell, maintaining both cell shape and the distribution of organelles in the cytosol. Furthermore, the cytoskeleton is ever changing and dynamic, allowing some cells to change shape and move around on their own. Unlike the bone skeleton of an adult human, which has fixed connections between bones, the cytoskeleton of the cell has many noncovalent associations between proteins that can break, re-form, and reshape the overall structure of the cell. The cytoskeleton is based on distinct systems of filaments, which include microtubules and actin filaments.

Microtubules support movement inside the cell

Microtubules are the thickest of the cytoskeleton filaments, with diameters of about 25 nanometers. Each microtubule is a helical polymer of the protein monomer

tubulin and has two distinct ends. Microtubules can grow and shrink in length by adding or losing tubulin monomers at either end. This ability allows microtubules to form dynamic structures capable of rapidly reorganizing the cell when necessary. The microtubules in most animal cells radiate out from the center of the cell and end at the inner face of the plasma membrane (Figure 6.11*a*). This radial pattern of microtubules serves as an internal scaffold that helps position organelles such as the ER and the Golgi apparatus.

Microtubules also define the paths along which vesicles are guided in their travels from one organelle to another or from organelles to the cell surface. The ability of microtubules to act as "railroad tracks" for vesicles depends on **motor proteins** that attach to both vesicles and microtubules. These specialized motor proteins convert the energy of ATP into mechanical movement, which allows them to move along microtubules in a specific direction, carrying an attached vesicle like cargo (Figure 6.11*b*).

Actin filaments allow cells to move

Actin filaments have the smallest diameters of the filament systems, but they are the most important when it comes to supporting cell movements. Each actin filament is a polymer of actin monomers, and, like microtubules, actin filaments can rapidly change length.

Perhaps the best example of rapid changes in complex actin structures can be found in a cell moving across a flat surface. When observed in a microscope, skin cells such as fibroblasts visibly crawl around on a flat surface.

Figure 6.12 Actin Filaments
Actin filaments help cells crawl on surfaces by allowing them to extend flattened projections called pseudopodia.

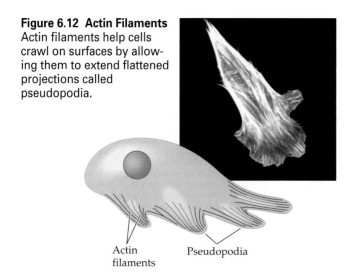

Actin filaments

Pseudopodia

They achieve this motion by extending flattened sheets of membrane called **pseudopodia** (*pseudo*, "false"; *podia*, "feet"; singular pseudopodium) and then pulling up their rear ends behind them (Figure 6.12). This ability to move is an important part of wound healing, since fibroblasts migrate into the area of the wound to assist in closing up the damaged edges.

Fibroblasts use different arrangements of actin filaments to alter the structure of the parts of the cell that are involved in extending the pseudopodia and retracting the rear end. Actin filaments in the protruding pseudopodia tend to point all in the same direction, toward the plasma membrane. When these well-organized filaments lengthen, they push on the plasma membrane and extend pseudopodia in the direction the cell is moving. At the same time on the other side of the cell, the actin filaments tend to run in all directions and shorten. This process loosens and breaks down the actin network at that end of the cell, resulting in retraction of the plasma membrane and what appears like the cell pulling up its rear end behind it (see Figure 6.12).

The example of the crawling fibroblast reveals that active changes in an actin network can result in move-

ment. Cells are accustomed to lengthening and shortening actin filaments as part of their daily activities, and this process is what the invading *Listeria* bacteria that we talked about at the beginning of the chapter take over and use. *Listeria* is able to use for its own purposes the actin proteins and natural processes of the host cell. In fact, biologists have identified proteins on the surface of *Listeria* that capture actin monomers and start the process of polymerization to form filaments.

The explosive propulsion created by this cometlike tail of growing actin filaments allows *Listeria* to move inside the infected cell and to jump from one cell to another cell, thus spreading the infection. As remarkable as this hijacking of the cell's machinery by *Listeria* might seem, other disease-causing bacteria, such as *Shigella*, also use actin-based propulsion to infect cells, causing dysentery.

■ The cytoskeleton provides eukaryotic cells with structural support and the ability to change shape and move. Protein filaments make up most cytoskeletal networks, each with a specific role in cellular processes. Microtubules are essential for the movement of vesicles and other organelles inside the cell, and actin filaments allow cells to change shape and move.

Highlight

The Evolution of Organelles

The use of the cell's own resources to further the agenda of an invading prokaryote like *Listeria* might seem terribly unfair. However, we eukaryotes may not have the right to pass judgment on *Listeria*'s behavior. In the distant evolutionary past, the ancestors of eukaryotic cells probably did a similar thing by capturing and using their prokaryotic neighbors (Figure 6.13).

Figure 6.13 How Primitive Eukaryotes May Have Acquired Mitochondria

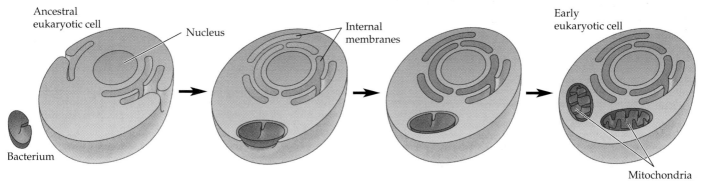

Ancestral eukaryotic cell

Nucleus

Internal membranes

Early eukaryotic cell

Bacterium

Mitochondria

Mitochondria and chloroplasts in eukaryotic cells bear a striking physical resemblance to primitive prokaryotes. Both organelles have their own DNA and are able to make some proteins. They also reproduce independently of the cell, by simply dividing in two. These striking characteristics imply that mitochondria and chloroplasts were once free-swimming prokaryotes that were engulfed by other cells, eventually forming a mutually beneficial relationship.

How did early single-celled eukaryotes benefit from capturing primitive prokaryotes? The answer to this question may lie in the environment of Earth more than 3.5 billion years ago. At that time, when the first cells are thought to have arisen, there was virtually no oxygen gas (O_2) in the atmosphere, and primitive prokaryotes could break down sugars in the absence of oxygen. However, as time passed and some prokaryotes evolved the ability to use the energy of sunlight to make organic compounds from carbon dioxide, the production of oxygen by these photosynthetic reactions slowly changed the atmosphere of Earth.

As oxygen gas accumulated in the early atmosphere, some prokaryotes evolved the means to use oxygen to break down sugars and release energy, giving them a significant advantage over earlier prokaryotes. At this turning point in evolution, about 2 billion years ago, a primitive single-celled eukaryote is thought to have captured a smaller prokaryote and gained the ability to use atmospheric oxygen in energy production.

The descendants of these captured prokaryotes are the mitochondria that we now find in every eukaryotic cell. Likewise, the ancestors of chloroplasts were probably primitive cyanobacteria that had evolved the ability to photosynthesize and thus were captured for it. The eukaryotic cells we know today all depend on mitochondria for their ability to use oxygen to produce energy by breaking down sugars. Likewise, plant cells depend on chloroplasts for photosynthesis. Thus we are reminded that cells are made up of very specialized compartments and chemical components that function together to support the processes of life. Since the division of labor seen in all eukaryotic cells may be based on the capture of useful organisms in the evolutionary past, it is unsurprising that the same principle can in turn be used by current invaders like *Listeria*. These phenomena confirm the old saying: What goes around comes around.

> ■ Mitochondria and chloroplasts are probably descendants of primitive prokaryotes that were engulfed by other cells billions of years ago.

Summary

Cells: The Simplest Units of Life

- ■ Cells are the basic units that make up all living organisms.
- ■ All multicellular organisms are made up of many different types of specialized cells.

The Plasma Membrane: Separating Cells from the Environment

- ■ Every cell is surrounded by a plasma membrane that separates the chemical reactions of life from the nonliving environment.
- ■ The plasma membrane is a highly fluid mixture of phospholipids and proteins that can move sideways in the plane of the membrane.
- ■ Proteins in the plasma membrane allow the cell to communicate and respond to the environment.

Comparing Prokaryotes and Eukaryotes

- ■ Living organisms can be classified as either prokaryotes or eukaryotes.
- ■ Prokaryotes are made up of cells that lack internal membrane-enclosed compartments; the cells of eukaryotes have internal compartments, such as the nucleus.
- ■ Prokaryotes were the first living organisms to arise in evolution. Today's prokaryotes are all single-celled bacteria.
- ■ Eukaryotic cells are approximately 100 times larger than prokaryotic cells and require internal compartments to concentrate and promote the chemical reactions of life.

The Specialized Internal Compartments of Eukaryotic Cells

- ■ The specialized membrane-enclosed compartments inside a eukaryotic cell are known as organelles, each of which makes a specific contribution to the life of the cell.
- ■ The nucleus is the most distinctive organelle in eukaryotic cells. It houses the DNA-encoded instructions that control every activity and structural feature of the cell.
- ■ The endoplasmic reticulum manufactures proteins and lipids for use by the cell or for export to the environment. The Golgi apparatus receives proteins and lipids from the endoplasmic reticulum and directs them to their final destinations either inside or outside of the cell.
- ■ Vesicles transport proteins and lipids between organelles and between an organelle and the plasma membrane.
- ■ Lysosomes break down organic macromolecules like proteins to simpler compounds that can be used by the cell. Vacuoles are similar to lysosomes but can also lend physical support to plant cells.
- ■ Mitochondria produce energy for use by both animal and plant cells. Chloroplasts harness the energy of sunlight for use by plant cells.

The Cytoskeleton: Providing Shape and Movement

- Eukaryotic cells depend on the cytoskeleton for structural support and the ability to change shape and move.

- The cytoskeleton consists of different types of protein filaments—including microtubules and actin filaments—each of which associates with a particular subset of other proteins in the cytosol.

- Microtubules and actin filaments change length frequently and rapidly. They are essential for the movement of organelles inside the cell, and for movement of the entire cell during locomotion.

Highlight: The Evolution of Organelles

- Mitochondria and chloroplasts are similar to simple prokaryotes.

- During evolution, eukaryotic cells probably acquired mitochondria and chloroplasts by engulfing simple prokaryotes.

Key Terms

actin filament p. 106	motor protein p. 106
cell p. 96	nuclear envelope p. 101
chlorophyll p. 105	nuclear pore p. 101
chloroplast p. 105	nucleus p. 99
cytoplasm p. 101	organelle p. 101
cytoskeleton p. 106	photosynthesis p. 105
cytosol p. 98	plasma membrane p. 98
endoplasmic reticulum (ER) p. 101	prokaryote p. 99
	pseudopodium p. 107
eukaryote p. 99	ribosome p. 102
fluid mosaic model p. 99	rough ER p. 102
Golgi apparatus p. 102	smooth ER p. 102
lumen p. 102	tubulin p. 106
lysosome p. 103	vacuole p. 103
microtubule p. 106	vesicle p. 102
mitochondrion p. 104	

Chapter Review

Self-Quiz

1. Which of the following statements about the plasma membrane is true?
   a. It is a solid layer of protein that protects the contents of the cell.
   b. The plasma membrane of a bacterium has none of the same components as the plasma membrane of an animal cell.
   c. It is a rigid and unmoving layer of phospholipids and protein.
   d. It allows selected molecules to pass into and out of the cell.

2. Which of the following cellular components can be used to distinguish a prokaryotic cell from a eukaryotic cell?
   a. a nucleus
   b. a plasma membrane
   c. DNA
   d. proteins

3. One key function of nuclear pores is to
   a. allow cells to communicate with each other.
   b. aid in the production of new nuclei.
   c. allow molecules like proteins to move in and out of the nucleus.
   d. form connections between different organelles.

4. Vesicles are essential for the normal functioning of the Golgi apparatus because
   a. they provide energy for chemical reactions.
   b. they move proteins and lipids between different parts of the organelle.
   c. they contribute to the structural integrity of the organelle.
   d. they produce the sugars that are added to proteins.

5. Which of the following statements is *not* true?
   a. Both mitochondria and chloroplasts provide energy to cells in the same way.
   b. Both mitochondria and chloroplasts have more than one membrane.
   c. Only chloroplasts contain the pigment chlorophyll.
   d. Both animal and plant cells contain mitochondria.

6. Actin filaments contribute to cell movement by
   a. providing energy in the form of ATP.
   b. lengthening and pushing against the plasma membrane.
   c. forming a stable and unchanging network inside the cell.
   d. allowing organelles to change position inside the cell.

Review Questions

1. Proteins embedded in the plasma membrane have several important functions in the life of the cell. Describe two of these functions and explain why they are important to the cell.

2. Vacuoles can have a wider variety of functions in plant cells than do lysosomes in animal cells. Describe one function that vacuoles perform in plant cells that lysosomes cannot in animal cells.

3. You have discovered a protein enzyme in the lumen of lysosomes that enables them to break down polysaccharides. Where in the cell do you think the ribosome that produced this enzyme is located—free in the cytosol or associated with the endoplasmic reticulum? Justify your answer.

4. Describe one common characteristic shared by microtubules and actin filaments. Relate this characteristic to the function of these filaments in the cell.

The Daily Globe

6

New Protein Provides Hope for Fatal Muscular Dystrophy

WASHINGTON, D.C. Researchers reported at a news conference today that mice with muscular dystrophy–like symptoms have decreased muscle cell damage when they produce more of a newly discovered protein known as utrophin. Researchers say the new study suggests that children with the fatal disease Duchenne's muscular dystrophy (DMD) might also benefit from drugs that increase their production of this protein. DMD is an inherited disease that afflicts more than 20,000 male infants each year worldwide. These children usually die from heart failure before reaching adulthood.

Children with DMD suffer from degenerating muscle cells due to the absence of a cytoskeletal protein called dystrophin. Biologists have been investigating DMD by altering the genetic profile of mice so that they, too, lack dystrophin and experience DMD symptoms. These mice become models for the disease and can be studied and subjected to experimental treatments that are too risky for humans.

"Based on what we see in DMD mice, a drug therapy that increases the production of utrophin could slow muscle degeneration and prolong these children's lives," said Dr. Sarah Benning, DMD researcher at Massachusetts Charitable Hospital. "This is potentially a huge breakthrough."

Evaluating "The News"

1. The similarity of utrophin to dystrophin and its potential ability to act as a substitute in affected cells implies that the cytoskeleton has built-in backup systems. What do you think might be the benefit of having backup proteins in a cell?

2. The development of mice that lack dystrophin was an important breakthrough in understanding DMD in humans. Why might studying these so-called model organisms be so useful in medically relevant research?

3. The genetic modification of animals such as mice to experience human diseases is now a common method used to study such diseases. Do the benefits of learning more about human diseases justify breeding animals that are doomed to suffer from those same diseases? When it comes to improving human health, do animals used in research have any rights?

7

Metabolism

Ben Shahn, *Helix and Crystal*, 1957.

"Take Two Aspirin and Call Me in the Morning"

If someone told you there was a wonder drug that reduced pain, fever, and inflammation and helped combat heart disease and cancer, would you believe them? As amazing as it may sound, just such a drug is about to celebrate its hundredth birthday; it is known to all of us as aspirin. Even more remarkable is that our understanding of how aspirin works and how it can have so many remarkable effects on the human body is still in its infancy.

Our knowledge of aspirin started with the laboratory synthesis of salicylic acid in 1860, which at the time was used as an antiseptic and fever-reducing agent. Unfortunately, salicylic acid also damages the stomach lining, so in 1899 a more palatable version was developed, which we know today as aspirin. In the years that followed, aspirin became an important means of lowering pain and reducing fevers and inflammation. For decades these therapeutic benefits were more than enough to make aspirin a staple in every doctor's bag and hospital around the world. It even spawned the well-known phrase "take two aspirin and call me in the morning," implying that this wonder drug could probably handle most medical problems at least overnight.

Today we know that aspirin could have an even more amazing impact on human health than its well-established use as a pain reliever. Well-publicized studies show that low doses of aspirin taken daily can reduce the risk of heart disease. In 1996, the importance of aspirin was further bolstered by studies showing that people who take aspirin regularly have lower rates of colon cancer. However, taking aspirin for prolonged periods does have negative side effects, including damage to the stomach and kidneys. So, given all the possible health benefits of taking aspirin, how does this wonder drug work, and can we make it better by reducing the side effects?

The first ideas about how aspirin works emerged almost 75 years after its development and focused on its ability to block the production of certain hormones. Today we know that the hormones in question are prostaglandins, and the ability of aspirin to affect so many processes in the body is directly related to its blocking of the chemical synthesis of prostaglandins. Ironically, these effects on chemical reactions in our bodies also account for the negative side effects of aspirin.

Armed with a growing knowledge of how aspirin affects metabolic processes in the body, researchers are now seeking to improve this old wonder drug by eliminating the negative aspects of its action. This chapter discusses the fundamental principles of metabolism that control every chemical reaction in our bodies,

and we will see how our understanding of metabolism will pave the way for the development of a new superaspirin.

An Apothecary's Kit from the Late Nineteenth Century

Key Concepts

1. Living organisms must obey the universal laws of energy conversion and chemical change.

2. Metabolism consists of chemical reactions that produce complex macromolecules such as sugars and proteins, which in turn are broken down to yield smaller molecules and usable energy.

3. The sun is the primary source of energy for living organisms. Photosynthetic cells in organisms such as plants capture energy from the sun and use it to synthesize sugars from carbon dioxide and water. All cells in turn break down sugars to release energy.

4. Enzymes control the speed of chemical reactions in cells. Metabolic pathways are successions of enzyme-controlled chemical reactions.

All living processes require energy that must be extracted from the environment. This energy is used by cells to manufacture and transform the various chemical compounds that make up the cell. Both the capture and the use of energy by living organisms involves thousands of chemical reactions that together are known as **metabolism**. The metabolic reactions that create complex molecules out of smaller compounds are described as **biosynthetic**; those that break down complex molecules to produce usable energy are described as **catabolic**.

All chemical reactions that occur in cells can be categorized into a surprisingly small number of metabolic pathways. In the same way that chemical building blocks fall into a limited number of categories, the reactions that allow the cell to manufacture and transform these building blocks are also limited in number. In this chapter, we examine the role played by energy in the chemical reactions that maintain living systems. We also discuss the role of specialized proteins that speed up chemical reactions that would otherwise take too long for us to survive.

The Role of Energy in Living Systems

The discussion of any chemical process in the cell is a discussion about energy. The idea that energy is behind every activity in the cell seems natural and unsurprising, since all of us are accustomed to thinking of energy as a form of fuel. However, energy is more than just fuel, since it also dictates which chemical reactions can occur and how molecules are organized into living systems.

The relationship between energy and the cell's activities is governed by the same physical laws that apply to everything else in the universe. These are the laws of thermodynamics, and they define the way in which cells transform chemical compounds and interact with the environment. The **first law of thermodynamics** states that the total energy of a system, such as a cell, and its surroundings always remains constant. In other words, energy is neither created nor destroyed, and the cell must use energy by converting it from one form to another (Figure 7.1a). For example, mitochondria convert the energy contained in small sugar molecules into the energy of covalent bonds in ATP (ADP + phosphate + energy → ATP) (see Chapter 5). Thus, mitochondria do not "produce" energy; that is, they do not create it from nothing. Instead, they convert energy from one form (sugars) to another (ATP), which can then be used by the cell. The way in which the energy stored in ATP is used to drive other chemical reactions is discussed later in this chapter.

The **second law of thermodynamics** determines how each cell relates to its environment and how it maintains a well-ordered internal organization. This law states that systems, such as a cell or even the whole universe, tend to become more disorderly. This statement may seem like a law written for an adolescent's room, which usually tends toward disorder and chaos, but it is true of all systems, including the internal organization of the cell (Figure 7.1b).

As we discussed in Chapters 5 and 6, the cell is made up of many chemical compounds assembled into complex structures. Such a high level of organization built out of small molecules flies in the face of disorder and must be compensated for. The tremendous structural complexity of organelles like the endoplasmic reticulum exists in a constant struggle against chaos. Thus, to counteract the natural tendency toward disorder, the cell must transfer some of this disorder elsewhere.

Living systems pass off or transfer disorder by releasing heat into the environment. Heat is a form of energy that causes the rapid and random movement of molecules, a condition that is highly disordered. Thus, when cells release heat, they increase the degree of disorder in the molecules of the environment, which in turn compensates for the increasing order inside the cell (Figure

7.1*c*). There is a direct connection between cell organization and the transfer of energy because the chemical processes used to build well-ordered structures like organelles are the same ones that produce the heat. Hence, the generation of order is directly coupled with the release of heat energy.

The flow of energy and the cycling of carbon connect living things with the environment

Where does the energy that creates order in the cell come from? We know from the first law of thermodynamics that the cell cannot create energy from nothing; thus it must come from outside of the cell. Energy must be transferred into the cell in some fashion. In the case of photosynthetic organisms such as plants, the energy comes from sunlight and is captured and converted into the chemical bonds used to form sugars. For nonphotosynthesizing cells, the energy comes from the chemical bonds in fuel molecules such as sugars and fats.

The two statements in the preceding paragraph reveal the relationship between photosynthetic producers and nonphotosynthetic consumers. For example, photosynthetic plants capture energy from sunlight, and animals in turn obtain this energy by consuming plants. Even in the case of animals that only eat other animals, at some point those "other" animals eat plants. This means that thanks to photosynthesis the sun is the primary energy source for all living organisms.

Now don't feel sorry for plants. They are more than consumable energy sources for nonphotosynthetic organisms like animals. First, plants also use the sugars made in photosynthesis, especially at night, when there

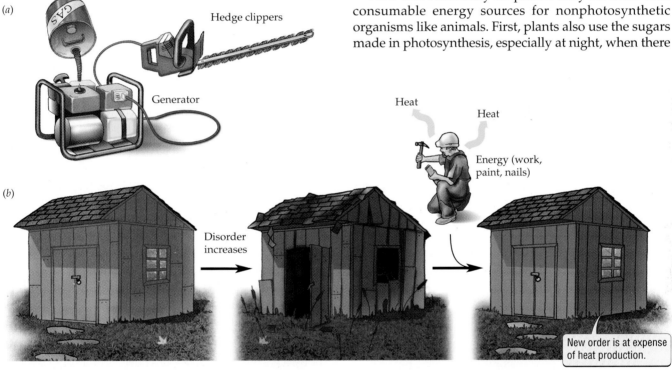

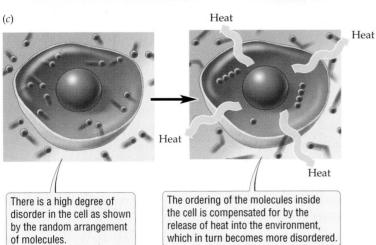

(*a*)

Hedge clippers

Generator

Heat

Heat

Energy (work, paint, nails)

(*b*)

Disorder increases

New order is at expense of heat production.

(*c*)

Heat

Heat

Heat

Heat

Heat

There is a high degree of disorder in the cell as shown by the random arrangement of molecules.

The ordering of the molecules inside the cell is compensated for by the release of heat into the environment, which in turn becomes more disordered.

Figure 7.1 Laws of Thermodynamics
(*a*) Energy is neither created nor destroyed. The energy contained in the covalent bonds of the gas molecules is converted into electrical energy by the generator. The electrical energy in turn is converted into the mechanical energy of motion by the hedge clippers. Neither the generator nor the hedge clippers creates or destroys energy. (*b*) The disorder of a system always tends to increase. Left unattended, all structures, such as this wooden shed, tend to lose their order and become disarrayed. The input of energy, by way of human effort in this case, maintains the careful order of the structure. (*c*) Cells maintain their organization through a continuous input of energy. Thus, they do obey the second law of thermodynamics, which states that the disorder of a system always increases unless there is an input of energy.

Figure 7.2 Carbon Cycling
Carbon atoms cycle among plants, animals, and the environment. Carbon atoms become parts of different kinds of molecules as they cycle between living organisms and the environment.

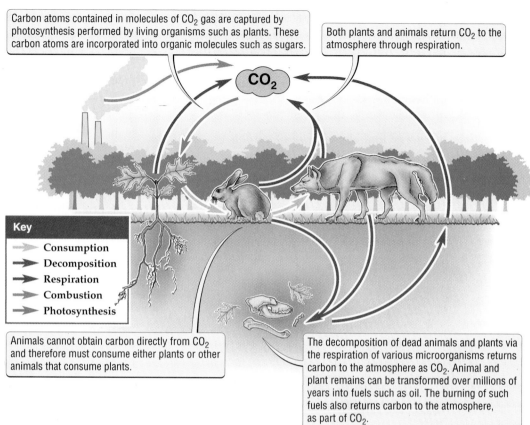

Carbon atoms contained in molecules of CO_2 gas are captured by photosynthesis performed by living organisms such as plants. These carbon atoms are incorporated into organic molecules such as sugars.

Both plants and animals return CO_2 to the atmosphere through respiration.

Key
→ Consumption
→ Decomposition
→ Respiration
→ Combustion
→ Photosynthesis

Animals cannot obtain carbon directly from CO_2 and therefore must consume either plants or other animals that consume plants.

The decomposition of dead animals and plants via the respiration of various microorganisms returns carbon to the atmosphere as CO_2. Animal and plant remains can be transformed over millions of years into fuels such as oil. The burning of such fuels also returns carbon to the atmosphere, as part of CO_2.

is no sunlight and no photosynthesis. Second, the carbon dioxide (CO_2) produced by a process called respiration (which will be discussed in Chapter 8) is a source of carbon for photosynthesizing plants. So during photosynthesis, a plant uses carbon dioxide as a carbon source in the production of sugars, which in turn are consumed either by the plant or by animals. Thus, carbon atoms are cycled from carbon dioxide in the atmosphere to sugars produced in plants and back to carbon dioxide released by respiring organisms (Figure 7.2). This recycling of atomic building blocks occurs not only for carbon but also for other atoms, such as nitrogen and phosphorus.

■ Living organisms obey the same laws of energy that apply to the physical world. In addition, organisms ultimately obtain energy from the nonliving environment and must convert it into more usable forms and structural components. Carbon atoms are recycled between living organisms and the environment.

Using Energy from the Controlled Burning of Food

Living systems obtain energy from food by burning organic molecules such as sugars to form carbon dioxide and water. If our cells were to burn food to carbon dioxide and water in a single chemical reaction, we would burst into flame like a lit match. Here's the chemical equation that describes what happens when the match burns:

$$\text{Wood} + O_2 \rightarrow CO_2 + H_2O + \text{energy (heat and light)}$$

This combustion reaction is similar to what occurs when our cells burn food, but fortunately for us, there are some important differences.

The energy released from the burning match is dispersed into the environment as heat and cannot be regained by the match in any fashion. In contrast, living systems need to capture the released energy, and they can do so only by controlling the combustion and breaking it down into a series of much smaller chemical reactions. This breakdown saves us from bursting into flame and gives our cells the opportunity to capture the smaller amounts of energy released at each step. In this section we review the key characteristics of these chemical reactions that release energy from food molecules for use by the cell.

Releasing energy from food requires the transfer of electrons

In the multiple chemical reactions that allow cells to capture energy from fuel, electrons are transferred from one molecule or atom to another. **Oxidation** is the loss of electrons from one molecule or atom to another; **reduction** is just the opposite, the gain of electrons by one molecule or atom from another.

Let's consider the carbon atom contained in the gas methane (Figure 7.3*a*). This carbon atom is bonded to hydrogen and is in a reduced, or electron-rich, state. If we compare the electron-rich carbon in methane to those in the CO_2 produced by combustion, we find that the oxygen atoms tend to attract electrons away from the carbon atoms, leaving them in an oxidized, or electron-poor, state. Thus, the combustion of organic compounds like methane to CO_2 is an oxidation reaction.

Biological oxidation takes place in a series of small steps, not all at once as in combustion. Instead of jumping from being part of the carbon backbone of a complex sugar to being part of the simple CO_2 molecule, the carbon atoms pass through several chemical reactions and intermediate compounds. Each intermediate compound is a little more oxidized than the preceding one (Figure 7.3*b*). This stepwise and controlled combustion allows the cell to couple each small energy-releasing oxidation reaction with other reactions, which store some of the released energy in newly formed chemical bonds. This transfer of energy from one compound to another is the basis of the reactions in metabolism.

Energy in a living system is transferred via the universal currency, ATP. The production of ATP from ADP and phosphate (see Figure 5.6) is an urgent priority for the human body, since each cell consumes its entire pool of ATP almost every minute. In fact, in nearly every chemical reaction in the cell, ATP is consumed or synthesized. The energy released from ATP when it is broken down to ADP and phosphate is used for such activities as moving molecules and ions between various cellular compartments, generating mechanical force in a crawling cell, and manufacturing complex macromolecules from simpler chemical compounds. This final activity comprises the biosynthetic reactions of metabolism.

Thus, the catabolic reactions in the cell that harness energy in the form of ATP are tightly coupled to the biosynthetic reactions, forming the two sides of metabolism: releasing energy by breaking down and using energy to build up. For an unexpected consequence of a high metabolism, see the Box on page 118.

Chemical reactions are governed by simple energy laws

How does the cell control such a powerful event as combustion and break it down into smaller, more manageable and useful steps? The answer to this question lies in the very nature of chemical reactions. Let's review some of the fundamental principles that govern chemical reactions, as represented by the following generic example:

$$A + B \rightarrow C + D$$

A and B are the starting materials, or reactants; C and D are the products that are formed by the reaction. A chemical reaction alters the composition and/or struc-

(*a*)

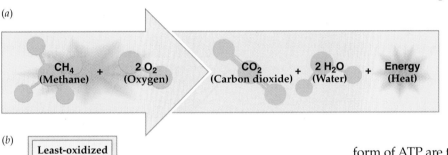

(*b*)

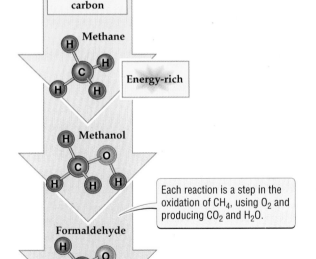

Figure 7.3 Oxidation of Methane
(*a*) When methane gas encounters a spark, it undergoes an explosive oxidation, or combustion, reaction. Energy is released to the environment as heat. (*b*) Methane can be oxidized in a series of smaller steps. The single carbon atom in each compound becomes increasingly oxidized with each step. That is, the carbon atom is gradually surrounded by more oxygen atoms that tend to attract its electrons, leaving it more oxidized.

Take It Easy, You Might Live Longer!

The idea of immortality has fascinated human beings throughout recorded history. In the past, people interested in extending their life span turned to alchemists or wizards. Today, researchers are discovering which biological factors limit our life span and how those factors might be controlled to let us live longer. One key factor that has emerged is the overall metabolic rate of an organism.

In general, smaller animals have faster metabolic rates and shorter life spans than larger animals. Furthermore, laboratory tests have shown that mice with a restricted diet and thus slower metabolism live longer than mice that eat as much as they like. The idea that a higher metabolic rate can shorten one's life span may seem contradictory, since metabolism includes all of the chemical reactions that maintain living organisms.

How might metabolism shorten the life span of an organism? The answer lies not in the reactions of metabolism themselves, but rather in the toxic chemical by-products that are sometimes produced. These chemical compounds, which are produced by accident, react with and damage cellular components like DNA. The gradual accumulation of this cellular damage is an important contributing factor to aging and ultimately death.

The link between slower metabolism and a longer life also holds true for a nematode worm named *Caenorhabditis elegans*. Abnormal worms that lack proteins responsible for higher metabolic rates live up to five times longer than they should. These worms also take longer to mature and display a slowing of specific behaviors. Similar phenomena have been observed in fruit flies and in yeasts, implying that a broad range of species are subject to these life-limiting metabolic events.

Given the supporting evidence gathered from several species, it is not surprising that the human life span may also be affected by metabolic rates. Women have slower metabolic rates than men, and they have a higher life expectancy (in the United States, 79 years versus 72 years in men). Even more striking is the fact that 9 out of 10 individuals 100 years or older are women. Thus, although genetic factors that affect overall health also have a role in determining life expectancy, the accumulated mistakes from a lifetime of metabolism clearly take their toll. Perhaps if we can limit the metabolic activities that run the highest risk of producing toxic by-products, we will finally achieve the goal of extending our life span.

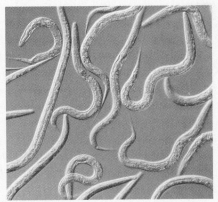

Studies of women, mice, and the nematode worm *C. elegans* show that a slower metabolism may lengthen an organism's life span.

ture of molecules. All chemical reactions tend to proceed in the direction that will result in greater stability and a lower energy state. Products in a lower energy state than their reactants have less energy stored in the form of well-ordered bonds and are therefore in a less ordered state, which is favored by the second law of thermodynamics. This tendency toward less order and a lower energy state favors a particular direction of change in chemical reactions. That is, the process of "going downhill" energetically from A + B to C + D is a good thing according to the laws of the physical world.

It is important to note that just because A and B are present together does not mean they will react. The reason is that all compounds such as A and B tend to be in a semistable state. Thus, they need to be destabilized, or activated ("jump-started"), before a chemical reaction can begin and proceed. The jump start that is required is the input of a small amount of energy, called the **activation energy** of the reaction. The activation energy is required to alter the chemical configuration of the reactants so that the reaction can take place. Once the reactants have overcome this activation energy barrier, the reaction proceeds and a product of lower energy is produced (Figure 7.4).

Where do chemical reactions obtain the activation energy required for them to proceed? Returning to the example of a burning match, the activation energy required to light the match can come from the friction generated by moving the head of the match across a rough surface. Chemical reactions in cells can acquire the activation energy they need from random collisions between molecules floating in the cytosol. These collisions are more frequent and energetic as the temperature increases and molecules in the cytosol move faster. But at the normal body temperatures of most organisms, these collisions do not release enough energy to drive all the reactions required for life. To compensate, cells contain a class of specialized proteins called enzymes, which directly promote chemical reactions in living organisms.

> ■ Living organisms capture and use the energy released from the stepwise oxidation of organic compounds. To proceed, each chemical reaction requires the input of a small amount of activation energy.

How Enzymes Speed Up Chemical Reactions

Chemical reactions are a crucial part of life. Some are involved in the production of macromolecules that make up the structures of the cell; while others allow the cell to obtain energy from fuel. These are just a few of the different chemical reactions that support life. Yet, if we were to depend on chemical reactions occurring unassisted, they would happen so slowly that life would not be possible.

To solve this problem, cells use a specialized class of proteins called **enzymes** to speed up the process. Nearly all chemical reactions that take place in living organisms require the assistance of enzymes. Our human cells contain several thousand different enzymes, each of which affects a specific chemical reaction. Since enzymes affect the rate at which reactions occur, but remain unchanged, they are called **catalysts**.

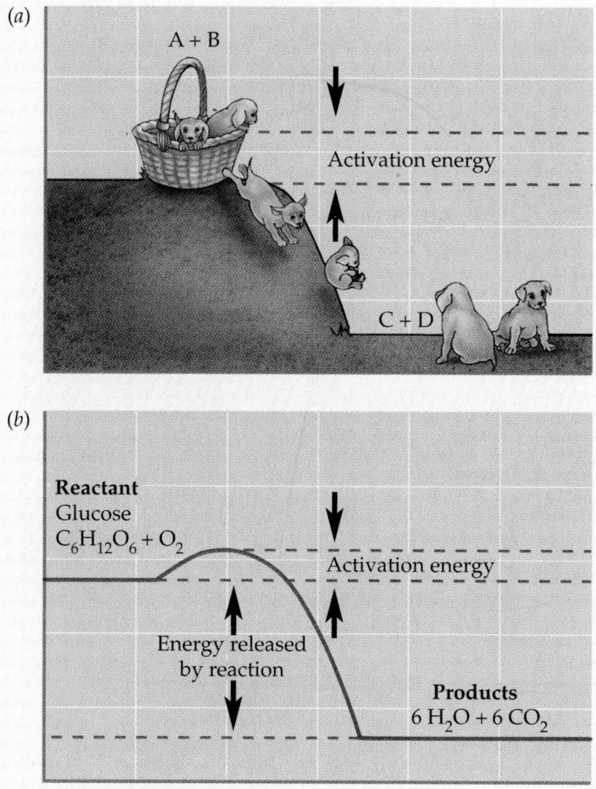

Figure 7.4 Getting Over the Activation Energy Barrier
(*a*) Imagine that the reactants of a chemical reaction (A + B) are a group of frisky puppies trying to get out of a basket sitting on a slope. In this analogy, the sides of the basket represent the activation energy barrier, and the end of the slope represents the lower energy state of the products. As each puppy tries to scramble over the edge of the basket, those that manage it are like the reactant molecules that receive enough energy from the environment to make it over the activation energy barrier. Once the successful puppies are at the bottom of the slope, it is more difficult for them to make it back into the basket; getting back in requires more effort. Thus, the preferred direction is for the puppies to move out of the basket and not back in. Likewise, the products of the chemical reaction (C + D) are in a lower energy state than the reactants, so a much larger input of energy is required to reverse the direction of the reaction. (*b*) The oxidation of glucose during respiration to produce water and carbon dioxide must overcome an activation energy barrier.

To increase the rate at which chemical reactions proceed, each enzyme binds to specific reactants called **substrates** and lowers the amount of activation energy that they require to react with each other. In fact, when an enzyme binds the reactants, it can bring them together in an orientation that favors the making or breaking of bonds required to form product. For a particular enzyme to bind the correct reactants and alter them appropriately for the chemical reaction, it must have a high degree of specificity for those reactants. In other words, each enzyme is specifically tailored to promote only one of the thousands of possible reactions that occur in the cell.

So, in the presence of many reactants, one enzyme rather than another can determine which chemical reaction takes place. However, an enzyme cannot make an impossible reaction happen. It cannot promote a particular reaction by changing the free energy associated with the reactants or the products. Instead, it can only affect the rate at which a reaction occurs by lowering the activation energy barrier, in a process called **catalysis**.

The control that enzymes exercise over the rate of chemical reactions is necessary for all living processes. For example, an essential process such as the removal of carbon dioxide from the cells in the human body depends on an enzyme in red blood cells. Before we can exhale carbon dioxide during breathing, the gas must first be transferred from our body's cells into the bloodstream and then into the lungs. For this transfer to occur, carbon dioxide must combine with water so that it can be transported in the blood as bicarbonate ions (HCO_3^-):

$$H_2O + CO_2 \rightarrow HCO_3^- + H^+$$

This simple reaction, described as the hydration of carbon dioxide, is necessary for normal respiration. Carbonic anhydrase is the enzyme that is responsible; it speeds up the hydration of carbon dioxide by a factor of nearly 10 million. In fact, a single carbonic anhydrase enzyme can hydrate more than 10,000 molecules of carbon dioxide in a single second. Needless to say, without carbonic anhydrase, the rate of carbon dioxide hydration would be so slow that we would not be able to rid our bodies of carbon dioxide fast enough. When circulating bicarbonate ions arrive at the lungs, they are converted back into carbon dioxide, which is then released as CO_2 gas.

The shape of an enzyme directly determines its activity

The specificity that enzymes have for their substrates depends on the three-dimensional shapes of both substrate and enzyme. In the same way that a particular lock accepts only a key with just the right shape, each enzyme has an **active site** that accepts only a substrate with the correct three-dimensional shape (Figure 7.5a). The matching of an enzyme shape to one or more substrates guarantees that a specific reaction will take place to yield the expected products. Indeed, which reaction occurs depends on the particular combination of enzyme with substrate molecules.

For example, carbonic anhydrase is able to bind molecules of both carbon dioxide and water in its active site. By bringing the two substrates ($H_2O + CO_2$) together, the active site of carbonic anhydrase promotes the hydration reaction (Figure 7.5b). All enzymes

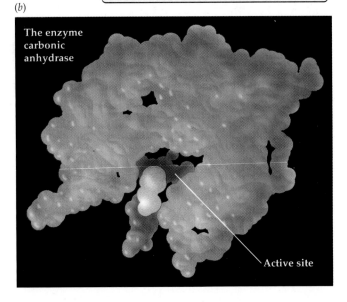

(a) The active site of the enzyme has the shape and chemical properties to bind to the reactants (substrates) and facilitate the reaction.

A   B
(Substrates)

Enzyme

The enzyme is not permanently changed by the reaction and can be recycled.

Substrate binds

Catalysis

Product released

C   D
(Products)

After the reaction is completed, the enzyme releases the products and is free to bind more substrate.

(b)

The enzyme carbonic anhydrase

Active site

Figure 7.5 Enzymes as Molecular Matchmakers
(*a*) An enzyme brings together two reactants (A and B) such that the chemical reaction proceeds to form the products (C and D). (*b*) Carbonic anhydrase catalyzes the combining of carbon dioxide and water to form bicarbonate.

have active sites that are specific for their substrates and will not bind other molecules.

The action of carbonic anhydrase demonstrates how specific binding of two substrates by an enzyme can push them together such that a chemical reaction takes place. Thus, we can think of enzymes as molecular matchmakers that bring the right substrates together. If there were no enzyme present, the two substrates would need to collide with each other in just the right way before the hydration reaction could take place. These sorts of molecular collisions do occur all the time, but not nearly as frequently as would be required for the continuous and rapid transfer of carbon dioxide from cells into the blood.

Enzyme chain reactions have energetic advantages

So far, we have discussed the activity of a single enzyme acting alone to promote a single chemical reaction, but this state of affairs is not so common in the cell. Instead, groups of enzymes usually catalyze the multiple steps in a sequence of chemical reactions known as a chemical **pathway**. This arrangement presents both advantages and challenges that illustrate important aspects of how enzymes usually behave in the cell.

Let's begin with a particularly noteworthy advantage granted by pathways. Being part of a multistep pathway permits the product formed by one enzyme to immediately become a reactant for the next reaction. In other words, the product of the first reaction is the immediate substrate for another enzyme and is rapidly consumed in a second catalyzed reaction. In general, a pathway of enzyme-catalyzed steps ensures a particular outcome from the sequence of chemical reactions. Such a sequence can be represented as follows:

$$A \xrightarrow{E1} B \xrightarrow{E2} C \xrightarrow{E3} D$$

where the enzyme E1 catalyzes the conversion of A to B, enzyme E2 catalyzes the conversion of B to C, and so on, ensuring that D will be produced in the end. Chemical pathways like this one produce most of the chemical building blocks of the cell, such as amino acids and nucleotides, and are necessary for the harnessing of energy from foodstuffs or sunlight.

The challenge faced by all enzyme-catalyzed pathways is the need for the product of each reaction to find the enzyme that catalyzes the next reaction in a timely fashion. Enzymes and their substrates do not actively swim after each other like sharks looking for prey. Instead they depend on random encounters or collisions inside the cell. If you have ever watched particles of dust float in a sunbeam, you have observed a phenomenon quite similar to what happens to enzymes and substrates inside the cell.

All free-floating molecules in the cell's watery cytoplasm experience a random or wandering motion that depends on the temperature. The warmer the cell, the faster these molecules move, just as dust particles move through the air on warm thermal drafts. So within the crowded environment of the cell, an enzyme collides with many molecules every second, but most of these molecules are not its substrates and do not fit into its active site, so nothing happens. Catalysis occurs only when the appropriate substrates are encountered that fit the active site of the enzyme. Thus, although molecular collisions do happen frequently in the cell, only some of these collisions result in enzyme-catalyzed reactions.

One way to increase the efficiency of a chemical pathway is to increase the frequency of molecular collisions between enzymes and their substrates. The multistep pathways common in metabolism accomplish this task by locating the enzymes for a particular pathway together. This physically close arrangement of enzymes means that the products of one reaction are closer to the next enzyme that uses them as substrates, increasing the likelihood of their collision and promoting the next chemical reaction. Both cells and molecules efficiently combine several enzymes in one location.

On the level of cellular organization, the enzymes necessary for a chemical pathway can be contained inside a specific organelle (Figure 7.6). As discussed in Chapter 6, organelles like mitochondria concentrate the proteins and chemical compounds required for specific life processes. For example, mitochondria are the sites where food substrates such as glucose are oxidized, generating most of the cell's ATP. The efficient production of ATP requires that the necessary enzymes and substrates be concentrated inside a small compartment. Several enzymes floating in the mitochondrial lumen participate in a series of reactions called the citric acid cycle, generating CO_2, while other enzymes are associated with the inner mitochondrial membrane, where ATP is produced from ADP and phosphate (see Chapter 8).

On the molecular level, several enzymes can be physically connected in a single giant multienzyme complex (see Figure 7.6). Many enzymes involved in the biosynthesis of cellular building blocks such as fatty acids and proteins function as large aggregates of multiple enzymes.

> ■ All chemical reactions in the cell require the assistance of enzymes to proceed at a rate that supports life. Enzymes speed up chemical reactions by lowering the amount of activation energy required to make and break bonds. Metabolism consists of many multistep pathways, each of which requires several different enzymes.

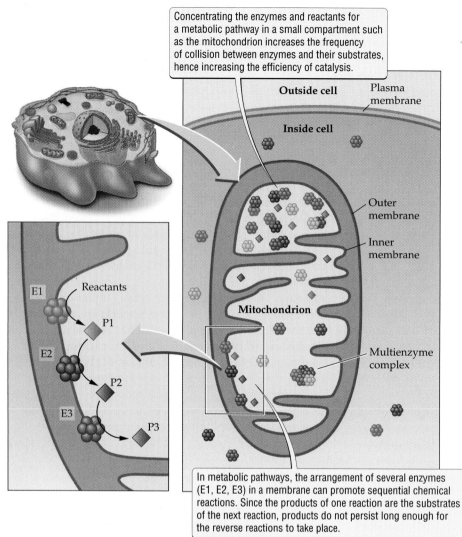

Concentrating the enzymes and reactants for a metabolic pathway in a small compartment such as the mitochondrion increases the frequency of collision between enzymes and their substrates, hence increasing the efficiency of catalysis.

Outside cell

Plasma membrane

Inside cell

Outer membrane

Inner membrane

Mitochondrion

Multienzyme complex

E1

Reactants

P1

E2

P2

E3

P3

In metabolic pathways, the arrangement of several enzymes (E1, E2, E3) in a membrane can promote sequential chemical reactions. Since the products of one reaction are the substrates of the next reaction, products do not persist long enough for the reverse reactions to take place.

Figure 7.6 Grouping of Enzymes in the Cell
Enzymes are arranged in the cell to promote multiple chemical reactions. These arrangements include localization of enzymes in organelle compartments, in membranes, and as parts of multienzyme complexes.

The Building of DNA

Enzyme-catalyzed chemical pathways are involved in the synthesis and breakdown of most complex molecules in the cell. The use of ATP by enzymes is a common feature of chemical pathways such as those involved in the biosynthesis of DNA. As discussed in Chapter 5, DNA is a polymer made up of nucleotides that are linked together by covalent bonds. Before cells can divide, their DNA must be duplicated so that each daughter cell can receive a complete version of the genetic material. Duplicating a cell's DNA requires the synthesis of new DNA polymers, using several enzymes. In this case, ATP is used to convert individual nucleotides to an activated form that can be added to a growing DNA chain.

Nucleotides can have one, two, or three phosphate groups bound to the pentose (five-carbon) sugar, and are described as monophosphate, diphosphate, or triphosphate nucleotides, respectively. The diphosphate and triphosphate nucleotides are commonly used to transfer energy in biosynthetic reactions. Energy-rich triphosphate nucleotides are used for DNA synthesis and are produced from monophosphate nucleotides with the help of two enzymes and ATP. These enzymes use two molecules of ATP to sequentially add two more phosphate groups to the one already bonded to the sugar group of the mononucleotide (Figure 7.7).

Since the nucleotides that make up DNA contain deoxyribose sugars, the triphosphate nucleotides produced are known as dATP, dCTP, dGTP and dTTP, where the "d" stands for deoxyribose; "A," "C," "G," and "T" stand for adenosine, cytosine, guanine, and thymidine, respectively; and "TP" stands for triphosphate. Since the high-energy phosphate bonds of the two ATPs were consumed in the reactions, the triphosphate nucleotide is now energy-rich and can form a covalent bond with another nucleotide. The chemical reaction that adds a triphosphate nucleotide to the end of a DNA chain is catalyzed by a third enzyme and results in the release of the two terminal phosphate groups (see Figure 7.7). In addition to DNA, other complex molecules such as sugars and fatty acids are produced using ATP and specific enzymes. These and other biosynthetic pathways create and maintain the complex structures found in every living cell.

> ■ Enzymes are necessary for the synthesis and breakdown of molecules in the cell. For example, the synthesis of DNA requires the action of at least three enzymes and energy from ATP.

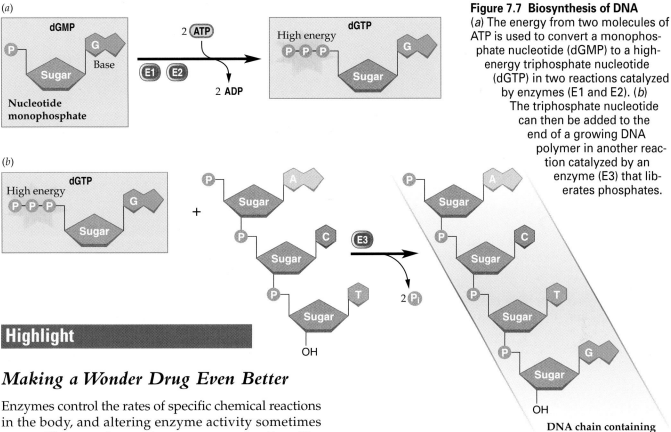

Figure 7.7 Biosynthesis of DNA
(*a*) The energy from two molecules of ATP is used to convert a monophosphate nucleotide (dGMP) to a high-energy triphosphate nucleotide (dGTP) in two reactions catalyzed by enzymes (E1 and E2). (*b*) The triphosphate nucleotide can then be added to the end of a growing DNA polymer in another reaction catalyzed by an enzyme (E3) that liberates phosphates.

DNA chain containing four nucleotides

Highlight

Making a Wonder Drug Even Better

Enzymes control the rates of specific chemical reactions in the body, and altering enzyme activity sometimes causes illness and sometimes promotes healing. The wondrous effects of aspirin, it turns out, are due to its blocking the activity of two important enzymes, COX-1 and COX-2. COX-1 is always produced in the body and catalyzes the biosynthesis of hormones that help maintain the lining of the stomach. In contrast, COX-2 is produced only when injury occurs, and it catalyzes the biosynthesis of different hormones that promote inflammation, fever, and the sensation of pain throughout the body. Although both enzymes are inhibited by aspirin, they participate in two different biosynthetic pathways and play different biological roles. It is clear that the therapeutic benefits of aspirin (reduction of pain, inflammation, and fever) are due to the blocking of COX-2, while the negative side effects (damage to the stomach lining) are due to the blocking of COX-1.

The blocking of COX-2 activity is probably also responsible for the effects of aspirin on colon cancer, which were mentioned at the beginning of the chapter. Some cancerous cells have an abnormally high level of COX-2, which may encourage the growth of blood vessels into the tumor, thereby feeding it and allowing it to grow into a more serious cancer. By blocking COX-2 activity, aspirin may limit the blood supply to tumors and reduce the spread of the cancer.

Although it has so many beneficial effects, how can we make aspirin better, with fewer negative side effects? The simple answer is to develop a drug that only blocks COX-2 activity and has little or no effect on COX-1. This challenge has been enthusiastically taken up by many research laboratories around the world, resulting in the recent development of a first generation of superaspirins.

The initial success in developing superaspirins was based on an understanding of the three-dimensional shape of COX-2. As we have already discussed, the shape of an enzyme helps define its catalytic activity, and knowing the shape of COX-2 allowed researchers to design inhibitors that bind only to COX-2 and not to COX-1. Since 1996, nearly a dozen new compounds have been developed that bind to COX-2 and block its activity, while having no significant effect on COX-1. This first generation of superaspirins is already being tested for anticancer properties, and although the results are pending, the development of ways to limit the metabolic activities of COX-2 is likely to yield significant health benefits.

■ Aspirin blocks the activities of both COX-1 and COX-2 enzymes. Superaspirins will have the beneficial effects of blocking COX-2 enzymes without the negative effects of blocking COX-1 enzymes.

Summary

The Role of Energy in Living Systems

■ Living organisms obey the same laws of energy that apply to the physical world.

■ The sun is the primary energy source for all living organisms.

■ The creation of biological order requires the transfer of disorder to the environment, most often in the form of heat.

■ Atoms such as carbon and nitrogen are recycled between various living organisms and the environment.

Using Energy from the Controlled Burning of Food

■ In oxidation, electrons are lost from one molecule or atom to another. In reduction, electrons are gained by one molecule or atom from another.

■ Catabolic reactions harness energy in the form of energy carriers like ATP and are tightly coupled to biosynthetic reactions that require energy.

■ All chemical reactions require the input of a small amount of activation energy to proceed.

How Enzymes Speed Up Chemical Reactions

■ Enzymes greatly increase the rate at which chemical reactions proceed by lowering the amount of activation energy they require.

■ All chemical reactions that support life require the assistance of enzymes.

■ The activity of enzymes is highly specific. Each enzyme binds to a specific set of substrates and catalyzes a specific chemical reaction.

■ The specificity of an enzyme depends on its three-dimensional shape and the chemical characteristics of its active site.

■ Several enzymes catalyze multiple steps in a chemical pathway. The harnessing of energy or the synthesis of macromolecules in metabolism depends on many different chemical pathways.

The Building of DNA

■ DNA consists of nucleotide monomers that are covalently bonded together to form a polymer.

■ Each monophosphate nucleotide must be converted to an energy-rich triphosphate form by the enzyme-catalyzed transfer of two phosphate groups from two molecules of ATP.

■ The production of triphosphate nucleotides and their addition to a growing DNA chain require the action of several enzymes.

Highlight: Making a Wonder Drug Even Better

■ Aspirin blocks the activity of both COX-1 and COX-2 enzymes.

■ The therapeutic benefits of aspirin (pain and fever relief) arise from the blocking of COX-2, while the negative side effects (damage to the stomach) arise from the blocking of COX-1.

■ New and improved superaspirins are chemical compounds that specifically block COX-2 without affecting COX-1, giving all the benefits of traditional aspirin without the negative side effects.

Key Terms

activation energy p. 119

active site p. 120

biosynthetic p. 114

catabolic p. 114

catalysis p. 120

catalyst p. 119

enzyme p. 119

first law of thermodynamics p. 114

metabolism p. 114

oxidation p. 116

pathway p. 121

reduction p. 116

second law of thermodynamics p. 114

substrate p. 120

Chapter Review

Self-Quiz

1. Which of the following statements is true?
   a. Cells are able to produce their own energy from nothing.
   b. Cells use energy only to generate heat and move molecules around.
   c. Cells obey the same physical laws of energy as the nonliving environment.
   d. Photosynthetic plants have no effect on the way animals obtain energy.

2. Living organisms use energy to
   a. organize chemical compounds into complex biological structures.
   b. decrease the disorder of the surrounding environment.
   c. cancel the laws of thermodynamics.
   d. cut themselves off from the nonliving environment.

3. The carbon atoms contained in biological molecules like proteins
   a. are manufactured by cells for use in the organism.
   b. are recycled from the nonliving environment.
   c. differ from those found in CO_2 gas.
   d. cannot be oxidized under any circumstances.

4. Oxidation is
   a. the removal of oxygen atoms from a chemical compound.
   b. the gain of electrons by an atom.

c. the loss of electrons by an atom.

d. the synthesis of complex molecules.

5. The small input of energy required before a chemical reaction can proceed
   a. is called the activation energy.
   b. is independent of the laws of thermodynamics.
   c. is provided by an enzyme.
   d. always takes the form of heat.

6. The active site of an enzyme
   a. has the same shape for all known enzymes.
   b. can bind both its substrate and other kinds of molecules.
   c. does not play a direct role in catalyzing the reaction.
   d. can bring molecules together such that a chemical reaction takes place.

Review Questions

1. Explain why it is important for cells to gradually oxidize food molecules in multiple steps instead of doing it all at once in a single reaction.

2. Why is the release of heat by cells so important for their organization?

3. For a chemical reaction to occur, the reactants must collide with each other. Compare the way in which higher temperature facilitates this process and speeds up the reaction with the way in which an enzyme does.

4. Cells use several methods to increase the efficiency of enzyme catalysis in chemical pathways. Describe two of these methods and how they apply to mitochondria.

5. Enzymes can be found in laundry detergents, where they assist in the removal of stains from clothing. How might this use of enzymes be similar to what they do in the cell?

7

The Daily Globe

Is Spinach Mightier than the Sword?

BETHESDA, MD. Everyone, from Popeye to vegetable-detesting children, knows about the benefits of eating spinach, a vegetable rich in vitamins and iron. Now, according to a new study in the journal *Advanced Military Science*, researchers report that this humble vegetable might also have the power to disarm explosives.

Spinach, researchers have discovered, contains a powerful enzyme known as nitroreductase, a substance that can neutralize dangerous explosives like TNT. The spinach enzyme reduces TNT to other compounds that can then be converted to CO_2 gas through additional chemical reactions.

Experts say this could be great news for the United States military, which has been struggling to find a safe way to dispose of more than half a million tons of stockpiled explosives. Researchers note that the by-product, CO_2 gas, is less harmful to the environment than the usual chemical by-products of TNT degradation.

A spokesman at DeArm, a company that specializes in products that defuse explosives, warned that their research indicates that spinach is not the environmentally friendly cure-all that the new research suggests. In fact, DeArm scientists claim it is more harmful to the environment than traditional methods, since it releases carbon dioxide, a gas implicated in global warming.

But many were undeterred by this potential flaw. "This is a fantastic finding," said Dr. John J. Blowemup, a chemist at Southern Michigan University. "Spinach may provide the most environmentally friendly method for getting rid of unwanted explosives. It's cheap, it's safe and it's absolutely silent. Who'd have guessed spinach could do all that?"

Evaluating "The News"

1. The possibility of using spinach enzymes to break down explosives has many advantages. Describe an advantage from each of the following viewpoints: the environment, the taxpayer, and the U.S. military.

2. The DeArm company claims that spinach is not an environmentally friendly solution, while this new study claims it is. What do you need to know to evaluate these two conflicting claims?

3. Why do you think a plant like spinach would have an enzyme with such reducing power?

chapter 8

Photosynthesis and Respiration

Paul Klee, *Around the Fish (Um den Fisch)*, 1926.

Food for Thought

The next time you feel hungry after skipping a meal, keep in mind that much of your body's demand for food is being made by your brain. One distinctive feature of the human brain is its size and need for energy. Although other animals, such as whales, certainly have larger brains by weight, the human brain is the largest when compared to the size of the human body. In other words, humans have the highest ratio of brain to body weight, which contributes to our status as the most intelligent of animals.

Photosynthesis and respiration are complementary chemical processes that govern the transfer of energy in cells.

A daily challenge of having such a relatively large brain is the urgent need to supply it with energy. Your brain consumes a large amount of energy while sending and receiving nerve impulses, a demand so high that more than half of the nourishment consumed by an infant is used up by the brain.

Given the tremendous energy demands made by your brain, how does your body manage to keep it satisfied? The answer ultimately lies in how energy is harvested from sunlight by photosynthetic organisms and used to manufacture sugars. These sugars are in turn used by all organisms, including humans, to provide the energy that supports life processes. In this chapter we discuss how energy is converted from sunlight into sugar, a form of energy that can satisfy the needs of all living organisms. Perhaps then we can answer the question of how we keep our brains running.

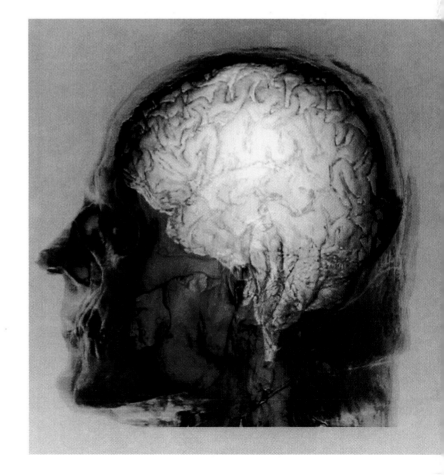

A 3-D Reconstruction of the Human Brain

Key Concepts

1. The temporary storage and transfer of energy in the cell requires the production of energy carrier molecules, including ATP, NADH, and NADPH.

2. Photosynthesis is a series of chemical reactions that use sunlight to manufacture sugar from the atmospheric gas carbon dioxide (CO_2) and water. The process also releases oxygen gas (O_2) into the environment.

3. The light-dependent reactions of photosynthesis capture solar energy using molecules embedded in the thylakoid membranes of chloroplasts. Light energy is converted to the movement of electrons, which indirectly generates ATP, NADPH, and O_2.

4. The reactions of photosynthesis that do not require light take place inside of chloroplasts and use the ATP and

NADPH generated by the light-dependent reactions to synthesize sugars from CO_2.

5. Glycolysis breaks down sugars in the cytosol of the cell, producing pyruvate and small amounts of ATP and NADH. In the absence of oxygen, the pyruvate remains in the cytosol and is converted into alcohol and CO_2, or lactic acid. In the presence of oxygen, the pyruvate is used by mitochondria to generate many additional molecules of ATP.

6. The citric acid cycle in the mitochondrial lumen uses pyruvate to produce NADH and CO_2. In oxidative phosphorylation, NADH donates electrons to the electron transport chain in the inner mitochondrial membrane, indirectly generating ATP. Oxygen is consumed in the process.

The harvesting of energy from the environment is one of the fundamental processes that support life. Given all the complex structures and chemical reactions of every living system, obviously some kind of fuel must power it all. Ultimately, the sun is the primary source of energy for all known life on Earth. Solar energy is used by plants and other photosynthetic organisms to make food such as sugars from carbon dioxide gas in the atmosphere and water. Animals that consume plants acquire this chemical energy, which in turn is passed on to animals that eat other animals (Figure 8.1).

The chemical reactions that transfer energy from one molecule to another and from one organism to another form the basis of metabolism. This chapter explores the biosynthetic processes of metabolism by discussing how plants capture solar energy and use it to form the chemical bonds found in food molecules such as sugars. In the case of the catabolic processes of metabolism, we will discuss how food is oxidized and broken down to produce usable forms of energy that function as fuel for the cell.

Before we can understand the chemical reactions of metabolism, we must first define how energy is transferred from one molecule to another. One method of energy transfer commonly found in the physical world depends on heat. When water in a kettle is boiled on a

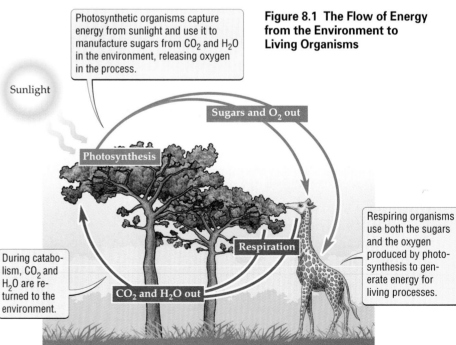

Photosynthetic organisms capture energy from sunlight and use it to manufacture sugars from CO_2 and H_2O in the environment, releasing oxygen in the process.

Sunlight

Sugars and O_2 out

Photosynthesis

During catabolism, CO_2 and H_2O are returned to the environment.

Respiration

CO_2 and H_2O out

Respiring organisms use both the sugars and the oxygen produced by photosynthesis to generate energy for living processes.

Figure 8.1 The Flow of Energy from the Environment to Living Organisms

stove, the energy from the flame is transferred in the form of heat to the metal of the kettle and then to the water molecules. Organisms generally cannot use such violent means and instead use specialized molecules that transfer the chemical energy stored in covalent bonds.

Energy Carriers: Powering All Activities of the Cell

The temporary storage and transfer of usable energy in the cell depends on several so-called **energy carrier** molecules. The most commonly used carrier of energy is ATP (which was introduced in Chapter 5). ATP stores energy in the form of covalent bonds between phosphate groups; an example of how it is used to synthesize a complex macromolecule like DNA is discussed in Chapter 7. In general, ATP contributes its stored energy to other molecules by transferring one of its phosphate groups to another molecule. This transfer energizes the recipient molecule, enabling it to change shape or chemically react with other molecules.

Chemical bond energy is not the only fuel that is donated by carrier molecules. Other important carriers—nicotinamide adenine dinucleotide phosphate ($NADP^+$) and nicotinamide adenine dinucleotide (NAD^+)—donate high-energy electrons to oxidation–reduction reactions. Both carriers pick up two high-energy electrons with a hydrogen ion (H^+), forming NADPH and NADH, respectively. The ability in turn to donate electrons means that NADPH and NADH can reduce other molecules. That is, they become oxidized by losing electrons while the compounds that accept the

electrons are reduced. (Oxidation and reduction were introduced in Chapter 7.)

Although NADPH and NADH have equal abilities as reducing agents, the difference of one phosphate group between them affects which target molecules they bind to and react with and which pathways they affect. NADPH is used in the biosynthetic reactions that manufacture sugar from carbon dioxide, and NADH is used in the catabolic reactions that produce ATP from the breakdown of sugars.

Two organelles—chloroplasts and mitochondria—produce the energy carriers that power the activities of the cell. These organelles are the power stations of the cell, and in this chapter we will see how their membrane arrangements facilitate the harvesting of usable energy. Chloroplasts harvest energy from sunlight and use it to synthesize sugars; mitochondria harvest energy from sugars and use it to synthesize energy carriers like ATP (Figure 8.2).

■ The chemical reactions that constitute life require energy. Energy is transferred within living organisms via specialized compounds called energy carriers. ATP is the most commonly used energy carrier and donates bond energy to chemical reactions. The energy carriers NADH and NADPH donate electrons to oxidation–reduction reactions.

Figure 8.2 The Exchange of Molecules between Chloroplasts and Mitochondria Produces Energy Carriers like ATP

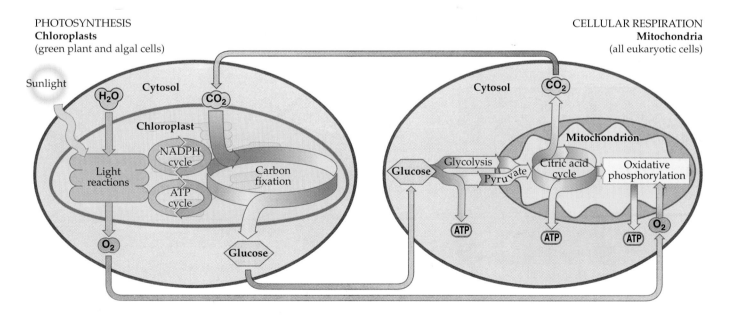

Photosynthesis: Harvesting Energy from Sunlight

The next time you walk outside, look at the plants around you and try to appreciate the critical role they play in supporting the web of life that includes human beings. Plants and other photosynthetic organisms, such as green algae and some bacteria, are able to capture the energy from sunlight in the form of chemical bonds. This process is **photosynthesis**, which uses solar energy to synthesize complex, energy-rich molecules such as sugars from carbon dioxide (CO_2) in the atmosphere and water (H_2O).

The process consists of a series of chemical reactions that also result in the chemical splitting of water and the release of oxygen gas (O_2) into the environment. The O_2 by-product of photosynthesis is essential for all oxygen-breathing life forms—another reason why plants are wor-thy of our respect. In other words, plants help support all animal life, including humans, since animals either indi-rectly or directly depend on plants for food and oxygen.

Chloroplasts are sites of photosynthesis

Photosynthesis takes place inside organelles called **chloroplasts**. These specialized membrane compartments were introduced in Chapter 6; here we focus on how the membranes and associated proteins of the chloroplast contribute to the chemical reactions of photosynthesis.

Chloroplasts are found in the cytosol of plant cells and contain chlorophyll molecules, which account for the green color of most plant foliage. These important organelles have a distinctive arrangement of membranes that is necessary for photosynthesis (Figure 8.3). Like mitochondria, chloroplasts are surrounded by a double membrane that divides the organelle into different com-

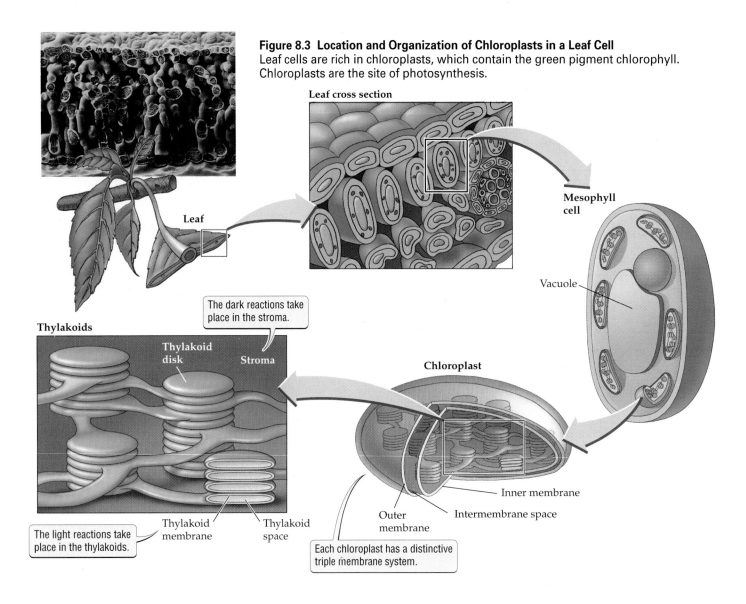

Figure 8.3 Location and Organization of Chloroplasts in a Leaf Cell
Leaf cells are rich in chloroplasts, which contain the green pigment chlorophyll. Chloroplasts are the site of photosynthesis.

Leaf cross section

Leaf

Mesophyll cell

Vacuole

The dark reactions take place in the stroma.

Thylakoids

Thylakoid disk

Stroma

Chloroplast

Inner membrane

Outer membrane

Intermembrane space

The light reactions take place in the thylakoids.

Thylakoid membrane

Thylakoid space

Each chloroplast has a distinctive triple membrane system.

partments. The first compartment, between the two membranes, is appropriately called the **intermembrane space**; the second compartment, enclosed by the inner membrane, is called the **stroma**. Inside the stroma, chloroplasts have yet another compartment in the form of flattened interconnected sacs made up of the **thylakoid membrane**. This third compartment, called the **thylakoid space**, is a structure that is not found in mitochondria.

Each internal compartment of the chloroplast has a specific role in the overall photosynthetic process. Both the thylakoid membrane and the thylakoid space house enzymes and other molecules needed to transfer solar energy to energy carriers such as ATP and NADPH. In contrast, the stroma contains enzymes that use these energy carriers to manufacture sugars from carbon dioxide and water. Thus, the chloroplast is a well-organized power plant and factory that segregates metabolic activities into specialized compartments.

The first series of reactions to consider take place at the thylakoid membrane and directly harness energy from sunlight. Since they require light, these reactions are collectively known as the **light reactions**. On the other hand, the reactions that directly use CO_2 to synthesize sugars are known as the **dark reactions**. (However, these reactions can also take place in the light.) In the following sections we discuss the light and dark reactions, emphasizing the key chemical events that characterize each group.

The light reactions harvest energy from sunlight

The capture of energy from sunlight is essential to photosynthesis and supports life throughout the biosphere. In this process solar energy is converted to the energy contained in the electrons of organic compounds. Specialized pigments, of which chlorophyll is the most common, carry out this process. Chlorophyll is a green pigment found embedded in the thylakoid membrane.

When exposed to light, the electrons associated with the covalent bonds of a chlorophyll molecule absorb energy from the light and become more energized. These energized electrons are often said to be "excited." To guarantee that the energy captured is not lost to the environment and wasted, hundreds of chlorophyll molecules are arranged together in a complex called the **antenna complex**, an appropriate name given its role in capturing solar energy (Figure 8.4). Excited electrons pass energy from one chlorophyll molecule to another in the antenna until ultimately they are passed to a group of electron-accepting proteins located nearby in the thylakoid membrane. These groups of proteins are collectively called **electron transport chains**, or **ETCs**, and together with the antenna systems they make up more than half of the thylakoid membrane.

Electron transport chains are arrangements of proteins embedded in membranes. As electrons are passed down an ETC from one protein component to another, small amounts of energy are released that are used to generate the energy carriers ATP and NADPH. The combination of an antenna with a neighboring ETC is called a **photosystem**, and there are two distinct types of photosystems in the thylakoid membrane.

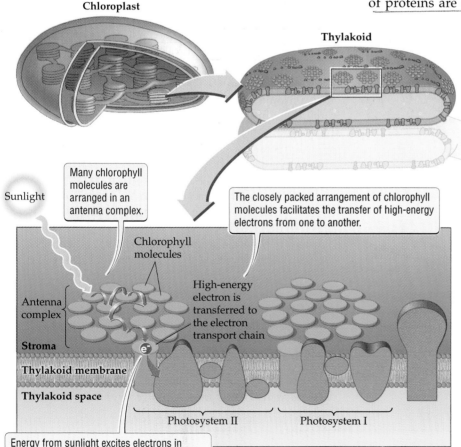

Chloroplast

Thylakoid

Sunlight

Many chlorophyll molecules are arranged in an antenna complex.

The closely packed arrangement of chlorophyll molecules facilitates the transfer of high-energy electrons from one to another.

Chlorophyll molecules

High-energy electron is transferred to the electron transport chain

Antenna complex

Stroma

Thylakoid membrane

Thylakoid space

Photosystem II

Photosystem I

Energy from sunlight excites electrons in chlorophyll molecules to a higher energy state.

Figure 8.4 Arrangement of Chlorophylls and Photosystems
The special arrangement of chlorophyll pigments in the thylakoid membrane facilitates the transfer of electrons from the initial antenna complex (a collection of chlorophylls) to the electron transport chains (colored purple) of photosystems I and II.

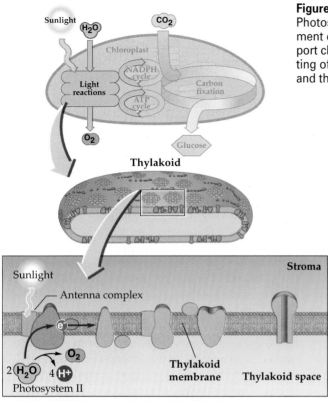

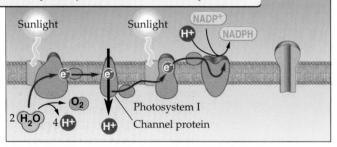

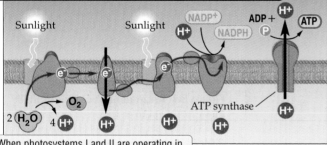

Figure 8.5 ATP Production and Photosynthesis

Photosynthesis produces ATP using the energy from light. The movement of electrons between two photosystems and their electron transport chains results in the production of NADPH and ATP, with the splitting of water molecules and the release of oxygen gas. Both the ATP and the NADPH produced are later used in carbon fixation.

They are called photosystem I and photosystem II, respectively, and each connects to a different ETC (see Figure 8.4).

Photosynthesis operates by integrating the activities of both photosystems. Photosystem I is primarily responsible for the production of the powerful reducing agent NADPH; the reactions of photosystem II lead to the important production of O_2. Let's consider how each photosystem contributes to photosynthesis as a single integrated process.

As Figure 8.5 shows, the journey of electrons along the ETC used by photosystem II includes transfer to and from a channel protein that spans the thylakoid membrane. The channel protein uses the energy of the electrons to pump protons (H^+) from the stroma into the thylakoid space. The protons accumulate inside the thylakoid space, causing an imbalance in the proton concentration across the thylakoid membrane. In other words, protons are abundant inside the thylakoid space and more scarce in the stroma.

This imbalance, called a **proton gradient**, is used to manufacture ATP by another protein that spans the thylakoid membrane, called ATP synthase. Since the thylakoid membrane will not allow protons to pass through on their own, the only way for them to return to the stroma is to move through the ATP synthase from a region of high concentration in the thylakoid space to one of low concentration in the stroma. This movement of protons releases the energy that powers the addition of phosphate to ADP to form ATP.

Photosystem I receives electrons from the last protein in the ETC used by photosystem II. Therefore, photosystem I uses the electrons transferred from photosystem II to replace the electrons lost from chlorophyll molecules in its own antenna. Photosystem I eventually transfers the electrons to an ETC protein that in turn transfers them to $NADP^+$. The transfer of electrons to $NADP^+$ gives it a negative charge, and it takes up H^+ from the stroma to form NADPH (see Figure 8.5). In this manner, the two photosystems work together to produce both ATP and NADPH, which are used in the dark reactions.

The dark reactions manufacture sugars

The energy carriers, ATP and NADPH, that are produced by the light reactions are used in the dark reac-

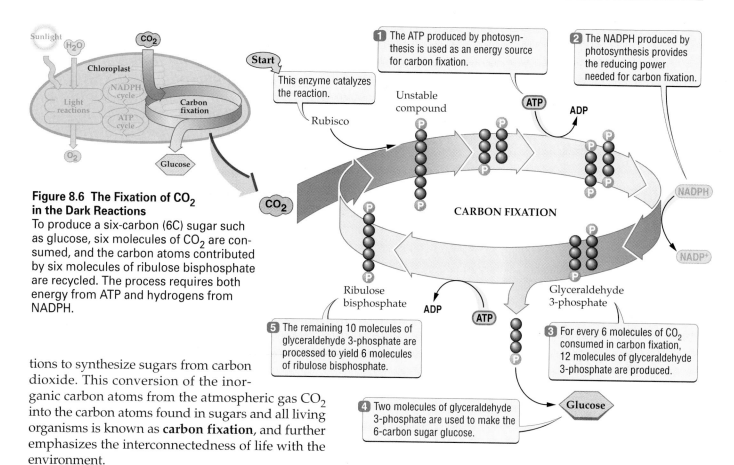

Figure 8.6 The Fixation of CO$_2$ in the Dark Reactions
To produce a six-carbon (6C) sugar such as glucose, six molecules of CO$_2$ are consumed, and the carbon atoms contributed by six molecules of ribulose bisphosphate are recycled. The process requires both energy from ATP and hydrogens from NADPH.

Labels within figure:

Sunlight   H$_2$O   CO$_2$ Chloroplast   NADPH cycle   Light reactions   ATP cycle   Carbon fixation   O$_2$ Glucose

Start — This enzyme catalyzes the reaction. — Rubisco

1 The ATP produced by photosynthesis is used as an energy source for carbon fixation.

2 The NADPH produced by photosynthesis provides the reducing power needed for carbon fixation.

Unstable compound   ATP   ADP   NADPH   NADP$^+$

CARBON FIXATION

CO$_2$

Ribulose bisphosphate   ADP   ATP   Glyceraldehyde 3-phosphate

5 The remaining 10 molecules of glyceraldehyde 3-phosphate are processed to yield 6 molecules of ribulose bisphosphate.

3 For every 6 molecules of CO$_2$ consumed in carbon fixation, 12 molecules of glyceraldehyde 3-phosphate are produced.

4 Two molecules of glyceraldehyde 3-phosphate are used to make the 6-carbon sugar glucose.

Glucose

---

tions to synthesize sugars from carbon dioxide. This conversion of the inorganic carbon atoms from the atmospheric gas CO$_2$ into the carbon atoms found in sugars and all living organisms is known as **carbon fixation**, and further emphasizes the interconnectedness of life with the environment.

As mentioned earlier, the dark reactions are catalyzed by enzymes that float freely in the stroma. The most important enzyme for these dark reactions is ribulose bisphosphate carboxylase, known as **rubisco**. Rubisco catalyzes the first reaction of carbon fixation, in which a molecule of the one-carbon (1C) compound CO$_2$ combines with a five-carbon (5C) compound called ribulose 1,5-bisphosphate to produce two three-carbon compounds ($1C + 5C = 2 \times 3C$). This first reaction is followed by a multistep cycle of many reactions designed both to manufacture sugars and to regenerate more of the 5C compound (ribulose 1,5-bisphosphate) to keep the dark reactions going. The entire process requires the input of both energy from ATP and hydrogens from NADPH.

Three turns of the carbon fixation cycle produces one molecule of the 3C sugar **glyceraldehyde 3-phosphate**. One can follow this process by tracking the number of carbon atoms as they are rearranged into different compounds at each step of the cycle (Figure 8.6). For every three molecules of CO$_2$ ($3 \times 1C = 3C$) combined with three ribulose 1,5-bisphosphates ($3 \times 5C = 15C$), six three-carbon compounds ($6 \times 3C = 18C$) are produced. These compounds eventually produce three ribulose 1,5-bisphosphates ($3 \times 5C = 15C$) and one molecule of glyceraldehyde 3-phosphate (3C). As the math indicates, it takes three

turns of the cycle to produce one three-carbon sugar, with the other carbon atoms being recycled to ribulose 1,5-bisphosphate. The formation of one molecule of glyceraldehyde 3-phosphate also requires the input of nine molecules of ATP and six molecules of NADPH.

Glyceraldehyde 3-phosphate is the chemical building block used to manufacture other sugars such as glucose and complex molecules needed by the cell. Most of the glyceraldehyde 3-phosphate is exported from the chloroplast and eventually consumed in chemical reactions that produce ATP. Molecules of glyceraldehyde 3-phosphate that are not immediately consumed in ATP synthesis are used to manufacture sucrose in the cytoplasm.

Sucrose is the simple sugar (glucose linked to fructose) most often used as a sweetener in desserts. It is an important food source for all the cells in a plant and is transported from the leaves where photosynthesis takes place to all other parts of the plant. Significant amounts of sucrose are stored in the cells of sugarcane and sugar beets, which is why these two plants are the major crops of the sugar industry worldwide.

Not all the glyceraldehyde 3-phosphate is exported from the chloroplasts; some of it is converted into starch by enzymes in the stroma. Starch is a polymer of glucose and is an important form of stored energy in plants. It tends to accumulate in chloroplasts during the day and is then broken down to simple sugars at night. This conservation of food to generate energy at night compensates for the lack of sunlight and hence photosynthesis. Plant seeds, roots, and tubers are also rich in starch to provide the energy required for germination and growth. Indeed, the energy-rich nature of these plant components explains why they are such an important food source for animals.

> ■ Photosynthesis harvests energy from sunlight and uses it to synthesize sugars from CO_2 and H_2O. It occurs in chloroplasts and produces as a by-product the O_2 that is essential for all oxygen-breathing organisms on Earth. The photosynthetic reactions that harvest energy from sunlight are called the light reactions and produce ATP and NADPH, which is used by the dark reactions to synthesize sugars.

Catabolism: Breaking Down Molecules for Energy

The conversion of food into useful energy requires the breakdown and gradual oxidation of food molecules. This is the foundation of **catabolism**, which is constantly occurring in our bodies. The first stage of catabolism is the breakdown of large, complex food molecules such as carbohydrates, proteins, and fats into their simpler components.

This is the digestive process that we all experience after every meal. It occurs in the stomach and intestines, and the simpler compounds that are released by digestion, such as amino acids and simple sugars, are absorbed by the intestine and passed on to other cells in the body via the bloodstream. Digestion is discussed in Chapter 28; here we are more concerned with how the simple sugars supplied by digestion are converted into fuel for use by the cell. This catabolic process consists of three major stages: glycolysis, the citric acid cycle, and oxidative phosphorylation.

Glycolysis is the first stage in the breakdown of sugar

The first stage of catabolism that occurs inside the cell is known as glycolysis (literally, "the splitting of sugar"). From an evolutionary standpoint, glycolysis was probably the earliest means of producing ATP from food molecules. In most organisms today, however, glycolysis is a necessary preparation for more efficient ways of producing ATP in mitochondria (Figure 8.7).

Figure 8.7 Glycolysis
In glycolysis, a single six-carbon glucose molecule is converted into two three-carbon molecules of pyruvate for fermentation or use by mitochondria. The process produces a relatively low net yield of two ATP molecules and two NADH molecules per glucose molecule. If oxygen is present, the NADH moves into the mitochondria for use in aerobic respiration. If oxygen is absent, the NADH is used in fermentation reactions.

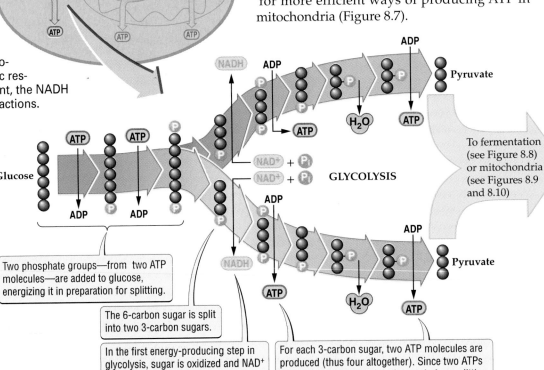

Two phosphate groups—from two ATP molecules—are added to glucose, energizing it in preparation for splitting.

The 6-carbon sugar is split into two 3-carbon sugars.

In the first energy-producing step in glycolysis, sugar is oxidized and NAD⁺ is reduced, producing NADH.

For each 3-carbon sugar, two ATP molecules are produced (thus four altogether). Since two ATPs were required to energize glucose before splitting, glycolysis has a net yield of two ATPs.

Specifically, **glycolysis** is a series of chemical reactions in the cytosol of the cell that provide mitochondria with the substrate molecules needed to generate ATP. A series of enzyme-catalyzed reactions partially oxidizes the simple 6C sugar glucose into pyruvate, which is then transported into mitochondria for further processing. Partway through the series of glycolytic reactions, a 6C sugar intermediate formed from glucose is split into two 3C sugars, which are then converted into two 3C molecules of pyruvate.

During glycolysis, for each molecule of glucose consumed, four molecules of ADP are phosphorylated to four molecules of ATP, and electrons are donated to two molecules of NAD^+, generating two molecules of NADH. Since the early steps of glycolysis consume two molecules of ATP per glucose molecule, a single glucose molecule produces a net yield of two ATP molecules and two NADH molecules (see Figure 8.7). Glycolysis does not require oxygen gas (O_2). For complete oxidation using O_2, the pyruvate molecules enter the mitochondria for use in the cell's most efficient ATP-producing system.

Fermentation sidesteps the need for oxygen

Since glycolysis does not require O_2 from the atmosphere, its reactions are described as **anaerobic**, and they were probably essential for early life forms in the oxygen-poor atmosphere of primitive Earth. This is in contrast to the **aerobic** reactions in the mitochondria, which require O_2 and dominate ATP production in most organisms living under Earth's present oxygen-rich atmosphere. However, some organisms still use glycolysis as their only means of generating ATP, in a process known as **fermentation**.

These organisms are anaerobic; that is, they can live without O_2. Yeast is a familiar anaerobic microorganism that has an essential role in the production of beer. Fermentation in yeast converts pyruvate into ethanol and CO_2 gas, giving beer its alcohol content and foamy effervescence (Figure 8.8). This production of CO_2 also explains the role of baker's yeast in making bread rise since the released gas expands the bread dough.

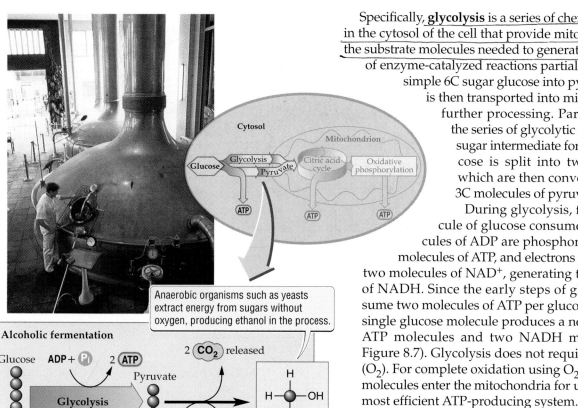

Anaerobic organisms such as yeasts extract energy from sugars without oxygen, producing ethanol in the process.

During strenuous activity, muscle cells resort to fermentation to produce extra ATP quickly. The process yields lactic acid instead of ethanol.

Glycolysis produces ATP whether or not O_2 is present, as long as the cell's supply of NAD^+ is replenished.

When O_2 is absent, NADH from glycolysis is used as a reducing agent to regenerate NAD^+, thus replenishing the cell's supply.

Figure 8.8 Fermentation Has a Variety of Uses
Fermentation is a glycolytic process that produces ATP without using oxygen. The production of ethanol by alcoholic fermentation is the reason that yeasts are used to brew alcoholic beverages like beer (top photo). A similar process occurs in strenuously exercised muscles (bottom photo), but instead of alcohol, lactic acid is produced.

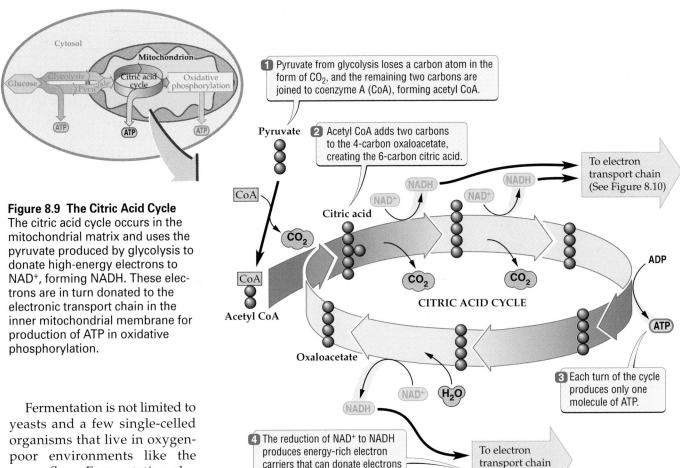

Figure 8.9 The Citric Acid Cycle
The citric acid cycle occurs in the mitochondrial matrix and uses the pyruvate produced by glycolysis to donate high-energy electrons to NAD$^+$, forming NADH. These electrons are in turn donated to the electronic transport chain in the inner mitochondrial membrane for production of ATP in oxidative phosphorylation.

Fermentation is not limited to yeasts and a few single-celled organisms that live in oxygen-poor environments like the ocean floor. Fermentation also takes place in the human body. When we exercise hard and push our muscles to the point of exhaustion, the pain we feel is due largely to fermentation processes in the muscles. A rapid burst of strenuous exercise can exhaust the ATP stores in muscle in a matter of seconds.

Both marathon runners and Olympic cyclists know from firsthand experience that to sustain strenuous physical activity, muscle cells must use both glycolysis and fermentation to generate more ATP. However, athletes' muscles do not produce alcohol and CO_2, as yeasts do; instead, pyruvate is converted into lactic acid, which causes the burning sensation in aching muscles.

Aerobic respiration produces the most ATP for the cell

The bulk of ATP produced in most organisms depends on those energy powerhouses, mitochondria (see Figure 6.9). They use both pyruvate and O_2 in a series of oxidation reactions that release energy. This energy is used to phosphorylate ADP to ATP. Mitochondria are therefore said to perform **aerobic respiration**, which produces many more molecules of ATP per glucose consumed than glycolysis does.

The important connections between oxygen, mitochondria, and ATP production turn up again and again in different cells in all living organisms. For example, the muscle cells of the human heart have an exceptionally large number of mitochondria to produce the enormous amounts of ATP needed to keep the heart beating. Blind mole rats that live underground in oxygen-poor burrows and must dig and dig every day have also optimized their means of using oxygen to generate ATP in their muscles. In comparison to white lab rats, blind mole rats have muscles that have 50% more mitochondria and 30% more blood capillaries. This means that their muscles are supplied with more blood to maximize the transfer of available oxygen, and have more mitochondria to maximize the resulting production of ATP.

Blind mole rat

The citric acid cycle produces NADH and CO_2

In aerobic eukaryotes such as humans, all the steps that follow the production of pyruvate by glycolysis take place inside mitochondria. Once pyruvate enters mitochondria, it is converted into CO_2 and an acetyl group that is attached to a large carrier molecule called coenzyme A. The high-energy compound that is formed, acetyl CoA, is an important substrate for a series of eight oxidation reactions called the **citric acid cycle**, all of which take place in the mitochondrial matrix (Figure 8.9). The name is based on the fact that citric acid is the product of the first reaction involving acetyl CoA. The major consequence of this oxidation cycle is the production of high-energy electrons stored in NADH, and CO_2 as waste. The molecules of NADH are then used in the next stage of ATP production.

Most of the oxidation reactions that take place in the cell are part of the citric acid cycle, emphasizing the importance of this process in energy production. In addition, stored fat can be broken down into fatty acids, which enter the mitochondria and are converted to acetyl CoA by a different set of reactions. Thus, the processes that produce energy from both sugars and fats come together at the beginning of the citric acid cycle.

Oxidative phosphorylation uses O_2 and NADH to produce ATP in quantity

The jackpot of ATP is generated in the third and last stage of aerobic respiration. This is also the stage at which the physical layout of mitochondria really comes into play. As discussed in Chapter 6, mitochondria have a double membrane that forms two separate compartments, the intermembrane space and the matrix. The NADH molecules produced from the citric acid cycle donate their high-energy electrons to an electron transport chain (ETC) embedded in the inner membrane of the mitochondrion. A component of the ETC in turn phosphorylates ADP to produce ATP. Since the electrons carried by the NADH molecules were gained by the oxidation of sugars in the citric acid cycle, and the components of the ETC are oxidized by the transfer of electrons, the whole process is appropriately called **oxidative phosphorylation** (Figure 8.10).

If this process seems familiar, the reason is that the ETC in the inner mitochondrial membrane has a function similar to that of the ETC found in the thylakoid membrane of chloroplasts. In mitochondria, the electrons donated by NADH are passed along a series of ETC components, releasing the energy used to pump protons from the matrix into the space between the two mitochondrial membranes. The resulting buildup of protons in the intermembrane space has the same effect as the buildup of protons in the thylakoid space: It forms a proton gradient

Figure 8.10 Oxidative Phosphorylation
Oxidative phosphorylation is the last stage of aerobic respiration and produces the most ATP of any process in the cell.

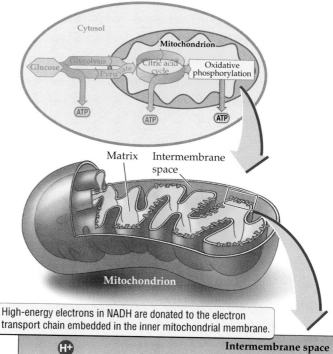

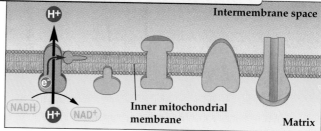

High-energy electrons in NADH are donated to the electron transport chain embedded in the inner mitochondrial membrane.

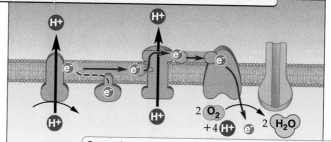

As electrons are passed down the electron transport chain, protons are pumped from the matrix into the intermembrane space, creating a proton imbalance across the inner mitochondrial membrane.

Oxygen is required as the final electron acceptor. In this final electron transfer, O_2 picks up protons and forms water.

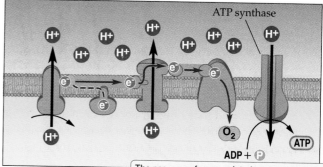

The passage of accumulated protons from the intermembrane space to the matrix through ATP synthase drives ATP synthase to produce ATP.

across the membrane. As is the case in in chloroplasts, the proton gradient causes protons to move through an ATP synthase in the inner mitochondrial membrane, catalyzing the phosphorylation of ADP to form ATP (see Figure 8.10).

Unlike what occurs in photosynthesis, in the final step of oxidative phosphorylation electrons are donated to O_2 that has diffused into the mitochondrion. When O_2 accepts these electrons, it also combines with H^+ in the matrix to form water (H_2O). Thus, O_2 is the last electron acceptor in the series of electron transfers and must be consumed in oxidative phosphorylation (see Figure 8.10).

If O_2 is not present, the incomplete oxidation seen in fermentation has a net production of only two ATP molecules per molecule of glucose consumed. In marked contrast, cellular respiration (glycolysis, citric acid cycle, and oxidative phosphorylation) has a net production of well over 30 ATP molecules per molecule of glucose. So, cellular respiration is clearly more efficient than fermentation at producing ATP, which may explain the need for the latter in larger multicellular organisms with higher energy demands.

> ■ Catabolism is a series of oxidation reactions that break down food molecules and produce ATP. The three stages of catabolism are glycolysis, the citric acid cycle, and oxidative phosphorylation. Glycolysis occurs in the cytosol and splits sugar molecules to eventually produce pyruvate and a small amount of ATP. The citric acid cycle and oxidative phosphorylation occur inside mitochondria and produce most of the cell's ATP.

Highlight

Solving the Brain Drain

Now that we know how energy is obtained from food, how do we meet the energy demands of our particularly large brains? One method would be to increase our resting rate of metabolism so that we consume more food and generate more ATP. However, humans do not generally have a higher rate of metabolism than other mammals that have far smaller brains, such as sheep. So the answer does not lie in a special alteration of the catabolic processes described in this chapter. Instead, to meet the high energy demands of the human brain, we evolved a way to redistribute energy to our various organ systems and make higher energy demands on our mothers early in life.

Approximately 70 percent of our resting metabolism supplies ATP to the heart, liver, kidney, gastrointestinal tract, and brain. For us to have such large brains relative to our body size, one or more of the other organs must either give up some of its precious energy resources or be smaller. In fact, when compared with the organs of other primates, the human gastrointestinal tract is nearly 40 percent smaller than expected. Thus, the large energy demands of our brains are compensated for by our smaller gastrointestinal tracts. During evolution, early humans switched from a strictly vegetarian to a more varied diet that included easily digested foods such as meat. Thus, there was no longer an advantage to having a large gastrointestinal tract for slowly digesting vegetation.

The evolution of a smaller human gastrointestinal tract was probably not the only change necessary to accommodate our large brains. The amount of energy that a mother contributes to her child during pregnancy and early infancy is equally important to the development of the large human brain. By the time a child reaches age 4, he or she has a brain that is 85 percent of the adult size. The growing human brain requires enormous amounts of energy during early development to support such rapid growth, and the only source for such metabolic resources is the mother.

During pregnancy the fetus greedily pulls nutrients from the mother to supply its growing energy needs, and this transfer of energy continues after birth with the intake of breast milk. Thus, we probably owe the development of our impressive brains to the evolutionary rearrangement of energy needs in the body and the generous investment of energy from our mothers. Thanks, Mom!

> ■ Satisfying the energy needs of a living organism involves more than the mere production of ATP. The ATP that is available to a multicellular organism must be carefully distributed to satisfy the varying needs of different cell types. Our large brains require a significant share of our bodies' ATP production, as well as a high investment of energy from our mothers early in life.

Summary

Energy Carriers: Powering All Activities of the Cell
- The chemical reactions that constitute life require energy.
- Energy is transferred within living organisms via specialized compounds called energy carriers.
- ATP is the most commonly used energy carrier; it donates bond energy to chemical reactions.
- The energy carriers NADH and NADPH donate electrons to oxidation–reduction reactions.

Photosynthesis: Harvesting Energy from Sunlight

- Photosynthesis harvests energy from sunlight and uses it to synthesize sugars from CO_2 and H_2O.

- Photosynthesis occurs in chloroplasts and produces as a by-product the O_2 that is essential for all oxygen-breathing organisms on Earth.

- The photosynthetic reactions that harvest energy from sunlight are called the light reactions. They produce ATP and NADPH.

- The ATP and NADPH produced by the light reactions are used in the dark reactions to synthesize sugars from CO_2.

Catabolism: Breaking Down Molecules for Energy

- Catabolism is a series of oxidation reactions that break down food molecules and produce ATP.

- The three stages of catabolism are glycolysis, the citric acid cycle, and oxidative phosphorylation.

- Glycolysis occurs in the cytosol and splits sugar molecules to eventually produce pyruvate and a small amount of ATP.

- The citric acid cycle occurs inside mitochondria and uses pyruvate to produce NADH and CO_2.

- Oxidative phosphorylation also occurs inside mitochondria and uses O_2 and NADH to produce most of the cell's ATP.

Highlight: Solving the Brain Drain

- Satisfying the energy needs of a living organism involves more than the mere production of ATP.

- Our large brains require a high investment of energy from our mothers early in life, as well as a significant share of our bodies' ATP production.

Key Terms

aerobic p. 135	glyceraldehyde 3-phosphate p. 133
aerobic respiration p. 136	glycolysis p. 135
anaerobic p. 135	intermembrane space p. 131
antenna complex p. 131	light reactions p. 131
carbon fixation p. 133	oxidative phosphorylation p. 137
catabolism p. 134	photosynthesis p. 130
chloroplast p. 130	photosystem p. 131
citric acid cycle p. 137	proton gradient p. 132
dark reactions p. 131	rubisco p. 133
electron transport chain (ETC) p. 131	stroma p. 131
energy carrier p. 129	thylakoid membrane p. 131
fermentation p. 135	thylakoid space p. 131

Chapter Review

Self-Quiz

1. Energy carriers like ATP
   a. transfer energy from the environment into cells.
   b. transfer energy to molecules inside the cell, allowing chemical reactions to occur.
   c. enable organisms to get rid of excess heat.
   d. are synthesized only by mitochondria.

2. Photosynthesis
   a. is found exclusively in plants.
   b. breaks down sugars with energy from sunlight.
   c. releases CO_2 into the atmosphere.
   d. releases O_2 into the atmosphere.

3. Which of the following compounds is produced by the light reactions and used by the dark reactions to synthesize sugars?
   a. ATP
   b. NaCl
   c. CO_2
   d. NADH

4. Which of the following statements is *not* true?
   a. Glycolysis produces most of the ATP required by aerobic organisms.
   b. Glycolysis produces pyruvate, which is consumed by the citric acid cycle.
   c. Glycolysis occurs in the cytosol of the cell.
   d. Glycolysis is the first stage of aerobic respiration.

5. Electron transport chains
   a. are found in both chloroplasts and mitochondria.
   b. are clumps of protein that float freely in the cytosol.
   c. pass protons from one ETC component to another.
   d. have no role in ATP production.

6. Which of the following is essential for oxidative phosphorylation?
   a. rubisco
   b. NADPH
   c. a proton gradient
   d. chlorophyll

Review Questions

1. The transfer of electrons down an ETC produces a similar chemical event in both chloroplasts and mitochondria. Describe this chemical event and how it contributes to the production of ATP by both organelles.

2. The cells in the root of a plant do not contain chloroplasts. How do these cells obtain energy without the benefit of direct photosynthesis?

3. Oxidative phosphorylation is the final stage of aerobic respiration. What important chemical reaction in oxidative phosphorylation directly consumes O_2?

4. Certain drugs allow protons to pass through the thylakoid membrane on their own and without the involvement of a channel. How you think these drugs affect ATP synthesis?

The Daily Globe

8

Tropical Rainforests May Be Source of Pollution

SAN JOSE, COSTA RICA. For years, the accepted scientific dogma has been that the greenhouse effect, global warming that is caused by increases in carbon dioxide in the atmosphere, has been combated by tropical rainforests, which consume such carbon dioxide. Now a series of new reports suggests that the rainforest in Costa Rica may have turned traitor, producing more carbon dioxide than it consumes, itself becoming a significant source of this greenhouse gas.

"This goes against everything we had thought," said Dr. Theodore Michois, ecologist at La Bruja Ecological Station in Costa Rica, who called the finding "alarming." "We must rethink the role of rainforests."

Rainforest plants photosynthesize during the day, consuming carbon dioxide and releasing oxygen. At night there is no photosynthesis and aerobic respiration dominates, consuming oxygen and releasing carbon dioxide. The findings at the La Bruja station suggest that rainforests may have become carbon dioxide producers, rather than consumers, due to global warming that has already occurred. Higher-than-normal temperatures at night caused by global warming may stress the trees, increasing their aerobic respiration and causing them to release more carbon dioxide. Meanwhile, levels of photosynthesis remain the same, tipping the balance sheet toward rainforests producing more carbon dioxide than they consume.

"If higher levels of carbon dioxide are driving the production of even more carbon dioxide in rainforests," said Dr. Will Helmina, climatologist at Purdue College, "then we are really in trouble. It could be the beginning of a truly vicious cycle."

Evaluating "The News"

1. Faced with the apparent increase in CO_2 levels caused by the Costa Rican rainforest, some people have suggested deforestation as a solution. Is this an acceptable solution? Will it work?

2. Trees in cooler climates seem to grow faster when levels of CO_2 in the atmosphere are higher. How might this phenomenon affect the global climate?

3. Does faster growth of trees in these regions make higher levels of CO_2 a plus?

9

How Cells Communicate

Ross Bleckner, *Healthy Spot*, 1996.

Sleeping with the Enemy

Since the dawn of recorded history, humans have waged wars both with each other and with deadly bacteria in the environment. Although the threat of nuclear war has decreased in the past decade with the end of the cold war between the United States and the Soviet Union, there has been an upsurge in smaller ethnic conflicts around the world and the threat of biological warfare. In an ironic twist of fate, our puzzling tendency to wage war with each other has now overlapped with our struggle against two types of bacteria, which cause the deadly diseases cholera and anthrax.

Recent ethnic conflicts in East Africa remind us of the tragic consequences of such events, including the creation of refugee camps to house fleeing civilians. As these camps swell with thousands of refugees, the crowded conditions and poor sanitation have led to serious epidemics of disease. An unfortunate consequence has been the resurgence of cholera.

Cholera is caused by the bacterium *Vibrio cholerae*, which produces a deadly toxin responsible for severe diarrhea and eventual death by dehydration. As alarming reports of antibiotic-resistant strains of cholera continue to flow from the refugee camps, it appears that human warfare has presented an opportunity for a deadly bacterium to spread. Perhaps even more disturbing is evidence that some individuals might even harness the deadly properties of this bacteria for use as a biological weapon of war.

In recent years, there has been increasing anxiety that international terrorists and misguided leaders might use biological weapons such as anthrax.

During the Gulf War in 1991, there were concerns that missile warheads armed with anthrax might be used on the battlefield. But what exactly is this deadly biological weapon?

Anthrax is caused by the bacterium *Bacillus anthracis*, which, like cholera, produces a deadly toxin that kills infected animals. In the case of anthrax infections in animals, the bacterium requires the death of its unfortunate host to release spores and spread from one individual to another. Fortunately, the passing of anthrax to humans is relatively rare under normal conditions, but the major concern today is how to counteract anthrax exposure caused by its use as a biological weapon.

To develop an effective countermeasure, we must understand how bacterial toxins damage and ultimately kill so many cells in the body of an infected individual. Thanks to intensive biological research, we now know that both the cholera and the anthrax toxins exercise their deadly effects by interfering with how eukaryotic cells communicate with each other. Multicellular organisms such as humans depend on the existence of proper communication between all the cells that make up our bodies. In this chapter we discuss how cells receive and respond to signals sent by other cells, and then how the cholera and anthrax toxins disrupt these essential processes, with deadly consequences.

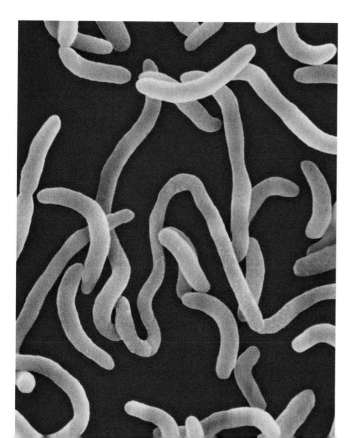

***Vibrio cholerae*, the Bacterium That Causes Cholera**

Key Concepts

1. Multicellular organisms require specialized cells to perform various living processes more efficiently. Specialization requires communication between cells.

2. Cells use a variety of different signaling molecules to communicate with each other, including proteins, fatty acid derivatives, and even gases.

3. To receive and interpret signals, cells contain specific receptor proteins that bind to signaling molecules and trigger certain chemical reactions inside the cell.

4. Some receptor proteins reside in the cytosol and bind to signaling molecules that pass through the plasma membrane; others reside in the plasma membrane and bind only to external signaling molecules.

5. One signaling molecule induces many chemical reactions that amplify the cell's response.

6. Different signaling molecules affect some of the same chemical reactions in the cell, resulting in the combining of different signals.

In the preceding chapters we have seen how chemical compounds are organized and function in living systems. We know how a limited selection of chemical components such as nucleic acids, sugars, amino acids, and lipids are arranged into the complex structures of the cell. We also have learned how energy is captured from the environment and used to power every living process. Armed with this understanding of the cell as a functioning unit, we can now discuss how cells communicate with each other and form large multicellular organisms like us humans.

Sometimes it is tempting and even useful to consider cells in isolation and to think of multicellular organisms as nothing more than a mere collection of cells with little interest in each other. However, this portrayal could not be further from the truth. Like any complex community, multicellular organisms benefit from specialized cells and must coordinate their activities in the body. During the course of evolution this simple need for coordination presented a tremendous challenge to living systems. It took almost 3 billion years before primitive single-celled life forms evolved the systems of communication and cooperation that led to the first multicellular organism. Clearly, cooperation is not an easy thing to achieve, especially when it involves millions upon millions of cells.

Specialization and Communication in the Community of Cells

Two organizational principles apply to every large multicellular organism: the specialization of cell function and the establishment of communication between cells. The principle of **cell specialization** means that not all the cells found in a multicellular organism are the same and have the same function. Indeed, with the increasing size of an organism, there is an urgent need for different types of cells with different structures and functions. Having specialized types of cells that are especially well suited to specific duties ensures that all the processes necessary for the life of the organism occur quickly and efficiently.

Specialized cells fulfill a wide variety of needs in multicellular organisms. For example, just as each cell needs the structural support provided by the cytoskeleton described in Chapter 6, every multicellular organism requires specialized cells that support, maintain, and strengthen its physical form. In the case of humans, the specialized cells that deposit bone in the body form the skeleton, which both supports the body and allows the movement of limbs. In another example, specialized cells in the roots, stems, and leaves of plants have a unique form that allows them to transport water between different parts of the plant.

Specialization does not mean that each type of cell functions in isolation. On the contrary, most cell types function in close cooperation with other cell types to fulfill particular functions for the body. A set of cooperating cell types is called a **tissue**. For example, the cells that make up our lungs are collectively known as lung tissue, since they work together to facilitate the exchange of carbon dioxide for oxygen that we need to live. However, not all the cells that make up lung tissue are the same. Some cells are very thin and thus are specialized to exchange the gases inside the lungs; other cells form the lining of the capillaries that bring blood to the lungs (Figure 9.1). The various cells that make up different tissues and how they work together are discussed in Unit 5, "Form and Function" (see also Figure 16.1).

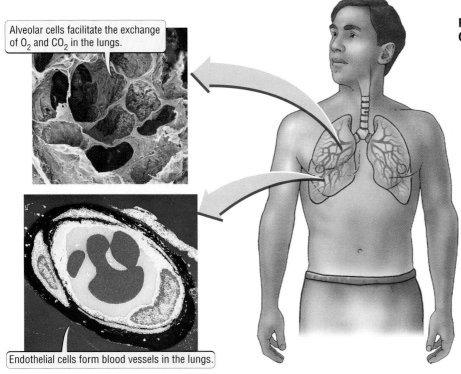

Alveolar cells facilitate the exchange of O_2 and CO_2 in the lungs.

Endothelial cells form blood vessels in the lungs.

Figure 9.1 Different Types of Specialized Cells Make up Lung Tissue

communication. Specialized nerve cells in the skin of the hand sense the heat and transmit signals to other nerve cells in the spinal cord (see Figure 9.2). The nerve cells in the spinal cord then transmit signals to other nerve cells, which instruct the muscles of the arm to contract, resulting in the quick jerk of the hand away from the heat. Remarkably, this entire series of cell-to-cell communications occurs without the brain or consciousness being involved in the decision. Such simple reflex responses are discussed further in Chapter 34.

Various nerve cells use small molecules called neurotransmitters to communicate with one another and with the muscles of the arm. These molecules are released by one cell and attach to another cell,

The second basic principle that applies to all multicellular organisms is **cell communication**. Given the possible diversity of cells and tissues found in a multicellular organism, they must have some way to communicate and hence coordinate with each other. The comparison to a corporate office building that was applied to the single cell in Chapter 6 also holds true for the multicellular organism, which must establish communication networks among its various cells and tissues. For example, when you touch an open flame, how is the perception of heat transferred to your spinal cord, which then tells your hand to jerk back quickly? The simple reflex of jerking back that we have all experienced is based on a rapid series of cell-to-cell communications (Figure 9.2).

The reflex just described uses several different specialized cells and means of

Figure 9.2 A Simple Reflex Requires Communication between Nerve Cells
The perception of heat results in the passing of nerve signals from the sensory nerve to the spinal cord, which in turn signals to muscle cells to contract via a motor nerve.

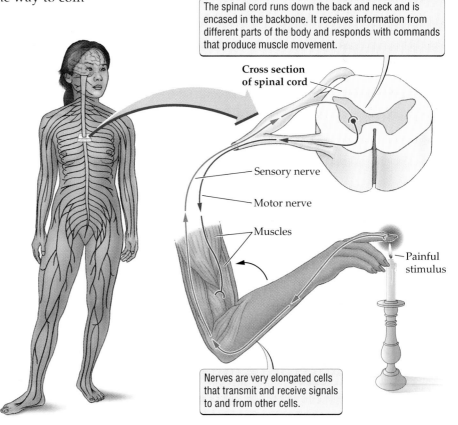

The spinal cord runs down the back and neck and is encased in the backbone. It receives information from different parts of the body and responds with commands that produce muscle movement.

Cross section of spinal cord

Sensory nerve

Motor nerve

Muscles

Painful stimulus

Nerves are very elongated cells that transmit and receive signals to and from other cells.

resulting in the transmission of a signal. The example of a simple reflex involves a very specific set of nerve cells, but the principle of using a small molecule to transmit signals between cells is widely applied in multicellular organisms. In general, communication between cells uses small proteins or other molecules that are released by one cell and received by another, which is called the **target cell**. These so-called **signaling molecules** form the language of cellular communication, as we'll discuss in the next section.

> ■ Multicellular organisms require specialized cells to perform specific functions in the body. The coordination of these functions depends on communication between different types of specialized cells. Cells communicate via small signaling molecules that are produced by specific cells and received by different target cells.

The Role of Signaling Molecules in Cell Communication

The passing of signaling molecules from one cell to another is an effective way for two cells to communicate.

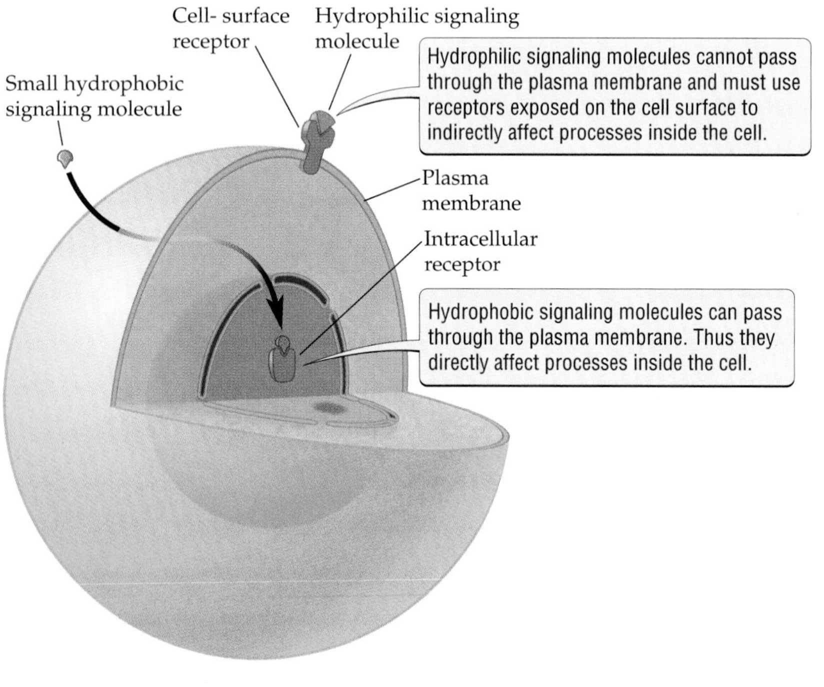

Figure 9.3 Signal Receptors
Two major classes of receptors bind signaling molecules. Cell surface receptors are embedded in the plasma membrane and bind to signaling molecules that cannot cross the membrane. Intracellular receptors reside in the cytosol and bind to signaling molecules that can cross the plasma membrane on their own.

However, the identity of the signaling molecule, and its effect on the target cell, are highly variable. The particular signaling molecule that is used often depends on the speed with which a particular signal must move and the distance that it must travel between cells.

Returning to the example of a corporate office building, a simple internal memo that has to travel only between persons within the same building can take the form of a single typewritten sheet of paper. In contrast, a letter that has to be mailed out to customers in different states must be placed in a protective envelope with a complete address label.

The same is true of signaling molecules that travel only between neighboring cells in a tissue versus those that circulate through the bloodstream to reach cells in different parts of the body. In the former situation, the signaling molecule can be relatively fragile and short-lived, much like the internal memo; in the latter situation, the signaling molecule must have a longer life span and be able to withstand environmental stresses. These issues of the physical structure and life span of signaling molecules are addressed in various ways by different cells and even within the same cell.

When a signaling molecule reaches a target cell, there is a specific means of receiving it and acting on its message. These tasks are the responsibility of a class of proteins called **receptors**. Since there are many kinds of signaling molecules, each of which has a specific effect on the cell, there must also be a matching diversity of receptors. Some receptors are located on the surface of the cell and encounter their matching signaling molecules on the outside (Figure 9.3). Other receptors are located either in the cytosol or inside the nucleus of the cell. To reach these receptors, their signaling molecules must cross the plasma membrane and perhaps the nuclear envelope.

In the sections that follow, we will identify and discuss some of the different kinds of molecules and receptors that cells use to communicate with one another, and the different effects that each can have on the cell. We'll start with a small signaling molecule, the gas nitric oxide.

Nitric oxide is a short-range signaling molecule

Signaling molecules often take the form of proteins or fatty acid derivatives. In the past decade, however, biologists have shown that there is a gas that functions as a signaling molecule inside the human body. This gas is **nitric oxide** (NO), and it has an important role in lowering blood pressure. How does a gas serve as a signal to lower blood pressure? The answer lies in how specific target cells respond to nitric oxide by changing their physical shape.

Blood pressure is directly related to the internal volume (capacity) of blood vessels. The larger the internal volume of a blood vessel, the lower will be the pressure of the blood inside it. Specific nerve signals induce endothelial cells in the walls of blood vessels to produce nitric oxide. Small molecules of NO gas rapidly spread into the neighboring muscle cells, causing them to relax. Relaxation of the muscle cells allows the blood vessel to expand, increasing the volume of blood that it can carry, and decreasing blood pressure. This process is known as **vasodilation** (*vaso*, "vessel"; *dilation*, "expansion") (Figure 9.4).

Although we only recently discovered the role of nitric oxide as a signaling molecule, we have long used it as a therapy without knowing it. Since the nineteenth century, nitroglycerin has been administered to patients suffering from angina (chronic chest pain), which is due in part to the narrowing of the blood vessels that deliver blood to the heart. Although nitroglycerin is often associated with dangerous explosives like TNT, the human body converts administered nitroglycerin into nitric oxide, resulting in increased blood flow to the heart and a reduction in angina pain.

The direct inhaling of nitric oxide has even been used to treat newborns with pulmonary hypertension. This serious disease is characterized by abnormally high blood pressure in the vessels that supply the lungs, usually caused by inherited heart defects. Before the identification of nitric oxide as a signaling molecule, customary therapies often lowered the overall blood pressure of the infant to dangerous levels. In contrast, inhaling nitric oxide affects only the pulmonary blood vessels of the lungs, causing them to expand and reducing the high blood pressure in that tissue alone.

The effect of inhaled nitric oxide is highly localized to the lungs because it acts only over short distances. Once nitric oxide diffuses out of the cell in which it is produced, it is degraded within a matter of seconds. Thus, only those target cells in the immediate vicinity are close enough to receive nitric oxide and respond to the signal. The inhaled nitric oxide does not travel far enough to affect blood vessels outside of the lungs, and thus it does not have the undesirable effect of lowering blood pressure throughout the body.

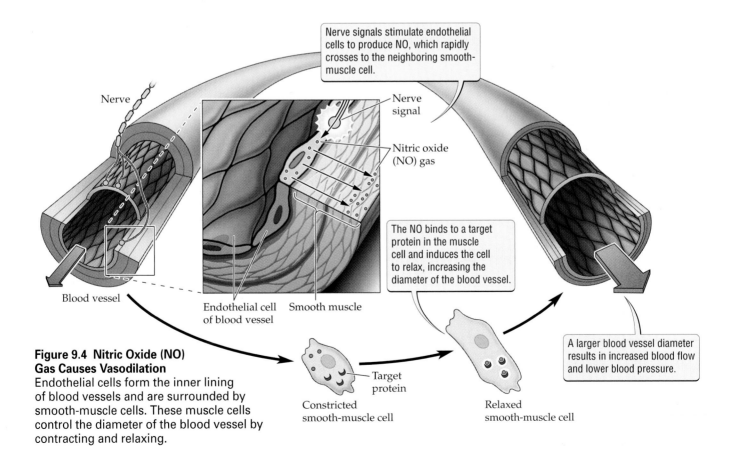

Nerve signals stimulate endothelial cells to produce NO, which rapidly crosses to the neighboring smooth-muscle cell.

Nerve

Nerve signal

Nitric oxide (NO) gas

The NO binds to a target protein in the muscle cell and induces the cell to relax, increasing the diameter of the blood vessel.

A larger blood vessel diameter results in increased blood flow and lower blood pressure.

Blood vessel

Endothelial cell of blood vessel

Smooth muscle

Target protein

Constricted smooth-muscle cell

Relaxed smooth-muscle cell

Figure 9.4 Nitric Oxide (NO) Gas Causes Vasodilation
Endothelial cells form the inner lining of blood vessels and are surrounded by smooth-muscle cells. These muscle cells control the diameter of the blood vessel by contracting and relaxing.

Hormones are long-range signaling molecules

The short life span of nitric oxide makes it an effective signaling molecule over short distances. But communication is not limited to cells that are near each other. Depending on the size of the organism, cells that are anywhere from centimeters to several meters apart must also be able to communicate with each other. Such long-distance communication requires a different class of signaling molecules called hormones.

Hormones are used by all multicellular organisms to coordinate the activities of different cells and tissues. In contrast to nitric oxide, hormones are produced by cells in one part of the body and are transported by fluids to target cells in another part of the body.

Transporting hormones from their site of production to their target cells often depends on the circulation of fluids inside the organism. In plants, some hormones are dissolved in the sap, which moves from the roots to the rest of the plant. In animals, hormones are dissolved in the blood, which circulates throughout the body, ensuring rapid distribution.

Steroid hormones cross cell membranes

Steroid hormones are an important class of signaling molecules that are essential for many growth processes, including the normal development of reproductive tissues in mammals. For example, the female sex hormone progesterone is produced by the ovary, circulates in the bloodstream, and ensures that the cells of the uterus are ready to support implantation of an embryo.

All steroid hormones are derived from a lipid called cholesterol and are hydrophobic. The hydrophobicity of steroid hormones plays an important role in their mode of action, by allowing them to pass easily through the hydrophobic core of the cell's plasma membrane and enter the cytosol. Their hydrophobic character also means that steroids must be packaged with proteins that help them dissolve in the watery environment of the body, and extends their life span in the bloodstream.

The ability of steroids to survive and be transported in the bloodstream for up to several days improves the likelihood that they will reach distant target cells. Survival in the blood allows the female steroid hormone progesterone to affect different cells in parts of the body very distant from the site of hormone synthesis. In addition to preparing the cells of the uterus for early pregnancy, progesterone can induce development of the more distant mammary glands for milk production.

Once steroids reach their target cells, they cross the plasma membrane and alter the production of specific proteins inside the cell (Figure 9.5). In order to do this, they must bind to receptors in the cytosol, forming an active complex that can enter the nucleus and affect specific genes. As discussed in Chapter 6, genes carry the information that dictates what proteins are synthesized in the cell. Therefore, the particular genes that are activated by a steroid such as progesterone will produce proteins that are necessary for the proper growth of the uterine lining. The action of a steroid hormone in effect coordinates the production of proteins necessary for specific changes in the cell.

Some hormones require the help of cell surface receptors

Not all signaling molecules enter the cell as steroids do. Other hormones use receptors on the cell surface to send their signals into the cell. Some of these hormones are small proteins; others are chemical derivatives of amino acids and fatty acids.

Adrenaline is one such hormone, whose effects we have all felt when startled or frightened. It is derived from the amino acid tyrosine. When an event or a situation frightens us, our

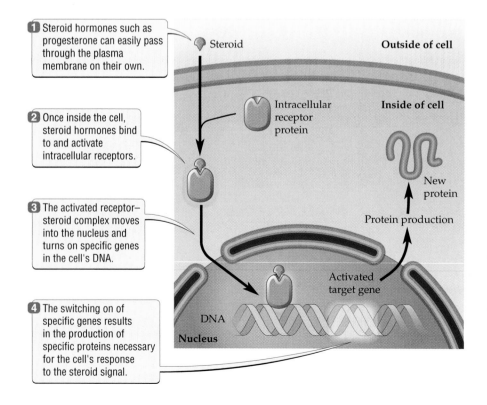

1 Steroid hormones such as progesterone can easily pass through the plasma membrane on their own.

2 Once inside the cell, steroid hormones bind to and activate intracellular receptors.

3 The activated receptor–steroid complex moves into the nucleus and turns on specific genes in the cell's DNA.

4 The switching on of specific genes results in the production of specific proteins necessary for the cell's response to the steroid signal.

Steroid

Outside of cell

Intracellular receptor protein

Inside of cell

New protein

Protein production

Activated target gene

DNA

Nucleus

Figure 9.5 A Cell's Response to a Steroid Hormone

adrenal glands release adrenaline into the bloodstream to enhance the body's ability to respond (Figure 9.6).

Adrenaline binds to receptors on the surfaces of cells in the liver and fatty tissues of the body. This binding results in changes in the metabolism of these cells, promoting the breakdown of glycogen in the liver and of fats in fatty tissue. The glucose and fatty acids that are produced can then be used as fuel by all of the cells in your body during times of stress.

Adrenaline has a different effect on muscle cells in the heart (see Figure 9.6). It increases the rate at which heart muscle contracts and relaxes, resulting in a faster heartbeat. When the heartbeat increases, blood circulates more rapidly in the body, delivering more oxygen and food to cells. Once again, the ability of the body to respond to the frightening stimulus is enhanced. Thus, adrenaline is a signaling molecule that has different but complementary effects on different types of cells.

Growth factors induce cell division

Growth factors are another class of signaling molecules that use receptors on the cell surface. Their name is based on their ability to induce cell growth and division. The proliferation of cells in the human body is controlled largely by growth factors, which both initiate and maintain the processes needed for growth and cell division. Scores of proteins exported by cells function as growth factors, and in most cases their effects are confined to neighboring cells. In this respect growth factors are like nitric oxide, but unlike NO, they do not enter the cell. Instead they depend on cell surface receptors to transmit a message into the cell and alter its function.

Growth factors were first discovered by biologists attempting to grow cells outside of the body. In the presence of the appropriate nutrients in a liquid medium, cells will live outside of the body in suitable culture dishes, but they will not divide and multiply. These cells will eventually die, thus presenting a problem for biologists interested in long-term studies. Eventually it was discovered that the addition of blood extracts to the liquid medium induced cells to divide, permitting the proliferation of cells outside of the body.

Further analysis of these blood extracts led to the discovery of the first growth factor that is produced by specialized blood cells known as platelets. Under normal conditions, platelets produce a growth factor appropriately called platelet-derived growth factor, or **PDGF**, which induces cell division at sites of tissue damage. PDGF therefore speeds up wound healing and is one example of how a growth factor helps coordinate a specific process in the body.

■ Signaling molecules affect target cells, but not other cells, by binding to protein receptors either on the cell surface or in the cytosol. Short-lived signaling molecules like nitric oxide allow rapid communication between neighboring cells. Long-range signaling molecules like hormones allow communication between cells that are much farther apart. Some hormones enter the cell directly; other hormones and growth factors use cell surface receptors.

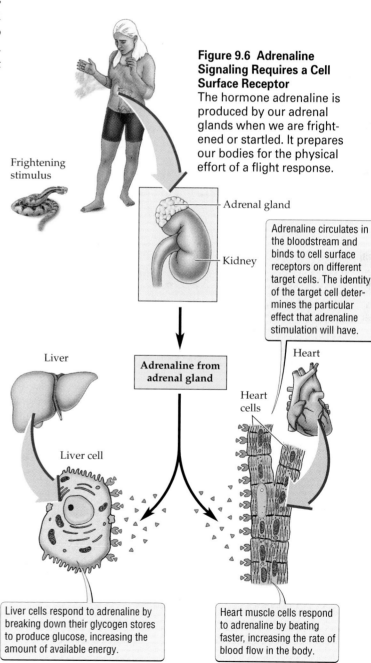

Figure 9.6 Adrenaline Signaling Requires a Cell Surface Receptor
The hormone adrenaline is produced by our adrenal glands when we are frightened or startled. It prepares our bodies for the physical effort of a flight response.

Frightening stimulus

Adrenal gland

Kidney

Adrenaline circulates in the bloodstream and binds to cell surface receptors on different target cells. The identity of the target cell determines the particular effect that adrenaline stimulation will have.

Liver

Adrenaline from adrenal gland

Heart

Heart cells

Liver cell

Liver cells respond to adrenaline by breaking down their glycogen stores to produce glucose, increasing the amount of available energy.

Heart muscle cells respond to adrenaline by beating faster, increasing the rate of blood flow in the body.

How Cell Surface Receptors Initiate Changes inside the Cell

Signaling molecules are important coordinators of a broad range of cellular processes (Table 9.1). Yet, as we have seen, many of these molecules affect events inside the cell without crossing the plasma membrane. They achieve these effects by using specialized receptors that reside in the plasma membrane, thereby connecting the exterior and the interior of the cell. These cell surface receptors are all membrane-spanning proteins that have one part exposed to the outside and another exposed to the cytosol. This arrangement in the membrane allows the receptor to bind the signaling molecule on the outside, change shape, and then induce a specific chemical reaction on the inside of the cell.

The triggering of a specific chemical reaction inside the cell in response to an external signal is known as **signal transduction**. As the name implies, the signal is transferred, or transduced, from the outside into the cell. Since the collection of external signals that affect cells is diverse, the range of receptors that bind to them with specific results inside the cell is equally diverse. In other words, each type of signaling molecule binds to a specific receptor on the cell surface, causing a specific signal transduction event.

The simple but effective process of transducing an external signal by triggering stepwise protein activations inside the cell is referred to as a **signal cascade**. However, the proteins that are activated depend on the nature of the external signal, which accounts for the wide range of responses that are caused by different signaling molecules.

In the case of hormones such as adrenaline, binding to a receptor causes a change in the shape of the receptor protein, converting it into an active form. The portion of the active receptor that is exposed to the inside of the cell is then able to bind and activate a G protein at the inner face of the plasma membrane (Figure 9.7).

G proteins are so named because they bind to and are regulated by guanine nucleotides: GDP (guanosine diphosphate) and GTP (guanosine triphosphate). When a G protein is bound to GDP, it is inactive. Then when it encounters an activated receptor, it is able to release GDP and bind GTP from the cytosol. This process of exchanging one guanine nucleotide for another alters the shape of the G protein and activates it. G proteins are just the first of several types of proteins that must be activated in order for the signal to be transduced inside the target cell.

9.1   *Examples of Signaling Molecules*

Name of molecule	Type of molecule	Site(s) of synthesis	Function(s)
Nitric oxide	Gas	Endothelial cells in blood vessel walls; nerve cells	Promotes relaxation of blood vessel walls
Testosterone	Steroid	Testes	Promotes the development of secondary male sex characteristics
Progesterone	Steroid	Ovaries	Prepares the uterus for implantation; promotes mammary gland development in females
Insulin	Protein	Beta cells of the pancreas	Promotes the uptake of glucose by cells
Adrenaline	Amino acid derivative	Adrenal glands	Promotes glycogen breakdown and increased heart rate
Thyroxin	Amino acid derivative	Thyroid gland	Promotes increased rate of metabolism
Platelet-derived growth factor (PDGF)	Protein	Many cell types	Promotes cell multiplication
Nerve growth factor (NGF)	Protein	Tissues richly supplied with nerves	Promotes nerve growth and survival
Auxin (plant hormone)	Acetic acid derivative	Most plant cells	Promotes root formation and stem elongation
Acetylcholine	Choline derivative	Nerve cells	Assists impulse transmission from nerves to many muscles

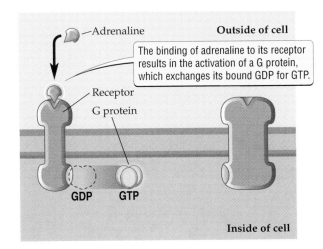

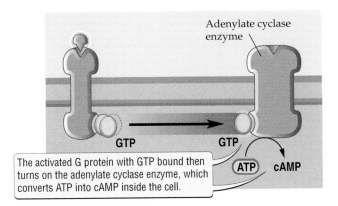

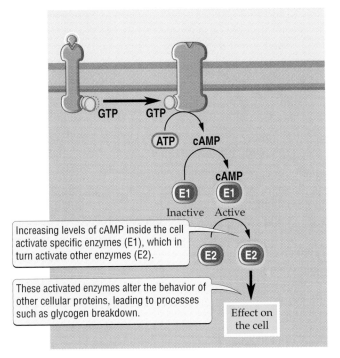

Figure 9.7 G Protein Signal Cascade

The active G protein in turn activates a cellular enzyme called **adenylate cyclase**, which catalyzes the conversion of ATP into the cyclic nucleotide cAMP (cyclic adenosine monophosphate, or cyclic AMP). As the amount of cAMP in the cytosol increases, several other enzymes and proteins in the cell bind cAMP and themselves become activated (see Figure 9.7). Among the enzymes that are activated by cAMP are those required for the adrenaline response. Thus, unlike steroid hormones, which directly activate genes and affect protein production when bound to their cytosolic receptor, adrenaline depends on a series of stepwise protein activations in the cytosol.

The proteins in a signal cascade can be activated in various ways, many of which do not depend on GTP or cAMP. For example, the binding of growth factors such as PDGF to the appropriate receptor results in the enzyme-catalyzed transfer of phosphate groups from ATP to several proteins inside the cell. This process is called **protein phosphorylation**, and the enzymes that catalyze these reactions are called **kinases** (Figure 9.8).

In the case of growth factor receptors, the part of the enzyme that binds growth factor is on the outside of the cell, and the part that catalyzes the phosphorylation reaction is on the inside. Therefore, binding of growth factor to the exposed portion of the receptor on the cell surface activates the enzymatic activity of the receptor that leads to multiple phosphorylation events on the inside of the cell.

The first protein to become phosphorylated in a growth factor signal cascade is the receptor itself. This process of self-, or auto-, phosphorylation targets specific tyrosine amino acids in the cytosolic portion of the receptor (see Figure 9.8). Once phosphorylated, these amino acids become suitable binding sites for other proteins involved in the growth factor signal cascade. Thus, instead of merely activating a target protein, the growth factor receptor creates phosphorylated binding sites on itself for other proteins. This process of assembling or concentrating proteins and enzymes necessary for the next steps of the signal cascade is yet another way of transducing a signal.

How does protein phosphorylation result in the transducing of an external signal? Upon phosphorylation, some enzymes are activated and catalyze reactions faster, while others are inhibited and catalyze reactions more slowly or not at all. Phosphorylation of proteins is well suited to achieving this aim, since the addition of the highly charged phosphate group (PO_4^{3-}) to one or more amino acids changes the three-dimensional shape and hence the activity of the protein (Figure 9.9).

As discussed in Chapter 5, the chemical characteristics of the amino acids that make up a given protein determine its folded three-dimensional shape. Thus,

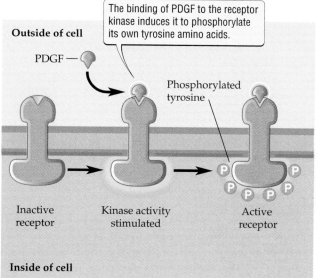

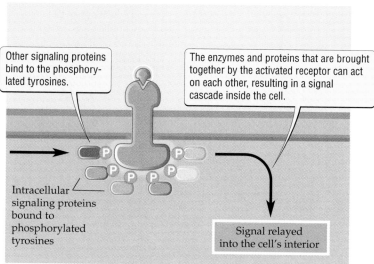

Figure 9.8 Receptor Kinase Signal Cascade

when a kinase adds a charged phosphate group to specific amino acids of a target protein, it dramatically changes their chemical characteristics, resulting in altered shape and activity of the protein.

▉ In signal transduction, specific chemical reactions are triggered inside a cell by an external signal, causing a series of stepwise protein activations known as a signal cascade. Cell surface receptors trigger signal cascades by binding external signaling molecules, changing shape, and then activating or inhibiting specific proteins inside the cell. A signal cascade may result in the modification or production of proteins required for a cellular process.

Amplifying and Combining Signals inside the Cell

Imagine a situation in which a single hormone or growth factor binds to a cell surface receptor. If this binding event resulted in the activation or phosphorylation of only one protein inside the cell, many binding events at the cell surface would be required in order to bring about a significant change in the cell. To get around such an inefficient arrangement, most signaling events initiate a signal cascade inside the cell that greatly amplifies the initial signal. A signal cascade is like a huge avalanche on a snow-covered mountain that is started by one small icicle falling off a tree at the top of the slope.

Signal amplification depends on a cascade of enzyme-catalyzed reactions

For the signaling molecules described so far, the binding of the hormone or growth factor to its cell surface receptor fulfills the role of the falling icicle. If we consider the case of a hormone like adrenaline, each activated receptor can activate several G proteins, each of which in turn can act on several enzymes.

Likewise, a single activated growth factor receptor leads to the phosphorylation and hence activation of many target proteins. This simple principle is apparent over and

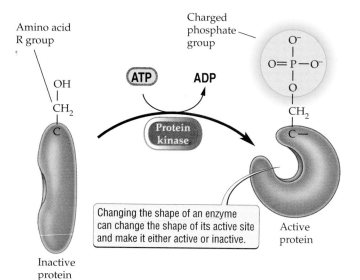

Figure 9.9 Phosphorylation Can Change a Protein's Activity
The addition of a highly charged phosphate group to an amino acid R group can change the overall chemical character of the protein. This addition in turn changes the three-dimensional shape of the protein.

over again in different signal cascades inside the cell, and explains why the binding of one signaling molecule to a single receptor has a much amplified effect in the cell.

The fact that a multitude of cellular processes are regulated by different external signals means that both the timing of the signal cascade and the identity of the target molecules can vary a great deal. So far we have discussed signal cascades that ultimately lead to either the activation of cellular enzymes or the turning on of genes and the production of specific proteins. In the former case, which is well illustrated by the example of adrenaline, the enzymes necessary for the breakdown of glycogen into glucose are activated in a matter of seconds after the hormone binds to its receptor (Figure 9.10*a*). Since adrenaline production is associated with a perception of danger and the possible need to flee, it makes sense that a rapid increase in the levels of available glucose in the bloodstream is necessary for an escape response.

In contrast, the promotion of cell division by PDGF depends on the turning on of genes and the production of proteins, which occur in a matter of hours rather than seconds (Figure 9.10*b*). This process does not have the immediate urgency of the adrenaline response, and given the importance of ensuring the appropriate timing and extent of cell division, it is not surprising that the body uses slower and more regulated signaling pathways to control the process.

Different signals can be combined inside the cell

In the examples of the previous section we addressed only a few of the signaling pathways and the processes they control. Almost every activity a cell undertakes involves some sort of regulatory input from external signals. Even the shapes that cells adopt are regulated by external signals, which lead to the phosphorylation and rearrangement of proteins in the cyto-

skeleton. Yet not all signal cascades that are started by particular signaling molecules stand alone. Indeed, several different signaling molecules can share some of the components of a single signal cascade in a manner that results in the cooperation of different signals. This process of **signal integration** means that one signal can either enhance or inhibit the effects of another signal.

Several related protein kinases, first discovered as part of the growth factor signal cascade, also participate in the transduction of signals by several other molecules including hormones. These kinases promote such diverse processes as cell growth in humans and sexual reproduction in yeasts. Therefore, the effect that activating such kinases can have on the cell must depend in part on the differing combinations of proteins that are

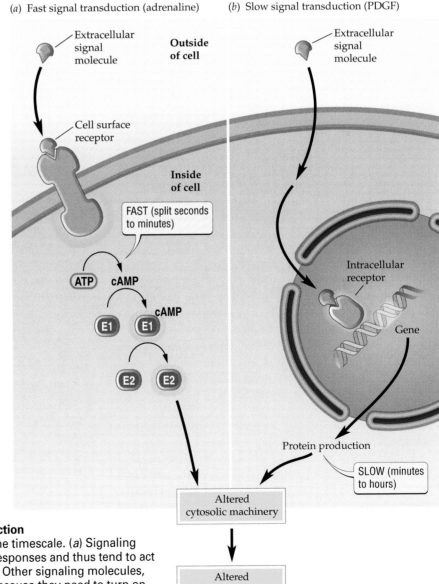

(a) Fast signal transduction (adrenaline)

(b) Slow signal transduction (PDGF)

Figure 9.10 The Timing of Signal Transduction
Not all signal cascades occur over the same timescale. (*a*) Signaling molecules like adrenaline produce rapid responses and thus tend to act directly on existing proteins in the cell. (*b*) Other signaling molecules, such as growth factors, act more slowly because they need to turn on genes, which in turn produce the necessary proteins.

found in different cells. The combining of different signals gives cells an increased range of possible responses to their environment.

> ■ The activation of many proteins at each step of a signal cascade greatly amplifies the original signal. Signal cascades that modify existing proteins inside the cell occur in a matter of seconds; those that activate genes to produce new proteins can take several hours. Certain proteins can participate in different signal cascades, allowing the cell to combine different external signals.

Highlight

Defeating Deadly Bacterial Toxins

The cell signaling events that we have discussed in this chapter are frequently the target of harmful bacteria. Returning to the severe dehydration associated with cholera infection (see the beginning of the chapter), we now know that the cholera toxin specifically affects a G protein found in the cells that line the intestine. Once the toxin enters an intestinal cell, it chemically modifies the G protein such that it becomes permanently bound to GTP. This means that the cholera toxin permanently activates the G protein, and the signal cascade in which it participates thus cannot be turned off.

This abnormal state of affairs leads to an excess of the enzymatic activities normally regulated by the G protein, which include the activation of membrane channels that facilitate the movement of water out of the bloodstream, through the intestinal cell, and into the lumen compartment of the intestine. The resulting excessive loss of water from the blood into the intestine causes the diarrhea and dehydration that are characteristic of cholera infections. As antibiotic-resistant strains of cholera bacteria appear, new drug therapies that limit the effect of the toxin on intestinal cells must be developed. Such drugs could even save tens of thousands of lives each year.

Instead of overactivating a signaling pathway as the cholera toxin does, the anthrax toxin blocks the action of a kinase pathway used by several signaling molecules. Once the kinase pathway is blocked, the cell can no longer respond to multiple external signals; the result is the deregulation of many normal cellular processes and ultimately cell death. One of the most difficult problems encountered in the treatment of anthrax is that by the time the bacteria are killed with antibiotics, there is already so much toxin released in the body that the host animal still dies.

The anthrax toxin is an enzyme that cuts up other proteins, making them inactive. These specialized enzymes are called proteases, and one target of the anthrax protease happens to be an important kinase. Now that scientists have uncovered the mechanism by which anthrax toxin kills cells, it may be possible to develop a drug that directly blocks the protease activity.

The idea of a drug inhibiting a protease is nothing new. One of the most effective therapies for HIV infection includes a drug that blocks a viral protease necessary for the HIV life cycle. In the case of anthrax exposure, such a therapy would prevent the toxin protease from cutting apart members of the kinase family, and it could be administered together with antibiotics both to neutralize the toxic effects and to clear the bacterial infection from the body.

> ■ Toxins produced by cholera and anthrax bacteria cause disease by affecting signal cascades inside the cell. Cholera toxin overactivates a G protein signal cascade, while anthrax toxin inhibits important kinases. New drug therapies are in development that will specifically block these effects.

Summary

Specialization and Communication in the Community of Cells

- Multicellular organisms require specialized cells to perform specific functions in the body.

- The coordination of activities in a multicellular organism depends on communication between different types of specialized cells.

- Communication between cells depends on small signaling molecules that are produced by specific cells and received by different target cells.

The Roles of Signaling Molecules in Cell Communication

- Signaling molecules affect target cells by binding to protein receptors either on the cell surface or in the cytoplasm.

- Short-lived signaling molecules like nitric oxide allow rapid communication between neighboring cells.

- Hormones are long-range signaling molecules that allow communication between cells that are far apart in a large multicellular organism.

- Some hormones (such as steroids) enter cells directly; others (such as adrenaline) use cell surface receptors to send their signals into cells.

- Growth factors are signaling molecules that induce the target cell to divide, leading to cell multiplication.

How Cell Surface Receptors Initiate Changes inside the Cell

■ In signal transduction, specific chemical reactions are triggered inside a cell by an external signal, causing a series of stepwise protein activations known as a signal cascade.

■ Cell surface receptors trigger signal cascades by binding external signaling molecules, changing shape, and then activating or inhibiting specific proteins inside the cell.

■ The cytosolic proteins that participate in a signal cascade include G proteins and specific enzymes.

■ Activation of proteins in a signal cascade often involves the enzyme-catalyzed addition of phosphate or the binding of specific nucleotides such as GTP.

■ A signal cascade may result in the production of proteins required for a cellular process like division or the enzyme-catalyzed modification of existing proteins inside the cell.

Amplifing and Combining Signals inside the Cell

■ The activation of many proteins at each step of a signal cascade greatly amplifies the original signal.

■ Signal cascades that modify existing proteins inside the cell occur in a matter of seconds; those that activate genes to produce new proteins can take several hours.

■ Certain proteins (such as kinase enzymes) can participate in different signal cascades, allowing the cell to combine different external signals.

Highlight: Defeating Deadly Bacterial Toxins

■ Knowing which signal cascades are affected by deadly bacterial toxins such as the cholera and anthrax toxins will allow biologists to design drug therapies that specifically block these effects.

Key Terms

adenylate cyclase p. 151	protein phosphorylation p. 151
adrenaline p. 148	receptor p. 146
cell communication p. 145	signal cascade p. 150
cell specialization p. 144	signal integration p. 153
G protein p. 150	signal transduction p. 150
growth factor p. 148	signaling molecule p. 146
hormone p. 148	steroid p. 148
kinase p. 151	target cell p. 146
nitric oxide p. 147	tissue p. 144
PDGF p. 149	vasodilation p. 147

Chapter Review

Self-Quiz

1. Multicellular organisms have different specialized cell types because
   a. cells are unable to coordinate their activities with other cells.
   b. only some cells will produce ATP for use by the entire organism.
   c. the products of cell division are variable.
   d. they are better able to perform specific functions required for the life of the organism.

2. Signaling molecules
   a. are found exclusively in animals.
   b. allow cells to communicate only with the nonliving environment.
   c. are always proteins.
   d. are produced by specific cells and received by target cells.

3. Which of the following statements is true?
   a. Signaling molecules bind to specific receptors that in turn affect chemical processes inside the cell.
   b. All receptors are located on the cell surface.
   c. G proteins are important receptors for hormone signaling molecules.
   d. Steroid hormones require cell surface receptors to affect target cells.

4. Which of the following statements is *not* true?
   a. Hormones tend to be long-range signaling molecules.
   b. Hormones are not found in plants.
   c. Hormones move from one part of the body to another via the bloodstream.
   d. Hormones can affect the production of specific proteins in a target cell.

5. Growth factors are signaling molecules that
   a. stimulate target cells to grow larger and not divide.
   b. were first discovered by biologists attempting to grow cells outside of the body.
   c. freely pass through the plasma membrane of the target cell.
   d. affect only cells in distant parts of the body.

6. Nitric oxide lowers blood pressure by
   a. causing the heart to beat more slowly.
   b. increasing water loss via the intestine.
   c. reducing the volume of blood circulating in the body.
   d. relaxing the muscle cells that line blood vessels.

Review Questions

1. The cellular location of a receptor depends on the chemical characteristics of the signaling molecule to which it binds. Where would you expect the receptor for a hydrophobic molecule to be located?

2. Signal cascades greatly amplify the effect of a single signaling molecule binding to a receptor. Why is signal amplification so important for a cell's ability to receive communications from other cells?

3. Why is phosphorylation by a kinase such an effective way to alter the activity of an enzyme?

4. G proteins and adenylate cyclase both participate in the signal cascade induced by adrenaline. How do these two classes of proteins use nucleotides in promoting the signal cascade?

9   𝕿𝖍𝖊 𝕭𝖆𝖎𝖑𝖞 𝕲𝖑𝖔𝖇𝖊

Equal Opportunity Viagra: Is What's Good for the Gander Good for the Goose?

PARAMUS, NJ. For decades, male erectile dysfunction has been the subject of veiled sneers and standup comedy. But that all changed when the U.S. Food and Drug Administration approved the sale of the drug Viagra in 1998. Today, erectile dysfunction is big business and the subject of vigorous public debate—and even television advertising.

Viagra relieves erectile dysfunction by increasing blood flow to the genital area. The drug achieves this effect by enhancing the signal cascade induced by nitric oxide. Since nitric oxide signals prompt blood vessels to expand, Viagra prolongs the resulting increase in blood flow.

Dr. Phyllis Hoffman of the New Jersey Medical Center firmly believes that Viagra can successfully treat female sexual dysfunction. "It is true that women are wired differently than men," Dr. Hoffman says. "But, increased blood circulation to the genital area is just as important in the female sexual response as it is in males."

While Viagra is still being tested in women, the attention brought to the biology of female sexuality has been applauded by many researchers. "Whether or not Viagra works in women," says Dr. Kris Carmichael of the Female Reproductive Biology Unit at Mt. Sinai Hospital, "these clinical studies will greatly improve our limited understanding of female sexual dysfunction."

Evaluating "The News"

1. Given what you know about nitric oxide as a signaling molecule, do you think that drugs like Viagra might have negative side effects?

2. The wide use of Viagra among men has exceeded even the most optimistic predictions of its manufacturer. The huge number of men using Viagra has led some physicians to fear that its use has become more recreational than therapeutic. Should the use of drugs like Viagra be regulated?

3. Why do you think human sexual dysfunction has received so little attention from the scientific community until now?

10

Cell Division

Hilma af Klint, *Group 4. The Ten Greatest no. 7 Manhood (Lnr 108)*, 1907.

An Army in Revolt

Your body is constantly under siege. With every breath you take, foreign invaders enter and must be repelled by your body's defenses. This continuous struggle against dangerous microorganisms in the environment is a normal part of your body's day-to-day existence. The tremendous organization and complex functioning of the body requires a great deal of effort to maintain and protect it. Given the different specialized cells that make up the body, it is not surprising that a specific group of cells is specialized to defend the body against foreign invaders. These cells make up the immune system: They are both the guards that protect exposed body surfaces and the soldiers that seek out and destroy any invaders that get inside.

Unfortunately, the immune system is not a perfect guardian. In fact, many dangerous organisms have developed ways not only to evade immune defenses, but also to use them to damage the body. One such organism is the bacterium *Borrelia burgdorferi*, which is the primary cause of Lyme disease. Tiny ticks that live on animals such as deer and sheep carry these microscopic invaders. Each year more than 16,000 persons are bitten by infected ticks

Main Message

Cell division is the means by which organisms grow and specific features are passed on from one generation to the next.

while camping or merely walking through tall grass. Just one bite from an infected tick can pass *Borrelia* to an unfortunate individual. The resulting Lyme disease begins with skin rashes and flulike symptoms and can lead to serious arthritic and neurological disorders.

Fortunately, *Borrelia* is easily killed by powerful antibiotics. However, some patients experience severe arthritic pain in their joints long after ridding the body of the bacteria. This mysterious prolonging of arthritic pain involves the inappropriate attacking of the joints by cells of the immune system. Infection by *Borrelia* bacteria allows abnormal immune cells that attack components of the body to divide rapidly, increase in number, and in effect turn your own immune system against you.

This alarming event emphasizes the importance of cell division in diverse biological processes and the need for controls that regulate it. In this chapter we will explore the mechanics of cell division and how it is involved in such diverse processes as the replacement of worn-out tissues and the defense of the body by the immune system.

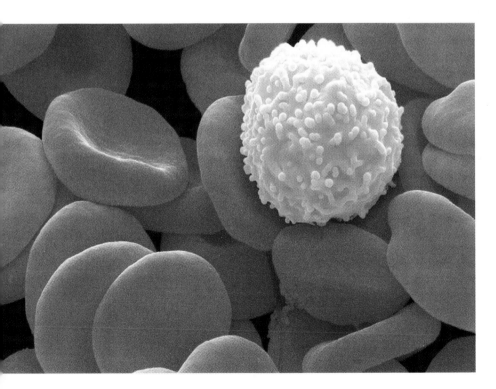

A Single Immune Cell among Many Red Blood Cells

Key Concepts

1. Cells divide to produce two daughter cells. This process is necessary for growth and development, the ability to respond to changes in the environment, and the passing on of genetic information.

2. Cell division has several distinct stages collectively known as the cell division cycle. Each stage is marked by physical changes in the cell that either prepare it for division or directly participate in the process.

3. Most of the events that prepare the cell for division occur during the interphase stage of the cell division

cycle. The physical event of cell division occurs during the stages of mitosis and meiosis.

4. Mitosis ensures that both daughter cells receive identical and complete sets of chromosomes. Each set of chromosomes is uniquely characteristic of each species. The cells that make up the body but are not involved in sexual reproduction all undergo mitosis.

5. Meiosis consists of two cycles of cell division that produce daughter cells with half the number of chromosomes of the parent cell. Meiosis occurs exclusively in cells that produce gametes—sperm and eggs.

As you sit and read this book, you are probably not aware of the millions of cells that are responding to growth signals and are dividing, replacing old cells with new ones. Cells proliferate via cell division when tissue is damaged or lost, as is the case when you fall and scrape your knee, and as a natural part of both repair and regeneration.

The production of new cells in the body is an essential activity for every multicellular organism. In Chapter 9 we discussed how cells communicate with each other by using signaling molecules. One of the most significant consequences of cell-to-cell communication is cell division. Why and when is cell division so important? The most obvious answer is during the growth and development of a multicellular organism. The increase in size and complexity of the body during growth and maturation requires not just more cells, but different kinds of cells. Thus, as a multicellular organism develops, a tremendous amount of cell division and proliferation takes place to expand existing tissues and create new types.

Once an organism has achieved its mature size, cells still divide to replace worn-out and damaged tissue. Skin, for example, must be continuously renewed as the dead cells on the surface are worn away. Simply put, the upper portion of

the skin consists of multiple layers of cells that gradually move to the surface as old cell surface layers are lost to wear and tear. In the process, the cells undergo dramatic physical changes such that by the time they reach the outside, they form dead, flattened scales of protein (Figure 10.1). As dead cells are lost from the skin surface, they are

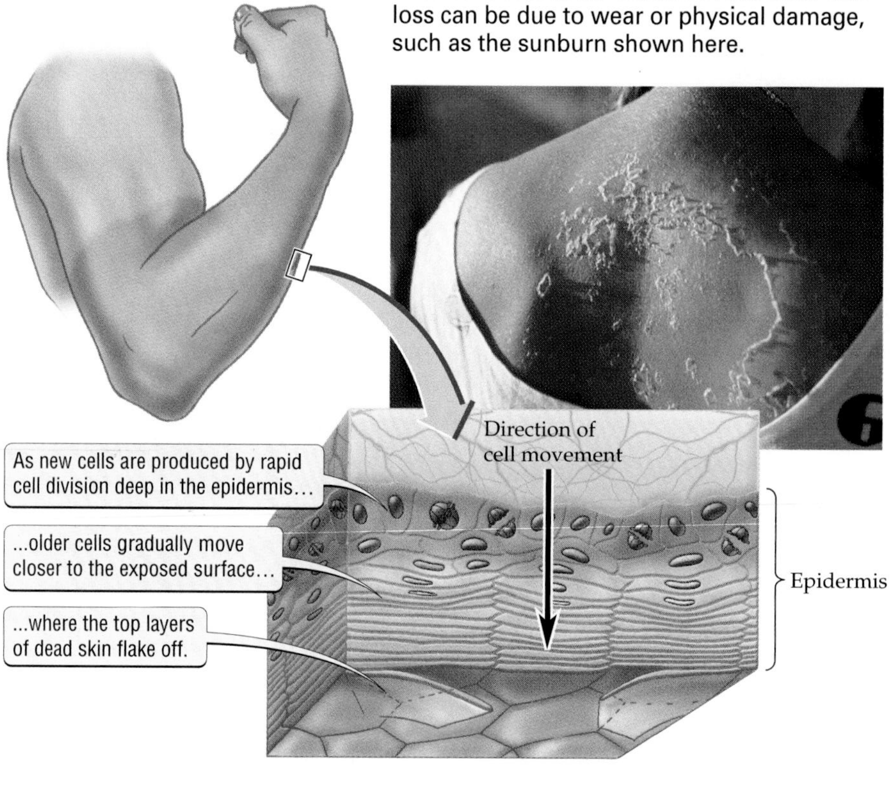

Figure 10.1 Cell Division Replenishes the Skin Rapid cell division in the deepest layer of the epidermis is necessary for the replacement of dead cells lost at the surface of the skin. This loss can be due to wear or physical damage, such as the sunburn shown here.

As new cells are produced by rapid cell division deep in the epidermis…

…older cells gradually move closer to the exposed surface…

…where the top layers of dead skin flake off.

Direction of cell movement

Epidermis

Cell Division and Regeneration of the Human Body

We have discussed how the skin must regenerate itself as layer after layer of cells is lost due to wear. This need for regeneration is not confined to the skin, as most tissues in the body also go through a related process. In other words, cells have a limited life span and must be replaced as they wear out, become less effective at performing their duties, or die. For example, the food-absorbing cells that line the small intestine live for less than a week and must therefore be continuously replaced with new cells.

Depending on the tissue, cells are regenerated in one of two ways. Some tissues, like the cells that form the inner walls of blood vessels, regenerate by simple cell division. Other tissues, like the skin and the small intestine, depend on both cell division and a special class of cells called **stem cells**.

The second type of regeneration involves two key stem cell features. First, stem cells are able to divide for the lifetime of the organism, making them an ideal source of new cells. Second, stem cells are unspecialized, lacking the specific characteristics of any particular tissue. They divide to produce the cells that then become specialized in response to external signals. This process of cell specialization is called **differentiation**.

Stem cells are also responsible for the production of specialized cells that circulate in the bloodstream. The bone marrow contains stem cells that divide and produce all the different cell types found in the blood, including red blood cells that carry oxygen and T cells that defend the body from foreign microorganisms. The discovery of bone marrow stem cells in 1991 was an important breakthrough for cancer patients undergoing radiation therapy. Since bone marrow is easily destroyed by radiation therapy, resulting in fewer blood cells and severe anemia, transplantation of donated stem cells into cancer patients enables their bodies to replace the lost blood cells.

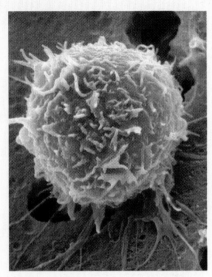

A Stem Cell

More recently, other kinds of stem cells have been discovered that can produce nearly any kind of tissue if given the right chemical signals. Transplantation of these stem cells may eventually let us regenerate any damaged or aged organ at will, creating our own fountain of youth.

---

replaced by cells produced by the division of specialized cells in the deepest layer of the skin (see the Box above).

This chapter focuses on how eukaryotic cells divide in the body of a multicellular organism.

Stages of the Cell Division Cycle

What exactly does cell division mean? How do we go from one cell to two cells? The simple answer is that a single cell divides to form two so-called daughter cells. But although the outcome of cell division is simple, the process requires a great deal of cellular preparation. For example, in order for the daughter cells to live and function normally with their complete set of proteins, both must receive the genetic material that contains the blueprint for all these proteins. In other words, both daughter cells must receive the full complement of DNA in the form of chromosomes that is characteristic of that organism. In addition, the parent cell must be large enough to divide in two and still contribute sufficient cytosol and organelles to each daughter cell.

Both requirements mean that before division takes place, key cellular components must be duplicated, including DNA, proteins, and lipids. These preparations are normally accomplished in a series of steps that make up the life cycle of every eukaryotic cell.

For a single eukaryotic cell, the life cycle both begins and ends with cell division and is therefore called the **cell division cycle**. In its simplest terms, the cell division cycle consists of two major stages that are very different from each other (Figure 10.2). The stage whose events are easily distinguished in the microscope is called **mitosis (M phase)**. Mitosis ends with the physical division of the parent cell into two daughter cells,

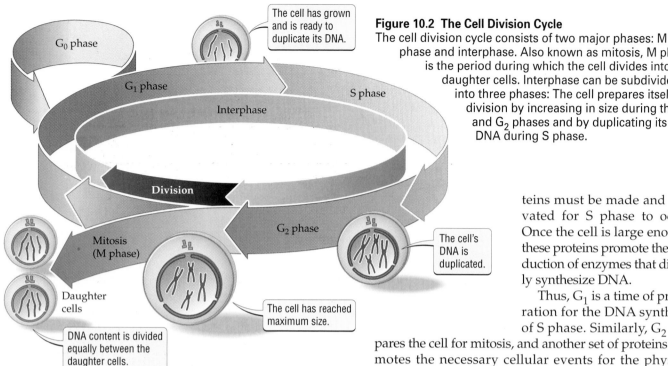

Figure 10.2 The Cell Division Cycle
The cell division cycle consists of two major phases: M phase and interphase. Also known as mitosis, M phase is the period during which the cell divides into two daughter cells. Interphase can be subdivided into three phases: The cell prepares itself for division by increasing in size during the G_1 and G_2 phases and by duplicating its DNA during S phase.

and lasts about an hour in most mammalian cells. The stage between two successive mitotic divisions is called **interphase**, which lasts about 10 to 14 hours in most actively dividing mammalian cells.

It was once thought that the relatively long period of interphase was uneventful for the cell. Today we know that interphase is an active stage during which the cell prepares itself for division.

Interphase prepares the cell for division

Interphase consists of three major stages: S, G_1, and G_2. The stages are defined by a key event that must occur for cell division to proceed. This event is the careful duplication of the entire DNA content of the nucleus in preparation for partitioning to the daughter cells during mitosis. Since duplication requires the synthesis of new DNA, this stage is called **S phase** (S for "synthesis") (see Figure 10.2).

In most cells, S phase does not come immediately before or after mitosis. Two other stages separate mitosis and S phase in the cell division cycle. The first stage, known as the **G_1 phase** (G for "gap"), occurs after mitosis but before S phase begins. The second stage, the **G_2 phase**, occurs between the end of S phase and the start of mitosis (see Figure 10.2). Although their names describe these stages as mere gaps between mitosis and S phase, many essential processes occur during both G_1 and G_2.

The G_1 and G_2 phases are important periods of growth, during which the size of the cell and its protein content increase. Furthermore, during the G_1 phase particular pro-

teins must be made and activated for S phase to occur. Once the cell is large enough, these proteins promote the production of enzymes that directly synthesize DNA.

Thus, G_1 is a time of preparation for the DNA synthesis of S phase. Similarly, G_2 prepares the cell for mitosis, and another set of proteins promotes the necessary cellular events for the physical dividing of the cell. These events include physical changes in the nucleus that are required before the parent cell DNA can be shared equally between the two daughter cells.

The time it takes to complete the cell division cycle depends on the type of cell and the life stage of the organism. Dividing cells in tissues that require frequent replenishing, such as the skin or the lining of the intestine, require about 12 hours to complete the cell division cycle. Most other actively dividing tissues in the human body require about 24 hours to complete the cycle. By contrast, a single-celled eukaryote like yeast can complete the cycle in only 90 minutes. However, the real speed champions of the cell division cycle are the cells that form the earliest stages of the frog embryo. These early cell divisions subdivide only the cytoplasm of the large egg cell and therefore do not require a growth period. Consequently, there are virtually no G_1 and G_2 phases, and the time between divisions is as little as 10 minutes.

Not all of the cells in your body go through the cell division cycle. If they did, it would be difficult to control the size of the body and its organs because many tissues do not require such rapid replenishing of cells. Instead, the cells that make up these tissues pause in the cell division cycle somewhere between mitosis and S phase for periods ranging from days to years. This resting period is called the **G_0 phase** and is easily distinguished from G_1 by the absence of preparations for DNA synthesis (see Figure 10.2).

Cells in G_0 have exited the cell division cycle. Liver cells remain in this resting state for up to one year before undergoing cell division. Other cells, such as those that form the lens of the eye, remain in G_0 for life, thus forming a nondividing tissue. Nerve cells that make up the brain also exist in this nondividing state, which explains why brain cells lost as a result of physical trauma or chemical damage are usually not replaced.

> ■ Cell proliferation requires the division of a single parent cell into two daughter cells. Cells divide in a carefully controlled cycle with two distinct stages: interphase and mitosis. Interphase prepares the cell for division and consists of three phases: S, G_1, and G_2. The cell division cycle ends in mitosis, the last step of which is the physical division of the parent cell into two identical daughter cells.

Mitosis: From One Cell to Two Identical Cells

Mitosis is the climax of the cell division cycle. The central event of mitosis is the equal distribution of the parent cell DNA to two daughter cells. This process, called **DNA segregation**, is a distinctly physical process that requires the coordinated actions of several structural proteins. But before discussing the details of DNA segregation during mitosis, we must consider the earlier preparation and packing of the DNA that occurs in the nucleus.

The DNA in the nucleus is not a random tangle of nucleic acid polymers, but rather is highly packed and organized into distinct individual structures. This packing is essential because about 1 to 2 meters of DNA must fit into the average nucleus, which has a diameter of less than 5 micrometers. DNA and proteins are packed to form thicker and more complex strands called **chromatin**. Chromatin in turn can be further looped and packed to form even more complex structures called **chromosomes** (Figure 10.3). The packing of chromatin into chromosomes is one of the earliest events of mitosis that can be identified in the microscope.

Each species has a distinctive karyotype

Every species has a characteristic number of chromosomes in each cell. Furthermore, each chromosome contains a single molecule of DNA with a defined set of genes. When chromosomes become visible during mitosis, they adopt characteristic shapes that allow them to be identified in the microscope (Figure 10.4a). The portrait formed by the number and shapes of chromosomes found in a species is known as that species' **karyotype**.

All cells in the human body, with the exception of the sex cells (eggs and sperm), contain 46 chromosomes each. In contrast, cells from horses have 64 chromosomes, and cells from corn have 20 chromosomes. However, the number of chromosomes has no particular significance other than being an identifying characteristic of the species.

To understand how the karyotype of a given species is duplicated and segregated during the cell division cycle, we must understand how chromosomes relate to each other. Returning to the example of the human karyotype, our 46 chromosomes can be arranged in 23 pairs, 22 of which are described as homologous (numbered 1 to 22), plus one pair of sex chromosomes (individually lettered X or Y). Each pair of **homologous chromosomes** consists of two of the same type, one received from the mother and the other from the

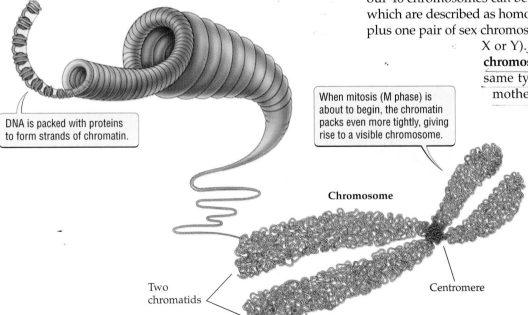

DNA is packed with proteins to form strands of chromatin.

When mitosis (M phase) is about to begin, the chromatin packs even more tightly, giving rise to a visible chromosome.

Chromosome

Two chromatids

Centromere

Figure 10.3 The Packing of DNA into a Chromosome

(a)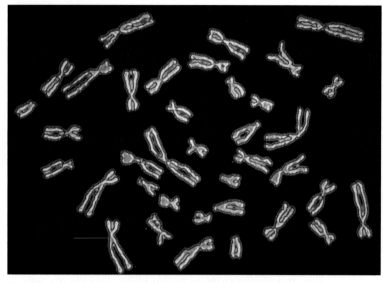

(b)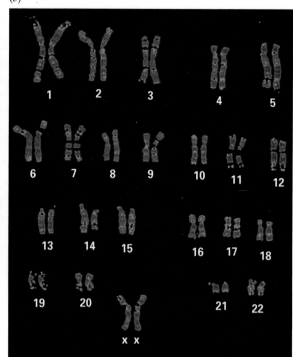

Figure 10.4 The Karyotype from a Human Female
(a) This typical preparation of mitotic human chromosomes shows the various shapes that are characteristic of each pair (except for the sex chromosomes). (b) In this preparation the chromosomes have been arranged in homologous pairs and are numbered.

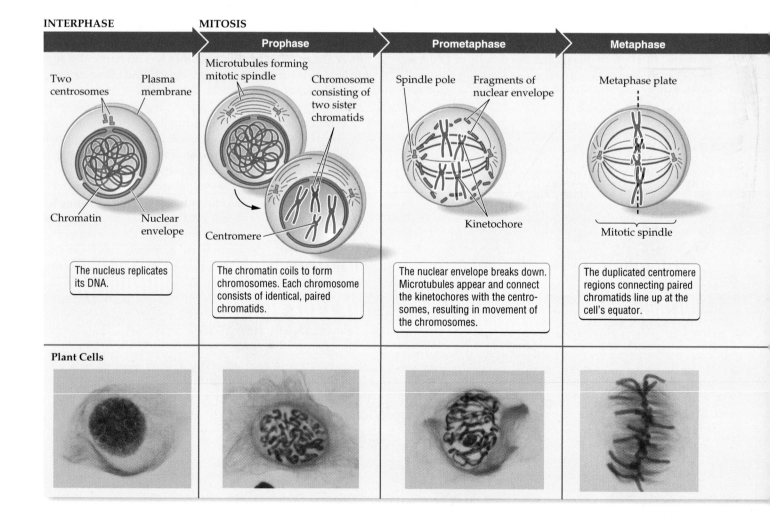

INTERPHASE

MITOSIS

| Prophase | Prometaphase | Metaphase |

Interphase

Two centrosomes · Plasma membrane · Chromatin · Nuclear envelope

The nucleus replicates its DNA.

Prophase

Microtubules forming mitotic spindle · Chromosome consisting of two sister chromatids · Centromere

The chromatin coils to form chromosomes. Each chromosome consists of identical, paired chromatids.

Prometaphase

Spindle pole · Fragments of nuclear envelope · Kinetochore

The nuclear envelope breaks down. Microtubules appear and connect the kinetochores with the centrosomes, resulting in movement of the chromosomes.

Metaphase

Metaphase plate · Mitotic spindle

The duplicated centromere regions connecting paired chromatids line up at the cell's equator.

Plant Cells

father. On the other hand, the two different types of **sex chromosomes** (X or Y) can form a nonhomologous pair (XY) or a homologous pair (XX). They are called sex chromosomes because humans with two X chromosomes are female (Figure 10.4*b*), and those with an X and a Y chromosome are male. The determination of gender on the basis of sex chromosomes and inheritance is discussed in Chapter 13.

Before cell division can proceed, the full karyotype of the parent cell must be duplicated so that each daughter cell receives a complete set of chromosomes. This duplication occurs during S phase and produces chromosomes made up of two identical, side-by-side strands called **chromatids**. Thus, at the beginning of mitosis the nucleus of a human cell contains twice the usual amount of DNA, since each of 46 chromosomes consists of a pair of identical sister chromatids held together at a constriction called the **centromere** (see Figure 10.3). The

important roles of chromatids and centromeres in mitosis will become obvious as we describe the stages of cell division: prophase, prometaphase, metaphase, anaphase, telophase, and cytokinesis.

Chromosomes become visible during prophase

Mitosis is divided into six stages, each with easily identifiable events that are visible in the light microscope (Figure 10.5). The first stage of mitosis, called **prophase** (*pro*, "before"; *phase*, "appearance"), is characterized by the first appearance of visible chromosomes. In an interphase cell, the chromatin is well dispersed throughout the nucleus, and specific chromosomes cannot be distinguished. As the cell moves from G$_2$ phase into prophase, the chromatin condenses and the chromosomes appear in the nucleus, looking like a tangled ball of spaghetti.

Important changes occur in the cytosol of the prophase cell as well. Two protein structures called **centrosomes** (*centro*, "center"; *some*, "body") begin to move around the nucleus, finally halting at opposite sides in the cell. As will be obvious later, this arrangement of centrosomes defines the opposite ends or poles of the cell that will eventually separate to form two daughter cells.

At the same time that the centrosomes are moving to the poles of the cell, microtubules grow outward from each centrosome. These filaments are the beginnings of a major cellular structure called the **mitotic spindle**, which will later guide the movement of chromosomes.

Chromosomes attach to the spindle during prometaphase

In the next stage of mitosis, **prometaphase**, the nuclear envelope breaks down (see Figure 10.5). In the process, the mitotic spindle radiating from the centrosome poles extends and enters the region of the cell that was once within the nucleus. The spindle microtubules then attach to chromosomes at their centromeres, effectively linking each chromosome to both centrosomes.

The physical structure of the centromere dictates how each chromosome will attach to the spindle microtubules. Each centromere has two placques of protein, called **kinetochores**, that are oriented on opposite sides of the constriction. Each kinetochore forms the site of attachment for a single microtubule such that each of the two chromatids that make up a chromosome end up being attached to opposite poles of the cell. This arrangement is essential for the later segregation of DNA.

Chromosomes line up in the middle of the cell during metaphase

Once each chromosome is attached to both spindle poles, the lengthening and shortening of its micro-

Figure 10.5 The Stages of Mitosis
Diagrams of a dividing animal cell (top) and microscope images of a dividing plant cell (bottom). Note the presence of the cell wall in the plant cell that still encloses the two daughter cells after cytokinesis (see also Figure 10.6).

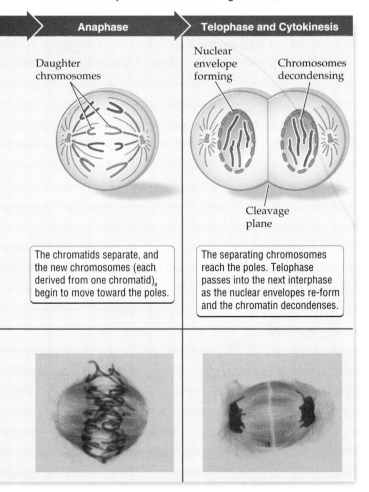

Anaphase — The chromatids separate, and the new chromosomes (each derived from one chromatid) begin to move toward the poles.

Telophase and Cytokinesis — The separating chromosomes reach the poles. Telophase passes into the next interphase as the nuclear envelopes re-form and the chromatin decondenses.

tubule attachments moves it toward the middle of the cell. There the chromosomes eventually line up in a single plane that is equally distant from both spindle poles. This stage of mitosis is called **metaphase** (*meta*, "after"), and the chromosomes are said to be arranged at the metaphase plate (see Figure 10.5). Metaphase is so visually distinctive that its appearance is used as an indicator of actively dividing cells in the examination of tissue in a microscope.

Chromatids separate during anaphase

During the next stage of mitosis, **anaphase** (*ana*, "up"), the chromosomes segregate. At the beginning of anaphase, the sister chromatids separate, each attached to a single spindle microtubule. Once separated, each chromatid is now considered to be a chromosome, and the gradual shortening of the microtubules pulls the newly separated chromosomes to opposite poles of the cell. This remarkable event results in the equal segregation of chromosomes between the two daughter cells.

Because the chromatin is duplicated during S phase, in a human cell each of the 46 chromosomes is duplicated, yielding 46 pairs of identical sister chromatids. Thus, when the chromatids separate at anaphase, identical sets of 46 chromosomes arrive at each spindle pole, ready to become a complete genetic blueprint for each daughter cell.

Figure 10.6 Mitosis in Higher Plants
The tough cell wall that surrounds the cells of higher plants requires a special kind of cytokinesis. Instead of pinching in two as animal cells do, plant cells divide by building a new cell wall down their middle. Vesicles filled with cell wall components accumulate in the middle of the cell at the start of telophase. As the vesicles fuse, the new cell wall then forms from the vesicles' contents, dividing the original cell into two new daughter cells at cytokinesis.

New nuclei form during telophase

Telophase (*telo*, "end") begins when a complete set of chromosomes arrives at a spindle pole, which marks the region of the parent cell that will form the cytosol of a new daughter cell. Major cytosolic changes also occur in preparation for division into two new cells. Once the respective sets of chromosomes arrive at the opposite poles of the parent cell, the spindle microtubules break down and nuclear envelopes begin to form around each set of chromosomes (see Figure 10.5). As the two new nuclei become increasingly distinct in the cell, the chromosomes within them start to unfold back into chromatin, making them less visible in the microscope.

In higher plants, additional changes occur during telophase that prepare the plant cell for the physical process of dividing in two (Figure 10.6). Vesicles containing cell wall components accumulate in the region previously occupied by the metaphase plate and begin to fuse with one another. As they fuse and share their contents, a new cell wall begins to form down the middle of the cell.

Cell division occurs during cytokinesis

The last stage of mitosis, **cytokinesis** (*cyto*, "hollow vessel"; *kinesis*, "movement"), features the division of the parent cell into two daughter cells (see Figure 10.5). The physical act of separation in animal cells is performed by a ring of actin filaments that lies against the inner face of the plasma membrane like a belt at the equator of the cell. When the actin ring contracts, it pinches the cytoplasm of the cell and divides it in two. Since the plane of constriction by the actin ring lies between the two newly formed nuclei, successful division results in two daughter cells, each with its own nucleus.

In higher plant cells, the new cell wall that began forming in telophase is completed, dividing the cell into two independent daughter cells (see Figure 10.6). Cyto-

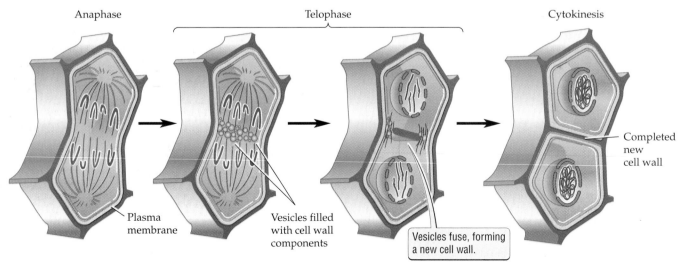

Anaphase   Telophase   Cytokinesis

Plasma membrane

Vesicles filled with cell wall components

Vesicles fuse, forming a new cell wall.

Completed new cell wall

kinesis marks the end of the cell division cycle, and once completed, the daughter cells are free to enter G_1 phase and start the process anew.

> ■ DNA in the nucleus is packed into chromatin using proteins. Mitosis requires that the cell's chromatin be further packaged into specific structures that are called chromosomes, which form a unique karyotype for each species. During mitosis the parent cell DNA is separated such that each daughter cell receives a complete karyotype.

Meiosis: Halving the Chromosome Number in Sex Cells

The remarkable process of mitosis occurs throughout the body during our entire lifetime. However, certain specialized cells in the body undergo a related but significantly different cell division cycle, called meiosis. **Meiosis** (*meio*, "less") produces daughter cells with half the number of chromosomes found in the parent cell. Human sex cells, or **gametes** (either sperm in males or eggs in females), are the only cells in the body produced by meiosis. What is it about these cells that requires them to undergo a different kind of division cycle? The answer lies in the role they play in sexual reproduction.

Gametes contain half the number of chromosomes

The creation of a new organism via sexual reproduction requires the fusion of two gametes in a process known as **fertilization** (see Chapter 36). A successful union of gametes forms a specialized single cell called a **zygote** (Figure 10.7). The zygote then undergoes multiple rounds of mitosis to form the embryo that will develop into a new organism.

However, if both the sperm and the egg contained a complete set of chromosomes (46 for humans), the zygote would have double that number (92 chromosomes for humans), and these would be duplicated and

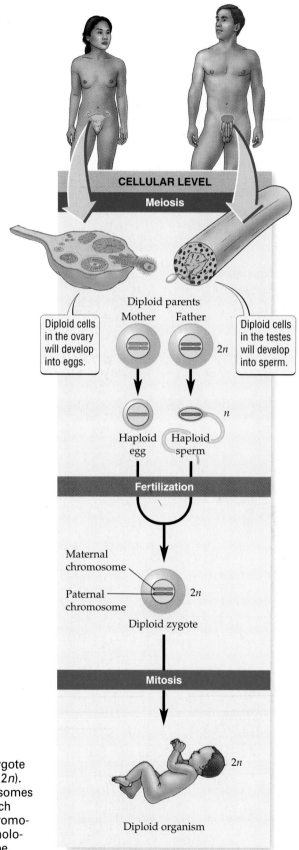

Figure 10.7 Sexual Reproduction Requires a Reduction in Chromosome Number
The fusion of a sperm and an egg at fertilization must produce a zygote with no more than the full number of chromosomes (diploid state, $2n$). Thus the two gametes must each have half the number of chromosomes found in the cells of the rest of the body. To solve this dilemma, each gamete receives only one member of each homologous pair of chromosomes (haploid state, n). Thus, when fertilization occurs, a full homologous pair re-forms in the zygote, which then has a diploid karyotype.

passed on in all the later cell divisions. The resulting embryos would have double the number of chromosomes, an abnormality that would be lethal. Therefore, if offspring are to have the same karyotype as their parents, the fusion of gametes must produce the usual number of chromosomes in the zygote after fertilization (see Figure 10.7).

The simple solution to this problem is for the gametes to contain half the number of chromosomes found in other cells. This arrangement is possible thanks to the organization of the karyotype into pairs of homologous chromosomes. In the case of humans, the 22 homologous pairs of chromosomes are found in all body cells except gametes. Each gamete contains only one homologous chromosome of each pair. Then, depending on the gender of the person, the female gametes (eggs) will all contain a single X chromosome, and the male gametes (sperm) will contain either an X or a Y chromosome. That is, each gamete contains one representative of all homologous chromosome types plus one sex chromosome.

Upon fertilization, the zygote will therefore contain a complete set of 22 homologous pairs of chromosomes and one pair of sex chromosomes—that is, a normal human karyotype. Furthermore, each pair of homologous chromosomes in the zygote will consist of one from the father and one from the mother. This equal contribution of chromosomes by each parent is the basis for genetic inheritance discussed in Chapter 13.

Gametes are haploid, while other cells are diploid

The difference in chromosome numbers between gametes and other cells in the body has led to alternative ways of describing the karyotype. Human gametes, which have a chromosome number of 23 (n), are said to be **haploid**. The cells that make up the rest of the human

Figure 10.8 Differences and Similarities between Meiosis and Mitosis
The major difference between meiosis and mitosis occurs during meiosis I. The homologous chromosomes pair during anaphase I through metaphase I, resulting in a separation of the homologues at the end of meiosis I. Meiosis I is the reduction division. In contrast, meiosis II is more similar to mitosis. For simplicity, not all stages are shown.

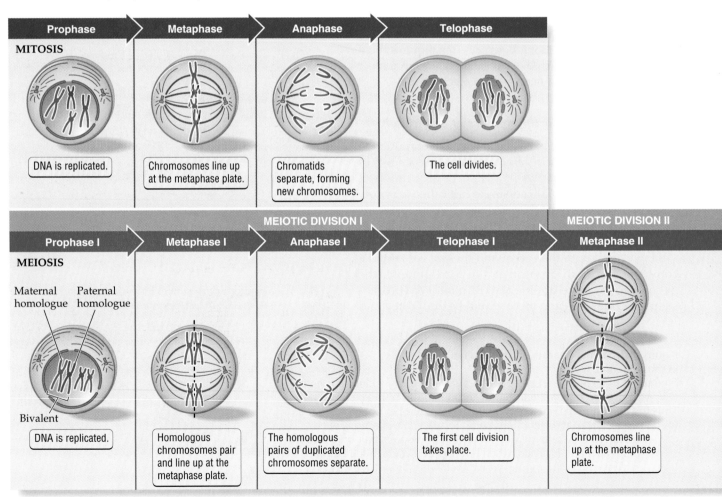

body, which have a chromosome number of 46 (2*n*), are said to be **diploid**. The symbol *n* designates the haploid state; 2*n* designates the diploid state.

Since gametes are the only cells that go on to form a new organism, the continued propagation of a species depends on the haploid state. On the other hand, the day-to-day processes that maintain the life of an individual organism depend on the diploid state. So for the sake of propagating the species, there must be a process to generate haploid gametes from diploid cells. This process is meiosis (Figure 10.8), the process of cell division in which a single diploid cell ultimately yields four haploid gametes.

The stages of meiosis are very similar to those of mitosis. However, unlike mitosis where a single nuclear division is sufficient, in meiosis the nucleus divides twice. These successive divisions are called meiosis I and meiosis II, and each has distinct roles in the important task of producing haploid cells from a diploid parent.

Meiosis I is the reduction division

Let's begin with the diploid cells that form the reproductive tissue responsible for the production of gametes. The first step in producing haploid gametes from these cells is the halving of the chromosome number. This reduction in chromosome number is achieved during **meiosis I** by a pairing of homologous chromosomes that is not seen in mitosis. Otherwise, meiosis I has all the same stages and overall appearance of mitosis that were discussed earlier. In fact, the stages of meiosis I have the same names as those of mitosis, except that the Roman numeral I is added to distinguish them from similar stages in meiosis II.

How does chromosome pairing during meiosis I reduce the number of chromosomes? The answer lies in both the preparation for and mechanics of chromosome segregation. As early as prophase, there are significant differences between meiosis I and mitosis. During prophase I, each homologous chromosome consisting of two chromatids pairs with its matching homologue. In other words, both copies of chromosome 6 pair up, as do both copies of chromosome 3, and so on. These pairs are called **bivalents** and have a total of four chromatids each. The formation of bivalents is an opportunity for genetic sequences to be exchanged between homologous chromosomes. (This important process is discussed in Chapter 13.)

An important consequence of the formation of bivalents appears during the later stages of prophase I, when spindle microtubules extend to meet the chromosomes. In contrast to mitotic prophase, in meiosis only one spindle microtubule attaches to each homologous chromosome, linking it to only one pole of the cell. Thus, when the bivalents move into position at the metaphase plate during metaphase I, each chromosome of a bivalent is attached to opposite spindle poles. When the microtubules shorten at anaphase I, they therefore pull apart the homologous chromosomes of each bivalent to opposite sides of the cell. This process is very different from what happens at anaphase of mitosis, in which the individual chromatids of each homologous chromosome are separated to form new chromosomes (see Figure 10.8).

After anaphase I of meiosis, the events of telophase I and cytokinesis follow the same patterns seen in mitosis and produce two daughter cells. Unlike those formed by mitosis, however, the daughter cells of meiosis I contain only half the number of chromosomes found in the parent cell. That is, each daughter cell has only one homologous chromosome of each type and is therefore haploid. Meiosis I is therefore called a reduction division because it reduces the chromosome number by half: Diploid (2*n*) becomes haploid (*n*).

Meiosis II is like mitosis

The two haploid cells formed in meiosis I go through a second round of cell division, called **meiosis II**. This time the stages of the division cycle are more comparable to those of mitosis. In particular, the chromatids separate at anaphase II, leading to an equal segregation of chromosomes to the daughter cells (see Figure 10.8). In this manner the two haploid cells produced in meiosis I ultimately give rise to a total of four haploid cells. These haploid cells are gametes and contain the appropriate set of chromosomes such that fusion of male and female gametes at fertilization will produce a diploid zygote with a complete karyotype.

Anaphase II

Telophase II

The second cell division takes place.

Four daughter cells form.

The remarkable reduction in chromosome number observed in meiosis I offsets the combining of chromosomes during fertilization and is an elegant way to maintain the constant chromosome number of a species during sexual reproduction.

> ■ Meiosis produces daughter cells with half the number of chromosomes found in the parent cell. The only cells in the human body that are produced by meiosis are gametes. Each gamete contains only one chromosome from each pair of homologous chromosomes and is haploid (*n*); other cells in the body contain a full karyotype and are diploid (2*n*). Meiosis has two nuclear divisions: meiosis I (the reduction division) and meiosis II (similar to mitosis).

Highlight

Immune Cell Proliferation and Lyme Disease

Cell division is essential for the replacement of worn-out cells and tissue. Less obvious is the role that mitosis plays in defending the body from foreign invaders such as harmful bacteria, viruses, and fungi. These undesirable guests are called **pathogens** (*patho*, "suffering"; *gen*, "producing") and differ from microorganisms that are helpful to the body, such as beneficial intestinal bacteria. Specialized cells collectively known as the immune system have the job of patrolling and defending the body against pathogens. Whereas human armies react to an attack on their country by recruiting new soldiers, these strong defenders of the body must undergo cell division to proliferate when needed. Thus, cell division is not confined to growth and tissue replacement; it also determines how an organism responds to challenges from the external environment.

Cytotoxic **T cells** are among the crucial soldiers of the immune system that are prime candidates for proliferation when a pathogen attacks. They are called T cells because they originate from the thymus, and they are described as cytotoxic (*cyto*, "hollow vessel"; *toxic*, "poison") because their duty is to kill other cells that are damaged by a pathogen. Their ability to recognize cells that are infected by a virus, for example, is based on the virus proteins that always appear on the surface of infected cells. Once these virus proteins are recognized as foreign, the T cell binds to them and kills the infected cell. This is how the body rids itself of infected cells and limits the illness caused by the pathogen.

This immune response depends on the body's ability to produce specific cytotoxic T cells that best recognize and destroy cells infected by a particular pathogen. These T cells are tailored to detect certain pathogen components usually during a previous exposure to the pathogen. After the first exposure, pathogen-specific T cells persist in the body as memory cells. When the pathogen reappears, the memory cells are stimulated to divide rapidly and create a large population that can both detect and destroy infected cells.

Normally, T cells are able to distinguish pathogen proteins from the body's own proteins, which they do not attack. This ability to ignore self proteins and bind only to foreign proteins is an essential characteristic of the immune system, known as self-tolerance. In fact, the body has protective mechanisms for quickly eliminating T cells that bind to self components, since every army must clearly distinguish friends from foes. However, certain pathogens, such as *Borrelia*, the bacteria that causes Lyme disease, are able to derail this normal process of self-tolerance.

As described at the start of this chapter, the painful arthritis associated with Lyme disease is linked to the fact that the immune system inappropriately attacks the body's joints. However, why does the arthritis continue in some patients long after the *Borrelia* bacteria have been killed by antibiotics? This mysteriously persistent arthritis may depend on a *Borrelia* protein that resembles a normal cellular protein. A protein found in the *Borrelia* cell wall has amino acid sequences similar to those found in a human protein that is present on the surface of many human cells.

T cells that recognize the foreign bacterial protein on infected cells are induced to divide in a normal immune response, but those same T cells also recognize the human protein on healthy cells. Thus, when T cells that recognize the *Borrelia* protein move into the body's joints to clear the bacterial infection, they also attack and destroy normal cells. The death of more and more cells in the joint causes inflammation and arthritis pain. Even after the bacteria have been eliminated, the earlier proliferation of these T cells along with their continued attack on normal body cells prolongs the arthritis. This painful consequence of inappropriate cell division reminds us that both the timing and the identities of the cells undergoing division must be carefully controlled in the body. Cancer, another dire consequence of inappropriate cell division, will be discussed in Chapter 11.

■ An effective immune response requires the division of specific T cells that recognize and destroy the pathogen. The similarity between a *Borrelia* protein and a human protein found on healthy cells causes T cells to recognize and destroy both infected cells and healthy cells. The inappropriate destruction of healthy cells in the joints causes persistent arthritis in patients even after *Borrelia* has been killed.

Summary

Stages of the Cell Division Cycle

■ Cell proliferation requires the division of a single parent cell into two daughter cells.

■ Cells divide in a carefully controlled cell division cycle with two distinct stages: interphase and mitosis.

■ Interphase prepares the cell for division and consists of three phases: S, G_1, and G_2.

■ The cell's DNA is duplicated during S phase, and the cell increases in size and produces specific proteins needed for division during the G_1 and G_2 phases.

■ Mitosis marks the end of the cell division cycle, itself ending with the physical division of the parent cell into two identical daughter cells.

Mitosis: From One Cell to Two Identical Cells

■ DNA in the nucleus is packed into chromatin with the assistance of proteins.

■ Mitosis requires that the cell's chromatin be further packed into chromosomes.

■ The specific set of chromosomes found in a species is known as that species' karyotype.

■ During mitosis the parent cell DNA is separated such that each daughter cell receives a complete karyotype.

■ The six stages of mitosis are prophase, prometaphase, metaphase, anaphase, telophase, and cytokinesis.

Meiosis: Halving the Chromosome Number in Sex Cells

■ Meiosis produces daughter cells with half the number of chromosomes found in the parent cell.

■ The only cells in the human body that are produced by meiosis are gametes.

■ Gametes are haploid (n), containing only one chromosome from each pair of homologous chromosomes; other cells in the body are diploid ($2n$), containing a complete karyotype.

■ Meiosis consists of two nuclear divisions: meiosis I, called the reduction division because it produces haploid daughter cells from a diploid parent cell, and meiosis II, which is similar to mitosis.

Highlight: Immune Cell Proliferation and Lyme Disease

■ An effective immune response requires the division of specific T cells that recognize and destroy the pathogen.

■ The bacterium *Borrelia* causes Lyme disease and persistent arthritis. The latter condition results because the similarity between a *Borrelia* protein and a human protein found on healthy cells causes T cells to recognize and destroy both infected cells and healthy cells.

■ The inappropriate destruction of healthy cells in the joints causes persistent arthritis in patients even after *Borrelia* has been killed.

Key Terms

anaphase p. 166	interphase p. 162
bivalent p. 169	karyotype p. 163
cell division cycle p. 161	kinetochore p. 165
centromere p. 165	meiosis p. 167
centrosome p. 165	meiosis I p. 169
chromatid p. 165	meiosis II p. 169
chromatin p. 163	metaphase p. 166
chromosome p. 163	mitosis (M phase) p. 161
cytokinesis p. 166	mitotic spindle p. 165
differentiation p. 161	pathogen p. 170
diploid p. 169	prometaphase p. 165
DNA segregation p. 163	prophase p. 165
fertilization p. 167	S phase p. 162
G_0 phase p. 162	sex chromosome p. 165
G_1 phase p. 162	stem cell p. 161
G_2 phase p. 162	T cell p. 170
gamete p. 167	telophase p. 166
haploid p. 169	zygote p. 167
homologous chromosome p. 163	

Chapter Review

Self-Quiz

1. Cell division is important for adult skin because
   a. it helps block the harmful effects of ultraviolet light.
   b. the dead cells lost at the surface because of wear and tear must be replaced.
   c. it allows the animal to grow larger.
   d. all cells must divide or die.

2. DNA segregation is an essential feature of mitosis because
   a. each daughter cell must receive a complete set of chromosomes.
   b. it is necessary for the physical separation of the daughter cells.
   c. only one daughter cell should receive the parental DNA.
   d. it ensures that the parent cell is large enough to divide.

3. Which of the following statements is true?
   a. The cell lies dormant during interphase of the cell division cycle.
   b. The key event of S phase is the synthesis of proteins required for mitosis.
   c. The cell increases in size during the G_0 phase.
   d. The cell increases in size during the G_1 and G_2 phases.

4. Which of the following statements is *not* true?
   a. DNA is packaged into chromatin with the help of proteins.
   b. Chromosomes are packaged into chromatin with the help of proteins.
   c. Chromosomes are visible in the microscope only during mitosis and meiosis.
   d. Each species is characterized by a particular number of chromosomes.

5. Which of the following correctly represents the order of phases in the cell division cycle?
   a. Mitosis, S phase, G_1 phase, G_2 phase
   b. G_2 phase, G_1 phase, mitosis, S phase
   c. S phase, mitosis, G_2 phase, G_1 phase
   d. G_1 phase, S phase, G_2 phase, mitosis

6. Cytokinesis occurs
   a. at the end of prophase.
   b. just before telophase.
   c. at the end of meiosis I.
   d. at the end of G_1 phase.

Review Questions

1. Horses have a karyotype of 64 chromosomes. How many chromosomes would a horse cell undergoing mitosis have? How many would a horse cell undergoing meiosis II have?

2. What essential role does the mitotic spindle play in mitosis?

3. Spindle microtubules attach to chromosomes in different ways during mitosis and meiosis I. Describe the effect of these differing attachments on the daughter cells produced by mitosis versus those produced by meiosis I.

4. Why is meiosis so important for the production of sex cells? How would the offspring of sexual reproduction be affected if sex cells underwent mitosis instead of meiosis?

10 𝔗𝔥𝔢 𝔇𝔞𝔦𝔩𝔶 𝔊𝔩𝔬𝔟𝔢

Spare Parts for Your Brain

Letter to the Editor:

Dear Madam,

The human brain is a delicate and precious part of the body. But the tendency of adult brain cells to not divide means that any brain cells that die are not replaced. When the brain is damaged, either by physical injury or the effects of old age, there is no way to replace the cells that are lost, so the chances of recovering full brain function are limited.

Today, several research laboratories, including my own, have isolated stem cells that can be induced to divide and form new brain cells if given the right chemical cues. We first discovered these stem cells in the brains of fetal mice and have tested for their ability to divide and repair brain tissue. These stem cells, when reintroduced into the brain, divide and form nerve cells that co-exist with the older brain cells of the mice.

This very exciting breakthrough means that we will probably be able to repair brain damage in the future by administering the appropriate stem cells to the site of injury. There they will respond to the normal chemical signals of the brain and divide to form new nerve cells that will replace those lost to injury. Since brain cells also die with advancing age, resulting in reduced cognitive ability, this approach may even allow us to maintain our mental faculties far into old age.

However, we can only learn so much using stem cells isolated from fetal mice. We must study stem cells from human fetal tissue to really understand how brain regeneration might occur in humans. A ban on research using human fetal tissue is therefore a major obstacle to our ultimate goal of limiting the ravages of disease and aging by regenerating essential tissues. Thus, I urge your readers to ask Congress not to ban research on human fetal tissue. Instead, support such research and help us develop the strict ethical guidelines that will allow us to improve human health with a clear conscience.

Yours sincerely,
Dr. Joseph Stemple
Division of Organ Transplantation
and Tissue Repair
New York Medical Institute

Evaluating "The News"

1. If you were to compare the brain cells of a human fetus with those of an adult, which would have more cells in G_0? How does this account for the potential importance of fetal tissue as a source for brain stem cells?

2. Some would argue that our pursuit of a prolonged life span has clouded our ethical judgement. In the hunt for human stem cells, do the possible benefits to our health justify research on tissues from aborted human fetuses?

3. If we do allow research on human fetal tissue, what restrictions do you think should be imposed?

chapter *11*

Cancer:
Cell Division
Out of Control

Eugene Von Bruenchenhein, *EVB 390 Untitled no. 659 (GS261)*, 1957.

Turning an Enemy Virus into a Cancer Treatment

ancer is the ultimate insult to the cooperative functioning of all multicellular organisms. Cancerous tumors are nothing more than stubborn rebel colonies that selfishly ignore the laws of cell-to-cell coordination that keep multicellular organisms alive. The cells that form a tumor divide with wild abandon, often failing to adopt the structures and behaviors required for the particular organ or tissue. At worst, these aggressive cells break out of the tumor and go on to establish other similarly rebellious tumors in other parts of the body.

This is the malignant form of cancer that everyone fears, and it accounts for more than 500,000 deaths in the United States each year. The National Cancer Institute estimates that the collective price tag for the various forms of cancer is $107 billion per year—$37 billion for direct medical costs, $11 billion for lost productivity, and $59 billion for costs due to individual deaths.

Main Message

The cell proliferation that is characteristic of cancer requires a loss of individual cell control and a failure in communication between neighboring cells.

Almost 25 years ago, President Richard Nixon declared a war on cancer by making anticancer research a high priority. Since then, some major victories have been won thanks to improvements in radiation and drug therapies. Whereas in the early twentieth century very few individuals survived bouts of cancer, today roughly 40 percent of patients are alive 5 years after treatment is begun. Nevertheless, the war against cancer is far from over, and the need for powerful new treatments to stop tumor growth and eliminate cancerous cells is as urgent as ever.

One of the most inventive methods of locating and destroying cancer cells depends on the assistance of an unlikely ally—a virus. The adenovirus, which infects mammals, takes over the enzymatic machinery of specific cells to produce more viruses. In the process the virus kills the infected cells, possibly causing respiratory tract disease. To harness this destructive power for the benefit of individuals with cancer, researchers have mutated the virus so that it successfully infects only cancer cells.

These mutant viruses are unable to multiply, but once they infect a cancer cell they are still able to destroy it. Thus, mutant viruses could be administered to tumor sites, where they would infect and kill only the cancer cells, leaving healthy cells untouched. Although it is still too early to tell whether this kind of therapy will work in patients, it represents a new potential solution to the problem of selectively eliminating cancerous cells.

This remarkable effort to tame a virus and turn it into an anticancer crusader depends on understanding how viruses take control of the cells they infect. New discoveries in this area make it clear that many viruses have ways to bypass the normal controls that limit cell division, revealing that some viruses play an important role in causing cancer. In this chapter we discuss some of the important ways in which cell division is controlled and how the loss of these controls leads to cancer. Then we look at how this knowledge has been used to develop anticancer viruses.

Adenoviruses, Which Can Cause Respiratory Disease

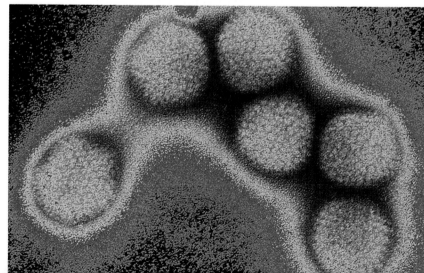

Key Concepts

1. Cancer is a group of diseases caused by rapid and inappropriate cell division. The resulting cell proliferation can form a tumor from which abnormal cells may spread, invading other tissues in the body.

2. All multicellular organisms have ways of regulating the proliferation of different types of cells. When a cell loses its ability to respond to external and internal regulatory signals, it divides uncontrollably, leading to cancer.

3. Genes that promote cell division in response to normal growth signals are called proto-oncogenes. Mutant, overactive versions of these genes can lead to excessive cell division and cancer.

4. Genes that inhibit cell division under normal conditions are called tumor suppressor genes. The complete loss of tumor suppressor activity can also lead to excessive cell division and cancer.

5. When and how often a cell divides depends on the balance between the promoting effects of proto-oncogenes and the inhibiting effects of tumor suppressors.

6. Most cancers are caused by a combination of environmental factors. A minority of cancers can be traced to inherited genetic defects or mutations.

To achieve a high level of organization, a multicellular organism must have a means of controlling and coordinating the behavior of individual cells. Any large community that does not have rules quickly falls into chaos, and the same is true of multicellular organisms. Therefore, both the metabolic activities of every cell and the frequency of cell division are closely regulated. As discussed in Chapter 9, different kinds of signaling molecules called growth factors promote cell division. Such positive growth signals achieve this task by activating cellular proteins required for division. In effect, positive growth signals define one set of rules that cells obey by promoting cell division when it is needed.

More than a decade ago, cell division was thought to be controlled exclusively by positive signals that promote the cell division cycle. Today we know that whether or not a cell divides is controlled not just by positive signals like growth factors. The proper functioning of a multicellular organism also depends on negative signals that counterbalance growth signals and halt the cell division cycle. The life of every cell is therefore managed by the delicate interplay of positive and negative signals, both of which directly affect multiple proteins inside the cell.

Because each multicellular organism is a cooperative community of cells, the failure of just one cell to maintain a delicate balance between opposing positive and negative signals can have serious consequences. One such consequence of uncontrolled cell division is cancer, since a cancerous tumor is a group of rapidly dividing cells that ignore the rules of cellular cooperation and behavior (Figure 11.1). In this chapter we discuss how

the rules that govern cell division are maintained inside each cell, and how the failure of these control systems results in cancer.

Positive Growth Regulators: Promoting Cell Division

Our understanding of how cells respond to signals that promote cell division is based largely on early observations of cancer in animals. Perhaps one of the best ways to study the effects of a particular control system is to discover what happens when it is no longer working. This principle was applied in the first decade of the twentieth century by the biologist Peyton Rous, who studied cancerous tumors called sarcomas in chickens.

Rous discovered that he could grind up tumors and extract an unidentified substance that, when injected into healthy chickens, caused cancer. He knew that the extract contained no bacteria because it was carefully filtered, so the cause of the cancer had to be something much smaller, which could thus pass through the filter. This finding led to the discovery of the first animal tumor virus, which was named the **Rous sarcoma virus** in honor of its discoverer and the type of tumor from which it was obtained (Figure 11.2).

Some viruses can cause cancer

Viruses are tiny assemblages of protein that surround either RNA or DNA. The nucleic acids found inside viruses contain genes that are necessary for the viral life cycle. However, viruses are more than 100 times small-

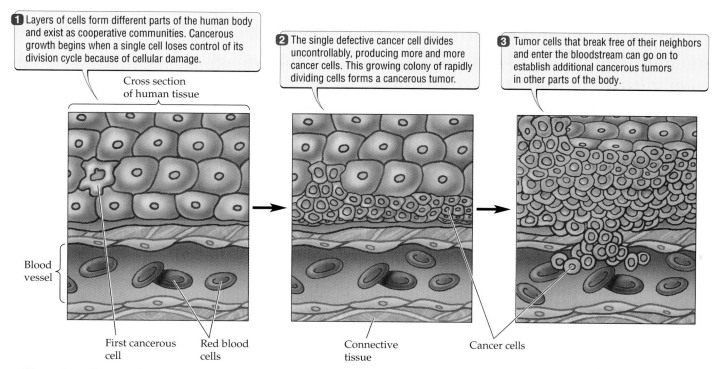

1 Layers of cells form different parts of the human body and exist as cooperative communities. Cancerous growth begins when a single cell loses control of its division cycle because of cellular damage.

2 The single defective cancer cell divides uncontrollably, producing more and more cancer cells. This growing colony of rapidly dividing cells forms a cancerous tumor.

3 Tumor cells that break free of their neighbors and enter the bloodstream can go on to establish additional cancerous tumors in other parts of the body.

Cross section of human tissue

Blood vessel

First cancerous cell

Red blood cells

Connective tissue

Cancer cells

Figure 11.1 Cancers Start with a Single Cell That Loses Control

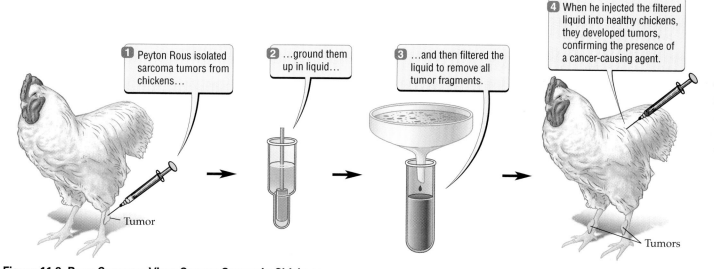

1 Peyton Rous isolated sarcoma tumors from chickens...

2 ...ground them up in liquid...

3 ...and then filtered the liquid to remove all tumor fragments.

4 When he injected the filtered liquid into healthy chickens, they developed tumors, confirming the presence of a cancer-causing agent.

Tumor

Tumors

Figure 11.2 Rous Sarcoma Virus Causes Cancer in Chickens

er than the average animal cell and are parasites that can multiply only by infecting cells and using the biochemical machinery of their unfortunate hosts. The discovery that a virus could cause cancer in animals was a major breakthrough in our understanding of this type of disease, but it took many more decades before we discovered how the Rous sarcoma virus derails the normal controls that regulate cell division.

The solution to this mystery came with the discovery of a particular Rous sarcoma virus that could multiply

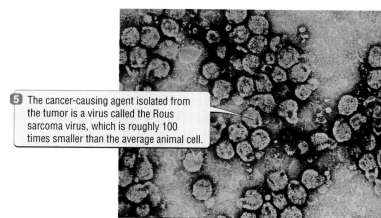

5 The cancer-causing agent isolated from the tumor is a virus called the Rous sarcoma virus, which is roughly 100 times smaller than the average animal cell.

in cells without causing them to divide rapidly like a cancerous tumor. Biologists could then compare the viral genes found in this virus with those found in the cancer-causing variety. The virus that did not cause cancerous cell division was missing a single gene. Further research showed that this viral gene was responsible for destroying the internal controls of the cell.

If you had to make an educated guess at the identity of the cancer-causing protein produced by the viral gene, what would you expect? As discussed in Chapter 9, growth factors activate many signaling proteins inside the cell, which in turn promote cell division. Perhaps the cancer-causing protein somehow affects these signaling events. This is in fact the case, since the cancer-causing agent produced by the viral gene is a very active protein kinase.

Protein kinases activate target proteins by adding phosphate groups to them (see Chapter 9). Under normal conditions, these activation events are counter-balanced by other enzymes that remove phosphates, effectively turning the target proteins off. When an over-active viral protein kinase happens to act on the same target proteins, there is no way to turn the signal cascade off. It is a clear case of too much of a good thing. The avalanche of enzymatic reactions that drive the cell toward mitosis spin out of control, leading to a cell that just keeps dividing. Since Rous sarcoma virus inserts all of its genes into the DNA of the infected cell, all the daughter cells also receive the cancer-causing gene. Eventually, the growing colony of cells that cannot stop dividing forms a tumor.

Oncogenes play an important role in cancer development

The overactive protein kinase gene that is found in Rous sarcoma virus is named *Src* in honor of the virus. (By convention, the names of genes are always given in italic type, and the names of proteins are given in regular type.) *Src* is just one of several so-called **oncogenes** (*onco*, "bulky mass"), or cancer-causing genes, found in other viruses. What came as a real surprise is the discovery that the *Src* oncogene is not unique to the Rous sarcoma virus. Instead, it is an altered version of a normal gene found in the genetic material of host cells.

Viruses like Rous sarcoma virus multiply by becoming part of the infected cell's DNA. At some point in history an altered protein kinase gene probably was acquired from an abnormal host cell and plugged into the genetic material of the virus. The alteration in question is a change in the DNA sequence of the gene known as a **mutation**. Mutations can alter the activity of the protein produced by the gene, either increasing or decreasing its activity or ability to function. In the case

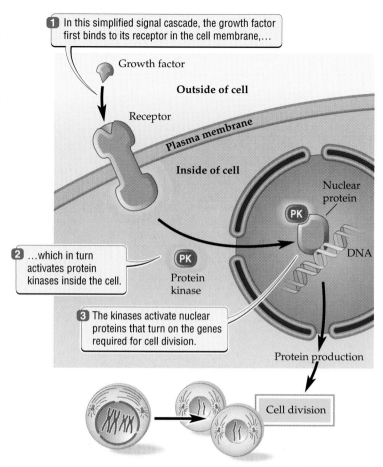

Figure 11.3 Proto-oncogene Proteins Are Found throughout the Cell
Most proteins in a signaling pathway such as the one shown here can cause cancer if their genes (proto-onco-genes) are mutated to become oncogenes. The conversion of a proto-oncogene into an oncogene often means that the protein it produces will be overactive, causing the cell to divide uncontrollably.

of *Src*, the mutation results in the production of an over-active enzyme that cannot be controlled like its normal counterpart in the cell.

The realization that the *Src* oncogene has a normal, controllable counterpart in the host cell was an important step in defining cellular genes that regulate growth. These normal cellular genes are called **proto-oncogenes** (*proto*, "first") because they are the predecessors of the viral oncogenes. Today scores of proto-oncogenes are known, most of which were first identified as oncogenic mutants in tumor viruses. Although most tumor viruses cause cancer only in animals like chickens, mice, and cats, all the proto-oncogenes identified in these animal host cells are also found in human cells.

We now know that oncogenes play a major role in human cancers. Instead of being brought into the cell by an infecting virus, human proto-oncogenes can be changed or mutated into cancer-causing oncogenes by environmental factors such as chemical pollutants. Well-studied examples of cancer-causing environmental pollutants include the aromatic amines present in cigarette smoke. Aromatic amines are organic compounds that can enter cells and bind to DNA, causing DNA damage or mistakes during DNA synthesis. Such chemical changes result in mutated DNA sequences. If such a mutation occurs in a proto-onco-gene, it may be converted into an oncogene. Such changes are just one way in which environmental pollutants can contribute to cancer.

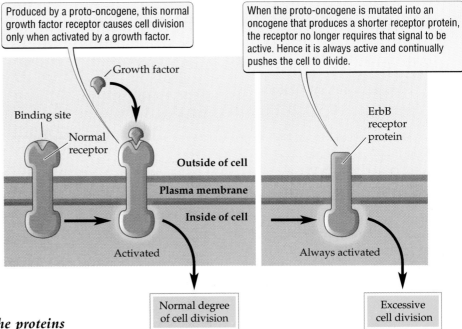

Produced by a proto-oncogene, this normal growth factor receptor causes cell division only when activated by a growth factor.

When the proto-oncogene is mutated into an oncogene that produces a shorter receptor protein, the receptor no longer requires that signal to be active. Hence it is always active and continually pushes the cell to divide.

Figure 11.4 The *ErbB* Oncogene Causes Cancer by Producing an Altered Receptor Protein
The *ErbB* oncogene produces a receptor protein that no longer has a complete extracellular portion and thus is always activated.

Proto-oncogenes produce many of the proteins in signal cascades

Not all proto-oncogenes produce protein kinases as *Src* does. Not surprisingly, nearly every gene that produces a protein used in a cell division signal cascade can be a proto-oncogene (Figure 11.3). In other words, mutant versions of genes for cell surface receptors, various kinases, and even signaling proteins like growth factors can function as oncogenes (Table 11.1). Since the outcome of increased cell division depends on a complex and interconnected cascade of protein activities, any player in this scheme that misbehaves and overactivates its target can cause uncontrolled cell division.

Let's consider an example of a cell surface receptor oncogene. Epidermal growth factor (EGF) promotes cell division by binding to a receptor embedded in the plasma membrane of the target cell. As described in Chapter 9, the binding of a growth factor to the external portion of the receptor activates its enzymatic activity on the inside of the cell. This enzymatic activity initiates the signal cascade of reactions that lead ultimately to cell division.

In one type of blood cell cancer found in chickens, a tumor virus produces a mutant form of the EGF receptor, called **ErbB**, that is missing the external portion needed to bind the growth factor. As a result, the receptor protein no longer needs the binding of EGF for activation and is therefore always turned on (Figure 11.4). Just like an overactive Src kinase, the overactive EGF receptor turns on too many target proteins over too long a time period, leading to uncontrolled cell division and cancer.

11.1	**Locations and Functions of Several Proto-oncogene Proteins**	
Proto-oncogene product (protein)	**Location**	**Type of protein**
Sis	Outside of cell	Growth factor
ErbB	Cell surface	Growth factor receptor
Neu	Cell surface	Growth factor receptor
ErbA	Cytosol	Steroid receptor
Src	Cytosol	Protein kinase
Abl	Cytosol	Protein kinase
Ras	Cytosol	G protein
Myc	Nucleus	Gene-activating factor
Fos	Nucleus	Gene-activating factor
Jun	Nucleus	Gene-activating factor

■ Viruses can multiply only by infecting cells and using the cells' biochemical machinery. The first onco-gene (*Src*) found to promote cell division was isolated from the Rous sarcoma virus. Oncogenes (overactive mutant versions of proto-oncogenes) promote excessive cell division, which can lead to cancer. Proto-oncogenes produce many of the protein components of growth factor signal cascades.

Negative Growth Regulators: Inhibiting Cell Division

The previous section might lead one to think that oncogenes are the villains responsible for the rampant cell growth that leads to cancer. Although this may be true in some cases, such as with *Src* in chickens, usually something else must also go wrong before cancer can occur. That is, normal cells have internal safeguards that must be overcome before the controls on cell division are totally removed. These safeguards consist of a family of proteins called **tumor suppressors** because their normal activities stop tumor growth. Tumor suppressor genes are therefore negative growth regulators that stop cells from dividing by opposing the action of proto-oncogenes.

Whether or not a normal cell divides depends on the activities of both proto-oncogenes and tumor suppressor genes. For a cell to divide, proto-oncogenes must be activated to promote the process, and tumor suppressor genes must be inactivated to allow the process to happen. Since controlling the timing and extent of cell division is so important, cells have these two counterbalancing control systems, which must be in agreement before the cells can divide.

How do tumor suppressors oppose the activity of proto-oncogenes under normal circumstances? The answer once again lies in the cascade of enzymatic events that lead to cell division.

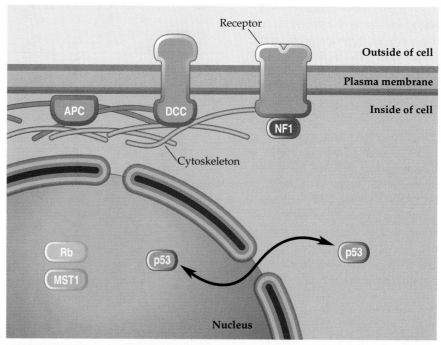

Tumor suppressor product (protein)	Location	Function
Rb	Nucleus	Blocks gene activation
MST1	Nucleus	Inhibits selected kinases
p53	Nucleus and cytoplasm	Turns on genes
NF1	Cytoplasm	Turns off G proteins
APC	Cytoplasm	Links cytoskeleton to membrane
DCC	Plasma Membrane	Assists cell adhesion

Figure 11.5 Locations and Activities of Several Tumor Suppressor Proteins

Tumor suppressors block specific steps of growth factor signal cascades

In the same way that oncogenes induce too much cell division by overactivating components of a growth factor signal cascade, tumor suppressors block cell division by inhibiting some of the same components (Figure 11.5). A well-known example of tumor suppressor activity was discovered during study of a rare childhood cancer called **retinoblastoma**. As the name indicates, retinoblastoma (*retino*, "net"; *blastoma*, "bud") is a cancer that consists of tumors in the retina of the eye that often lead to blindness (Figure 11.6).

Retinoblastoma strikes one in every 15,000 children born in the United States and accounts for about 4 percent of childhood cancers. A major breakthrough in our understanding of the disease came with the observation that some patients have visible abnormalities in chromosome 13. This important discovery was made by examination of the karyotype of cancer cells removed from retinoblastoma patients.

Every species has a very characteristic set of chromosomes, termed the karyotype, that can be viewed in the microscope (see Chapter 10). In the case of some retinoblastoma patients, a portion of chromosome 13 appeared to be missing, hinting that the cancer might be caused by the absence of a particular gene. Today we know that the gene in question normally produces a protein called **Rb**. Thus, a lack of Rb protein results in retinoblastoma.

What can we conclude from the observation that the absence of the *Rb* gene leads to cancer? For one thing, this is not what you would expect for an oncogene. An oncogene causes cancer by producing an excessive protein activity that pushes the cell to divide. *Rb* is the exact opposite, since its *absence* promotes cell division. A simple explanation is that the Rb protein normally inhibits a process required for cell division. When it is missing, the brakes on cell division no longer work and cells divide uncontrollably.

As predicted, the Rb protein inhibits a key process in the cell's preparations for division. It binds to and inhibits the activities of proteins that are required for the cell's response to growth factor signals. Thus, when cells are stimulated to divide by growth factors under normal conditions, the signal cascade involves not just the activation of proto-oncogenes, but also the inactivation of tumor suppressors like Rb. Given what is known about cell signaling, it should not be surprising that the Rb protein is inhibited by a protein kinase, which is activated by the growth factor signal cascade.

True to its enzymatic identity, the kinase phosphorylates the Rb protein, causing it to change shape and release its protein targets, which can then activate the genes needed for cell division (Figure 11.7). This example shows that the phosphorylation events resulting from growth factor signals act both to turn on some proteins and turn off others, thus confirming the balance of positive and negative regulatory controls that come into play before a cell can divide.

Unlike oncogenes, both copies of a tumor suppressor gene must be mutated in cancer

The differences in how oncogenes and tumor suppressor genes function highlight differences in the kinds of mutations that can lead to cancer. Since there are two copies of each type of chromosome (see Chapter 10), there are also two copies of each gene—one contributed by each parent. For an oncogene to promote cancer, only one copy of the matching proto-oncogene must be mutated to an oncogenic form. As long as one copy of the gene is mutated to produce an overactive protein, this excessive activity can push the cell to divide.

In contrast, for a tumor suppressor gene to promote cancer, both copies of the gene must be mutated to an inactive form. In other words, if only one copy is inactivated, there might still be enough tumor suppressor protein made by the remaining copy to stop cell division. Therefore, complete loss of this negative control, meaning that no tumor suppressor is being made, requires both copies to be inactive (Figure 11.8).

Roughly 40 percent of retinoblastoma cases are the result of inherited mutations in the *Rb* gene. The remaining noninherited cases appear to be caused by random mutation events linked to environmental factors such as

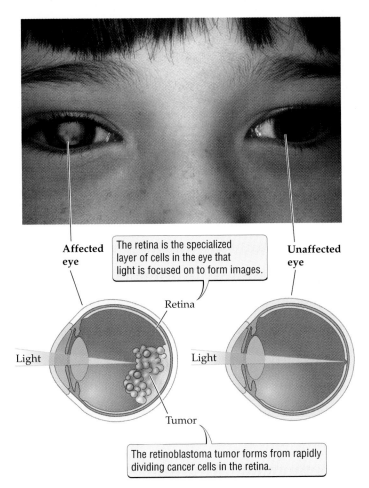

Affected eye

The retina is the specialized layer of cells in the eye that light is focused on to form images.

Unaffected eye

Retina

Light

Light

Tumor

The retinoblastoma tumor forms from rapidly dividing cancer cells in the retina.

Figure 11.6   Retinoblastoma
This child has a visible retinoblastoma in her right eye. The retinoblastoma both blocks the light and destroys the ability of retinal cells to respond to light, frequently resulting in blindness.

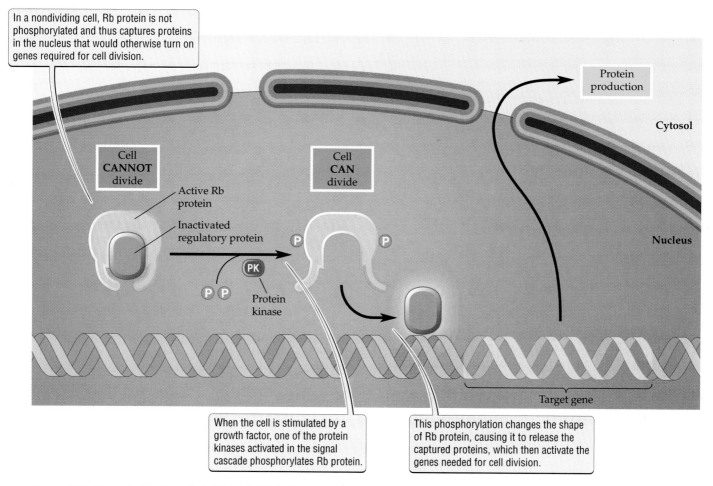

In a nondividing cell, Rb protein is not phosphorylated and thus captures proteins in the nucleus that would otherwise turn on genes required for cell division.

Protein production

Cytosol

Cell **CANNOT** divide

Active Rb protein

Inactivated regulatory protein

Nucleus

PK Protein kinase

Target gene

When the cell is stimulated by a growth factor, one of the protein kinases activated in the signal cascade phosphorylates Rb protein.

This phosphorylation changes the shape of Rb protein, causing it to release the captured proteins, which then activate the genes needed for cell division.

Figure 11.7 How the Rb Protein Inhibits Cell Division

exposure to chemical pollutants or radiation. A comparison of how the cancer progresses in inherited versus noninherited cases confirms the prediction that both copies of *Rb* must be inactive for cancer to occur.

Inherited cases of retinoblastoma are often more severe than noninherited cases and involve multiple tumors in both eyes. Since most patients with inherited retinoblastoma receive a single mutated copy of the *Rb* gene from one parent, every cell in both retinas has only one working copy of the gene from birth. Therefore, when environmental factors and other random events cause the accidental mutation of the remaining *Rb* gene in any cell of the retina, *Rb* activity in that cell is completely lost and the cell no longer has a block against inappropriate cell division. Because these events require only one random event in each cell, there is a greater chance of such mutations occurring multiple times in both eyes.

Noninherited cases of retinoblastoma tend to be milder and usually involve a single tumor in one eye. Since chil-

Figure 11.8 The Control of Cell Division by Proto-oncogenes and Tumor Suppressors   ▶

Whether or not a cell divides depends on the balance between proto-oncogene and tumor suppressor activity. (*a*) A normal cell that is not dividing can be compared to a cart attached to two ponies pulling in opposite directions. Because the ponies are of equal size and strength, the cart remains stationary. Likewise, the activities of proto-oncogenes, which promote cell division, are counterbalanced by the activities of tumor suppressors, which inhibit cell division: No cell proliferation occurs. (*b*) The mutation of one copy of a proto-oncogene to an overactive oncogene is like substituting a workhorse for one of the ponies. The workhorse is larger and stronger than the pony; hence it pulls the cart to one side. In a cell, the result is inappropriate cell division. (*c*) When both copies of a tumor suppressor gene are inactivated, the result is similar to completely eliminating one pony from the cart: The remaining pony can pull the cart to one side. Again, the result in a cell is inappropriate cell division.

dren in these cases do not inherit a defective copy of the *Rb* gene from their parents, the complete loss of *Rb* function depends on two random events mutating both copies of the gene in a single cell. The likelihood of these two random events occurring in one cell is quite low, explaining the lower frequency of tumors in children with the non-inherited form of the disease. These children are also not as prone to other types of cancer as those who are unfortunate enough to have inherited a defective *Rb* gene.

> ■ Tumor suppressor genes stop cells from dividing by opposing the action of proto-oncogenes. When cells divide, proto-oncogenes must be activated to promote the process, and tumor suppressors must be turned off to allow the process to happen. Whereas overactivation of just one copy of a proto-oncogene can cause inappropriate cell division, both copies of a tumor suppressor gene must be inactivated to have the same effect. Mutations that inactivate the *Rb* tumor suppressor gene cause the eye cancer retinoblastoma.

A Series of Chance Events Can Cause Cancer

Most human cancers involve more than the overactivation of one oncogene or the complete inactivation of one tumor suppressor gene. Indeed, the complex series of events that leads to cancer means that both genetic and environmental factors can come into play. Only about 1 to 5 percent of all cancer cases can be traced exclusively to an inherited genetic defect. The remaining majority of cases involve either a combination of genetic and environmental factors or environment alone. In some ways this is good news, since we have the power to try to limit our exposure to environmental factors that cause cancer.

We must try both to reduce the likelihood of dangerous mutations accumulating in the DNA of our cells and to understand the genetic characteristics that may lead to cancer later in life. In this section we will see just how important it is to lower the likelihood of mutations occurring in different genes.

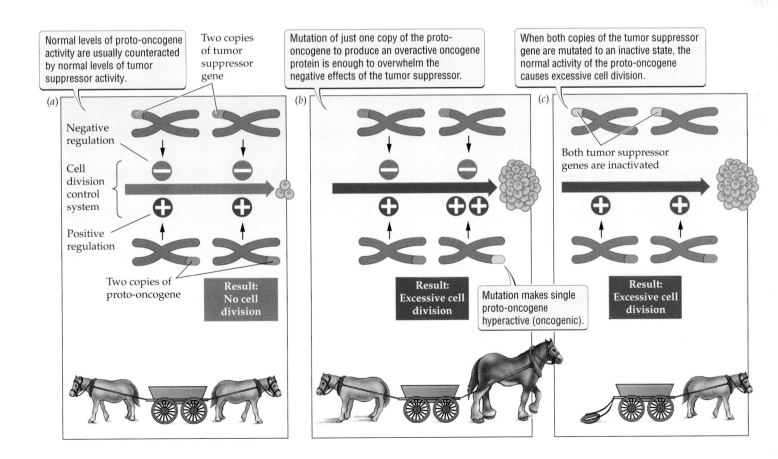

Cancer is a multistep process

Cancer is a group of diseases (Table 11.2) that is likely to affect all our lives eventually. During his lifetime, an American male has a 50 percent chance of developing cancer or dying from it. American women fare slightly better, with a 33 percent chance of developing cancer or dying from it.

In the United States, one in four deaths are due to cancer, and more than 8 million Americans alive today have been diagnosed with cancer and are either cured or are undergoing treatment. Given such a high incidence of cancer, you might think that only one or two mutations are sufficient to cause the disease. However, careful study of human cancers shows that not a few, but several, cellular safeguards would have to fail before a malignant tumor can form. The unlucky string of failures that produces a malignant tumor includes both the over-activation of oncogenes and the loss of tumor suppressor activity.

Consider cancer of the colon (large intestine), which is diagnosed in more than 94,000 individuals each year in the United States. In many cases of colon cancer, tumor cells contain at least one overactive oncogene and several completely inactive tumor suppressor genes. In fact, since the mutations in different genes that lead to colon cancer occur over a period of years, the gradual accumulation of these mutations can be linked with the stepwise progression of the cancer.

Let's look at the step-by-step sequence of chance mutations that might lead to malignant colon cancer. In most cases the first sign of colon cancer is a relatively harmless or **benign** growth described as a polyp (Figure 11.9). The cells that make up the polyp are undergoing division at an inappropriate rate. These cells are the descendants of a single abnormal cell in the lining of the colon.

In many large polyps the cells contain mutations that inactivate both copies of a tumor suppressor gene called *APC*, and a single mutation that creates an overactive oncogene named *Ras*. The complete loss of one tumor suppressor's activity combined with the overactivation of one oncogene is therefore enough to allow inappropriate cell division. However, most such polyps are not harmful, in that they have not spread to other tissues and can be safely removed surgically.

The progression from a benign polyp to a **malignant** tumor that spreads throughout the body with life-threatening consequences depends on the inactivation of additional tumor suppressor genes. In many colon tumors, the start of true malignancy coincides with the loss of a part of chromosome 18 that contains at least two important tumor suppressor genes. This complete loss of two additional tumor suppressors results in a far more aggressive and rapid multiplication of the cancer cells, greatly increasing the chance that they will spread to other tissues.

One of the last key events in the unlucky slide to full malignancy is the complete inactivation of yet another tumor suppressor gene, named *p53*. For reasons that are not entirely clear, the loss of p53 function seems to remove all controls on cell division, allowing the cancer cells to break free of the original tumor and travel through the bloodstream to other parts of the body. At this point the cancer cells are entirely resistant to both regulatory and defensive signals from the body, and the worst possible scenario—a malignant tumor—has come true (see Figure 11.9).

11.2 Selected Human Cancers in the United States		
Type of cancer	**New cases in 1999**[a]	**Deaths in 1999**[a]
Breast cancer The second leading cause of cancer deaths in women	175,000	43,300
Colon cancer The number of new cases is declining as a result of early detection and polyp removal	94,700	47,900
Leukemia Often thought of as a childhood disease, it affects more than 10 times as many adults each year	30,200	22,100
Lung cancer Accounts for 28 percent of all cancer deaths and kills more women than breast cancer does	171,600	158,900
Ovarian cancer Accounts for 4 percent of all cancers in women	25,200	14,500
Prostate cancer The second leading cause of cancer deaths in men	179,300	37,000
Skin cancer Melanoma accounts for 80 percent of skin cancer deaths and has been on the rise since 1973	1,000,000	9,200

[a] Estimated numbers courtesy of the National Center for Health Statistics (NCHS) at the Centers for Disease Control and Prevention, with additional information from the American Cancer Society.

■ Most cases of cancer involve either a combination of genetic and environmental factors or environment alone. Colon cancer requires mutations that alter the activities of several different genes, resulting in at least one overactive oncogene and several completely inactive tumor suppressor genes.

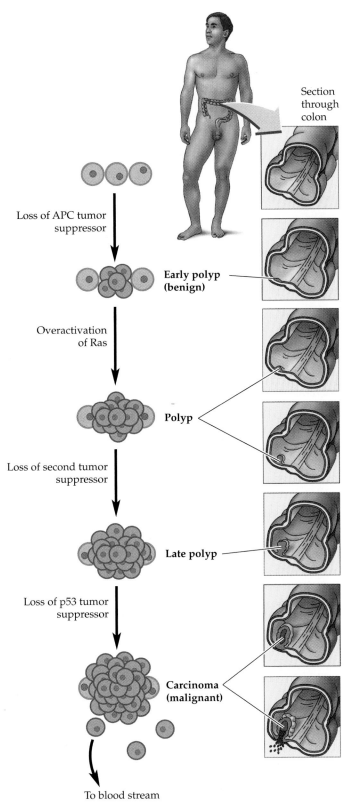

Loss of APC tumor suppressor

Early polyp (benign)

Overactivation of Ras

Polyp

Loss of second tumor suppressor

Late polyp

Loss of p53 tumor suppressor

Carcinoma (malignant)

Section through colon

To blood stream

Figure 11.9 Colon Cancer Is a Multistep Process
The sequential mutation of several growth regulatory genes coincides with the progression from a benign polyp to a malignant cancer.

Highlight

Making the Most of Losing p53

Although the connection between loss of p53 activity and fully malignant colon cancer emphasizes the importance of this tumor suppressor, it only hints at the broad range of different p53 activities. The tumor suppressor protein p53 is perhaps the most famous because of its multiple roles in guarding the integrity of the cell. It not only prevents the cell from dividing at inappropriate times; it also halts cell division when there is evidence of DNA damage that could result in harmful mutations. This protection gives cells the opportunity to repair the damage. If the repair process fails, p53 then goes so far as to induce a cascade of enzymatic reactions that kills the cell. In other words, if the cell's DNA is too badly damaged to repair, the cell commits suicide rather than passing on mutations that could potentially harm the entire organism.

Given the important guardian functions fulfilled by p53, it is not surprising that more than half of all cancers include a complete loss of p53 activity in tumor cells. The number goes as high as 80 percent in some types of cancer, such as colon cancer. Although the loss of cell division control in each case of cancer involves mutations in a variable roster of several genes, the loss of p53 activity is a factor in most cancers. Thus the absence of p53 activity might be used as a recognizable characteristic of cancer cells.

Returning to the cancer therapy introduced at the start of this chapter, the first hint that the adenovirus could be used against cancer cells came with the discovery that the virus must inactivate p53 in order to multiply. The same mechanism that enables p53 to halt cell division also stops the adenovirus DNA in an infected cell from being used to make viral proteins. To avoid this defensive measure, an adenovirus gene produces a protein that binds to and disables the p53 protein, thereby allowing the virus to use the cell to make components for new viruses. In other words, the adenovirus can function effectively only in cells that have no p53 activity. Since p53 also happens to be the tumor suppressor gene that is most often defective in cancer cells, an important connection can be forged between the virus and cancer cells that lack p53 activity.

In a clever turn of events, biologists have mutated adenovirus such that it no longer produces the protein that disables p53. This mutant virus can multiply only in cells that already lack p53 activity, resulting in the selective killing of these cells. Since the only cells in the body that are likely to lack p53 activity are cancer cells, this virus works like a smart bomb that seeks out its target and destroys it. Early tests in humans have shown that injection of the mutant virus can reduce the size of some tumors, but only time will tell if this inventive application of our understanding of cancer will help those with the disease.

> ■ Adenoviruses successfully infect cells by inactivating the p53 tumor suppressor protein. Mutant adenoviruses that cannot inactivate p53 can only infect cells that already lack p53. These mutant adenoviruses can therefore be used to selectively infect and destroy malignant cancer cells, which usually lack p53 activity.

Summary

Positive Growth Regulators: Promoting Cell Division

■ Viruses can multiply only by infecting cells and using the cells' biochemical machinery.

■ The first gene found to promote cell division was isolated from the Rous sarcoma virus. This gene, the *Src* oncogene, causes cancer in chickens.

■ Oncogenes (overactive mutant versions of proto-oncogenes) promote excessive cell division, which leads to cancer.

■ Proto-oncogenes produce many of the protein components of growth factor signal cascades.

Negative Growth Regulators: Inhibiting Cell Division

■ Tumor suppressor genes stop cells from dividing by opposing the action of proto-oncogenes.

■ For a cell to divide, proto-oncogenes must be activated to promote the process, and tumor suppressor genes must be inactivated to allow the process to happen.

■ The Rb tumor suppressor protein inhibits proteins required for growth factor signal cascades. A complete lack of Rb activity can cause the eye cancer retinoblastoma.

■ Overactivation of one copy of an oncogene is sufficient to cause inappropriate cell division. In contrast, both copies of a tumor suppressor gene must be inactivated to cause inappropriate cell division.

■ Inherited cases of retinoblastoma are more severe than noninherited cases because there is a greater chance in inherited cases that both copies of the *Rb* gene will be inactivated by mutation.

A Series of Chance Events Can Cause Cancer

■ Most cases of cancer involve either a combination of genetic and environmental factors or environment alone.

■ Cancer usually requires mutations that alter the activities of several different genes. In colon cancer, tumor cells contain at least one overactive oncogene and several completely inactive tumor suppressor genes.

■ The inactivation of both copies of the *p53* tumor suppressor gene is a key step in the transition from benign growth to malignant cancer.

Highlight: Making the Most of Losing p53

■ The tumor suppressor gene *p53* has several roles in guarding the integrity of the cell. It prevents the cell from dividing at inappropriate times, such as when DNA damage could result in harmful mutations.

■ The adenovirus must inactivate the p53 protein in order to multiply in cells. Biologists have mutated adenovirus such that it no longer disables p53, and it can only multiply in cells that already lack p53 activity. Since cancer cells frequently lack p53 activity, the mutant virus multiplies only in these cells, thus selectively killing cancer cells.

Key Terms

APC p. 184	proto-oncogene p. 178
benign p. 184	*Ras* p. 184
ErbB p. 179	Rb p. 181
malignant p. 184	retinoblastoma p. 180
mutation p. 178	Rous sarcoma virus p. 176
oncogene p. 178	tumor suppressor p. 180
p53 p. 184	virus p. 176

Chapter Review

Self-Quiz

1. Viruses are different from cells because
   a. they do not contain nucleic acids.
   b. they do not have genes.
   c. they cannot multiply on their own.
   d. they are larger than most cells.

2. The *Src* oncogene
   a. was first discovered in a virus that causes cancer in chickens.
   b. is found in all healthy cells.
   c. inhibits cell division.
   d. can be found only in cancer-causing viruses.

3. Proto-oncogenes
   a. are not related to oncogenes.
   b. can be mutated to become oncogenes.
   c. inhibit signal cascades.
   d. are found only in viruses.

4. Which of the following statements is correct?
   a. Tumor suppressors are able to halt cell division only in cancer cells.
   b. Tumor suppressors inhibit proteins that promote cell division.
   c. Only one type of tumor suppressor is known.
   d. Proto-oncogenes and tumor suppressors have similar effects on the cell.

5. Which of the following proteins is least likely to be produced by a proto-oncogene?
   a. actin cytoskeletal protein
   b. growth factor receptor
   c. protein kinase
   d. growth factor

6. A benign tumor is more likely to become malignant when
   a. the tumor cells no longer respond to growth factor signals.
   b. the rate of cell division slows down.
   c. it reaches a certain size.
   d. several tumor suppressor genes are inactivated by mutations.

Review Questions

1. Describe one way in which p53 prevents the passing on of harmful mutations from one cell to its offspring.

2. When Peyton Rous isolated the first cancer-causing virus, why was it so important for him to filter the extract made from the cancerous tumors?

3. On the basis of what you know about how malignant cancers develop, why is it so important to reduce your long-term exposure to chemicals that damage DNA?

4. How does the differing severity of inherited versus non-inherited retinoblastoma confirm the involvement of a tumor suppressor and not an oncogene in this cancer?

11 The Daily Globe

Cell Phones as a Health Risk?

MELBOURNE, AUSTRALIA. While millions of Americans work and socialize using increasingly popular cell phones, a new study on mice suggests that these devices, which many people already cannot live without, may in fact increase a person's chances of developing cancer.

Researchers in Australia reported today in the journal *Medical Research* that mice exposed to high-frequency microwaves like those produced by cell phones developed twice as many cancers as mice that were not exposed to the radiation. Furthermore, the mice that did receive radiation were exposed to the equivalent of only 1 hour on a cell phone per day.

ATIO, the country's largest cell phone producer, called the new study ridiculous, stating: "There have been and will be no scientific studies to indicate that cell phones are a hazard to human health."

Dr. Wade Garland, geneticist at Melbourne Provincial College in Australia, said the new study indicates that the high-frequency microwaves produced by cell phones might be causing cancer by damaging cell's DNA. Damaging DNA is how ultraviolet light and X-ray radiation cause cancer. Dr. Janice

Owen of The University of Boston, however, argued that cell phones probably cannot damage DNA, since the microwaves they produce lack sufficient energy. However, other researchers interviewed argued that these microwaves might make the cell's DNA vibrate, making that DNA more prone to damage by other factors; or microwaves might even alter the physical properties of plasma membranes, thereby changing the behavior of cells.

"There's still more work to do," said Dr. Garland. "But until we know more, I'm not going to use my cell phone again."

Evaluating "The News"

1. Why does DNA damage increase the risk of cancer?

2. Given the unknown risks of extended cell phone use, do you think telecommunications companies should encourage using a cell phone as your primary means of calling?

3. Even if cell phone use has only a small risk of causing cancer, would this risk make you less likely to make cell phone calls?

4. Some people argue that so many factors in the environment contribute to cancer, it doesn't matter if cell phone use is added to the list. Given what you know about the multistep process that leads to cancer, would you agree with this lack of concern?

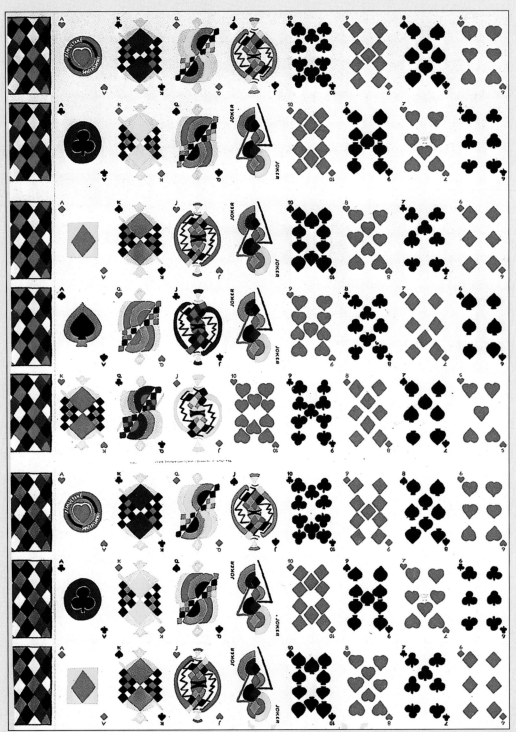

Sonia Delaunay, *Playing Cards (Jeu de Cartes)*, 1959.

UNIT 3
Genetics

c h a p t e r ## 12

Patterns of Inheritance

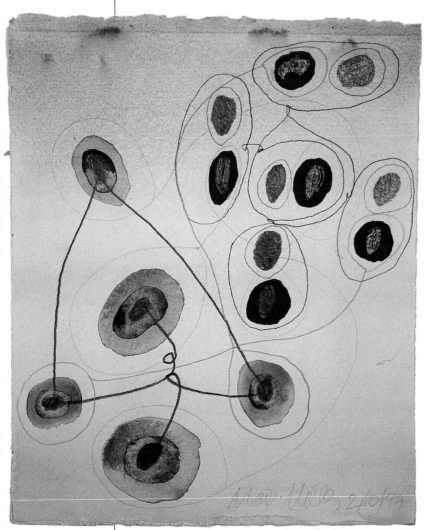

Max Miller, *Cause and Because*, 1997.

Who's the Father?

*I*magine you are the proud owner of a young female cat, a beautiful black tabby ("tabby" refers to a pattern of striping like the one in the photo). You got her from the animal shelter, and, as is usually the case in such adoptions, you don't know anything about your cat's breeding history. But you are so fond of her and you like her color and pattern of striping so much, that you decide to breed your cat with another black tabby, hoping to produce an entire litter of black tabby kittens.

You know someone with an attractive male black tabby, and his owner tells you that the cat is a "pure" black tabby. After carefully examining the certificates that back up the owners claim, you pay the owner a fee, the male and your cat mate, and you wait expectantly for the birth of the kittens.

Then comes the bad news: You wanted a female black tabby, but only two of the kittens are black tabbies and, as chance would have

190

it, they are both males. The rest of the kittens are tabbies, but they have a chocolate or chestnut brown color that does not appeal to you.

At first you blame the owner of the black male tabby. Perhaps, you fume to yourself, the male cat was not a pure black tabby after all. But the certificates seemed authentic, so you start to wonder what happened. Assuming the owner's claims were true, could a "pure" black tabby have produced chestnut brown kittens when mated with your cat? If not, did that mean that another, unknown cat actually was the father of some or all of your kittens?

These questions go to the heart of a major branch of genetics, the branch concerned with the transmission of traits from parents to offspring. What determines the fur color of kittens? How are characteristics like fur color inherited? Why do offspring—whether kittens or people or pea plants—often resemble their parents, but sometimes resemble their grandparents instead? To answer these and many other questions about inherited characteristics, we must learn about genetics, the scientific study of genes.

A Black Tabby Cat

Key Concepts

1. Genetics is the study of genes, which are the basic units of heredity. Alternative versions of genes are called alleles. Alleles cause variation in inherited characteristics.

2. Organisms contain two copies of each gene, one inherited from each parent. If the two copies of a gene are identical, the individual is homozygous for that gene. If the two copies of a gene are different, the individual is heterozygous for the gene.

3. An allele is dominant if it determines the observable characteristics of an organism when paired with a different (recessive) allele. An allele is recessive if it has no observable effect when paired with a dominant allele.

4. During meiosis, alleles separate equally into egg (in a female) or sperm (in a male) cells. With some exceptions (see Chapter 13), the separation of alleles for one gene is independent of the separation of alleles for other genes.

5. The genetic makeup of an organism is its genotype. The phenotype is the observable characteristics of an organism, such as its physical appearance, behavior, and pattern of development.

6. For most characteristics, the phenotype is determined by many genes that interact with each other and with the environment.

Genetics is an extremely useful field of biology. For example, the past few years have witnessed an impressive series of discoveries about the genetic basis of human disease. These discoveries have the potential to improve our understanding and treatment of conditions ranging from cancer to Alzheimer's Disease to diabetes, diseases that touch all our lives. In addition, genetic principles are used routinely in plant and animal breeding, which lead ultimately to the foods we eat. It is also now common to use genetic techniques in some criminal investigations and in paternity lawsuits. Genetic methods have even been used to solve a long-standing mystery concerning the princess Anastasia, who many thought had escaped the 1918 execution of the Russian royal family (she did not escape).

Throughout this unit, we will explore many practical applications of genetics. To do so we will need to understand how genetic traits are passed from one generation to the next. The transmission of genetic traits is the subject of this chapter, which focuses on basic principles of inheritance and important extensions of those basic principles. We begin our journey into the study of inheritance with an overview of the field of genetics.

Genetics: An Overview

Humans have used principles of inheritance for thousands of years. For example, knowing that offspring resemble their parents, humans have improved crop species by allowing only those organisms with desirable traits, like larger grains of wheat, to reproduce. Over time, we used this method to domesticate animals and to develop agricultural crops from wild plant species.

As a field of science, however, genetics did not begin until 1866, the year that Gregor Mendel (Figure 12.1) published his landmark paper on inheritance in pea plants. Prior to Mendel's work, many facts about heredity were known, but no one had organized those facts by describing and testing a hypothesis that could explain how traits are passed from parent to offspring.

Mendel changed all that. His experiments led him to propose that characteristics like fur color are controlled by hereditary factors—now known as genes—and that one factor for each characteristic is inherited from each parent. Although he did not use the word "gene," Mendel was the first to propose the concept of the gene as the basic unit of heredity. The emphasis that Mendel placed on genes continues today. In fact, we define **genetics** as the study of genes.

In the more than 100 years since Mendel, we have learned a great deal about genes. As described throughout this unit, most genes contain instructions for the synthesis of a single protein product. Genes are located on chromosomes. Structurally each gene is a length of DNA within the DNA of the chromosome. Finally, most of the trillions of cells in our body contain two copies of the same set of genes (Figure 12.2); the exception to this rule is our sperm and egg cells, which have only one of the usual two copies of each gene.

> ■ Genetics is the study of genes. Inherited characteristics are controlled by genes. One gene for each characteristic is inherited from each parent. Most genes contain instructions for a single protein product. Genes are made of DNA and are located on chromosomes.

Figure 12.1 Mendel and the Monastery Where He Performed His Experiments
Mendel was a monk at the monastery shown in this photograph. For many years people thought Mendel had performed his experiments behind the fence located in front of the monastery. Staff members at a museum devoted to Mendel recently discovered that Mendel's garden actually was located in the foreground of this photograph.

Essential Definitions in Genetics

By the end of this chapter, you should understand how genetic traits are inherited. You will be able to use that information to answer the hypothetical question posed in the introduction: Which male cat fathered the litter of kittens? You will also be able to predict the chance that a newborn baby will inherit certain traits from its parents, such as an inherited disease. To understand the material in this chapter, however, you must first become very familiar with key genetic terms, which we introduce here.

Organisms differ in many traits, or **characters**, which are features of an organism such as its height, flower color, or chemical structure of a protein. Many characters are determined at least in part by **genes**, which are separate units of genetic information for specific characters. The gene is the basic unit of heredity. With the exception of the gametes (sperm and egg cells), cells in

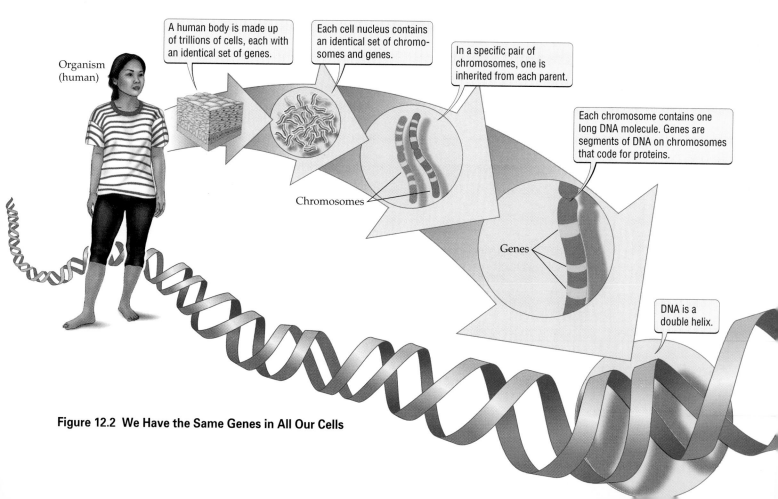

Figure 12.2 We Have the Same Genes in All Our Cells

plants and animals contain two copies of each gene, one inherited from each parent. Alternative versions of a gene are called **alleles**. The different alleles of a gene are often denoted by upper- and lowercase letters—for example, *A* and *a*. No individual has more than two alleles for a particular gene, but within a population of individuals of the same species a single gene can be represented by many more than two alleles.

An individual that carries two copies of the same allele, such as an *AA* or an *aa* individual, is called a **homozygote**. An individual that carries one copy of each of two different alleles, such as an *Aa* individual, is called a **heterozygote**. The genetic makeup of an organism is called its **genotype**; for example, a heterozygote has genotype *Aa*. The **phenotype** of an organism is its observable physical characteristics. Aspects of an organism's phenotype include its appearance (for example, flower color), behavior (for example, courtship displays in birds), and biochemistry (for example, the amount of the protein product of a gene in the body). Two individuals with the same phenotype can have different genotypes (Figure 12.3).

An allele that determines the phenotype of an organism when it is paired with a different allele is referred to as a **dominant** allele. Dominant alleles are often denoted by uppercase letters, such as *A*. An allele that does not express its phenotypic effect when paired with a dominant allele is said to be **recessive**. Recessive alleles are often denoted by lowercase letters, such as *a*.

A **genetic cross**, or cross for short, is a controlled mating experiment performed to examine the inheritance of a particular character. "Cross" can also be used as a verb, as in "individuals of genotype *AA* were crossed with

individuals of genotype *aa*." The parent generation of a genetic cross is called the **P generation**. The first generation of offspring in a genetic cross is called the **F₁ generation** ("F" is for "filial," a word that refers to a son or daughter). The second generation of a cross is called the **F₂ generation**.

Definitions of these important genetic terms are collected in Table 12.1. Study these terms carefully, and refer to them as needed throughout the chapter.

> ■ The following are some key genetic terms: character, gene, allele, homozygote, heterozygote, genotype, phenotype, dominant, recessive, genetic cross, P generation, F_1 generation, F_2 generation.

Gene Mutations: The Source of New Alleles

In the previous section we learned that alleles are different versions of a gene. Different alleles of a gene are responsible for hereditary differences among organisms. Although a single individual has at most two different alleles for any given gene, when the genotypes of many individuals are examined, a particular gene may be found to have many different alleles. For example, in human populations many proteins have three or more forms. The different forms of a protein are produced by different alleles of the gene that makes the protein.

New alleles arise by **mutation**, which we define here as a change to the DNA that makes up the gene (see Chapter 15 for a more complete definition). Many mutations are harmful, and many have little impact on the organism. A few are beneficial. Even when they are beneficial, mutations occur at random with respect to their usefulness. There is no evidence that specific mutations occur because they are needed.

Mutations can happen at any time and in any cell of the body. In multicellular organisms, however, only the mutations that occur in cells that lead to the formation of gametes or in the gametes themselves can be passed on to offspring.

> ■ In a population of many individuals, a particular gene may have one to many alleles. Different alleles of a gene cause hereditary differences among organisms. New alleles arise by mutation, a change to the DNA that makes up the gene. Many mutations are harmful, many are of little impact, and a few are beneficial.

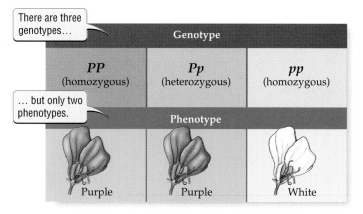

Figure 12.3 Genotype and Phenotype
A character like flower color in peas is controlled by a gene with two alleles (*P* and *p*), resulting in three genotypes (*PP*, *Pp*, and *pp*) and two phenotypes (purple and white flowers).

12.1   *Key Genetic Terms*

Term	Definition
Character	Feature of an organism, such as height, flower color, or the chemical structure of a protein.
Gene	Individual unit of genetic information for a specific character. Genes are located on chromosomes and are the basic functional unit of heredity.
Allele	Alternative versions of a gene. Different alleles are often denoted by upper- and lowercase italic letters; for example, two alleles for a gene might be denoted *A* and *a*.
Homozygote	An individual that carries two copies of the same allele (for example, an *AA* or an *aa* individual).
Heterozygote	An individual that carries one copy of each of two different alleles (for example, an *Aa* individual).
Genotype	Genetic makeup of the organism.
Phenotype	Physical characteristics of an organism.
Dominant allele	Allele that determines the phenotype of an organism when paired with a different (recessive) allele.
Recessive allele	Allele that does not have a phenotypic effect when paired with a dominant allele.
Genetic cross	Controlled mating experiment, usually performed to examine the inheritance of a particular character.
P generation	Parent generation of a genetic cross.
F_1 generation	First generation of offspring in a genetic cross.
F_2 generation	Second generation of offspring in a genetic cross.

Basic Patterns of Inheritance

Now that we've defined the key genetic terms and discussed how mutations produce new alleles, we are ready to explore how genes are transmitted from parents to offspring. Prior to Mendel, many people argued that the traits of both parents were blended in the offspring, much as paint colors blend when mixed together. According to this theory, which was known as blending inheritance, offspring are intermediate in form to the two parents, and it should not be possible to recover traits from previous generations. Thus, if a white-flowered plant were mated with a red-flowered plant, the offspring should have pink flowers and later generations should not return to the original flower colors of white and red.

In contrast to these predictions, the features of offspring often are not intermediate to those of their parents, and it is common for traits to skip a generation (for example, for a child to have blue eyes like its grandparent, but unlike its brown-eyed parents). How can such observations be explained? Gregor Mendel answered this question with a series of experiments on pea plants.

Mendel's experiments

During 8 years of investigation, Mendel conducted experiments on inheritance in peas. His results led him to reject the blending theory of inheritance. Mendel proposed instead that for each characteristic, offspring inherit two separate units of genetic information (genes), one from each parent.

Peas are an excellent organism for studying inheritance. Ordinarily, peas self-fertilize; that is, a given plant contains both male and female reproductive parts, and it fertilizes itself. But because peas also can be mated (crossed) experimentally, Mendel was able to perform carefully controlled experiments. In addition, peas have varieties that breed true for easy-to-measure characters, such as the color and shape of seeds. When a plant of a **true-breeding variety** is self-fertilized, all of its offspring are of the same variety as the parent. For example, when self-fertilized, a variety that breeds true for yellow seeds produces only peas with yellow seeds. Mendel used only true-breeding varieties.

In his experiments, Mendel observed hereditary characters for three generations. For example, he crossed plants that bred true for purple flowers with plants that bred true for white flowers (Figure 12.4). Mendel then allowed the F_1 plants (the first generation of offspring) to self-fertilize, thereby producing the F_2 generation.

Inherited characteristics are determined by genes

According to the theory of blending inheritance, the cross shown in Figure 12.4 should yield F_1 generation plants bearing flowers of intermediate color. Instead, all F_1 plants had purple flowers. Furthermore, when the F_1

Figure 12.4 Three Generations in One of Mendel's Experiments

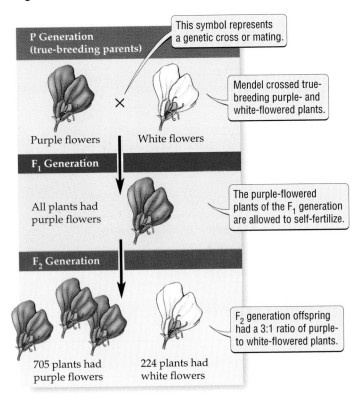

P Generation
(true-breeding parents)

This symbol represents a genetic cross or mating.

×

Purple flowers   White flowers

Mendel crossed true-breeding purple- and white-flowered plants.

F₁ Generation

All plants had purple flowers

The purple-flowered plants of the F₁ generation are allowed to self-fertilize.

F₂ Generation

705 plants had purple flowers   224 plants had white flowers

F₂ generation offspring had a 3:1 ratio of purple- to white-flowered plants.

ers). Mendel was right: Most cells in the adult organism contain one maternal and one paternal copy of each of their many genes (reexamine Figure 12.2). The only exceptions to this rule are the gametes—the egg or sperm cells produced by meiosis. Gametes contain only one copy of each gene. Two copies of each gene are restored when two gametes, an egg and a sperm, fuse to form a zygote, the first cell of a new offspring.

3. *An allele is dominant if it determines the phenotype of an organism when paired with a different (recessive) allele. An allele is recessive if it has no phenotypic effect when paired with a dominant allele.* For example, let's call the allele for purple flower color P and the allele for white flower color p. Plants that breed true for purple flowers have two copies of the P allele (otherwise they would produce white flowers occasionally). Similarly, plants that breed true for white flowers have two copies of the p allele. Thus, the F₁ plants in Figure 12.4 received a P allele from the parent with purple flowers, and they received a p allele from the parent with white flowers. Since all F₁ plants had purple flowers, the P allele is dominant and the p allele is recessive.

4. *The two copies of a gene separate during meiosis and end up in different gametes.* As already mentioned, each gamete receives only one copy of each gene. If an organism has two identical alleles for a particular character, like Mendel's homozygous true-breeding varieties, all the gametes will contain that allele. However, if the organism has two different alleles, like the heterozygous F₁ plants in Mendel's crosses, then 50% of the gametes will receive one of the alleles and 50% of the gametes will receive the other allele.

5. *Gametes fuse without regard to which allele they carry.* When gametes fuse to form a zygote, they do so randomly with respect to the alleles they carry for a particular gene. As we'll see, this element of randomness allows us to use a simple method to determine the chance the offspring will have a particular genotype.

plants self-fertilized, about 25% of the F₂ offspring had white flowers. Thus, the white-flowered trait skipped a generation, something that should not happen under blending inheritance.

Mendel studied seven characters in peas. His results for each of the different characters were similar to those shown in Figure 12.4, leading him to propose a new, particulate theory of inheritance. Mendel's theory is referred to as a "particulate" theory because in it genes behave like separate units or particles, not like colors of paints that blend together. Using modern terminology, Mendel's theory can be summarized as follows:

1. *Alternative versions of genes cause variation in inherited characters.* For example, pea plants have one version of a gene that causes flowers to be purple, and another version of the same gene that causes flowers to be white. The alternative versions of a gene are known as alleles.

2. *Offspring inherit one copy of a gene from each parent.* In his analysis of crosses like that of Figure 12.4, Mendel reasoned that for the white-flower trait to reappear in the F₂ generation, the F₁ plants must have had two copies of the flower color gene (one copy that caused white flowers and one copy that caused purple flow-

■ Mendel's experiments on pea plants led him to propose a new, particulate theory of inheritance, which states that (1) alternate versions of genes (alleles) cause variation in inherited characters, (2) offspring inherit one copy of a gene from each parent, (3) alleles can be dominant or recessive, (4) the two copies of a gene separate into gametes, and (5) gametes fuse without regard to which allele they carry.

Mendel's Laws: Equal Segregation and Independent Assortment

Mendel summarized the results of his experiments in two laws: the law of equal segregation and the law of independent assortment. Let's take a look at each of Mendel's laws and how he developed them.

Mendel's first law: Equal segregation

The **law of equal segregation** states that the two copies of a gene separate during meiosis and end up in different gametes. This law can be used to predict how a single character is inherited. As illustration, we will revisit the experiment shown in Figure 12.4. In that experiment, Mendel crossed individuals that bred true for purple flowers (genotype PP) with individuals that bred true for white flowers (genotype pp). This cross produced an F_1 generation composed entirely of heterozygotes (individuals with genotype Pp). According to the law of segregation, when the F_1 plants reproduced, 50% of the pollen (sperm) should have contained the P allele, and the other 50% the p allele. The same is true for the eggs.

We can represent the equal separation of alleles by the Punnett square method, first used in 1905 by the British geneticist Reginald Punnett (Figure 12.5). In a **Punnett square**, all possible male gametes are listed on one side of the square, and all possible female gametes are listed on an adjacent side of the square. Regardless of whether it has a P or a p allele, each sperm has an equal chance of fusing with an egg that has a P allele or an egg that has a p allele. Thus, the four combinations of alleles shown within the Punnett square are all equally likely.

Reginald Punnett

Using the Punnett square method, we predict that on average ¼ of the F_2 generation should have genotype PP, ½ should have genotype Pp, and ¼ should have genotype pp. Because the allele for purple flowers (P) is dominant, plants with PP or Pp genotypes have purple flowers, while pp genotypes have white flowers. Thus, we predict that on average ¾ (75%) of the F_2 generation should have purple flowers and ¼ (25%) should have white flowers. This prediction is very close to Mendel's actual results for the F_2 generation: 76% had purple flowers and 24% had white flowers.

Mendel's second law: Independent assortment

Mendel also performed experiments in which he simultaneously tracked the inheritance of two characteristics of pea plants. For example, pea seeds can have a round or wrinkled shape, and they can be yellow or green. Two different genes control these aspects of the plant's phenotype. With respect to seed shape, Mendel determined that the allele for round seeds (denoted R) was dominant to the allele for wrinkled seeds (r). With respect to seed color, he determined that the allele for yellow seeds (Y) was dominant to the allele for green seeds (y).

What would happen if round, yellow-seeded individuals of genotype $RRYY$ were crossed with wrinkled, green-seeded individuals of genotype $rryy$? As might be expected, when Mendel performed this experiment all of the resulting F_1 plants had genotype $RrYy$ (Figure 12.6).

Next, Mendel crossed $RrYy$ plants with each other. He obtained the following results in the F_2 generation: $\frac{9}{16}$ of the seeds were round and yellow, $\frac{3}{16}$ were round and green, $\frac{3}{16}$ were wrinkled and yellow, and $\frac{1}{16}$ were wrinkled and green (a 9:3:3:1 ratio). As shown in Figure 12.6, these results indicate that the alleles for these two

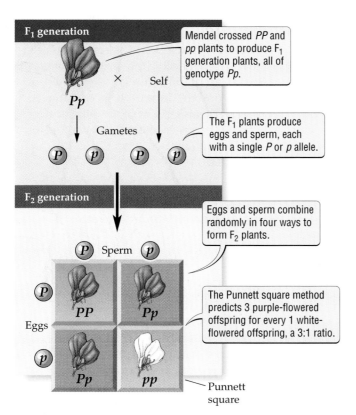

Figure 12.5 The Punnett Square Method
The Punnett square method can be used to represent the equal separation of alleles and to predict the outcome of a genetic cross.

characters are inherited independently of one another: The ratios that Mendel observed are exactly the ratios predicted by that hypothesis.

Mendel made similar crosses for various combinations of the seven characters he studied. His results led him to propose the **law of independent assortment**, which states that when gametes form, the separation of alleles for one gene is independent of the separation of alleles for other genes.

Finally, when developing both his first and second laws, it was important that Mendel observed the results of his genetic crosses in large numbers of offspring, for reasons we explore in the Box on page 199.

> ■ Mendel used basic principles of probability to summarize the results of his experiments in two laws, the law of equal segregation and the law of independent assortment. The law of equal segregation states that the two copies of a gene separate into gametes. The law of independent assortment states that when gametes form, the separation of alleles for one gene is independent of the separation of alleles for other genes. The Punnett square method considers all the possible combinations of gametes when predicting the outcome of a genetic cross.

Extensions of Mendel's Laws

Mendel's laws describe how genes are passed from parents to offspring. In some cases—like the seven characters that Mendel studied—these laws allow accurate prediction of patterns of inheritance. In other cases, however, they do not. To account for these special cases, extensions of Mendel's laws have been developed. These extensions supplement rather than invalidate Mendel's laws. Even when Mendel's laws do not accurately predict *observed* patterns of inheritance, the genes in question are inherited according to Mendel's laws. As we'll see, what differs is how the genes affect the phenotype of the organism, not how the genes are inherited.

Many alleles do not show complete dominance

For dominance to be complete, a single copy of the dominant allele must be enough to produce the maximum phenotypic effect (for example, one *P* allele ensures that even a *Pp* individual has purple flowers). But often dominance is not complete. There are many examples of a lack of complete dominance in plants. In snapdragons, for example, when a true-breeding variety with red flowers (*AA*) is crossed with a true-breeding variety with white flowers (*aa*), the heterozygote offspring (*Aa*) have pink flowers. Animals also

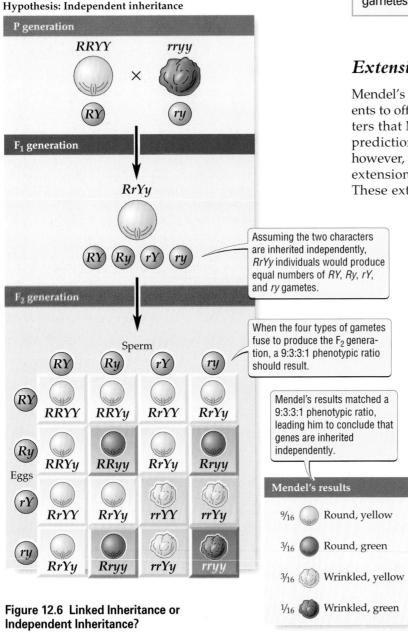

Hypothesis: Independent inheritance

Assuming the two characters are inherited independently, *RrYy* individuals would produce equal numbers of *RY*, *Ry*, *rY*, and *ry* gametes.

When the four types of gametes fuse to produce the F₂ generation, a 9:3:3:1 phenotypic ratio should result.

Mendel's results matched a 9:3:3:1 phenotypic ratio, leading him to conclude that genes are inherited independently.

Mendel's results

⁹⁄₁₆ Round, yellow

³⁄₁₆ Round, green

³⁄₁₆ Wrinkled, yellow

¹⁄₁₆ Wrinkled, green

Figure 12.6 Linked Inheritance or Independent Inheritance?

Snapdragon

The Scientific Process

Tossing Coins and Crossing Plants: Probability and Mendel's Experiments

Many offspring were produced in Mendel's experiments on peas. The fact that there were many offspring was important, for it allowed Mendel to identify the simple rules that determine the inheritance of flower color and other characteristics in peas.

The basic principles of probability explain why Mendel had to produce large numbers of offspring. The probability of an event refers to the chance that the event will occur. For example, there is a probability of 0.5 that a fair coin will turn up "heads" when the coin is tossed. A probability of 0.5 is the same thing as a 50% chance.

As illustration, consider a hypothetical coin-tossing experiment. If a fair coin is tossed only a few times, the observed percentage of heads can differ greatly from 50%. For example, if you tossed a coin only 10 times, it would not be unusual to get 70% (7) heads. However, if you tossed a coin 10,000 times, it would be very unusual to get 70% (7,000) heads. If you got such a result, you would (and should) suspect that the coin was not fair after all.

Each toss of a coin is an independent event, in the sense that the outcome of one coin toss does not alter the outcome of another coin toss. For a series of independent events, we can estimate the probability of the event from the results. For example, assume that we tossed a coin 10,000 times and got 5,092 heads. From these results, it would be reasonable to estimate the chance of getting heads as 50%, the percentage we expect from a fair coin. When only a small number of events are observed, our estimates of the underlying probabilities are likely to be inaccurate, as when we toss a coin only a few times.

How does all this relate to Mendel's experiments? When Mendel crossed heterozygous plants (for example, *Pp* individuals) with each other, the offspring always had a phenotypic ratio close to 3:1. The fundamental reason for this consistent ratio is that the chances of getting *PP*, *Pp*, and *pp* individuals are 25%, 50%, and 25%, respectively (see Figure 12.5). Because the 25% *PP* individuals and the 50% *Pp* individuals have the same phenotype, 75% of the individuals should look like each other, thus giving a 3:1 phenotypic ratio.

The Punnett square method predicts the percentages of *PP*, *Pp*, and *pp* offspring that Mendel should have observed. The method assumes that all sperm cells and all egg cells have an equal chance of achieving fertilization. But in reality, some sperm or egg cells do not achieve fertilization. With large numbers of offspring, the assumption that all sperm and egg cells have an equal chance of achieving fertilization is not too far off, because on average the success or failure of the different types of gametes tends to balance one another. But if Mendel had recorded results from small numbers of offspring in his experiments, his results probably would have differed greatly from a 3:1 ratio. In that case, he might not have discovered his two fundamental laws: the law of equal segregation and the law of independent assortment.

In general, it is important to understand that the chance of obtaining a particular type of offspring is just that, a chance. If there is a ¼ chance that an offspring will be a homozygote recessive (*pp*) individual, that means that on average, 25% of the offspring will have genotype *pp*. But it does not mean that if four offspring are produced, one will always be of genotype *pp*. That could happen, but often, in fact, none of the offspring would have genotype *pp*,

spring (*Aa*) have pink flowers. Animals also show a lack of complete dominance, as in the coat color of horses (Figure 12.7).

The colors of snapdragons and horses illustrate **incomplete dominance**: The heterozygotes are intermediate in form between the two homozygotes. Although incomplete dominance superficially resembles blending inheritance, it is really just an extension of Mendelian inheritance. For example, if two heterozygote snapdragons (*Aa*) are crossed, on average ¼ of the off-

spring will have red flowers (genotype *AA*), ½ will have pink flowers (genotype *Aa*), and ¼ will have white flowers (genotype *aa*).

Work this out for yourself using the Punnett square method. You will see that Mendel's laws still apply; the main difference is that the heterozygotes (*Aa*) look different from *AA* individuals. Because later generations return to the original flower colors of red and white—something that cannot occur under blending inheritance—this example shows that incomplete dominance

Figure 12.7 Incomplete Dominance in Horses

Palominos have genotype *Cc* and are intermediate in color to genotypes *CC* and *cc*.

Chestnut (sorrel), genotype *CC*

Palomino, genotype *Cc*

Cremello, genotype *cc*

Alleles for one gene can alter the effects of another gene

A particular phenotype often depends on more than one gene. In such cases the genes are said to interact because their phenotypic effect depends partly on their own function, and partly on the function of other genes. Gene interactions are very common.

Coat color in mammals provides an example of gene interactions. In mice and many other mammals, for example, a gene that controls pigment has a dominant allele (*B*) that produces black fur and a recessive allele (*b*) that produces brown fur. But the effects of the pigment alleles (*B* and *b*) can be eliminated completely, depending on which alleles are present at another gene that interacts with the pigment gene.

To illustrate this point, if a mouse has genotype *cc* at the gene that interacts with the pigment gene, it produces no pigment, regardless of which alleles it has for the pigment gene (Figure 12.8). A lack of pigment causes mice to have white fur. Thus, although we would expect *BB* mice to be black and *bb* mice to be brown, both *BBcc* and *bbcc* mice actually have white fur because they have the *cc* genotype at the gene that interacts with the pigment gene.

The environment can alter the effect of a gene

The effects of many genes depend on environmental conditions, such as the temperature, amount of sunlight, or concentration of salt. For example, an allele for coat color in Siamese cats (Figure 12.9) is sensitive to temperature: This allele leads to the production of dark pigment only at low temperatures. Because the extremities are colder than the rest of the body, the paws, nose, ears and tails of Siamese cats tend to be dark. If a patch of light fur were shaved from the body of a Siamese cat and covered with an ice pack, when the fur grew back it would be dark. Similarly, if dark fur were shaved from the tail and allowed to grow back under warm conditions, it would be light-colored.

Chemicals, nutrition, sunlight, and many other environmental factors also can alter the effects of genes. In plants, for example, genetically identical individuals (clones) grown in different environments often differ greatly in many aspects of their phenotype, including their height and the number of flowers they produce. Thus, plants on a windswept mountainside may be short and have few flowers while clones of the same plants grown in a warm, protected valley are tall with many flowers.

Ordinarily, mice of genotype *BB* or *Bb* for a pigment gene are black and mice of genotype *bb* are brown. However, mice with two *c* alleles at a gene that interacts with the pigment gene produce no pigment and are white in color, regardless of their genotype for the pigment gene (*BB*, *Bb*, or *bb*.)

Figure 12.8 Gene Interactions
Gene interactions are very common, as illustrated here by the effects of the *c* allele of a gene that interacts with a pigment gene in mice.

Figure 12.9 The Environment Can Alter the Effect of Genes
Coat color in Siamese cats is controlled by an allele that produces dark pigment (as on the nose, tail, paws, and ears) only at low temperatures.

Many other characters, however, are influenced by sets of genes that interact with each other and with the environment. For such characters, the relation between genotype and phenotype is more complex: A given gene does not act in isolation; rather, its effect depends on its own function, the function of other genes with which it interacts, and the environment (Figure 12.10).

For example, many human diseases, including heart disease, cancer, alcoholism, and diabetes, are strongly influenced by multiple genes and by many different environmental influences, such as smoking, diet, and overall mental and physical health. To predict the phenotypes of offspring for such characters requires a detailed understanding of how genes and the environment influence the final product of the genes, the phenotype. Such prediction is a challenging and important task. For example, we could reduce the death rate from heart disease and cancer if we knew how specific genes interacted with the environment to cause

Most characters are determined by multiple genes

Mendel studied characters that were under simple genetic control: A single gene determined patterns of inheritance for each trait. Most characters, however, are determined by the action of more than one gene. For example, skin color, running speed, and body size are controlled by multiple genes in humans, as are height, flowering time, and seed number in plants. Let's look at one of these examples, the inheritance of skin color in humans, in more detail.

A dark pigment, melanin, determines a person's skin color. Many of the differences among people for the amount of melanin in the skin are controlled by three or four genes. None of these genes are dominant to each other. The skin colors that result from these three or four genes vary considerably. Differences between genotypes are then smoothed over by suntans, causing the skin color of humans to vary nearly continuously from light to dark.

Putting it all together

Patterns of inheritance are determined by genes that are passed from parent to offspring according to the simple rules summarized in Mendel's laws. Some characters are controlled by one gene and are little affected by environmental conditions. For such characters, such as seed shape and flower color in peas, it is possible to predict the phenotypes of offspring just from knowledge of the parent alleles for a single gene.

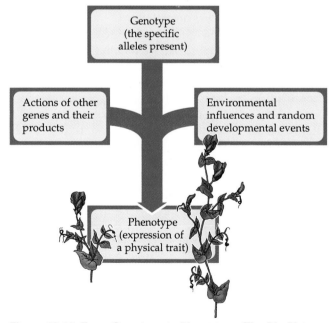

Figure 12.10 From Genotype to Phenotype: The Big Picture
The effect of a gene on the phenotype depends on its own function, the function of other genes with which it interacts, and the environment.

these diseases. Yet, despite years of effort, we still have much to learn about how such interactions cause heart disease and cancer.

> ■ For some characters, Mendel's laws do not predict patterns of inheritance accurately. The reasons for such exceptions include: (1) Many alleles do not show complete dominance, (2) alleles for one gene can alter the effects of another gene, (3) the environment can alter the effect of a gene, and (4) most characters are determined by multiple genes. Even when Mendel's laws do not accurately predict observed patterns of inheritance, the genes in question are inherited according to Mendel's laws.

Highlight

Predicting Fur Color in Kittens

Fur color in cats is controlled by at least nine genes, including genes that influence such traits as whether the color of the cat is solid or striped (tabby) and dense or light-colored. Some of the fur color genes interact. For example, as we saw for coat color in mice (see Figure 12.8), cats can have a white coat color when alleles at one gene mask the effects of a pigment gene. Overall, the many coat color genes in cats can produce fur with a wide range of colors and patterns.

Let's return to the question described in the chapter opener: Could a "pure" black tabby male have fathered chestnut brown kittens? Black coat color in cats is controlled by a dominant allele (*B*) of a pigment gene. A "pure" black tabby would be homozygous for the dominant allele—that is, it would have genotype *BB*. In addition, a "pure" black tabby would carry dominant alleles at other coat color genes that prevented the effect of the *B* alleles from being masked (as when a cat is white despite the fact that it carries a dominant *B* allele). Thus, if the male was a "pure" black tabby, then no, he could not have fathered chestnut brown kittens.

How then were the chestnut brown kittens produced? If more than one male mates with a female, the female's kittens can have different fathers. Thus, one answer to how the chestnut brown kittens were produced could be that the female actually mated with more than one male. More genetic information would be needed to discover who fathered the kittens, but we can offer the owner of the black tabby female some simple advice: To have a litter that consists entirely of black tabby kittens, breed the black tabby female with a "pure" black tabby male and then make absolutely sure that the female does not mate with any other males.

> ■ Fur color in cats is controlled by at least nine genes, several of which interact.

Summary

Genetics: An Overview

■ Genetics is the study of genes.

■ Inherited characteristics are controlled by pairs of genes. One gene for each characteristic is inherited from each parent.

■ Most genes contain instructions for a single protein product.

■ Genes are made of DNA and are located on chromosomes.

Essential Definitions in Genetics

■ Key genetic terms include the following: character, gene, allele, homozygote, heterozygote, genotype, phenotype, dominant, recessive, genetic cross, P generation, F_1 generation, F_2 generation.

Gene Mutations: The Source of New Alleles

■ In a population of many individuals, a particular gene may have one to many alleles.

■ Different alleles of a gene cause hereditary differences among organisms.

■ New alleles arise by mutation.

■ Mutations occur at random with respect to their usefulness.

■ Many mutations are harmful, many have little impact, and a few are beneficial.

Basic Patterns of Inheritance

■ Mendel's experiments on pea plants led him to propose a new, particulate theory of inheritance, summarized as follows: (1) Alternative versions of genes (alleles) cause variation in inherited characters. (2) Offspring inherit one copy of a gene from each parent. (3) Alleles can be dominant or recessive. (4) The two copies of a gene separate equally into gametes. (5) Gametes fuse without regard to which allele they carry.

Mendel's Laws: Equal Segregation and Independent Assortment

■ Mendel used basic principles of probability to summarize the results of his experiments in two laws: the law of equal segregation and the law of independent assortment.

■ The law of equal segregation states that the two copies of a gene separate equally into gametes.

■ The law of independent assortment states that when gametes form, the separation of alleles for one gene is independent of the separation of alleles for other genes.

■ The Punnett square method considers all possible combinations of gametes when predicting the outcome of a genetic cross.

Extensions of Mendel's Laws

■ For some characters, Mendel's laws do not predict patterns of inheritance accurately.

■ Reasons for departures from Mendelian inheritance patterns are as follows: (1) Many alleles do not show complete dominance. (2) Alleles for one gene can alter the effects of another gene. (3) The environment can alter the effect of a gene. (4) Most characters are determined by multiple genes.

■ Even when Mendel's laws do not accurately predict observed patterns of inheritance, the genes in question are inherited according to Mendel's laws. What differs is how the genes affect the phenotype, not how they are inherited.

Highlight: Predicting Fur Color in Kittens

■ Fur color in cats is controlled by at least nine genes, several of which interact.

Key Terms

allele p. 194

character p. 193

dominant p. 194

F₁ generation p. 194

F₂ generation p. 194

gene p. 193

genetic cross p. 194

genetics p. 192

genotype p. 194

heterozygote p. 194

homozygote p. 194

incomplete dominance p. 199

law of equal segregation p. 197

law of independent assortment p. 198

mutation p. 194

P generation p. 194

phenotype p. 194

Punnett square p. 197

recessive p. 194

true-breeding variety p. 195

Chapter Review

Self-Quiz

1. Alternative versions of a gene for a given character are called
   a. alleles.
   b. heterozygotes.
   c. genotypes.
   d. phenotypes.

2. If long hair (*L*) is dominant to short hair (*l*), then a cross of *Ll* × *ll* should yield
   a. ¼ short-hair offspring.
   b. ¾ short-hair offspring.
   c. ½ short-hair offspring.
   d. all offspring with intermediate hair length

3. Individuals of genotype *Aa* are
   a. homozygous.
   b. heterozygous.

   c. dominant.
   d. recessive.

4. Genes
   a. are the basic units of heredity.
   b. are located on chromosomes and composed of DNA.
   c. usually contain instructions for a protein product.
   d. all of the above

5. Coat color in horses shows incomplete dominance. *CC* individuals have a chestnut color, *Cc* individuals have a palomino color, and *cc* individuals have a cremello color. What is the predicted phenotypic ratio of chestnut: palomino:cremello if *Cc* individuals are crossed with other *Cc* individuals?
   a. 3:1
   b. 2:1:1
   c. 9:3:1
   d. 1:2:1

Review Questions

1. For flower color in peas, the allele for purple flowers (*P*) is dominant to the allele for white flowers (*p*). A purple-flowered plant, therefore, could be of genotype *PP* or *Pp*. What genetic cross could you make to determine the genotype of a purple-flowered plant? Explain how your cross enables you to figure out the correct genotype of the purple-flowered plant.

2. Many lethal human genetic disorders are caused by a recessive allele, whereas relatively few are caused by a dominant allele. Why might dominant alleles for lethal human diseases be uncommon? (*Hint:* Solve Sample Genetics Problems 4 and 5 below and use the results to guide your answer to this question.)

3. Referring to the Box on page 199, explain in your own words why it was important that Mendel observed large numbers of offspring in his genetic crosses.

Sample Genetics Problems

1. One gene has alleles *A* and *a*, a second gene has alleles *B* and *b*, and a third gene has alleles *C* and *c*. List the possible gametes that can be formed from the following genotypes: (a) *Aa*, (b) *BbCc*, (c) *AAcc*, (d) *AaBbCc*, (e) *aaBBCc*.

2. For the same three genes described in problem 1, what are the predicted genotype and phenotype ratios of the following genetic crosses (assume that alleles written in uppercase letters are dominant to alleles written in lowercase letters): (a) *Aa* × *aa*, (b) *BB* × *bb*, (c) *AABb* × *aabb*, (d) *BbCc* × *BbCC*, (e) *AaBbCc* × *AAbbCc*.

3. Sickle-cell anemia is inherited as a recessive genetic disorder in humans. That means that in terms of disease onset, the normal hemoglobin allele (*S*) is dominant to the sickle-cell allele (*s*). If two people of genotype *Ss* have children, construct a Punnett square to predict the possible genotypes and phenotypes (does or does not have the disease) of their children. Also list the genotype and phenotype ratios. Each time two *Ss* individuals have a child together, what is the chance that the child will have sickle-cell anemia?

4. For any human genetic disorder caused by a recessive allele, call the allele that causes the disease *n*, and the normal allele *N* (the capital "*N*" is for "normal" individuals). (a) What are the phenotypes of *NN*, *Nn*, and *nn* individuals? (b) Predict the outcome of a genetic cross between two *Nn* individuals. List the genotype and phenotype ratios that would result from such a cross. (c) Predict the outcome of a genetic cross between an *Nn* and an *NN* individual. List the genotype and phenotype ratios that would result from such a cross.

5. For any human genetic disorder caused by a dominant allele, call the allele that causes the disease *D*, and the normal allele *d* (the capital "*D*" is for "disease"). (a) What are the phenotypes of *DD*, *Dd*, and *dd* individuals? (b) Predict the outcome of a genetic cross between two *Dd* individuals. List the genotype and phenotype ratios that would result from such a cross. (c) Predict the outcome of a genetic cross between a *Dd* and a *DD* individual. List the genotype and phenotype ratios that would result from such a cross.

6. If purple flower color (*A*) is dominant to white flower color (*a*), what are the *genotypes for the parents* in the following genetic cross: purple flower × white flower yields only purple-flowered offspring?

7. In one of his experiments, Mendel crossed plants that bred true for yellow seeds with plants that bred true for green seeds. All seeds in the F_1 generation were yellow. Which allele is dominant—the one for green or the one for yellow? Explain why.

12   𝕿𝖍𝖊 𝕯𝖆𝖎𝖑𝖞 𝕲𝖑𝖔𝖇𝖊

A New Era Begins

Letter to the Editor:

Recent advances in genetic research have allowed us the opportunity—if we take it—to rid ourselves of the suffering caused by human genetic disease. Scientists are now able to identify disease-causing genes, and we have a moral obligation to make use of this technology through mandatory genetic testing for all individuals who wish to have children.

Such testing would ensure embryos were free of genetic diseases. If testing revealed disease-causing genes in either the potential mother or father, the couple could still have kids, but they would have to mix their sperm and eggs in a test tube and then pick the developing embryo that did not have the disease gene. The embryo (which would just be a ball of a few cells) could then be placed into the mother and she would have a normal child. We're not talking science fiction here; this kind of thing has already been done. In one or two generations, the human race would emerge with a clean bill of health. Finally, for the first time in human history, we would be truly free.

Name witheld

Evaluating "The News"

1. Do you think it's reasonable to force people to be tested for disease genes? What are the advantages and disadvantages of such testing? If not forced to do so, should people be allowed to be tested for disease genes and to use the results to make sure their children don't have a genetic disease?

2. In some states, couples that wish to marry must be tested for HIV, the virus that causes AIDS. Do you think the suggestion in this letter to the editor is similar to or very different from current laws that force people to be tested for HIV? Is it different to regulate the act of having children?

3. Should people be allowed to test for genes that don't affect human genetic diseases? That is, should parents be allowed to test for many different genes and to "design" the genetic makeup of their child? Would you have wanted your parents to decide your genetic makeup?

4. Mutation regularly generates new alleles of genes, including genes that cause disease. Could we ever really be "free" of disease genes?

13

Chromosomes and Human Genetics

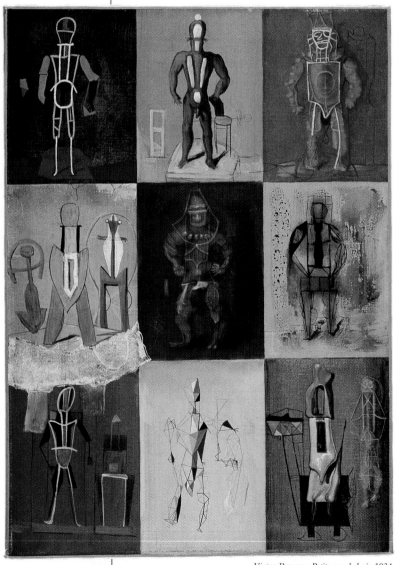

Victor Brauner, *Petite morphologie*, 1934.

A Horrible Dance

As a child in the mid-1800s, George Huntington went with his father, a medical doctor, as he visited patients in rural Long Island, New York. On one such trip, the young boy and his father saw two women by the roadside, a mother and daughter, who were bowing and twisting uncontrollably, their faces contorted in a series of strange grimaces. Huntington's father paused to speak with them, then left, continuing on his rounds.

For any child, such an encounter would be unforgettable. For the young Huntington, it was that and more. Like his father and grandfather before him, George Huntington became a doctor, and in 1872 he described and named the disease that had plagued the two women. He called it "hereditary chorea" (the word "chorea" comes from the Greek word for "dance").

Huntington described hereditary chorea as an inherited disease, one that destroyed the nervous system and caused jerky, involuntary movements of the body and face. There was no cure. Eventually, it killed its victims, but first it reduced them to a quivering wreck of their former selves. The disease also caused memory loss, severe depression, mood shifts, personality changes, and intellectual deterioration.

If a parent had the disease, it did not necessarily strike all the children of that parent. The children who did get it usually showed no symptoms until they were in their 30s, 40s, or 50s. Thus, hereditary chorea was like a genetic time bomb. In Huntington's words, the combination of the terrible symptoms of the disease and its late and uncertain onset caused "those in whose veins the seeds of the disease are known to exist [to speak of it] with a kind of horror."

Hereditary chorea is now known as Huntington's disease, in honor of George Huntington. There still is no cure, but researchers have identified a mutation in a gene on chromosome 4 that causes the disease, and they have isolated the gene. With the gene in hand, the quest continues for further understanding and, ideally, an effective treatment.

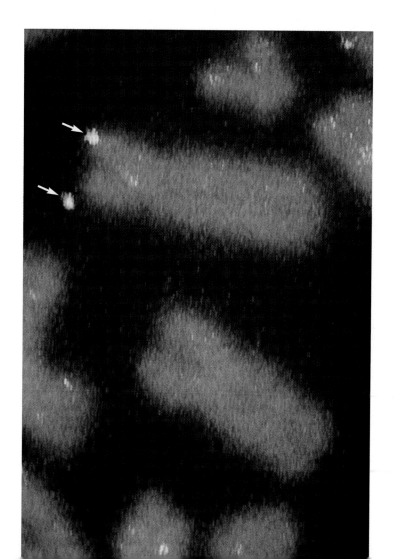

The Gene for Huntington's Disease
This gene is located at one end of chromosome 4 (shown by the white arrows).

Key Concepts

1. Chromosomes contain a single DNA molecule. A gene is a region of DNA within the DNA of the chromosome. Each gene has a specific location on the chromosome.

2. In humans, males have one X and one Y chromosome, and females have two X chromosomes. A specific gene on the Y chromosome is required for human embryos to develop as males.

3. Unless they are located far from one another, genes on the same chromosome tend to be inherited together. Genes that are inherited together are said to be linked.

4. The two copies of a chromosome that pair during meiosis (one inherited from each parent) can exchange genes in a process called crossing-over.

5. Genetic recombination produces genotypes in offspring that are different from that of either parent. Genetic recombination is caused by crossing-over and by the random distribution of maternal and paternal chromosomes into gametes.

6. Many inherited genetic diseases in humans are caused by mutations to single genes. A far smaller number of human genetic disorders are caused by abnormalities in chromosome number or structure.

Humans are afflicted by many types of genetic disorders, with effects that range in severity from mild to deadly. For some of these conditions, such as some forms of breast cancer, an understanding of the genetic basis of the disorder has contributed to effective means of treatment. For others, such as Huntington's disease, the search for successful treatments is still under way. Although we emphasize human genetic disorders in this chapter, first we continue our discussion of basic genetic principles.

In Chapter 12, we described Mendel's discovery that inherited characteristics are determined by genes. We begin this chapter with a second foundation of modern genetics, the chromosomal theory of inheritance. We then explain how gender (sex) is determined in humans and other organisms, and how new nonparental combinations of genes can occur in offspring. The information about chromosomes, gender determination, and new gene combinations provides helpful background material for our discussion of human genetic disorders.

The Role of Chromosomes in Inheritance

By 1882, studies using microscopes revealed that threadlike structures, the chromosomes, exist inside of dividing cells. The German biologist August Weismann hypothesized that the number of chromosomes was first reduced by half during the formation of sperm and egg cells, then restored to its full number during fertilization. In 1887, meiosis was discovered (see Chapter 10), thus confirming Weismann's hypothesis. Weismann also suggested that the hereditary material was located on chromosomes, but no experimental data supported or refuted this idea.

Genes are located on chromosomes

The idea that genes are located on chromosomes is known as the **chromosome theory of inheritance**. Much experimental evidence now indicates that genes are indeed located on chromosomes. In fact, modern genetic techniques allow us to pinpoint where genes are located on a chromosome, as in the photograph on page 207 of the gene for Huntington's disease.

How are chromosomes, DNA, and genes related? Chromosomes are made of a single DNA molecule and many proteins. Each gene on a chromosome is a region of DNA within the long strand of DNA in the chromosome. The physical location of a gene on a chromosome is called a **locus** (plural loci) (Figure 13.1). Chromosomes that pair during meiosis and that have the same genetic loci and structure are called **homologous chromosomes**. At each gene locus there are two alleles, each one inherited from a different parent and situated on a different chromosome of the pair of homologous chromosomes. Chromosomes contain many gene loci, and these loci have a physical relationship to each other: Some genes are located near each other on the chromosome, others far from each other (see Figure 13.1).

■ The chromosome theory of inheritance states that genes are located on chromosomes. Chromosomes are composed of a single DNA molecule and many proteins. A gene is a region of DNA within the DNA of the chromosome. The physical location of a gene on a chromosome is called a locus.

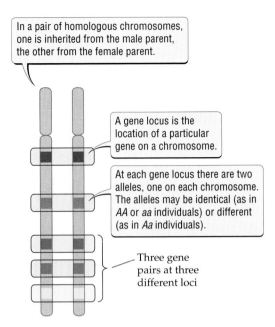

In a pair of homologous chromosomes, one is inherited from the male parent, the other from the female parent.

A gene locus is the location of a particular gene on a chromosome.

At each gene locus there are two alleles, one on each chromosome. The alleles may be identical (as in *AA* or *aa* individuals) or different (as in *Aa* individuals).

Three gene pairs at three different loci

Figure 13.1 Genes Are on Chromosomes
The genes shown here take up a larger portion of the chromosome than they would if they were drawn to scale.

Autosomes and Sex Chromosomes

For most chromosomes, the maternal and paternal copies of the chromosome are exactly alike in terms of length, shape, and the sequence of genes they carry. But in humans and many other organisms, this is not true for the chromosomes (X and Y) that determine the gender (sex) of the organism. In humans, for example, males have one X chromosome and one Y chromosome, whereas females have two X chromosomes (Figure 13.2). The Y chromosome in humans is much smaller than the X chromosome, and it does not contain the same set of genes. In other organisms, such as birds, butterflies, and some fish, the males have two identical chromosomes (which we denote ZZ), whereas the females have one Z chromosome and one W chromosome. Chromosomes that determine gender are called **sex chromosomes**; all other chromosomes are called **autosomes**.

Sex determination in humans

Because human females have two copies of the X chromosome, all egg cells produced by a woman contain one X chromosome. In males, however, half of the sperm contain an X chromosome and half contain a Y chromosome. The sex chromosome carried by the sperm therefore determines the gender of the child: If a sperm carrying an X chromosome fertilizes the egg, the child will

be a girl, but if a sperm carrying a Y chromosome fertilizes the egg, the child will be a boy (see Figure 13.2).

The Y chromosome has few genes. It does, however, carry one very important gene: a gene that functions as a master switch, committing the gender of the developing embryo to "male." In the absence of this gene, human embryos develop as females, but when this gene is present, the child is a male. Occasionally, XY individuals develop as females, and XX individuals develop as males. In most cases, this occurs because the XY females are missing the portion of their Y chromosome that contains the sex-determining gene (and hence they cannot develop as a male), or similarly, because the XX males have the same portion of the Y chromosome attached to

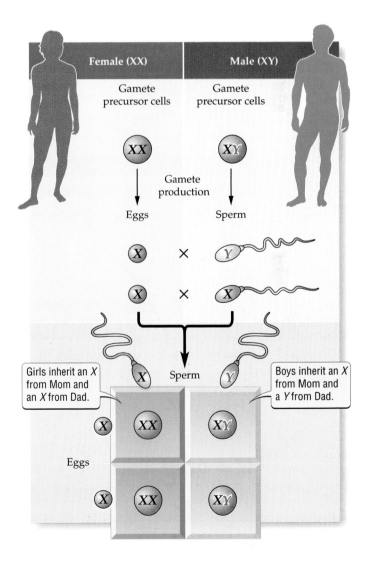

Figure 13.2 Sex Determination in Humans
Human females have two X chromosomes, while human males have one X and one Y chromosome.

one of their X chromosomes (and hence they develop as a male).

> ■ Chromosomes that determine gender are called sex chromosomes. All other chromosomes are called autosomes. Human males have one X and one Y chromosome. Human females have two X chromosomes. A specific gene on the Y chromosome is required for human embryos to develop as males.

Linkage and Crossing-Over

As described in Chapter 12, Mendel's results indicated that genes are inherited independently of one another. This conclusion led him to propose the law of independent assortment, which stated that the alleles for one gene separate into gametes independently of alleles for other genes. Early in the twentieth century, however, results from several laboratories indicated that genes were often inherited together, and thus that their behavior contradicted the law of independent assortment.

Thomas Hunt Morgan

For example, Thomas Hunt Morgan discovered genes that were inherited together in his research on fruit flies, which began in 1909 at Columbia University in New York City. In one such experiment, Morgan crossed a pure-breeding variety of fruit fly that had a gray body (*G*) and wings of normal length (*W*) to a pure-breeding variety that had a black body (*g*) and wings that were greatly reduced in length (*w*). That is, he crossed *GGWW* flies with *ggww* flies, to obtain flies with a genotype of *GgWw* in the F$_1$ generation. He then mated the *GgWw* flies with *ggww* flies, as Figure 13.3 shows. Morgan's results were very different from the results he expected based on the law of independent assortment. What had happened?

From his data Morgan concluded that the genes for body color and wing length were located on the same chromosome.

Because they were on the same chromosome, they did not assort independently during meiosis. Instead, they were inherited together. Genes that are located on the same chromosome and that do not assort independently are **genetically linked**. All genes on the same chromosome are members of the same **linkage group**.

Figure 13.3 Some Alleles Do Not Assort Independently
Thomas Hunt Morgan found that the gene for body color (dominant allele *G* for gray, recessive allele *g* for black) was linked to the gene for wing length (dominant allele *W* for normal length, recessive allele *w* for greatly reduced length). This linkage occurred because the two genes were on the same chromosome.

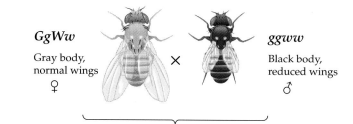

GgWw
Gray body, normal wings
♀

×

ggww
Black body, reduced wings
♂

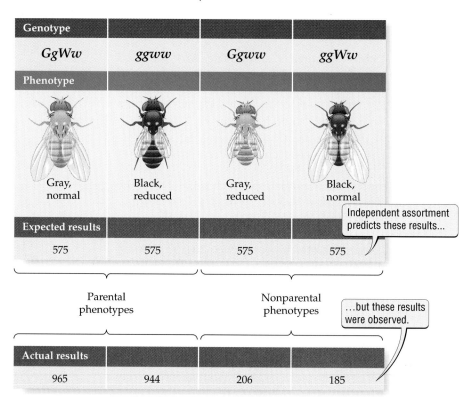

Genotype			
GgWw	*ggww*	*Ggww*	*ggWw*
Phenotype			
Gray, normal	Black, reduced	Gray, reduced	Black, normal
Expected results			
575	575	575	575

Independent assortment predicts these results...

Parental phenotypes | Nonparental phenotypes

...but these results were observed.

Actual results			
965	944	206	185

Conclusion: These two genes do not assort independently. They are linked together on the same chromosome.

Because organisms have many more genes than chromosomes, many of their genes are inherited together or linked. For example, humans have an estimated 100,000 genes, which are located on 23 pairs of homologous chromosomes. Thus, on average our 23 chromosomes have 100,000/23, or 4348, genes per chromosome.

Crossing-over disrupts linked genes

If the linkage between two genes on a chromosome is complete, all offspring should be of a parental type; that is, all of them should have a genotype that matches that of one of their parents. For example, in the cross shown in Figure 13.3, if the genes were completely linked, half of the offspring should have had genotype *GgWw*, and the other half should have had genotype *ggww* (Figure 13.4). Since many of the offspring did have these two parental genotypes, the two genes clearly were linked. But how could the appearance of nonparental—*Ggww* and *ggWw*—genotypes be explained?

To explain the appearance of nonparental genotypes, Morgan suggested that genes were physically exchanged between chromosomes during meiosis. The exchange of genes between homologous chromosomes is called **crossing-over**. To make this concept more concrete, imagine that the two chromosomes illustrated in Figure 13.5 come from one of your cells. You inherited one of these chromosomes from your father, the other from your mother.

In crossing-over, part of a chromosome inherited from one parent is replaced with the corresponding genetic material inherited from the other parent. By physically exchanging pieces of one homologous chromosome with another, crossing-over combines alleles inherited from one parent with those inherited from the other. This exchange makes possible the formation of nonparental genotypes, such as the *Ggww* and *ggWw* genotypes found in Figure 13.3.

Crossing-over can be compared to the cutting of a string. Two points that are far apart on the string will be separated from each other in most cuts of the string, whereas points that are very close to each other will rarely be separated. Similarly, genes that are far from each other on a chromosome are more likely to be separated by crossing-over than are genes that are close to each other. In fact, two genes on the same chromosome

that are very far from each other may be separated by crossing-over so often that they are no longer effectively linked. Such genes assort independently even though they are located on the same chromosome. Among the traits that Mendel studied, we now know that the genes for flower color and seed color are on the same chromosome but are so far apart that they are not effectively linked. Thus, the law of independent assortment holds for these genes.

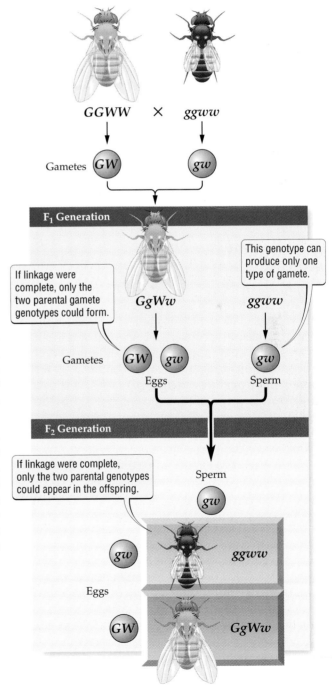

Figure 13.4 Consequences of Complete Linkage
If the linkage between genes on a chromosome were complete, all F$_2$ offspring would have a parental genotype. Morgan's experiment (see Figure 13.3) showed that complete linkage does not occur. Gametes are indicated by circles.

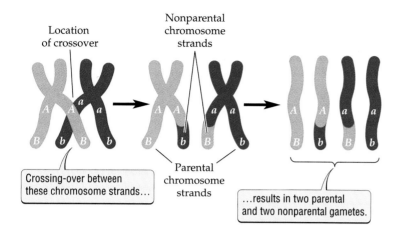

Location of crossover

Nonparental chromosome strands

Parental chromosome strands

Crossing-over between these chromosome strands...

...results in two parental and two nonparental gametes.

Figure 13.5 Crossing-Over Disrupts the Linkage between Genes
In the case shown here, crossing-over occurs between two genes, causing half of the gametes to have a parental genotype (either *AB* or *ab*), and the other half to have a nonparental genotype (either *Ab* or *aB*).

> ■ Genes that are located on the same chromosome and that do not assort independently are genetically linked. All genes on the same chromosome are members of the same linkage group. Crossing-over, the exchange of genes between chromosomes, disrupts the linkage between genes.

Genetic Recombination

In broad overview, inheritance is both a stable and a variable process. It is stable in that genetic information is transmitted accurately from one generation to the next. It is variable in that offspring are never exact replicas of parents, even in species, like bacteria and dandelions, that reproduce by asexual cloning. In this section we describe how offspring receive new combinations of genes that differ from those in their parents. As we will discuss in Unit 4, new combinations of genes provide the genetic variation on which evolution can act.

How do offspring receive new combinations of genes? First, new alleles arise by mutation. Once formed by mutation, alleles are shuffled or arranged in new ways by crossing-over, independent assortment of chromosomes, and fertilization. Here we focus on crossing-over and the independent assortment of chromosomes. The **independent assortment of chromosomes** refers to the random distribution of maternal and paternal chromosomes into gametes during meiosis.

We mention fertilization only to point out that when a particular sperm unites with a particular egg cell, a unique individual with a unique set of alleles is formed. If, for example, a different one of your father's sperm cells had fertilized your mother's egg cell, someone different from you—with a different set of alleles—would have been formed.

Crossing-over and independent assortment of chromosomes cause **genetic recombination**, which is the formation of nonparental genotypes. For example, every time meiosis occurs, crossing-over produces some "new" chromosomes. These chromosomes are new in the sense that they contain some alleles inherited from one parent and other alleles inherited from the other parent. By exchanging alleles between chromosomes, crossing-over allows the formation of nonparental genotypes, as seen in Figure 13.5 and in the results shown in Figure 13.3.

The independent assortment of chromosomes does not alter individual chromosomes. Instead, it shuffles the chromosomes, randomly combining maternal and paternal chromosomes in the gametes. Independent assortment happens because the orientation of the maternal and paternal chromosomes varies at random when the chromosomes line up on the spindle microtubules during meiosis. In humans, for example, the 23 pairs of homologous chromosomes can be arranged on the spindle microtubules in 2^{23}, or 8,388,608, different ways. Of these 8,388,608 ways of arranging the chromosomes, only two are the combination originally inherited from the parents. Thus, like crossing-over, the independent assortment of chromosomes can cause the formation of nonparental genotypes.

> ■ The independent assortment of chromosomes is the random distribution of maternal and paternal chromosomes into gametes during meiosis. Crossing-over and the independent assortment of chromosomes cause the formation of nonparental genotypes. The formation of nonparental genotypes, called genetic recombination, provides genetic variation on which evolution can act.

Human Genetic Disorders

The topics covered thus far in this chapter have included the chromosome theory of inheritance, how gender is determined in humans and other organisms, linkage and crossing-over, and the independent assortment of

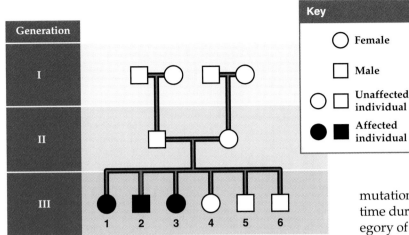

Generation	
I	
II	
III	1 2 3 4 5 6

Key

○ Female

□ Male

○ □ Unaffected individual

● ■ Affected individual

Figure 13.6 Human Pedigree Analysis
This pedigree of cystic fibrosis, a recessive genetic disorder in humans, illustrates symbols commonly used by geneticists. The Roman numerals at left identify different generations. Numbers listed below symbols are used to identify individuals of a given generation.

chromosomes. We will now apply this information to the study of genetic disorders in humans. The application of genetic principles to human genetic disorders has been a powerful and useful approach, one that has led to greater understanding and improved treatment of human genetic diseases.

Each of us knows someone who has suffered from a human genetic disorder, such as a hereditary form of cancer, heart disease, or one of the many other diseases caused by gene mutations. Because human genetic disorders are so widespread, it is important to study them, seeking a prevention or cure for much human suffering. But the study of human genetic disorders faces daunting problems. Unlike fruit flies and other organisms studied by geneticists, we humans have a long generation time, we select our own mates, and we decide when and if to have children. In addition, understanding the inheritance of human genetic disorders can be difficult because our families are so small (see the Box in Chapter 12, page 199).

One approach to these problems is to analyze pedigrees. A **pedigree** is a chart that shows genetic relationships among family members over two or more generations of a family's history (Figure 13.6). Pedigrees provide geneticists with a way to combine information from many families so as to learn about the genetic control of a particular disease. For example, the pedigree shown in Figure 13.6 is for the lung disease cystic fibrosis. Cystic fibrosis usually causes death before the age of 30. Individuals 1, 2, and 3 in generation III had cystic fibrosis, but their parents did not. The pedigree in Figure 13.6 indicates that the allele that causes cystic fibrosis is not dominant, for if it were, one of the parents of the affected individuals would have had the disease.

Humans suffer from a variety of genetic disorders (Figure 13.7). Some of these disorders result from new

mutations that occur in the cells of an individual sometime during his or her life. Most cancers fall into this category of genetic disorder (see Chapter 11). Other genetic disorders are passed down from parent to child. These inherited genetic disorders can be caused by mutations of individual genes or by abnormalities in chromosome number or structure.

Often the tendency to develop diseases, like heart disease, diabetes, and some cancers, is caused by interactions among multiple genes and the environment. However, the identity of these genes and how they lead to disease is poorly understood. Consequently, we will focus on genetic disorders caused by mutations to a single gene or by chromosome abnormalities. We will organize our discussion of single-gene genetic disorders by whether the gene is located on an autosome or a sex chromosome.

> ■ Humans suffer from a variety of genetic disorders, including those caused by mutations to a single gene and those caused by abnormalities in chromosome number or structure. Pedigrees, which show genetic relationships among family members over two or more generations of a family's history, provide a useful way to study human genetic disorders.

Autosomal Inheritance of Single-Gene Mutations

Several thousand human genetic disorders are inherited as recessive characteristics. Most of these, such as cystic fibrosis, Tay-Sachs disease, and PKU (phenylketonuria) (see Figure 13.7), are caused by recessive mutations to genes located on autosomes. Recessive genetic disorders range in severity from those that are lethal to those with relatively mild effects. For example, Tay-Sachs disease is a recessive genetic disorder in which the brain and spinal cord begin to deteriorate during a child's first year of life, causing death within a few years. At the

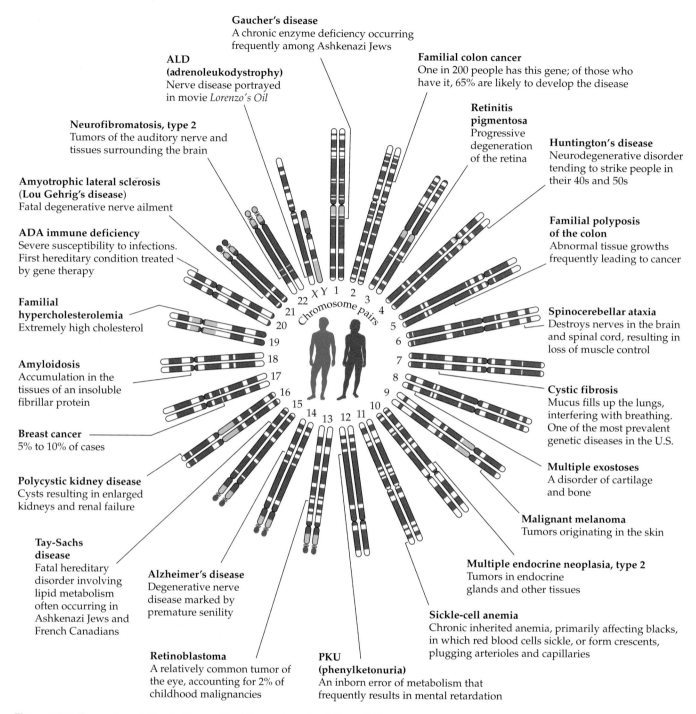

Gaucher's disease
A chronic enzyme deficiency occurring frequently among Ashkenazi Jews

ALD (adrenoleukodystrophy)
Nerve disease portrayed in movie *Lorenzo's Oil*

Neurofibromatosis, type 2
Tumors of the auditory nerve and tissues surrounding the brain

Amyotrophic lateral sclerosis (Lou Gehrig's disease)
Fatal degenerative nerve ailment

ADA immune deficiency
Severe susceptibility to infections. First hereditary condition treated by gene therapy

Familial hypercholesterolemia
Extremely high cholesterol

Amyloidosis
Accumulation in the tissues of an insoluble fibrillar protein

Breast cancer
5% to 10% of cases

Polycystic kidney disease
Cysts resulting in enlarged kidneys and renal failure

Tay-Sachs disease
Fatal hereditary disorder involving lipid metabolism often occurring in Ashkenazi Jews and French Canadians

Alzheimer's disease
Degenerative nerve disease marked by premature senility

Retinoblastoma
A relatively common tumor of the eye, accounting for 2% of childhood malignancies

PKU (phenylketonuria)
An inborn error of metabolism that frequently results in mental retardation

Familial colon cancer
One in 200 people has this gene; of those who have it, 65% are likely to develop the disease

Retinitis pigmentosa
Progressive degeneration of the retina

Huntington's disease
Neurodegenerative disorder tending to strike people in their 40s and 50s

Familial polyposis of the colon
Abnormal tissue growths frequently leading to cancer

Spinocerebellar ataxia
Destroys nerves in the brain and spinal cord, resulting in loss of muscle control

Cystic fibrosis
Mucus fills up the lungs, interfering with breathing. One of the most prevalent genetic diseases in the U.S.

Multiple exostoses
A disorder of cartilage and bone

Malignant melanoma
Tumors originating in the skin

Multiple endocrine neoplasia, type 2
Tumors in endocrine glands and other tissues

Sickle-cell anemia
Chronic inherited anemia, primarily affecting blacks, in which red blood cells sickle, or form crescents, plugging arterioles and capillaries

Chromosome pairs

Figure 13.7 Examples of Single Genes That Cause Inherited Genetic Diseases
Mutations of single genes that cause genetic disorders are found on the X chromosome and on each of the 22 autosomes in humans.

other end of the severity spectrum, albino skin color in humans is controlled by a complex set of genes, including a recessive allele similar to that which produces a white coat color in mice, cats, and other mammals (see Figure 12.8).

For genetic disorders caused by a recessive allele (*a*), the only individuals that actually get the disease are those that have two copies of the disease-causing allele (*aa*). Most commonly, when a child inherits a recessive genetic disorder, both parents are heterozygous for the disorder. That is, the parents have genotype *Aa*, where the *A* allele is dominant and does not cause the disease, and the *a* allele is recessive and does cause the disease. In such a case, the parents are said to be **carriers**: They carry the disease-causing allele but do not get the disease.

If two carriers of a recessive genetic disease have children, the patterns of inheritance are the same as for any recessive characteristic: on average, ¼ of the children have genotype *AA*, ½ have genotype *Aa*, and ¼ have genotype *aa*. Thus, each child has a 25% chance of not carrying the disease-causing allele, a 50% chance of being a carrier, and a 25% chance of actually getting the disease (Figure 13.8). These percentages identify one way in which lethal recessive disorders like Tay-Sachs disease can persist in the human population: Although homozygous recessive individuals (with genotype *aa*) die long before they reproduce, carriers (with genotype *Aa*) are not harmed, and hence they can pass the disease-causing allele to half of their children. In a sense, then, the *a* alleles can "hide" in the heterozygous carriers. Recessive genetic disorders also remain in the human population because new mutations generate new copies of the disease-causing alleles.

A dominant allele (*A*) that causes a genetic disorder cannot "hide" in the same way that a recessive allele can: *AA* and *Aa* individuals get the genetic disorder; only *aa* individuals are symptom-free. For a dominant genetic disorder with serious negative effects, individuals that have the allele tend to survive or reproduce poorly; hence few of them pass the allele on to their children. In particular, lethal dominant genetic disorders often are not common and remain in the population primarily because new disease-causing alleles are generated by mutation. Huntington's disease, which was described on page 206, illustrates another way in which dominant, lethal alleles persist. Huntington's disease begins relatively late in life, often after victims of the disease have had children. Because the allele that causes the disease can be passed on to the next generation before the victim dies, it is more common than it would be if it persisted by mutation alone.

■ For a genetic disorder caused by a recessive allele (*a*) of a gene on an autosome, only homozygous *aa* individuals get the disease; heterozygous (*Aa*) individuals are merely carriers. For dominant genetic disorders, both *AA* and *Aa* individuals suffer from the disorder. Lethal dominant genetic disorders can remain in the population because of late onset of disease symptoms or because new disease-causing alleles are generated by mutation.

Sex-Linked Inheritance of Single-Gene Mutations

Roughly 4000 of the estimated 100,000 human genes are located on sex chromosomes. Such genes are said to be **sex-linked**. In humans, the Y chromosome contains few genes, and there are no well-documented cases of disease genes located on it. X chromosomes, however, do contain genes that cause inherited genetic disorders (see Figure 13.7). Genes on the X chromosome, whether or not they cause a genetic disorder, are said to be **X-linked**.

Because males inherit only one X chromosome, genes on sex chromosomes have different patterns of inheritance than do genes on autosomes. For example, consider how an X-linked, recessive allele that causes a human genetic disorder is inherited (Figure 13.9). We label the recessive, disease-causing allele *a*, and in the Punnett square (see Figure 12.5) we write this allele as X^a to emphasize the fact that the allele is on the X chro-

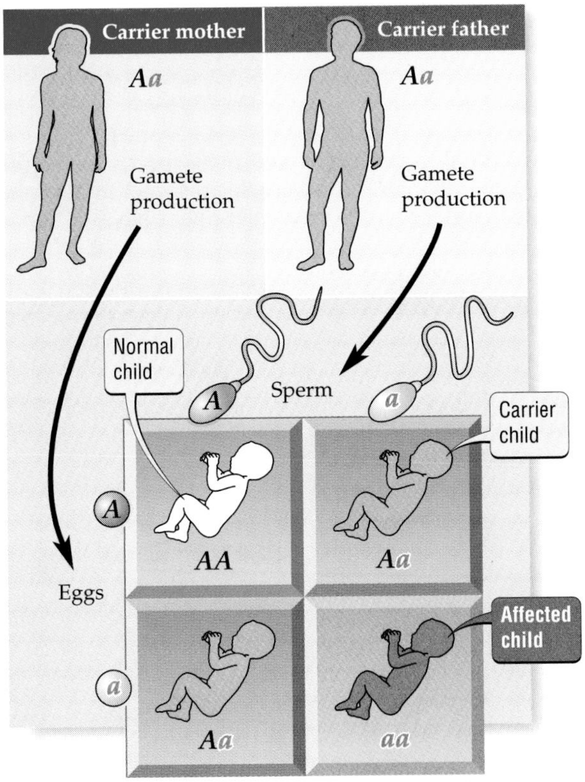

Figure 13.8 Autosomal Recessive Inheritance
Recessive, disease-causing alleles are denoted by *a*. Dominant, normal alleles are denoted by *A*. Here, a carrier male (genotype *Aa*) mates with a carrier female (genotype *Aa*).

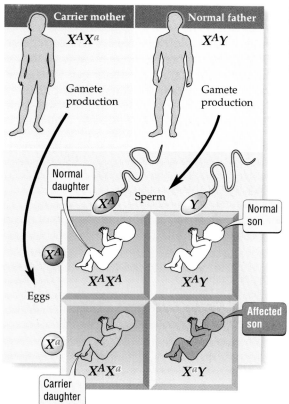

Carrier mother
$X^A X^a$

Normal father
$X^A Y$

Gamete production

Gamete production

Normal daughter

X^A Sperm Y

Normal son

Eggs

X^A

$X^A X^A$ $X^A Y$

X^a

$X^A X^a$ $X^a Y$

Affected son

Carrier daughter

mosome. Similarly, the dominant allele is labeled *A* and is written as X^A in the Punnett square. If a carrier female (with genotype $X^A X^a$) has children with a normal male (with genotype $X^A Y$), on average 50% of the male children will get the disease. This result differs greatly from what would happen if the same disease-causing allele (*a*) were on an autosome: In that case, none of the children, male or female, would get the disease.

Figure 13.9 X-Linked Recessive Inheritance

The recessive, disease-causing allele (*a*) is on the X chromosome and so is denoted by X^a. The dominant, normal allele (*A*) is also on the X chromosome and is denoted by X^A. Here, a carrier female (genotype $X^A X^a$) mates with a normal male (genotype $X^A Y$).

For X-linked genetic disorders, males of genotype $X^a Y$ get the disease because the Y chromosome does not have a copy of the gene, and hence a dominant *A* allele cannot mask the effects of the *a* allele. In general, males are much more likely than females to become afflicted with an X-linked disorder (see Figure 13.9). The reason is that to get the disorder, males have to inherit only a single copy of the disease-causing allele, whereas females must inherit two copies of the allele.

X-linked genetic disorders in humans include hemophilia, a serious disorder in which minor cuts and bruises can cause a person to bleed to death, and Duchenne muscular dystrophy, a lethal disorder that causes the muscles to waste away, usually leading to death in the early 20s. Both of these X-linked disorders are caused by recessive alleles. An example of a dominant, X-linked disorder is shown in Figure 13.10. Congenital generalized hypertrichosis (CGH) is a very rare genetic disorder that causes extreme hairiness on the face and upper body. Study the pedigree in Figure 13.10 carefully. Do you understand why researchers concluded that CGH is a dominant, X-linked disorder?

Figure 13.10 Congenital Generalized Hypertrichosis (CGH)

CGH is a dominant, X-linked genetic disorder that causes extreme hairiness on the face and upper body.

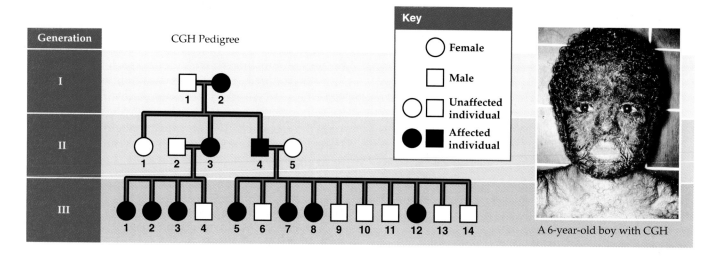

A 6-year-old boy with CGH

■ Because males inherit only one X chromosome, genes on sex chromosomes have different patterns of inheritance than do genes on autosomes. Males are much more likely than females to get an X-linked genetic disorder, whereas both sexes are equally likely to get an autosomal genetic disorder.

Inherited Chromosomal Abnormalities

Relatively few human genetic disorders are caused by inherited chromosomal abnormalities, probably because most large changes to the chromosomes kill the developing embryo. However, two main types of changes to chromosomes occur in humans and other organisms: changes to the structure (for example, the length) of an individual chromosome and changes to the overall number of chromosomes.

Changes to the structure of chromosomes have dramatic effects. As we discussed earlier in this chapter, some XY individuals develop as females because their Y chromosome is missing a region that contains the sex-determining gene. Likewise, some XX individuals develop as males because a similar region of a Y chromosome has become attached to one of their X chromosomes. Another example is cri du chat syndrome, which develops when a child inherits a chromosome 5 that is missing a particular region (Figure 13.11). *Cri du chat* is French for "cry of the cat," a name that describes the characteristic mewing sound made by infants suffering from this condition. Individuals with this syndrome grow slowly and tend to have severe mental retardation, small heads, and low-set ears.

Most changes to the number of chromosomes result in death of the embryo. It is estimated that at least 20% of all pregnancies spontaneously abort, largely as a result of changes in chromosome number. Unusual numbers of chromosomes—such as one or three copies instead of the usual two—occur when chromosomes fail to separate properly during meiosis (see Chapter 10). Down's syndrome is the only nonfatal disorder in which a person inherits too many or too few autosomes. This condition occurs when an individual has three copies of chromosome 21, the smallest autosome in humans (see Figure 18.6). Individuals with Down's syndrome tend to be short and mentally retarded, and to have defects of the heart, kidneys, and digestive tract. They also tend to be cheerful and affectionate, to enjoy music and dance, and to have a flair for mimicry. Live births also can result when an infant has three copies of chromosome 13, 15, or 18. However, in each of these cases there

are severe birth defects and the children rarely live beyond their first year.

Compared to autosomes, changes to the number of sex chromosomes can have relatively minor effects. For example, XXY males have normal life spans and normal intelligence, and tend to be tall. However, they also have reduced fertility, and some have feminine characteristics, such as enlarged breasts.

■ Chromosome abnormalities include changes in the structure of an individual chromosome and changes in the overall number of chromosomes. Changes to the structure of an individual chromosome can have profound effects. In humans, most changes to the number of autosomes result in the death of the developing embryo. Down's syndrome, which occurs in individuals with three copies of chromosome 21, is an exception to this rule. Changes to the number of sex chromosomes in humans can have relatively minor effects.

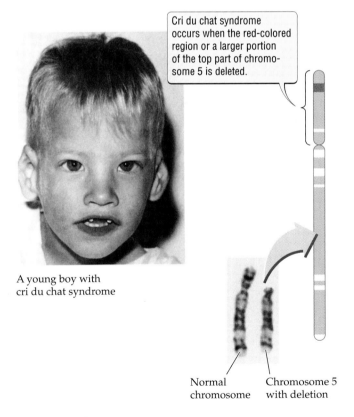

Cri du chat syndrome occurs when the red-colored region or a larger portion of the top part of chromosome 5 is deleted.

A young boy with cri du chat syndrome

Normal chromosome

Chromosome 5 with deletion

Figure 13.11 Cri du Chat Syndrome
Cri du chat syndrome is caused by the deletion of a portion of the top part of chromosome 5.

Uncovering the Genetics of Huntington's Disease

George Huntington wrote the classic paper describing Huntington's disease in 1872. From then until 1983, there was little hope for progress in treatment of the disease. It was known that the gene for the disease is on an autosome, and that the disease-causing allele is dominant (*A*). But there was no cure, and little helpful information could be given to potential victims of the disease. For example, by constructing a pedigree a geneticist might learn that a person's father (who had the disease) was of genotype *Aa*. If the mother did not have the disease, and hence was of genotype *aa*, all the geneticist could say was that the person had a 50% chance of developing the disease (see Problem 3 in the sample genetics problems at the end of the chapter).

In 1983, however, this situation changed dramatically when researchers discovered that the gene for Huntington's disease is located on a portion of chromosome 4. This discovery was possible because pedigree analyses indicated that the gene for Huntington's disease (HD) was linked to other genes on chromosome 4. The discovery that the HD gene was carried on chromosome 4 set off an intense effort to isolate the gene. Ten years later, the gene was found. To find the gene, researchers carefully determined which chromosome 4 genes were linked most closely to the HD gene, a process that eventually allowed them to pinpoint the gene's location on chromosome 4 and then isolate the gene itself.

By isolating the gene, scientists learned how the proteins of individuals with Huntington's disease differ from those of individuals without the disease. (We will return to the general topic of how this is done in Chapter 15.) The key point is that from these analyses, researchers learned that the gene for Huntington's disease causes the production of an abnormal protein. This protein forms clumps in the brains of people with the disease. Some scientists think that these clumps interfere with normal brain function and cause the disease. Others think that the clumps are a symptom of the disease rather than the cause. In either case, the new results have started an effort to understand the molecular cause of the disease and, ultimately, to use that information to design effective treatments.

Isolation of the gene for Huntington's disease had another, immediate effect: It allowed scientists to design a diagnostic genetic test for the disease. With this test, a person whose parent had the disease can now know with near certainty whether or not he or she also will get the disease. In some cases, the genetic test offers hope. For example, if a person is at risk and wants to have a family without the fear of passing a terrible disease to their children, they can take the test and make an informed decision about having children. But the test poses an agonizing choice for these and all other individuals at risk: If they take the genetic test, they may experience tremendous relief if the results show that they will not get the disease. Alternatively, they may experience a crushing burden and loss of hope if they find out they are doomed to die a horrible death. Given these alternatives, would you take the test?

> ■ Analysis of human pedigrees indicated that the gene that causes Huntington's disease (HD) is located on a portion of chromosome 4. By determining which chromosome 4 genes were most closely linked to the HD gene, researchers pinpointed the location on chromosome 4 of the HD gene and then isolated the gene itself. The isolation of the HD gene has improved our understanding of the disease and has led to the development of a diagnostic test for the disease.

Summary

The Role of Chromosomes in Inheritance

■ The chromosome theory of inheritance states that genes are located on chromosomes.

■ Chromosomes are composed of a single DNA molecule and many proteins.

■ A gene is a region of DNA within the DNA of the chromosome.

■ The physical location of a gene on a chromosome is called a locus.

Autosomes and Sex Chromosomes

■ Chromosomes that determine gender are called sex chromosomes; all other chromosomes are called autosomes.

■ Human males have one X and one Y chromosome. Human females have two X chromosomes.

■ A specific gene on the Y chromosome is required for human embryos to develop as males.

Linkage and Crossing-Over

■ Genes that are located on the same chromosome and that do not assort independently are genetically linked.

■ All genes on the same chromosome are members of the same linkage group.

■ Crossing-over, the exchange of genes between chromosomes, disrupts the linkage between genes.

Genetic Recombination

■ In the independent assortment of chromosomes, maternal and paternal chromosomes are randomly distributed into gametes during meiosis.

■ Crossing-over and the independent assortment of chromosomes cause the formation of nonparental genotypes.

■ The formation of nonparental genotypes (genetic recombination) provides genetic variation on which evolution can act.

Human Genetic Disorders

■ Humans suffer from a variety of genetic disorders, including those caused by mutations to a single gene, and those caused by abnormalities in chromosome number or structure.

■ Pedigrees provide a useful way to study human genetic disorders.

Autosomal Inheritance of Single-Gene Mutations

■ For a genetic disorder caused by a recessive allele (*a*) of a gene on an autosome, only homozygous *aa* individuals get the disease. Heterozygous (*Aa*) individuals are merely carriers of the disease.

■ For dominant genetic disorders, both *AA* and *Aa* individuals suffer from the disorder.

■ Lethal dominant genetic disorders can remain in the population because of late onset of disease symptoms as in Huntington's disease, or because new disease-causing alleles are generated by mutation.

Sex-Linked Inheritance of Single-Gene Mutations

■ Because males inherit only one X chromosome, genes on sex chromosomes have different patterns of inheritance than do genes on autosomes.

■ Males are much more likely than females to get an X-linked genetic disorder; both sexes are equally likely to get an autosomal genetic disorder.

Inherited Chromosomal Abnormalities

■ Chromosome abnormalities include changes to chromosome structure and overall number.

■ Changes to the structure of an individual chromosome can have profound effects.

■ Changes to the number of autosomes in humans are usually fatal. Down's syndrome, which occurs in individuals with three copies of chromosome 21, is an exception to this rule.

■ Changes to the number of sex chromosomes in humans can have relatively minor effects.

Highlight: Uncovering the Genetics of Huntington's Disease

■ Analysis of human pedigrees indicated that the gene that causes Huntington's disease is located on a portion of chromosome 4.

■ By determining which chromosome 4 genes were most closely linked to the HD gene, researchers pinpointed the location on chromosome 4 of the HD gene and then isolated the gene itself.

■ Isolation of the HD gene has improved our understanding of the disease and has led to the development of a diagnostic test for the disease.

Key Terms

autosome p. 209
carrier p. 215
chromosome theory of inheritance p. 208
crossing-over p. 211
genetic linkage p. 210
genetic recombination p. 212
homologous chromosome p. 208
independent assortment of chromosomes p. 212
linkage group p. 210
locus p. 208
pedigree p. 213
sex chromosome p. 209
sex-linked p. 215
X-linked p. 215

Chapter Review

Self-Quiz

1. Genes are
   a. located on chromosomes.
   b. composed of DNA.
   c. composed of both protein and DNA.
   d. both a and b

2. Which of the following human genetic disorders is an autosomal, dominant disorder that has late onset and destroys the nervous system, resulting in death?
   a. Tay-Sachs disease
   b. Huntington's disease
   c. Down's syndrome
   d. Cri du chat syndrome

3. Crossing-over is
   a. more likely between genes that are close together on a chromosome.
   b. more likely between genes that are on different chromosomes.
   c. more likely between genes that are far apart on a chromosome.
   d. not related to the distance between genes.

4. Comparatively few genetic disorders are caused by chromosome abnormalities. One reason is that
   a. most chromosome abnormalities have little effect.
   b. it is difficult to detect changes in the number or length of chromosomes.
   c. most chromosome abnormalities result in spontaneous abortion of the embryo.
   d. it is not possible to change the length or number of chromosomes.

5. Genetic recombination can be caused by
   a. crossing-over and the independent assortment of chromosomes.
   b. linkage.
   c. autosomes.
   d. sex chromosomes.

Review Questions

1. Explain how nonparental genotypes are formed.

2. Do you think alleles that cause lethal, dominant genetic disorders are likely to be as common as alleles that cause lethal, recessive genetic disorders? Explain your answer.

3. Huntington's disease is a genetic disorder that begins in middle age and destroys the nervous system, eventually killing its victims. Although at present no cure is available, the gene that causes the disease has been isolated and a genetic test has been developed that can tell potential victims with near certainty whether or not they will get the disease. If you were at risk for the disease, would you take the genetic test? More generally, what are the advantages and disadvantages of genetic tests that identify whether or not a person has a particular genetic disorder?

Sample Genetics Problems

1. Human females have two X chromosomes; human males have one X chromosome and one Y chromosome.
   (a) Do males inherit their X chromosome from their mother or their father?
   (b) If a female has one copy of an X-linked recessive allele that causes a genetic disorder, does she have the disorder?
   (c) If a male has one copy of an X-linked recessive allele that causes a genetic disorder, does he have the disorder?
   (d) A female is a carrier of an X-linked, recessive disorder. With respect to the disease-causing allele, how many types of gametes can the female produce?
   (e) Assume that a male with an X-linked, recessive genetic disorder has children with a female who does not carry the disease-causing allele. Could any of their children have the genetic disorder? Could any of their children be carriers for the disorder?

2. Cystic fibrosis is a genetic disorder caused by a recessive allele, *a*; the disease-causing allele is located on an autosome. What are the chances that parents with the following genotypes will have a child with the disorder? (a) *aa* × *Aa*, (b) *Aa* × *AA*, (c) *Aa* × *Aa*, (d) *aa* × *AA*.

3. Huntington's disease is a genetic disorder caused by a dominant allele, *A*; the disease-causing allele is located on an autosome. What are the chances that parents with the following genotypes will have a child with Huntington's disease? (a) *aa* × *Aa*, (b) *Aa* × *AA*, (c) *Aa* × *Aa*, (d) *aa* × *AA*.

4. Hemophilia is a genetic disorder caused by a recessive allele, *a*; the disease-causing allele is located on the X chromosome. What are the chances that parents with the following genotypes will have a child with hemophilia?
   (a) $X^A X^A \times X^a Y$
   (b) $X^A X^a \times X^a Y$
   (c) $X^A X^a \times X^A Y$
   (d) $X^a X^a \times X^A Y$
   Do male and female children have the same chance of getting the disease?

5. Explain why the terms "homozygous" and "heterozygous" do not apply to X-linked traits in males.

6. Study the pedigree shown below. Is the disease-causing allele dominant or recessive? Is the disease-causing allele located on an autosome or on the X chromosome? What are the genotypes of individuals 1 and 2 in generation I?

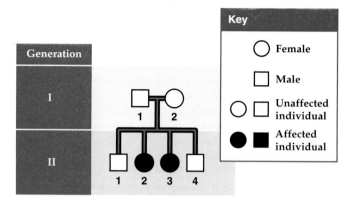

7. Study the pedigree shown below. Is the disease-causing allele dominant or recessive? Is the disease-causing allele located on an autosome or on the X chromosome? To answer this question, assume that individual 1 in generation I and individuals 1 and 6 in generation II do not carry the disease-causing allele.

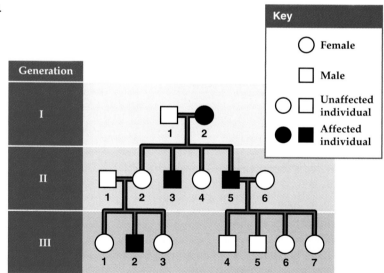

8. Consider the inheritance of two genes, one with alleles *A* or *a*, the other with alleles *B* or *b*. *AABB* individuals are crossed with *aabb* individuals to produce F_1 offspring, all of which have genotype *AaBb*. If these *AaBb* F_1 offspring are then crossed with *aaBB* individuals, construct Punnett squares and list the possible offspring genotypes that would you expect if (a) the two genes were completely linked or (b) the two genes were on different chromosomes.

The Daily Globe

Alcoholism Gene Questioned

NEW YORK, NY. Six months ago, there was a flurry of news coverage about the discovery of a gene for alcoholism. Now it seems that in their rush to win the race for isolating an alcoholism gene, the scientists who reported that discovery may not have isolated the right gene after all.

Alcoholism is one of several addictive behaviors that is thought to be at least partially under genetic control. Because alcoholism is so common and has such a heavy social cost, a lot of media hoopla accompanied the discovery of an alcoholism gene. But new studies question whether the isolated gene really does influence alcoholism.

A second group of scientists, writing in the journal *Natural Science*, report that the gene discovered 6 months ago is linked to the alcoholism gene, but is not the actual alcoholism gene. The scientists who made the original discovery are sticking to their guns, so only time will tell which group of reseachers is right.

Evaluating "The News"

1. In a court case, a man accused of killing his family while under the influence of drugs and alcohol was found to carry the putative alcoholism gene. He claimed he was genetically fated to be an alcoholic and therefore, should not be held accountable for behavior caused by this disease. He pleaded innocent to the murders. If you were on the jury, would you vote innocent or guilty? Why?

2. Given the latest news on the alcoholism gene, what scientific information would you as a jury member want to have to decide whether genes and alcoholism could have made him commit his crime?

3. To generalize from the alcoholism gene, assume that a gene has been isolated that is thought to control a particular trait. This gene might actually control the trait in question. Alternatively, the gene might be closely linked to a gene that controls the trait in question but does not itself cause the trait. How might one distinguish between these two possibilities? What role would any of this play in your decision as a jury member?

chapter 14

DNA

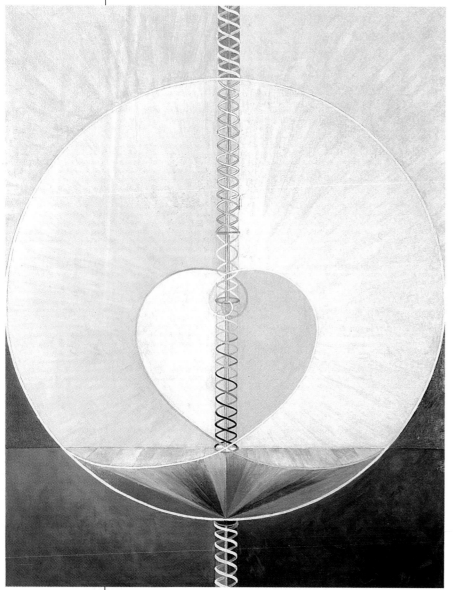

Hilma af Klint, *Group 9. Series UW. Dove no. 25*, 1915.

The Library of Life

The book you are reading has more than a million letters in it. How long do you think it would take you to copy these letters by hand, one by one? How many mistakes do you think you would make? Would you check your work for mistakes, and if so, how long do you think that would take?

Difficult as the job of copying the letters of this book would be, it pales in significance compared to the job that your cells do each time they divide. When a cell divides, it has to make a copy of all its genetic information. That information is stored in deoxyribonucleic acid, or DNA, the substance of which genes are made.

DNA is the hereditary material, its three-dimensional structure consisting of two strands of nucleotides twisted into a double helix.

The amount of information stored in your DNA is mind-boggling: Whereas this book contains roughly one million letters, the DNA in each of your cells has the equivalent of 3,300,000,000 letters. If it were printed with the same font and the same size as the letters on this page, the information that is in your DNA would fill thousands of books similar in length to this one.

Your cells copy the phenomenal amount of information in your DNA in a matter of hours. Yet despite how fast they work, on average they make only one mistake for every billion "letters." How do they do this? What are the "letters" that make up the DNA molecule? More broadly, what is the structure of the DNA molecule, and what implications does this structure have for life?

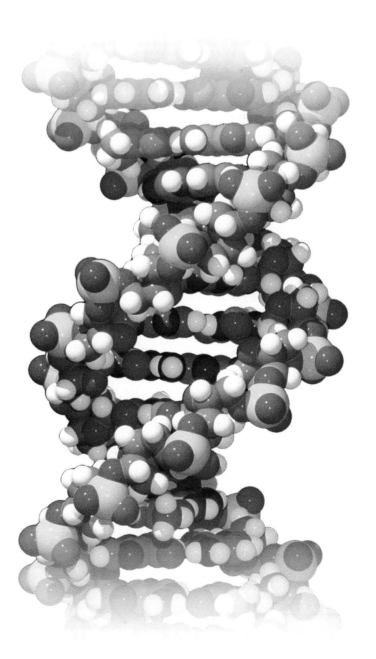

A Model of the DNA Molecule

1. Genes are composed of DNA.

2. Four nucleotides are the building blocks of DNA. Each nucleotide contains one of four nitrogen-containing molecules, or "bases": adenine, cytosine, guanine, and thymine.

3. DNA consists of two twisted strands of nucleotides. The two strands are held together by hydrogen bonds that form between adenine and thymine, and between cytosine and guanine.

4. Each strand of DNA has an enormous number of bases arranged one after another. The sequence of bases in

DNA differs among species and among individuals within a species. These differences are the basis of inherited variation; for example, different alleles of a gene have different DNA sequences.

5. Because ordinarily adenine pairs only with thymine, and cytosine pairs only with guanine, each strand of DNA can serve as a template with which to duplicate the other strand.

6. DNA in cells is damaged constantly by factors like heat energy and UV light. If this damage were not repaired, the organism would die.

We have discussed genes in several different ways in this unit. In Chapter 12, for example, genes were treated as abstract entities that control the inheritance of characters. In Chapter 13, we described how genes are physically located on chromosomes. Knowing that genes are located on chromosomes makes them less abstract, but this knowledge leaves unanswered many fundamental questions, such as, What are genes made of? When a cell divides to form two daughter cells, how is the information in the genes copied? How are errors in copying corrected, and how is damage to the cell's hereditary material repaired?

To answer such questions, geneticists had to discover the substance of which genes are made, and they had to learn the physical structure of this substance. As they began this search, they were guided by three basic biological facts about the nature of the hereditary material.

First, the hereditary material had to contain the information necessary for life. For example, it had to contain the information needed to build the body of the organism and to control the complex metabolic reactions on which life depends. Second, the hereditary material had to be composed of a substance that could be replicated (copied) accurately. If this could not be done, reliable genetic information could not be passed from one generation to the next. Finally, the hereditary material had to be variable. If the genetic material were not variable, there would be no genetic differences within or among species, something that scientists knew was not true.

Parallel to the search for the chemical composition and physical structure of genes was a search for the function of genes: At the molecular level, how did genes

produce their effects? In Chapter 5, we identified DNA as the hereditary material of genes. In this chapter we describe how scientists discovered that genes are composed of DNA. We also discuss the physical structure of genes and how the genetic material is copied and repaired. In Chapter 15 we will discuss how genes produce their effects.

The Search for the Hereditary Material

By the early 1900s, geneticists knew that genes control the inheritance of characters, that genes are located on chromosomes, and that chromosomes are composed of DNA and protein. With this knowledge in hand, the first step in the quest to understand the physical structure of genes was to determine whether DNA or protein was the hereditary material.

Initially, most geneticists thought that protein was the more likely candidate. Proteins are large, complex molecules, and it was not hard to imagine that they could store the tremendous amount of information needed to govern the lives of cells. Proteins also vary considerably within and among species; hence it was reasonable to assume that they caused the inherited variation observed within and among species.

DNA, on the other hand, was initially judged a poor candidate for the hereditary material, mainly because DNA was incorrectly thought to be a small, simple molecule whose composition varied little among different species. Over time, these ideas about DNA were discovered to be wrong. In fact, DNA molecules are large and vary tremendously in their nucleotide composition within and among species. Still, as we will see, the variation contained in DNA is more subtle than the variation in shape, electrical charge, and function shown by proteins, so it is not surprising that most researchers initially favored proteins as the hereditary material.

Over a period of roughly 25 years (1928–1952), geneticists became convinced that DNA, not protein, was the hereditary material. Three key experiments helped to cause this shift of opinion.

Harmless bacteria can be transformed into deadly bacteria

In 1928, a British medical officer named Frederick Griffith published an important paper on *Streptococcus pneumoniae*, a bacterium that causes pneumonia in humans and other mammals. Griffith was studying two genetic varieties, or strains, of *Streptococcus* in order to find a cure for pneumonia, a common cause of death at that time. The two strains, called strain S and strain R, were named after differences in appearance. When grown on a dish, strain S produced colonies that appeared smooth, while strain R produced colonies that appeared rough.

Griffith conducted four experiments on these bacteria (Figure 14.1). First, when he injected bacteria of type R into mice, the mice lived and did not develop pneumonia. In contrast, when he injected bacteria of type S into mice, the mice developed pneumonia and died. In the third experiment, he injected heat-killed strain S bacteria into mice, and once again the mice lived. His plan was to test mice from the third experiment to see if they were resistant to later exposure to strain S.

None of these results were particularly unusual: Griffith had shown simply that there were two strains of bacteria, one of which (strain S) killed mice and was itself killed and rendered harmless by heat. In the fourth experiment, however, Griffith mixed heat-killed bacteria of strain S with live bacteria of strain R. On the basis of the results from the first three experiments, he expected the mice to live. Instead, the mice died, and Griffith recovered live bacteria of strain S from the blood of the dead mice.

In Griffith's fourth experiment, something caused harmless, strain R bacteria to change into deadly, strain S bacteria. Griffith showed that the change was genetic: When they reproduced, the altered strain R bacteria produced strain S bacteria. Overall, the results of Griffith's fourth experiment

suggested that genetic material from heat-killed strain S bacteria somehow changed living strain R bacteria into strain S bacteria.

This remarkable and unexpected result stimulated an intensive hunt to identify the material that caused this change. We now know that the strain R bacteria had

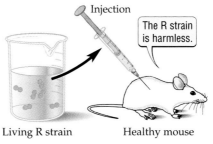
Injection

The R strain is harmless.

Living R strain   Healthy mouse

Figure 14.1 Genetic Transformation of Bacteria
Harmless strain R bacteria can be transformed into deadly strain S bacteria.

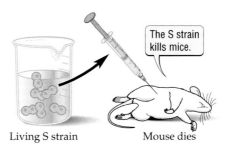
The S strain kills mice.

Living S strain   Mouse dies

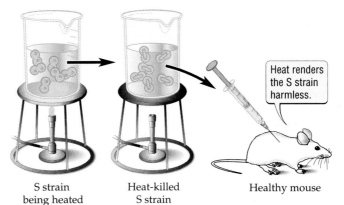
Heat renders the S strain harmless.

S strain being heated   Heat-killed S strain   Healthy mouse

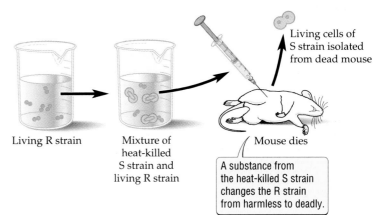
Living cells of S strain isolated from dead mouse

Living R strain   Mixture of heat-killed S strain and living R strain   Mouse dies

A substance from the heat-killed S strain changes the R strain from harmless to deadly.

absorbed DNA from the heat-killed strain S bacteria, causing the genotype of the strain R bacteria to change. A change in the genotype of a cell or organism after exposure to the DNA of another genotype is called **transformation**.

DNA can transform bacteria

For 10 years, the American Oswald Avery and his colleagues at the Rockefeller University in New York struggled to identify the genetic material that had transformed the bacteria in Griffith's experiments. They isolated and tested different compounds from the bacteria. Only DNA was able to transform harmless strain R bacteria into deadly strain S bacteria. In 1944, Avery, Colin MacLeod, and Maclyn McCarty published a landmark paper that summarized their results. The paper created quite a stir.

In addition to showing that DNA transforms bacteria, Avery, MacLeod, and McCarty's paper led many biologists to a broader conclusion: that DNA, not protein, was the hereditary material. As a leading DNA researcher later remarked, the paper stimulated an "avalanche" of new research on DNA. Some biologists remained skeptical, arguing, for example, that DNA was too simple a molecule to be the hereditary material. However, the tide was definitely turning in favor of DNA.

Hereditary instructions in viruses are contained in DNA

In Griffith's experiments, heat killed the strain S bacteria but did not destroy its genetic material. Since most proteins are destroyed by heat, this result suggested that proteins were not the transforming material. Then the work by Avery, MacLeod, and McCarty provided very strong evidence that DNA was the hereditary material. Additional proof came in 1952, when Alfred Hershey and Martha Chase published an elegant study on the genetic material of viruses.

Hershey and Chase studied a virus that attacks bacteria and consists only of DNA and protein (Figure 14.2). To reproduce, this virus attaches to the wall of a bacterium and injects its genetic material into the bacterium. The genetic material of the virus then takes over the

Figure 14.2 DNA Is the Hereditary Material
The Hershey–Chase experiments on viruses that infect bacteria showed that DNA, not protein, is the hereditary material. In their experiments, Hershey and Chase grew viruses in a solution that contained either radioactive sulfur (^{35}S) or radioactive phosphorus (^{32}P). Sulfur is a common element of the virus's proteins, but it is absent from DNA. Viruses grown in radioactive sulfur make proteins but not DNA with ^{35}S, thus providing a radioactive tag, called a label, for the proteins. For phosphorus, the reverse is true: It labels DNA, not protein.

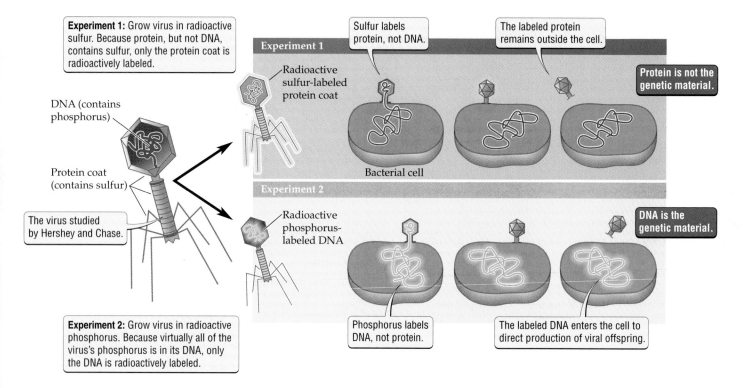

bacterial cell, eventually killing it, but first causing it to produce many new viruses. Because the virus is composed of only protein and DNA, it was an excellent organism in which to test whether DNA or protein was the genetic material.

Hershey and Chase demonstrated that only the DNA portion of the virus was injected into the bacterium (see Figure 14.2). This result indicated that DNA was responsible for taking over the bacterial cell and causing the production of new viruses. These experiments convinced most remaining skeptics that DNA, not protein, was the hereditary material.

■ Geneticists initially thought that protein was the genetic material. Three key experiments showed, however, that DNA, not protein, is the genetic material.

The Three-Dimensional Structure of DNA

By the early 1950s, the stage was set for the discovery of the three-dimensional structure of genes. By then, genes were known to be composed of DNA, so the next big step was to figure out the structure of DNA and hence the structure of genes.

In 1951, Linus Pauling became the first person to figure out the three-dimensional structure of a protein. Pauling's success suggested that determining the three-dimensional structure of DNA should also be possible. Leading laboratories from around the world, including Pauling's, devoted great effort toward reaching that goal. This effort was a race to unlock some of the greatest mysteries of life: How was the cell's genetic material duplicated so that it could be passed from parent to offspring? How was genetic information stored in DNA?

It was hoped that knowledge of the physical structure of DNA would provide clues to these fundamental questions. As we will see, this hope was fulfilled in a dramatic way: The discovery of the structure of DNA immediately suggested answers to these and other important biological questions.

DNA is a double helix

Working at Cambridge University in England, the American James Watson and the Englishman Francis Crick won the race to determine the physical structure of DNA (see also the Box on page 228). In a two-page paper published in 1953, they proposed that DNA was a **double helix**, a structure that can be thought of as a ladder twisted into a spiral coil (Figure 14.3).

This "ladder" is more accurately described as two long strands of covalently bonded nucleotides, with the two strands held together by hydrogen bonds and twisted into a helix. As discussed in Chapter 5, **nucleotides** are composed of a sugar–phosphate backbone and one of four nitrogen-containing molecules, called **bases**: adenine (A), cytosine (C), guanine (G), and thymine (T). The sugar–phosphate backbones of the two nucleotide

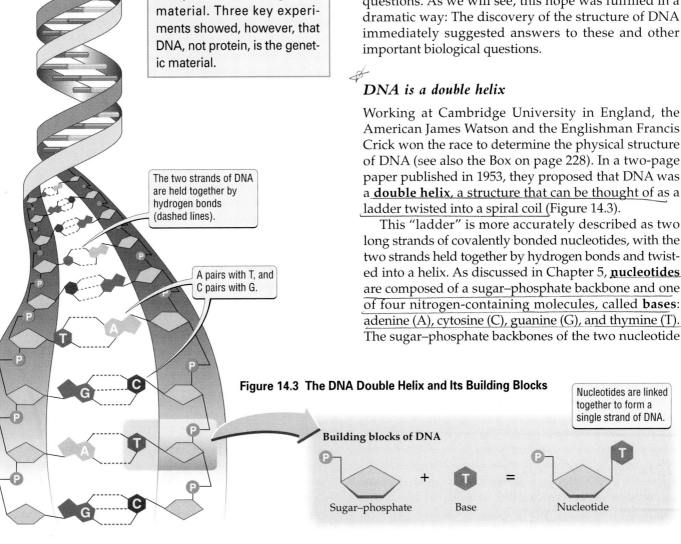

The two strands of DNA are held together by hydrogen bonds (dashed lines).

A pairs with T, and C pairs with G.

Figure 14.3 The DNA Double Helix and Its Building Blocks

Nucleotides are linked together to form a single strand of DNA.

Building blocks of DNA

P Sugar–phosphate + T Base = P T Nucleotide

Rosalind Franklin: Crucial Contributor to the Discovery of DNA's Structure

In 1962, James Watson and Francis Crick received the Nobel prize in physiology or medicine for determining the three-dimensional structure of DNA. They shared the prize with Maurice Wilkins, who also had worked on the structure of DNA. Missing from the 1962 Nobel ceremony was Rosalind Franklin, a gifted young scientist whose research provided critical data that Watson and Crick used to determine the structure of DNA.

Rosalind Franklin was born in London in 1920. At St. Paul's Girls' School and Cambridge University, she studied physics and chemistry. After leaving Cambridge University in 1942, she worked in industry for several years. From 1947 to 1950 she worked in Paris, learning how to use X-rays to produce photographs of crystals. In 1951, she returned to Cambridge University to work on DNA. By that time, she was a master at taking X-ray photographs of crystals.

Rosalind Franklin's research at Cambridge provided several important clues for Watson and Crick. First, she took very clear X-ray photographs of the DNA fibers prepared by Wilkins. As described by Watson in his 1968 book, *The Double Helix*, these photographs provided essential information to Watson and Crick as they sought to determine the molecular structure of DNA. Rosalind Franklin also demonstrated that phosphate groups are located on the outside of the DNA molecule, a key feature of Watson and Crick's description of the three-dimensional structure of DNA.

Rosalind Franklin died of cancer in 1958 at the age of 37. Nobel prizes cannot be awarded after the recipient's death, so no one will ever know whether or not Franklin would have been awarded a Nobel prize for her critical contributions to determining the structure of DNA. Unfortunately, at times during her scientific career, her work was not given the high regard that it fully deserved. In *The Double Helix*, Watson commented that he realized years too late what a struggle it was for a woman like Rosalind Franklin to be accepted by a scientific community that often failed to take female scientists seriously. Increasingly recognized for her scientific accomplishments, Rosalind Franklin, because of her achievements and her courage in the face of such skepticism, is an excellent role model for scientists today, both male and female.

Rosalind Franklin

strands are the "sides" of the ladder; the bases are the "rungs" of the ladder. The fact that DNA consists of twisted strands explains why DNA is called a double helix.

Watson and Crick's paper contained two key realizations. First, they recognized that there were two strands of nucleotides in DNA. Second, they realized that only certain combinations of bases could pair with each other. Watson and Crick proposed that adenine could pair only with thymine, and that cytosine could pair only with guanine. These pairing rules had an important consequence: If the sequence of bases on one strand of the DNA molecule was known, the sequence of bases on the other strand of the molecule was automatically known.

For example, if one strand consisted of the sequence

A-C-C-T-A-G-G-G

the other strand had to have the sequence

T-G-G-A-T-C-C-C

Any other sequence would violate the restriction that A pairs only with T, and C pairs only with G.

We now know that the physical structure of DNA proposed by Watson and Crick is correct in all essential elements. This structure has great explanatory power. For example, as will be described in the following section, the fact that adenine can pair only with thymine, and that cytosine can pair only with guanine, immediately suggested a simple way in which the DNA molecule could be replicated: The original strands could serve as templates on which to build the new strands.

Knowledge of the three-dimensional structure of DNA also indicated that DNA could be viewed as a long string of the four bases: A, C, G, and T. Although A had to pair with T, and C had to pair with G, the four bases could be arranged in any order along a strand of the DNA molecule. The fact that each strand of DNA is composed of millions of these bases suggests that a tremendous amount of information can be contained in how the bases are ordered along the DNA molecule (see Chapter 15).

The sequence of bases in DNA differs among species and among individuals within a species (Figure 14.4). Differences in DNA sequence are the basis of inherited variation: Different alleles of a gene have different DNA sequences. For example, people with a genetic disorder like Huntington's disease or cystic fibrosis (see Chapter 13) inherit particular alleles that cause the disease. At the molecular level, one allele causes a disease and another allele does not because the two alleles have a different sequence of bases.

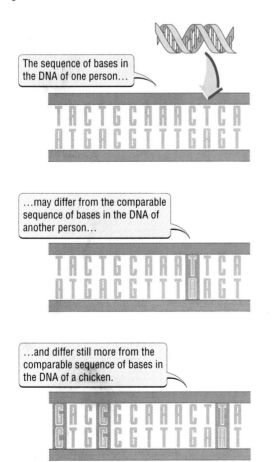

The sequence of bases in the DNA of one person…

…may differ from the comparable sequence of bases in the DNA of another person…

…and differ still more from the comparable sequence of bases in the DNA of a chicken.

> ■ DNA is a double helix consisting of two long strands of covalently bonded nucleotides held together by hydrogen bonds between the nitrogen-containing bases adenine, cytosine, guanine, and thymine. Adenine pairs only with thymine; cytosine pairs only with guanine. The specific base-pairing rules indicate that each strand can serve as a template from which to copy the other strand. The sequence of bases in DNA, which differs by species and by individual, is the basis of inherited variation.

How DNA Is Replicated

As Watson and Crick noted in their historic 1953 paper, the structure of the DNA molecule suggested a simple way that the genetic material could be copied. They elaborated on this suggestion in a second paper, also published in 1953. Because A pairs only with T, and C pairs only with G, each strand of DNA contains the information needed to duplicate the other strand. For this reason, Watson and Crick suggested that **DNA replication**, the duplication of a DNA molecule, could work in the following way (Figure 14.5):

1. The hydrogen bonds connecting the two strands could be broken.

2. Breaking the hydrogen bonds would cause the two strands to unwind and separate.

3. Each strand could then be used as a template for the construction of a new strand of DNA.

4. When this process was completed, there would be two copies of the DNA molecule, each composed of one "old" strand of DNA (from the parent DNA molecule), and one newly synthesized strand of DNA.

Five years later, other researchers confirmed that DNA replication produces DNA molecules composed of one old strand and one new strand, as predicted by Watson and Crick.

The Watson–Crick model of DNA replication is elegant and simple, but the mechanics of copying DNA are far from simple. More than a dozen enzymes and proteins

Figure 14.4 Variation in the Sequence of Bases in DNA
The sequence of bases in DNA differs among species and among individuals within a species. Here the sequence of bases for a hypothetical region of DNA in two humans and a chicken is compared. Base pairs highlighted in red are different from the corresponding base pairs of the first human.

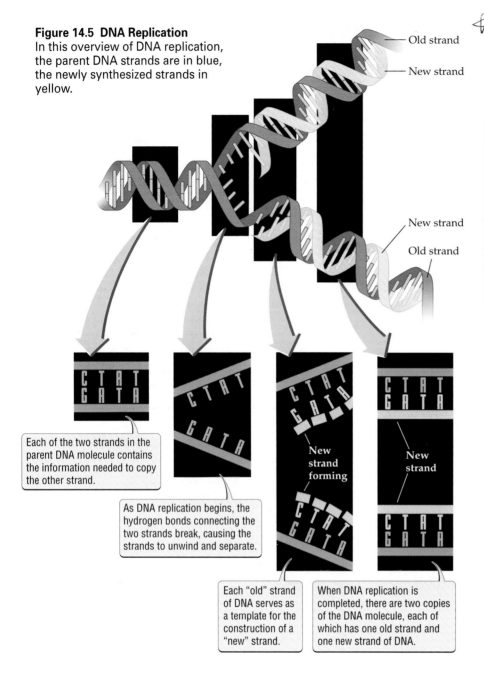

Figure 14.5 DNA Replication
In this overview of DNA replication, the parent DNA strands are in blue, the newly synthesized strands in yellow.

Old strand

New strand

New strand

Old strand

Each of the two strands in the parent DNA molecule contains the information needed to copy the other strand.

As DNA replication begins, the hydrogen bonds connecting the two strands break, causing the strands to unwind and separate.

New strand forming

New strand

Each "old" strand of DNA serves as a template for the construction of a "new" strand.

When DNA replication is completed, there are two copies of the DNA molecule, each of which has one old strand and one new strand of DNA.

■ Because A pairs only with T, and C pairs only with G, each strand of DNA contains the information needed to duplicate the other strand. To replicate DNA, proteins break the hydrogen bonds connecting the two nucleotide strands, which then unwind and separate. Each of these strands is then used as a template for building a new strand of DNA. DNA replication produces two copies of the DNA molecule, each composed of one old strand of DNA (from the parent DNA molecule) and one newly synthesized strand.

Repairing Damaged DNA

When DNA is copied, there are many opportunities for mistakes to be made. In humans, for example, more than 3 billion nucleotides must be copied each time a cell divides, so there are more than 3 billion opportunities for mistakes. In addition, the DNA in cells is constantly being damaged by various sources, including radiation, collisions with other molecules in the cell, and chemicals, many of which are produced by the cell itself. A failure to repair copying errors and damage to the DNA would be disastrous for the organism: Normal cell functions would grind to a halt if essential genes were damaged or replicated incorrectly. This damage would lead to cell death and, ultimately, to the death of the organism.

are needed to unwind the DNA, to stabilize the separated strands of DNA, to start the replication process, to attach nucleotides to the correct positions on the template, and to join partly replicated fragments of DNA to each other.

Although DNA replication is a complex task, cells can copy DNA molecules that contain billions of nucleotides in a matter of hours. This speed is achieved in part by starting the replication of the DNA molecule at thousands of different places at once. Cells make few mistakes when they copy their DNA. As we discuss in the following section, when a mistake does occur, DNA repair operations usually correct it.

Few mistakes are made in DNA replication

The enzymes that replicate DNA sometimes insert an incorrect base in the newly synthesized strand of DNA. For example, if a cytosine (C) were inserted across from an adenine (A) located on the template strand, an unusual C–A pair bond would form instead of the correct T–A pair bond (Figure 14.6). Such mistakes are made about once in every 10,000 bases. Depending on the organism, one or more enzymes checks or "proofreads" the pair

Figure 14.6 Mistakes Can Be Made in DNA Replication
Here, a cytosine (C) has incorrectly been inserted opposite an adenine (A). DNA repair enzymes almost always fix such problems before the DNA is replicated again.

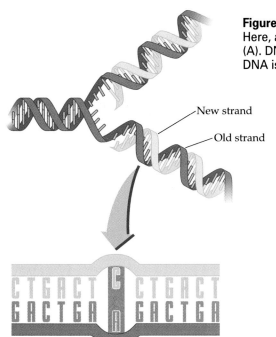

New strand

Old strand

ple, the DNA in each of our cells is damaged thousands of times by heat energy and random chemical accidents. The vast majority of the damage to DNA is fixed by a complex set of repair proteins. Single-celled organisms like yeasts have more than 50 different repair proteins, and humans probably have even more.

There are three steps in the repair of damage to DNA, or **DNA repair**: The damaged DNA must be recognized, removed, and replaced (Figure 14.7). Different sets of repair proteins specialize in recognizing and removing different types of DNA damage. Once these first two steps have been accomplished, the final step is to add the correct sequence of bases to the damaged strand of

bonds as they form, reducing the error rate to about one in every 10 million bases.

When an incorrect base is added and escapes the proofreading mechanism, a **mismatch error** is said to occur. Mismatch errors are so named because the bases are not correctly matched with each other. Cells contain an additional set of repair proteins that specialize in fixing mismatch errors. These repair proteins fix 99% of the remaining errors, reducing the overall chance of an error to the incredibly low rate of one mistake in every billion bases.

On the rare occasions when a mistake is not corrected, the DNA sequence changes. A change in the sequence of bases in DNA is called a **mutation**. Thus, when a mistake in the copying process is not corrected, a mutation has occurred. Mutations cause the formation of new alleles, including those that cause cancer and other human genetic disorders, such as sickle-cell anemia and Huntington's disease.

Normal gene function depends on DNA repair

The DNA in cells constantly suffers from mechanical, chemical, and radiation damage. Every day, for exam-

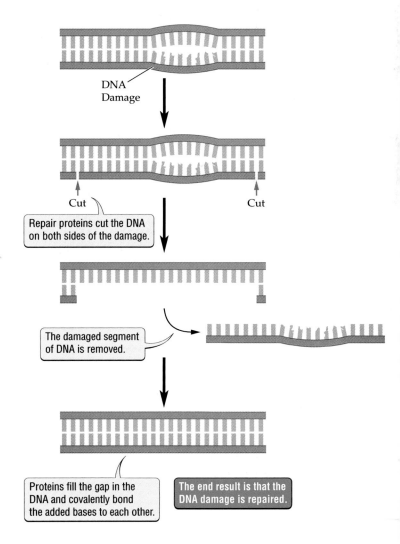

DNA Damage

Cut   Cut

Repair proteins cut the DNA on both sides of the damage.

The damaged segment of DNA is removed.

Proteins fill the gap in the DNA and covalently bond the added bases to each other.

The end result is that the DNA damage is repaired.

Figure 14.7 DNA Repair
DNA repair proteins fix many types of damage to DNA by cutting out the damaged DNA and replacing it with newly synthesized DNA.

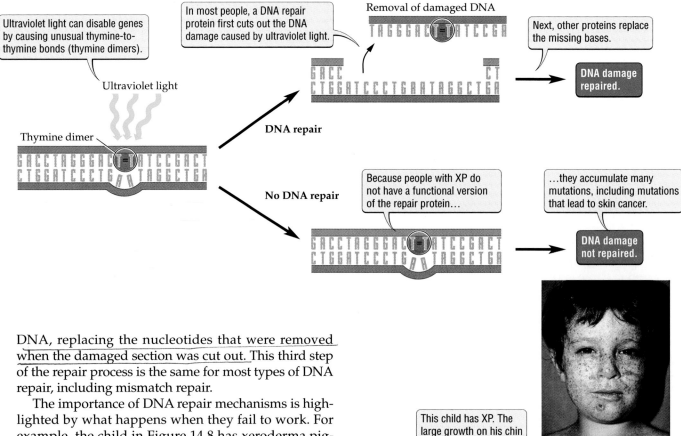

Ultraviolet light can disable genes by causing unusual thymine-to-thymine bonds (thymine dimers).

Ultraviolet light

Thymine dimer

In most people, a DNA repair protein first cuts out the DNA damage caused by ultraviolet light.

Removal of damaged DNA

Next, other proteins replace the missing bases.

DNA damage repaired.

DNA repair

No DNA repair

Because people with XP do not have a functional version of the repair protein...

...they accumulate many mutations, including mutations that lead to skin cancer.

DNA damage not repaired.

This child has XP. The large growth on his chin is a skin cancer.

DNA, replacing the nucleotides that were removed when the damaged section was cut out. This third step of the repair process is the same for most types of DNA repair, including mismatch repair.

The importance of DNA repair mechanisms is highlighted by what happens when they fail to work. For example, the child in Figure 14.8 has xeroderma pigmentosum (XP), a recessive genetic disorder in which even brief exposure to sunlight causes painful blisters. In XP, the disease-causing allele (*a*) produces a nonfunctional version of one of our many DNA repair proteins. The normal form of the protein that is disabled in XP repairs damage to DNA caused by ultraviolet light. The lack of this DNA repair protein makes individuals with XP highly susceptible to skin cancer. Several inherited tendencies to develop cancer, including types of breast and colon cancer, also appear to be caused by defective versions of genes that control DNA repair.

Figure 14.8 Impact of DNA Repair Mechanisms
The failure of DNA repair mechanisms to work properly has severe consequences. This result is illustrated by the high frequency of skin cancers in people who have xeroderma pigmentosum (XP), a recessive genetic disorder whose victims cannot produce a protein that is used to repair damage to DNA from ultraviolet light.

■ On rare occasions, mistakes are made when DNA is copied. Mistakes in the copying process introduce mutations, which are a change in the sequence of bases in DNA. In addition, the DNA in each of our cells is altered thousands of times every day by mechanical, chemical, and radiation damage. If DNA damage in an organism's cells were not repaired, the cells and ultimately the organism would die. Mutations produced by errors in copying DNA or by damage to DNA are fixed by DNA repair proteins.

Highlight

Sunburns and Parsnips

The medical literature is filled with reports of individuals whose skin blistered and turned red, somewhat resembling a sunburn, but who had not been overexposed to sunlight. Often doctors made the wrong diagnosis when they examined such patients, initially blaming the skin damage on factors like child abuse, jellyfish stings, or a possible malignant form of cancer. Sometimes the skin damage cases appear in batches, as when workers at a particular grocery or farm all develop similar symptoms.

Other times they are isolated events: a child who was playing with limes at the beach, a woman who drank a home remedy of celery broth (which, ironically, was meant to improve a skin condition).

The common thread to these examples is a class of chemicals called psoralens (pronounced SORE-uh-luns).

These chemicals are found in certain plant species, including celery, limes, parsnips, and figs. Many plants contain chemicals that can cause rashes or other problems when they contact human skin. But psoralens are a skin irritant with a twist: If the plant were handled or eaten at night or in a room with no sunlight, it would have no effect.

Parsnips

Why do psoralens cause problems only in sunlight? The answer concerns how they work. Sunlight contains a type of light known as ultraviolet, or UV, light. When in the presence of UV light, psoralens can insert themselves into the DNA double helix, bridging the two strands of the DNA molecule and disrupting its normal function. This damage can kill the cells whose DNA is under assault, causing blistering, changes in skin color, and considerable pain. (Psoralens are by no means all bad: They have been used in Egypt and India for thousands of years to treat various skin problems, and they continue to be used today to treat skin and other medical problems.)

The attack of psoralens on DNA mimics the damage that UV light causes on its own. UV light damages DNA by causing unusual chemical bonds to form within the DNA molecule. If the damage is severe enough, the cell dies. Sunburns result from massive damage to DNA caused by UV light. The damage to the DNA causes the death and removal (by peeling) of many skin cells. Cells that survive try to repair the damage caused by UV light, but some damage remains, resulting in mutations to the DNA. If a mutation occurs in certain genes, such as a gene that controls DNA repair or suppresses the growth of tumors, a sunburn can be the first step on a path that leads ultimately to skin cancer (Figure 14.9).

Knowledge of what UV light does to your DNA suggests cautionary actions: While enjoying the outdoors, you should always protect yourself from overexposure to the sun. This protection is especially important if you have a higher-than-average risk of developing skin cancer. For example, you should use special caution if you have any of the following characteristics: (1) fair skin; (2) a family history of cancer; (3) patches of skin that remain sore, dry, or scaly; or (4) moles that are frequently subject to rubbing or other forms of irritation. Skin cancer can strike people of all ages, so if you find a suspicious growth on your skin, such as those shown in Figure 14.9, consult a doctor immediately. Prompt action can save your life.

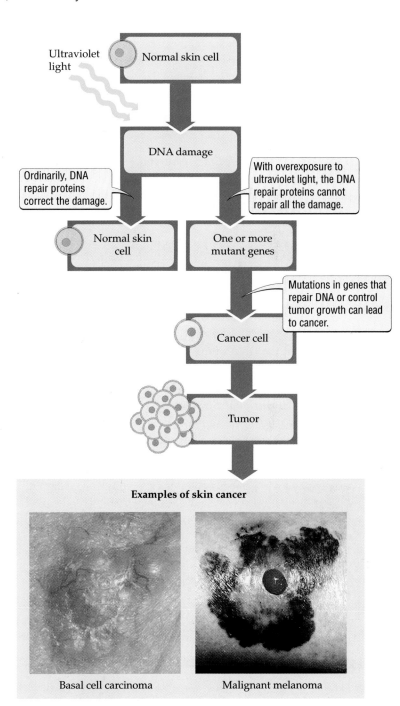

Figure 14.9 Sunburn Can Cause Cancer
Overexposure to ultraviolet light can cause DNA mutations that transform normal cells into cancer cells.

■ Chemicals called psoralens, found in certain plant species, can cause painful blisters and damage DNA if the plant is handled or eaten in the presence of UV light. Acting on its own, UV light also can damage DNA. Over-exposure to psoralens or UV light can damage DNA so extensively that the cell's DNA repair mechanisms fail and the cell dies. In the case of UV light, the mutations to DNA that result from sunburn can be the first step on the path to skin cancer.

Summary

The Search for the Hereditary Material

■ Geneticists initially thought that protein was the genetic material. Three landmark experiments showed that this initial view was wrong and that DNA, not protein, is the genetic material.

■ The first experiment showed that harmless strain R bacteria could be transformed into deadly strain S bacteria when exposed to heat-killed strain S bacteria.

■ The second experiment showed that only the DNA from heat-killed strain S bacteria was able to transform strain R bacteria into strain S bacteria.

■ The third experiment used a virus that attacks bacteria and showed that the DNA of the virus, not the protein, was responsible for taking over the bacterial cell and producing new viruses.

The Three-Dimensional Structure of DNA

■ In 1953, James Watson and Francis Crick determined that DNA is a double helix.

■ The double helix has two long strands of covalently bonded nucleotides. The two strands of nucleotides are held together by hydrogen bonds between the nitrogen-containing bases: adenine, cytosine, guanine, and thymine.

■ An adenine on one strand pairs only with a thymine on the other strand. Similarly, a cytosine on one strand pairs only with a guanine on the other strand.

■ The specific base-pairing rules indicate that each strand can serve as a template from which the other strand can be copied.

■ The sequence of bases in DNA, which differs among species and among individuals within a species, is the basis of inherited variation.

How DNA Is Replicated

■ Because A pairs only with T, and C pairs only with G, each strand of DNA contains the information needed to duplicate the other strand.

■ A complex set of proteins guides the replication of DNA. To replicate DNA, these proteins must first break the hydrogen bonds connecting the two nucleotide strands.

■ Breaking the hydrogen bonds causes the two strands to unwind and separate. Each of these strands is then used as a template from which to build a new strand of DNA.

■ DNA replication produces two copies of the DNA molecule, each composed of one old strand of DNA (from the parent DNA molecule) and one newly synthesized strand of DNA.

Repairing Damaged DNA

■ On rare occasions, mistakes occur when DNA is copied. Mistakes in the copying process introduce mutations, which are a change in the sequence of bases in DNA.

■ The DNA in each of our cells is altered thousands of times every day by mechanical, chemical, and radiation damage. If DNA damage in an organism's cells were not repaired, the cells and ultimately the organism would die.

■ Mutations produced by errors in copying DNA or by damage to DNA are fixed by DNA repair proteins.

■ Several inherited genetic disorders result from the failure of DNA repair proteins to work properly.

Highlight: Sunburns and Parsnips

■ Certain plant species contain psoralens, which can cause painful blisters if the plant is handled or eaten in the presence of UV light.

■ In the presence of UV light, psoralens damage DNA. Acting on its own, UV light also can damage DNA. Overexposure to psoralens or UV light can damage DNA so extensively that the cell's DNA repair mechanisms fail and the cell dies.

■ In the case of UV light, the mutations of DNA that result from sunburn can be the first step on the path to skin cancer.

Key Terms

base p. 227	mismatch error p. 231
DNA repair p. 231	mutation p. 231
DNA replication p. 229	nucleotide p. 227
double helix p. 227	transformation p. 226

Chapter Review

Self-Quiz

1. The base-pairing rules in DNA are
   a. any combination of bases is allowed.
   b. T–C, A–G.
   c. A–T, C–G.
   d. C–A, T–G.

2. DNA replication results in
   a. two DNA molecules, one with two old strands and one with two new strands.
   b. two DNA molecules, each of which has two new strands.
   c. two DNA molecules, each of which has one old strand and one new strand.
   d. none of the above

3. Experiments performed by Oswald Avery and colleagues showed that
   a. protein, not DNA, transformed bacteria.
   b. DNA, not protein, transformed bacteria.
   c. carbohydrates, not protein, transformed bacteria.
   d. both DNA and protein were able to transform bacteria.

4. The DNA of cells is damaged
   a. thousands of times per day.
   b. by collisions with other molecules, chemical accidents, and radiation.
   c. not very often and only by radiation.
   d. both a and b

5. The DNA of different species differs in terms of
   a. the sequence of bases.

   b. the base-pairing rules.
   c. the number of nucleotide strands.
   d. the location of the sugar–phosphate backbone on the DNA molecule.

Review Questions

1. Explain why the structure of DNA proposed by Watson and Crick immediately suggested a way in which DNA could be replicated.

2. Explain how the following three things are related: (a) the sequence of bases in DNA, (b) mutation, and (c) alleles that cause human genetic disorders.

3. Explain how a mutation that disables a DNA repair protein can lead to cancer.

---

14   **The Daily Globe**

Saving Our Children from Cancer

Dear Editor:

When surveyed, most people say they fear cancer. So why have Americans chosen to ignore the biggest news of all about this frightening disease? According to scientists, the vast majority of cancers are not inherited but are caused by mutations to your DNA that occur over the course of your lifetime, and many of these mutations can be avoided by simple changes in diet.

Yet despite this good news, Americans continue to eat a deadly diet—high in animal fat and salt, and low in fiber, fruits and vegetables. The result? In North America and Europe alone, nearly 400,000 people die each year from cancers caused by what is or isn't on their plates, and many of the victims are children.

One look at any public school cafeteria's offerings reveals a cancer-causing diet: hamburgers, french fries, pepperoni pizza. And with all the desserts available— cakes, cookies, and candy bars from the vending machine—what chance is there that these children will eat the lifesaving fruit and vegetables that they are served?

It's time for America to face the facts: We are using tax dollars to serve up sickness every day at our nation's schools. The Americans for Healthy Living are calling for the immediate removal of all meat, nonfruit desserts, and other dangerous foods from public schools. We remove asbestos from our schools. It's time for the hamburgers to go, too. What is your child's school serving for lunch today?

The Americans for Healthy Living

Evaluating "The News"

1. Is there a difference between removing asbestos and removing hamburgers from a school?

2. There is overwhelming evidence that the link between diet and cancer is real. However, scientists still don't know exactly how what you eat translates into the danger of or protection from cancer. Should that

uncertainty alter whether or not public policy—like what children can eat at school—is changed?

3. Smoking causes 30 percent of all cancer deaths. The drinking of very hot beverages and overexposure to sunlight contribute a few more percentage points each to the total number of cancer deaths. Yet,

drinking hot beverages and sitting in the bright sun are allowed at schools across the country. Many high schools provide areas where students can smoke as well. Should these activities be banned for sake of the health of the nation's children?

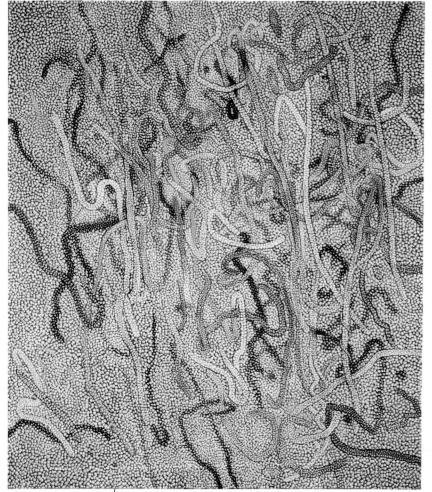

chapter 15

From Gene to Protein

Finding the Messenger and Breaking the Code

We live in a global economy. The headquarters of a company may be located in one country—say, Germany—but the company's factories may be located elsewhere—say, the United States. Immediately there is a problem: Decisions made in Germany need to be communicated to employees in the United States. This problem is easy to solve: A message is sent from one location to the other. In addition, however, the message must be translated from German, the language in which the decision was made, to English, the language in which the decision must be implemented.

Eukaryotic cells face similar challenges. Genes work by controlling the production of proteins. Whereas genes are located in the nucleus of the cell, their protein products are made on ribosomes, which are located outside of the nucleus, in the cytoplasm. How does a gene control from a distance how a ribosome constructs a protein? Like our imaginary corporate head-

Ross Bleckner, *In Replication*, 1998.

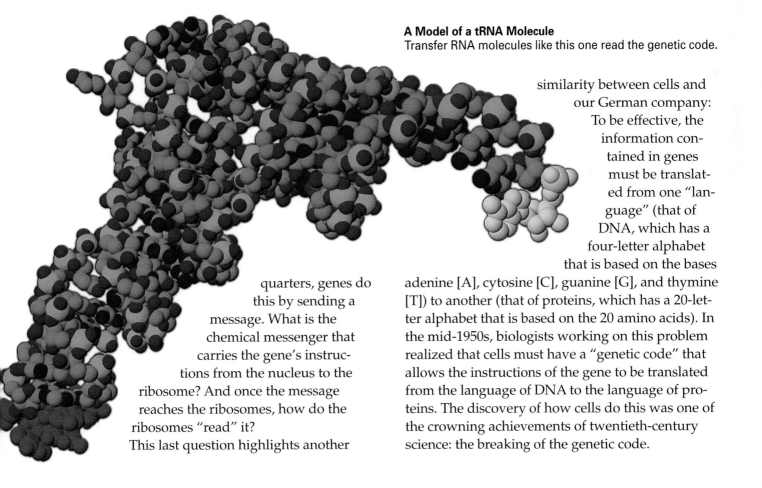

A Model of a tRNA Molecule
Transfer RNA molecules like this one read the genetic code.

quarters, genes do this by sending a message. What is the chemical messenger that carries the gene's instructions from the nucleus to the ribosome? And once the message reaches the ribosomes, how do the ribosomes "read" it?

This last question highlights another similarity between cells and our German company: To be effective, the information contained in genes must be translated from one "language" (that of DNA, which has a four-letter alphabet that is based on the bases adenine [A], cytosine [C], guanine [G], and thymine [T]) to another (that of proteins, which has a 20-letter alphabet that is based on the 20 amino acids). In the mid-1950s, biologists working on this problem realized that cells must have a "genetic code" that allows the instructions of the gene to be translated from the language of DNA to the language of proteins. The discovery of how cells do this was one of the crowning achievements of twentieth-century science: the breaking of the genetic code.

Key Concepts

1. Most genes specify how to build proteins. The DNA sequence of the gene codes for the amino acid sequence of its protein product. A few genes specify how to make RNA molecules that assist in protein synthesis.

2. For genes that specify a protein, two steps are required to go from gene to protein: transcription and translation.

3. Transcription occurs in the nucleus and produces an RNA copy of the information stored in the gene. In transcription of a gene that specifies a protein, a single-stranded messenger RNA (mRNA) molecule is copied from the DNA sequence of the gene.

4. Translation occurs in the cytoplasm and converts the sequence of bases in an mRNA molecule to the sequence of amino acids in a protein. In translation, mRNA and two other types of RNA are used to make the protein product specified by a gene.

5. Gene mutations can alter the sequence of amino acids in a gene's protein product, and this change, in turn, can alter the protein's function. Although changes in protein function are usually harmful, occasionally they benefit the organism.

Chapters 12 through 14 described how genes are inherited, where they are located (on chromosomes), and what they are made of (DNA). But we have yet to cover several major issues related to gene function. First, we must describe how genes work. At the molecular level, how do genes store the "information" needed to build their final products, proteins? How does the cell access that information? Knowing how genes work helps us understand how mutations produce new phenotypes, including disease phenotypes. We begin this chapter by discussing how genetic information is encoded in genes and how the cell uses that information to build proteins. We then discuss how a change to a gene can change an organism's phenotype.

How Genes Work

Genes work by controlling the production of proteins. For most genes, the relationship between gene and protein is direct: The gene contains instructions for how to build a particular protein. Some genes have an indirect relationship to proteins. Rather than encoding a particular protein, these genes specify how to build ribonucleic acid (RNA) molecules that help the cell construct proteins.

Proteins are essential to life. They provide cells and organisms with structural support, means of transporting materials, defense against disease-causing organisms, and the ability to move. In addition, the many chemical reactions on which life depends are controlled by a crucial group of proteins, the enzymes. Enzymes and other proteins influence so many features of the organism that they, along with the environment, determine the organism's phenotype.

Early clues that genes work by controlling the production of proteins came in the beginning of the twentieth century from the work of British physician Archibald Garrod. Garrod studied several inherited, human metabolic disorders. In 1902, he argued that human metabolic disorders were caused by an inability of the body to produce specific enzymes. For example, Garrod was interested in alkaptonuria, a condition in which the urine of otherwise healthy infants turns black when exposed to air. He proposed that infants with alkaptonuria had a defective version of an enzyme that ordinarily would break down the substance that caused urine to turn black. Garrod did not stop there; he and his collaborator, William Bateson, went on to suggest that in general, genes worked by controlling the production of enzymes.

Genes contain information for the synthesis of RNA molecules

Garrod and Bateson were on the right track but were not entirely correct: Genes can control the production of other proteins, not just enzymes. In addition, some genes do not directly specify proteins. Rather, some genes specify as their final product one of several RNA molecules used in the construction of proteins. Thus, directly and indirectly, genes control the production of proteins.

As we will see shortly, even genes that specify proteins make an RNA molecule as their initial gene product. Thus, modifying our statement in Chapter 12, we redefine **gene** as a sequence of DNA that contains information for the synthesis of one of several types of RNA molecules used to make protein.

Gen fu = 6+2

■ Genes work by controlling the production of proteins, which are essential to life. Most genes contain information for how to build a protein product. A few genes specify as their final product one of several RNA molecules used to synthesize proteins.

15.1   *RNA Molecules and Their Functions*

Type of RNA	Function
Ribosomal RNA (rRNA)	Component of ribosomes, the molecular machines that make the covalent bonds that link amino acids together in a protein
Messenger RNA (mRNA)	Specifies the order of amino acids in a protein
Transfer RNA (tRNA)	Transfers the correct amino acid to the ribosome, on the basis of information from the gene that is encoded in the mRNA

Molecules Involved in the Production of Proteins

The nucleic acids DNA and RNA (see Chapter 5) play key roles in the construction of proteins. Several types of RNA and many enzymes and other proteins are required for the cell to make proteins. As already described, DNA controls the production of all these essential molecules, so DNA controls all aspects of protein production.

Cells use three types of RNA molecules to construct proteins, **ribosomal RNA (abbreviated rRNA), messenger RNA (mRNA)**, and **transfer RNA (tRNA)**. The function of each kind of RNA is defined in Table 15.1 and described in more detail in the section that follows. But first we describe several general differences between the structure of RNA and the structure of DNA.

Whereas DNA molecules are double-stranded, most RNA molecules are single-stranded. Overall, the structure of a single strand of RNA is similar to the structure of a single strand of DNA. RNA is composed of a long string of nucleotides, each of which in turn is composed of a sugar–phosphate backbone and one of four nitrogen-containing bases (Figure 15.1). However, the nucleotides in RNA and DNA differ in two respects: (1) RNA uses the sugar ribose, whereas DNA uses the sugar deoxyribose; and (2) in RNA, the base uracil (U) replaces the DNA base thymine (T). The other three bases (A, C, and G) are the same in RNA and DNA.

■ Three types of RNA (rRNA, mRNA, and tRNA) and many enzymes and other proteins are required for the cell to make proteins. RNA consists of a single strand of nucleotides, each composed of a sugar–phosphate backbone and one of four nitrogen-containing bases: A, C, G, or U.

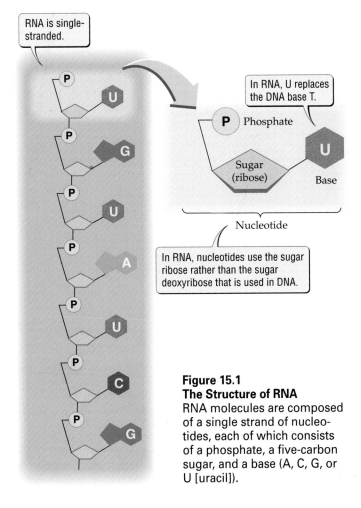

RNA is single-stranded.

In RNA, U replaces the DNA base T.

P Phosphate

Sugar (ribose)

Base

Nucleotide

In RNA, nucleotides use the sugar ribose rather than the sugar deoxyribose that is used in DNA.

Figure 15.1
The Structure of RNA
RNA molecules are composed of a single strand of nucleotides, each of which consists of a phosphate, a five-carbon sugar, and a base (A, C, G, or U [uracil]).

How Genes Control the Production of Proteins

In both prokaryotes and eukaryotes, genes specify the production of proteins in two steps: transcription and translation. In transcription, an RNA molecule is made from the DNA sequence of a gene. Most genes specify the production of mRNA, which in turn specifies the amino acid sequence of a protein (see Table 15.1). Some genes specify the production of rRNA and tRNA, molecules that aid in protein synthesis. In translation, rRNA, mRNA, and tRNA molecules direct the synthesis of a gene's protein product. We discuss transcription and translation in detail later in the chapter.

Before we discuss the details of transcription and translation, let's consider how genes work from the perspective of information flow. We'll look here at the flow of genetic information in eukaryotes. Events are similar in prokaryotes except that, because they lack a nucleus, both genes and ribosomes are located in the cytoplasm.

To make the protein specified by a gene, the information in the gene must be sent from the gene, located in the nucleus, to the site of protein synthesis, the ribosome, which is located in the cytoplasm (Figure 15.2). The information in the gene is transferred from the nucleus to the ribosome by mRNA. This transfer of information is made possible by transcription, in which the sequence of bases in mRNA is copied directly from the DNA sequence of a gene. Because it is a direct copy of the gene's DNA sequence, mRNA provides the ribosome with all the information that is in the gene.

Once the mRNA molecule arrives at the ribosome, the information on how to build the protein must be translated from the language of DNA (nucleotides) to the language of proteins (amino acids) (see Figure 15.2). The information is translated at the ribosomes by tRNA molecules. Transfer RNA molecules can translate genetic information because one portion of the molecule binds to specific amino acids, while another portion binds to specific sequences of mRNA. The specificity of these binding rules allows the message in mRNA to be translated into the exact sequence of amino acids that is called for by the gene.

■ In both prokaryotes and eukaryotes, two steps are required for the synthesis of proteins: transcription and translation. In transcription, an RNA molecule is made using the DNA sequence of the gene. In translation, rRNA, mRNA, and tRNA molecules direct protein synthesis. Information for the synthesis of a protein flows from the gene, located in the nucleus, to the site of protein synthesis, the ribosome, located in the cytoplasm.

Transcription: From DNA to RNA

Transcription is the synthesis of RNA from a DNA template. It is somewhat similar to DNA replication in that one strand of DNA is used as a template against which a new strand—in this case RNA—is formed. However, transcription differs from DNA replication in three important ways. First, a different enzyme guides the process: Whereas the enzyme used in DNA replication is DNA polymerase, the key enzyme in transcription is RNA polymerase. Second, whereas the entire DNA molecule is duplicated in DNA replication, in transcription only the small portion of the DNA molecule that includes the gene is transcribed into RNA. Finally, whereas the process of DNA replication produces a double-stranded DNA molecule, the process of transcription produces a single-stranded RNA molecule.

Transcription begins when the RNA polymerase enzyme binds to a region of DNA that is called a **promoter**. Promoters contain specific sequences of bases to which RNA polymerase can bind. Once bound to the promoter, the RNA polymerase enzyme unwinds the DNA double helix and separates the two strands. Then the enzyme begins to construct an RNA molecule (Figure 15.3).

As discussed earlier, RNA molecules consist of a sequence of four bases: adenine (A), cytosine (C), guanine (G), and uracil (U). These four bases pair with each other according to specific rules: A in

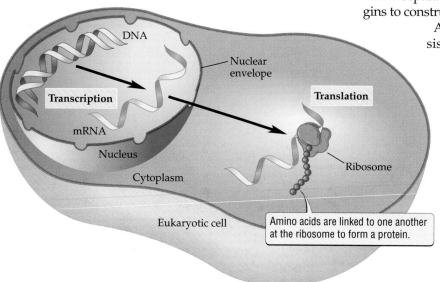

Figure 15.2 The Flow of Genetic Information
Genetic information flows from DNA to RNA to protein in two steps, transcription and translation. For a gene that encodes a protein, transcription produces an mRNA molecule that then moves to the ribosome, where translation occurs and the protein is made. This diagram shows the flow of information in a eukaryotic cell.

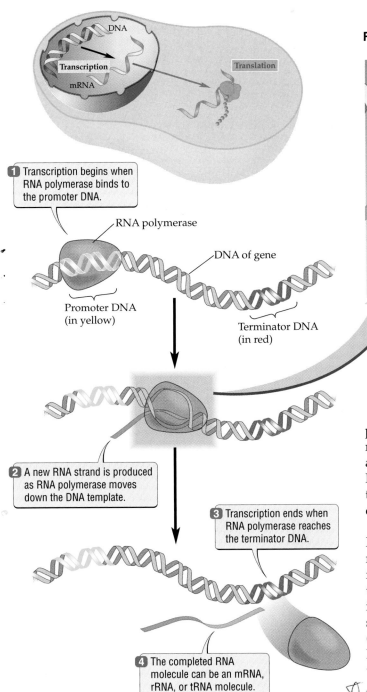

1 Transcription begins when RNA polymerase binds to the promoter DNA.

RNA polymerase

DNA of gene

Promoter DNA (in yellow)

Terminator DNA (in red)

2 A new RNA strand is produced as RNA polymerase moves down the DNA template.

3 Transcription ends when RNA polymerase reaches the terminator DNA.

4 The completed RNA molecule can be an mRNA, rRNA, or tRNA molecule.

DNA pairs with U in RNA, A in RNA pairs with T in DNA, and C pairs with G. These base-pairing rules determine the sequence of bases in an RNA molecule that is made from a DNA template. For example, if the DNA sequence were TTATGGCACCG, an mRNA molecule synthesized from this DNA template would have the sequence AAUACCGUGGC.

Figure 15.3 Overview of Transcription

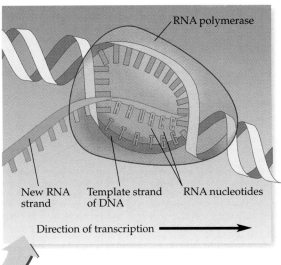

New RNA strand   Template strand of DNA   RNA nucleotides

Direction of transcription ⟶

Synthesis of an RNA molecule from the DNA template continues until the RNA polymerase enzyme reaches a sequence of bases called the terminator DNA, at which point transcription ends and the newly formed RNA molecule separates from its DNA template. The two strands of the DNA template then bond back to each other, ready to be used again when needed by the cell.

Transcription produces each of the three types of RNA involved in the synthesis of proteins: rRNA, mRNA, and tRNA. When transcription ends, the newly formed RNA products often must be modified before they are used. For example, in eukaryotes, genes contain internal sequences of bases called **introns** that do not specify part of their final protein or RNA product (Figure 15.4). Introns must be removed from the initial RNA product if the final product of the gene is to function properly.

■ In transcription, an mRNA, rRNA, or tRNA molecule is synthesized from a gene's DNA template. The RNA molecule is constructed from the DNA sequence of the gene according to specific base-pairing rules. In eukaryotes, genes contain internal sequences of DNA called introns that must be removed from the initial RNA product if the final protein or RNA product of the gene is to function properly.

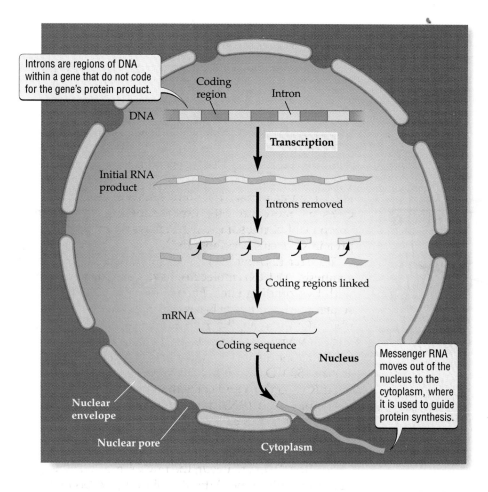

Introns are regions of DNA within a gene that do not code for the gene's protein product.

DNA

Coding region   Intron

Transcription

Initial RNA product

Introns removed

Coding regions linked

mRNA

Coding sequence

Nucleus

Nuclear envelope

Nuclear pore

Cytoplasm

Messenger RNA moves out of the nucleus to the cytoplasm, where it is used to guide protein synthesis.

Figure 15.4 Removal of Introns by Eukaryotic Cells
Before RNA molecules can be used in eukaryotes, enzymes in the nucleus must remove noncoding sequences (introns) from mRNA (shown here), tRNA, or rRNA.

The Genetic Code

At the molecular level, genes consist of a sequence of the four bases adenine, cytosine, guanine, and thymine. For genes that specify a particular protein, the sequence of bases in a gene must somehow code for, or specify, the sequence of amino acids in a protein. From the late 1950s to the mid-1960s, geneticists worked feverishly to figure out the code that accomplishes this task. By 1966, the genetic code had been broken (Figure 15.5).

In the **genetic code**, an amino acid is specified by a nonoverlapping sequence of three nucleotide bases (a triplet) in an mRNA molecule. For example, as Figure 15.6 shows, if a portion of an mRNA molecule consisted of the sequence UUCACUCAG, the first triplet (UUC) would specify one amino acid (phenylalanine), the next triplet (ACU) would specify a second amino acid (threonine), and the last triplet (CAG) would specify a third amino acid (glutamine).

Each group of three bases in mRNA is called a **codon**. There are four possible bases at each of the three positions of a triplet, so there are a total of 64 possible codons ($4 \times 4 \times 4 = 64$).

When reading the code, the cell begins at a fixed starting point, called a start codon (usually the codon AUG), and ends at one of several stop codons, such as UGA or UAA (see Figure 15.5). By beginning at a fixed point, the cell ensures that the message from the gene does not become scrambled. (Use Figure 15.5 to determine the amino acid sequence that results if

Figure 15.5 The Genetic Code

UAA, UAG, and UGA do not code for an amino acid. Translation stops when these codons are reached.

Like arginine, most amino acids are encoded by more than one codon.

Second letter

	U	C	A	G	
U	UUU UUC Phenylalanine UUA UUG Leucine	UCU UCC UCA UCG Serine	UAU UAC Tyrosine UAA Stop codon UAG Stop codon	UGU UGC Cysteine UGA Stop codon UGG Tryptophan	U C A G
C	CUU CUC CUA CUG Leucine	CCU CCC CCA CCG Proline	CAU CAC Histidine CAA CAG Glutamine	CGU CGC CGA CGG Arginine	U C A G
A	AUU AUC Isoleucine AUA AUG Methionine; start codon	ACU ACC ACA ACG Threonine	AAU AAC Asparagine AAA AAG Lysine	AGU AGC Serine AGA AGG Arginine	U C A G
G	GUU GUC GUA GUG Valine	GCU GCC GCA GCG Alanine	GAU GAC Aspartate GAA GAG Glutamate	GGU GGC GGA GGG Glycine	U C A G

First letter

Third letter

Figure 15.6 How Cells Use the Genetic Code

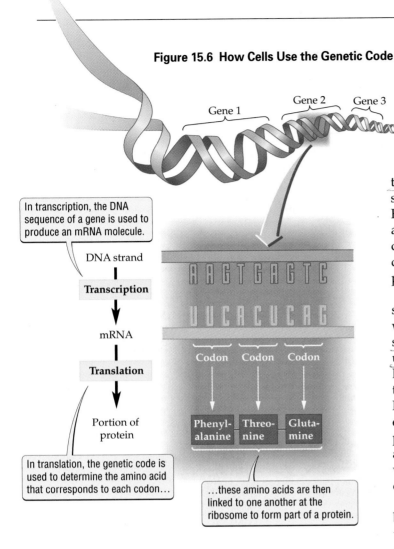

In transcription, the DNA sequence of a gene is used to produce an mRNA molecule.

DNA strand

Transcription

mRNA

Translation

Portion of protein

In translation, the genetic code is used to determine the amino acid that corresponds to each codon...

Gene 1   Gene 2   Gene 3

A A G T G A G T C

U U C A C U C A G

Codon   Codon   Codon

Phenyl-alanine   Threo-nine   Gluta-mine

...these amino acids are then linked to one another at the ribosome to form part of a protein.

Translation: From mRNA to Protein

The genetic code provides the cell with the equivalent of a dictionary with which to translate the language of DNA into the language of proteins. The conversion of a sequence of bases in mRNA to a sequence of amino acids in a protein is called **translation**. Translation is the second major step in the process by which genes specify proteins (see Figure 15.2). Translation occurs at the ribosomes, which are composed of several different sizes of rRNA molecules and more than 50 different proteins. As a major component of ribosomes, rRNA plays a central role in protein synthesis (see Table 15.1).

Transfer RNA molecules are also crucial to protein synthesis. All tRNA molecules have a similar structure with two binding sites. Each tRNA has a particular sequence of nucleotide bases that can bind to a particular mRNA codon, the **anticodon** site in Figure 15.7. Each tRNA molecule also has a site that binds to a particular amino acid (the "amino acid attachment site" in Figure 15.7). Different tRNA molecules bind to different mRNA codons and carry different amino acids. In particular, each tRNA molecule carries the specific amino acid that is called for by the mRNA codon to which it can bind. If this were not the case, the genetic code would not work.

For translation to occur, the mRNA molecule must bind to the ribosome. After this binding has occurred, translation begins at the AUG start codon that is nearest to the region where mRNA is bound to the ribosome. Here's how the amino acid chain of the protein is built.

the sequence UUCACUCAG from Figure 15.6 is read in triplets that begin with the second U, not the first.)

The information in a gene is encoded in its sequence of bases. In the genetic code, an amino acid is specified by a non-overlapping sequence of three nucleotide bases in an mRNA molecule. When reading the genetic code, the cell begins at a fixed starting point, thus ensuring that the message from the gene does not become scrambled.

Amino acid

Amino acid attachment site

Serine

tRNA

Anticodon (pairs with a specific codon on mRNA)

Codon (on an mRNA molecule)

U C G

A G C

mRNA

Figure 15.7 Transfer RNA (tRNA)

Translation is accomplished by tRNA molecules, which bind to specific mRNA codons and specific amino acids. A model of a tRNA molecule is shown above at left. The site at which tRNA binds to an mRNA codon is called the anticodon.

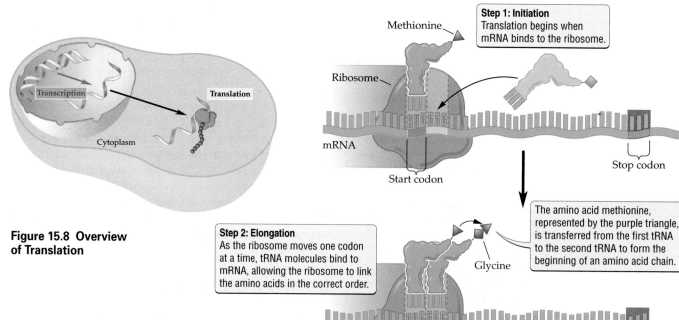

Figure 15.8 Overview of Translation

Step 1: Initiation
Translation begins when mRNA binds to the ribosome.

Methionine
Ribosome
mRNA
Start codon
Stop codon

The amino acid methionine, represented by the purple triangle, is transferred from the first tRNA to the second tRNA to form the beginning of an amino acid chain.

Step 2: Elongation
As the ribosome moves one codon at a time, tRNA molecules bind to mRNA, allowing the ribosome to link the amino acids in the correct order.

Glycine
mRNA
Start codon
Stop codon

Later

Completed amino acid chain

Step 3: Termination
When the ribosome reaches a stop codon, the mRNA and the completed amino acid chain both separate from the ribosome.

mRNA
Start codon
Stop codon

First, a tRNA molecule binds to the AUG codon, bringing with it the amino acid methionine (Figure 15.8). The ribosome then moves to the next codon on the mRNA molecule (in the example shown in Figure 15.8, GGG). Next, a different tRNA molecule carrying the appropriate amino acid (glycine) binds to the GGG codon. The ribosome then forms a covalent bond between the first amino acid (methionine) and the second amino acid (glycine). The ribosome then moves to the next mRNA codon, and the first tRNA—the one bound to AUG—is released.

Once the first tRNA is released, a tRNA molecule that can bind to the third codon does so, bringing with it the third amino acid of the growing amino acid chain. The first two amino acids are then attached by the ribosome to the third one, and the second tRNA is released. This process continues until a stop codon (see Figure 15.5) is reached, at which point the mRNA molecule and the completed amino acid chain both separate from the ribosome. The protein then folds to its compact, specific three-dimensional shape.

■ In translation, which occurs at the ribosomes, a sequence of bases in mRNA is converted into a sequence of amino acids in a protein. Ribosomes are molecular machines that make the covalent bonds that link amino acids together in a protein. Transfer RNA molecules translate the genetic information in mRNA by carrying the specific amino acid called for by the mRNA.

Impact of Mutations on Protein Synthesis

In Chapter 13 we discussed two types of mutations: those that affect individual genes and those that alter the number or structure of chromosomes. Both types of mutations alter the DNA of a cell. In general, we define **mutation** as a change to either the sequence or the amount of DNA found in a cell. Mutations range in extent from a change in the identity of a single base pair in the DNA sequence of a gene to the addition or deletion of one or more chromosomes.

In this section we describe how gene mutations affect protein synthesis, emphasizing mutations that occur in portions of a gene that code for proteins rather than mutations that affect entire chromosomes or that occur in introns (see Figure 15.4 to review introns).

Many mutations alter a single base pair

Many gene mutations are changes in a single base pair of the gene's DNA sequence. There are three major types of such mutations: substitution, insertion, and deletion mutations. In a **substitution mutation**, one base is changed to another at a single position in the DNA sequence of the gene. In the substitution mutation of Figure 15.9, for example, the sequence of a normal gene is changed when a thymine (T) is replaced by a cytosine (C). As the figure shows, this particular change causes the substitution of one amino acid for another because the mRNA triplets made from the DNA sequences CCG and TCG code for different amino acids (see Figure 15.5).

Not all substitution mutations lead to changes in the amino acid sequence of the protein. For example, although a change in the DNA sequence from GGG to GGA alters the mRNA sequence from CCC to CCU, since both CCC and CCU code for the same amino acid (see Figure 15.5) this change does not alter the amino acid sequence of the protein. In such cases, the substitution mutation is said to be "silent" because it produces no change in the structure of the protein, and no change in phenotype.

Insertion or **deletion mutations** occur when a single base is inserted or deleted. Such mutations alter the identity of many of the amino acids for the protein under construction (see Figure 15.9), and this change usually prevents the protein from functioning properly. When an insertion or deletion mutation occurs, it is said to cause a **frameshift**. A genetic frameshift is similar to what happens if you accidentally record the answer to a question twice on the answer sheet of a multiple choice test: All the answers from that point forward are likely to be wrong, since each is an answer to the previous question.

Insertions or deletions of a series of bases are common

Mutations can alter more than one base pair of the DNA sequence of a gene. For example, from two to thousands of bases can be inserted within the DNA sequence of a gene. Unless the number of bases that are inserted is a multiple of three (the length of a complete codon), such an insertion will cause a frameshift, usually with disastrous results. Even if no frameshift occurs, the insertion of extra bases into a gene causes extra amino acids to be added to the protein being built, often destroying protein function and hence harming the organism.

Mutations can cause a change in protein function

Gene mutations alter the DNA sequence of a gene, which in turn alters the sequence of bases in an mRNA molecule made from the gene. Changes to the sequence of bases in mRNA can have a wide range of effects. If the change is to a single base and does not result in the substitution of one amino acid for another, the structure and function of the protein are not changed. Even if one or a few amino acids of the protein are changed, there may be little or no effect. For example, changes that do not alter the active site of the protein may not affect protein function.

However, amino acid changes that alter the active site of a protein usually do change how the protein works. These changes often are harmful to the organism because they decrease or destroy protein function. On rare occasions, changes to an active site of a protein benefit the

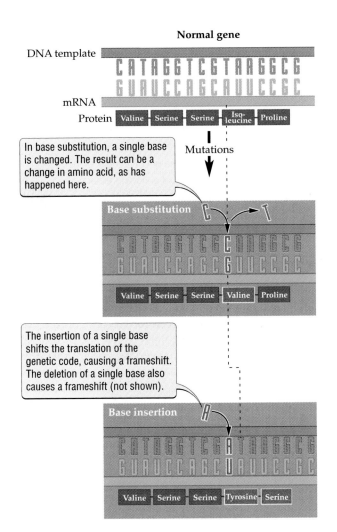

Figure 15.9 Impact of DNA Mutations
Changes to the sequence of bases in a gene can change the sequence of amino acids in the gene's protein product. DNA bases, RNA bases, and amino acids that change as a result of a mutation are shown in red.

organism by improving how the protein works, or by causing the protein to take on a new and useful function.

> ■ Many mutations are caused by the substitution, insertion, or deletion of a single base pair of a gene's DNA sequence. Insertions or deletions of a series of bases into or from the DNA sequence of a gene are also common. Mutations can change the function of a gene's protein product.

Putting It Together: From Gene to Phenotype

Humans have approximately 100,000 genes. More than 99% of these genes code for the amino acid sequence of proteins. Here we review the major steps in how cells go from gene to protein to phenotype. It is important to understand that the first step in the process that leads from gene to protein, transcription, is similar in all genes, including the small percentage that specify tRNA and rRNA molecules. However, translation does not occur for genes that specify tRNA or rRNA, because the

tRNA and rRNA molecules produced by such genes are the gene's final product.

Chromosomes contain many genes. Each gene on a chromosome is composed of DNA and consists of a sequence of the four bases adenosine (A), cytosine (C), guanine (G), and thymine (T). For genes that code for proteins, the particular sequence of bases in the DNA of a gene specifies the amino acid sequence of the gene's protein product.

As we have discussed in this chapter, the two major steps from gene to protein are transcription and translation (see Figure 15.2). In transcription, the sequence of bases in the DNA of a gene is used as a template to produce an RNA molecule. For genes that code for proteins, the mRNA molecule so produced contains all the information in the gene. The cell transports mRNA molecules to the ribosome, where translation occurs. In translation, the genetic code is used to synthesize the gene's protein product from the sequence of bases in the mRNA molecule.

The proteins produced by genes are essential to life. A mutation in a gene can alter the sequence of amino acids in the gene's protein product, and this change can disable or otherwise alter the function of the protein. When a critical protein is disabled, the entire organism may be harmed. For example, in people that suffer from the disease sickle-cell anemia, a single amino acid is altered in hemoglobin, a protein that functions in the transport of oxygen (Figure 15.10). People with sickle-cell anemia suffer from heart and kidney failure, among many other serious effects.

In sickle-cell anemia, a gene mutation alters the gene's protein product, which in turn produces a change in the organism's phenotype. A similar chain of events occurs for other genes. Overall, the phenotype of an organism is determined by the organism's enzymes and other proteins, and by the environment (see Chapter 12). Because genes control the production of proteins, genes play an important role in determining the phenotype of the organism.

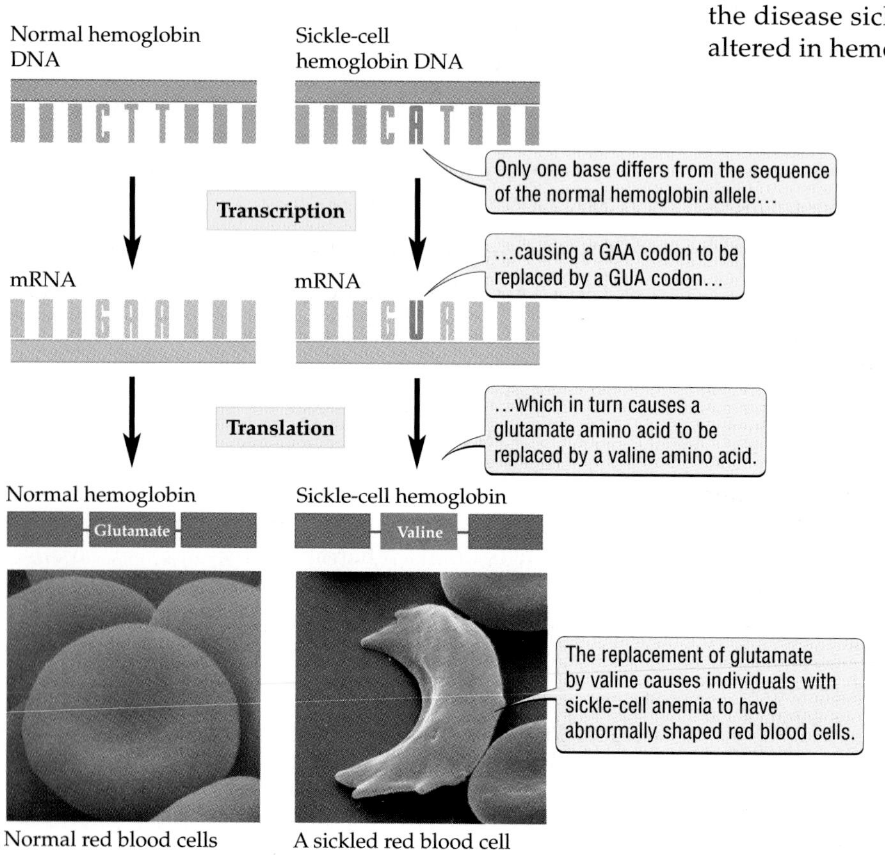

Figure 15.10 A Small Genetic Change Can Have a Large Effect
Sickle-cell anemia is caused by mutation of a single base. People with sickle-cell anemia usually die before they reach childbearing age.

■ More than 99% of genes specify a protein product. Proteins are essential to life. In conjunction with the environment, proteins determine an organism's phenotype. Because genes control the production of proteins, genes play a key role in determining an organism's phenotype.

Highlight

The Evolving Genetic Code

The genetic code lies at the heart of the process by which genes make proteins: It provides the "dictionary" that cells use to translate the language of DNA into the language of proteins. As we have learned in this chapter, the genetic information stored in genes is first transcribed into mRNA molecules, then translated by tRNA molecules. Transfer RNA molecules are able to translate the genetic code because one portion of their structure binds to specific mRNA triplets (codons) while another portion binds to the specific amino acid called for by that mRNA triplet.

The consequences of a change to the genetic code would seem disastrous. For example, consider what would happen if a particular type of tRNA molecule always carried the wrong amino acid. For every protein made by every cell in the body, whenever the codon recognized by that tRNA was translated, the wrong amino acid would be inserted into the protein that was being built. As a result, the function of many proteins likely would be destroyed, which would kill or severely harm the organism.

The line of reasoning just given suggests that once the genetic code reached its current form, early in the history of life, it should have remained "frozen" in place because changes to it would be so harmful. This strong emphasis on the unchanging nature of the genetic code explains why it is often called a universal code, common to all life.

All organisms do have a similar genetic code, and many use the exact same genetic code (the code shown in Figure 15.5). But the genetic code is not really "frozen." Instead, it has slowly evolved or changed over time. For example, in six species of yeasts, CUG codes for serine instead of leucine. In the mitochondria of many organisms, AUA codes for methionine instead of isoleucine. Departures from the genetic code are not limited to yeasts and mitochondria; bacteria, protozoans, algae, and even humans violate portions of the code. In total, 15 of the 64 codons—almost 25%—have changed their meaning at least once throughout the history of life.

How did the genetic code evolve without killing the organisms in which the changes occurred? One key to answering this question is that the genetic code is "redundant": Different codons have the same meaning; that is, they call for the same amino acid. In some cases, organisms do not use one of the redundant codons at all, nor do they produce the tRNA molecules that are usually associated with the unused codon. If an organism no longer uses a particular codon or produces its tRNA molecule, a codon can change its meaning without damaging the organism.

Such changes appear to have happened repeatedly, albeit very slowly, during the history of life. Early in the history of life, all organisms probably did have the same genetic code. But since that time the genetic code, like all other aspects of life, has evolved and continues to evolve.

■ The genetic code has evolved slowly over time. The genetic code can change without killing the organism because different codons call for the same amino acid, and in some cases, particular codons are not used.

Summary

How Genes Work

- Genes work by controlling the production of proteins.
- Most genes contain information for how to build a protein product.
- A few genes specify as their final product one of several RNA molecules used to synthesize proteins.

Molecules Involved in the Production of Proteins

- Three types of RNA (rRNA, mRNA, and tRNA) and many enzymes and other proteins are required for the cell to make proteins.
- RNA consists of a single strand of nucleotides, each composed of a sugar–phosphate backbone and one of four nitrogen-containing bases: adenine (A), cytosine (C), guanine (G), or uracil (U).

How Genes Control the Production of Proteins

- In both prokaryotes and eukaryotes, two steps are required for the synthesis of proteins: transcription and translation.
- In transcription, an RNA molecule is made using the DNA sequence of the gene.
- In translation, rRNA, mRNA, and tRNA molecules direct protein synthesis.
- In eukaryotes, information for the synthesis of a protein flows from the gene, located in the nucleus, to the site of protein synthesis, the ribosome, located in the cytoplasm.

Transcription: From DNA to RNA

■ In transcription, an mRNA, rRNA, or tRNA molecule is synthesized from a gene's DNA template.

■ The RNA molecule is constructed using the DNA sequence of the gene according to specific base-pairing rules: A in DNA pairs with U in RNA, A in RNA pairs with T in DNA, and C pairs with G.

■ In eukaryotes, genes contain internal sequences of DNA (introns) that must be removed from the initial RNA product if the final protein or RNA product of the gene is to function properly.

The Genetic Code

■ The information in a gene is encoded in the sequence of bases of the gene.

■ In the genetic code, an amino acid is specified by a nonoverlapping sequence of three nucleotide bases in an mRNA molecule.

■ When reading the genetic code, the cell begins at a fixed starting point, thus ensuring that the message from the gene does not become scrambled.

Translation: From mRNA to Protein

■ In translation, a sequence of bases in mRNA is converted into a sequence of amino acids in a protein.

■ Translation occurs at the ribosomes, which are composed of rRNA and more than 50 different proteins.

■ Ribosomes are molecular machines that make the covalent bonds that link amino acids together in a protein.

■ Transfer RNA molecules translate the genetic information in mRNA by carrying the specific amino acids called for by the mRNA.

Impact of Mutations on Protein Synthesis

■ Many mutations are caused by the substitution, insertion, or deletion of a single base pair of a gene's DNA sequence.

■ Insertion or deletion of a single base pair causes a genetic frameshift, which usually destroys the function of the gene's protein or RNA product.

■ Insertions or deletions of a series of bases into or from the DNA sequence of a gene are common.

■ Mutations can change the function of a gene's protein product.

Putting It Together: From Gene to Phenotype

■ More than 99% of genes specify a protein product.

■ Proteins are essential to life. In conjunction with the environment, proteins determine an organism's phenotype.

■ Because genes control the production of proteins, genes play a key role in determining an organism's phenotype.

Highlight: The Evolving Genetic Code

■ The genetic code has evolved slowly over time.

■ The genetic code can change without killing the organism because different codons call for the same amino acid and, in some cases, particular codons are not used.

Key Terms

anticodon p. 243

codon p. 242

deletion mutation p. 245

frameshift p. 245

gene p. 238

genetic code p. 242

insertion mutation p. 245

intron p. 241

messenger RNA (mRNA)
   p. 239

mutation p. 244

promoter p. 240

ribosomal RNA (rRNA) p. 239

substitution mutation p. 244

transcription p. 240

transfer RNA (tRNA) p. 239

translation p. 243

Chapter Review

Self-Quiz

1. For genes that specify a protein, what molecule carries information from the gene to the ribosome?
   a. DNA
   b. mRNA
   c. tRNA
   d. rRNA

2. During translation, the nucleotide bases in mRNA are read _____ at a time to produce a protein.
   a. one
   b. two
   c. three
   d. four

3. Which molecule carries the amino acid called for by mRNA to the ribosome?
   a. rRNA
   b. tRNA
   c. codon
   d. DNA

4. In transcription, which of the following molecules is produced?
   a. mRNA
   b. rRNA
   c. tRNA
   d. all of the above

5. A portion of the DNA sequence of a gene has the nucleotide bases CGGATAGGGTAT. What is the sequence of amino acids specified by this DNA sequence?
   a. alanine-tyrosine-proline-isoleucine
   b. arginine-tyrosine-tryptophan-isoleucine
   c. arginine-isoleucine-glycine-tyrosine
   d. none of the above

6. What molecular machine makes the covalent bonds that link the amino acids of a protein together?
   a. tRNA
   b. mRNA
   c. rRNA
   d. ribosome

Review Questions

1. What is a gene?

2. What are the functions of the three types of RNA used by cells to make proteins: rRNA, tRNA, and mRNA?

3. Summarize the key steps in transcription and translation.

4. Why is it essential that tRNA be able to bind to both an amino acid and an mRNA codon?

5. Describe the flow of genetic information from gene to phenotype.

15

The Daily Globe

OPINION

Human Genome Project Is Not Science

In the past 50 years, science has revealed how genes control the production of proteins. Building on this effort and seeking to improve our understanding of the genetic basis of disease, the Human Genome Project (HGP) was launched in 1990, a multi-billion-dollar undertaking to sequence the 3 billion nucleotide bases that make up human DNA.

Although it is supported by many scientists and politicians, the HGP is fundamentally flawed and is a colossal waste of money. First, the HGP is not real science: no hypotheses are being tested

and no set of scientific questions underlies and justifies its enormous expense. It is, basically, a multi-billion-dollar fishing expedition.

Second, the HGP has yielded few useful results, and for many years will continue to do so. Virtually all we have achieved by spending billions of dollars on the HGP are new diagnostic tests for particular disease genes. Since we can't cure most of these diseases, the only real "treatment" that these tests may lead to is the abortion of embryos that have the disease genes. Is this how we as a society want to spend our money?

Finally, large-scale science projects such as the HGP are useful only when a scientific discipline is in a very mature stage of development. That is not the case for genetics. We know little about how genes and the environment interact to produce the onset of human disease and other aspects of our phenotype. We would be much better off spending our money to understand fundamental scientific principles, not throwing it away in the scientific equivalent of blindly hunting for a needle in a haystack.

Evaluating "The News"

1. Do you think specific hypotheses must be tested for a scientific project to have merit?

2. The Human Genome Project is an example of the large-scale collection of observations on nature. To date, governmental and other funding agencies have not been

willing to pay comparable billions of dollars for, say, observations on the number of species on Earth. Is this discrepancy in the funding for the collection of scientific observations reasonable? What are the reasons for the discrepancy in funding levels?

3. Genetic testing has led to more abortions of embryos with genetic disorders. What are the ethical implications of this unintended side effect of genetic testing?

16

Control of Gene Expression

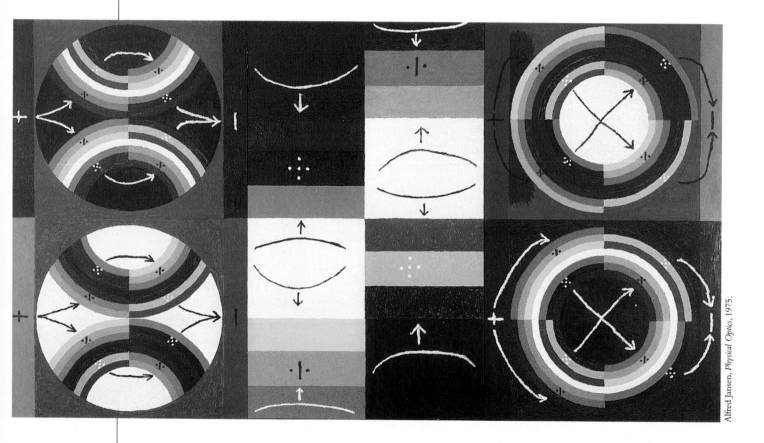

Alfred Jansen, *Physical Optics*, 1975.

Greek Legends and One-Eyed Lambs

Among his many adventures, the Greek mythological hero Odysseus encountered (and outwitted) a Cyclops, a gigantic, humanlike creature with great strength and a single, large eye. The Cyclopes of legend have characteristics that resemble characteristics of some rare genetic or developmental disorders. Lambs, mice, and humans occasionally are born with a single, large eye, along with other abnormalities of the brain and face. Such individuals die soon after birth.

Organisms control when, where, and how much of each gene's product is made.

Cyclopean Lamb

tain instructions for the synthesis of a protein or an RNA molecule. A gene is expressed, or "turned on," when the cell makes this protein or RNA product.

Some one-eyed individuals have a defective homeotic gene. The defect prevents the homeotic gene from turning on other genes that control the normal development of the brain and face. Other one-eyed individuals have a functional homeotic gene, but as embryos they were exposed to chemicals that prevented the homeotic gene's protein product from having its usual effect (which is to turn on the other crucial genes).

Deciphering the causes of developing a single eye brings us to one of the most exciting areas of modern genetics: the control of gene expression. To function normally, organisms must express the right genes at the right place and time, producing just the right amount of each gene's product. This is a task of bewildering complexity, but each of us does it, many times, every day. How organisms control gene expression is the subject of this chapter.

What causes an animal to be born with one eye? Two causes are known, both of which relate to the function of a "homeotic" gene. Homeotic genes are master-switch genes that guide the development of an organism by "turning on" a series of other, crucial genes. As discussed in Chapter 15, genes con-

Key Concepts

1. Eukaryotic DNA is organized by a complex packing system that allows cells to store an enormous amount of information in a small space.

2. Prokaryotes have relatively little DNA, most of which encodes protein or RNA. Eukaryotes have far more DNA than prokaryotes because they have more genes and because, unlike prokaryotes, eukaryotes have large amounts of DNA that does not encode protein or RNA.

3. Organisms turn genes on and off in response to short-term changes in food availability or other features of the environment. Organisms also control genes over long periods of time, as when different genes are expressed at different times during development.

4. In multicellular organisms, different cell types express different genes, both as developing embryos and as adults.

5. The main way in which cells control gene expression is to regulate transcription. Cells also control gene expression in other ways, such as by preventing the translation of mRNA molecules.

6. Genetics is moving from the study of single genes to the study of interactions among large numbers of genes. This shift has the potential to revolutionize our understanding of genes and gene expression, leading to dramatic improvements in the practice of medicine.

Figure 16.1 Different Cells Have the Same Genes
The nucleus of each cell type has the same DNA. Although they have the same genes, cells can differ greatly in structure and function because different genes are active in different types of cells.

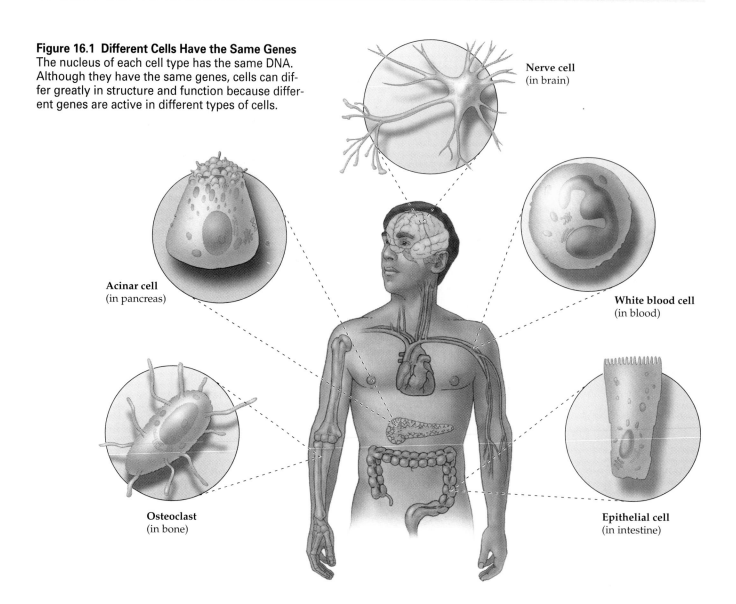

Nerve cell
(in brain)

Acinar cell
(in pancreas)

White blood cell
(in blood)

Osteoclast
(in bone)

Epithelial cell
(in intestine)

Even though they have the same genes, the various cells of a multicellular organism differ greatly in structure and function (Figure 16.1). Since genes provide cells with the "blueprint" of life, how can cells with the same genes be so different? The answer lies in how the genes are used. Cells differ in which of their genes are active, and as we will see, these differences relate to function.

For different cells to use different genes, organisms must have ways to control which genes are expressed, and indeed they do. Organisms control where, when, and how much of a given gene's product is made. Factors that influence gene expression include cell type, the chemical environment of the cell, signals received from other cells in the organism, and signals received from the external environment. In this chapter we describe how cells control their genes.

Before describing control of gene action, we need to describe how DNA is organized in chromosomes. To be expressed, the information in a gene must first be transcribed to an RNA product. For transcription to occur, the proteins that guide transcription must be able to reach the gene of interest. This task may sound simple, but it involves what may be the ultimate storage problem: how to store an enormous amount of information in a small space, yet still be able to retrieve each single piece of that information precisely when it is needed.

DNA Packing in Eukaryotes

Each chromosome in each of our cells contains one DNA molecule. These molecules hold a vast amount of genetic information. For example, the haploid number of chromosomes in humans is 23, which together contain about 3.3 billion base pairs of DNA. If the DNA from all 46 chromosomes in a human cell were stretched to its full length, it would be more than 2 meters long (taller than most of us). That is a huge amount of DNA, especially considering that it is packed into a nucleus that is only 0.000006 meter in diameter. In total, the amount of DNA in our bodies is staggering: The human body has about 10^{13} cells, each of which contains roughly 2 meters of DNA. Therefore, each of us has 2×10^{13} meters of DNA in his or her body, a length that is more than 130 times the distance from Earth to the sun.

How can our cells pack such an enormous amount of DNA into such small spaces? The reason is that the DNA molecule is very narrow (only 2 nanometers wide, where a nanometer is 1 billionth of a meter), and it is highly organized in a complex **DNA packing** system (Figure 16.2). At the first level of packing, the DNA is wound around a "spool" composed of proteins called histones. These histone spools are then packed together tightly, to form a fiber that is 30 nanometers wide. These fibers, in turn, form loops, which are themselves tightly packed together, giving the chromosome its characteristic shape.

The packing scheme we've just described holds for eukaryotes during mitosis or meiosis, when the chromosomes are most condensed. The DNA of eukaryotes is much less tightly packed during interphase (see Chapter 10), when most gene expression occurs. During interphase, much of the DNA is folded into loops of the

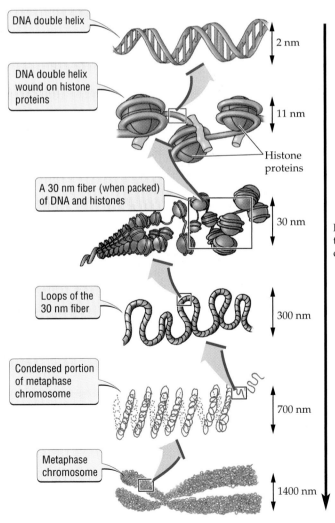

Figure 16.2 DNA Packing in Eukaryotes
The DNA of eukaryotes is highly organized in a complex packing system.

30-nanometer fibers, but some remains as tightly packed loops. Genes in the tightly packed regions are not expressed; the DNA is so tightly packed that the proteins that guide transcription cannot reach them. Compared to eukaryotes, prokaryotic cells have much less DNA per cell, and their DNA packing system is less complex.

> ■ The DNA molecule is very narrow and in eukaryotes is highly organized in a complex packing system. Together, these two features allow cells to pack an enormous amount of DNA into a very small space. During interphase, when most gene expression occurs, the tightness of the DNA packing system is partially reduced. Genes in DNA regions that are tightly packed cannot be expressed, because the proteins that guide transcription cannot reach them.

Functional Organization of DNA

We've just seen how the massive amount of DNA in a eukaryotic cell must be packaged to fit into the nucleus. Now let's turn to several related topics, such as: How much DNA do different organisms have? What are the functions of different portions of an organism's DNA?

Compared to eukaryotes, prokaryotes have little DNA. A typical bacterium has several million base pairs of DNA. Humans have 3.3 billion base pairs, and some plants have more than 100 billion base pairs. In addition, prokaryotic DNA is organized by function in a straightforward way: The different genes that are needed for a given metabolic pathway are often grouped together in the DNA. In contrast, eukaryotic genes with related functions usually are not grouped near one another on a chromosome. Finally, most of the DNA in prokaryotes codes for a protein or an RNA product, and prokaryotic genes rarely contain introns (see Chapter 15). Overall, then, prokaryotic DNA is streamlined and organized by function in a simple, direct way.

The situation is much more complicated in eukaryotes. Eukaryotes are more complex organisms than prokaryotes; hence eukaryotes need more genes to run their metabolic machinery. A typical prokaryote has about 2000 genes. Among the eukaryotes studied to date, the single-celled yeast *Saccharomyces cerevisiae* has 6300 genes, a nematode worm has 19,100 genes, and humans have an estimated 100,000 genes. Finally, in most eukaryotes, only a small percentage of the DNA actually consists of genes. Instead, most eukaryotic DNA is composed of introns and other noncoding sequences, as described in the section that follows.

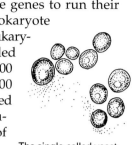

The single-celled yeast
Saccharomyces cerevisiae

Thus differences in the amount of DNA in prokaryotes and eukaryotes reflect not only more genes in eukaryotes but also the presence in eukaryotes of additional DNA that does not encode RNA.

Most of the DNA in eukaryotes is noncoding DNA

Scientists estimate that genes make up less than 4 percent of human DNA. The rest is **noncoding DNA**—that is, DNA that does not encode protein or RNA. There are several types of noncoding DNA, including introns and spacer DNA. **Spacer DNA** consists of stretches of DNA that separate genes (Figure 16.3). Eukaryotic DNA also contains many **transposons**, the mobile genetic elements discussed in the Box on page 255. Transposons make up at least 10 percent of human DNA and more than 50 percent of the 5.4 billion base pairs in corn.

> ■ Compared to eukaryotes, prokaryotes have little DNA. Most prokaryotic DNA encodes protein or RNA, and functionally related genes in prokaryotes are grouped together in the DNA. Eukaryotes have more genes than prokaryotes, and most eukaryotic DNA is noncoding DNA. Eukaryotic genes with related functions usually are not near each other.

Figure 16.3 The Composition of Eukaryotic DNA
Eukaryotic genes are surrounded by spacer DNA and transposons, as shown here for five human genes that produce hemoglobin proteins. Each of the two different types of transposons shown here is found in many copies throughout human DNA.

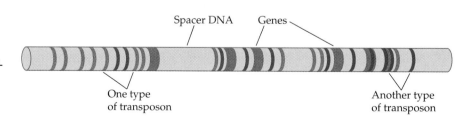

Spacer DNA   Genes

One type of transposon

Another type of transposon

The Scientific Process

Barbara McClintock and Jumping Genes

Transposons, known informally as "jumping genes," are DNA sequences that are capable of moving from one position on a chromosome to another, or from one chromosome to another. Transposons were discovered by Barbara McClintock in 1948, a discovery that earned her the 1983 Nobel prize in physiology or medicine. Originally discovered in corn by McClintock, transposons have since been found in many other organisms, ranging from bacteria to peas to humans.

Born in 1902, Barbara McClintock (see figure) completed her undergraduate and graduate studies at Cornell University, where she was awarded a Ph.D. in 1927. Working at Cornell as an instructor, in 1931 McClintock published an important paper proving that genetic recombination is due to crossing-over. This discovery established McClintock's reputation as one of the world's leading geneticists.

Despite the great respect in which McClintock was held, her discovery of transposons in 1948 was very slow to be appreciated, in part because geneticists thought that genes had fixed locations on chromosomes. It took roughly 20 years for the field of genetics to catch up with McClintock, but by 1967 she had begun to receive a series of awards for her work on transposons, culminating in the 1983 Nobel prize. McClintock continued to do research until her death in 1992, a few months after her ninetieth birthday.

What role do transposons play in organisms? Many scientists view transposons as genetic parasites that spread through an organism's DNA. When a transposon moves from one location in the DNA to another, it may insert itself into a gene, causing a mutation that changes gene function. As with any gene mutation, mutations caused by transposons can be harmful. In contrast, some scientists argue that transposons may be advantageous. For example, transposons alter patterns of gene expression in corn and many other species. McClintock viewed the transposon-caused changes in gene expression as a mechanism that organisms use to cope with changing environmental conditions.

Whether transposons are genetic parasites or are beneficial to organisms, their discovery helped to change how scientists think about DNA. Biologists used to think of genes as unchanging entities, like beads on a string. The discovery of transposons changed that thinking forever. DNA can move from one location to another, organisms change patterns of gene expression quickly, and over the course of evolution, genes take on new functions. The world of genes is exciting and dynamic, and Barbara McClintock's discoveries played a major role in showing biologists just how dynamic that world can be.

Barbara McClintock

Patterns of Gene Expression

Gene expression is the synthesis of the gene's protein or RNA product, the process by which a gene influences the cell or organism. At any given point in time, roughly 5 percent of the genes in a typical human cell are actively being expressed. The rest of the genes are not in use. The particular genes that are used by a cell can change over time, and different cells use different sets of genes. How do organisms control which of their genes are expressed at a particular time or in a particular type of cell?

Figure 16.4 Bacteria Express Different Genes as Food Sources Change

In prokaryotes, genes are turned on and off as a direct, short-term response to changing environmental conditions. Gene control in eukaryotes is more complex. Eukaryotes have more genes, their DNA is more highly packed, and they must control genes over both short and long periods of time. This section provides a broad overview of patterns of gene expression. In the following section, "How Cells Control Gene Expression," we discuss how cells turn genes on and off.

Organisms turn genes on and off in response to short-term environmental changes

Single-celled organisms such as bacteria face a big challenge from the environment: They are directly exposed to the environment, and they have no specialized cells to help them deal with changes in the external environment. One way that they meet this challenge is to express different genes as conditions change (Figure 16.4).

For example, if the nutrient lactose (a sugar found in milk) is given to *E. coli* bacteria, within a matter of minutes the bacteria will begin to produce the enzymes needed to metabolize lactose. When the lactose is gone, the bacteria stop producing these enzymes. In effect, the bacteria specialize temporarily on an available resource. When that resource runs out, they switch to the next resource that becomes available (in Figure 16.4, the sugar arabinose). By producing the enzymes to process a particular food only when that food is available, bacteria do not waste energy and cellular resources making enzymes that are not needed.

Multicellular organisms also change which genes they use in response to short-term changes in the environment. For example, we humans change the genes we express when our blood sugar or blood pH levels change, allowing us to keep these quantities from becoming too high or too low. Similarly, when exposed to high temperatures, humans, plants, and many other organisms turn on cer-

Head of a normal fruit fly

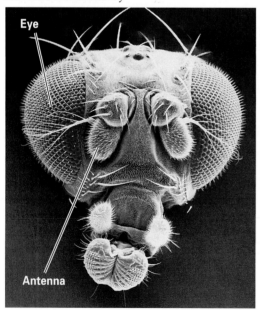

Head of a developmental mutant

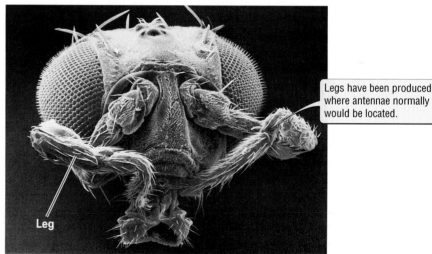

Figure 16.5 A Developmental Mutation with Bizarre Results
A mutation in a homeotic gene that controls development in fruit flies produces legs where antennae should be.

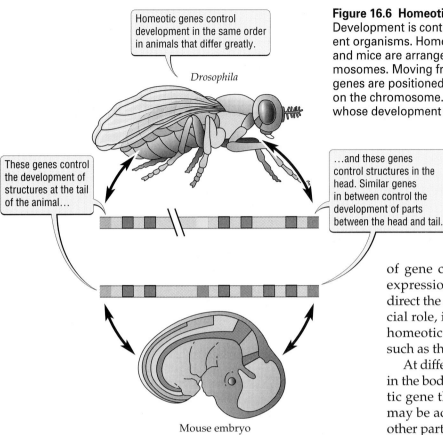

Homeotic genes control development in the same order in animals that differ greatly.

Drosophila

These genes control the development of structures at the tail of the animal…

…and these genes control structures in the head. Similar genes in between control the development of parts between the head and tail.

Mouse embryo

Figure 16.6 Homeotic Genes in Different Organisms Are Similar
Development is controlled by similar homeotic genes in greatly different organisms. Homeotic genes that control development in fruit flies and mice are arranged in a similar order on fruit fly and mouse chromosomes. Moving from tail to head, body parts affected by these genes are positioned in the same order that the genes are arranged on the chromosome. Similar homeotic genes and the structures whose development they control are matched here by color.

whose protein products alter the structure and function of the cells, allowing cells to specialize for particular tasks.

The "master-switch," or homeotic, genes described at the beginning of this chapter play a central role in the control of gene cascades. Each **homeotic gene** controls the expression of a series of other genes whose proteins direct the development of the organism. Given this crucial role, it is not surprising that defective versions of homeotic genes can have striking phenotypic effects, such as those shown in Figure 16.5.

At different times, different homeotic genes are active in the body's different cell types. For example, a homeotic gene that coordinates the development of the eye may be active in cells that will give rise to the eyes. In other parts of the body, this gene is not in use, but other homeotic genes are in use. Finally, as the body changes during development, the homeotic genes used by cells also change.

Recently it was discovered that similar homeotic genes control development in organisms as different as flies, mice, and humans (Figure 16.6). This similarity indicates that these genes are ancient. Homeotic genes first evolved hundreds of millions of years ago, and since then they have been used in similar ways by a wide variety of organisms.

tain genes. Such heat-induced genes produce proteins that protect against heat damage.

Different genes are expressed at different times during development

Turning the correct genes on and off in response to changing environmental conditions is a challenging task. But multicellular organisms also coordinate an even more difficult operation: the timing of gene activity during embryonic development. The control of gene expression during development is a task of great complexity, and errors in the control process can result in death or deformity (Figure 16.5; see also the lamb on page 251).

Organisms accomplish gene control in development through what are called cascades of gene expression. In a **gene cascade**, the protein products of different genes interact with each other and with signals from the environment, thereby turning on other sets of genes in some cells, but not in other cells. This process continues: The products from the newly turned-on genes interact with each other and with the environment to turn on still more genes, and so on. Eventually, genes are expressed

Different cells express different genes

We have just described how different cells use different homeotic genes during development. The same holds true for many other genes: In general, different types of cells express different genes, in both developing embryos and adults.

Whether a cell expresses a particular gene depends on the function of the gene's product. Not surprisingly, a gene for a specialized protein will be expressed only in cells that need the protein or in cells whose function is to produce the protein. For example, red blood cells are the only cells in the human body that need the oxygen transport protein hemoglobin. Developing red blood cells thus are the only cells that express the gene

Figure 16.7 Different Types of Cells Express Different Genes
Some genes, such as hemoglobin, crystallin, and insulin genes, are active
only in cells that require or produce the gene's protein product; other genes,
such as the rRNA gene, are active in most types of cells. "OFF" signifies
inactive genes; "ON," active genes.

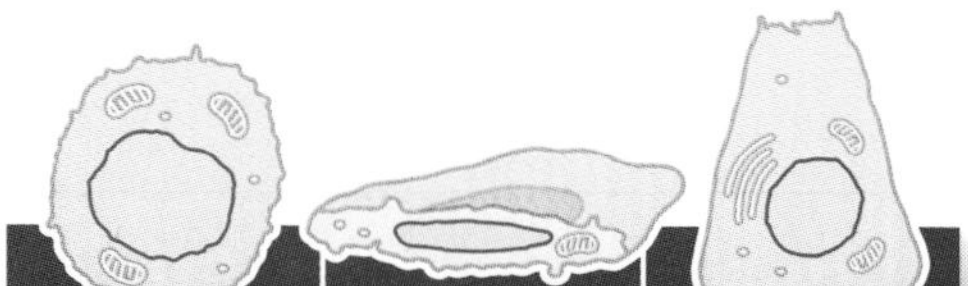

	Developing red blood cell	Eye lens cell (in embryo)	Pancreas cell
Hemoglobin gene	ON	OFF	OFF
Crystallin gene	OFF	ON	OFF
Insulin gene	OFF	OFF	ON
rRNA gene	ON	ON	ON

for this protein (Figure 16.7). Similarly, the gene for crystallin (a protein needed in the lens of the eye) is expressed only in developing eye lens cells. Finally, the gene for insulin, a hormone produced in the pancreas and used elsewhere in the body, is expressed only in pancreas cells.

Genes known as **housekeeping genes** have an essential role in the maintenance of cellular activities and are expressed by most cells in the body. For example, genes for ribosomal RNA (rRNA) are expressed by most cells (see Figure 16.7). This is not surprising, since rRNA is a key component of ribosomes, which are the sites at which proteins are synthesized. Thus, if a cell is to make proteins, it must express the genes that allow it to make rRNA, and hence to make ribosomes.

■ Both prokaryotes and eukaryotes turn genes on and off in response to short-term changes in environmental conditions. Multicellular eukaryotes also must regulate genes over long periods of time, as when they express different genes at different times during development. Organisms control genes in development through gene cascades, which are regulated by homeotic genes. Finally, the different cells of multicellular organisms express different genes, in both developing embryos and adults.

How Cells Control Gene Expression

There are two essential aspects to how cells control genes: The cell must "decide" which genes are needed at a given point in time, and the cell must then implement its decision by turning the appropriate genes on and off. After briefly discussing how cells "decide" which genes to express, in this section we focus on how cells turn genes on or off.

Cells receive signals that influence which genes they express. Such signals include those sent from one cell to another, as when one cell releases a molecule that alters gene expression in another cell. Cells also receive signals from the organism's internal environment (for example, blood sugar level) and external environment (for example, sunlight).

Overall, cells combine information from a variety of signals and use that information to "decide" which genes to express. Once the decision has been made concerning which genes to express, the cell must actually turn the appropriate genes on or off. In the discussion that follows, we describe how cells turn genes on and off.

Cells control most genes by controlling transcription

The most common way in which cells control gene expression is to turn the transcription of particular genes on or off, depending on whether the genes' products are needed. For example, the bacterium *E. coli* needs the amino acid tryptophan. If this amino acid is available in the environment, the bacterium absorbs it from the environment and does not waste cellular resources making it. But if tryptophan is not readily available, the bacterium expresses a series of five genes that code for enzymes used to make tryptophan.

E. coli controls these five genes in the following way. When tryptophan is present, a molecule of the amino acid binds to a **repressor protein**, so called because it represses the expression of the tryptophan genes. When bound to tryptophan, the repressor protein can bind to an operator. An **operator** is a sequence of DNA whose function is to control the transcription of a gene or group of genes—in this case, the five genes required to make tryptophan.

When bound to the operator, the repressor protein prevents RNA polymerase from binding to the operator, thus

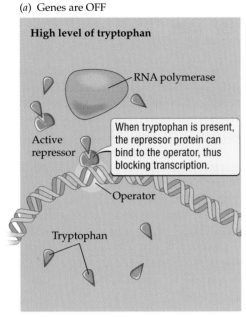

(a) Genes are OFF

High level of tryptophan

RNA polymerase

Active repressor

When tryptophan is present, the repressor protein can bind to the operator, thus blocking transcription.

Operator

Tryptophan

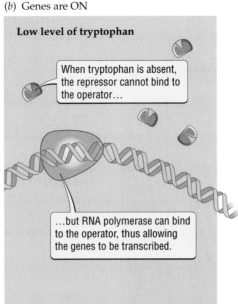

(b) Genes are ON

Low level of tryptophan

When tryptophan is absent, the repressor cannot bind to the operator…

…but RNA polymerase can bind to the operator, thus allowing the genes to be transcribed.

Figure 16.8 Repressor Proteins Turn Genes Off
In prokaryotes, a repressor protein binds to an operator to control the transcription of a gene or group of genes. (a) When tryptophan is present, it binds to the repressor protein, which then turns the genes off by binding to the operator. (b) When tryptophan is absent, the repressor protein cannot bind to the operator; thus the genes used to make tryptophan are turned on.

blocking transcription of the five genes needed to produce tryptophan (Figure 16.8a). In the absence of tryptophan, the repressor protein cannot bind to the operator, and transcription occurs (Figure 16.8b). Thus, cells do not make tryptophan when it is already present, but they do make it when tryptophan levels are low. This control of gene expression ensures that the cell does not waste precious resources producing tryptophan when tryptophan is readily available.

A few genes, such as those in *E. coli* that code for the tryptophan repressor protein, are always expressed at a low level; their transcription is not regulated. But nearly all genes in prokaryotes and eukaryotes are regulated. In general, the control of transcription has two essential elements, both of which are illustrated by tryptophan synthesis in *E. coli*.

First, there are **regulatory DNA sequences**, such as the tryptophan operator, that can switch a gene on and off. Second, to switch genes on and off, regulatory DNA sequences must interact with **regulatory proteins** that signal whether a gene should be expressed. The repressor protein that binds to the tryptophan operator when tryptophan is present is an example of a regulatory protein. Together, regulatory DNA sequences and regulatory proteins turn genes on and off in both prokaryotes and eukaryotes.

In eukaryotes, the situation is considerably more complex than in prokaryotes: More proteins are needed, and often dozens of regulatory DNA sequences are involved. Despite this added complexity, the basic concepts are the same: Regulatory proteins bind to regulatory DNA sequences that, in turn, promote or inhibit transcription.

Cells also control gene expression in other ways

In addition to directly controlling transcription, cells control other key steps in the path from gene to protein (Figure 16.9):

- *Tightly packed DNA is not expressed.* Most gene expression occurs during the interphase portion of the cell cycle. At that time, the chromosomes are long and narrow, and their DNA is relatively loosely packed. Even during interphase, however, some DNA remains densely packed. Genes in this densely packed DNA are not transcribed, in part because the transcription machinery simply cannot reach them.

- *Cells regulate how quickly messenger RNA molecules are broken down.* Most mRNA molecules are broken down by cells a few minutes or hours after being made, although a few persist for days or weeks. When it takes longer to break down an mRNA molecule, more protein can be made from the mRNA molecule. In some cases, cells alter how long mRNA molecules persist by chemically modifying the mRNA.

- *Cells can inhibit translation.* Proteins can bind to mRNA molecules and prevent translation. This method of control is especially important for some long-lived mRNA molecules. It allows the cell to deactivate an mRNA molecule that otherwise might continue to produce a protein product that is no longer needed.

- *Proteins can be regulated after translation.* Cells often must modify or transport proteins before they are used; both modification and transport can be used to regulate the availability of a protein. Cells also can target certain proteins for destruction, thus controlling gene expression at the final step in the chain from gene to protein.

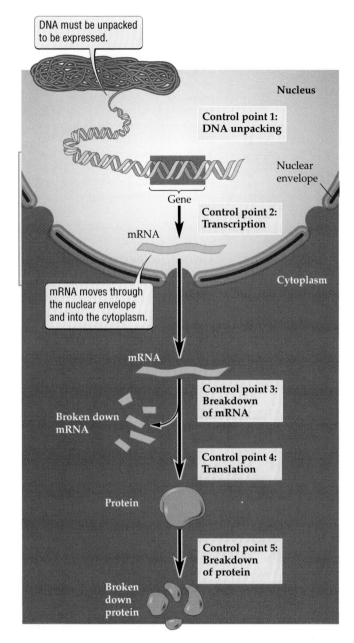

DNA must be unpacked to be expressed.

Nucleus

Control point 1: DNA unpacking

Nuclear envelope

Gene

Control point 2: Transcription

mRNA

Cytoplasm

mRNA moves through the nuclear envelope and into the cytoplasm.

mRNA

Control point 3: Breakdown of mRNA

Broken down mRNA

Control point 4: Translation

Protein

Control point 5: Breakdown of protein

Broken down protein

Figure 16.9 Control of Gene Expression in Eukaryotes
Eukaryotes can control gene expression in many ways. Each control point on the pipeline from genes to protein represents a point at which cells can regulate the production of proteins.

■ Cells control most genes by controlling transcription. Transcription is controlled by regulatory DNA sequences that can switch genes on and off. To switch genes on and off, regulatory DNA sequences interact with regulatory proteins that signal whether a gene should be expressed. Cells also can control gene expression by other means.

Highlight

Genome Biology

A revolution is brewing in the field of genetics—one that, in its impact on society, may ultimately rival the computer revolution. Like the computer revolution, technological advances are pushing this revolution forward (see Chapter 17). However, the revolution in genetics is not centered on a particular new type of machine or technology. Instead, it is due to a shift in perspective, from the study of single genes to the study of whole genomes.

A **genome** is all the DNA of an organism, including its genes. Thus, the human genome consists of all the DNA in our 46 chromosomes. Genome biologists are currently striving to meet two major goals: (1) determining the DNA sequence of entire genomes and (2) understanding the expression and function of large numbers of genes.

How much progress has been made in determining the DNA sequence of entire genomes? As of this writing, the entire DNA sequences of 19 species of bacteria and two eukaryotes (the budding yeast *Saccharomyces cerevisiae* and the nematode worm *Caenorhabditis elegans*) have been determined. The human genome is on schedule to be completely sequenced by the year 2003. Many other genome sequencing projects are nearing completion: By the year 2001 it is estimated that at least another 22 genomes will be completely determined, many of which are for disease-causing organisms.

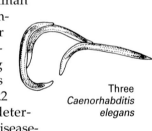

Three *Caenorhabditis elegans*

What can all this DNA sequence data—long lists of millions of the bases adenine, cytosine, guanine, and thymine—tell us? We'll illustrate the answer with several examples. In 1996, the DNA sequence of the yeast genome was published. There are approximately 6300 genes in yeast, which is a simple, single-celled eukaryote. Once the sequence of these genes was known, scientists placed each of the yeast genes on a small glass surface (roughly 2 centimeters by 2 centimeters), producing a "DNA chip" packed with genetic information. Using methods that will be described in Chapter 17, geneticists can use DNA chips to monitor the activity of all the yeast genes at once (Figure 16.10). This approach lets biologists observe how a particular stage of development or a particular set of environmental conditions influences the activity of *all* genes of an organism. Geneticists have long realized that many genes influence the metabolism and

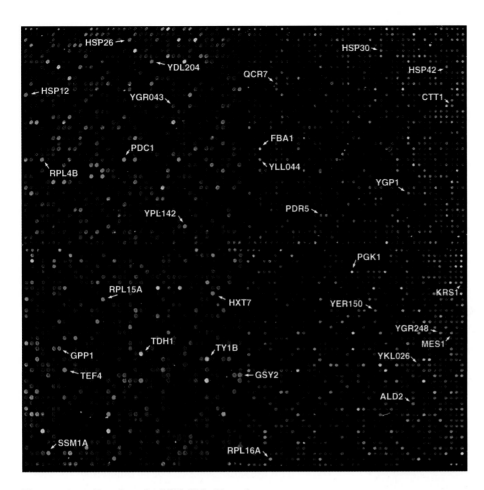

Figure 16.10 Results of a DNA Chip Experiment
In this experiment, yeast cells were provided nutrients that caused them to switch from an anaerobic (fermentation) to an aerobic (respiration) metabolism. Each colored dot represents a particular gene. Genes colored in red were turned on when the yeast cells switched to respiration. Genes colored in green were turned off when the yeast cells switched to respiration. Genes colored in yellow were expressed at roughly equal levels during fermentation and respiration.

hence should not be prescribed by the physician).

In the examples just given, DNA chips would remove the guesswork from the diagnosis and treatment of both cancer and routine disease. By identifying the correct treatment, DNA chips have the potential to save lives and improve medical care for millions of people.

> ■ The science of genetics is changing from the study of single genes to the study of whole genomes. New technological advances allow geneticists to monitor the activity of all genes of an organism simultaneously, making it possible to study how large numbers of genes influence metabolism and phenotype. Genome biology also has the potential to revolutionize the diagnosis and treatment of human disease.

phenotype of an organism; now, for the first time, they can begin to study how.

In a more applied setting, genome sequence data could be used to design DNA chips that alter the practice of medicine. For example, chips could be used to determine whether the cells of a breast cancer have a particular mutation: If the answer is yes, the most aggressive treatments should be used to treat the tumor; if the answer is no, less traumatic treatments could be used with equal effectiveness. Similarly, DNA chips could be mass-produced and used to identify the exact strain of bacteria causing a particular individual's sore throat, as well as a list of antibiotics to which the bacteria are resistant (and

Summary

DNA Packing in Eukaryotes

- The DNA molecule is very narrow and in eukaryotes is highly organized in a complex packing system. Together, these two features allow cells to pack an enormous amount of DNA into a very small space.

- The tightness of the DNA packing system is partially reduced during interphase, when most gene expression occurs.

- Genes in DNA regions that are tightly packed cannot be expressed, because the proteins that guide transcription cannot reach them.

Functional Organization of DNA

- Compared to eukaryotes, prokaryotes have little DNA. Most prokaryotic DNA encodes protein or RNA, and functionally related genes in prokaryotes are grouped together in the DNA.

- Eukaryotes have more genes than prokaryotes, and most eukaryotic DNA is noncoding DNA.

- Eukaryotic genes with related functions usually are not located near each other.

Patterns of Gene Expression

■ Both prokaryotes and eukaryotes turn genes on and off in response to short-term changes in environmental conditions.

■ Multicellular eukaryotes also must regulate genes over long periods of time during development.

■ Organisms control genes in development through gene cascades, which are regulated by homeotic genes.

■ The different cells of multicellular organisms express different genes, in both developing embryos and adults.

How Cells Control Gene Expression

■ Cells control most genes by controlling transcription.

■ Transcription is controlled by regulatory DNA sequences that can switch genes on and off.

■ To switch genes on and off, regulatory DNA sequences interact with regulatory proteins that signal whether a gene should be expressed.

■ Cells also can control gene expression in other ways, such as by preventing the translation of an mRNA or by targeting certain proteins for destruction.

Highlight: Genome Biology

■ The science of genetics is changing from the study of single genes to the study of whole genomes.

■ New technological advances allow geneticists to monitor the activity of all genes of an organism simultaneously, making it possible to study how large numbers of genes influence metabolism and phenotype.

■ Genome biology has the potential to revolutionize the diagnosis and treatment of human disease.

Key Terms

DNA packing p. 253

gene cascade p. 257

gene expression p. 255

genome p. 260

homeotic gene p. 257

housekeeping gene p. 258

noncoding DNA p. 254

operator p. 258

regulatory DNA sequence p. 259

regulatory protein p. 259

repressor protein p. 258

spacer DNA p. 254

transposon p. 254

Chapter Review

Self-Quiz

1. All the DNA of an organism is called the _____ of the organism.
   a. genes
   b. spacer DNA
   c. gene frequency
   d. genome

2. In prokaryotes and eukaryotes, gene expression is most often controlled by regulation of which of the following?
   a. the destruction of a gene's protein product
   b. how long mRNA remains intact
   c. transcription
   d. translation

3. Which of the following is a regulatory DNA sequence?
   a. repressor protein
   b. operator
   c. intron
   d. housekeeping gene

4. During development, different cells express different genes, leading to
   a. the formation of different cell types.
   b. gene mutation.
   c. DNA packing.
   d. the formation of gene families.

5. Assume that an organism has 40,000 genes and a large quantity of DNA, most of which is noncoding DNA. What kind of organism is it most likely to be?
   a. insect
   b. plant
   c. bacterium
   d. either a or b

Review Questions

1. Cell types in multicellular organisms often differ considerably in structure and in the metabolic tasks they perform (see Figure 16.1), yet each cell of a multicellular organism has the same set of genes. (a) Explain how cell types with the same genes can be so different in structure. (b) Explain how cell types with the same genes can be so different in the metabolic tasks they perform.

2. Genes in eukaryotes are often separated by large amounts of noncoding DNA. How would you expect long stretches of noncoding DNA located between genes to influence the frequency of gene recombination?

3. Describe the tryptophan operator in *E. coli*. Relate how this operator works to the general way in which gene expression most often is controlled in prokaryotes and eukaryotes.

4. As outlined in the section "Highlight: Genome Biology" (see page 260), there is an ongoing shift in genetics from the study of single genes to the study of interactions among large numbers of genes. What are the advantages and disadvantages of this approach?

The Daily Globe

New Genetic Technique Threatens Monarch Butterflies

LINCOLN, NE. In recent years, geneticists have developed ways to alter the genes of organisms ranging from bacteria to humans. In particular, much effort has been directed toward altering the genes of crop plants to improve their resistance to insect pests. One such success story was the insertion into corn plants of a toxin-producing bacterial gene, known as the *Bt* gene. The toxin produced by the *Bt* gene protected corn from attack by the corn borer, an insect pest that causes millions of dollars of damage each year.

Now it seems that the *Bt* gene, while promising, has a dark side: The toxin protects corn from the corn borer, but it also gets into corn pollen. Corn pollen spreads by wind from corn to nearby plants, including the milkweed plants that are eaten by the caterpillar stage of monarch butterflies. In lab experiments, 50 percent of monarch caterpillars that ate pollen with the *Bt* gene died. If these effects are observed in the wild, natural populations of these beautiful butterflies could be hit hard.

The *Bt* gene has already been approved for use in crop plants; when testing for unwanted side effects, no one considered how the *Bt* toxin might spread via pollen and harm other species. Some environmentalists have been quick to call for a stop in the use of the *Bt* gene until its potential negative effects are more carefully examined. Researchers at the MonPont Company, which produced the corn with the *Bt* gene, have countered that corn pollen does not spread very far and thus is not likely to cause much of a threat.

Evaluating "The News"

1. Do you agree with environmentalists who argue that more careful testing is needed before genetically modified crops are used by farmers?

2. If the MonPont Company researcher is correct that corn pollen does not spread very far, should farmers use corn that has the *Bt* gene? Which do you think is more likely to have unintended negative effects—the use of corn with the *Bt* gene or the spraying (often by airplane) of pesticides on crop fields?

3. The study that showed how the *Bt* toxin harms monarch caterpillars was conducted in a laboratory. How would you conduct a more realistic test of the impact of *Bt* corn pollen on monarch caterpillars?

4. In principle, how could an understanding of the control of expression of the *Bt* gene be used to remove the risk that the *Bt* toxin will spread in pollen and harm monarch caterpillars?

17

DNA
Technology

Glowing Plants
and Deadly Bacteria

*T*he blinking of firefly lights on warm, summer nights has long fascinated children and adults. Recently these lights have made scientific headlines: The gene that enables the production of light in fireflies has been isolated, transferred to other species, and used in a variety of laboratory studies and medical applications. For example, in 1986 the gene for the enzyme luciferase, which causes the production of light in fireflies, was inserted into tobacco plants. The plants expressed the firefly gene and glowed in the dark! This experiment provided a dramatic example of how genes can be transferred from one species to another—in this case, from an insect to a plant.

Since 1986, the luciferase gene has been transferred to a variety of other species, including frogs, fish, and the bacterium that causes tuberculosis (TB). An airborne disease that destroys the lungs, TB was a common cause of death in the nineteenth and early twentieth centuries. Although the incidence of TB in the United States had been in decline since 1882, it began to increase in 1985. This rise in the incidence

Arthur Tress, *Fish Tank Sonata, Corn Farmer*, 1990.

Main Message

DNA technology allows us to isolate genes, produce many copies of them, determine their sequence, and insert them back into organisms.

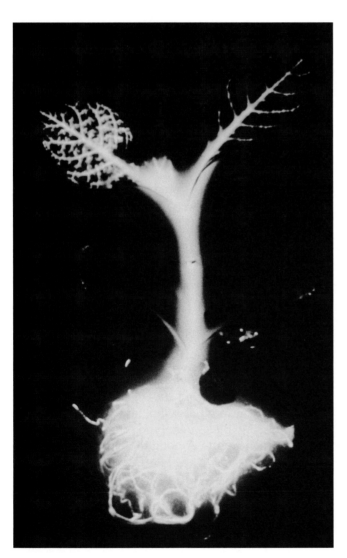

A Glowing Tobacco Plant

of TB was due in part to the appearance of new strains of the bacteria, which were resistant to one or more antibiotics. Resistance to antibiotics made the TB bacterium a major threat to public health once again.

As unlikely as it might seem, one way that we are meeting this threat is with the firefly luciferase gene. Because the TB bacterium grows very slowly outside the human body, it took 2 to 3 months to test which antibiotics would be effective for a particular patient. By inserting the firefly gene into bacteria taken from a TB patient, we can now tell whether a patient is infected with resistant bacteria in 2 to 3 days. If the bacteria are resistant, they express the firefly gene and glow in the dark. If the bacteria are not resistant, they sicken or die and don't glow in the dark. This rapid turnaround time of the firefly gene test lets doctors prescribe the appropriate antibiotic quickly, improving a patient's chance for a speedy recovery and lowering the chance that the disease will spread to new victims.

Insertion of the firefly gene into the TB bacterium was made possible by new innovations in DNA technology, the set of techniques used to manipulate DNA. These new techniques have led to many medical and commercial applications, including the isolation of disease-causing genes and the development of genetically modified crop species. In this chapter we describe the techniques scientists use to manipulate DNA, and we discuss applications and risks of these techniques.

Key Concepts

1. Recent innovations in the laboratory techniques used to manipulate DNA have greatly increased our ability to isolate and study genes and to alter the DNA of organisms.

2. Restriction enzymes cut DNA molecules at specific target sequences. When used with gel electrophoresis, a technique that sorts the chopped pieces of DNA by size, restriction enzymes provide a powerful way to examine DNA sequence differences.

3. A gene is said to be cloned if geneticists can isolate and produce many copies of it. Once a gene is cloned, automated sequencing machines can quickly determine its DNA sequence.

4. Cloning and sequencing a gene can provide vital clues about gene function, making these techniques critical to the study of genes that cause inherited genetic disorders.

5. In genetic engineering, a gene is isolated, modified, and inserted back into the same species or into a different organism. Expression of the transferred gene changes the performance of the genetically modified organism.

6. DNA technology provides many benefits, but its use also raises ethical dilemmas and poses risks to human society.

Human beings have been changing the DNA of other organisms for thousands of years. This fact is well illustrated by the many differences we have sculpted between domesticated species and their wild ancestors. For example, because of genetic changes brought about through selective breeding, dogs differ greatly from each other and from their wild ancestor, the wolf (see the Box in Chapter 21, page 333).

Although we have a long history of altering the DNA of other organisms, the past 30 years have witnessed a huge increase in the power, precision, and speed with which we make such changes. For the first time, we can now select a particular gene, produce many copies of it, and insert it back into living organisms. In doing so, we rapidly alter DNA in ways that would never happen naturally, as when the firefly luciferase gene was inserted into plants and TB bacteria.

This chapter discusses **DNA technology**, the set of techniques with which scientists manipulate DNA. We will see how scientists locate a gene, analyze its sequence, and insert it back into living organisms. DNA technology is increasingly a subject of the news, whether because of the millions of dollars made or lost in biotechnology stocks, the identification of new disease-causing genes, or the use of DNA fingerprinting in murder and rape trials. We begin the chapter by describing methods used in DNA technology, then we'll turn to practical applications, ethical issues, and risks of DNA technology.

Working with DNA: Techniques for DNA Manipulation

The genetic material of most organisms is composed of DNA (a few viruses use RNA). Although the sequence of DNA varies greatly among species, the molecule itself has the same chemical structure—a double helix—in all species. This consistency in the structure of DNA means that similar techniques can be used to analyze DNA from organisms that are as different as bacteria and humans.

In this section we describe basic methods of DNA technology that scientists are currently able to use: key enzymes, gel electrophoresis, DNA hybridization, and DNA sequencing and synthesis. As we describe each of these methods, we will discuss how they can be applied to sickle-cell anemia, a lethal, recessive genetic disorder in humans.

Sickle-cell anemia is caused by a mutation that alters a single amino acid of hemoglobin, a protein that is involved in the transport of oxygen in the blood (see Figure 15.10). Individuals who have two copies of the disease-causing allele (genotype *ss*) suffer from many serious complications. Among the worst such complications are damage to the heart, lungs, kidneys, and brain. Most individuals with sickle-cell anemia die before they have children. Individuals of genotype *Ss*, who carry the sickle-cell allele but usually have few or no symptoms, can pass the disease gene on to the next generation.

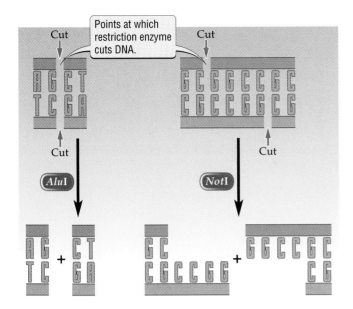

Figure 17.1 Restriction Enzymes Cut DNA at Specific Places
The two restriction enzymes *Alu*I and *Not*I cut the specific DNA sequences shown here.

Key enzymes of DNA technology

Humans have 6.6 billion base pairs of DNA on the 46 chromosomes located in their cells. Each chromosome contains a DNA molecule so large that it is difficult for scientists to work with. Thus, after DNA has been extracted from a person's cells, it must be broken into smaller pieces.

DNA is broken into small pieces by **restriction enzymes**, which cut DNA at highly specific sites. For example, the restriction enzyme *Alu*I cuts DNA everywhere that its target sequence (AGCT) occurs, but nowhere else (Figure 17.1). When restriction enzymes are used to chop up a person's DNA in a test tube, the same results are obtained at different times or from the DNA of different tissues (for example, skin and hair). Because restriction enzymes work on the DNA of all organisms, they are used in virtually all applications of DNA technology.

Ligases and DNA polymerases are two other very important enzymes used in DNA technology. **Ligases** are used in laboratory experiments to connect two DNA fragments to each other, a procedure that is critical to our ability to insert a gene from one species into the DNA of another species. **DNA polymerases**, the key enzymes that cells use to replicate their DNA, can be used to make many copies of a gene or other DNA sequence in a test tube.

Gel electrophoresis

Once a DNA sample is cut into fragments by a restriction enzyme, researchers often use gel electrophoresis to help them see and analyze the fragments. In **gel electrophoresis**, DNA that has been chopped up by a restriction enzyme is placed into a depression (a "well") in a gelatin-like substance (a "gel") through which DNA can move (Figure 17.2). The gel is then subjected to an electrical current. Since DNA has a negative charge (see Chapter 5), the electrical current causes the DNA to move through the gel toward the positive end of the gel.

Large pieces of DNA (those with more base pairs) pass through the gel with more difficulty than small pieces, causing them to move more slowly. Because they move more slowly, large fragments of DNA do not travel as far as small fragments (see Figure 17.2). Thus,

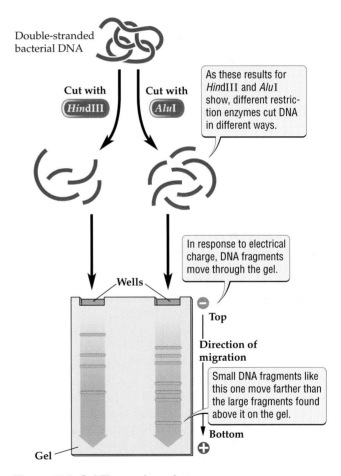

Figure 17.2 Gel Electrophoresis
Under an electrical charge, DNA fragments move through the gel. Fragments toward the bottom of the gel are smaller than fragments toward the top of the gel.

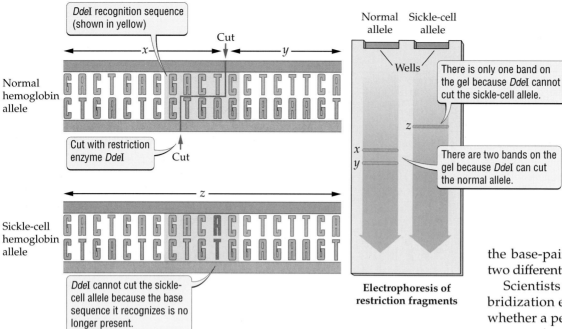

Figure 17.3 A Restriction Enzyme Identifies the Sickle-Cell Allele
The restriction enzyme *Dde*I cuts DNA every time it encounters the sequence GACTC. The mutation that causes the disease sickle-cell anemia changes the DNA sequence of part of the hemoglobin gene from the normal sequence of GACTC to GACAC, thus preventing *Dde*I from cutting the DNA at this location. As a result, the normal hemoglobin allele has two bands on the gel, whereas the sickle-cell hemoglobin allele has only one. The mutation that causes sickle-cell anemia is highlighted in red.

rapidly moving, small fragments are located toward the bottom of the gel, while slowly moving, large fragments are located toward the top. The different-sized fragments are invisible to the human eye, so to be seen they must be stained or labeled by one of various methods.

By using restriction enzymes and gel electrophoresis together, we can examine differences in DNA sequences. For example, the restriction enzyme *Dde*I cuts the normal hemoglobin allele into two pieces but cannot cut the sickle-cell allele, providing a simple test for the disease allele (Figure 17.3).

DNA hybridization

Another way of testing for the sickle-cell allele is to use DNA probes. A **DNA probe** is a short sequence of DNA, usually tens to hundreds of bases long, that can base-pair with a particular gene or region of DNA. DNA probes are used in DNA hybridization experiments (Figure 17.4). **DNA hybridization** is the base-pairing of DNA from two different sources.

Scientists can use DNA hybridization experiments to test whether a person has one, two, or no copies of the allele (s) that causes sickle-cell anemia. In such experiments, two DNA probes are used: one probe that can bind only to the sickle-cell allele, and a second probe that can bind only to the normal allele. If only the probe for the normal allele can bind to a person's DNA, the person must have two copies of the normal allele (genotype *SS*). Similarly, if only the probe for the sickle-cell allele can bind, the person must have two copies of the sickle-cell allele (genotype *ss*). If both probes can bind, the person must be a carrier with genotype *Ss*.

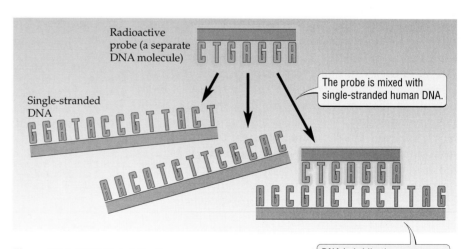

Figure 17.4 DNA Hybridization
DNA hybridization occurs when DNA from two different sources can form complementary base pairs. Here, a DNA probe forms complementary base pairs with a single-stranded segment of human DNA.

DNA sequencing and synthesis

Both basic research and practical applications of modern genetics often depend on knowing the sequence of bases in a DNA fragment, a gene, or even an entire genome. DNA sequences can be determined by several methods, the most efficient of which rely on automated sequencing machines (Figure 17.5). One of these machines can sequence many tens of thousands of base pairs per day, thus making it possible to determine the sequence of a single gene quickly. DNA can also be sequenced without a sequencing machine, by means of relatively slow but still highly effective methods.

The synthesis of probes and other DNA fragments is also automated: DNA synthesis machines can rapidly produce DNA segments hundreds of bases long. For example, in less than an hour a DNA synthesis machine can produce the two probes used to test for sickle-cell anemia, each of which is 21 bases long.

> ■ Scientists manipulate DNA extracted from cells with a variety of laboratory techniques: (1) Restriction enzymes are used to break DNA into small pieces. (2) Gel electrophoresis is used to separate DNA fragments by size. (3) DNA probes are used in DNA hybridization experiments to test for the presence of a particular allele or gene. (4) DNA is sequenced and synthesized by automated machines. The methods of DNA technology can be used to study the DNA of a wide range of organisms.

Producing Many Copies of a Gene: DNA Cloning

A gene is **cloned** if geneticists have isolated the gene and can make many copies of it. A single copy of a gene is difficult to study, but once a gene is cloned, the gene can be sequenced, transferred to other cells or organisms, and used in DNA hybridization experiments. For example, cloning and sequencing a gene enable researchers to gain vital clues about its function, making DNA cloning a key step in the study of genes that cause inherited genetic disorders or cancer. As described in the sections that follow, genes are cloned via the construction of DNA libraries or through the use of the polymerase chain reaction (PCR).

DNA libraries

A **DNA library** is a collection of an organism's DNA fragments that are stored in a host organism. For humans, a complete DNA library would contain millions of DNA fragments, which collectively would include all the 3.3 billion bases in the human genome.

The concepts behind the formation of a DNA library are simple. First, the DNA is broken into pieces by a restriction enzyme. These fragments are then inserted into a **vector**, which is a piece of DNA that is used to transfer a gene or other DNA fragment from one species to another. Common types of vectors include DNA from viruses and **plasmids**, which are small, circular segments of DNA that are found naturally in bacteria (Figure 17.6).

If, for example, plasmids were used to construct a human DNA library, fragments of human DNA would be inserted into a plasmid vector and then mixed with bacteria under conditions that cause the bacteria to take up a single plasmid (Figure 17.7). Once the bacteria had taken up the plasmids, our library would be formed: We would have millions of bacteria, each containing a single piece of human DNA. Collectively, these bacteria would contain many fragments of the 3.3 billion base pairs in the human genome.

Vectors move DNA fragments into a host organism, such as a bacterium. Bacteria reproduce rapidly, and as they reproduce they make new copies of the inserted DNA fragments. Thus, we can use the bacteria in a DNA library to make many copies of a gene. In this process, a few bacterial cells from the library are grown on a small dish. Each bacterial cell forms a mass of cells, called a

Figure 17.5 A DNA Sequencing Machine
Automated DNA sequencing machines can determine DNA sequences rapidly.

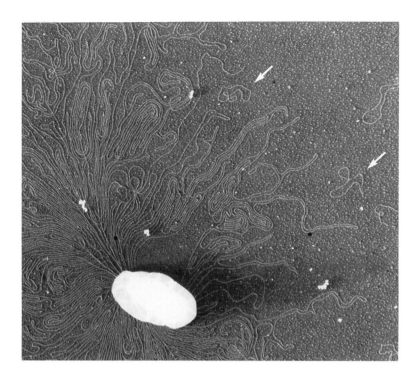

Figure 17.6 Plasmids
Plasmids are circular segments of DNA found naturally in bacteria. Plasmids are not part of and are much smaller than the main chromosome of the bacterium. Here, a ruptured *E. coli* bacterium spills out its chromosome and several plasmids, two of which are indicated with arrows.

colony. Each cell in the colony contains a copy of the gene inserted into the bacterium that started the colony.

After the bacterial colonies are established, we must determine which of them carry the gene of interest. Therefore, the colonies are tested, or "screened," to see if they can base-pair with a DNA probe for the gene of interest. Colonies whose DNA can base-pair with the probe contain the gene of interest. Bacteria from a colony containing the gene can then be grown in a liquid broth, producing billions of bacterial cells. Each of these cells contains a plasmid that has all or part of the gene of interest in it. Thus, by screening a DNA library we can isolate a particular gene from among millions of human DNA fragments, and then produce many copies of the gene.

The polymerase chain reaction

Many genes are cloned by use of DNA libraries. In some cases, DNA can be cloned through use of the polymerase chain reaction. The **polymerase chain reaction (PCR)** is a method that uses the DNA polymerase enzyme to make billions of copies of a targeted sequence of DNA in just a few hours (Figure 17.8).

When using PCR to clone a particular gene, two short segments of synthetic DNA called primers must be used. Each primer is designed to base-pair with one of the two ends of the gene of interest. The DNA polymerase enzyme then produces many copies of the sequence of DNA that is between the primers; that is, it produces many copies of the gene. To use PCR in DNA cloning, scientists must know the DNA sequence of both ends of the gene; without this knowledge, they could not synthesize the specific primers that will base-pair with the ends of the gene.

■ A gene is cloned if it has been isolated and many copies of it can be made. After being cloned, a gene can be sequenced, transferred to other organisms, or used in DNA hybridization experiments. Genes can be cloned via construction of a DNA library or by the polymerase chain reaction.

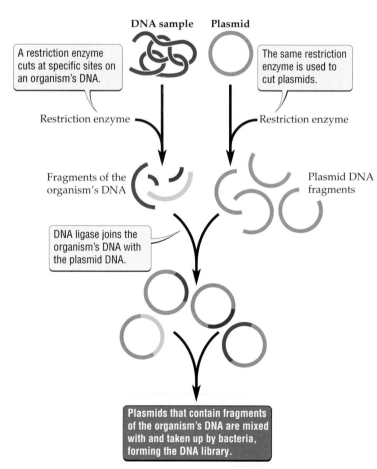

A restriction enzyme cuts at specific sites on an organism's DNA.

The same restriction enzyme is used to cut plasmids.

Restriction enzyme — Restriction enzyme

Fragments of the organism's DNA — Plasmid DNA fragments

DNA ligase joins the organism's DNA with the plasmid DNA.

Plasmids that contain fragments of the organism's DNA are mixed with and taken up by bacteria, forming the DNA library.

Figure 17.7 Construction of a DNA Library

Figure 17.8 Polymerase Chain Reaction

In PCR, short primers that can base-pair with an organism's DNA are mixed in a machine with the organism's DNA, the enzyme DNA polymerase, and nucleotides (containing the bases A, C, G, or T). The machine then goes through a repeated series of three steps, as indicated. Billions of copies of a targeted DNA sequence can be made in this way in a few hours. The targeted DNA sequence (blue) and the primers (red) are color-coded here for reasons of clarity only; the colors do not indicate any biological differences in the DNA sequences.

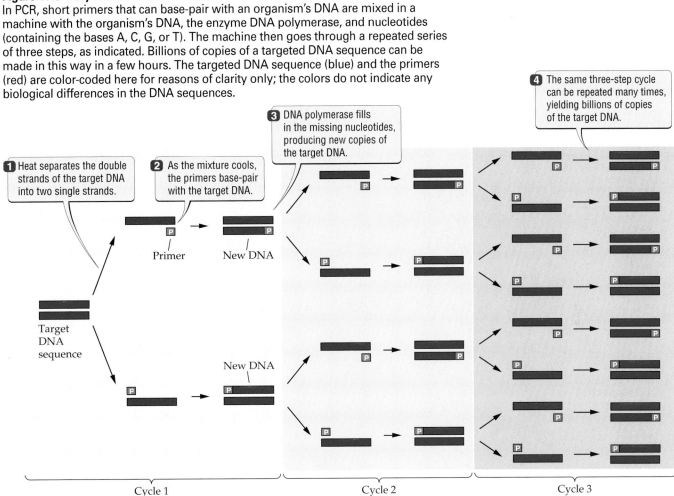

1 Heat separates the double strands of the target DNA into two single strands.

2 As the mixture cools, the primers base-pair with the target DNA.

3 DNA polymerase fills in the missing nucleotides, producing new copies of the target DNA.

4 The same three-step cycle can be repeated many times, yielding billions of copies of the target DNA.

Primer   New DNA

New DNA

Target DNA sequence

Cycle 1   Cycle 2   Cycle 3

Applications of DNA Technology

There are many important applications of DNA technology. DNA fingerprinting, prenatal screening, and tests designed to reveal the presence of a particular disease-causing organism, such as the virus that causes AIDS, are recent and important applications of DNA technology (see Chapter 18). Other examples include gene therapy, a topic addressed at the close of this chapter, and the cloning of medically important genes, such as the gene for Huntington's disease, as described in Chapter 13. In this section we'll focus on genetic engineering, an approach that includes a wide range of applications of DNA technology.

Genetic engineering

All organisms share a similar, often identical, genetic code. For this reason, if a gene can be transferred from one species to another, it often can make a functional protein product in the new species. For example, the luciferase gene of the firefly has been transferred to and expressed in organisms as different as plants, mice, and bacteria. The deliberate transfer of a gene from one species to another is an example of genetic engineering. **Genetic engineering** is a three-step process in which a DNA sequence (often a gene) is isolated, modified, and inserted back into the same species or into a different species.

Firefly

Several techniques can be used to insert genes into organisms. We have already discussed how plasmids can be used to transfer a gene from humans or other organisms to bacteria. Plasmids also can be used to transfer genes to plant or animal cells (Figure 17.9). For some species, including many plants, and recently mammals such as sheep, cows, and mice, genetically

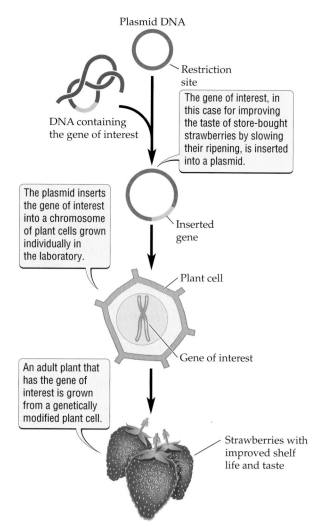

Plasmid DNA

DNA containing
the gene of interest

Restriction
site

The gene of interest, in
this case for improving
the taste of store-bought
strawberries by slowing
their ripening, is inserted
into a plasmid.

The plasmid inserts
the gene of interest
into a chromosome
of plant cells grown
individually in
the laboratory.

Inserted
gene

Plant cell

Gene of interest

An adult plant that
has the gene of
interest is grown
from a genetically
modified plant cell.

Strawberries with
improved shelf
life and taste

Figure 17.9 Genetic Engineering of Plants via Plasmids
Plasmids can be used to insert a gene of interest into plant cells.
For many plant species, adult plants can then be grown from the
genetically modified cells, thus producing a genetically engineered plant.

amounts of the cell surface proteins of particular disease-
causing organisms. When injected into our body, these cell
surface proteins stimulate our immune system to recognize the organism that normally carries the cell surface
protein. Hence, they can be used as a vaccine to protect us
against future attack by that organism. Currently, a test is
under way to evaluate the effectiveness of an AIDS vaccine that uses a genetically engineered cell surface protein
from the AIDS virus.

■ In genetic engineering, a DNA sequence (often a
gene) is isolated, modified, and inserted back into the
same species or into a different organism. Genetic engineering is used to alter the performance of the genetically modified organism and to produce many copies of
a DNA fragment, a gene, or a gene's protein product.
Genetic engineering works because all organisms share
a similar, often identical genetic code.

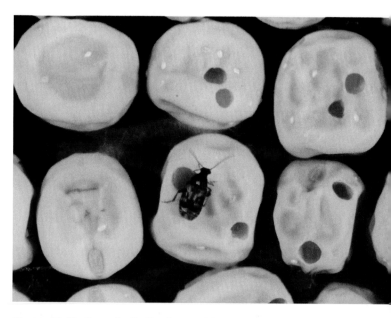

**Figure 17.10 Genetically Engineered Protection
from Insect Attack**
The pea seeds on the left, which were genetically engineered to be protected against insect attack, are intact. The
seeds on the right have holes caused by damage from wee-
vils. A weevil is shown here.

modified adults can be generated from these altered
cells. Other techniques for gene transfer include the use
of viruses that infect cells with genes from other species
and "gene guns" that fire small pellets coated with the
gene of interest into target cells.

Genetic engineering is commonly used to alter the
performance of the genetically modified organism. Pea
seeds have been engineered to be protected against
insect attack (Figure 17.10). Other crop plants have been
genetically engineered for disease resistance, frost tol-
erance, and herbicide resistance (to allow crops to survive weed-killing chemicals).

Genetic engineering is also commonly used to produce
many copies of a DNA sequence, a gene, or a gene's protein product (Table 17.1). For example, insulin, a human
hormone used to treat millions of people suffering from
diabetes, is mass-produced by bacteria that have been
engineered to contain the human insulin gene. In addition,
bacteria have been genetically engineered to produce large

17.1   *Products of Genetic Engineering*

Product	Method of production	Use
Protein		
Taxol	*E. coli*	Treatment of ovarian cancer
Human insulin	*E. coli*	Treatment of diabetes
Hepatitis B vaccine	Yeast cells	Prevention of hepatitis B
Human factor VIII	Mammalian cells	Treatment of hemophilia
DNA sequence		
Sickle-cell probe	DNA synthesis machine	Testing for sickle-cell anemia
BRCA1 probe	DNA synthesis machine	Testing for breast cancer mutations
M13 probe	*E. coli*, PCR	DNA fingerprinting in plants
33.6 and other probes	*E. coli*, PCR	DNA fingerprinting in humans
Gene		
Luciferase gene	Bacterial cells	Testing for antibiotic resistance
ADA gene	Human cells	Treatment of ADA deficiency
HD gene	*E. coli*	Testing for Huntington's disease

Ethical Issues and Risks of DNA Technology

The immense power and scope of genetic engineering and other aspects of DNA technology raise ethical dilemmas and pose a variety of risks. For example, at a most basic level, what gives us the right to alter the DNA of other species? We typically do so for our own advantage, but is this an ethical thing to do? And if, say, most people could agree that there was no ethical conflict in altering the DNA of a bacterium or a virus, does that mean there is also no dilemma associated with altering the DNA of a plant, a dog, a chimpanzee, or a human?

With respect to altering our own DNA, how do we distinguish between acceptable and unacceptable cases of genetic engineering? If it is ethical to genetically engineer a human to prevent a horrible disease, is it also acceptable to make less critical changes? For example, if it were possible to do so before birth, would it be ethical to genetically alter the future intelligence, personality, looks, or sexual orientation of our children? According to a March of Dimes survey, more than 40 percent of Americans would make such modifications if given the chance, but is it fair for parents to make such decisions on behalf of their children? We will return to these and similar issues in Chapter 18, but be forewarned: Often there are no easy answers to questions like these.

The use of DNA technology also involves risks. For example, many crop plants mate and produce offspring with closely related species. If a crop plant is genetically engineered to be resistant to an herbicide, the potential exists that the resistance gene will be transferred (by mating) from the crop to the wild species. Thus, there is a risk that by engineering our crops to resist herbicides, we will unintentionally create "superweeds" that are resistant to the same herbicides. Similar risks exist for most efforts to alter the performance of a genetically engineered organism. In essence, such risks boil down to the problem that a gene that is good for humans when it is in one species (or one set of circumstances, like an agricultural field), may be very bad for us if it is in another species (or another set of circumstances, like a more natural field environment).

Other sources of risk also exist. For example, environmental or social costs may be associated with genetic engineering. Engineering crops to be resistant to herbicides might promote the increased use of herbicides, many of which are harmful to the environment. Alternatively, a product might be environmentally safe yet still entail social costs. Consider bovine growth hormone (BGH). This hormone is mass-produced by genetically engineered bacteria. Among other effects, BGH increases milk production in cattle. Before the introduction of genetically engineered BGH in the 1980s, milk surpluses were already common. The use of BGH by large, corporate milk producers has further increased the amount of milk available. The resulting drop in milk prices threatens to drive small producers of milk—the traditional family farms—out of business. Are lower milk prices for consumers worth the social cost of driving small dairy farms into bankruptcy?

■ The use of DNA technology raises ethical questions and poses risks. There is a fine line between acceptable and unacceptable changes to the DNA of humans and other species. Benefits and risks of DNA technology must be considered carefully in evaluations of the potential impact to human society of a particular use of DNA technology.

Highlight

Human Gene Therapy

On September 14, 1990, 4-year-old Ashanthi DeSilva made medical history when she sat in a hospital bed and received intravenous fluid that contained genetically modified versions of her own white blood cells. She suffered from adenosine deaminase (ADA) deficiency, a lethal genetic disorder that severely limits the ability of the body to fight disease. This disorder is caused by a mutation to a single gene that is expressed in white blood cells, which play a central role in our ability to combat disease. Earlier, doctors had removed some of Ashanthi's white blood cells and added the normal ADA gene to them (Figure 17.11). Thus, her cells were genetically engineered in an attempt to fix a lethal genetic defect. A few months later, a similar experiment was performed on 8-year-old Cynthia Cutshall, who also suffered from ADA deficiency.

Since white blood cells do not reproduce, for several years these two girls were given genetically engineered white blood cells once every few months. More than 6 years after they last received any engineered cells, they continue to have white blood cells that express the normal version of the ADA gene. As of this writing, both Ashanthi and Cynthia are doing very well and lead essentially normal lives.

The treatments received by Ashanthi DeSilva and Cynthia Cutshall were the first clinical gene therapy experiments ever performed. Human **gene therapy** seeks to correct genetic disorders by fixing the genes that cause them. It is a bold and captivating prospect, the goal being to cure even the worst of genetic diseases by reaching into our cells and repairing the mutation that caused the disease.

As such, gene therapy has attracted much media attention—some of it, unfortunately, bordering on hype. Take the cases of Ashanthi and Cynthia. In addition to gene therapy, both girls received other treatments for ADA deficiency. Hence, contrary to what some reports

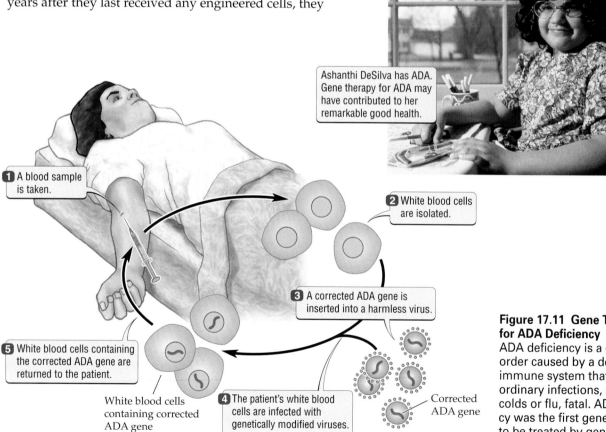

Ashanthi DeSilva has ADA. Gene therapy for ADA may have contributed to her remarkable good health.

1 A blood sample is taken.

2 White blood cells are isolated.

3 A corrected ADA gene is inserted into a harmless virus.

5 White blood cells containing the corrected ADA gene are returned to the patient.

White blood cells containing corrected ADA gene

4 The patient's white blood cells are infected with genetically modified viruses.

Corrected ADA gene

Figure 17.11 Gene Therapy for ADA Deficiency
ADA deficiency is a genetic disorder caused by a defect in the immune system that can make ordinary infections, such as colds or flu, fatal. ADA deficiency was the first genetic disorder to be treated by gene therapy.

in the media might have us believe, their remarkable good health cannot be attributed to gene therapy alone. Overall, although worldwide there are now more than 300 gene therapy experiments in progress, not a single one provides a clear success story. Why? Has gene therapy been "oversold"?

At one level, scientists know that gene therapy works: There are examples in which the corrected genes have been transferred and expressed. But whether gene therapy can achieve its ultimate goal, that of permanently fixing a genetic disease, is not yet known. Formidable hurdles remain to be cleared—perhaps most importantly, with how to deliver the engineered gene to the cells where it is needed. Harmless viruses are often used for this purpose, but to date viral delivery methods have not been very successful. In part, the reason for the low rate of success is that the human body defends itself so well against viruses that the viruses often are destroyed before they deliver the corrected gene to where it is needed.

Viral methods for delivering engineered genes continue to be improved, and recently a novel approach that uses an artificial "hybrid" molecule (part DNA, part RNA) was astonishingly effective at delivering an engineered gene to the liver cells of live laboratory rats. In this work, up to 60 percent of the rat liver cells received the engineered gene—far more than would be needed to cure many genetic disorders. Thus, although many challenges lie ahead, there is cause for excitement and hope.

> ■ Human gene therapy seeks to correct genetic disorders by fixing the genes that cause them. In some gene therapy experiments, the corrected gene has been transferred and expressed in the patient. Although results to date are encouraging, there are no clear examples in which gene therapy has permanently fixed a genetic disease.

Summary

Working with DNA: Techniques for DNA Manipulation

■ Scientists manipulate DNA using a variety of laboratory techniques that work on a wide range of organisms.

■ Restriction enzymes break DNA into small pieces, and gel electrophoresis is used to separate the resulting DNA fragments by size.

■ DNA probes are used in DNA hybridization experiments to test for the presence of a particular allele or gene.

■ DNA is sequenced and synthesized by automated machines.

Producing Many Copies of a Gene: DNA Cloning

■ A gene is cloned if it has been isolated and many copies of it can be made.

■ After being cloned, a gene can be sequenced, transferred to other organisms, or used in DNA hybridization experiments.

■ Genes can be cloned by constructing a DNA library or by using the polymerase chain reaction.

Applications of DNA Technology

■ In genetic engineering, a DNA sequence (often a gene) is isolated, modified, and inserted back into the same species or into a different organism.

■ Genetic engineering is used to alter the performance of the genetically modified organism and to produce many copies of a DNA fragment, a gene, or a gene's protein product.

■ Genetic engineering works because all organisms share a similar, often identical genetic code.

Ethical Issues and Risks of DNA Technology

■ The use of DNA technology raises ethical questions and poses risks. There is a fine line between acceptable and unacceptable changes to the DNA of humans and other species.

■ Benefits and risks of DNA technology must be considered carefully in evaluations of the potential impact to human society of a particular use of DNA technology.

Highlight: Human Gene Therapy

■ Human gene therapy seeks to correct genetic disorders by fixing the genes that cause them.

■ In some gene therapy experiments, the corrected gene has been transferred and expressed in the patient.

■ Although results to date are encouraging, there are no clear examples in which gene therapy has permanently fixed a genetic disease.

Key Terms

clone (of a gene) p. 269
DNA hybridization p. 268
DNA library p. 269
DNA polymerase p. 267
DNA probe p. 268
DNA technology p. 266
gel electrophoresis p. 267
gene therapy p. 274

genetic engineering p. 271
ligase p. 267
plasmid p. 269
polymerase chain reaction (PCR) p. 270
restriction enzyme p. 267
vector p. 269

Chapter Review

Self-Quiz

1. Which of the following cuts DNA at highly specific target sequences?
   a. ligase
   b. DNA polymerase
   c. restriction enzyme
   d. RNA polymerase

2. A collection of an organism's DNA fragments that are stored in a host organism is called a
   a. DNA library.
   b. DNA restriction site.
   c. plasmid.
   d. DNA clone.

3. The base pairing of DNA from two different sources is called
   a. DNA replication.
   b. DNA hybridization.
   c. DNA probing.
   d. DNA cloning.

4. Genetic engineering
   a. can be used to make copies of a DNA sequence, a gene, or a gene's protein product.
   b. can be used to alter the performance of the genetically modified organism.
   c. raises ethical questions and poses risks to society.
   d. all of the above

5. When DNA fragments are subjected to an electrical current on a gel, the _____ fragments move the farthest.
   a. smallest
   b. largest
   c. tRNA
   d. DNA library

Review Questions

1. (a) Define DNA cloning and describe how it is done.
   (b) Discuss advantages of gene cloning.

2. When a DNA library is made, millions of fragments of an organism's DNA are stored in a host organism. How is the library screened so that scientists can find a particular gene of interest?

3. Is it ethical to genetically alter the DNA of a bacterium? A single-celled yeast? A worm? A plant? A cat? A human? Give reasons for your answers.

4. If it were possible to do so, should it be legal to alter the DNA of a person to cure a genetic disorder, like Huntington's disease, that causes great suffering and kills its victims? Are some changes to the DNA of humans not acceptable? Assuming you think some changes are not acceptable, what criteria would you use to draw the line between acceptable and unacceptable changes to the DNA of humans?

The Daily Globe

"Terminator Gene" Causes Riots in Developing World

LINCOLN, NE. Seed companies spend money to develop new genetic varieties of crop plants. They "own" these new genetic varieties, so when seed companies sell their seed to farmers, the farmers must agree not to replant any of the seed they harvest. Thus, each year the farmers must start from scratch and buy more seed from the seed companies.

Despite this agreement, until now the possibility remained that some farmers could "cheat" by planting some of the seed they harvested. But the MonPont Company is now marketing a new, genetically engineered variety of corn that gives the seed companies the final say: MonPont researchers have inserted a "terminator gene" that makes the next generation of corn plants sterile. Thus, if a farmer "cheated" and planted seed that he harvested, the plants produced from that seed would be sterile and would not produce any corn.

When word of the terminator gene spread, farmers in developing countries were outraged. They feared that the terminator gene would spread from the large fields farmed by agribusiness companies to the small plots on which they depended for food, making their crop plants sterile. Their fear and outrage over the terminator gene led them to burn fields owned by the MonPont Company, fields that contained genetically engineered plants but not the terminator gene. As one upset, local farmer put it, "We are hungry enough already. We don't need MonPont to genetically engineer our crops to be sterile." A spokesperson for MonPont assured local farmers that the gene could not spread to their crop plants, but few farmers appeared to be in a mood to listen.

Evaluating "The News"

1. When seed companies sell seed for genetic varieties that cost a lot of money to develop, is it reasonable for them to insist that farmers not replant any of the seed they harvest?

2. Farmers work hard to grow their crops. When the harvest is complete, do you think farmers should have the right to do whatever they want with the seeds they harvest, including saving some of them to plant as crops for the next year?

3. Nearly a billion of Earth's 6 billion people are malnourished. Given that so many people lack food, do you think the "terminator gene" is a wise use of genetic engineering?

chapter 18

Genetic Screening

Sandy Skoglund, *Babies at Paradise Pond*, 1996.

Brave New World

Pregnancy has always been a time of mystery—of waiting, anticipating, and wondering about the growing fetus hidden deep in a woman's womb. Today much of that mystery has been removed by genetic screening. Parents can know, long before a baby is born, whether the baby will be male or

female, whether it has Down's syndrome, a disease that causes mental retardation, or whether it will develop any of the hundreds of disabling or deadly genetic diseases that can now be diagnosed in the uterus. This shockingly clear peek into the future has provided valuable and beneficial information to many parents. But at the same time, genetic screening is providing parents with a bewildering new set of choices.

These genetic tests are presenting some parents with the agonizing choice of whether to terminate a pregnancy or proceed with the birth of a child that

they know will develop a particular disease. Or in the case of a fetus with a clean bill of health for the tested diseases, some parents are pondering whether they will continue a pregnancy with the male fetus now developing, or try again in the hope of having the baby girl they want. Striking out into entirely new moral territory, parents are making decisions with information never before available in the history of parenthood or medicine.

Meanwhile the Human Genome Project, a huge effort that is under way to sequence the entire human genome, is moving quickly forward, with new genes found and beginning to be understood every day. In addition, in research laboratories around the country, scientists are specifically seeking genes that affect not only very straightforward genetic diseases but also characteristics like intelligence, tendencies to alcoholism and schizophrenia, violent behavior, shyness, and sexual orientation. If and when such genes are discovered, the ability to identify them in individuals is sure to offer even more perplexing choices and consequences not only for parents but for the rest of society as well.

As the data pour in, people are being presented with a wealth of information about their own genetic predispositions, as well as those of other people. Are we ready for a world in which people know so much about their own and one another's future?

Scientists Put Their Fingers on the Future

Key Concepts

1. Scientists are now able to examine specific regions of a person's DNA in a process known as genetic screening, identifying genes that reveal an increasing number of things about that person and that person's future.

2. Genetic screening can be used to determine the likelihood that a person will contract a particular disease. This information is already being put to widespread and sometimes controversial use by potential parents, health insurance companies, patients, and others.

3. DNA fingerprinting, a form of genetic screening, can match a blood, semen, or body tissue sample to the person from whom it came. It has been used to convict criminals, to prove innocence, and to determine blood relationships.

4. Genetic screening raises a host of ethical, medical, and social concerns. Can privacy be maintained in a world in which a drop of blood can reveal so much about a person's possible future? Is it ethical to choose which children will be born on the basis of their genetic potential? Is it useful to know so much about one's own genetic tendencies?

DNA, the genetic blueprint we carry in every one of our cells, holds a wealth of information. Passed down from generation to generation, it provides the information required for a single cell to develop into a functioning, human body. But our DNA can also harbor the seeds of genetic diseases, programmed with tendencies for cancer, heart disease, and many other ailments. Scientists are continuing to learn more about how different genes produce different phenotypes, including disease phenotypes. On the basis of the presence of a specific DNA fragment, they can predict the likelihood that a particular phenotype will be expressed. As a result, genetics is entering an age in which researchers may soon be able to predict a person's potential not only for disease, but also for such characteristics as high intelligence or a quick temper, just on the basis of looking at that person's DNA. **Genetic screening** is the process of looking for such genes in order, for example, to assess a person's future health risks.

With hundreds of tests for genetic diseases already available, people are able to learn more about their children, their families, their coworkers, and themselves than they ever thought possible. Done for an increasing number of different reasons, genetic screening is an extremely powerful tool that has saved lives and convicted violent criminals. But at the same time, genetic screening has raised difficult, ethical questions. What do we want to know about ourselves and others? What do we have the right to know? And what do other people have the right to know about us? In this chapter we examine several types of genetic screening, looking at their power to help people, as well as the potential for harm and the diversity of ethical issues they raise.

How Genetic Screening Works

The techniques for screening directly for a particular disease mutation are surprisingly simple (Figure 18.1). All researchers need is a sample donated from the patient that contains cells with DNA. Those cells can be, for example, from blood, skin, hair, semen, or saliva. After researchers isolate the DNA from these samples, they can then use a variety of techniques to screen for genes that cause disease. For example, researchers can perform the polymerase chain reaction (PCR; see Figure 17.8) using primers that are designed to attach only to a particular disease-causing mutation. If an allele carrying the disease-causing mutation is present, then the targeted region of DNA will amplify—that is, many copies of it will be made. If the person does not carry the mutation of interest, nothing will amplify. Researchers can also amplify the entire gene of interest and use different methods, including direct DNA sequencing, to test for the presence of the disease gene mutation or other mutations of interest.

Researchers can also screen for genetic diseases that are caused by chromosomal abnormalities. To test for Down's syndrome, they count the chromosomes from individual cells in a developing fetus. Down's syndrome is caused by the presence of three rather than the normal two copies of chromosome 21—hence the disease's other name, trisomy 21.

Finally, researchers can test indirectly for genetic diseases. This can be done by looking for products in the body or by looking for certain aspects of a body's metabolism that indicate a person carries a genetic defect that will result in a disease in that person or in his or her children.

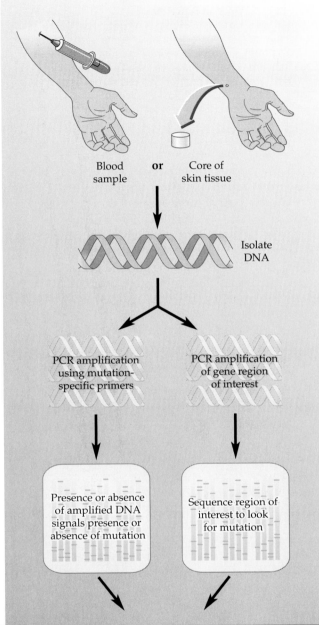

Figure 18.1 From Tissue to Test Results
Using cells from tissue or body fluids, researchers can isolate a person's DNA and use a variety of different methods to screen for a particular disease form of a gene. In the two methods charted here, PCR amplification (the polymerase chain reaction; see Figure 17.8) is followed by analysis of the presence or absence of DNA fragments or by sequencing of DNA fragments. Results gathered are presented to patients in the form of genetic counseling, a service that helps patients decide how they want to use the information they have been given.

Blood sample   **or**   Core of skin tissue

Isolate DNA

PCR amplification using mutation-specific primers

PCR amplification of gene region of interest

Presence or absence of amplified DNA signals presence or absence of mutation

Sequence region of interest to look for mutation

Genetic counseling

> ■ Using a variety of techniques, researchers can screen the genome of an individual—a grown person or even a fetus—to learn his or her genetic tendencies for certain conditions and diseases.

Retinoblastoma: Screening, Treatment, Prevention

In many cases, genetic screening is always beneficial. Screening for **retinoblastoma**, a rare childhood cancer of the retina, is a case in point. Susceptibility to retinoblastoma can be passed along in the form of defective copies of what's known as the retinoblastoma, or *Rb*, gene, a tumor suppressor gene (see Chapter 11). *Rb* keeps tumors in check by acting as a gatekeeper on cell division, producing a protein that allows cell division only when conditions are correct. If a person carries two defective copies of this gatekeeping gene, cells can divide out of control, resulting in a tumor and destroying the retina and potentially spreading cancer throughout the body. When young children or infants who have inherited one defective copy of the *Rb* gene develop a spontaneous mutation in their one good copy of the gene, they can develop retinoblastoma.

In the past, those who developed retinoblastoma often lost their eyes, and sometimes their lives. Today researchers can largely prevent the cancer by screening for mutations in the retinoblastoma gene of newborns, even fetuses, in families with a history of the disease. If the child does not carry the gene, then he or she is at no greater risk than anyone in the general population and parents can breathe easy. If a defective *Rb* gene is detected, the child can be carefully examined every 6 months (Figure 18.2) until reaching age 4, at which point there is no longer a risk of developing retinoblastoma. If cancerous cells are detected by this process, they typically can be destroyed by laser surgery without damage to

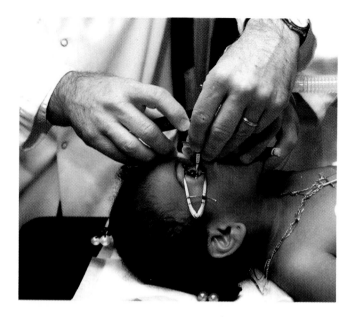

Figure 18.2 A Win-Win Situation
This child with a genetic tendency for retinoblastoma is being examined by doctors for early signs of tumor formation in the retina. By intensively examining such children, physicians are usually able to recognize the beginnings of the cancer, even at the one- or two-cell stage. They then destroy the tiny tumor with laser surgery.

includes a region with a string of three nucleotides (CAG) repeated again and again. Each triplet of CAG codes for the amino acid glutamine in the resulting protein. Different people carry different numbers of repeats in their genes and, as a result, different numbers of glutamines in the protein the gene produces. Anyone who has more than 42 repeats will almost certainly develop Huntington's disease. In some affected people, the gene has mutated such that there are more than 100 CAG repeats in a row.

the growing eye. An ideal case, *Rb* testing leads either to a diagnosis of no risk or to the prescription of a simple, reliable treatment that can usually prevent the disease from progressing.

> ■ Retinoblastoma is an example of a purely beneficial form of genetic screening, raising no difficult ethical issues. Babies, even fetuses, can be screened to see if they are at risk of developing the cancer. Children at high risk can be thoroughly and regularly examined for signs of cancer. Many lives and eyes have been saved by this powerful new screening procedure.

The Dilemma of Screening for Untreatable Diseases

Not every genetic test leads to the simple resolution described in the previous section for retinoblastoma. There is growing controversy among researchers, doctors, patients, and ethicists about the benefits of screening healthy people for diseases for which there is no cure. Two such incurable diseases are **Huntington's disease**, a deadly, degenerative disease of the nervous system, and Duchenne's dystrophy, which causes the muscles to waste away and leads to death in the early twenties.

As discussed in Chapter 13, Huntington's disease proceeds from mild difficulty with walking to complete loss of the ability to walk, stand, and speak, ending with dementia and death (Figure 18.3). Since isolating the gene for this disease, scientists have discovered that it

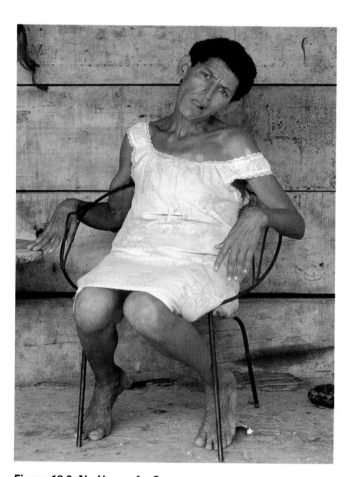

Figure 18.3 No Hope of a Cure
This woman is suffering through the late stages of Huntington's disease—a horrible and deadly degenerative disease of the nervous system for which there is no cure.

As explained in Chapter 13, Huntington's disease is transmitted as an autosomal dominant trait. As a result, children who have one afflicted parent (it would be extremely unlikely for a child to have two parents with the rare Huntington's disease) have a fifty-fifty chance of receiving the defective gene. If they receive the oversized gene, they will almost certainly develop Huntington's disease.

Before testing was available, people had no way of knowing whether they would be free of the disease until they reached old age without developing symptoms. For those who have seen relatives succumb to Huntington's, the threat that they may have this horrible disease is a constant anxiety. In the past, without knowledge of their genetic status, many potentially afflicted people chose not to have children, even though half of those people would not have carried the gene for Huntington's disease.

Now people can choose to learn their fate by having their DNA tested for the number of CAG repeats. For those who turn out not to be at risk of Huntington's, a lifetime of dread can be replaced by great peace of mind and, if they choose, the freedom to have a family without fear of passing on this disease.

But while the test provides great benefits to some, there are those patients, doctors, and ethicists who question whether people should have access to such tests. In the case of Huntington's disease, as well as other diseases for which there remains no cure, if a tested person turns out to carry the disease gene, is it really a good thing to know? For such people, hope is replaced by the wait for the inevitable descent into sickness without any chance of a cure.

Some say the knowledge provides a chance to plan for the future. Others argue that only the very wealthiest can plan for a future in which one will become completely disabled at a relatively young age. And researchers say that even those who discover they are free of the disease can suffer intense feelings of guilt at having escaped the fate that others in their family suffer. Like other genetic tests for diseases without cures, the tests for Huntington's raises questions for which there are no easy answers.

> ■ For those with a family history of Huntington's disease, fear of the unknown has been replaced by the dilemma of knowledge—deciding whether or not they want to know their fate through genetic screening. Some argue that knowing whether you will have an incurable disease can be useful, regardless of the outcome of the test. Others argue that learning that you will have such a disease provides no real benefits.

The Controversy over Cancer Screening

In some ways the cases we have described thus far in this chapter are the least controversial because the prediction of the outcome and the treatments are clear. Even with Huntington's disease, the available treatment, which is to do nothing, is obvious. The knottiest problems come from genetic diseases for which doctors cannot determine for sure that a person will develop the disease but can only determine the *likelihood* of that happening, as is the case with some cancer screening.

Supporters of genetic screening for such cancer risks say that people have the right to be tested and to know their genetic tendencies, now that screens are available. But others argue that genetic screens for tendencies to cancer, while very easily done, are much less easily interpreted. These tests, they say, present patients with the difficult dilemma of deciding how to play the percentages, how to play the odds on their health, sometimes using data that even cancer researchers are still struggling to understand fully. Moreover, there is concern about the kind and quality of counseling that is provided in this rapidly growing field. Breast cancer screens illustrate the dilemmas well.

The discovery of breast cancer genes led quickly to genetic testing

Breast cancer strikes one in every 10 women in the United States and kills about 50,000 women each year. Most cancers are caused by what are known as **sporadic mutations**, mutations that are not inherited from a parent but arise first in the stricken individual's genome. A small percentage of breast cancers, about 5 percent, are caused by inherited mutations. For some inherited mutations, the genetic tendency to breast cancer can be quite high, with generations of women in a single family succumbing to the disease. So it was with great excitement that in 1994, researchers reported the discovery of *BRCA1*, or breast cancer gene 1. Fifteen months later the gene *BRCA2* was discovered. Together, mutations in these two genes account for most of the cases of inherited breast cancer.

Thus far, scientists have the best information about *BRCA1* mutations. With a mutation in the *BRCA1* gene and a family tendency to the cancer, a woman has an 87 percent likelihood of developing breast cancer in her lifetime and a 64 percent chance of developing ovarian cancer. Scientists suspect that both *BRCA* genes are involved with DNA repair, keeping mutations at bay and suppressing tumors. Although much remains to be learned about these genes, commercial testers are making screening for them widely available.

The implications of the test results are often unclear

The great advantage of the tests for *BRCA* mutations is that a woman whose family has a tendency to breast cancer because of a known *BRCA* mutation can be screened and discover that she does not have that particular *BRCA* mutation. The worst fears of these women are thus put to rest. Upon learning their test results, they describe an extreme elation and relief.

But scientists worry that a woman with a family history of breast cancer who tests negative for some of the more than 200 mutations in the two *BRCA* genes could easily and falsely conclude that she is one of the lucky few not at risk for breast cancer in her family. In fact, she would be safe only if she knew that she did not carry the *particular* mutation that caused the women in her family to develop breast cancer. If the family cancers were caused by *BRCA* mutations for which she had not been tested or if they were caused by mutations outside of the *BRCA* genes, then the negative test results would be misleading. If such a woman were to decrease her vigilance against cancer (by performing breast exams and undergoing mammography less frequently), she could be making a deadly mistake (Figure 18.4).

For those who test positive for a *BRCA* mutation, there are still more complications. The 200 different *BRCA* mutations may well confer different levels of risk for breast and ovarian cancer, differences that remain largely unstudied. And each of those mutations may act differently depending on what other genes are active in a particular woman's genome. As discussed in Chapter 12, the presence of one gene can modify or alter the effect of another gene. There may well be genes that can increase or decrease a woman's risk of cancer given that she carries a particular *BRCA* mutation. These subtleties also remain poorly understood. Finally, researchers cannot predict the risk of cancer for women with *BRCA* mutations who are from families *without* a history of breast cancer.

Genetic screening for breast cancer has obvious and tangible benefits; lives have been and are being saved as a result of screening for breast cancer genes. But at the same time, for many women these genetic tests are producing a great deal of information that is unclear and difficult to interpret. Critics challenge the ability of counseling to prepare women to understand and make decisions about data that are often still poorly understood by cutting-edge researchers.

In the midst of these controversies, many women who screen positively for *BRCA* mutations—but for whom the risk of breast cancer is unclear—are taking the drastic measure of having their currently healthy breasts and sometimes ovaries surgically removed. More worrisome still, the benefits of such surgeries have yet to be clearly quantified; in some cases small amounts of breast or ovarian tissue remain that can become cancerous.

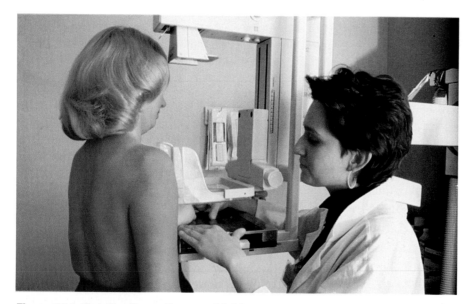
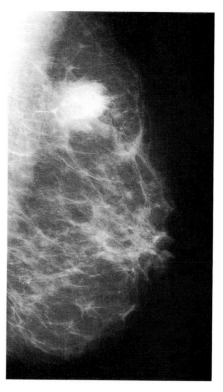

Figure 18.4 Fighting Breast Cancer with Mammograms
This woman is getting a mammogram, an X-ray of the breast tissue, which physicians use to visualize tumors and to diagnose breast cancers before they become lumps that can be felt. In the mammogram (right), the distinct white spot is a cancerous lump. For women with a genetic tendency to cancer, vigilance in the form of mammograms and other diagnostic procedures can be lifesaving.

■ The most complex ethical issues arise from genetic screening, like that for many cancers, in which having a particular allele doesn't guarantee either good health or a cancer, but only indicates a certain risk of the disease. Breast cancer screening is a good example of the difficulties of testing for genes whose function and mutations are still not fully understood. Screening for breast cancer genes has saved lives. But some argue that because of the unclear nature of the data these screens produce, they may cause women to make drastic mistakes with their health.

Prenatal Screening for Genetic Defects

Genetic screening has also become widespread for developing fetuses, a process known as **prenatal** (*pre,* "before"; *natal,* "birth") **screening**. One of the most controversial forms of genetic screening, prenatal screening provides parents-to-be with the information and opportunity to make decisions—either to continue a pregnancy or to end one.

While a woman is pregnant, doctors can carry out numerous tests to reveal genetic characteristics of the fetus. There are already more than 200 available genetic tests for fetuses. Depending on the outcome, parents may or may not terminate a pregnancy. Such decisions are already altering what babies are being born. In the United States, the number of Down's syndrome babies has decreased, apparently as a result of genetic screening. According to some reports, abortion rates of Down's fetuses range from 50 percent to 90 percent.

One study in Italy reported a 99 percent abortion rate of fetuses found to test positive for a particularly severe form of a genetic blood disease known as thalassemia. In Asia, where

in many cultures boys are more highly valued than girls, female fetuses are being aborted with a greater frequency than male fetuses. Scientists are already seeing a changing sex ratio in Asia and are predicting that the ratio of men to women will become highly skewed for the generations that will come of age in the next decades.

Scientists use a variety of methods in prenatal screening

In the most direct approach for prenatal screening, a sample of the fetus's genetic material is taken. Such samples can be obtained in several ways. One method is **amniocentesis** (Figure 18.5*a*), a procedure in which a needle is inserted directly into the uterus to extract some of the fluid, known as amniotic fluid, which surrounds the fetus. This fluid contains fetal cells from which chromosomes can be counted or that can be used for DNA testing. Another method is a **fetal biopsy** (Figure 18.5*b*), a procedure in which tissue is extracted directly from the developing fetus's body. These sampling methods have fairly high risks. For example, amniocentesis has a spontaneous abortion rate of one in 250 or more. After the sample is collected, DNA is extracted from the fetal tissue, and any genetic screening can be done as shown in Figure 18.1.

Prenatal screening presents parents with ethical dilemmas

What should prospective parents do when they discover that their fetus has Down's syndrome and is destined to be mentally retarded and to suffer other health problems, like heart disease (Figure 18.6)? Caring for a Down's child can be extremely difficult for a family—financially, emotionally, and physically. For many the answer is that the pregnancy should be mercifully terminated. But others argue that Down's children are among the most joyful of people, characteristically pleasant in their disposition, impaired yet able to enjoy many of the pleasures of life.

What about cystic fibrosis, another genetic ailment for which testing is available? Children with cystic fibrosis often have great difficulty breathing and typically die in their twenties. In order for the thick sub-

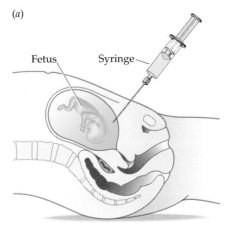

(*a*)

Fetus   Syringe

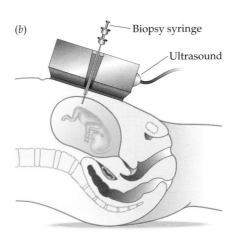

(*b*)   Biopsy syringe

Ultrasound

Figure 18.5 Peering into the Future
(*a*) In amniocentesis, amniotic fluid, which contains fetal cells, is extracted from the uterus. (*b*) Fetal biopsy can also be used to obtain larger quantities of fetal cells, which can be used to screen for DNA sequence as well as chromosomal defects.

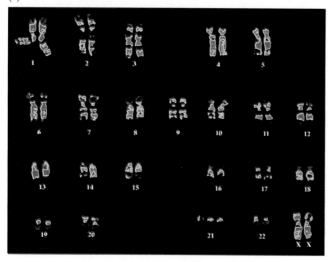

Figure 18.6 A Disappearing Condition?
(*a*) This child shows the typical facial symptoms of Down's syndrome, or trisomy 21. The major effect of the extra chromosome is mental retardation but there are often other health problems as well. The number of Down's syndrome babies being born is on the decline, a trend attributed to prenatal screening for the condition. (*b*) Down's individuals carry two copies of all chromosomes except chromosome 21, for which they carry three copies, as shown.

with such choices, how do we measure the value of these lives? And can we or should we?

And what of the many other genes for which genetic screening tests exist, but for which doctors can give no more than a probability that the person will develop the disease? What should parents do with the knowledge that their fetus has a 40 percent chance of developing Alzheimer's or breast cancer or colorectal cancer? Should they terminate the pregnancy or accept the risks for their child?

Although most people use prenatal screening as a way to decide whether to continue or terminate a pregnancy, some people use the information simply as a way to prepare for the future. They look for good news as a way to relieve their anxiety about the state of their child's genetic health. They accept bad news as a first step toward preparing financially and emotionally for the birth of their child. As one doctor put it, if parents are prepared with the knowledge that their child has an impairment—for example, Down's syndrome—they can be ready to joyously celebrate instead of suffering disappointment at the moment of the child's birth.

Parents can now pick and choose among embryos

For a variety of medical reasons, some women become pregnant by **implantation**, a procedure in which eggs are fertilized outside a woman's body and implanted into her uterus. Parents with such test tube babies can choose the genetic makeup of the fetus before a pregnancy has even begun. As more genes are discovered, the possibilities will become even greater as parents may be able, one day, to screen not only for severe diseases but for intelligence, shyness, or athletic ability.

The increasing availability of knowledge about developing fetuses raises a host of difficult questions. Do we need to prevent parents from trying for the perfect designer baby? Do parents have a moral responsibility to prevent the birth of children who will suffer from terrible, predictable genetic ailments? And who will have access to such technologies? Will only the very rich have access, or should health insurance pay for everyone to know the genetics of their developing fetuses?

stances produced in their lungs to be loosened and their risk of infection and pneumonia to be reduced, they must endure painful, daily treatments. They often suffer digestive problems and have trouble growing. On the other hand, though babies with cystic fibrosis once were unlikely to survive childhood, some have survived into their fifties.

Although some people see the termination of such pregnancies as immoral, others see the continuation of such pregnancies as immoral and irresponsible. Faced

■ Prenatal screening is one of the most controversial forms of genetic screening. Parents can now learn a great deal about developing fetuses and are making decisions to continue or terminate pregnancies based on such information. Some say that preventing the birth of children who will suffer painful diseases is the moral choice. Others say that the access to prenatal genetic information promotes a world in which parents will be working toward creating a perfect, designer baby.

The Right to Know and the Right to Privacy

One of the dilemmas in all this newfound knowledge is the question of who has the right to know and who has the right to privacy. Within families, if one person in a family has undergone genetic screening, it often means that he or she has information about that of other people in the family as well. In an extreme example, a pair of identical twins were at risk for Huntington's. Only one wanted to be tested. She promised not to reveal her results to anyone in her family. She was found to have a very high probability of having Huntington's disease. Very quickly her entire family—including her twin—knew. Although she had exercised her right to know, her sister had simultaneously lost the right to privacy and the right, in this case, not to know that she was almost certain to succumb to Huntington's disease.

> ■ Because families share genes, it can sometimes be very difficult to protect the right of one person to know about his or her genetic future while simultaneously protecting the right of a related person not to know.

DNA Fingerprinting

Another form of genetic screening, put to entirely different uses, is **DNA fingerprinting**, in which researchers develop a profile of DNA characteristics so specific that they can be used to identify an individual, just as a traditional fingerprint can. Laboratories can take a biological sample of unknown origin, like blood, tissue, or semen from a crime scene, and develop a DNA fingerprint or profile for the person from whom the sample came. That profile can then be compared with other profiles—for example, from a criminal suspect—to see if the two profiles match. A match indicates that a crime scene sample (for example, a drop of blood on a carpet) came from the tested suspect. Widely used in criminal cases, DNA fingerprints have been used to convict criminals of murder and rape and to prove people's innocence, sometimes after years of wrongful imprisonment.

Of great concern to many is the fact that huge databases of such DNA profiles are now being compiled. In many states, convicted sex offenders are required to pro-

vide samples that can be used for DNA profiling. In South Dakota, DNA samples are taken upon arrest, before any conviction, just as fingerprints are. In Virginia, law enforcement officials take samples from all convicted felons and some juveniles. In Minnesota, DNA testing is done at all crime scenes, even those of burglaries. In the U.S. military, everyone is required to provide a tissue sample that can also be used for DNA fingerprinting for purposes of identification, similar to dental records, after death. In addition, different commercial DNA-fingerprinting companies are maintaining records. Some parents pay such companies to compile their children's DNA fingerprints, in case they are ever needed for identification purposes later.

Many of these databases are being used. Numerous states have now pooled their DNA profiles over the Internet, allowing the DNA identification process to span large parts of the country, instantly. In late 1997, after eight states linked their databases, within minutes they had their first "hit": A convicted sex offender in Illinois appeared to be the same person who had committed rape and attempted murder in Wisconsin (Figure 18.7).

DNA profiling and the growing DNA databases are providing law enforcement agencies with a previously unimagined power to track down and convict people who are guilty of crimes. At the same time, many are

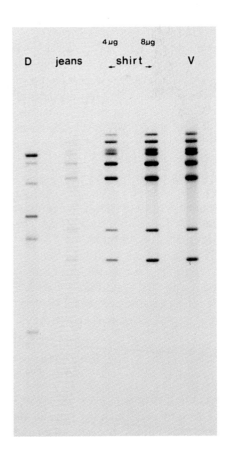

Figure 18.7 Catching the Bad Guys Using Their Own DNA
The DNA profile on the left comes from a sample taken from the defendant (D) on trial for murdering the victim (V). The defendant's jeans and shirt were splattered with blood—the DNA from which clearly matches the victim's. Courtesy of Cellmark Diagnostics, Germantown, MD.

concerned about who is going to have access to these criminal and noncriminal DNA databases and samples. Can the rights of individuals be protected?

Critics point out that the information in these databases may not always be accurate. Critics have also expressed particular concern over the kind and amount of training provided to those creating and maintaining these databases, databases that can falsely implicate a person with a simple mislabeling of a tube or mistyping of a piece of data. The concern is particularly great because, as a general rule, juries find DNA evidence extremely convincing, even when it does not agree with other evidence.

Are we as a society willing to give up, or restrict, the power of such databases to take violent criminals off the streets? Or are we willing to accept the risks involved in having detailed genetic information of many individuals available to so many different people?

Scientists can use a variety of methods to create DNA fingerprints

DNA fingerprinting takes advantage of the fact that all individuals (except identical twins) are genetically unique. In order to distinguish between different people, researchers look for highly variable regions of the genome to compare, often screening what is known as "junk" DNA, sequences that do not code for any proteins and that can be highly variable in size and base composition between different people.

Fingerprinting can be done in various ways, and the methods continue to evolve and improve. Currently the major commercial DNA fingerprinters use one of two methods: restriction fragment length polymorphism (RFLP) analysis or PCR amplification. In RFLP analysis, technicians document the size of the DNA fragments produced when a particular region of a person's DNA is broken down into small pieces by what is called a restriction enzyme. In order to produce a unique fingerprint, technicians document the size of many such fragments in regions of the genome which vary greatly between individuals. The unique pattern of DNA fragments in a number of different gene regions makes up a person's DNA profile.

Similarly, PCR amplification can be carried out using primers from highly variable regions of the genome. By looking at a number of these highly variable amplified regions, technicians can assemble a DNA profile, which can then be compared to other such profiles.

The applications for DNA fingerprinting continue to grow

Law enforcement officials have found an increasing number of uses for profiling in criminal cases. When plant seeds were found in the back of a truck of a man

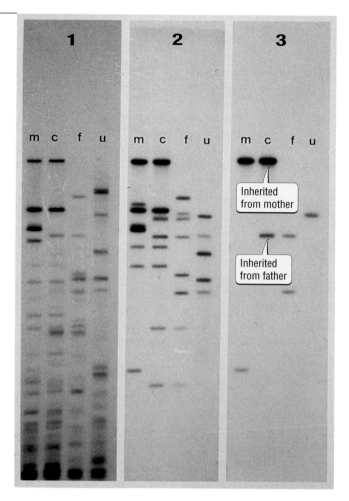

Figure 18.8 Mommy and Daddy and Baby Make Three
Here a child's paternity and maternity can be clearly seen written in its DNA profile using three different restriction enzymes. Half of the child's (c) markers come from its mother (m), and half from its father (f). An unrelated individual is shown in the last lane (u). Courtesy of Cellmark Diagnostics, Germantown, MD.

suspected of murdering a woman in the Arizona desert, prosecutors were able to link the truck to the murder scene by producing a DNA profile of the seeds, which matched the DNA profile of a particular tree growing near where the woman's body was found. DNA profiling is even being used to prosecute poachers. Poachers confessed to taking a trophy buck from actor Clint Eastwood's California ranch when DNA profiling linked their carcass to the entrails of the animal that they left behind and that were found near the ranch.

Sometimes, as in paternity cases, it is important to determine whether and in what way two people are related. DNA fingerprinting allows researchers to do this. For example, because we receive half our genes from our mother and half from our father, we should share half the fragments of our DNA profile with each parent (Figure 18.8).

In another application, DNA fingerprinting was used to identify a military man who had died in Vietnam and lay for years in the Tomb of the Unknown Soldier, a man now recognized as Lieutenant Michael J. Blassie. With all military personnel now providing a sample that can be genetically screened, every one of the military dead is theoretically knowable, making Lieutenant Blassie the last of the unknown soldiers.

> ■ DNA fingerprinting is being used for a wide variety of purposes, the most common of which are convicting criminals and freeing the innocent. Growing databases compiling DNA fingerprints are becoming highly controversial, as critics question whether individual privacy rights can be maintained and whether such databases are accurate enough.

Highlight

Health Insurance and the New Genetics

We've already seen that human genetic screening for disease tendencies can be used to try to prevent disease, to gain peace of mind, and to decide how to proceed with pregnancies and marriages. But what is to stop the information gained through genetic screening from being used against people in a form of "genetic discrimination"? Genetic discrimination already has cost some people their health insurance.

For example, when Theresa Morelli, a young single attorney, tried to get disability insurance, the company discovered that her father had Huntington's disease. The company would only agree to give her insurance beginning at age 50 and only if she had no symptoms of Huntington's disease. So while she was still healthy and earning a good living, but had the chance of becoming ill—exactly the time when a person might want to buy health insurance—she was declared uninsurable because of her genetic makeup. The company was willing to give her the insurance only if they had good evidence that she was unlikely to develop the costly Huntington's disease.

Studies have documented numerous such discrimination cases, even for easily treated diseases like phenylketonuria (PKU), a genetic disease in which the body cannot metabolize the amino acid phenylalanine. The disease can lead to severe retardation if untreated. However, a simple, effective treatment is available: a lifelong diet low in phenylalanines. Nevertheless, some health insurers have declined to cover people because they have PKU.

For health insurers, information about genetics, complete with detailed predictions about what diseases their potential customers will contract, presents a wealth of possibilities. Health insurance companies are in the business of assessing health risks—for example, charging smokers and nonsmokers different premiums. With genetic screening such companies can now have a very powerful tool to weed out people likely to succumb to diseases with expensive treatments. Some argue, however, that such practices are immoral, preventing those most likely to need the assistance of medical insurance from getting it. Others counter that health insurers are a business like any other, not a charitable organization for supporting society's most ill.

As with other forms of genetic testing, testing that can affect health insurance raises difficult issues. Genetic screening is quickly changing our world, and we must all begin to seek answers to these difficult questions as society tries to find its way in what remains largely uncharted territory.

> ■ As more genetic information about patients becomes accessible and available, health insurance companies are using this information. Some say that genetic screens are just another form of risk assessment, such as asking if a person is a smoker or not. Others argue that studying a prospective customer's DNA is an invasion of privacy and an immoral way of guaranteeing that the very sick are not insured.

Summary

How Genetic Screening Works

- DNA isolated from cells can be screened for particular forms of genes—for example, disease-causing mutations.

- Knowing whether a person carries particular alleles, researchers can make predictions about the person—for example, the likelihood that that person will contract a particular disease.

Retinoblastoma: Screening, Treatment, Prevention

- For retinoblastoma, genetic screening is purely beneficial. Accurate testing is available and straightforward, and successful treatments are also available.

- When infants are screened and found to be at high risk of retinoblastoma, their eyes are carefully monitored and treated with lasers if the cancer does develop.

The Dilemma of Screening for Untreatable Diseases

- Researchers can now screen people for genetic tendencies to develop certain untreatable diseases, like Huntington's disease.

■ For those who discover they are not at risk of developing an incurable disease, dread is replaced by relief; for those who discover they will succumb to the disease, their worst fears are confirmed.

■ Some say that screening for such diseases is beneficial, but others question the benefits of knowing that you have a disease for which there is no hope of a cure.

The Controversy over Cancer Screening

■ Cancer screening is an example of some of the most difficult and complex genetic screening. Screening that reveals only that a person has a certain level of risk, but not certainty, that the disease will develop.

■ Screening for breast cancer provides solid, easy-to-understand information for some, and information that is unclear or difficult to understand to others.

■ Despite the difficulties and complexities, women are making critical decisions—for example, to keep or remove breasts and ovaries—on the basis of these tests.

Prenatal Screening for Genetic Defects

■ Prenatal screening is one of the most controversial forms of genetic screening. Often parents make decisions to continue or terminate pregancies based on the information they receive.

■ Some argue that preventing the birth of children who will suffer severe diseases is the only moral choice; others argue that parents are treading on shaky moral ground when they choose their pregnancies by the fetuses' genetics.

The Right to Know and the Right to Privacy

■ Genetic screening can create dilemmas by making it impossible for some individuals to retain their right to privacy while others retain their right to learn more about their genetic fate or tendencies.

DNA Fingerprinting

■ In DNA fingerprinting, a DNA profile is developed that, like a traditional fingerprint, is specific to a particular individual, making it possible to match samples of skin or other tissue or fluids to the person from whom they came.

■ Law enforcement officials are building huge, nationwide databases of DNA profiles, enabling them to search the country for criminals. Other, noncriminal databases are being created as well.

■ Critics worry that these databases could be inaccurate and that they could be misused, convicting the innocent or violating privacy rights.

Highlight: Health Insurance and the New Genetics

■ Genetic screening is being used not just by patients, but by insurance companies, raising questions about the ethics of using genetic information from patients to deny them health insurance or restrict their coverage.

Key Terms

amniocentesis p. 285

BRCA1 p. 283

DNA fingerprinting p. 287

fetal biopsy p. 285

genetic screening p. 280

Huntington's disease p. 282

implantation p. 286

prenatal screening p. 285

retinoblastoma p. 281

sporadic mutation p. 283

Chapter Review

Self-Quiz

1. Currently, genetic screening cannot be used to
   a. connect individuals to blood or semen left at a crime scene.
   b. predict whether an individual will contract Huntington's disease.
   c. tell parents-to-be whether their infant will be of average or above-average intelligence.
   d. tell parents-to-be whether their infant may suffer from particular forms of mental retardation.

2. Retinoblastoma is
   a. a childhood cancer.
   b. not treatable.
   c. an example of a highly controversial form of genetic screening.
   d. a degenerative nerve disease.

3. A person's tendency to develop Huntington's disease is determined by whether he or she has
   a. any CAG nucleotide repeats in the disease gene.
   b. too few CAG nucleotide repeats in the disease gene.
   c. too many CAG nucleotide repeats in the disease gene.
   d. no CAG nucleotide repeats in the disease gene.

4. Which of the following is *not* a reason for controversy over genetic screening?
   a. It is too difficult to perform.
   b. It risks invasions of personal privacy.
   c. It causes parents to make decisions that some find questionable about pregnancies.
   d. It is not available to the very poor.

5. Critics of genetic screening are concerned about breast cancer screening because
   a. the test's results can be difficult to interpret.
   b. women are having breasts removed as a result of their screening.
   c. only 5 percent of breast cancer cases are inherited.
   d. there are more than 200 known mutations in the two *BRCA* genes.

6. Prenatal genetic screening can be done by
   a. amniocentesis.
   b. ultrasound.
   c. genetic screening of parents.
   d. embryo implantation.

Review Questions

1. If you were a parent-to-be and were offered genetic testing that would reveal whether your fetus was destined to suffer a painful disease that could drastically alter the quality of his or her life, would you want to know? Why or why not?

2. Some people who have learned that they have the gene for Huntington's disease think they would have been better off not knowing and that the test should not be provided to patients. Should the government prohibit testing for incurable diseases?

3. Geneticists are discovering genes for a wide variety of human traits. Are there any traits that people—employers, insurance agencies, parents—should not be allowed to screen for? Intelligence? Kindness? Violent tendencies?

---

18   **The Daily Globe**

Should Doctors Genetically Screen All Their Patients?

ALTONIA, LOUISIANA. Last month, when Jason Shultze, a 6-year-old boy suffering from leukemia, was given a standard leukemia treatment—a drug known as a thiopurine—he nearly died. Jason nearly died not because of his leukemia, but because he had a genetic tendency that made the drug deadly for him to take—a tendency for which his doctors could have, but did not, test him.

"All this could have been avoided," said Jason's father, Mitch Schultze, "if they'd done a simple test that would've cost less than 200 dollars."

In fact, every year 2.2 million Americans have adverse reactions to drugs, and 100,000 of these people die as a result. And the number of genetic screening tests available that can identify people who are genetically at risk of harmful or nonbeneficial reactions to drugs is growing. There are now genetic screening tests for the asthma medication albuterol (trade name Ventolin), the painkiller codeine, and the antidepressant Prozac, as well as others.

Jason is one of the one in 300 people who carry a gene that makes them unable to metabolize thiopurines. Rather than helping cure his leukemia, the drugs his doctors gave him accumulated in his body, reaching toxic levels and destroying his bone marrow. An emergency bone marrow transplant saved Jason's life.

Despite their potential usefulness, genetic screening tests for drug reactions are still very uncommon and, pharmaceutical industry representatives say, rightly so. "These kinds of adverse reactions are extremely rare," said John Smith, spokesman for United Pharmaceutical Industrial Corporations Inc., a trade association of pharmaceutical companies. "There is no reason to do widespread testing for these genetic defects or to frighten people into thinking that medicines that have been used safely for decades pose a high risk."

Evaluating "The News"

1. Should doctors begin genetically screening their patients before issuing them drug treatments? Why or why not? Are there any ethical issues to be considered?

2. Why might a pharmaceutical trade industry be against widespread testing?

3. As researchers discover more and more genes that predict other tendencies—for example, tendencies to obesity, to high blood pressure, to depression—should doctors be screening their patients for some or all of these genes to help them adopt the healthiest lifestyles possible? Why or why not?

Philip Taaffe, *Passage II*, 1998.

UNIT $\Big|$ 4

Evolution

chapter 19

How Evolution Works

Franz de Hamilton, *Concert of Birds*, c. 1682.

A Journey Begins

The Galápagos Islands are isolated, encrusted with lava, and home to bizarre creatures found nowhere else on Earth. Life here offers odd twists on the usual: tortoises that reach giant size, land-dwelling lizards that take to the sea, and vampire finches that suck the blood of other birds rather than eating seeds, as most finches do.

The unusual species on the Galápagos Islands have long fascinated biologists. One such biologist was the young Charles Darwin, who visited the Galápagos as part of his remarkable 5-year journey on the ship the *Beagle*. The *Beagle* left England in 1831, sailed to South America, from there to the Galápagos Islands, and eventually back to England in 1836. Throughout this long voyage, Darwin collected specimens and made careful observations of the regions he visited and the organisms he found there.

Upon his return to England, Darwin consulted other scholars and thought deeply about what he had seen on his journey. Early in 1837, Darwin learned from taxonomists that many of the specimens he had collected in the Galápagos were entirely new species, often confined to a single island. This fact, along with other observations, led the young Darwin to a bold conclusion: Species were not, as he and everyone else of his time had been taught, the unchanging result of separate creations by God. Instead,

they had descended with modification from ancestor species; that is, they had evolved. Darwin's conviction in 1837 that species had evolved was a key step in an intellectual journey that would blossom, 22 years later, into the publication of a book that shook the world: *The Origin of Species*.

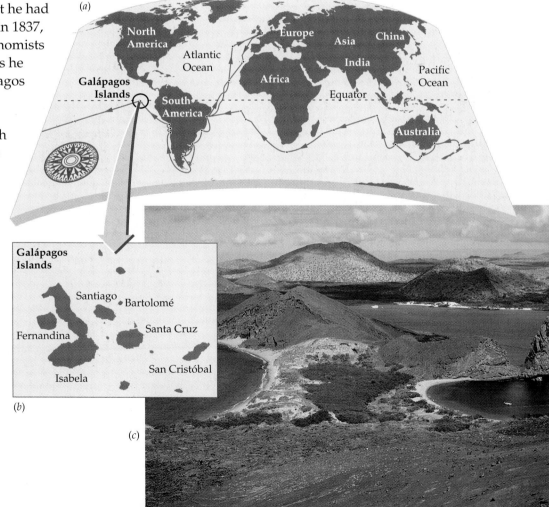

Charles Darwin's Voyage
(*a*) The course sailed by the *Beagle*. (*b*) The Galápagos Islands, located 1000 kilometers to the west of Ecuador. (*c*) Bartolomé, one of the Galápagos Islands.

Key Concepts

1. Biological evolution is change in the inherited characteristics of populations of organisms over generations. For evolution to occur, there must be heritable differences among the individuals in a population.

2. Populations evolve when some of their individuals leave more offspring than other individuals. The result is that heritable characteristics of the favored individuals are more common in the following generation.

3. Adaptations are features of an organism that improve its performance in its environment. Adaptations are the products of natural selection, which is the process in which individuals with particular, heritable characteristics survive and reproduce at a higher rate than other individuals.

4. The great diversity of life on Earth has resulted from the repeated splitting of one species into two or more species.

5. Similarities among organisms are due to common descent: When one species splits into two, the two species that result share similar features because they evolved from a common ancestor.

6. There is overwhelming evidence that evolution has occurred. Some of the strongest evidence comes from the fossil record, which allows biologists to reconstruct the history of life on Earth and shows how new species arose from previous species.

Earth teems with organisms, many of which are exquisitely matched to their environment. The soaring flight of a hawk, the beauty and practicality of a flower, and the stunning camouflage of a caterpillar each provide a glimpse into the remarkable design of organisms. How did organisms come to be as they are, seemingly engineered to match their surroundings (Figure 19.1)? What has caused the amazing diversity of life? And within this diversity, why do organisms share so many characteristics? Scientists who study evolution seek to answer questions such as these.

Defined broadly, evolution is biological change over time. "Biological change" can refer to changes in the genetic characteristics of populations, or to changes in the kinds of species that have lived on Earth at different points in time. In this chapter we provide an overview of evolution, the evidence for it, and its consequences for life on Earth.

cuss throughout this chapter, no "designer" guides biological evolution in nature, though humans can, and do, direct the course of evolution in some species.

Biological evolution can be defined in several ways. We define **biological evolution** as change in the genetic characteristics of populations of organisms over time. We take this approach to emphasize how evolution occurs—through changes in the heritable characteristics of organisms. Biological evolution also can be defined in terms of the pattern of changes that occur over time. From this perspective, evolution is history; specifically, it is the history of the formation and extinction of species over time. Focusing on the pattern of evolution provides us with a grand view of the history of life on Earth, a view to which we will return in Chapter 22.

■ Biological evolution is change in the genetic characteristics of populations of organisms over time.

Biological Evolution: The Sum of Genetic Changes

Evolution is descent with modification, often with an increase in the variety of forms. The term can be applied to organisms, cars, computers, or hats. In each of these cases, new items represent modified versions of previous items, and often several varieties arise where one existed before. But there is an important and fundamental difference between biological evolution and, say, the evolution of hats. Hats change over time because of deliberate decisions made by their designers. As we dis-

Mechanisms of Evolution

Evolutionary change requires that heritable differences exist among the individuals within a population. Heritable differences among individuals are generated by gene mutations. As we learned in Unit 3 ("Genetics"), gene mutations occur at random and are not directed toward a goal. In many organisms, the variation produced by mutations is then increased further by a shuffling of the arrangement of genetic information, as occurs when chromosomes exchange genes during crossing-over.

Figure 19.1 The Match between an Organism and Its Environment
This frog has evolved to hide from predators by resembling the leaves of its Malaysian rainforest home.

The net effect of gene mutations, crossing-over, and other processes that rearrange the genetic information is that individuals within populations differ greatly for many inherited characteristics, including aspects of their form and structure (morphology), biochemistry, and behavior. These differences in the inherited characteristics of individuals in a population provide the raw material upon which evolution acts.

One way that populations evolve is when some individuals within the population survive and reproduce at a higher rate than other individuals. This causes the heritable characteristics of the favored individuals to be more common in the following generation. There are two main ways by which some individuals are favored over others: natural selection and genetic drift.

There are consistent, nonrandom differences in the survival and reproduction of individuals within a population. The process in which individuals that possess particular, heritable characteristics survive and reproduce at a higher rate than other individuals is called **natural selection** (Figure 19.2). Over time, natural selection can cause a population to evolve such that more and more individuals possess the characteristics that enhance survival and reproduction.

Chance events also can also cause some individuals to reproduce at a higher rate than others (Figure 19.3). For example, by chance alone, some individuals are crushed and others are not when a windstorm causes trees to fall amid a population of forest wildflowers. Such chance events can lead to **genetic drift**, a process in which the genetic makeup of the population drifts at random over time rather than being shaped in a nonrandom way by natural selection. As with natural selection, genetic drift can cause populations to evolve.

Although it is a simplification, we can summarize how evolution works as follows: Mutations occur at random, thus generating genetically based differences in the characteristics of individuals, and these differences are then sorted by natural selection or by genetic drift to produce evolutionary change. In Chapter 20 we will examine other mechanisms besides natural selection and genetic drift that also influence the evolution of populations.

> ■ Evolution can be summarized as a three-step process: (1) mutations occur at random, (2) these mutations generate inherited differences in the characteristics of individuals in populations, and (3) these genetic differences are then sorted by natural selection or genetic drift to produce evolutionary change.

Soot from industrial pollution darkens the bark of trees.

Dark moths are harder for birds to find against the dark tree bark.

Figure 19.2 Natural Selection
The peppered moth has two color forms, a light form and a dark form, both shown here. Moth color is an inherited characteristic. Many peppered moths are eaten by birds, which usually select the moths that differ most in color from the trees on which the moths rest. Thus, in regions where soot darkens the bark of trees, moth populations evolve to consist mostly of dark moths.

Consequences of Evolution for Life on Earth

Life on Earth is distinguished by a match between organisms and their environment (adaptations), by a great diversity of species, and by many puzzling examples of dissimilar organisms that share characteristics. Evolutionary biology seeks to explain these features of life on

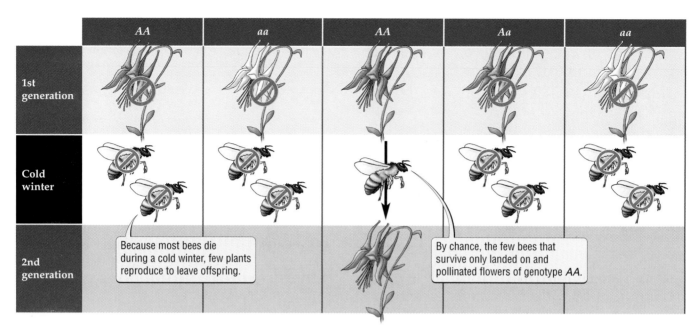

Figure 19.3 Genetic Drift
Chance events can determine which individuals reproduce to leave offspring. Here, a cold winter has caused most of the insect pollinators of this plant to die. By chance alone, the few pollinators that are present fertilize only *AA* individuals, causing the frequency of the *A* allele to change from 50% to 100% in a single generation.

Earth. This attempt to understand why the living world is as it is motivates the three great themes of evolutionary biology: adaptation, the diversity of life, and the shared characteristics of life. Here we describe consequences of evolution for life on Earth, focusing on these three themes.

Adaptations result from natural selection

Some of the most striking features of the natural world concern the complex design of living things and the often remarkable way in which organisms are suited to their environment (see Figure 19.1). These aspects of design in life are called **adaptations**, which are characteristics of an organism that improve the performance (reproductive success) of the organism in its environment.

Adaptations are the product of natural selection. Here's the reasoning behind this statement: If not restricted in some way, all species can reproduce well enough that their populations could quickly outstrip the limited resources available to them. Because organisms produce more offspring than can survive, individuals in a population must struggle for existence. During this struggle individuals whose inherited characteristics provide the best match to their environment tend to leave more offspring than other individuals. This is natural selection in action. Over time natural selection leads to

the accumulation of favorable characteristics—adaptations—within the population.

The diversity of life results from the splitting of one species into two or more species

A major focus of evolutionary biology is to explain the great diversity of life on Earth. Here, too, evolution provides a simple, clear explanation: The diversity of life has been caused by the repeated splitting of one species into two or more species, a process called **speciation**. Speciation can result from a variety of processes (see Chapter 21), one of the most important of which is adaptation to different environments.

For example, consider two populations of a species that live in different environments and that are isolated from each other, as by a mountain or other barrier that prevents individuals from moving between the populations. Over time, natural selection may cause each population to become better adapted to its environment, leading to changes in the genetic makeup of both populations. Eventually, these genetic changes may be so great that individuals from the two populations can no longer reproduce with each other. As we learned in Chapter 1, biological species are often defined in terms of reproduction: A species is a group of interbreeding populations whose members cannot reproduce with members of other such groups. Thus, evolution by natural selection can lead to the formation of new species.

Shared characteristics of life are due to common descent

The natural world is filled with puzzling examples of organisms that share characteristics. For example, the wing of a bat, the arm of a human, and the flipper of a whale all have five digits and the same kind of bones (Figure 19.4*a*). Why do limbs that look so different and have such different functions share the same set of bones? Surely if the best wing and the best arm were designed from scratch, they would not be so similar. Why do we humans have reduced tailbones and the remnants of muscles for moving a tail? Why do some snakes have rudimentary leg bones, but no legs (Figure 19.4*b*)?

Evolution explains these and many other questions about shared characteristics of life: Many similarities among organisms are due to common descent—that is, to the fact that the organisms share a common ancestor. When one species splits into two, the two species that result share similar features because they evolved from a common ancestor. Some snakes have reduced leg bones because they evolved from reptiles with legs, and humans have rudimentary bones and muscles for a tail because we evolved from organisms that had tails.

> ■ Evolution explains how organisms are matched with their environment (natural selection), how the diversity of species is created (speciation), and why dissimilar organisms share characteristics (descent from a common ancestor).

Evolution Is a Fact

> [E]volution is fact, not theory. . . . Birds arose from nonbirds and humans from nonhumans. No person who pretends to any understanding of the natural world can deny these facts any more than she or he can deny that the earth is round, rotates on its axis, and revolves around the sun.
>
> — Richard C. Lewontin, 1981

A 1994 survey showed that 46 percent of adults in the United States did not think that humans had evolved from earlier species of animals, and an additional 9 percent of those surveyed were not sure. The results of this and other similar surveys are startling because evolution has been a settled issue in science for nearly 150 years.

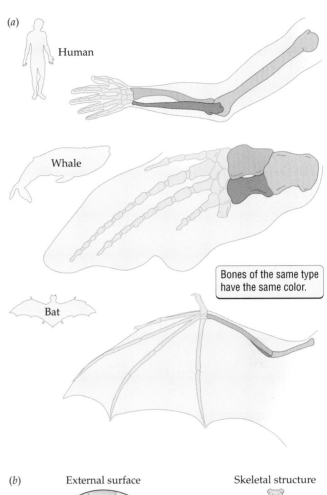

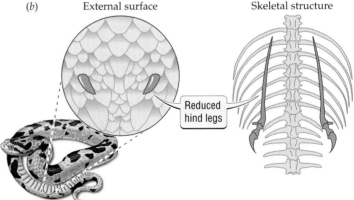

Figure 19.4 Shared Characteristics
(*a*) The human arm, whale flipper, and bat wing all have five digits and the same set of bones. (*b*) Rudimentary hind legs of a python snake, as seen in the skeletal structure and from the external surface of the snake.

Scientists from all nations, races, and creeds regard evolution as a fact. In his landmark book, *The Origin of Species*, Charles Darwin argued convincingly that organisms had descended with modification from common

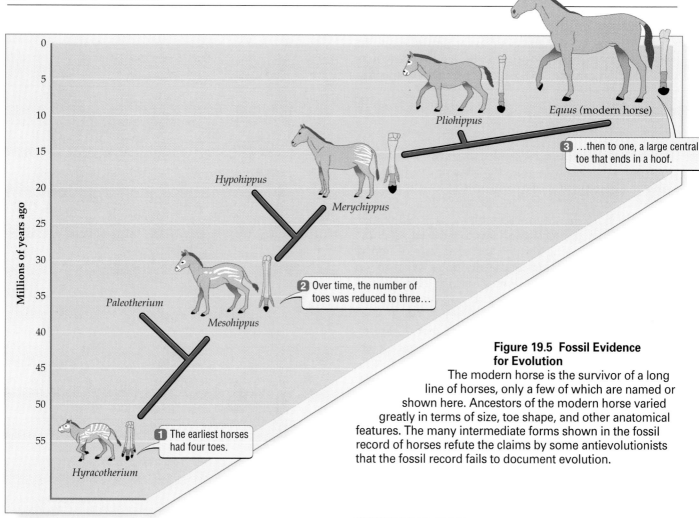

3 ...then to one, a large central toe that ends in a hoof.

2 Over time, the number of toes was reduced to three...

1 The earliest horses had four toes.

Figure 19.5 Fossil Evidence for Evolution
The modern horse is the survivor of a long line of horses, only a few of which are named or shown here. Ancestors of the modern horse varied greatly in terms of size, toe shape, and other anatomical features. The many intermediate forms shown in the fossil record of horses refute the claims by some antievolutionists that the fossil record fails to document evolution.

ancestors—that is, that evolution had occurred. The scientific issue today is not *whether* evolution occurs, but *how.* To this question Darwin also offered an answer: He argued that natural selection is the principal cause of evolutionary change.

On the issue of natural selection Darwin was less successful convincing other scientists, in part because at that time no one understood the underlying cause of heredity. For 60 years after the 1859 publication of *The Origin of Species,* many scientists thought Darwin was wrong to place so strong an emphasis on natural selection. However, the rediscovery of Mendelian genetics and its application to the genetics of populations made it clear that natural selection could cause significant evolutionary change and, hence, that Darwin was at least partially correct.

Biologists still argue about the relative importance of natural selection and other mechanisms that cause evolution, but they do not dispute the fact that evolution occurs. This debate about causes of evolution is similar to a dispute over what caused World War I to progress as it did: Although we might argue over the causes, we all recognize that the war did indeed happen.

■ Evolution is a fact.

The Evidence for Evolution

Why do scientists consider evolution to be a fact? In this section we look at five compelling lines of evidence: fossils, traces of evolutionary history in existing organisms, continental drift, direct observations of change, and the experimental production of new species.

The fossil record strongly supports evolution

Fossils are the preserved remains of former living organisms. The fossil record provides some of the strongest evidence that species evolved over time (Figure 19.5). The fossil record also contains excellent examples of how major new groups of organisms arose from previously existing organisms. We will discuss one of these examples, the evolution of mammals from reptiles, in Chapter 23. Fossils showing how new organisms evolved from previous organisms also exist for microorganisms, fish, amphibians, reptiles, birds, and humans.

A fossil

Another important line of evidence for evolution is that the time at which organisms appear in the fossil record matches predictions based on evolutionary patterns of descent. For example, on the basis of anatomical data we predict that horses evolved relatively recently (about 5 million years ago). This evidence suggests that horse fossils should not be found in very old rocks, and thus far they have not.

Organisms contain evidence of their evolutionary history

A major prediction of evolution is that organisms should carry within themselves evidence of their evolutionary past—and they do. We described examples of such evidence earlier in this chapter: namely, the reduced versions of organs used by ancestor species (for example, the "legs" of a snake and the "tail" of a human) and the remarkable similarity in design of limbs that differ greatly in function (for example, the bat wing and the human arm).

Patterns of embryonic development in organisms provide an additional, and similar, line of evidence. For example, anteaters and some whales do not have teeth as adults, but as embryos they do. Why should the embryos of these organisms produce teeth and then reabsorb them? In general, why do the embryos of organisms as different as humans and fish look alike? These and many other similarities in patterns of development are caused by common descent: Anteater and whale embryos have teeth because anteaters and whales evolved from organisms with teeth. Similarly, human and fish embryos resemble each other because humans and fish share a common ancestor.

The evolutionary relationships among organisms—their pattern of descent from a common ancestor—often can be determined from anatomical data. These patterns of descent can then be used to make predictions about the similarity of the molecules within different species, such as the structure of the species' proteins or the sequence of the species' DNA. Biologists have correctly predicted that the proteins and DNA of organisms that share a recent ancestor should be more similar than the proteins and DNA of organisms that do not share a recent ancestor (Figure 19.6). Because molecular data confirm predictions based on the evolutionary relationships of organisms, these data provide strong evidence for evolution.

Figure 19.6 Evolution in Animals
The evolutionary relationships shown here are based on the number of mutations required to change the amino acid sequence of cytochrome *c* from that found in humans to that found in other organisms. Cytochrome *c* is an enzyme found in all eukaryotes that functions in cellular respiration. The evolutionary relationships shown here, such as humans being most closely related to rhesus monkeys and least closely related to moths, match the relationships based on anatomical data.

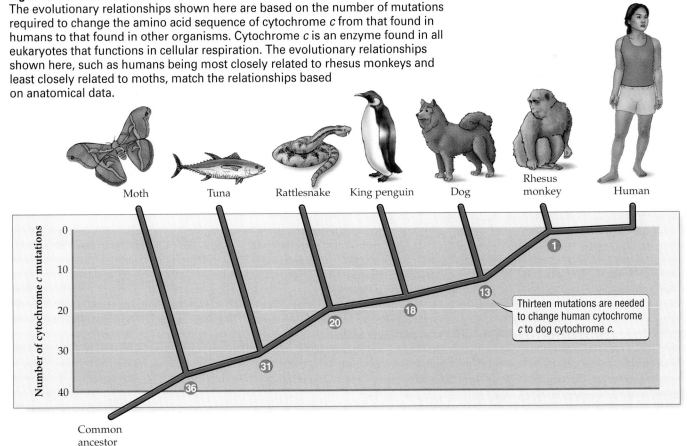

Continental drift explains some geographic trends in evolution

Earth's continents move over time, a process called **continental drift**. Each year, for example, the distance between South America and Africa increases by about 3 centimeters. But 240 million years ago the land masses of Earth drifted together to form one giant continent, called Pangea. Beginning about 200 million years ago, Pangea split up to form the continents we know today (see Figure 22.6).

We can use knowledge about evolution and continental drift to make predictions about the geographic location of fossils. For example, organisms that evolved when Pangea was intact could move relatively easily between what later became widely separated regions, such as Antarctica and India. Because such organisms could move across large geographic regions, we predict that their fossils should be found on most or all continents. In contrast, the fossils of species that evolved after the breakup of Pangea should be found on one or a few continents (for example, the continent on which they evolved and any connected or nearby continents). These predictions have been proven correct, and they provide another important line of evidence for evolution.

Direct observation reveals evolutionary changes within a species

In thousands of studies, researchers observing populations in the wild or in the laboratory have seen them change genetically over time. Such observations provide direct, concrete evidence for evolution. Consider how humans have altered the crop species *Brassica oleracea* (Figure 19.7). By allowing only individuals with certain characteristics to breed, a process called **artificial selection**, humans have crafted enormous evolutionary changes within this species. The tremendous variety that humans have produced within dogs, ornamental flowers, and many other species illustrates the power of artificial selection to produce evolutionary change. Similarly, natural selection can produce large evolutionary changes, as shown by the often striking match of organisms to their environment.

The formation of new species can be produced experimentally and observed in nature

Biologists have directly observed the formation of new species from previously existing species. The first experiment in which a new species was formed occurred in the early 1900s, when the primrose *Primula kewensis* was produced. Scientists have also observed the formation of new species in nature. For example, two new species of salsify plants were discovered in Idaho and eastern Washington in 1950. Neither of the new species were found in this region or anywhere else in the world in 1920. Thus both of the new species appear to have evolved from previously existing species sometime between 1920 and 1950. The two new species continue to thrive, and one of them has become common since its discovery in 1950.

Primrose

> ■ Evolution is viewed as fact because of (1) evidence from the fossil record; (2) the many examples of organisms that share characteristics because they share a common ancestor; (3) evidence from continental drift; (4) direct observations of biological change in populations; and (5) the experimental and natural formation of new species.

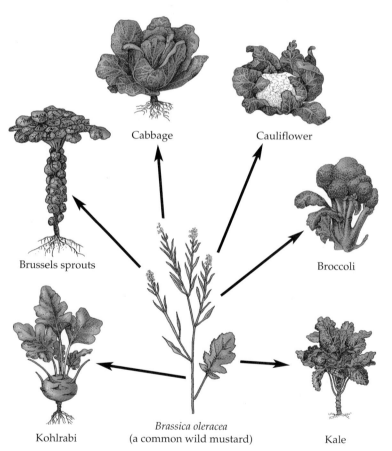

Figure 19.7 Artificial Selection
Humans directed the evolution of *Brassica oleracea*, the wild mustard shown in the center of the figure, to produce such different crop plants as brussels sprouts, cabbage, and broccoli. Despite their obvious differences, all the crop plants shown here are members of the same species, *B. oleracea*.

The Impact of Evolutionary Thought

Before the concept of evolution was developed, adaptations were taken as evidence for the existence of a creator. The clear examples of organisms that seem well designed for their environments implied to many people that there must be a supernatural designer—a God—much as the existence of a watch implies the existence of a watchmaker. In addition, there was a long tradition, dating from Plato, Aristotle, and other Greek philosophers, in which species were viewed as unchanging through time. Darwinian evolution shook these ideas to their very foundation: No longer could species be viewed as unchanging, nor could apparent design in nature be offered as logical proof that God existed, since evolution by natural selection provided an alternative, scientific explanation for the design of organisms.

The evolution of species and, even more, the argument that the design of organisms could be explained by natural selection, were radical ideas in the mid-nineteenth century. These ideas not only revolutionized biology, but they had a large impact on other fields, ranging from poetry to psychology to economics.

The idea of Darwinian evolution also had a large impact on religion. Evolution was viewed initially as a direct attack on Judeo-Christian religion, and this presumed attack in turn prompted a spirited counterattack by many prominent members of the clergy. Today, however, most religious leaders and most scientists do not view evolution and religion as incompatible (but see the Box on this page). For example, in a letter transmitted to the Pontifical Academy of Sciences in 1996, Pope John Paul II reaffirmed that evolution has occurred. For their part, most scientists recognize that religious beliefs are up to the individual, and that science cannot answer questions regarding the existence of God or other matters of religious import.

The emergence of evolutionary thought has also had a great impact on applied aspects of human society. For example, an understanding of evolution has proven essential when farmers and researchers have sought to prevent or slow insects and weeds from evolving resistance to pesticides. In business, information about the evolutionary relationships among organisms can be used to increase the efficiency of searching for new antibiotics and other pharmaceuticals, food additives, pigments, and many other valuable products.

Biology in Our Lives

Creationism: A Challenge to Evolution

Even though most scientists and religious leaders no longer see evolution and religion as in conflict, a minority of Christian fundamentalists remain opposed to evolutionary biology. They exert pressure on the governments of many states not to allow evolution to be taught in secondary schools, or if it is taught, to give equal time to what they call creation science.

Creation science states that all species were created by God roughly 10,000 years ago and that they have not evolved since. As scientific issues, we know that these assertions are false: The scientific evidence indicates that (1) Earth is more than 4 billion years old, (2) life began about 3.5 billion years ago, and (3) evolution has occurred and continues to occur, and is responsible for the great diversity of life.

Creation science is built on faith, not science. Since its "scientific" statements are false, scientists do not think creation science has a place in biology classrooms. The reason is that it can be confusing, even dangerous, to mix science with nonscientific beliefs when teaching science. For example, if medical doctors had no understanding of evolution, they would not realize that overuse of antibiotics has the disastrous effect of causing bacteria to evolve resistance to the antibiotics (see the highlight in Chapter 20, page 321). Although antibiotics have been overused, medical doctors are aware of the situation, they know why bacteria evolve resistance, and they are taking steps to fix the problem. If doctors had no understanding of evolution, the very serious problem that we already face with respect to the emergence of antibiotic-resistant bacteria would be much worse.

If we prevent our secondary school students from learning what science has to offer, we run the risk that they will not be able to compete effectively in college classrooms or in today's global economy. When scientific understanding and nonscientific beliefs come into conflict, as illustrated by the conflict between evolutionary biology and creation science, the debates that result are interesting and are appropriately presented in social science classrooms. Such debates, however, should not be used as an excuse to water down how we teach science.

■ The concept of evolution has had a large impact on the sciences, philosophy, religion, agriculture, and many other aspects of human society.

Highlight

Evolution in Action

The Galápagos Islands greatly influenced Darwin's thinking about evolution. The islands have remained important in evolutionary biology because of ongoing studies on the species that live there. These studies confirm that evolution is not just something that happened in the past; it continues to happen, all around us, every day.

In the Galápagos, it is usually hot and relatively wet from January to May, and cooler and dry for the rest of the year. In 1977 the wet season never arrived: Very little rain fell the entire year. On Daphne Major, a small island near the center of the Galápagos Islands, the lack of rain had a large impact on the plants that lived there (Figure 19.8). Soon the effects were also felt by a seed-eating bird, the medium ground finch: During the drought the number of these birds on Daphne Major plummeted from 1200 to 180.

The lack of rain not only caused many birds to die, it also set the wheels of evolution in motion. One effect of the drought was that the seeds available to the medium ground finch were larger than normal, a condition that prevailed until the wet season of 1978. Large seeds can be difficult or impossible for birds with small beaks to crack open and eat. Thus finches with large beaks had an edge, and many more large-beak than small-beak finches survived the drought to contribute offspring to future generations. As a result, the beak size of the medium ground finch population evolved toward a larger size.

As the research on the medium ground finch shows, we live in an ever-changing world in which species are constantly being shaped by evolutionary forces. In particular, this research shows how natural selection can cause bill size to vary over time within a species. Natural selection and other evolutionary forces also can have much greater effects, such as causing entirely new species to form.

New species have evolved many times on the Galápagos Islands. These islands are home to many unique and unusual species, including 13 species of finches (Figure 19.9). The Galápagos finches are closely related to each other, they are found nowhere else on Earth, and they exhibit many behaviors that are unusual for a finch. With respect to behavior, for example, whereas finches usually eat seeds, Galápagos finches feed on everything from seeds to insects to green leaves to blood.

Figure 19.8 A Drought Quickly Sets Evolution in Motion
The 1977 drought on Daphne Major in the Galápagos Islands greatly affected the plant life there, thus speeding the mechanisms of natural selection.

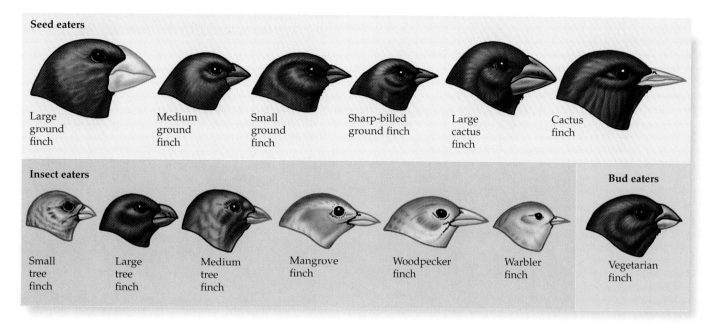

Figure 19.9 The Galápagos Finches
Thirteen unique species of finches evolved in the
Galápagos Islands.

Why do these small islands harbor so many unique but closely related species of finches? The answer seems to be that all the finches on the Galápagos Islands descended from a single species, most likely a species of finch that migrated from the nearest mainland, South America. Upon its arrival, this species found itself in a place where many birds, such as insect-feeding woodpeckers and warblers, were absent. Over time, natural selection favored birds that developed new ways (for a finch) to live in their new home. The end result of this process was that new species of finches evolved from the single species that originally colonized the islands. The odd behaviors of these newly evolved finches make sense when viewed from the perspective of "evolution in action." For example, while it is unusual for finches to feed on insects, the Galápagos finches that do so have evolved to fill an ecological role that is usually taken by other birds, such as the absent woodpeckers and warblers.

■ The Galápagos Islands represent a biological laboratory in which both small (changes in beak size over a few generations) and large (the formation of new species) evolutionary changes have been documented.

Summary

Biological Evolution: The Sum of Genetic Changes

■ Biological evolution can be defined in different ways. Here we emphasize that biological evolution is change in the genetic characteristics of organisms over time.

■ Biological evolution also can be defined as history—namely, the history of the formation and extinction of species over time.

Mechanisms of Evolution

■ Gene mutations occur at random and are not directed toward a goal.

■ Mutations and processes that rearrange genetic information cause individuals to differ for inherited morphological, behavioral, and biochemical traits.

■ Genetic differences among individuals in a population are sorted by natural selection or genetic drift to produce evolutionary change.

Consequences of Evolution for Life on Earth

■ Life on Earth is characterized by the match between organisms and their environment (adaptations), by the great diversity of species, and by many puzzling examples in which dissimilar organisms share characteristics for no clear reason. Evolution explains these features of life on Earth.

■ Adaptations are characteristics of an organism that improve its performance (reproductive success) in its environment. Adaptations result from natural selection.

- The diversity of life on Earth has resulted from the repeated splitting of one species into two or more species.

- The shared characteristics of organisms are due to descent from a common ancestor.

Evolution Is a Fact

- Organisms descend with modification from common ancestors; that is, they evolve over time.

- Scientists from all nations, races, and creeds regard evolution as a fact.

The Evidence for Evolution

- The fossil record provides clear evidence of the evolution of species over time. It also documents the evolution of major new groups of organisms from previously existing organisms.

- The extent to which organisms share characteristics is consistent with our understanding of evolution. For example, the proteins and DNA of organisms that share a recent ancestor are more similar than the proteins and DNA of organisms that do not share a recent ancestor.

- As predicted by our understanding of evolution and continental drift, fossils of organisms that evolved when the continents were still connected to each other have a wider geographic distribution than do fossils of more recently evolved organisms.

- In thousands of studies, researchers have observed genetic changes in populations over time, providing direct evidence of small evolutionary changes.

- Biologists have observed the evolution of new species from previously existing species.

The Impact of Evolutionary Thought

- Darwin's ideas on evolution and natural selection revolutionized biology and had a large impact on many other fields, including poetry, economics, and religion.

- Evolutionary biology influences many applied aspects of human society.

Highlight: Evolution in Action

- The Galápagos Islands provide a natural laboratory in which scientists have studied evolution.

- On the Galápagos Islands, natural selection causes rapid changes in the size of the beak of a seed-eating bird, the medium ground finch.

- Large evolutionary changes have also taken place on the Galápagos Islands, including the evolution of many new and unusual plant and animal species.

Key Terms

adaptation p. 298
artificial selection p. 302
biological evolution p. 296
continental drift p. 302

fossil p. 300
genetic drift p. 297
natural selection p. 297
speciation p. 298

Chapter Review

Self-Quiz

1. Which of the following provides evidence for evolution?
   a. direct observations of genetic changes in populations
   b. shared characteristics of organisms
   c. the fossil record
   d. all of the above

2. In natural selection,
   a. the genetic composition of the population changes at random over time.
   b. new mutations are generated over time.
   c. all individuals in a population are equally likely to contribute offspring to the next generation.
   d. individuals that possess particular heritable characteristics survive and reproduce at a higher rate than other individuals.

3. Adaptations
   a. are features of the organism that hinder its performance in its environment.
   b. are not common.
   c. result from natural selection.
   d. result from genetic drift.

4. The first mammals evolved 220 million years ago. The supercontinent Pangea began to break apart 200 million years ago. Therefore, fossils of the first mammals should be found
   a. on most or all of the current continents.
   b. only in Antarctica.
   c. only on one or a few continents.
   d. none of the above

5. The fact that the flipper of a whale and the arm of a human both have five digits and the same kind of bones illustrates that
   a. genetic drift can cause the evolution of populations.
   b. organisms can share characteristics simply because they share a common ancestor.
   c. whales evolved from humans.
   d. humans evolved from whales.

Review Questions

1. How does evolution explain (a) adaptations, (b) the great diversity of species, and (c) the many examples in which dissimilar organisms share characteristics?

2. Why do scientists regard evolution as a fact? Consider the five lines of evidence discussed in this chapter.

3. Although biologists agree that evolution occurs, they debate which mechanisms are most important in causing evolutionary change. Does this mean that the "theory" of evolution is wrong?

4. Genetic drift occurs when chance events cause some individuals in a population to contribute more offspring to the next generation than other individuals. Are such chance events likely to have a greater effect in small or in large populations? (*Hint:* Examine Figure 19.3. Consider whether the percentage of the *A* allele would be as likely to change if there were 1000 individuals in the population instead of 10.)

ℭhe 𝔅aily 𝔊lobe

Genetically Engineered Corn Could Aid Farmers

AMES, IA. In a discovery that some are saying could save farmers millions of dollars in pesticide use and lost crops, researchers have produced a genetically engineered variety of corn that can knock out one of its most troublesome pests, the corn mealybug. This pest, which has evolved resistance to most pesticides and attacks millions of acres of crops each year, has been troublesome to farmers.

"We think this is going to revolutionize corn farming," said Ms. Carol Barnes, public relations director at the MonPont Company. The genetically engineered corn contains a protein that prevents the larva, or immature form, of the mealybug from developing normally. Thus the mealybugs are unable to grow and reproduce on the engineered corn.

But Dr. Joseph Purpurata, evolutionary biologist at Western Iowa University, urged caution. "The mealybug has proven a tough foe in the past. If farmers plant only the engineered corn, the mealybug will evolve resistance in a few years, and then they'll be right back where they started." MonPont spokesperson Barnes agrees and points out that her company routinely asks farmers not to plant all of their fields with the resistant corn.

Many farmers, however, are not convinced by these calls for restraint. As farmer Joe Henderson of Straight, Iowa puts it, "It's fine to worry about the future, but I need to feed my family now. I'm planting all my fields with the engineered corn."

Evaluating "The News"

1. The protein present in the genetically engineered corn is also widely used by organic farmers who depend heavily on it, as it is one of a few powerful, naturally produced toxins available for fighting insect attack. As a result, organic farmers have a big stake in seeing that insects do *not* develop resistance to this protein. Organic farmers use the protein only when large numbers of mealybugs threaten crop production, whereas the engineered plant produces the protein continuously. Which method is more likely to cause the insect to evolve resistance to the protein? (Hint: The evolution of resistance is due to natural selection. Consider whether it would be easier to select for resistance if the protein was always in use or was only used sporadically.)

2. Organic farmers are insisting that traditional farmers do more to keep insects from becoming resistant to this potent protein. Does one set of farmers have the right to dictate methods that might alter yields and profits of another set of farmers? Why or why not?

3. If farmers followed Dr. Purpurata's advice, the vast majority of mealybugs that successfully produced offspring probably would do so on the non-resistant corn. If farmers set aside 20 percent of their fields in nonresistant corn, do you think the mealybugs would evolve resistance to the engineered corn? Why or why not? Given the losses farmers will sustain in yields in the 20 percent of their fields unprotected by the genetically engineered crop, is this a fair thing to ask farmers to do?

4. It costs a lot of money to develop a genetically engineered crop. If farmers decide to plant all of their fields with the engineered corn and the mealybug rapidly evolves resistance, do you think MonPont or other companies would continue to develop crops that are genetically engineered to resist insect attack? Is there any way to coordinate the activities of individual farmers?

chapter 20

Evolution of Populations

Betty La Duke, *Africa: Osun's Children*, 1990.

Evolution and AIDS

In 1996, spectacular advances in the battle against HIV (human immunod-eficiency virus), the virus that causes AIDS, were heralded in the media as breakthroughs that might lead to the end of that dread disease. The new work included exciting advances in clinical treatment and great strides in basic research. With respect to clinical practice, new therapies reduced the blood concen-tration of HIV to undetectable levels in a majority of patients.

These new treatments relied on "triple drug cocktails," so called because they contained three different drugs. Two of these drugs inhibited the ability of HIV to transcribe its genetic material, an early stage of viral attack. The third drug inhib-ited the ability of HIV to assemble off-spring viruses, a late stage of viral attack.

Complementing the development of treatments that reduced the concentration of HIV in the blood, new discoveries in basic research shed light on a mystery related to how HIV first invades human cells. Researchers had known for more than 10 years that HIV must bind to a par-ticular cell surface protein, called CD4, before it can enter target cells. But they also knew that additional (but unidenti-

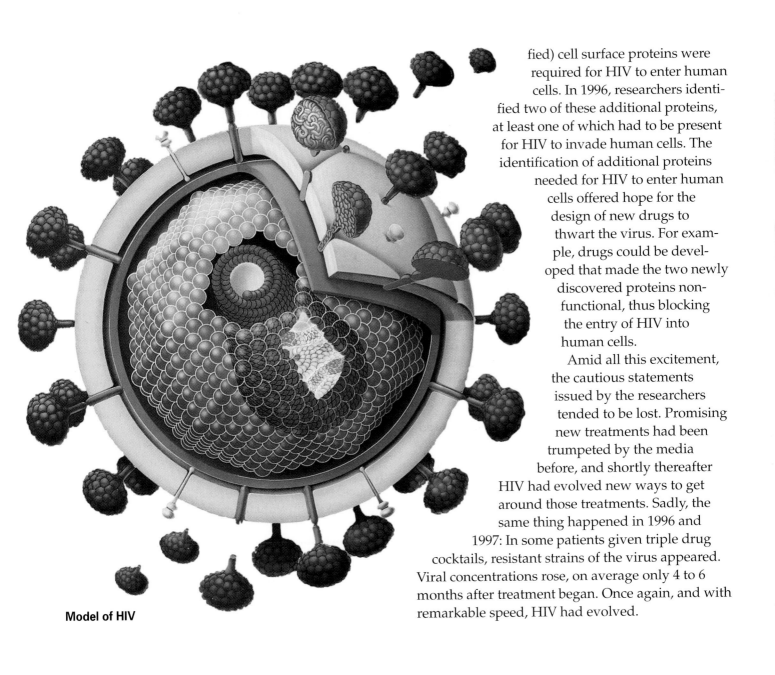

Model of HIV

fied) cell surface proteins were required for HIV to enter human cells. In 1996, researchers identified two of these additional proteins, at least one of which had to be present for HIV to invade human cells. The identification of additional proteins needed for HIV to enter human cells offered hope for the design of new drugs to thwart the virus. For example, drugs could be developed that made the two newly discovered proteins nonfunctional, thus blocking the entry of HIV into human cells.

Amid all this excitement, the cautious statements issued by the researchers tended to be lost. Promising new treatments had been trumpeted by the media before, and shortly thereafter HIV had evolved new ways to get around those treatments. Sadly, the same thing happened in 1996 and 1997: In some patients given triple drug cocktails, resistant strains of the virus appeared. Viral concentrations rose, on average only 4 to 6 months after treatment began. Once again, and with remarkable speed, HIV had evolved.

1. Individuals within populations differ genetically for behavioral, morphological, physiological, and biochemical characteristics. This genetic variation provides the raw material on which evolution can work.

2. Mutations are the original source of genetic variation within populations. The genetic variation produced by mutations is then increased by recombination.

3. Populations evolve when allele or genotype frequencies change from generation to generation.

4. Five factors can cause the evolution of populations: mutation, nonrandom mating, gene flow, genetic drift, and natural selection.

5. Populations can evolve very rapidly (months to years), depending in part on the generation time of the organism. This ability for rapid evolution has serious implications for the evolution of pesticide resistance by insects and antibiotic resistance by bacteria.

Charles Darwin wrote that evolution occurs too slowly to observe, but now thousands of studies document the evolution of populations. Collectively, these studies indicate that Darwin had the basic ideas right, but he was wrong about the time required for evolution to occur. The new studies show that populations can evolve very rapidly (within a few generations, or in months to years), and even that new species can form in short periods of time (one year to a few thousand years).

The **evolution of populations** is defined as a change in allele or genotype frequencies (percentages) over generations. For example, the percentage of a particular genotype in a population might change from 5% to 15% in several generations. Such changes in allele or genotype frequencies result from a two-step process. First, mutation and recombination create genetic variation among the individuals in the population. Second, the pattern of genetic variation in the population changes over time, as when an allele that once was common is replaced by another allele.

Changes in allele or genotype frequencies in a population over time are referred to as **microevolution**, so called because they represent the smallest scale at which evolution occurs. Evolution also occurs on a much larger scale, as when new species or entire new groups of organisms evolve (macroevolution).

We will discuss the formation of new species in Chapter 21 and macroevolution in Chapter 22. In this chapter we describe factors that cause allele and genotype frequencies in a population to change from generation to generation. We begin by defining several essential terms: gene pool, genotype frequency, and allele frequency.

Key Definitions: Gene Pool, Genotype, and Allele Frequencies

The **gene pool** of a population consists of all alleles at all genetic loci in all the individuals of a population. In practice, however, this term is often used to describe the alleles of a restricted set of genetic loci. For example, if two loci were being studied, an investigator might use the term "gene pool" to refer only to all alleles at those two loci, not to all alleles at all genetic loci.

Genotype frequency and **allele frequency** refer, respectively, to the proportion of a genotype or an allele in a population. For example, assume that wing color in a population of 1000 moths is caused by a single gene, with two alleles: W, for orange wing color, and w, for white wing color. If the population contains 160 WW individuals, 480 Ww individuals, and 360 ww individuals, then we obtain the frequencies for the three genotypes (WW, Ww, and ww) by dividing their number by the total number of individuals in the population (1000). Thus, the genotype frequencies for WW, Ww, and ww are 0.16, 0.48, and 0.36, respectively.

Allele frequencies are computed as follows. There are 1000 individuals in the moth population, each of which has two alleles at the wing color locus. Thus, the total number of alleles in the population equals 2000. There are 160 WW individuals, each of which carries two W alleles, for a total of 320. Ww individuals have one W allele each, for a total of 480, and ww individuals have no W alleles. Thus, there are 800 (320 + 480 + 0) W alleles in the population. Finally, we calculate the frequency of the W allele by dividing the number of W alleles by the total number of alleles: (800 W alleles)/(2000 total alleles) = 0.4. Since the gene for wing color has only two alleles, *W* and *w*, the sum of their frequencies must equal 1.0. Hence, the frequency of the *w* allele is 1.0 – 0.4 = 0.6. We can check this value by performing a calculation for the *w* allele similar to the calculation we made for the *W* allele.

With definitions for gene pool, genotype frequency, and allele frequency in hand, we are ready to begin our study of the evolution of populations. We start with a discussion of genetic variation, the raw material on which evolution acts.

> ■ The gene pool of a population consists of all alleles at all genetic loci in all individuals of a population. Genotype frequency and allele frequency are, respectively, the proportion of a genotype or an allele in a population.

Genetic Variation: The Raw Material of Evolution

Genetic variation refers to genetic differences among the individuals of a population. Within a population, the individuals of a species often differ in morphological (Figure 20.1), physiological, and behavioral traits. Much of this variation is under genetic control. Organisms also vary greatly for biochemical characteristics that are under direct genetic control, such as the amino acid sequences of their proteins. The underlying cause of all genetic differences among individuals within a population is that individuals within populations differ in their DNA sequences.

The overall message is clear: Individuals within populations differ for many important characteristics, and much of this variation is under genetic control. The genetic variation among individuals within populations is important, because it provides the raw material on which evolution can work.

Mutation and recombination create genetic variation

Mutations are changes in the sequence of an organism's DNA. New alleles arise only by mutation. Thus, mutations are the original source of all genetic variation. In addition, mutations are thought to occur at random in the sense that (1) they are not directed toward a particular goal (that is, genes do not "know" when and how

it is good for them to mutate), and (2) we cannot predict which copy of a gene in a population will mutate.

Mutations occur regularly in all organisms, even though the mechanisms that correct mistakes in DNA fix most of the errors (see Chapter 14). For example, humans have 100,000 genes, each of which occurs in two copies (one copy from each of our parents). On average, two or three of these 200,000 gene copies have mutations that differ from our parents.

The genetic variation that is initially present as a result of mutations is increased when recombination (for example, fertilization, crossing-over, independent assortment) reshuffles alleles within and among individuals (see Chapter 13).

Many mutations have little effect, many are harmful, and a few are beneficial. The impact of a mutation often depends on the environment in which it occurs. For example, mutations that provide houseflies with resistance to the pesticide DDT also reduce their growth rate. In the absence of DDT, such mutations are harmful, but when DDT is sprayed, these mutations provide an advantage great enough to offset the disadvantage of slow growth, and hence the mutant alleles spread throughout housefly populations.

Figure 20.1 Morphological Variation
Individuals within a population often vary greatly for many morphological traits, such as the color patterns of the starfish shown here.

■ Individuals within populations differ for many important characteristics, and much of this variation is under genetic control. The genetic variation among individuals in a population provides the raw material on which evolution can work.

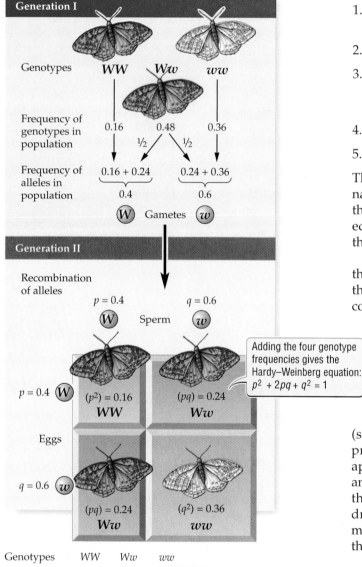

Genotypes   WW   Ww   ww

Frequency of genotypes in population   0.16   0.48   0.36   ½   ½

Frequency of alleles in population   0.16 + 0.24   0.24 + 0.36

0.4   0.6

W   Gametes   w

Generation II

Recombination of alleles

$p = 0.4$ $q = 0.6$

W   Sperm   w

$p = 0.4$ W $(p^2) = 0.16$ $(pq) = 0.24$
                WW   Ww

Eggs

$q = 0.6$ w $(pq) = 0.24$ $(q^2) = 0.36$
                Ww   ww

Adding the four genotype frequencies gives the Hardy–Weinberg equation: $p^2 + 2pq + q^2 = 1$

Genotypes   WW   Ww   ww
Frequency of genotypes in population: 0.16   0.48   0.36

Conclusion:
Generation II genotype and allele frequencies = frequencies of generation I.

When Populations Do Not Evolve

In this section we turn to a discussion of conditions under which populations do not evolve. Specifically, we discuss the Hardy–Weinberg equation, a simple formula that allows us to predict genotype frequencies in a nonevolving population. This calculation is useful, for it provides a baseline against which real populations can be compared.

A population does not evolve when the following five conditions hold:

1. There is no net change in allele frequencies due to mutation.

2. Members of the population mate randomly.

3. New alleles do not enter the population via immigrating individuals, seeds, or gametes (for example, pollen).

4. The population contains many individuals.

5. Natural selection does not occur.

These five conditions do not hold completely in most natural populations. However, populations often meet these conditions well enough that the Hardy–Weinberg equation is approximately correct, at least for some of the genetic loci within the population.

To derive this equation, we will return to the hypothetical population of 1000 moths introduced earlier. In that example, the dominant allele (W) for orange wing color had a frequency of 0.4, and the recessive allele (w) for white wings had a frequency of 0.6. What we seek to do now is predict the frequencies of WW, Ww, and ww genotypes in the next generation.

If mating among the individuals in the population is random, and if natural selection and other evolutionary forces do not alter allele or genotype frequencies, we can use a Punnett square approach (see Chapter 12) to predict the offspring that will be produced in the next generation (Figure 20.2). This approach is similar to mixing all the gametes in a bag and then randomly drawing two at a time to determine the genotype of each offspring. With such random drawing, the allele and genotype frequencies in our moth population do not change from one generation to the next (see Figure 20.2).

Because the WW, Ww, and ww genotypes are the only three types of zygotes that can be formed, the sum of

Figure 20.2 The Hardy–Weinberg Equation
Sexual reproduction alone does not change allele or genotype frequencies. p = frequency of the W allele, q = frequency of the w allele.

Highlight

The Crisis in Antibiotic Resistance

Thirty years ago, the U.S. surgeon general confidently testified to Congress that it was time to "close the book on infectious diseases." He was sadly mistaken. For example, one or more species of bacteria are resistant to every antibiotic now in use. Similarly, most viruses, fungi, and parasites that cause disease have evolved resistance to clinical treatments. Thus, HIV (see the Box on page 315) is not the exception, but rather the rule: Most disease agents have evolved resistance to our best efforts to destroy them, a compelling illustration of directional selection at work.

With respect to bacterial infections, during the past 60 years the bacteria that cause rheumatic fever, staph infections, pneumonia, strep throat, tuberculosis, typhoid fever, dysentery, gonorrhea, and meningitis all have evolved resistance to multiple antibiotics. The widespread resistance of bacteria to antibiotics has serious implications. Some biologists worry that we may enter a "postantibiotic" era, in which bacterial diseases ravage human populations on a scale not seen since the widespread introduction of antibiotics in the late 1930s. Before antibiotics were in common use, even a seemingly mild infection, such as a boil on the face, was potentially lethal.

We are not in a postantibiotic era yet, but there are early signs of a return to just such a nightmarish situation. For example, in 1941, patients could be cured in several days from pneumonia if they took 40,000 units of penicillin per day; today, however, despite excellent medical care and the administration of 24 million units of penicillin per day, the patient could die from complications of the disease.

Resistance to antibiotics evolves so rapidly in bacteria partly because bacteria can transfer resistance genes within and across the species boundary with great ease. The rapid movement of resistance genes from one species to another is especially troubling: Genes for resistance that evolve in a relatively harmless species of bacteria can be transferred to a highly pathogenic species, thus having the potential to create a public health disaster. To prevent this sort of doomsday scenario, what should we, as a society, do? Consider the costs and benefits of the following actions:

- Devoting greater resources to the study of the biology of disease agents, thereby improving our ability to design drugs or pest management strategies that attack weak points in their life cycle.

- Insisting on better and more prudent use of antibiotics in human, plant, and animal health care. Medical doctors and agriculturists frequently use antibiotics inappropriately. For example, one U.S. study found that doctors prescribed antibiotics for 51 to 66 percent of adult patients with colds or bronchitis. Antibiotics do little for either of these conditions because they are usually caused by viruses, which are not affected by antibiotics. The indiscriminate use of antibiotics encourages the evolution of antibiotic resistance in the many species of bacteria that are normally found in our bodies. As we have seen, these resistant, nonpathogenic bacteria can then transfer genes for antibiotic resistance to other, harmful species of bacteria.

- Improving sanitation, thus decreasing the spread of resistant bacteria from one patient to another. This action is of critical importance in hospitals, where the abundant use of antibiotics has led to the emergence of highly resistant strains of bacteria that can cause a variety of "hospital diseases," some of which can be fatal.

> ■ Bacteria evolve resistance to antibiotics rapidly. One or more species of bacteria are resistant to every antibiotic now in use. Genes for resistance that evolve in harmless bacteria can be transferred to highly pathogenic species, having the potential to create a public health disaster. This makes it essential that antibiotics be used only when needed.

Summary

Key Definitions: Gene Pool, Genotype, and Allele Frequencies

- The gene pool of a population consists of all alleles at all genetic loci in all individuals of a population.
- Genotype frequency and allele frequency are, respectively, the proportion of a genotype or an allele in a population.

Genetic Variation: The Raw Material of Evolution

- Individuals within populations differ for biochemical, physiological, morphological, and behavioral characteristics, many of which are under genetic control.
- Mutation and recombination create genetic variation.
- Genetic variation provides the raw material on which evolution can work.

When Populations Do Not Evolve

■ A population does not evolve when (1) mutation does not change allele frequencies, (2) mating is random, (3) the movement of individuals does not bring new alleles into the population, (4) the population is large, and (5) natural selection does not occur.

■ When these five conditions hold, the Hardy–Weinberg equation can be used to calculate genotype frequencies in a population.

Mutation: The Source of Genetic Variation

■ Mutation creates new alleles. Hence all evolutionary change depends ultimately on mutations.

■ Mutations cause little direct change in allele frequencies over time.

■ New mutations can stimulate the rapid evolution of populations by providing new genetic variation on which evolution can work.

Nonrandom Mating: Changing Genotype Frequencies

■ Nonrandom mating does not alter allele frequencies directly but does alter genotype frequencies.

■ Nonrandom mating leads to an increase in the frequency of homozygotes.

Gene Flow: Exchanging Genes between Populations

■ Gene flow makes the gene pools of populations more similar.

■ Gene flow can introduce new alleles to a population, providing new genetic variation on which evolution can work.

Genetic Drift: The Effects of Chance

■ Genetic drift causes random changes in allele frequencies over time.

■ Genetic drift can cause small populations to lose genetic variation.

■ Genetic drift can cause the fixation of harmful, neutral, or beneficial alleles.

■ In a genetic bottleneck, a drop in population size reduces genetic variation or causes the fixation of harmful alleles.

Natural Selection: The Effects of Advantageous Alleles

■ There are three forms of natural selection: directional, stabilizing, and disruptive selection.

■ In all types of natural selection, individuals that possess certain forms of a heritable characteristic tend to survive better or produce more offspring than individuals that possess other forms of the heritable characteristic.

■ Natural selection is the only evolutionary mechanism that consistently favors alleles that improve the performance (for example, reproductive success) of the organism in its environment.

Highlight: The Crisis in Antibiotic Resistance

■ Bacteria evolve resistance to antibiotics rapidly. One or more species of bacteria are resistant to every antibiotic now in use.

■ The indiscriminate use of antibiotics encourages the evolution of resistance in the many species of bacteria normally found in our bodies.

■ Genes for resistance that evolve in harmless bacteria can be transferred to highly pathogenic species, having the potential to create a public health disaster. Thus, it is essential that antibiotics be used only when needed.

Key Terms

allele frequency p. 310

directional selection p. 318

disruptive selection p. 318

evolution of populations p. 310

gene flow p. 314

gene pool p. 310

genetic bottleneck p. 317

genetic drift p. 316

genetic variation p. 311

genotype frequency p. 310

Hardy–Weinberg equation p. 313

heterozygote advantage p. 319

microevolution p. 310

mutation p. 311

natural selection p. 318

nonrandom mating p. 313

stabilizing selection p. 319

Chapter Review

Self-Quiz

1. A population of birds has roughly 15 individuals each year. If allele frequencies were observed to change in a random way from year to year, which of the following would be the most likely cause of the observed changes in gene frequency?
   a. stabilizing selection
   b. disruptive selection
   c. genetic drift
   d. mutation

2. A study of a population of plants finds that large individuals consistently survive at a higher rate than small individuals. Most likely, the evolutionary mechanism at work here is
   a. heterozygote advantage.
   b. directional selection.
   c. stabilizing selection.
   d. genetic drift.

3. Two large populations of a species that are found in nearby but different environments are observed to become genetically more similar over time. Which evolutionary mechanism is the most likely cause of this trend?
   a. gene flow
   b. nonrandom mating
   c. natural selection
   d. genetic drift

4. Which of the following terms describes the situation in which individuals within a subset of the population are more likely to mate with each other than with individuals selected at random from the population at large?
   a. gene flow
   b. preferential mate choice

c. nonrandom mating

d. random mating

5. Assume that individuals of genotype *Aa* are intermediate in size and that they leave more offspring than either *AA* or *aa* individuals. This situation is an example of
   a. heterozygote advantage.
   b. disruptive selection.
   c. stabilizing selection.
   d. both a and c

6. A population of toads has 280 individuals of genotype *AA*, 80 individuals of genotype *Aa*, and 60 individuals of genotype *aa*. What is the frequency of the *a* allele?
   a. 0.24
   b. 0.33
   c. 0.14
   d. 0.07

Review Questions

1. Using your own words, define the following terms: nonrandom mating, gene flow, genetic drift, and natural selection.

2. To prevent a small population of a plant or animal species from going extinct, some individuals from a large population of the same species can be moved from the large to the small population. In terms of the evolutionary mechanisms discussed in this chapter, what are potential benefits and drawbacks of transferring individuals from one population to another? Do you think biologists and concerned citizens should take such actions?

3. Reconsider the toads in Question 6 of the Self-Quiz. How do the numbers of toads with genotypes *AA*, *Aa*, and *aa* compare to the numbers you would expect based on the Hardy–Weinberg equation? Discuss factors that could cause any differences you find.

20

The Daily Globe

Patients Are Demanding, Doctors Are Spineless, a New Study Finds

CHICAGO, IL. What do you want from your doctors? Do you expect them to give you the treatment *you* want? Or do you want them to treat you according to what they think is the best and most appropriate medical treatment?

Many patients, it seems, want the former. And, perhaps not surprisingly, their doctors give in to them. A new study found that doctors altered their diagnosis of a child's cough, depending on whether parents wanted their child to receive an antibiotic. If the parent wanted the child to receive an antibiotic, the doctor was twice as likely to diagnose the cough as caused by bronchitis, not a cold. Roughly 90 percent of the children diagnosed with bronchitis were then treated with antibiotics. Although in children bronchitis is not treated effectively with antibiotics, bronchitis can be treated effectively in some adults who are heavy smokers or who have chronic lung infections. For parents who want their child to receive an antibiotic, doctors apparently prefer to diagnose bronchitis as the cause of the cough because they (incorrectly) assume that a treatment occasionally effective in adults will also be effective in children.

Evaluating "The News"

1. Do you think doctors should make diagnoses according to whether their patients want antibiotics?

2. Have you ever tried to convince your doctor to give you antibiotics? Have you ever given subtle, nonverbal hints that you wanted an antibiotic for a particular illness?

3. Bronchitis is usually caused by a virus. Viruses are not affected by antibiotics, which are designed to treat bacterial infections. Should patients challenge their doctors if their doctors prescribe antibiotics for bronchitis?

4. If a patient has bronchitis, antibiotics usually won't do any good. But antibiotics could help if the patient had an unusual bacterial infection. How could a doctor determine if a particular case of bronchitis was caused by a bacterium rather than a virus?

21

Adaptation and Speciation

Attracting Mates and Avoiding Predators

Richard Ross, *Museum National D'histoire Naturelle, Paris, France, 1982.*

The color patterns of guppies that live in the forested mountain streams of Trinidad and Venezuela are so variable that no two males look exactly alike (in the wild, only adult males express the color genes). Their bright and variable colors play an important role in mating behavior: Females prefer colorful males as mates. But the bright colors that attract females also make the males easier for visually hunting predators to find. The males, then, appear to be caught between an evolutionary rock and a hard place: They must have colors bright enough to attract mates, yet not so bright as to attract predators.

How have these conflicting pressures affected evolution? As in many other stream systems, the mountain streams of Trinidad and Venezuela contain fewer predators at high elevations than at low elevations. In high mountain streams, where few predators lurk, the guppies have bright colors. Proceeding downstream, the number of predators increases and the guppies become more drab, which makes them harder to spot against the gravel stream bottoms. Thus, field observations suggest that the color pattern of guppies that live in a particular stream reflects a compromise between natural selection to hide from predators and natural selection to be attractive to mates.

A skeptic could argue, however, that the changes in the color patterns of the guppies observed in the field resulted from other, unmeasured factors. To convince such skeptics otherwise, researchers conducted a series of field and laboratory experiments. The results were clear: When the guppies were transferred to a new environment, their color patterns evolved to match their new conditions within 10 to 15 generations (14 to 23 months). For example, when guppies were moved from streams with many predators to streams with few predators, as predicted the descendants of these guppies evolved to have brighter and more conspicuous color patterns.

As the guppy example shows, natural selection can improve the match of organisms to their environment, and it can do so over short periods of time. In addition, as we'll see in this chapter, evolution by natural selection is one of several factors that can cause one species to split into two species, thereby helping to generate the great diversity of life.

A Low Predation Site in a Mountain Stream in Trinidad

Key Concepts

1. Adaptations are features of an organism that improve the performance of the organism in its environment. Adaptations result from natural selection.

2. The process by which natural selection improves the quality of an adaptation over time is called adaptive evolution. Adaptive evolution allows organisms to adjust to environmental change, sometimes over short periods of time (months to years).

3. A biological species is a group of interbreeding natural populations that is reproductively isolated from other such groups.

4. Speciation, the process by which one species splits to form two or more species, is usually an accidental by-product of genetic differences between populations that are caused by other factors (for example, natural selection or genetic drift).

5. Speciation usually occurs when populations of a species become geographically isolated. Such isolation limits gene flow between the populations, which makes the evolution of reproductive isolation more likely.

The three great themes of evolutionary biology are adaptation, the diversity of life, and the shared characteristics of life (see Chapter 19). In this chapter we return to two of these themes, adaptation and diversity. We will examine the characteristics of adaptations and discuss how they are shaped by natural selection. Then we will focus on speciation, the process that generates the great diversity of life.

Adaptation: Adjusting to New Challenges

An **adaptation** is a feature of an organism that improves the performance of the organism in its environment. Many adaptations are remarkable, resulting in what appears to be a well-designed match between the organism and its environment. As we discussed in Chapter 19, however, adaptations do not result from a "designer" as such. Instead, adaptations result from natural selection: Individuals with favorable characteristics replace those with less favorable characteristics, and this process, over time, improves the match between organisms and their environment. This process, in which natural selection causes the quality of an adaptation to improve over time, is called **adaptive evolution**.

There are many different types of adaptations

Weaver ants construct nests of living leaves by the concerted actions of many weaver ants, some of which draw the edges of leaves together while others weave them in place by moving silk-spinning larvae (immature ants) back and forth over the seam of the two leaves. These actions are not the result of conscious planning on the part of the ants; rather they illustrate how a simple evolutionary mechanism—natural selection—can produce a complex behavioral adaptation (the building of a nest).

Weaver ants

In another example of an adaptation, caterpillars of a species of moth have a different shape depending on whether they feed on the flowers or leaves of their food plants, oak trees. Caterpillars that feed on flowers resemble oak flowers; those that feed on leaves resemble oak twigs (Figure 21.1). These changes in the development of the larvae cause the moths to match whatever background they feed on, making them more difficult for predators to locate.

Natural selection has also shaped astonishing adaptations that facilitate reproduction. For example, some species of orchids attract bee or wasp pollinators by fooling the males into attempting to mate with their flowers. In these attempts to mate, the insects become coated with pollen, which they then transfer from one plant to another during their repeated attempts to mate with flowers (Figure 21.2).

All adaptations share certain key characteristics

Although there are millions of examples of adaptations, the few examples given in the previous section illustrate the most important characteristics of adaptations:

1. They have the appearance of having been designed for a close match between the organism and its environment.

(a)

(b)

Figure 21.1 Caterpillars That Match Their Environments
Caterpillars of the moth *Nemoria arizonaria* have a different shape depending on their diet. (*a*) Caterpillars that hatch in the spring resemble the oak flowers on which they feed. (*b*) Caterpillars that hatch in the summer eat leaves and resemble oak twigs. Experiments demonstrated that chemicals in the leaves control the switch that determines whether caterpillars mimic flowers or mimic twigs.

2. They are often complex, as exemplified by the nest-building behavior of weaver ants.

3. They help the organism accomplish important functions, such as food capture, regulation of body chemistry, defense against predators, reproduction, and dispersal.

Populations can adjust rapidly to environmental change

The guppies described in the introduction to this chapter evolved new color patterns in less than 2 years. Similarly, viruses, bacteria, and insects evolve resistance to our best efforts to kill them in only a few months or a few years (see Chapter 20). Finally, beak sizes in the medium ground finch evolve from year to year to match the size of the seeds on which they depend for food (see Chapter 19). Collectively, these examples illustrate an important point: Evolution by natural selection can improve the adaptations of organisms in short periods of time.

> ■ Adaptations cause an apparent match between an organism and its environment, they are often complex, and they help organisms accomplish important functions, such as food capture, reproduction, and defense against predators. Natural selection can improve the quality of adaptations in short periods of time (months to years).

Adaptation Does Not Craft Perfect Organisms

Organisms do not match their environment perfectly, in part because genetic or developmental limitations and ecological trade-offs prevent further improvements in

Figure 21.2 A Plant Fools an Insect
The flowers of this orchid use chemical attractants and appearance to mimic female wasps, thereby attracting male wasps. In their attempts to mate with the flowers, males transfer pollen from one plant to another, but they do not benefit from this process.

an organism's adaptation. In this section we'll look at these barriers to perfection.

Genetic limitations

For the quality of an adaptation to increase over time, there must be genetic variation for the traits that enhance the match between the organism and its environment. In some cases, the absence of such genetic variation places a direct limit on the ability of natural selection to result in adaptive evolution. For example, the mosquito *Culex pipiens* is now resistant to organophosphate pesticides, but this resistance is based on a single mutation that occurred in the 1960s (see Chapter 20). Before this mutation occurred, billions of these mosquitoes were killed because their populations lacked genetic variation for resistance to the pesticides.

Developmental limitations

A change in the developmental program of an organism often influences more than one part of the phenotype. Thus, changes to genes that control development may have many effects, some of which may be advantageous and others of which may harm or kill the organism. The many different effects of developmental genes can limit the ability of the organism to evolve in certain directions, which in turn limits what can be achieved by adaptive evolution. For example, the larval stage of some insects, such as beetles and moths, never have wings or well-developed eyes, two important adaptations that the adult forms of these insects have (Figure 21.3). Beetle and moth larvae have a wide range of lifestyles, so wings or

well-developed eyes probably would benefit many of them. Since the adult and larval forms of these insects have exactly the same genetic instructions, the lack of wings and eyes in beetle and moth larvae is probably due to developmental limitations.

Ecological trade-offs

Organisms must perform many functions, such as finding food and mates, avoiding predators, and surviving the challenges posed by the physical environment. Within the realm of what is genetically and developmentally possible, natural selection increases the overall ability of the organism to survive and reproduce. However, the many and often conflicting demands that organisms face causes trade-offs or compromises in the ability of organisms to perform important functions.

For example, high levels of reproduction often are associated with an increased death rate (Figure 21.4). In general, the widespread existence of trade-offs ensures that organisms are not perfect, for the simple reason that it is not possible to be the best at all things at once.

> ■ Adaptation does not craft perfect organisms. Adaptive evolution can be limited by genetic constraints, developmental constraints, and ecological trade-offs.

Figure 21.3 An Apparent Developmental Limitation
The immature form of this moth (*a*) does not have wings or eyes, two important adaptations that the adult form (*b*) has.

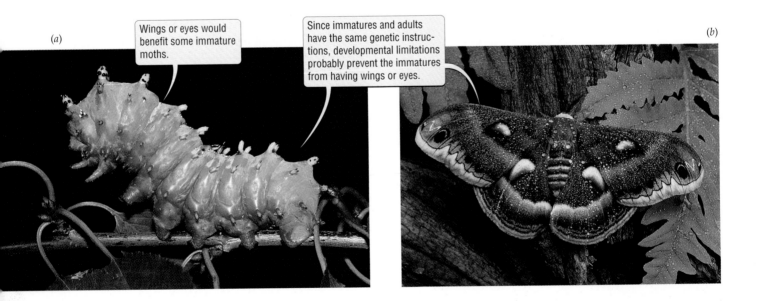

(*a*) Wings or eyes would benefit some immature moths.

Since immatures and adults have the same genetic instructions, developmental limitations probably prevent the immatures from having wings or eyes.

(*b*)

Figure 21.4 A Cost of Reproduction
In female deer, the performance of one important biological function (reproduction) reduces the ability to perform another important function (survive the winter).

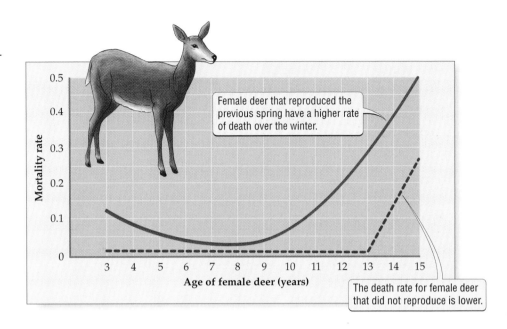

Female deer that reproduced the previous spring have a higher rate of death over the winter.

The death rate for female deer that did not reproduce is lower.

What Are Species?

Before we discuss speciation, we must first define what a species is. In a practical sense, species are usually defined in terms of morphology; that is, two groups of organisms are classified as members of different species if they look sufficiently different. All of us use such a definition: For example, we distinguish red-winged blackbirds from other birds by how they look (Figure 21.5), and most species can be identified by morphological characteristics. However, a morphological definition of species does not tell us what makes one species different from another, nor does it always work well. What holds a species together and causes it to be phenotypically different from other species?

How we answer this question can be important, for the species definition that we use can have large practical effects. For example, the definition we choose can determine whether or not a particular organism is classified (in a legal sense) as a rare or endangered species. A classification as "rare" can have major effects: For the organism, it can make the difference between recovery and extinction; for people, it can limit the rate or extent of economic development.

Species are reproductively isolated from each other

In many cases, barriers prevent reproduction between members of different species. When barriers to reproduction exist between species, the species are said to be **reproductively isolated** from one another. Reproductive isolation can occur in a variety of ways (Table 21.1), but it always has the same net effect: Few or no genes are exchanged between species. This restriction ensures that the members of a species share a common and exclusive

Figure 21.5 Members of a Species Look Alike
Red-winged blackbirds that live in Montana (*a*) look the same as red-winged blackbirds that live in New Jersey (*b*). Although these birds live far apart, they remain phenotypically similar.

(*a*)

(*b*)

21.1 Barriers That Can Reproductively Isolate Two Species in the Same Geographic Region

Type of barrier	Example	Effect
Prezygotic		
Ecological isolation	The two species breed in different portions of their habitat at different seasons or different times of the day	Mating is prevented
Behavioral isolation	The two species respond poorly to each other's courtship displays	Mating is prevented
Gametic isolation	The gametes of the two species are incompatible	Fertilization is prevented
Postzygotic		
Zygote death	Zygotes fail to develop properly, and die before birth	No offspring are produced
Hybrid performance	Hybrids survive poorly or reproduce poorly	Hybrids are not successful

Note: Barriers to reproduction can occur before (prezygotic) or after (postzygotic) the formation of a zygote.

gene pool, which in turn ensures that they remain phenotypically similar to each other but different from other species.

Species, then, can be defined in terms of reproductive isolation: A **species** is a group of interbreeding natural populations that is reproductively isolated from other such groups. The phrase "interbreeding natural populations" is meant to include populations that could interbreed if they were in contact with one another, but do not because they have no opportunity to do so (for example, they are geographically isolated from one another).

A definition of species based on reproductive isolation has important limitations. For example, it is of no use when defining fossil species, since no information can be obtained about whether or not two fossil forms were reproductively isolated from one another. Instead, fossil species are named on the basis of morphology. Such definition also does not apply to organisms, such as bacteria and dandelions, that reproduce mainly by asexual means (see Chapter 10).

Our definition of species also does not work well for the many plant and animal species which mate in nature to form fertile offspring. Species that interbreed in nature are said to **hybridize**, and their offspring are called **hybrids**. Although they can reproduce with each other, species that hybridize often look different (Figure 21.6) or are distinct ecologically (for example, they are usually found in different environments, or they differ in how they perform important biological functions, such as obtaining food).

Despite the limitations of such a definition, most biologists define species on the basis of reproductive isolation, and this is the definition we will use. Many alternative definitions of species exist, but there are problems with these as well. It is perhaps best to think of our def-inition of species as a simple conceptual definition, but to recognize that the reality of a species in nature can be considerably more complicated.

> ■ A species is a group of interbreeding natural populations that is reproductively isolated from other such groups.

Speciation: Generating Diversity

The tremendous diversity of life is caused by **speciation**, the process in which one species splits to form two or more species that are reproductively isolated from each other. The study of speciation is fundamental to understanding a key component of life on Earth: How did the current diversity of life arise?

The crucial event in the formation of new species is the evolution of reproductive isolation. Thus a challenging question arises: How does reproductive isolation develop within a species, whose members share a common gene pool? Populations within a species are connected by gene flow, which tends to keep the populations similar to one another. Throughout this section, we discuss how obstacles to speciation posed by gene flow are overcome, thus allowing reproductive isolation to occur.

Speciation can be explained by mechanisms that cause the evolution of populations

Speciation is usually considered a secondary consequence of the evolutionary divergence of populations. In essence, populations evolve genetic differences from one another—for whatever reason—and some of these

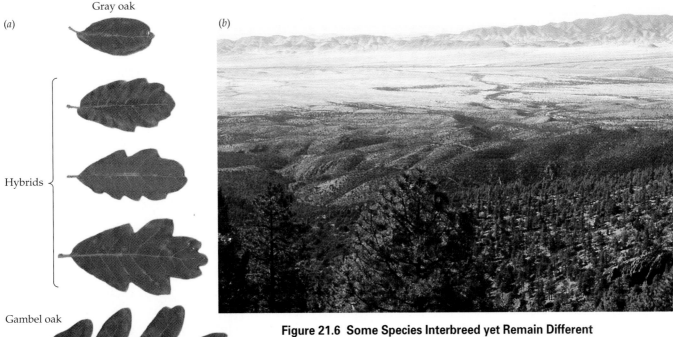

(a)

Gray oak

Hybrids

Gambel oak

(b)

Figure 21.6 Some Species Interbreed yet Remain Different
(a) The Gambel oak and gray oak mate to produce fertile hybrids. However, the species remain phenotypically different, as is evident from their leaves. (b) The two oak species and their hybrids are found throughout the region pictured in the foreground of this photograph.

genetic differences have the accidental by-product of causing partial or total reproductive isolation.

For example, natural selection can cause populations to diverge genetically, as when populations located in different environments face different selection pressures. Over time, these different selection pressures can cause the populations to differ genetically from one another. But the divergence of populations does not have to be due to natural selection. Populations can also diverge as a result of mutation, nonrandom mating, and genetic drift. Gene flow always operates to *prevent* the genetic divergence of populations. Thus, for populations to accumulate genetic differences, factors that promote divergence must have a greater effect than the amount of ongoing gene flow has.

Speciation usually results from geographic isolation

Most speciation events occur when populations of a single species become separated or **geographically isolated** from one another. This process could begin, for example, when a newly formed geographic barrier, such as a river or a mountain chain, isolates two populations of a single species. Alternatively, geographic isolation can occur when a few members of a species colonize a

region that is difficult to reach, such as an island located far outside the usual range of the species.

The distance required for geographic isolation varies tremendously with the species in question: Populations of squirrels and other rodents that live on opposite sides of the Grand Canyon have diverged considerably, whereas populations of birds—whose members can cross the canyon relatively easily—have not. In general, geographic isolation is said to occur whenever populations are separated by a distance that is great enough to limit gene flow.

However they arise, geographically isolated populations are connected by little or no gene flow. For this reason, mutation, nonrandom mating, genetic drift, and natural selection can more easily cause the populations to diverge genetically from one another. If the populations remain isolated for long enough periods of time, they can evolve into new species (Figure 21.7).

Much evidence indicates that geographic isolation can lead to speciation. For example, in many groups of organisms the number of species increases in regions where strong geographic barriers increase the potential for geographic isolation. Examples include species that live in mountainous regions, on island chains, or in a series of small, isolated lakes. In addition, when tested in the laboratory, geographically isolated populations of a single species often show considerable reproductive isolation from one another. In some cases, individuals from populations that live at the extreme ends of a species' geographic range cannot reproduce with each other, even though they can mate with individuals from intermediate portions of the species' range.

Speciation can occur without geographic isolation

In populations whose ranges overlap or are adjacent to one another, there is a greater potential for gene flow than in populations that are geographically isolated. Thus it can be difficult for speciation to occur in the absence of geographic isolation. Nevertheless, it has long been known that plants can form new species in the absence of geographic isolation, and recent work has provided convincing evidence that animals can as well.

In plants, rapid chromosomal change can cause new species to form without geographic isolation. One important way for speciation to occur in this situation is by **polyploidy**, a condition in which an organism has more than two entire sets of chromosomes. Polyploidy can occur within a single species, usually as a result of the failure of chromosomes to separate during meiosis. Polyploidy also can occur when a hybrid spontaneously doubles its chromosome number, thereby forming a new species. A doubling of the chromosomes by the hybrid leads to reproductive isolation because the chromosome numbers in the gametes of the new species no longer match those of either parent species. Although relatively few species originate directly in this way, polyploidy has had a large impact on life on Earth: More than half of all plant species alive today descended from species that originated by polyploidy.

Evidence is mounting that new species of animals also can arise from populations that are not geographically isolated. For example, there is compelling evidence that new species of fish can form in the absence of geographic isolation. In one such case, genetic data indicate that 9 and 11 species of cichlid fishes originated, respectively, within the confines of Lake Bermin and Lake Barombi Mbo, two very small lakes in East Africa.

Cichlid fish

Researchers also have made an excellent case that North American populations of the picture-wing fruit fly, *Rhagoletis pomonella*, are in the process of diverging into new species, even though their ranges overlap. Historically, this species usually ate native hawthorn fruits, but in the mid-nineteenth century these fruit flies were first recorded as pests on introduced apples.

Figure 21.7 Geographic Speciation
New species can form when a population is divided by a geographic barrier, such as a rising sea.

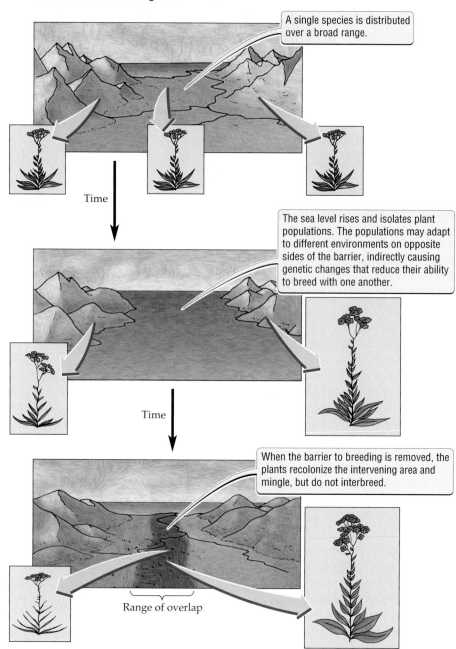

A single species is distributed over a broad range.

Time

The sea level rises and isolates plant populations. The populations may adapt to different environments on opposite sides of the barrier, indirectly causing genetic changes that reduce their ability to breed with one another.

Time

When the barrier to breeding is removed, the plants recolonize the intervening area and mingle, but do not interbreed.

Range of overlap

The Origin of Dogs, Corn, and Cows

Humans can—and do—direct the course of evolution. For thousands of years we have domesticated wild species and controlled their breeding, thus molding their evolution to suit our own ends. For example, wheat, rice, corn, cows, chickens, and dogs all were derived from wild species. Among animal species, the evidence suggests that cattle were domesticated three times, all from the now-extinct wild ox, while chickens were domesticated once (from a southern Asian jungle fowl). Dogs also were domesticated only a few times, from wolves. Thus, the remarkable diversity of dogs represents the effects of artificial selection on a small number of lineages of domesticated wolves (see the figure).

Among plant species, wheat and corn illustrate two different ways in which crop species have evolved. Wheat evolved from two polyploid speciation events, the first leading to the production of emmer wheat (*Triticum turgidum*), the second to the formation of bread wheat (*T. aestivum*).

Corn, on the other hand, evolved directly from a single plant species, teosinte. Teosinte and corn have very different forms: Whereas teosinte has long side branches, each tipped with male reproductive structures, corn has short side branches, each tipped with female reproductive structures (see Figure 1.10). Recent molecular genetic studies indicate that most of these changes in the form of the plants are controlled by changes in the regulation of a single gene. Early Central American farmers probably spotted a teosinte plant with mutations that altered the regulation of this gene, and, by selectively breeding this mutant, they went on to guide the evolution of corn from teosinte.

Wolf
(Common ancestor)

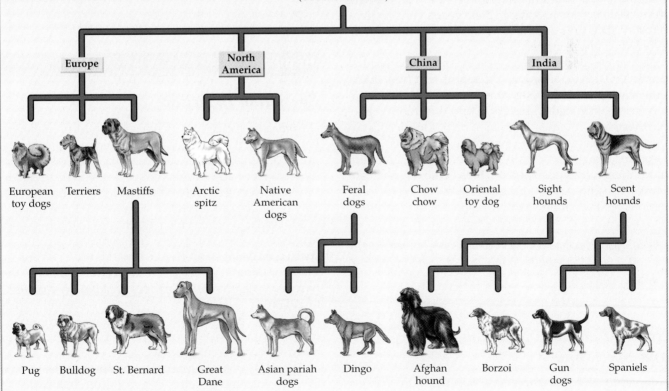

Europe			North America		China			India	
European toy dogs	Terriers	Mastiffs	Arctic spitz	Native American dogs	Feral dogs	Chow chow	Oriental toy dog	Sight hounds	Scent hounds
Pug	Bulldog	St. Bernard	Great Dane	Asian pariah dogs	Dingo	Afghan hound	Borzoi	Gun dogs	Spaniels

Hawthorn and apple *Rhagoletis* populations are now genetically distinct, and members of these populations mate at different times and usually lay their eggs only on the fruit of their particular plant species. Thus, there is little gene flow between the apple and hawthorn populations. Over time, the ongoing research on *Rhagoletis* may well provide a dramatic case history of speciation without geographic isolation.

Picture-wing fruit fly

> ■ Speciation usually is an accidental by-product of the evolutionary divergence of populations. Most speciation events occur in geographic isolation, but new species also can form in the absence of geographic isolation.

Rates of Speciation

When speciation is caused by polyploidy or other types of rapid chromosomal change, new species form in a single generation. New species also appear to have formed with extraordinary speed in the case of some cichlid fishes. For example, genetic analyses indicate that 300 species of cichlids descended from a single common ancestor within Lake Victoria, a large, East African lake. Recent estimates for the length of time during which the lake has remained filled with water range from 200,000 years to a mere 12,400 years. Thus, in perhaps as little time as 12,400 years, 300 species of fish evolved from one species.

In many—perhaps most—cases, speciation occurs more slowly. In organisms as different as fruit flies, snapping shrimp, and birds, the time required for speciation has been estimated to be 600,000 to 3 million years. Furthermore, in cases such as North American and European sycamore trees, populations have been separated for more than 20 million years, yet they remain morphologically similar and can breed with one another. Thus, populations can be geographically isolated for long periods of time without evolving reproductive isolation.

> ■ Speciation occurs rapidly in some cases, but requires hundreds of thousands to millions of years in other cases.

Implications of Adaptation and Speciation

Adaptation and speciation are, respectively, the means by which organisms adjust to the challenges posed by new environments, and the means by which the diversity of life has come into being. Both, therefore, are critical to understanding how evolution works.

Adaptation and speciation are also very important from an applied perspective. For example, to combat rapidly evolving disease organisms—such as HIV, the virus that causes AIDS—we must have a detailed understanding of the new adaptations that enable them to overcome our best efforts at treatment. Speciation has long been of practical importance to humans, as the development of domesticated species readily attests (see the Box on page 333).

In addition, the often relatively slow pace of speciation is a strong reason to stop the ongoing extinction of species (see Chapters 4 and 45). Speciation can require hundreds of thousands to millions of years, yet we are driving species extinct in decades to hundreds of years. If we continue to drive species extinct at the present rate, it will take millions of years before the speciation process can replace the large number of species that are currently being lost.

> ■ Adaptation allows organisms to adjust to the challenges posed by new environments. Speciation is the means by which the diversity of life has come into being. Adaptation and speciation also influence such practical matters as how we fight disease and develop new crop species.

Highlight

Speciation in Action

Helianthus anomalus is a very unusual species of sunflower. For example, it has a thick, waxy covering on the leaves, unlike any other sunflower. It also has the most distinctive habitat of any sunflower species: It is restricted to sand dunes in Arizona and Utah and sometimes is the only plant that can survive in such areas (Figure 21.8*a*).

Molecular genetic evidence indicates that this species originated as a hybrid between two other sunflower species, *H. annuus* and *H. petiolaris* (Figure 21.8*b* and *c*). The two parent species grow on wetter soils than the parched soils that support *H. anomalus*. *H. annuus* is found on wet, clay soils, and *H. petiolaris* is found on dry, sandy soils, but not on true sand dunes. Although it originally arose as a hybrid, *H. anomalus* now cannot breed with its parent species.

The study of these sunflowers has provided a unique glimpse into the genetics of speciation. In a series of greenhouse experiments, researchers created hybrids

(a)

(b)

Figure 21.8 Speciation in Sunflowers
The sunflower *Helianthus anomalus* (*a*) lives in dry, sand dune
environments. *H. anomalus* evolved from *H. annuus* (*b*) and
H. petiolaris (*c*), both of which live in moister environments.

(c)

by natural selection). Finally, gene combinations found
in the three experimental hybrid lineages resembled
those found in naturally occurring *H. anomalus* plants.

This final result provides dramatic evidence that the
researchers understood and could duplicate at least the
first steps of a natural speciation event. The similarity
between evolution in the greenhouse and evolution in
the wild also suggests that natural selection favored certain combinations of genes because of their effects on
fertility (which is very low in first-generation hybrids),
not because of any advantages they provided for life in
harsh, sand dune environments.

> ■ Research on speciation in sunflowers provides a
> dramatic example in which researchers understood and
> could duplicate at least the first steps of a natural speci-
> ation event.

between *H. annuus* and *H. petiolaris* and then used these
hybrids to form three independent hybrid lineages.
After five generations, they compared the DNA of the
three hybrid lineages to each other and to naturally
occurring *H. anomalus* plants.

The results were remarkable. First, the DNAs of the
three independent hybrid lineages resembled each other
closely, indicating that natural selection consistently
favored certain gene combinations. Second, 5 to 6% of
the genes from *H. petiolaris* appeared to provide an
advantage in the hybrid lineages, even though the
genetic background of the hybrids came mainly from
H. annuus. This result indicates that genes from one
species can be advantageous in the genetic background
of another species, which in turn suggests that hybridization between species may promote adaptive evolution (by creating novel combinations of genes favored

Summary

Adaptation: Adjusting to New Challenges

- Adaptations cause an apparent match between organisms
  and their environment.

- Adaptations are often complex.

- Adaptations help organisms accomplish important functions, such as mate attraction, dispersal, and predator avoid-
  ance.

- The quality of an adaptation can be improved in short periods of time (months to years).

Adaptation Does Not Craft Perfect Organisms

- Adaptive evolution can be limited by genetic constraints, developmental constraints, and ecological trade-offs.

What Are Species?

- A species is a group of interbreeding natural populations that is reproductively isolated from other such groups.

- A definition of species in terms of reproductive isolation, like the one we use, has important limitations. It does not apply to fossil species, to organisms that reproduce mainly by asexual means, or to organisms that hybridize extensively in nature.

Speciation: Generating Diversity

- The crucial event in the formation of a new species is the evolution of reproductive isolation.

- Speciation usually occurs as an accidental by-product when natural selection, genetic drift, or mutation cause populations to diverge genetically from one another.

- Speciation usually occurs when populations are geographically isolated from one another.

- Speciation can occur without geographic isolation.

Rates of Speciation

- Speciation occurs rapidly in some cases, but it requires hundreds of thousands to millions of years in other cases.

Implications of Adaptation and Speciation

- Adaptation allows organisms to adjust to the challenges posed by new environments.

- Speciation is the means by which the diversity of life has come into being.

- Adaptation and speciation influence such practical matters as how we fight disease and develop new crop species.

Highlight: Speciation in Action

- Genes from one species can be advantageous in other species.

- Some of the steps in natural speciation events can be duplicated in the laboratory.

Key Terms

adaptation p. 326

adaptive evolution p. 326

geographic isolation p. 331

hybrid p. 330

hybridize p. 330

polyploidy p. 332

reproductive isolation p. 329

speciation p. 330

species p. 330

Chapter Review

Self-Quiz

1. Species that do not interbreed in nature are said to be
   a. geographically isolated.
   b. reproductively isolated.
   c. influenced by genetic drift.
   d. hybrids.

2. Which of the following evolutionary mechanisms acts to slow down or prevent the evolution of reproductive isolation?
   a. natural selection
   b. gene flow
   c. mutation
   d. genetic drift

3. Speciation usually occurs
   a. when populations are not geographically isolated.
   b. by genetic drift.
   c. when populations are geographically isolated.
   d. suddenly.

4. The time required for populations to diverge to form new species
   a. varies from a single generation to millions of years.
   b. is always greater in plants than in animals.
   c. is never less than 100,000 years.
   d. is rarely more than 1000 years.

5. Adaptations
   a. have the appearance of a close match between the organism and its environment.
   b. are often complex.
   c. help the organism accomplish important functions.
   d. all of the above

Review Questions

1. Should species that look different and are ecologically distinct, such as the oaks in Figure 21.6, be called one species or two? These oak species hybridize in nature. Should species that hybridize in nature be called one species or two?

2. Hundreds of new species of cichlid fish evolved within the confines of Lake Victoria, but some of these species live in different habitats within the lake and rarely encounter one another. In this example of speciation, did the species evolve with or without geographic isolation?

3. A species that is legally classified as rare and endangered is discovered to hybridize with a more common species. Since the two species interbreed in nature, should they be considered a single species? Since one of the two species is common, should the rare species no longer be legally classified as rare and endangered?

21

The Daily Globe

Protected Plant Causes Development Project to Grind to a Halt

RED MOUNTAIN, NEW MEXICO. Two months ago, things were looking good for this small town. Construction was about to begin on a new tourist resort, the town's first. The construction jobs would have provided badly needed work, and once built, the resort was expected to attract tourist dollars to this scenic, but economically poor area.

Those hopes were dashed when environmentalists announced that a small population of indigo trumpet flowers lived on the site of the future resort. This small wildflower is protected as a rare and endangered species by the federal Endangered Species Act (ESA). Private landowners cannot harm species protected by the ESA. Since the trumpet flower probably would be harmed by development at the resort site, as environmentalist Steven Cooper says, "Either the resort or the plant had to go. We're glad that we've saved the plant."

Cooper thinks the plant evolved at the site very recently. As he explains, "We surveyed that area several times in the past 15 years. We found the indigo trumpet flower for the first time a few years ago, and it is not known to live anywhere else. So it seems to have evolved there very recently. Knowing exactly when and where the species originated makes it unique." But Cooper's enthusiasm is not shared by many of the locals. As one woman, who preferred to remain anonymous, said, "I only wish we had known the plants were there. We could have dug all of them up and saved our town. What right do these outsiders have to ruin our future?"

Evaluating "The News"

1. Should we protect rare species? Why or why not?

2. The indigo trumpet flower is thought to have evolved from a common species of trumpet flower, the crested trumpet flower. The two species are very similar to each other. Since the indigo trumpet flower is not all that different from another plant species, is it really worth the effort to save it?

3. Should an endangered species be valued more than the economic well-being of a town? Why or why not?

4. Are environmental and development interests necessarily in conflict? Could the town of Red Mountain turn the discovery of the indigo trumpet flower into an economic benefit? How might the town be able to protect the trumpet flower, yet still build the resort?

chapter 22

The Evolutionary History of Life

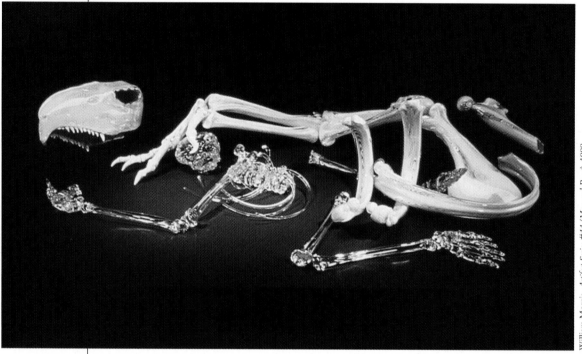

William Morris, *Artifact Series #11 (Man and Beast)*, 1988.

Puzzling Fossils in a Frozen Wasteland

Antarctica is a crystal desert, a land in which heat and liquid water are very scarce. Few organisms can survive the extreme cold and lack of available water, and most of those that can are small and live near the sea. The entire continent has only two species of flowering plants (bundle grass and Antarctic pink), and its largest terrestrial animal is a fly 2 millimeters long.

In the interior of the continent, the organisms are even smaller: Often the only living things are microscopic bacteria, algae, and protists, some of which survive in a state of suspended animation (frozen but alive in the ice). Some of the interior valleys are mostly free of ice—hence they seem a little less for-

338

bidding—but they are so dry and cold that they support no visible life. There, the only living things are photosynthetic bacteria and lichens that spend their entire lives in a narrow zone just under the translucent surface of certain types of rocks.

Despite the nearly lifeless appearance of this continent, fossils reveal that Antarctica used to differ greatly from today's frozen landscape. Where life now maintains an uncertain foothold, the land once was bordered by tropical reefs and later was covered with forests. At different points in time, ferns, freshwater fish, large amphibians, aquatic beetles, and trees as tall as 22 meters thrived in what is now the harshest of environments. Dinosaurs roamed these lands, and millions of years later mammals and reptiles were pursued by the terrorbird, a fast, flightless bird that stood 3.5 meters tall.

Early explorers and scientists were amazed by the fossils that revealed these ancient life forms. They were left with a simple question: What happened?

Life in a Harsh Environment
Bundle grass, one of only two flowering plants in Antarctica, is shown in the foreground of the inset photo.

Key Concepts

1. The fossil record documents the history of life on Earth and provides clear evidence of evolution.

2. Early photosynthetic organisms released oxygen to the atmosphere as a waste product, thereby setting the stage for the evolution of the first eukaryotes, followed later by multicellular organisms.

3. An astonishing increase in animal diversity occurred 530 million years ago, when large forms of most of the major living phyla of animals appeared suddenly in the fossil record.

4. The colonization of land by the first plants (descendants of green algae) and animals (millipedes and spiders) marked the beginning of another major increase in the diversity of life.

5. The history of life is summarized by the rise and fall of major groups of protists, plants, and animals. This history has been greatly influenced by continental drift, mass extinctions, and evolutionary radiations.

*E*arth abounds with life. About 1.5 million species have been described, and millions more await discovery (estimates for the total number of species range from 5 to 50 million). Although numerous, the species alive today represent far less than 1 percent of all the species that have ever lived.

In previous chapters of this unit we discussed how the diversity of life arose, focusing on mechanisms that drive the evolution of populations and lead to the formation of new species (speciation). The evolution of populations is referred to as microevolution, the smallest scale at which evolution occurs. In this chapter, we broaden our scope to discuss large-scale evolutionary changes, or macroevolution.

Macroevolution refers to the rise and fall of major groups of organisms—that is, to the evolutionary expansions that bring new groups to prominence and the large-scale extinction events that greatly alter the diversity of life on Earth. Macroevolution emphasizes the pattern of evolutionary change over time. As we discussed in Chapter 19, an emphasis on pattern leads us to define evolution as history, specifically, the history of the formation and extinction of species over time. Let's look at how that history is documented in the fossil record.

The Fossil Record: Guide to the Past

Fossils are the preserved remains or impressions of past organisms (Figure 22.1). They are usually found in sedimentary rock (rock that consists of layers of hardened sediments), but also can be found in a few other situations. For example, insects have been found in amber, the fossilized sap of a tree (see Figure 22.1*e*), and many mammals, including mammoths and a 5000-year-old man (see the photograph on page 45), have been found in melting glaciers.

The fossil record documents the history of life and is central to the study of evolution. Historically, fossils provided the first compelling evidence of the fact that past organisms were unlike living forms, that many forms had disappeared completely from Earth, and that life had evolved through time.

The order in which organisms appear in the fossil record agrees with our understanding of evolution, thus providing strong support for evolution. Analyses of morphology, DNA sequences, and other characteristics indicate that bony fishes gave rise to amphibians, which later gave rise to reptiles, which still later gave rise to mammals. This is exactly the order in which fossils from these groups appear in the fossil record. The fossil record also provides excellent examples of the evolution of major new groups of organisms, such as the evolution of mammals from reptiles (see Chapter 23).

Although many fossils have been found, the fossil record is not complete. Most organisms decompose rapidly after death; hence, very few form fossils. Even if an organism is preserved initially as a fossil, a variety of common geologic processes (for example, erosion, extreme heat or pressure) can destroy the rock in which it is embedded. Finally, fossils can be difficult to find. Given the unusual circumstances that must occur for a fossil to form, remain intact, and be discovered, a species could evolve, thrive for millions of years, and become extinct without leaving evidence of its existence in the fossil record.

Although it is not complete, the fossil record shows clearly that there have been large changes in the dominant groups of organisms over time. These changes have been caused by the extinction of some groups and the expansion of other groups. In the discussion that follows we describe broad patterns in the history of life revealed by the fossil record, and we discuss factors that cause these patterns.

(a)

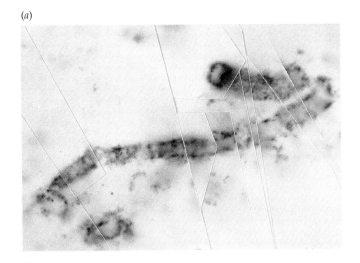

(b)

(c)

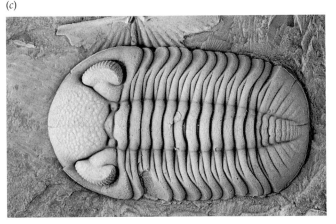

(d)

Figure 22.1 Fossils through the Ages

(a) The oldest fossils are of bacteria, such as this 3.5-billion-year-old fossil found in Western Australia. (b) Soft-bodied animals such as the ones that left these fossils dominated life on Earth 600 million years ago (mya). (c) A fossil of a trilobite that lived in the Devonian period (410 to 355 mya). Note the rows of lenses on each eye. (d) This leaf of a 300-million-year-old seed fern was found near Washington, D.C. The fossil formed during the Carboniferous period (355 to 290 mya), the great forests of which led to the formation of the fossil fuels we use today as sources of energy (for example, oil, coal, and natural gas). (e) These 40-million-year-old insects are preserved in amber, the fossilized sap of a tree.

(e)

■ The fossil record documents the history of life on Earth and provides strong evidence that life has evolved through time. The fossil record shows that past organisms were unlike living organisms, that many species have gone extinct, and that there have been large changes in the dominant groups of organisms over time.

The History of Life on Earth

Figure 22.2 provides a sweeping overview of the history of life on Earth; study it carefully. The sections that follow focus on three of the main events in the history of life: the origin of cellular organisms, the beginning of multicellular life, and the colonization of land.

The first single-celled organisms arose 3.5 billion years ago

Our solar system and Earth formed 4.6 billion years ago. The oldest known rocks (3.8 billion years old) contain carbon deposits that hint at life. The first solid evidence for life, however, comes from 3.5-billion-year-old fossils that resemble present-day bacteria (see Figure 22.1a). These fossils are of fully formed, cellular organisms, suggesting that the earliest forms of life arose earlier, probably between 4 and 3.5 billion years ago. (See Chapter 1 for more on the origin of life.)

After the origin of prokaryotes 3.5 billion years ago, it took roughly 2 billion years for the first eukaryotes to evolve. During this long period of time, the evolution of eukaryotes may have been limited by low levels of oxygen in the atmosphere. Initially, Earth's atmosphere contained almost no oxygen. Shortly after life began, some groups of bacteria evolved the ability to conduct photosynthesis, which releases oxygen as a waste product. Over time, photosynthesis

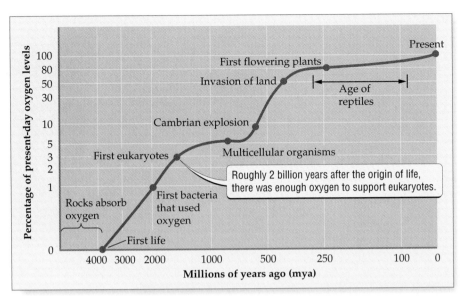

Figure 22.3 Oxygen on the Rise
The release of oxygen as a waste product by photosynthetic organisms has caused its concentration in Earth's atmosphere to increase greatly in the last 3 to 4 billion years.

caused the oxygen concentration in the atmosphere to increase (Figure 22.3).

Early single-celled prokaryotes that used oxygen absorbed it across their cell membranes. These oxygen-

Figure 22.2 The History of Life on Earth

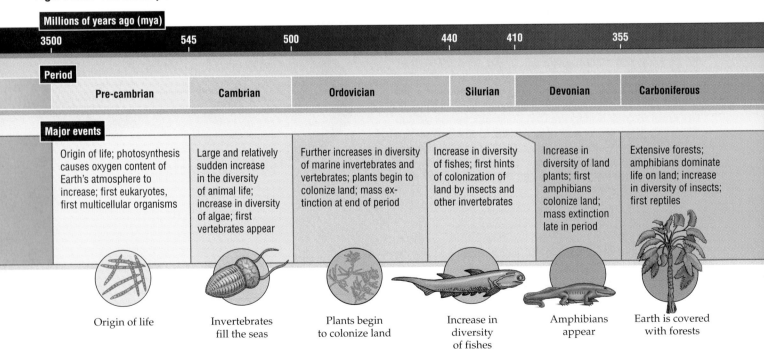

using organisms began to thrive when the concentration of oxygen in the atmosphere reached 1% of current levels. Oxygen absorbed across a cell membrane spreads more slowly through large cells than through small cells. Because they are larger than prokaryotic cells, eukaryotic cells cannot get enough oxygen unless the concentration of oxygen is at least 2–3% of current levels. Once these levels were reached, the first eukaryotes—single-celled algae—evolved 1.5 billion years ago (see Figure 22.3). As oxygen levels continued to increase, the evolution of larger and more complex multicellular organisms became possible.

Oxygen was toxic to many of the early forms of life. Thus, as the oxygen concentration increased, many early prokaryotes went extinct or were restricted to environments that lacked oxygen. Because it drove many early organisms extinct while simultaneously setting the stage for the origin of multicellular eukaryotes, the biologically driven increase in the oxygen concentration of the atmosphere was one of the most important events in the history of life on Earth.

Multicellular life evolved about 650 million years ago

All early forms of life evolved in water. Among these early life forms, there was an increase in the number of organisms appearing in the fossil record 650 million years ago (mya). At that time, much of Earth was covered by shallow seas, which were filled with protists, small multicellular animals, and algae.

By 600 mya, larger soft-bodied animals had evolved (see Figure 22.1*b*). These animals were flat and appear to have crawled or stood upright on the seafloor, probably feeding on living plankton or their remains. No evidence indicates that any of them preyed on the others. Many of these early multicellular animals may have belonged to groups of organisms no longer found on Earth.

The early to middle Cambrian period (530 mya) witnessed an astonishing burst of evolutionary activity. In an increase in the diversity of life which is known as the **Cambrian explosion**, large forms of most of the major living animal phyla, as well as other phyla that have since become extinct, appeared suddenly in the fossil record. The Cambrian explosion lasted only 5 to 10 million years, a blink of the eye in geologic terms (compare this time span to the 2 billion years it took for eukaryotes to evolve).

The Cambrian explosion was one of the most spectacular events in the evolutionary history of life. It changed the face of life on Earth: From a world of relatively simple, soft-bodied scavengers and herbivores, suddenly there emerged a world that was filled with large, mobile predators and well-defended herbivores (Figure 22.4).

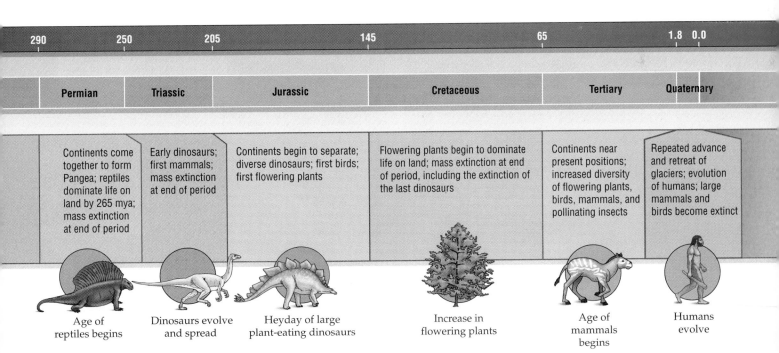

290	250	205	145	65	1.8 0.0
Permian	Triassic	Jurassic	Cretaceous	Tertiary	Quaternary
Continents come together to form Pangea; reptiles dominate life on land by 265 mya; mass extinction at end of period	Early dinosaurs; first mammals; mass extinction at end of period	Continents begin to separate; diverse dinosaurs; first birds; first flowering plants	Flowering plants begin to dominate life on land; mass extinction at end of period, including the extinction of the last dinosaurs	Continents near present positions; increased diversity of flowering plants, birds, mammals, and pollinating insects	Repeated advance and retreat of glaciers; evolution of humans; large mammals and birds become extinct
Age of reptiles begins	Dinosaurs evolve and spread	Heyday of large plant-eating dinosaurs	Increase in flowering plants	Age of mammals begins	Humans evolve

Colonization of land followed the Cambrian explosion

Land was probably first colonized by crusts of bacteria living close to the water's edge. A more extensive colonization of land did not begin until the late Ordovician (about 450 mya), at which time plant spores and burrows that have been interpreted as formed by millipedes appear in the fossil record.

Life arose in water, and the colonization of land posed enormous challenges. As will be described in Unit 5, many of the functions basic to life are very different on land and in water, including support, movement, reproduction, and the conservation of ions, water, and heat. Among plants, descendants of green algae were the first to meet these challenges. These early colonists had few cells and a simple body plan, but from them land plants diversified greatly.

As new groups of land plants arose, they evolved a series of key innovations, including waterproofing, stems with efficient transport mechanisms, structural tissues (wood), leaves and roots of various kinds, seeds, the tree growth form, and specialized sexual organs. These and other important changes allowed plants to cope with the difficulties of life on land. For example, waterproofing, stems with efficient transport mechanisms, and roots are important features that help plants acquire and conserve water while living on land.

The key innovations that made life on land possible for plants took roughly 120 million years to evolve. Taken together, these evolutionary changes represent an unparalleled episode in the history of plant life: Nothing like it has occurred before or since. By the end of the Devonian period (345 mya), Earth was covered with plants. As is true today, the plants of the Devonian included low-lying spreading species, short upright species, shrubs, and trees.

Although there are hints of land animals in the late Ordovician, the first definite fossils of terrestrial animals are of spiders and millipedes that date from about 410 mya. Many of the early animal colonists on land were predatory; others, such as millipedes, fed on plant material and decaying matter. Insects, which are now the most diverse group of terrestrial animals, first appeared roughly 400 mya, and they played a dominant role on land by 350 mya.

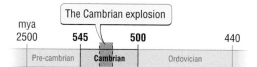
Millipede

The first vertebrates to colonize land were amphibians, the earliest fossils of which date to about 365 mya. Early amphibians resembled, and probably descended from, lobe-finned fish (Figure 22.5). Amphibians were the dom-

mya
2500   **545**   **500**   440

| | The Cambrian explosion | |
| Pre-cambrian | Cambrian | Ordovician |

Figure 22.4 Before and After the Cambrian Explosion
The Cambrian explosion greatly altered the history of life on Earth.

Before the Cambrian explosion, Earth was dominated by soft-bodied scavengers and plant-eaters (herbivores).

After the Cambrian explosion, life was dominated by more complex animals, including well-defended herbivores and animals that kill other animals for food (predators).

Millions of years ago (mya)

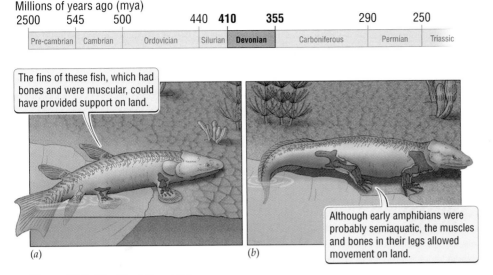

Figure 22.5 The First Amphibians
(*a*) Amphibians probably descended from a lobe-finned fish, such as that shown here. (*b*) This early amphibian was reconstructed from a 365-million-year-old (late-Devonian) fossil.

inant organisms on land for about 100 million years. In the late Permian, the reptiles, which had evolved from a group of reptilelike amphibians, rose to be the dominant group. Reptiles were the first group of vertebrates that could complete their entire life cycles without returning to open water (for example, to lay eggs). Thus reptiles could fully exploit the opportunities for life on land.

Reptiles dominated life on land for 200 million years (265 mya to 65 mya) and remain important today. Mammals, the group that currently dominates life on land, evolved from reptiles roughly 220 mya (see Chapter 23). The origin of mammals from reptiles is beautifully documented in the fossil record and provides an excellent example of macroevolution. Since the time the dinosaurs went extinct, 65 mya, the mammals have diversified greatly. The increase in the diversity of mammals has been influenced in part by continental drift, the subject we turn to next.

■ Three major events in the history of life were (1) the evolution of the first single-celled prokaryotes (3.5 billion years ago) and eukaryotes (1.5 billion years ago), (2) the great expansion of multicellular organisms that took place in the Cambrian explosion (530 to 525 mya), and (3) the colonization of land by plants and invertebrates (450 to 410 mya), followed later by the vertebrates (365 mya).

The Impact of Continental Drift

The enormous size of the continents causes us to think of them as immovable. But this notion is not correct. The continents move slowly relative to one another, and over hundreds of millions of years they travel considerable distances (Figure 22.6). This movement of the continents over time is called **continental drift**. The continents can be thought of as plates of solid matter that "float" on the surface of Earth's mantle, a hot layer of semisolid rock.

Two main forces cause the continental plates to move: (1) Hot plumes of liquid rock can rise to the surface and push the continents apart (Figure 22.7). This process can cause the seafloor to spread, as between North America and Europe, which are separating at a rate of 2.5 centimeters per year. This process also can cause bodies of land to break apart, as is currently happening in Iceland and East Africa. (2) Where two continental plates collide, one can sink below the other (see Figure 22.7). This sinking action pulls the rest of the continental plate along with it, eventually causing the sinking plate to melt.

Continental drift, most notably the breakup of the supercontinent Pangea, has had a tremendous impact on the history of life. Pangea began to break apart early in the Jurassic period (about 200 mya), ultimately separating into the continents we know today (see Figure 22.6). As the continents drifted apart, populations that once were connected by land became isolated from each other.

As we learned in Chapter 21, geographic isolation promotes speciation. The separation of the continents was geographic isolation on a grand scale, leading to the formation of many new species. Among mammals, for example, kangaroos, koalas, and other marsupials evolved in geographic isolation in Australia, which broke apart from Antarctica and South America about 40 mya.

Continental drift also affects the climate, which has a great impact on organisms. For example, shifts in the position of the continents alter ocean currents, which have a large influence on the global climate. At various points in time, changes in the global climate caused by the movement of the continents has led to the extinction of many species. In the next section we'll look at the evolutionary impact of such large-scale extinction events, which are known as mass extinctions.

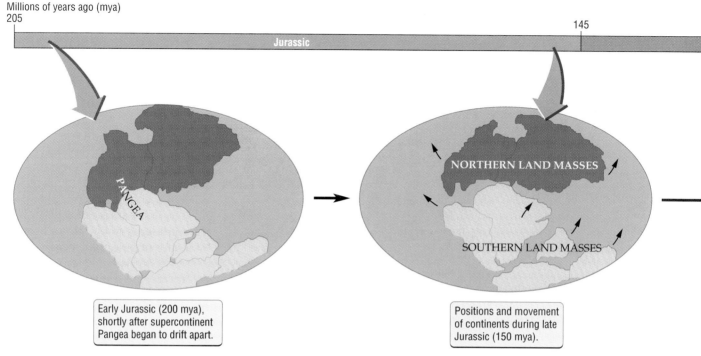

Millions of years ago (mya)

Jurassic

205   145

PANGEA

NORTHERN LAND MASSES

SOUTHERN LAND MASSES

Early Jurassic (200 mya), shortly after supercontinent Pangea began to drift apart.

Positions and movement of continents during late Jurassic (150 mya).

Figure 22.6 Movement of the Continents over Time
The continents have moved over time, as these snapshots of history show.

Figure 22.7 The Causes of Continental Drift
Two forces cause continental drift: hot plumes of liquid rock that push apart continental plates, and collisions that force one plate under another.

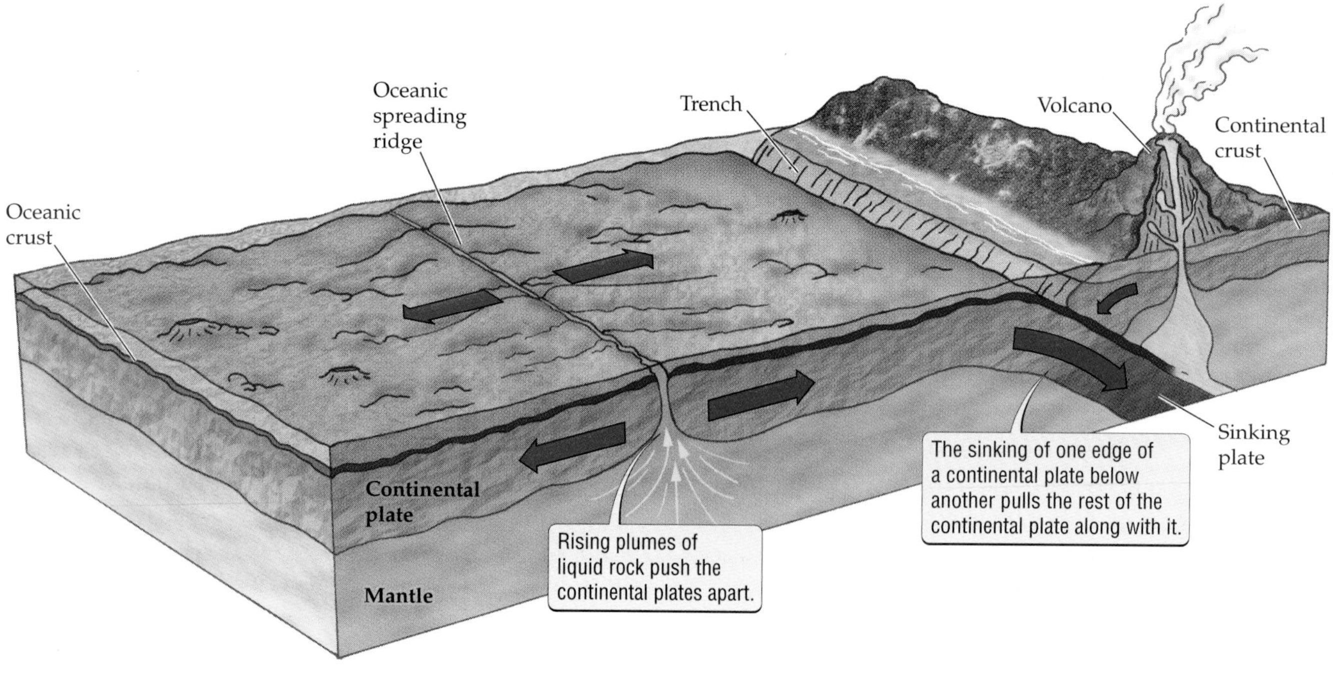

■ Continental drift has had a great impact on the history of life on Earth. The separation of the continents during the past 200 million years led to geographic isolation on a grand scale, promoting the evolution of many new species. In addition, at various points in time changes in the climate caused by the movement of the continents led to the extinction of many species.

Oceanic spreading ridge

Trench

Volcano

Continental crust

Oceanic crust

Continental plate

Mantle

Rising plumes of liquid rock push the continental plates apart.

The sinking of one edge of a continental plate below another pulls the rest of the continental plate along with it.

Sinking plate

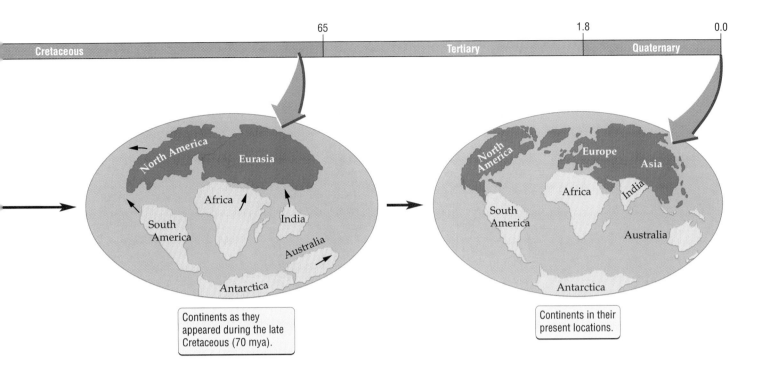

Continents as they appeared during the late Cretaceous (70 mya).

Continents in their present locations.

Mass Extinctions: Worldwide Losses of Species

The fossil record shows that species have become extinct throughout the long history of life. The rate at which species go extinct appears to vary continuously, from low to very high rates. At the upper end of this scale there have been five **mass extinctions**, periods of time during which great numbers of species became extinct throughout most of Earth. Each of these five biological disasters left a permanent mark on the history of life (Figure 22.8). In addition to the five mass extinctions revealed by the fossil record, we may be entering a sixth, human-caused mass extinction (see Chapters 4 and 45).

Of the five previous mass extinctions, the largest occurred during the Permian, 250 mya. This mass extinction radically altered life in the oceans. For example, among marine invertebrates, 50 to 63 percent of the families, 82 percent of the genera, and 95 percent of the species became extinct. The Permian mass extinction was also highly destructive on land, removing 62 percent of the families, bringing the reign of the amphibians to a close, and causing the only major extinction of insects in their 390-million-year history (8 of the 27 orders of insects became extinct). Although not as severe as the Permian extinction, each of the other mass extinctions also had a profound impact on life on Earth (see Figure 22.8).

The best-studied mass extinction occurred at the end of the Cretaceous, 65 mya. At that time, half of the marine invertebrate species perished, as did many families of terrestrial plants and animals, including the dinosaurs. Unlike the first four mass extinctions, which were caused by climate changes, drops in sea levels, and other factors, the Cretaceous mass extinction was probably caused at least in part by the impact of an asteroid. A 65-million-year-old crater of an asteroid 10 kilometers wide was recently found buried in sediments off the Yucatán coast of Mexico. An asteroid of this size could have caused great clouds of dust to hurtle into the atmosphere, blocking sunlight around the globe for months to years, thus causing temperatures to drop drastically and driving many species extinct.

The effects of mass extinctions on the diversity of life are twofold. First, entire groups of organisms perish in a mass extinction. The extinction of some groups—but not others—greatly alters the later course of evolution and the history of life. For example, if the dinosaurs had been spared and instead our early primate ancestors had become extinct 65 mya, humans would not exist and the world would certainly be a very different place.

Second, the extinction of a dominant group of organisms can provide new ecological and evolutionary opportunities for groups of organisms that previously were of relatively minor importance, thus dramatically altering the course of evolution. We discuss this second effect of mass extinctions in the following section.

Figure 22.8 The Five Mass Extinctions of History

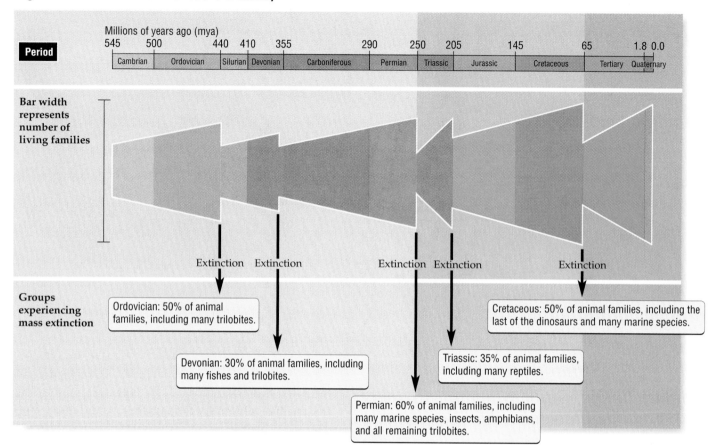

■ There have been five mass extinctions during the history of life on Earth. Mass extinctions have two large effects on the diversity of life. First, the extinction of some groups but not others greatly alters the later course of evolution. Second, the extinction of a dominant group of organisms can provide new opportunities for organisms that were of relatively minor importance, thus greatly changing the shape of life on Earth.

Evolutionary Radiations: Increases in the Diversity of Life

After each of the five mass extinctions, other groups of organisms diversified to replace those that had become extinct. These bursts of evolutionary activity lasted 1 to 7 million years each and were just as important for the future course of evolution as the extinction events themselves. Regardless of whether it occurs after a mass extinction or at other times, the rapid diversification of a group of organisms to form new species and higher taxonomic groups is called an **evolutionary radiation**. In the sections that follow we'll look at four factors that promote evolutionary radiations: release from competition, key evolutionary innovations, ecological interactions, and low amounts of gene flow.

Release from competition

Evolutionary radiations can be caused when species are released from competition. For example, the first mammals evolved in the Triassic period, about 220 mya. Fossil and molecular genetic evidence suggests that several orders of living mammals diverged from each other 100 to 85 mya, well before the extinction of the dinosaurs. But major radiations within these and other groups of mammals occurred mostly after the extinction of the dinosaurs (65 mya). Many dinosaurs were large and fierce; thus competition with them may have prevented the mammals from expanding to fill new ecological roles (such as those of large herbivores or large predators).

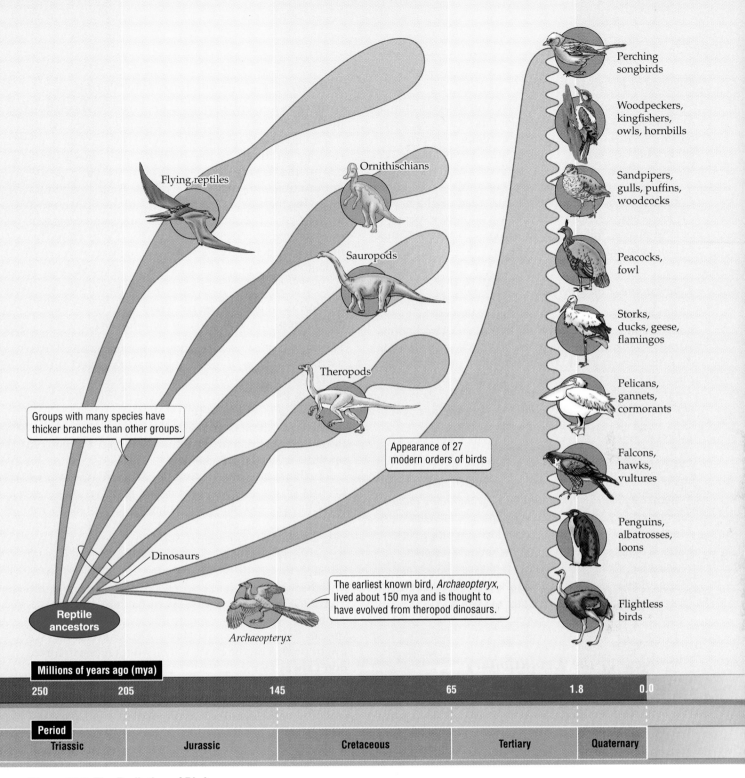

Perching songbirds

Woodpeckers, kingfishers, owls, hornbills

Sandpipers, gulls, puffins, woodcocks

Peacocks, fowl

Storks, ducks, geese, flamingos

Pelicans, gannets, cormorants

Falcons, hawks, vultures

Penguins, albatrosses, loons

Flightless birds

Flying reptiles

Ornithischians

Sauropods

Theropods

Groups with many species have thicker branches than other groups.

Appearance of 27 modern orders of birds

Dinosaurs

Reptile ancestors

The earliest known bird, *Archaeopteryx*, lived about 150 mya and is thought to have evolved from theropod dinosaurs.

Archaeopteryx

Millions of years ago (mya)

| 250 | 205 | 145 | 65 | 1.8 | 0.0 |

Period

| Triassic | Jurassic | Cretaceous | Tertiary | Quaternary |

Figure 22.9 The Radiation of Birds
Birds have radiated greatly since they first evolved flight. Representatives of the nine largest living bird orders are shown here.

On a much shorter time scale, the spectacular radiations of organisms that colonize newly formed islands (see Figure 19.9) or lakes also illustrate how migration to a place free from the usual competitor species can promote evolutionary radiations.

Key evolutionary innovations

A group may diversify greatly if it acquires a key adaptation that lets it use its environment in new ways. There are many examples of such diversification in the fossil record, as in the Cambrian explosion, the radiation of land plants, and the radiations that followed the evolution of flight in insects, birds (Figure 22.9), and mammals (bats).

Ecological interactions

Species compete with each other and depend on one another for food and services. For this reason, diversification within one group of organisms can promote diversification within other groups. For example, if a group radiates because it has evolved a new way to use its environment, this process can stimulate the evolutionary radiation of organisms that feed on the newly expanded group. Such was the case with the expansion of land plants, followed by the expansion of insects that fed on land plants.

Low amounts of gene flow

Low amounts of gene flow (the exchange of genes between populations) promote evolutionary radiations because when there is little gene flow, populations diverge more rapidly, a condition that favors the formation of new species. On a broad geographic scale, the separation of the continents in the past 200 million years isolated many previously connected populations, causing a large increase in the diversity of life.

> ■ After each of the five mass extinctions, other groups of organisms diversified greatly to replace those that had become extinct. Each of these great evolutionary radiations took 1 to 7 million years and forever altered the history of life on Earth. Regardless of whether they occur after a mass extinction or at other times, evolutionary radiations can be promoted by (1) release from competition, (2) key evolutionary innovations, (3) ecological interactions, and (4) low amounts of gene flow.

Overview of the Evolutionary History of Life

The history of life on Earth can be summarized by the rise and fall of major groups of protists, plants, and animals (see Figure 22.2). These broad patterns in the history of life are caused by the extinction or decline of some groups, and the origin or expansion of other groups. Taken together, mass extinctions and evolutionary radiations have largely been responsible for shaping evolution above the species level.

In addition to offering this broad view of the evolutionary history life, we want to emphasize two important related concepts: the increase in diversity over time, and the difference between the evolution of populations and evolution above the species level. We'll close the chapter by briefly summarizing these two points.

The diversity of life has increased over time

Despite the severity of mass extinctions, the diversity of life has increased over time, especially during the last 250 million years (Figure 22.10). One major cause of this increase in diversity has been the movement of the continents during the past 200 million years. Other important factors include the release from competition following mass extinctions, the evolution of key innovations (such as the ability to live on land), and ecological interactions that promote diversity (for example, an increase in diversity can cause a further increase in diversity).

Evolution above the species level differs from the evolution of populations

The often exquisite adaptations of organisms result from natural selection. Evolution by natural selection is a short-term process: Adaptations are shaped by selection to match the organism's current environment. Natural selection cannot predict future changes to the environment; hence there is no reason to suppose that adaptations that are currently advantageous will remain so once the environment changes.

Furthermore, mass extinctions can remove entire groups of organisms—seemingly at random—even those that possess highly advantageous adaptations. For example, a group of predatory gastropods (snails and their relatives) went extinct in the Triassic mass extinction shortly after they had evolved the ability to drill through the shells of other gastropods. The ability to drill through shells had opened a major new way of life and, if the shell-drilling gastropods had not gone extinct, probably would have led to the evolution of many species with such an adaptation. Thus, species that we might predict on the basis of their adaptations to thrive don't always win.

As this discussion illustrates, broad patterns in the history of life cannot be predicted solely from an understanding of the evolution of populations. To have a full understanding of the history of life on Earth, we must understand not only genetic changes within populations, but also factors such as mass extinctions, evolutionary radiations, and continental drift, all of which can have a large impact on evolution above the species level.

> ■ Evolutionary radiations and mass extinctions have largely been responsible for shaping the rise and fall of major groups of organisms. In addition, the diversity of life has increased over time, and evolution above the species level differs from the evolution of populations.

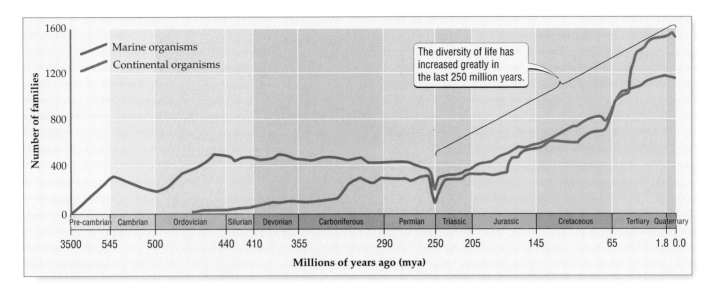

Figure 22.10 Increased Diversity of Life
The diversity of life in the oceans and on land has increased dramatically over time, most notably during the past 250 million years.

Highlight

When Antarctica Was Green

Antarctic fossils of tropical organisms, dinosaurs, and forests are vivid testimony to the fact that we live in a highly changeable world. These fossils reveal large changes over time, ranging as they do from Cambrian marine organisms to early land plants to birds and mammals. Changes in the dominant organisms of Antarctica at different points in time reflect the broad changes in the history of life described in this chapter, such as the Cambrian explosion, the colonization of land, and the respective periods of domination by amphibians, reptiles, and mammals.

The Antarctic fossils also show a striking contrast between the diversity of life forms that once lived in Antarctica and the few that live there today. The low number of organisms that now live in Antarctica is a consequence of continental drift. First, as the continents broke apart, the climate of Earth grew colder. This process was heightened in Antarctica, which experienced an ever-colder climate as it moved toward its present position over the South Pole. Once Antarctica separated from Australia and South America, about 40 mya, the organisms on Antarctica were trapped there. Thus, as the climate on Antarctica became increasingly cold, most species perished.

The movement of Antarctica to its present position may have contained within it not only the seeds of destruction, but those of creation as well. As Antarctica and the rest of the continents drifted apart during the last 40 million years, they altered the flow of ocean currents. The rerouting of ocean currents contributed to the formation of the Antarctic ice cap and produced the largest differences in the temperature between the poles and the tropics that Earth has ever known. The wide range of new habitats that these temperature differences created, along with the ensuing period of repeated glaciations, helped set the stage for evolutionary radiations in many organisms, including humans.

■ Antarctic fossils include Cambrian marine organisms, early land plants, dinosaurs, birds, and mammals. These fossils illustrate the large changes in dominant groups of organisms that have occurred in the history of life on Earth. The low number of species that currently live on Antarctica is a consequence of continental drift. Continental drift isolated Antarctica from other continents and caused most species that lived there to perish as the climate became increasingly cold.

Summary

The Fossil Record: Guide to the Past

■ The fossil record documents the history of life on Earth.

■ Fossils reveal that (1) past organisms were unlike living organisms, (2) many species have gone extinct, and (3) there have been large changes in the dominant groups of organisms over time.

- The order in which organisms appear in the fossil record is consistent with our understanding of evolution.

- Although the fossil record is not complete, it provides excellent examples of the evolution of major new groups of organisms, such as the evolution of mammals from reptiles.

The History of Life on Earth

- The first single-celled organisms resembled bacteria and evolved 3.5 billion years ago.

- Shortly after life began, some groups of bacteria evolved the ability to conduct photosynthesis, which releases oxygen as a waste product.

- The release of oxygen by bacteria caused oxygen concentrations in the atmosphere to increase. As oxygen concentrations continued to increase, the evolution of single-celled eukaryotes (1.5 billion years ago) became possible.

- Life on Earth changed dramatically during the Cambrian explosion, with the sudden appearance of large predators and well-defended herbivores.

- The colonization of land by plants and invertebrates (450 to 400 mya), followed later by the vertebrates (365 mya), stimulated a series of evolutionary radiations that greatly increased the diversity of life.

The Impact of Continental Drift

- Continental drift has had a great impact on the history of life on Earth.

- The separation of the continents in the past 200 million years led to geographic isolation on a grand scale, promoting the evolution of many new species.

- At different points in time, changes in the climate caused by the movement of the continents led to the extinction of many species.

Mass Extinctions: Worldwide Losses of Species

- There have been five mass extinctions during the history of life on Earth, each of which had two large effects on the diversity of life.

- First, the extinction of some groups but not others greatly alters the later course of evolution.

- Second, the extinction of a dominant group of organisms can provide new opportunities for organisms that were of relatively minor importance, thus greatly changing the shape of life on Earth.

Evolutionary Radiations: Increases in the Diversity of Life

- After each of the five mass extinctions, other groups of organisms diversified greatly to replace those that had become extinct.

- Each of these great evolutionary radiations took 1 to 7 million years and forever altered the history of life on Earth.

- Evolutionary radiations are promoted by (1) release from competition, (2) key evolutionary innovations, (3) ecological interactions, and (4) low amounts of gene flow.

Overview of the Evolutionary History of Life

- Evolutionary radiations and mass extinctions are primarily responsible for the rise and fall of major groups of organisms that characterize the history of life on Earth.

- The diversity of life has increased over time due to continental drift, release from competition after mass extinctions, the evolution of key innovations, and ecological interactions.

- Evolution above the species level differs from the evolution of populations. Mass extinctions, evolutionary radiations, and continental drift have all had large impacts on evolution above the species level.

Highlight: When Antarctica Was Green

- The diversity of organisms that once lived in Antarctica, as indicated by fossils, contrasts sharply with the few organisms that live there today.

- Large changes in the organisms that lived in Antarctica at different points in time reflect broad changes in groups of organisms that dominated life on Earth over time.

- The low number of species that currently live on Antarctica is a consequence of continental drift, which isolated Antarctica from other continents and caused most species that lived there to perish as the climate became increasingly cold.

Key Terms

Cambrian explosion p. 343	fossil p. 340
continental drift p. 345	macroevolution p. 340
evolutionary radiation p. 348	mass extinction p. 347

Chapter Review

Self-Quiz

1. Continental drift
   a. can occur when liquid rock rises to the surface and pushes the continents apart.
   b. no longer occurs today.
   c. has led to the geographic isolation of many populations, thus promoting speciation.
   d. both a and c

2. The fossil record
   a. documents the history of life.
   b. provides examples of the evolution of major new groups of organisms.
   c. is not complete.
   d. all of the above

3. Mass extinction events
   a. of the past were all caused by the impact of asteroids.
   b. are periods of time in which many species become extinct throughout Earth.
   c. have little lasting impact on the history of life.
   d. are usually recovered from within 100,000 years or so.

4. The Cambrian explosion
   a. caused a spectacular increase in the diversity and complexity of animal life.

b. caused a mass extinction event.

c. was the time during which all living animal phyla suddenly appeared.

d. had few consequences for the later evolution of life.

5. The history of life shows that

a. the diversity of life has remained constant for about 400 million years.

b. extinctions have little effect on the diversity of life.

c. macroevolution is greatly influenced by mass extinctions and evolutionary radiations.

d. macroevolution can be understand solely in terms of the evolution of populations.

Review Questions

1. Mass extinction events can remove entire groups of organisms, seemingly at random—even groups that possess wonderful adaptations. How can this be?

2. Evidence from the fossil record indicates that it usually takes 1 to 7 million years for an evolutionary radiation to replace the species lost during a mass extinction. Discuss this observation in light of your understanding of the speciation process (see Chapter 21). What are the consequences of this length of time for the current loss of species?

3. Is macroevolution fundamentally different from microevolution? Can macroevolutionary patterns be explained solely in terms of microevolutionary processes? Does any evolutionary mechanism that we have studied link macroevolution and microevolution?

The Daily Globe

22

Fish Diversity Crashes, Thousands of Years of Evolution Lost

LAKE VICTORIA, EAST AFRICA. Lake Victoria has seen it all. First, it was home to a spectacular rise in a group of fish called cichlids. Hundreds of new species evolved in the lake in the past 12,400 years. And now these species are going extinct at an alarming rate.

The extinctions seem to have two causes. First, a large predator, the Nile perch, was introduced (as a source of food for people) to the lake several decades ago. The Nile perch wiped out the populations of many cichlid species, which had evolved in the absence of such a powerful predator.

Second, the water in the lake is becoming more cloudy as a result of pollution and other human activities. Why is cloudy water driving large numbers of species extinct? The reason relates to the fact that although the cichlids can interbreed without loss of fertility, they usually choose members of their own species as mates. These choosy fish tell each other apart by differences in their color patterns. When the water is cloudy, the fish cannot tell each other apart; thus they interbreed so extensively that many species have disappeared.

Efforts are under way to clean up the water and thus prevent further extinctions. But unless the water can be cleaned up quickly, more species will go extinct, and thousands of years of evolution will continue to be lost.

Evaluating "The News"

1. What do you think will happen to the diversity of cichlids in Lake Victoria if the water is cleaned up and the numbers of Nile perch are kept low? Will species that are extinct reappear?

2. Assuming that clean water and a reduction in the numbers of Nile perch cause cichlid diversity to increase, how long do you think such increases in diversity will take?

3. The cichlids in Lake Victoria are not alone; many other freshwater fish species are going extinct throughout the world. What can we learn from events in Lake Victoria that would help us prevent the extinction of entire groups of species in other parts of the world?

4. Past mass extinctions are one of the factors that have shaped the current diversity of life on Earth. Since mass extinctions are a normal part of the evolutionary process, why should we worry about the current extinction of species?

chapter 23

Human Evolution

Rosamund Purcell, *Homo sapiens, Gorilla gorilla, Ourangutan*, 1992.

The Neanderthals

Neanderthals loom large in our imagination, having been the stuff of many movies, books, and magazine articles. Their name alone calls to mind words like "brutish," "primitive," "subhuman," and "caveman." These words are not accurate. Neanderthals were not subhuman; they may even have been early members of our own species, *Homo sapiens*. However, if a group of them were to stroll down the street, they certainly would attract attention.

Neanderthals had large, arching ridges above their eyes, a low, sloping forehead, no chin, and a face that, compared to ours, looked as if it were pulled forward. Without shirts, they would be even more striking. Slightly shorter than humans of today, Neanderthals had thick necks, were heavily boned, and as indicated by muscle markings on their skeletons, were very strong.

Neanderthals take their name from the Neander Valley, in eastern Germany ("thal" comes from the German for "valley"). Fossilized bones discovered there in 1856 were

354

thought by some people to be those of a bear. Others, however, argued that they were the remains of an ancient human, not a bear. Known as the Neanderthal Man, these bones became the subject of great debate: Were the bones really human? Finally, after similar fossils were found in other places, scientists were convinced that the bones were indeed of human origin. Collectively, the Neanderthal fossils shook our understanding of ourselves, for they provided dramatic proof that different forms of humans once existed.

The Neanderthals hold a central place in the study of human evolution. Nevertheless, their relationship to modern humans has long been a mystery. In both physical appearance and culture, they were like us, yet not like us. Neanderthals first appeared in Europe 165,000 years ago, where they thrived until they were suddenly replaced by anatomically modern humans, about 35,000 years ago.

What caused the rapid disappearance of the Neanderthals? Were they our direct ancestors? Or were they an evolutionary dead end, an experiment that failed and contributed little or nothing to the gene pool of modern humans?

An Artist's Interpretation of a Neanderthal's Appearance

Key Concepts

1. Humans are primates, which in turn are a type of mammal. Among primates, gibbons and the apes (including humans) are hominoids. Humans and our humanlike ancestors are hominids.

2. Mammals evolved from reptiles 220 million years ago. The first primates split off from other mammals more than 65 million years ago. Hominids diverged from other primates 5 to 6 million years ago.

3. Early hominids walked upright but resembled the earlier apes from which they evolved. Over time, hominid brain size increased greatly.

4. The human evolutionary tree is "bushy," consisting of multiple species and many side branches at different points in time.

5. Early members of our genus, *Homo*, used stone tools 2.5 million years ago. Toolmaking technology improved slowly until 400,000 years ago, increased somewhat at that time, then took a great leap forward 40,000 years ago.

6. Modern humans probably arose in Africa 100,000 to 200,000 years ago.

Human beings—*Homo sapiens*—are animals, classified as members of the chordate phylum, the mammal class, the primate order, and the hominid family. Humans share with all other mammals certain distinguishing characteristics, including body hair (which provides insulation in many species) and the feeding of their young with milk that is produced by mammary glands. As we'll describe in this chapter, humans also share more specific characteristics with other primates and with other members of the hominid family (which contains humans and our humanlike ancestors).

As animals, we evolved from earlier animals. Most recently, the evolutionary line leading to humans split from the line leading to chimpanzees 5 to 6 million years ago. Although chimpanzees are our closest living relatives, we did not evolve directly from them, as is sometimes misleadingly stated. Rather, the evolutionary lines leading to humans and to chimpanzees both originated from a common ape ancestor. In this chapter, we describe the origin of mammals and primates, then focus on the evolution of humans and our immediate ancestors.

Evolution of Mammals: From Subordinance to Dominance

Mammals evolved from reptiles. More specifically, mammals arose from the cynodonts, the end of a long line of mammal-like reptiles. Living mammals differ from living reptiles in many respects, including the way they move (Figure 23.1), the nature of their teeth, and the structure of their jaws. In the fossil record, however, it is difficult to draw the line between mammals and reptiles. Some fossil species form excellent intermediates between what we now call reptiles and mammals, thus providing a beautiful illustration of the evolutionary shift from one major group of organisms (the reptiles) to another (the mammals).

Mammal-like reptiles enjoyed great success in the early Triassic (245 million years ago, or mya). By 200 mya, however, the mammal-like reptiles had declined as other reptiles, most notably, the dinosaurs, came to dominate. The mammal-like reptiles left behind the first

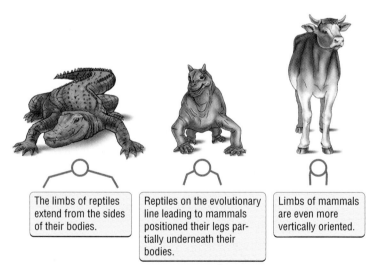

The limbs of reptiles extend from the sides of their bodies.

Reptiles on the evolutionary line leading to mammals positioned their legs partially underneath their bodies.

Limbs of mammals are even more vertically oriented.

Figure 23.1 From Reptiles to Mammals
Over time, the legs of mammal-like reptiles became positioned under the body, leading eventually to the vertical orientation of the legs in living mammals. Today, the legs of most reptiles stick out to the side of their bodies, causing them to have a sprawling gait as compared to the upright gait of mammals.

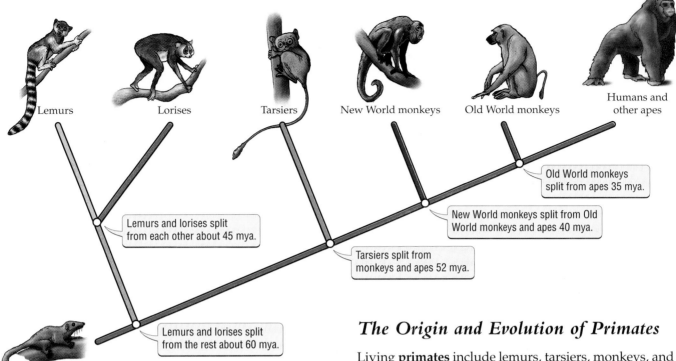

Figure 23.2 Evolutionary Tree of the Living Primates

Lemurs and lorises split from each other about 45 mya.

Lemurs and lorises split from the rest about 60 mya.

Tarsiers split from monkeys and apes 52 mya.

New World monkeys split from Old World monkeys and apes 40 mya.

Old World monkeys split from apes 35 mya.

Lemurs

Lorises

Tarsiers

New World monkeys

Old World monkeys

Humans and other apes

Early primate

mammals as their descendants. Early mammals were small, rodent-sized organisms that evolved at about the same time as the dinosaurs, 220 mya.

Throughout the long reign of the dinosaurs, the mammals remained small and appear to have been active at night, or nocturnal (activity at night is assumed because of the large size of their eye sockets). By being nocturnal and small, early mammals may have been to dinosaurs what a mouse is to a lion: hard to notice and too small to eat. Following the extinction of the dinosaurs 65 mya, the mammals branched out greatly to include many forms that were large and active by day.

A fundamental event in the evolution of mammals was an increase in the speed and duration of brain growth during development. The resulting increase in adult brain size had many effects. The tendency toward increased brain size was especially prominent in the order to which humans belong, the primates.

■ Mammals evolved from reptiles about 220 million years ago. Throughout the long reign of the dinosaurs, mammals remained small and nocturnal. Following the extinction of the dinosaurs, the mammals branched out greatly to include many species that were large and active by day.

The Origin and Evolution of Primates

Living **primates** include lemurs, tarsiers, monkeys, and humans and other apes (Figure 23.2). Primates are thought to have originated more than 65 mya, from small, nocturnal mammals, similar to tree shrews, that ate insects and lived in trees. However, the fossil evidence on primate origins is sketchy, and the first definite primate fossils are roughly 50 million years old. These early primates resembled modern lemurs and lorises (see Figure 23.2).

Tree shrew

The first higher primates, a group that includes New World monkeys, Old World monkeys, and humans and other apes, arose 37 to 45 mya. They had small bodies and probably ate insects or fruit. By 35 mya, the group had diversified greatly. Early higher primates gave rise to the **hominoids**, a group of primates whose living members include gibbons and the great apes (orangutans, gorillas, chimpanzees, and humans).

The hominoids evolved to form many new species and higher groups of organisms. From 23 to 5 mya, there were 30 genera of hominoids, most of which are now extinct. The hominoids also gave rise to the **hominids**—that is, humans and our now extinct, humanlike ancestors.

Evolutionary trends in primates

Primates share numerous characteristics, including flexible shoulder and elbow joints, five functional fingers and toes, thumbs and big toes that are opposable (that is, they can be placed opposite other fingers or toes), flat nails (instead of claws), forward-facing eyes, short snouts, and brains that are large in relation to body size. Many of these traits suit primates to life in trees. Three characteristics became increasingly well developed during primate evolution: limb mobility and grasping ability, daytime vision, and brain size.

- *Enhanced limb mobility and grasping ability.* The limbs of animals that run on four legs often evolve for greater speed and stability, as in horses. Primates, on the other hand, evolved greater limb mobility, as in the development of flexible shoulder and elbow joints. These flexible joints improved the ability of primates to climb and swing from branch to branch. Primates also evolved greater grasping ability in their hands and feet, largely because of their opposable big toes and thumbs. The freedom of movement of their limbs and the grasping ability of their feet and hands were essential to primates as they moved through trees. Humans no longer have opposable big toes, and unlike other living primates, we walk upright.

- *Improved daytime vision.* Early primates were nocturnal. They had relatively large snouts, as well as eyes located on the sides of their heads. During primate evolution, the position of the eyes moved forward, causing greater overlap in their fields of vision and hence improving depth perception. Simultaneously, the snout shortened, further improving the forward-facing vision of primates. Primates also evolved an increased ability to distinguish the color and brightness of light, both of which were useful for daytime vision. In general, the evolutionary changes in primate vision benefited organisms that lived in trees. For example, improved depth perception made it less likely that a primate moving from branch to branch would miscalculate and fall.

- *Increased brain size.* Perhaps the most notable feature of primate evolution is the tendency toward large brain size. In many primate lines, parents raised fewer offspring but placed more effort into each of them. For such primates, the period of time in which offspring depended on and learned from their parents increased. Increases in brain size were linked to this greater emphasis on learning, as well as to the more complex social behaviors shown by higher primates.

> ■ Primates originated more than 65 mya. Primates have flexible elbow and shoulder joints, forward-facing eyes, opposable big thumbs, and (with the exception of humans) opposable big toes. These traits make them well adapted for life in trees. Three characteristics became increasingly well developed during primate evolution: limb mobility and grasping ability, daytime vision, and brain size.

Hominid Evolution: The Switch to Walking Upright

Hominids—humans and our now extinct, humanlike ancestors—are characterized by large brains, upright locomotion (on two feet), and toolmaking behaviors. Of these traits, our intelligence, toolmaking abilities, and associated culture are central to what it means to be human. In an evolutionary sense, however, the increases in our intelligence and toolmaking abilities were secondary changes: They occurred relatively late in our evolutionary history and resulted from a general trend toward large brain size in primates.

The first big step in human evolution was the switch from moving on four legs to walking upright on two legs. This switch required a drastic reorganization of primate anatomy, most especially of the hip bones. Walking upright brought other important changes, such as a shift in how the skull is oriented on the spinal cord, a change in the position of the big toe, a loss of opposability in the big toe, the development of a pronounced heel and arch on the foot, a lengthening of the legs, and a change in the angle of weight support from the hip to the knee (Figure 23.3).

It is not necessary—or even feasible—to walk upright in a tree, and for an organism that lived primarily in trees, the loss of an opposable big toe would be a handicap. On the ground, however, walking upright would have provided several advantages, including freeing the hands for carrying objects or using tools and improving the line of sight (that is, being able to see over nearby objects). Thus, it is likely that the evolution of an upright posture was linked to a switch from life in trees to life on the ground, a switch that probably occurred between 8 and 5 mya.

The switch to life on the ground was probably not sudden or complete. The skeletal structure of the oldest fossil hominids (from 4.4 mya) indicates that they walked upright. However, footbones and fossilized footprints from 3 to 3.5 mya show that early hominids had partially opposable big toes (Figure 23.4), suggesting they may have continued to use trees somewhat.

The earliest known hominids are ***Ardipithecus ramidus*** (4.4 mya), ***Australopithecus anamensis*** (4.2 to 3.9 mya), and ***Australopithecus afarensis*** (3.9 to 3.0 mya). Although each of these organisms is thought to have walked upright, their brains were relatively small, and their skulls and teeth are more similar to other apes than to humans (compare *A. afarensis* to *Homo sapiens* in Figure 23.5).

Australopithecus species probably used simple tools, as do living chimpanzees and other animals, but no direct evidence to support this assumption has been found. This lack of evidence is not surprising; the tools used by early hominids probably consisted of items like sticks, blades of grass, or rocks, none of which can be identified as "tools" in the fossil record.

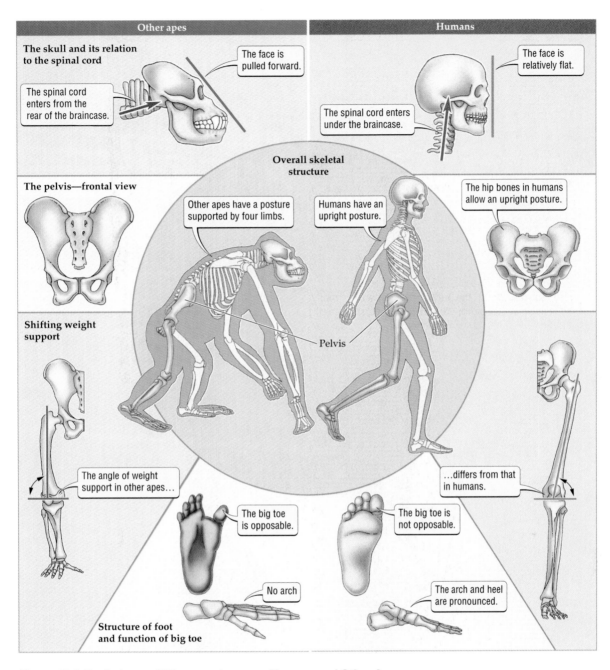

Figure 23.3 Evolutionary Differences between Humans and Other Apes

■ The first big step in human evolution was the switch to walking upright, which was linked to a change from life in trees to life on the ground. Hominids or their ancestors probably adapted to living on the ground between 8 and 5 mya. Fossil evidence indicates that early hominids walked upright but had relatively small brains, as well as jaws and skulls that differed greatly from those of living humans.

Evolution in the Genus **Homo**

The first members of the genus *Homo*, to which modern humans belong, originated in Africa 2 to 3 million years ago. The earliest *Homo* fossil fragments date from 2.4 mya. More complete early *Homo* fossils exist for the period from 1.9 to 1.6 mya; these fossils are known as **Homo habilis**. The oldest *H. habilis* fossils resemble *Australopithecus africanus*, the species from which *H. habilis* may

Figure 23.4 Early Hominids Had Partially Opposable Big Toes
(*a*) Footprints of two early hominids (3 to 3.5 million years old)
walking upright, side by side. (*b*) Chimpanzees.

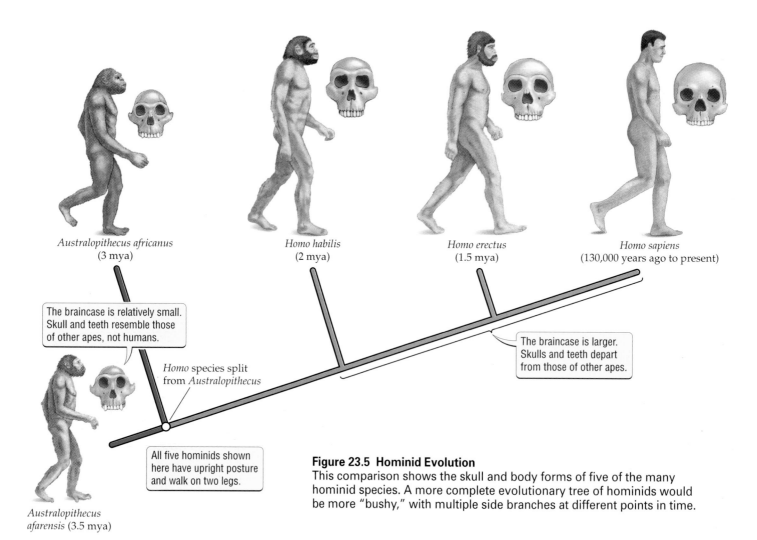

Figure 23.5 Hominid Evolution
This comparison shows the skull and body forms of five of the many
hominid species. A more complete evolutionary tree of hominids would
be more "bushy," with multiple side branches at different points in time.

have evolved. In more recent *H. habilis* fossils, the face is not pulled forward as much and the skull is more rounded. In these and other ways, more recent *H. habilis* specimens have features that are intermediate between those of *A. africanus* and *Homo erectus*, a species that evolved after *H. habilis*. Thus, *H. habilis* fossils provide an excellent example of the evolutionary shift from ancestral (*Australopithecus*) to more recent (*H. erectus*) characteristics.

H. habilis used stone tools, such as those shown in Figure 23.6a. Tools of this type are known as the Oldowan technology, named after Olduvai Gorge, Tanzania, a site at which many important fossils have been found. Oldowan tools were first made 2.5 mya. Oldowan tools were replaced in Africa and southwestern Asia by more advanced tools about 1.4 mya, but they persisted in Europe and eastern Asia until 1 mya. The more advanced tools were probably made by *H. erectus*, a species that first appeared 1.8 mya.

Taller and more robust than *H. habilis*, **Homo erectus** also had a larger brain and a face more like that of the modern human (see Figure 23.5). It is likely that by 500,000 years ago *H. erectus* could use, but not necessarily make, fire. With respect to hunting, our ancestors probably hunted large species of game for hundreds of thousands of years. The evidence to support this conclusion includes the remarkable discovery in Germany of three 400,000-year-old spears, each around 2 meters long and designed for throwing with a forward center of gravity, as a modern javelin is.

It was long thought that *H. habilis* gave rise to *H. erectus*, which then spread from Africa about 1 mya and later evolved into *Homo sapiens*. This simple picture has become more complicated with recent fossil discoveries. Some evidence now suggests that the *H. habilis* fossils are from two different species, and there is debate over which of these species gave rise to *H. erectus*. In addition, it now appears that *H. erectus* or an earlier form of *Homo* migrated from Africa much earlier than previously thought. *Homo* fossils dating from 1.9 to 1.7 mya have been found in Java, the central Asian republic of Georgia, and China.

Overall, the research on *H. habilis*, *H. erectus*, and other early *Homo* species indicates that there were more species of *Homo* than once thought, and that several of these species existed in the same places and times. More research will be necessary before general agreement is reached regarding the number of early *Homo* species and their evolutionary relationships.

Figure 23.6 A Gallery of Tools and Art
Toolmaking technology changed relatively little from 2.5 million to 400,000 years ago, but it has changed drastically in the past 40,000 years.

(*a*) Oldowan stone tools of *Homo habilis* (2.5–1.4 mya)

(*b*) Stone tools of *Homo erectus* (1 million–400,000 years ago)

(*c*) Neanderthal decorative art (45,000 years ago)

(*d*) Spear head and figurine of a bison (40,000–10,000 years ago)

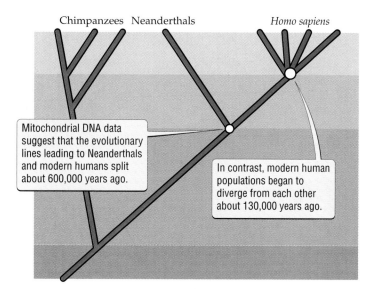

Chimpanzees   Neanderthals   *Homo sapiens*

Mitochondrial DNA data suggest that the evolutionary lines leading to Neanderthals and modern humans split about 600,000 years ago.

In contrast, modern human populations began to diverge from each other about 130,000 years ago.

Figure 23.7 DNA Evidence from Fossils
Mitochondrial DNA data from humans and from a Neanderthal suggest that Neanderthals were not the direct ancestors of modern humans.

■ The first members of the genus *Homo*, to which modern humans belong, originated in Africa 2 to 3 mya. By 400,000 to 500,000 years ago, our ancestors used fire and hunted large game species. The number of early *Homo* species and their evolutionary relationships are the subject of ongoing debate.

The Origin and Spread of Modern Humans

Fossils with a mix of features from *Homo erectus* and *Homo sapiens* appear in the fossil record 400,000 to 130,000 years ago. Known as archaic (meaning "old" or "early") *H. sapiens*, these fossils occur in Africa, China, Java, and Europe. These ancestors of living humans developed new tools and new ways of making tools, used new foods, built more complex shelters, and controlled the use of fire.

Populations of archaic *H. sapiens* gave rise both to the Neanderthals and to anatomically modern humans. The Neanderthals first evolved in Europe about 165,000 years ago, spreading from there to western Asia. DNA extracted from Neanderthal fossils suggests that Neanderthals were not direct ancestors of modern humans (Figure 23.7). The oldest fossils of anatomically modern humans have been found in Africa, dating from 130,000 years

ago. More recent fossils of anatomically modern humans have been found in such places as Israel (115,000 years ago), China (50,000 years ago), Australia (40,000 years ago), and the Americas (12,000 years ago).

There has been considerable controversy over the origin of anatomically modern humans. Two extreme hypotheses have been proposed: the out-of-Africa model and the multiregional model. According to the **out-of-Africa model** (Figure 23.8*a*), modern humans first evolved in Africa within the past 200,000 years. They then spread from Africa to the rest of the world, completely replacing archaic *H. sapiens*, including advanced forms such as the Neanderthals. In contrast, the **multiregional model** (Figure 23.8*b*) proposes that modern humans evolved over time from *H. erectus* populations located throughout the world. According to this hypothesis, regional differences among human populations developed early, but worldwide gene flow caused these different populations to evolve modern characteristics simultaneously and to remain a single species.

Which of these hypotheses is correct? Let's consider some of the evidence. According to the multiregional model, when different types of early humans came into contact, extensive gene flow should have caused them to evolve toward one another. Thus, we would not expect different types of early humans to coexist in the same area, yet remain different for long periods of time. But in fact Neanderthals and more modern humans coexisted in western Asia for about 80,000 years, calling into question the extensive gene flow assumed by the multiregional model.

The best fossil evidence for the shift from archaic to modern *H. sapiens* comes from Africa, providing some support for the out-of-Africa hypothesis. However, the "complete replacement" part of the model appears not to be correct. Genetic studies indicate that there was enough breeding with ancient peoples outside Africa to cause some of their genes to enter the modern gene pool. Overall, the evidence suggests that anatomically modern humans arose in Africa, but that they had limited breeding with and did not completely replace more ancient *Homo* populations.

■ Fossils with a mix of features from *Homo erectus* and *Homo sapiens* appear in the fossil record 130,000 to 400,000 years ago. The oldest fossils of anatomically modern humans have been found in Africa, dating from 130,000 years ago. Genetic and fossil evidence suggests that anatomically modern humans arose in Africa and had limited breeding with more ancient *Homo* populations.

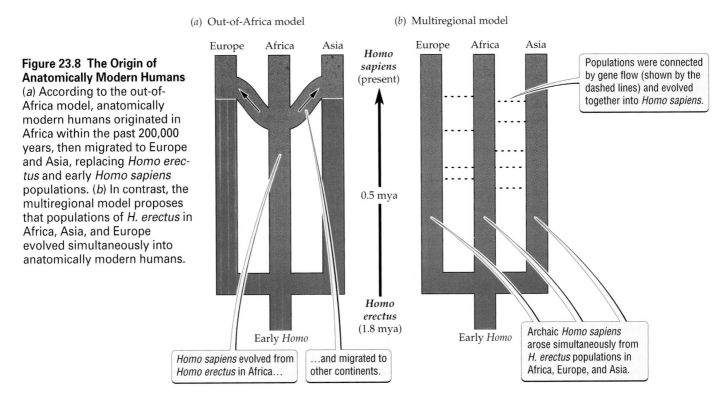

Figure 23.8 The Origin of Anatomically Modern Humans
(*a*) According to the out-of-Africa model, anatomically modern humans originated in Africa within the past 200,000 years, then migrated to Europe and Asia, replacing *Homo erectus* and early *Homo sapiens* populations. (*b*) In contrast, the multiregional model proposes that populations of *H. erectus* in Africa, Asia, and Europe evolved simultaneously into anatomically modern humans.

(*a*) Out-of-Africa model

(*b*) Multiregional model

Europe   Africa   Asia

Homo sapiens (present)

0.5 mya

Homo erectus (1.8 mya)

Early *Homo*

Populations were connected by gene flow (shown by the dashed lines) and evolved together into *Homo sapiens*.

Homo sapiens evolved from *Homo erectus* in Africa…

…and migrated to other continents.

Archaic *Homo sapiens* arose simultaneously from *H. erectus* populations in Africa, Europe, and Asia.

Why Are Humans So Different from Our Closest Relatives?

Humans are very similar genetically to both chimpanzees and gorillas. For example, 99 percent of our DNA is identical to chimpanzee DNA. Chimpanzees, our closest relatives, share many characteristics with us, including the use of tools, a capacity for symbolic language (chimpanzees have been taught hundreds of words, including those for abstract concepts), and, it seems, a sense of self-awareness and the performance of deliberate acts of deception.

Despite these similarities, humans also differ in many and obvious ways from chimpanzees. The extent to which humans make tools, solve problems, communicate, exterminate members of our own and other species, and create art, music, and other forms of culture are all unique to our species. How can the 1 percent difference in DNA have caused this enormous gulf between us and chimpanzees?

This question is difficult to answer. Many of the genetic differences between humans and chimpanzees probably relate to differences in brain size and to differences in how the two species walk and look. But our ancestors had large brains and differed from other apes in how they walked and looked long before our culture was particularly advanced. What, then, does cause the enormous difference in our problem-solving abilities and culture?

The answer may lie in the evolution of a modern capacity for spoken language. Before the last 400,000 years, technology improved slowly. The pace of technological development then began to increase, taking a great leap forward roughly 40,000 years ago. The technological advances shown by archaic and modern *Homo sapiens* may have been due to genetic changes that improved spoken language. This theory seems especially plausible when we consider what an organism with a large brain but no complex language skills might be like. What would it mean to be intelligent in the absence of all but the most basic language skills? To what extent could even a large-brained early *Homo* develop new technologies if it could not express its ideas, or share them with others?

■ Given that humans and chimpanzees differ by only 1 percent of their DNA, why are we so different? The evolution of morphological differences between humans and chimpanzees probably involved many genetic changes. However, these morphological differences had relatively little impact on our culture or success as a species. Differences in the intelligence, toolmaking abilities, and cultures of humans and chimpanzees may have been due to genetic changes that improved our capacity for spoken language.

Overview of Human Evolution

The fossil record covers 4 to 5 million years of hominid evolution, and provides evidence of the following points:

1. Living humans evolved from ape ancestors through many intermediate steps. From 4 million to 30,000 years ago, there was usually more than one hominid species alive. Thus, hominid evolution did not consist of a single evolutionary line that began with *Ardipithecus ramidus* and ended with anatomically modern humans. Rather, the human evolutionary tree is "bushy," consisting of multiple species and many side branches at different points in time.

2. Some hominid traits evolved toward a modern condition more rapidly than others. For example, hominids were walking on two feet by 4.4 mya but did not evolve large brains until much later (Table 23.1).

3. Brain size (volume of the braincase) increased greatly from early hominids to *Homo sapiens*, then decreased slightly in recent times (see Table 23.1). Comparing among species, the size of the brain relative to the size of the body provides a crude measure of intelligence.

4. Early *Homo* first made tools 2.5 mya. Toolmaking technology improved slowly for more than 2 million years, improved more rapidly as of 400,000 years ago, and finally, took a great leap forward 40,000 years ago. The increased pace of technological development may have been associated with the evolution of modern language skills.

Highlight

The Evolutionary Future of Humans

What is our evolutionary future? Will we go the way of science fiction stories and evolve toward beings with huge brains? Such a direction appears unlikely, since for the last 75,000 years brain size in humans appears to have been decreasing, not increasing (see Table 23.1). More realistically, what changes can we expect as a result of genetic drift, gene flow, and natural selection? (To review these concepts, see Chapter 20.)

Before the development of agriculture about 11,000 years ago, human populations were small, widely scattered, and dependent on hunting and gathering (collecting edible plants, insects, and small vertebrates). On a large geographic scale, these populations were isolated from one another by geographic barriers. For example, 20,000 years ago the chance of an African meeting an Australian was virtually nil. Such isolation was probably true on a much smaller geographic scale as well, as illustrated by the fact that before the twentieth century, people from nearby valleys in New Guinea spoke many different languages and had little contact with each other.

Translated into biological terms, the conditions in early human populations were exactly those under which genetic drift should be important: Population sizes were small, and there probably was little gene flow among them. Thus, we would predict that genetic drift has played a major role in causing genetic differences among human populations.

Some evidence supports this claim: Analyses of the rate of evolution of the skulls and teeth of modern

23.1 *Body Weight and Brain Size in Hominids, Chimpanzees, and Gorillas*

Species	Dates (thousands of years ago)	Body weight (kg)	Brain size (cubic cm)	Relative brain size (EQ)[a]
Homo sapiens	Present	58	1349	5.3
H. sapiens	35–10	65	1492	5.4
Neanderthal	75–35	76	1498	4.8
Late *Homo erectus*	600–400	68	1090	3.8
Early *H. erectus*	1800–600	60	885	3.4
Homo habilis	2400–1600	42	631	3.3
Australopithecus africanus	3000–2300	36	470	2.7
Australopithecus afarensis	4000–2800	37	420	2.4
Chimpanzee	Present	45	395	2.0
Gorilla	Present	105	505	1.7

[a]The encephalization quotient, EQ, is the ratio of the actual brain size to the size expected for a mammal of that body weight.

Figure 23.9 Gene Flow in Our Future
Gene flow among populations could reduce some of the features that distinguish different groups of people. This computer composite image was formed from photographs of eight Afro-Caribbean models, eight Caucasian models, and eight Japanese models.

humans indicate that genetic drift was a more important factor than natural selection. However, because the human population is now large and mobile, drift is less likely to play a major role in future human evolution. Instead, gene flow among populations could reduce their differences, as highlighted in Figure 23.9.

What about the role of natural selection? Our technological advancements have removed many of the selection pressures faced by our ancestors. Thus, for many characteristics—such as body size, eyesight, and treatable metabolic disorders like phenylketonuria (PKU, a disorder that causes retardation but which can be prevented by diet)—there is now little selection against individuals who might have been greatly disadvantaged at previous times. This does not mean that natural selection will be powerless in the future. For example, infectious disease takes a terrible toll each year in human death and suffering; hence strong selection pressures exist for the evolution of increased disease resistance in human populations.

In many ways, however, humans have now stepped outside the evolutionary framework described in this unit. In a sense, we have taken control of our own evolutionary future. We constantly modify both ourselves (at present, mostly in an indirect way through the use of machines) and our environment to suit our needs. We transmit our culture—our technology, languages, institutions, and traditions—within and between genera-

tions, changing it rapidly to meet new challenges. Cultural change proceeds at a far more rapid pace than biological evolution, and this fast pace provides us with both hope and peril: Will we direct our future wisely?

> ■ Early human populations were small and had little gene flow among them, conditions that favor evolution by genetic drift. Because the human population is now large and mobile, gene flow could lead to a blurring of existing differences among populations. Technological developments have removed many of the selection pressures faced by our ancestors. To a large extent, we now rely on rapid cultural change to meet new challenges posed by our environment.

Summary

Evolution of Mammals: From Subordinance to Dominance

- Humans are apes, which are a type of primate. Primates are a type of mammal.
- Mammals evolved from reptiles about 220 mya.
- Throughout the long reign of the dinosaurs, mammals remained small and nocturnal.
- Following the extinction of the dinosaurs, the mammals branched out greatly to include many species that were large and active by day.

The Origin and Evolution of Primates

- Primates are thought to have originated more than 65 mya.
- Primates have flexible elbow and shoulder joints, opposable big thumbs and big toes, and forward-facing eyes.
- Humans are the only primates that walk upright and that have lost the opposability of the big toe.
- With the exception of humans, primates are well adapted for life in trees.
- Three characteristics became increasingly well developed during primate evolution: limb mobility and grasping ability, daytime vision, and brain size.

Hominid Evolution: The Switch to Walking Upright

- The first big step in human evolution was the switch to walking upright.
- The evolution of an upright posture probably was linked to a change from life in trees to life on the ground, a switch that probably occurred between 8 and 5 mya.
- Though they walked upright, early hominids had relatively small brains, and their jaws and skulls were more similar to those of nonhuman apes than to those of modern humans.

Evolution in the Genus Homo

- The first members of the genus *Homo*, to which modern humans belong, originated in Africa 2 to 3 mya.

- *H. habilis* fossils (from 1.9 to 1.6 mya) provide an excellent example of the evolutionary shift from ancestral (*Australopithecus*) to more recent (*Homo erectus*) characteristics.

- By 400,000–500,000 years ago, our ancestors were using fire and hunting large game species.

- The number of early *Homo* species and their evolutionary relationships are the subject of ongoing debate.

The Origin and Spread of Modern Humans

- Fossils with a mix of features from *Homo erectus* and *Homo sapiens* appear in the fossil record from 400,000 to 130,000 years ago.

- The oldest fossils of anatomically modern humans have been found in Africa, dating from 130,000 years ago.

- Although the origin of anatomically modern humans is still the subject of a lively debate, genetic and fossil evidence suggests that they arose in Africa and had limited breeding with more ancient *Homo* populations.

Why Are Humans So Different from Our Closest Relatives?

- Despite the many and obvious differences between the two species, the DNA of chimpanzees and humans differ by only 1 percent.

- Many of the genetic differences between humans and chimpanzees probably relate to differences in brain size and to differences in how we walk and look. These genetic differences were in place long before human culture outstripped those of other apes.

- Differences in the intelligence, toolmaking abilities, and culture of humans and chimpanzees may have been due to genetic changes that improved our capacity for spoken language.

Overview of Human Evolution

- The human evolutionary tree is "bushy," consisting of multiple species and many side branches at different points in time.

- Some hominid traits (for example, upright locomotion) evolved toward a modern condition more rapidly than others (for example, large brain size).

- Brain size (volume of the braincase) increased greatly from early hominids to *Homo sapiens*.

- Early *Homo* first made tools 2.5 mya. Toolmaking technology improved slowly for more than 2 million years, improved more rapidly as of 400,000 years ago, and finally, took a great leap forward 40,000 years ago.

Highlight: The Evolutionary Future of Humans

- Early human populations were small and had little gene flow among them, conditions that favor evolution by genetic drift.

- Because the human population is now large and mobile, gene flow could lead to a blurring of existing differences among populations.

- With respect to natural selection, technological developments have removed many of the selection pressures faced by our ancestors.

- To a large extent, we now rely on rapid cultural change to meet new challenges posed by our environment.

Key Terms

Ardipithecus ramidus p. 358
Australopithecus afarensis p. 358
Australopithecus anamensis p. 358
hominid p. 357
hominoid p. 357
Homo erectus p. 361
Homo habilis p. 359
multiregional model p. 362
out-of-Africa model p. 362
primate p. 357

Chapter Review

Self-Quiz

1. An early (about 5 to 8 mya) and crucial step in human evolution was
   a. the development of large brains.
   b. sudden and dramatic improvement in toolmaking technology.
   c. the switch to walking upright.
   d. genetic changes that improved spoken language.

2. Which of the following is *not* correct?
   a. A single evolutionary line led from *Ardipithecus ramidus* to modern humans.
   b. Some hominid traits evolved more rapidly than others.
   c. Brain size increased greatly from early hominids to *Homo sapiens*.
   d. Toolmaking technology took a great leap forward 40,000 years ago.

3. Fossils of anatomically modern humans and Neanderthals coexisted for thousands of years in many parts of the world, yet remained different. Such data are not what you would expect for the
   a. out-of-Africa model.
   b. multiregional model.
   c. genetic drift model.
   d. Oldowan model.

4. Which of the following features do humans lack that other primates have?
   a. forward-facing eyes
   b. short snouts
   c. flexible shoulder and elbow joints
   d. opposable big toes

5. Five to 6 mya,
   a. humans evolved directly from chimpanzees.
   b. the evolutionary line leading to humans split from that leading to monkeys.

c. the evolutionary lines leading to humans and chimpanzees both originated from a common ape ancestor.

d. primates first diverged from other mammals.

Review Questions

1. Gene flow has the potential to make living human populations far more similar than they are now. Would this change represent a net loss or gain for human societies? What factors operate to prevent gene flow from making human populations more similar? (*Hint:* Refer to Chapter 20.)

2. Humans are so closely related to chimpanzees that some have referred to modern humans as the "fourth chimpanzee." Why do you think we label ourselves members of the genus *Homo*, instead of calling ourselves members of the genus *Pan*, along with the other three species of chimpanzees?

3. Researchers are currently trying to identify the genes that cause crucial differences between humans and chimpanzees. If they are successful, what should we do with this information? Would it be ethical for us to use these genes to create "designer chimps"—that is, chimpanzees that become ever more humanlike?

23

The Daily Globe

Evolution Explains Proper Roles

To the Editor:
Society today is in bad shape. The signs are everywhere: Children do not respect their parents like they used to, television continues to sink to new lows depicting sex and violence, drug use is up, young people are piercing their body parts, and our society is so violent that we are not surprised when even 10-year-old children commit horrible murders.

These and many other possible examples show that we are a society spinning out of control, a species that has lost touch with itself.

That's the problem in a nutshell, and identifying the problem leads to the solution. To regain control of our society we must return to our roots. Evolution teaches us that women must stay at home and take care of the children, because that is how we evolved. And men must go out and get jobs because men evolved to be hunters that provide for the family.

Once we return to our roots, other problems will be solved. Women will no longer divide their lives in ways that shatter their sense of self. Men will no longer struggle

with being what they are not, caregivers and nurturers. And a new generation of children will grow up with the guidance and support they need, leading them to respect their parents, their society, and themselves. Once we return to who we really are, we won't tolerate poor television shows, we won't have need for drugs, and the respect we gain for ourselves and our society will curb the many forms of violence that are destroying the social fabric of our country today.

Jack Numsley
New York, NY

Evaluating "The News"

1. In many hunting and gathering societies, much of the food that people eat comes from the gathering activities of women. What are the implications of women as food providers for the line of argument in this letter to the editor?

2. Given your understanding of evolution in general, and of human evolution in particular, is there likely to be any basis for the views presented in this letter to the editor? Why or why not?

3. Some psychologists think that the most important nongenetic influence on the development of children comes not from parents, but from children's interactions with

their peers. If these psychologists are right, are the societal roles of men and women all that important?

4. What factors do you think are causing the problems that human societies face today? Where do you think the solutions to these problems lie?

Robert Kushner, *Chrysanthemum Brocade,* 1994.

Form and Function

chapter 24

An Overview of the Form and Function of Life

Remedios Varos, *Planta insumisa*, 1961.

The Life of a Lawn

Lawns surround us. We find them in city parks, where they offer welcome relief from the surrounding concrete. We find them around suburban homes, where we spend hours of our lives pushing lawn mowers. We sunbathe and we play on lawns. Rarely, however, do we think of the lives that make lawns possible.

The life of a lawn depends on an amazing diversity of organisms. Most visible of these are plants like the grasses and the weedy dandelions that sneak in. The plants use photosynthetic tissues in their leaves to capture the sun's energy as sugars. They combine these sugars with minerals absorbed from the soil by their roots to form the proteins essential to their survival.

The success of the grasses and weeds depends on the activities of hundreds of unseen organisms that live in the soil. For example, we see earthworms only when heavy rains force them to the soil surface. Earthworms spend most of their time, however, burrowing through the soil feeding on fragments of dead

Main Message

All life must perform a shared set of functions to survive, including the ability to regulate what enters and leaves cells.

plant and animal tissue. As they burrow, they loosen the soil, allowing air to reach the plant roots. The thousands of fungi and single-celled bacteria and protists in each handful of soil, organisms that we cannot see without the aid of a microscope, decompose dead plant and animal tissues into minerals that plant roots can absorb.

Invisible though these organisms may be, we see the workings of earthworms and microbes in the quiet disappearance of lawn clippings from one mowing to the next. Animals like rabbits and robins make their living on lawns by feeding on other lawn dwellers. Whereas rabbits feed on lawn plants, robins roam the lawns listening for the worms that make up their meals.

All these organisms contribute in different ways to the function of a healthy lawn. Each species of lawn organism has a unique appearance and way of making its living. A grass plant seems to have little in

common with the earthworms and microbes on which it depends, or with the rabbits to which it falls victim. In this and the other chapters in this unit, however, we will explore some of the common features that, as living organisms, unite all these lawn inhabitants.

Lawns Provide a Familiar Example of the Diversity of Life

Key Concepts

1. The basic unit of life, the cell, is arranged in an organized fashion to form complex, multicellular plants and animals.

2. The survival of any organism depends on its ability to carry out basic functions that include most conspicuously the exchange of chemicals with its environment and reproduction.

3. The exchange of chemicals between living organisms and their environment depends on a combination of passive transport, which requires no energy input, and active transport, which does require energy.

4. The plasma membrane of the cell acts as a selective filter that determines which chemicals enter and leave a cell.

As the description of lawn life at the beginning of this chapter shows, organisms can consist of one cell or many and can make their living in many different ways. In the face of such great differences, it might seem possible to study the form and function of living things only by considering each unique species separately. In this chapter, however, we focus on some of the features that unite all living organisms, and that reflect their shared evolutionary origins: (1) All groups of organisms, no matter how simple or complex, depend on the cell as the basic unit of organization. (2) All species, regardless of how they make their living, must carry out a common set of functions to survive. (3) Finally, all organisms transfer materials into and out of their cells, a process that underlies much of the discussion in the later chapters in this unit.

Life's Organization

All life consists of basic building blocks called cells (see Chapter 6). The human body consists of trillions of cells, each of which shares certain essential features with the cells of all other living organisms. Each cell contains DNA that carries the genetic blueprint for the entire organism, has a characteristic lipid-based plasma membrane that separates the inside of the cell from its surroundings, and can closely control what crosses the plasma membrane to maintain an internal environment that differs from that outside the cell. All functions essential to life, which we will introduce later in this chapter, take place in cells.

Single-celled organisms lack the specialization of multicellular organisms

In single-celled organisms, all functions essential to life occur in one, all-purpose cell (Figure 24.1a). Each of the vast number of cells in multicellular plants and animals has the same design and size as those of single-celled organisms from which they evolved. In multicellular organisms, however, each cell is specialized to perform just a fraction of the many functions needed to keep the organism alive. As a result, the cells of plants and animals can often perform their particular specialties more effectively than the all-purpose cell of single-celled organisms can. On the other hand, because individual plant or animal cells depend on the other cells in the body to carry out the functions that they cannot, they could not survive as individuals.

The millions of cells that make up the body of a typical plant or animal are highly organized (see Figure 24.1b and c). Think of the human body: We are not random masses of specialized cells. Our bodies contain recognizable **organs** like eyes and lungs, each consisting of collections of cells that work together to perform a shared function.

Perhaps the most surprising feature of the organization of plant and animal bodies is how few basically different cell types form the many different specialized organs that make up a functioning organism. Animals consist of four distinctly different cell types, plants of 10.

Cells of the same basic type are gathered into **tissues**, which consist of many similar cells. The cells in a tissue produce a **matrix** of chemicals, which surrounds them and glues them together. Each organ, in turn, consists of a characteristic arrangement of several tissues that together perform a specific task. Thus our intestines consist of the four animal tissues (see Figure 24.1c) working together to absorb nutrients from food that we have digested, and plant leaves consist of the three plant tissues (see Figure 24.1b) working together to convert the sun's energy into sugars.

Animals have four basic cell types

Although more than 200 different kinds of animal cells have been identified, they fall into just four distinct classes that share the same basic function and appearance (see Figure 24.1c). **Epithelial cells** form the surface of animals and line their body cavities. A matrix called the basal membrane holds epithelial cells tightly togeth-

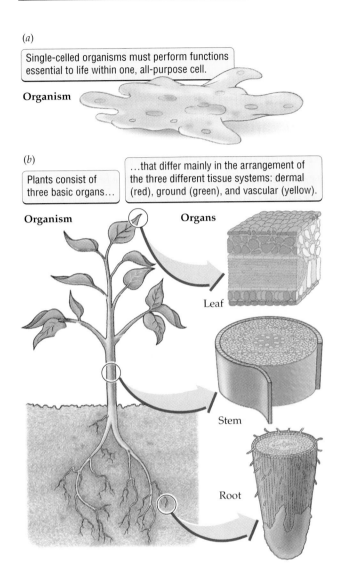

(a)

Single-celled organisms must perform functions essential to life within one, all-purpose cell.

Organism

(b)

Plants consist of three basic organs...

...that differ mainly in the arrangement of the three different tissue systems: dermal (red), ground (green), and vascular (yellow).

Organism **Organs**

Leaf

Stem

Root

(c)

The organs of animals, which are devoted to specialized functions...

Organism **Organ**

...are formed from tissues consisting of similar cells. Animal cells lack rigid cell walls.

System

Tissue

Epithelial

Connective

Muscle

Nerve

Figure 24.1 The Organization of Life
(*a*) Single-celled organisms like bacteria and protists must perform all basic functions of life within a single, general-purpose cell. Multicellular plants (*b*) and animals (*c*) have cells that are specialized for particular functions arranged into organs that work together, each carrying out one or a few functions.

er to form epithelial tissue. The epithelial tissues that form the surface of the skin act as a barrier between the inside of the body and the outside environment. In contrast, the epithelial tissues that line lungs and intestines control how gases and nutrients enter and leave the body.

Connective cells perform many functions, including most prominently the connecting and supporting of other cells in the body. Unlike tightly packed epithelial cells, connective cells typically lie loosely packed in a matrix that occupies more space than the cells themselves. Connective cells form the bones that support the body and provide the elastic connections that allow the skin to bend and stretch. Both epithelial and connective cells replace themselves by dividing throughout the life of the organism.

In contrast, more narrowly specialized muscle and nerve cells stop dividing shortly after birth. **Muscle cells** contract to give animals their unique ability to move. **Nerve cells** are highly specialized to transmit nerve signals from one part of the body to another, allowing an animal to coordinate the function of its various body parts.

Plants have 10 basic cell types arranged into seven tissues

Plant biologists have identified 10 structurally and functionally similar cell types combined into seven tissue arrangements that differ somewhat from those of animals (Table 24.1). Whereas animal tissues consist of a single cell type, plant tissues consist of various combinations of the 10 basic plant cell types. As is true of animal tissues, however, each plant tissue fulfills a fairly well defined function.

The seven plant tissues can be further grouped into three **tissue systems**, which consist of tissues that serve related functions (see Table 24.1). The **dermal tissue system** forms the outer surface of the plant. Like the epithelial tissue of animals, the dermal tissue system of plants protects the plant, as on the aboveground stem and leaves, and absorbs nutrients, as on the belowground roots. The **vascular tissue system** transports water and mineral nutrients from the roots to the leaves, and sugars produced photosynthetically in the leaves to other parts of the plant. The **ground tissue system** plays an important role in supporting the plant, and it houses the sites of photosynthesis and nutrient storage.

24.1	**The Organization of Plant Cell Types into Tissues and Tissue Systems**			
Cell types		**Tissue**	**Tissue system**	**Function**
Parenchyma, fiber, sclereid		Epidermis	Dermal	Protection, absorption
Parenchyma, fiber, sclereid		Periderm	Dermal	Protection
Tracheid, vessel, paren- chyma, fiber, sclereid		Xylem	Vascular	Water and nutrient transport, support
Sieve, sieve tube, albuminous, companion, parenchyma, fiber, sclereid		Phloem	Vascular	Transport of photosyn- thetic products
Parenchyma		Parenchyma	Ground	Photosynthesis, storage, respiration
Collenchyma		Collenchyma	Ground	Support in young tissues
Fiber, sclereid		Sclerenchyma	Ground	Support in mature tissues

The concept of tissue systems helps us understand plant structure because each of the three basic plant organs—the roots, the stems, and the leaves—represent three different arrangements of the three tissue systems (see Figure 24.1*b*).

> ■ The cell is the basic functional unit of all organisms. In multicellular organisms the specialized cells show clear levels of organization: Cells of similar types are collected into tissues containing many cells, and the various tissues, in turn, are arranged into organs that perform particular functions.

The Functions Essential to Life

Although organisms make their living in an overwhelming diversity of ways, we can identify certain functions that are essential to the success of all organisms (Figure 24.2). In this unit we sometimes consider a function as separate from the others, but keep in mind that the various functions depend heavily on one another. Here we merely introduce the various functions; the remaining chapters in this unit cover each function in much greater detail.

The transfer of chemicals into and out of organisms makes life possible

Much of what any organism does to accumulate the resources needed to stay alive and reproduce depends on its ability to transfer the chemicals it needs into its body and the wastes it produces out of its body. Nutrients are the chemicals that are taken up in the form of molecules dissolved in water (see Chapter 28), including the sugars and amino acids we get from our food and the mineral nutrients that plants extract from soil water.

Other essential chemicals, such as oxygen and carbon dioxide, move in and out of the body as gases and therefore require different exchange structures (see Chapter 29). The constant transfer of materials into and out of organisms works well only if organisms can maintain a relatively constant and appropriate environment inside their cells (see Chapter 31).

Reproduction measures an organism's success

Individuals that accumulate enough resources can eventually reproduce—that is, make more genetic copies of themselves (see Chapter 36). Reproduction measures how successfully individuals accumulate resources: Individuals that accumulate resources leave behind more offspring more quickly than do those that accumulate resources slowly. Thinking back to the discussion of natural selection (see Chapter 20), recall that individuals that produce many offspring quickly generally contribute the most genes to the next generation. Thus reproduction connects individual success with evolutionary processes.

Many common functions facilitate the transfer of chemicals

Several functions make it much easier to accumulate resources. Most, but not all, organisms carry out these functions. Most organisms can support their bodies against the pull of gravity, allowing them to maintain shapes that determine how they interact with their environment (see Chapter 25). Many organisms can move

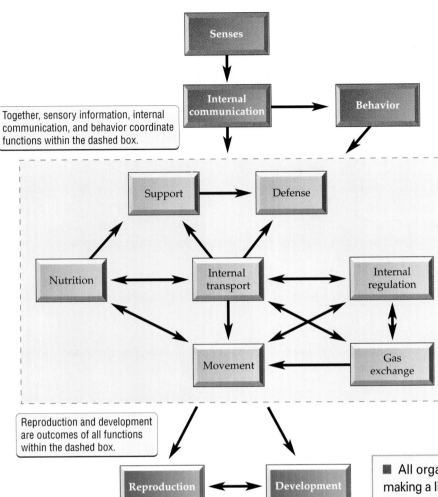

Together, sensory information, internal communication, and behavior coordinate functions within the dashed box.

Reproduction and development are outcomes of all functions within the dashed box.

Figure 24.2 The Functions Basic to Life
All organisms must carry out several inter-acting functions to live. The functions listed here correspond to the topics discussed in the remaining chapters of this unit.

Having many cells also creates two important problems not faced by sin-gle-celled organisms. Once food or gas, for example, enters the body of a multicellular organism, it has not nec-essarily reached the cells that need it. Thus most multicellular organisms have a way of rapidly carrying mate-rial from the place where it enters the body to the cells that use it (see Chapter 30). In addition, multicellular organisms must coordinate the func-tions of individual cells that carry out the chemistry basic to life. This co-ordination requires internal commu-nication systems that allow the ex-change of information among cells throughout the body (see Chapter 33).

■ All organisms, regardless of their forms or ways of making a living, must carry out a series of essential func-tions. Centering on the transfer of chemicals into and out of the organism, these functions interact closely with one another.

from one place to another, an ability that gives them some control over where they feed (see Chapter 27).

An essential part of survival is a defense against being eaten or becoming diseased (see Chapter 32). Finally, to coordinate the various functions so that they take place in the right way and under the right condi-tions, organisms must have a way of sensing what is happening in their environment (see Chapter 34) and of responding to this information (see Chapter 35).

Multicellular organisms face some challenges not faced by single-celled organisms

Large, multicellular plants, animals, and fungi face some special problems that are not faced by single-celled bac-teria and protists. After reproduction, multicellular organisms must undergo development, in which single-celled, fertilized eggs transform into adults consisting of many different kinds of cells organized into tissues and organs (see Chapter 37).

What Makes Chemicals Move into and out of Organisms?

The most basic functions of life depend on the highly selective transfer of chemicals into and out of organisms (Figure 24.3). Thus all organisms must have a means of moving chemicals in and out, as well as a way of con-trolling which chemicals can enter or leave.

To understand how materials move into and out of organisms, keep in mind that (1) chemicals move down **concentration gradients**, or from areas of high concen-tration to areas of low concentration, **passively** (with no energy input), and (2) chemicals move up concentration gradients, or from areas of low concentration to areas of high concentration, **actively** (with energy input) (Figure 24.4). We can use the example of a ball moving down or up a hill, in which the ball represents a chemical and the hill represents a concentration gradient: The ball rolls

Figure 24.3 Life Depends on Exchange between Organisms and Their Environment
Marathon runners lose water, salt, and heat to the environment by sweating and gain water by drinking. Every breath brings in oxygen and takes away carbon dioxide.

downhill on its own, but the ball cannot roll uphill unless the energy needed to push it is provided.

Try placing some orange juice concentrate in a pitcher of water without stirring. Over time, you can see the chemicals that give orange juice its color spreading outward from the concentrate without any work on your part. The color will continue to **diffuse**, or spread passively, until evenly distributed throughout the pitcher. Once the color is distributed uniformly, the juice reaches an equilibrium in which the concentration differences that make diffusion possible have disappeared.

In contrast, orange juice concentrate will not form spontaneously in a pitcher of orange juice, no matter how long you watch, because the formation of concentrate requires that the chemicals in orange juice move up a concentration gradient from relatively low concentrations in the orange juice to high concentrations in the concentrate. Orange juice concentrate forms only with the input of energy to remove water.

Both passive and active movement are biologically important

Organisms rely heavily on both passive and active transport to take up and remove nutrients, gases, and wastes. As we will see in the next section, organisms depend on active transport mech-

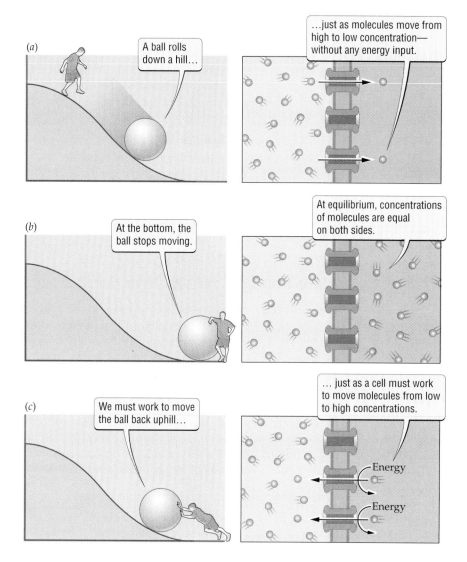

(a) A ball rolls down a hill...
...just as molecules move from high to low concentration—without any energy input.

(b) At the bottom, the ball stops moving.
At equilibrium, concentrations of molecules are equal on both sides.

(c) We must work to move the ball back uphill...
... just as a cell must work to move molecules from low to high concentrations.
Energy
Energy

Figure 24.4 Active versus Passive Movement of Chemicals
Materials move into and out of organisms either passively (without an input of energy) or actively (with an input of energy). Materials can move passively from areas of high concentration to areas of low concentration (*a*). When the concentration of molecules is equal on both sides of a membrane, there is no tendency of molecules to move (*b*). However, it takes energy to move materials from areas of low concentration to areas of high concentration (*c*).

anisms to move chemicals up concentration gradients. The same simple, passive process that allows orange juice concentrate to diffuse through a pitcher of water, however, also plays an essential role in the transfer of some of the most important chemicals into and out of organisms. Water, oxygen, and carbon dioxide, for example, move into or out of organisms only by diffusion.

> ■ All life relies on both passive transport, which requires no energy but can move chemicals only down concentration gradients, and active transport, which can move materials up concentration gradients at an energetic cost, to transfer materials into and out of the body.

Biological Membranes Control What Enters and Leaves an Organism

Passive and active transfer are the mechanisms that move materials into and out of organisms, but biological membranes select which chemicals can enter or leave

organisms. In this section, we briefly review the structure of plasma membranes found in cells of all living things (introduced in Chapter 6) and then consider how the plasma membrane assists in the selective transfer of chemicals.

The plasma membrane consists of a phospholipid bilayer studded with proteins

The simple structure of the plasma membrane is as universal a feature of life on Earth as is the DNA-based genetic code. A double layer of lipids, or **phospholipid bilayer**, provides the framework for the membrane (Figure 24.5). Although some essential chemicals cross the plasma membrane through the phospholipid bilayer, most biologically important chemicals cannot. Embedded in the phospholipid bilayer we find many proteins, which together typically make up more than half the weight of the membrane. Many of these proteins span the plasma membrane and provide a path by which various essential chemicals can cross into or out of cells.

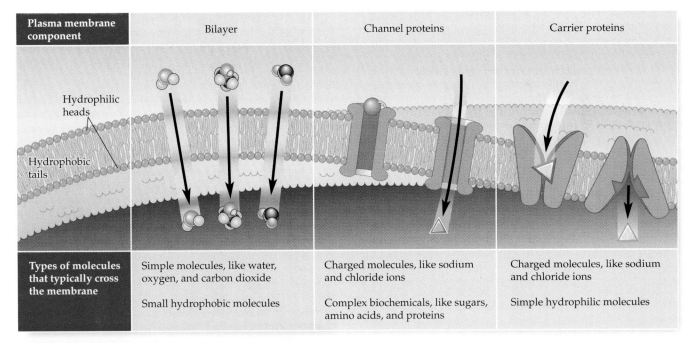

Plasma membrane component	Bilayer	Channel proteins	Carrier proteins
Hydrophilic heads / Hydrophobic tails			
Types of molecules that typically cross the membrane	Simple molecules, like water, oxygen, and carbon dioxide Small hydrophobic molecules	Charged molecules, like sodium and chloride ions Complex biochemicals, like sugars, amino acids, and proteins	Charged molecules, like sodium and chloride ions Simple hydrophilic molecules

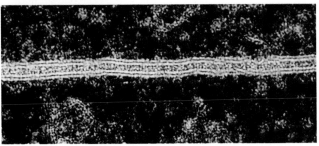

Figure 24.5 The Organization of the Plasma Membrane
The plasma membrane determines which materials leave and enter a cell. The lipid portion of the plasma membrane consists of a double layer of phospholipid molecules with water-soluble "heads" and fat-soluble "tails." This phospholipid bilayer is clearly visible in the electron micrograph at left. Proteins that span the plasma membrane play an important role in the movement of materials into and out of cells. Different kinds of molecules cross the membrane by different routes.

Small molecules can cross the phospholipid bilayer passively by diffusion

Materials that cross the phospholipid bilayer of the plasma membrane do so strictly by passively diffusing down concentration gradients. Some of the molecules of most importance biologically, such as water, oxygen, and carbon dioxide, often enter or leave organisms in this way. All these molecules consist of just a few atoms each. In addition, some simple **hydrophobic** (*hydro*, "water"; *phobic*, "fearing") molecules can pass through the hydrophobic core of the phospholipid bilayer, including many of the early pesticides, like DDT, that worked effectively precisely because they could enter the cells of the insect pests they were meant to control.

At the same time, the phospholipid bilayer acts as a barrier to the movement of most molecules of biological importance. Various **ions**, which are small atoms or molecules carrying a positive or negative charge, are too **hydrophilic** (*philic*, "loving") to penetrate the hydrophobic core of the phospholipid bilayer. Even the simplest sugars and amino acids consist of more than 20 atoms each and are both too large and too hydrophilic to diffuse through the phospholipid bilayer. As we will see, the plasma membrane proteins provide the only path by which these molecules can cross plasma membranes.

Channel and passive carrier proteins allow molecules to cross the plasma membrane passively

Two types of plasma membrane proteins allow large and hydrophilic molecules to cross the plasma membrane passively. **Channel proteins** form openings through the phospholipid bilayer that allow hydrophilic molecules of the right size and charge to cross the plasma membrane down a concentration gradient (see Figure 24.5). **Passive carrier proteins** change shape in such a way when they bind to a molecule that fits into the protein that they transfer the molecule from one side of the plasma membrane to the other (see Figure 24.5). Passive carrier proteins release the molecule being transferred on the other side of the membrane only if its concentration there is relatively low.

Only active carrier proteins can carry materials up a concentration gradient

Molecules can cross a plasma membrane up a concentration gradient only by active transport. **Active carrier proteins** change shape in the way needed to transfer a molecule that fits the protein across a plasma membrane only when they also bind to an energy storage molecule like ATP (Figure 24.6).

Like passive carrier proteins, active carrier proteins bind only to molecules having a shape that fits into the folds of the protein. Unlike passive carrier proteins, the energy-induced shape change of active carrier proteins forcibly releases the molecule being transferred, regardless of the concentration of that molecule near the site of release.

The cost to an organism of active transport can be substantial. Plants, for example, spend one energy-rich ATP molecule for active transport of one hydrogen ion. Of the energy that the human body uses at rest, for example, 30 to 40 percent fuels active transport across plasma membranes.

Active and passive transport often work together

Active carrier proteins often work together with passive carrier proteins to move molecules up concentration gradients. In these situations, the active carrier typically moves one kind of positively charged ions—sodium ions in animals and hydrogen ions in all other organisms—against a concentration gradient. As a result, ion concentrations build up outside the cell.

The passive carrier proteins involved in this type of cooperative transfer differ from the simple ones already described in that they simultaneously bind to two different molecules: the positively charged ion and another molecule. Such carrier proteins usually bind first to the positively charged ion, which changes the shape of the carrier protein so that it can bind the second molecule. Binding to the second molecule transfers both molecules to the opposite side of the membrane, where the carrier protein

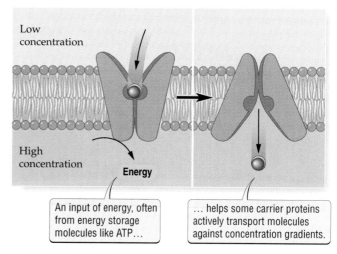

Figure 24.6 Active Carrier Proteins
Active carrier proteins move materials from areas of low concentration to areas of high concentration through the release of energy, often from the conversion of ATP to ADP.

releases the ion readily because its concentration there is low. Upon release of the ion, the carrier protein returns to its original shape, forcibly ejecting the second molecule.

Although the ion always moves down a concentration gradient, the other molecule can move against a gradient to an area of higher concentration. This indirect active transport allows a single power source, a concentration gradient in one kind of positively charged ion, to move many different molecules into or out of cells against concentration gradients.

> ■ The plasma membrane surrounding cells acts as a selective filter that controls which chemicals can enter and leave organisms. Although some essential chemicals diffuse directly through the phospholipid bilayer of the plasma membrane, most enter or leave cells through various proteins embedded in the plasma membrane.

Highlight

Transport Mechanisms and the Cure for Cholera

Cholera is a disease that has ravaged all parts of the world in a series of seven global epidemics, the most recent of which afflicted hundreds of thousands of victims in Southeast Asia and South America in the early 1990s. Cholera victims suffer from severe diarrhea (producing an astonishing 10 to 20 liters per hour in severe cases), which can quickly kill its victims by draining them of water and nutrients. *Vibrio cholerae*, the bacterium that causes cholera, wreaks this havoc not by physically damaging tissues, but by producing a toxic protein that disrupts the function of a single type of carrier protein in the cells lining the small intestine.

Although cholera toxin acts on just one kind of carrier protein, it inter-

feres with two kinds of epithelial cells that are critical to the normal functioning of the small intestine (Figure 24.7). The epithelial tissue lining the small intestine contains specialized epithelial cells, called secretory cells, that extract sodium and chloride ions from the body and

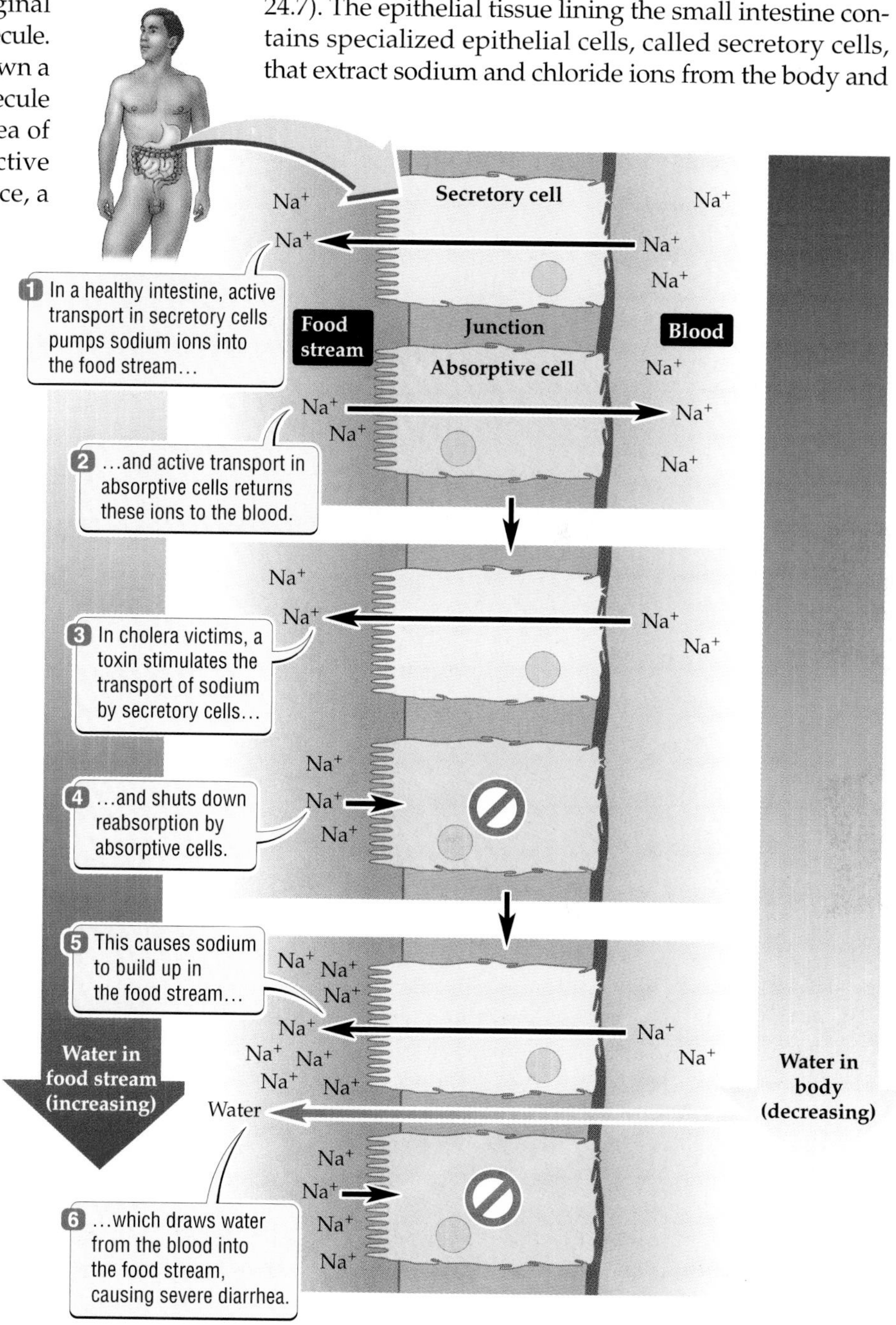

1 In a healthy intestine, active transport in secretory cells pumps sodium ions into the food stream...

2 ...and active transport in absorptive cells returns these ions to the blood.

3 In cholera victims, a toxin stimulates the transport of sodium by secretory cells...

4 ...and shuts down reabsorption by absorptive cells.

5 This causes sodium to build up in the food stream...

6 ...which draws water from the blood into the food stream, causing severe diarrhea.

Figure 24.7 The Causes of Cholera
The bacteria that cause the epidemic disease cholera kill their victims by specifically affecting one passive carrier protein in two different types of epithelial cells (secretory and absorptive cells) that line the small intestine. The toxin turns on carrier proteins that transport sodium (Na^+) ions into the cells from the body. At the same time, the toxin turns off the carrier proteins that carry Na^+ from the gut into the epithelial cells. The combined effect of these two actions leads to a buildup of Na^+ in the gut that draws water passively out of the body and into the food stream. The severe diarrhea that ultimately results can quickly drain the victim of water and nutrients.

(Figure 24.8). Whereas more than 50 percent of cholera victims in early epidemics perished, the death rate of treated victims today has fallen to below 1 percent. Currently, cholera victims are treated effectively by being given a salt solution laced with a sugar. The sodium and the chloride help replenish the salt pumped into the intestine by the poisoned epithelial cells. The sugar stimulates the activity of a second carrier protein in the absorptive cells of the small intestine that simultaneously carries sugar and sodium from the food stream into the body.

> ■ The deadly disease cholera results when a toxin produced by a bacterium affects transport across the plasma membrane in the small intestine, leading to severe diarrhea.

Figure 24.8 Curing Cholera
Cholera victims receive a salt and sugar solution at a refugee camp in the African country of Sudan.

release them into the digested food passing through the intestine. Intestinal epithelial cells of a second type, called absorptive cells, absorb sodium and chloride ions from the food stream and return them to the body. In a healthy small intestine, absorptive epithelial cells remove sodium and chloride from the stream of digested food as quickly as the secretory epithelial cells release them.

Cholera toxin affects a carrier protein that transports sodium and chloride across the plasma membranes of both secretory and absorptive cells in opposite ways—with disastrous results. Cholera toxin stimulates the carrier protein in secretory cells, causing them to release more sodium and chloride into the digested food in the intestine, and it shuts down the same carrier protein in absorptive cells, preventing them from absorbing the excess chloride and sodium ions released by the secretory cells. As a result the digested food becomes very salty (sodium and chloride together make table salt). Water diffuses passively out of our body into the gut as a way of equalizing the salt concentration on both sides of the epithelium of the intestine. This movement of water into the food stream causes diarrhea and can rapidly lead to fatal dehydration of the victim.

Once scientists understood that cholera kills by disrupting sodium and chloride transport in the gut, they could devise relatively simple, but effective, treatments

Summary

Life's Organization

■ The cell is the basic unit of life.

■ Large, multicellular organisms consist of cells organized into a hierarchy: Cells are grouped into tissues, and tissues are grouped into organs that serve specific functions.

■ All animal organs consist of various arrangements of four basic cell types; all plant organs consist of various arrangements of 10 basic cell types.

The Functions Essential to Life

■ All organisms, regardless of their forms or ways of making a living, carry out a series of closely interacting essential functions.

■ Functions related to the transfer of chemicals into and out of organisms make life possible.

■ Reproduction links the success of individuals to evolutionary processes.

■ Multicellular organisms like plants and animals carry out various functions related to the development of complexity, and to the transport of chemicals and communication among many cells.

What Makes Chemicals Move into and out of Organisms?

■ Passive movement from areas of high concentration to areas of low concentration provides an energy-free way of moving chemicals.

- Active transport makes it possible to move chemicals from areas of low concentration to areas of high concentration, but it requires a lot of energy.

Biological Membranes Control What Enters and Leaves an Organism

- The plasma membrane consists of a lipid bilayer studded with proteins.

- Small, uncharged molecules like oxygen and carbon dioxide cross the phospholipid bilayer passively by diffusion.

- Channel and passive carrier proteins selectively allow charged ions and larger molecules like sugars and amino acids to cross the plasma membrane passively.

- Only active carrier proteins can transport materials up a concentration gradient, but they consume a great deal of energy in doing so.

- Active transport of positively charged ions—sodium in animals and hydrogen in all other organisms—often acts together with passive transport to move biologically important molecules indirectly against a concentration gradient.

Highlight: Transport Mechanisms and the Cure for Cholera

- A toxin produced by the bacterium that causes the devastating epidemic disease cholera upsets a single carrier protein in the epithelial cells that line the human small intestine to cause often fatal diarrhea.

- Replacing the nutrients and the water lost in diarrhea, and stimulating another carrier protein that removes sodium ions from the watery mess in the small intestine, easily and quickly reduce the danger posed by a cholera infection.

Key Terms

active carrier protein p. 378	ion p. 378
active movement p. 375	matrix p. 372
channel protein p. 378	muscle cell p. 373
concentration gradient p. 375	nerve cell p. 373
connective cell p. 373	organ p. 372
dermal tissue system p. 374	passive carrier protein p. 378
diffusion p. 376	passive movement p. 375
epithelial cell p. 372	phospholipid bilayer p. 377
ground tissue system p. 374	tissue p. 372
hydrophilic p. 378	tissue system p. 374
hydrophobic p. 378	vascular tissue system p. 374

Chapter Review

Self-Quiz

1. The single cell of a bacterium
   a. typically carries out many more functions than each cell in the human body does.
   b. can survive on its own, whereas the individual cells of a human being cannot.
   c. lacks a plasma membrane.
   d. both a and b

2. Which of the following gives reproduction a unique place among the basic functions of life?
   a. It is the only function carried out by all organisms.
   b. It is the only function without which an organism could not survive.
   c. It connects what happens to individuals and evolutionary processes.
   d. It doesn't depend on any of the other basic functions.

3. Passive movement of a chemical involves which of the following?
   a. movement down a concentration gradient
   b. the expenditure of a lot of energy
   c. movement only of chemicals unimportant to living organisms
   d. movement from areas where the chemical is at low concentration to areas where the chemical is at high concentration

4. The plasma membrane of the cell
   a. plays an important role in determining which chemicals an organism can absorb.
   b. is found only in the cells of multicellular plants and animals.
   c. consists of almost pure lipid arranged in a characteristic bilayer.
   d. all of the above

5. Active carrier proteins do which of the following?
   a. carry molecules up concentration gradients
   b. work together with passive carrier proteins to transport many molecules across plasma membranes
   c. require an interaction with energy-rich molecules like ATP to transport molecules across plasma membranes
   d. all of the above

Review Questions

1. When we wash with soap, we kill bacteria on our hands by breaking open their plasma membranes. Why does destruction of the plasma membrane kill a bacterium?

2. One of the main points of this chapter is that despite their outward differences, all living things have many features in common. In light of what you learned about evolution in Unit 4, do the shared features of life come as a surprise? Why or why not?

3. The various functions that make life possible all interact closely with one another. Although we will discuss some of these interactions in the chapters to follow, can you think of any examples from your own experience of interactions between the functions described in this chapter?

4. Consider how it is possible to create an organism as complex as a tree from a very small number of different kinds of plant cells.

5. Compare the challenges faced by the single cell of a bacterium to those faced by the individual cells in the human body.

24   𝕿𝖍𝖊 𝕯𝖆𝖎𝖑𝖞 𝕲𝖑𝖔𝖇𝖊

Improper Formula Preparation Puts Babies at Risk

PORTLAND, OR. The Portland Regional Health Authority unveiled a campaign yesterday that will warn new parents of the risks associated with improper preparation of baby formula.

Spokesperson Vera Dulay noted that a recent spate of hospitalizations of infants highlights the importance of following manufacturers instructions in preparing formula. According to Ms. Dulay, five babies have been hospitalized for severe dehydration in the past month alone after having been fed concentrated baby formula at full strength.

"Concentrated baby formula must be diluted with water before being fed to infants," warned Dr. Beverly Ridge of the Breastfeeding Clinic at Stoneyview Hospital.

"When a baby is fed undiluted formula concentrate, the infant actually *loses* water from its body as a result of its attempts to dilute its meal."

"We were shocked," said Mrs. Beverly Frampton, mother of an infant who became extremely de-hydrated due to drinking concentrated formula. "We never thought that giving our baby a liquid formula would cause her to become dehydrated and so sick."

Most cases of dehydration occur when parents unwittingly switch from ready-to-serve formula to the concentrate, which often comes in very similar packaging. Ms. Dulay emphasized the importance of checking the label carefully to make sure of the contents before feeding it to an infant with potentially tragic results.

Evaluating "The News"

1. The parent interviewed in the column expressed shock that an improperly mixed baby formula could lead to dehydration. How can babies lose water while drinking a liquid baby formula that consists mostly of water?

2. We use many concentrated or powdered drink mixes not only for infants, but also for adults for weight loss and for rehydration during exercise. Explain why it is important that we make these drinks the correct strength.

3. Rats and mice are often used as test animals in nutritional studies. Both of these animals can survive on diets containing much less water than is necessary in the diet of human babies. Would tests with rats or mice reveal the hazards of using a too-concentrated infant formula? Explain why or why not.

25

Size and Complexity

Charles Ray, *Artist With Fall '91*, 1992.

Movie Monsters

O n a remote, fog-enshrouded island in the Pacific Ocean lives a Stone Age tribe that worships a giant ape. They keep the ape from harming them by occasionally sacrificing a virgin to him. An adventurous film crew from New York City intrudes into this peaceful setting and decides that a giant ape is just the thing needed to make them rich. With considerable effort, they capture the ape and return to civilization.

This is the idea behind the famous 1933 movie *King Kong*. Kong (the ape) doesn't particularly like the idea of being put on public display, so he escapes, and the cast and the audience quickly discover that an angry, 7.2-meter-tall ape can have a wild night on the town even by New York City standards: Subways derail, people die, and the city ordinance against climbing up skyscrapers is broken.

Kong belongs to a long tradition of mythical creatures based on real animals that become monstrous by virtue of huge size and foul tempers. Large size makes the monsters terrifying because with large size comes great strength. Thus, as a giant gorilla, Kong knocks out dinosaurs, lays waste to an entire village, and trashes the New York City public transportation system. Moreover, weapons like spears and guns that work at a human scale do little more than annoy something as big as Kong.

This fearful and shallow view of the effect of increasing size on monsters like Kong ignores the complex way in which size affects the biology of organisms. As we will see in this chapter, large organisms are not simply magnified versions of small organisms. By understanding how biology changes with size, we can get an idea of how Kong might function if he existed.

**King Kong atop
the Empire State Building**

Key Concepts

1. Living things span a tremendous range of size and complexity.

2. As the dimensions of an organism increase, volume increases faster than surface area.

3. Large organisms have less surface area relative to mass than small organisms have, and this relationship affects how they exchange materials with their environment.

4. Large size stems mainly from an increase in the number of cells, rather than from an increase in the size of cells.

5. Complexity probably evolved by the specialization of cells or organs present in multiple copies.

6. Specialization of cells increases the efficiency with which they can carry out their functions, but it also increases their dependence on other cells in the organism.

Adult humans typically weigh between 50 and 100 kilograms, stand between 1.5 and 2 meters tall, and as we saw in Chapter 24, they consist of a vast number of cells arranged into tissues and organs that are specialized to carry out well-defined functions. Amid the incredible diversity of life, humans stand out as large, complex organisms. As we will see, simply understanding the implications of our size and complexity can tell us a great deal about what we are and how we function.

In this chapter we consider some of the implications of size and complexity. After surveying the range in size and complexity of life, we consider how size affects biology and how scientists use the relationship between size and function to understand better the biology of various organisms. Finally, we consider the relationship between size and complexity in an attempt to understand the evolution of large, complex organisms from small, simple ones.

An Overview of Size and Complexity

The variation in the size and complexity of living things is astonishing (Table 25.1). For example, it would take an incredible 100,000,000,000,000,000 individuals of *Mycoplasma*, one of the smallest bacteria known, to weigh as much as a single 100-kilogram person. In turn, it would take 10,000 100-kilogram people to weigh as much as a giant sequoia tree, one of the largest land plants. Whereas we need a pow-

Fungus
A. bulbosa

25.1 *Some of the Smallest and Largest Living Things*		
Organism	**Type of organism**	**Weight (g)**
Mycoplasma	Bacterium	0.000000000001
Giant squid	Mollusk	450,000
King Kong	Movie character	13,950,000
Baluchitherium	Extinct land mammal	14,250,000
Blue whale	Aquatic mammal	100,000,000
Armillaria	Fungus	110,000,000
Giant sequoia	Plant	1,000,000,000

erful microscope to see *Mycoplasma*, some individuals of the fungus *Armillaria bulbosa* can spread over an area larger than 10 city blocks.

Life encompasses an equally impressive range of complexity. Whereas bacteria consist of a single cell with a relatively simple internal structure, complex animals and plants consist of many millions of cells, each containing well-defined organelles and organized into highly structured tissues and organs.

Although size and complexity often go hand in hand, they do not mean the same thing (Figure 25.1). Large organisms consist of many cells, and complex organisms consist of many different kinds of cells organized into tissues and organs. As we will see later in this chapter, multicellular organisms have had more opportunities for the evolution of complexity than have single-celled organisms. Nonetheless, large organisms may be simple and small organisms may be complex.

For example, impressively large marine sponges may consist of several cell types, but these cells are not arranged into tissues or organs, and sponges function essentially as a collection of independent cells. In contrast, the microscopic aquatic animals called rotifers con-

**Figure 25.1 Living Things Vary Tremendously
in Size and Complexity**

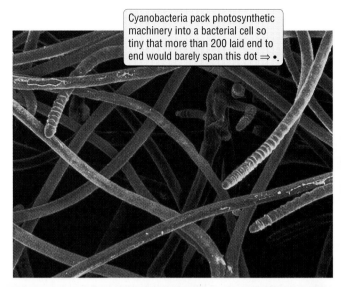

Cyanobacteria pack photosynthetic machinery into a bacterial cell so tiny that more than 200 laid end to end would barely span this dot ⇒ •.

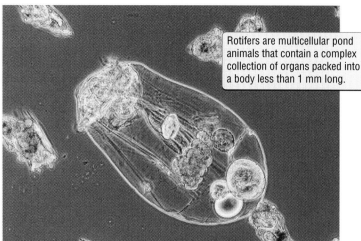

Rotifers are multicellular pond animals that contain a complex collection of organs packed into a body less than 1 mm long.

Large sponges may contain just a few cell types and lack tissues.

Individual aspen trees can cover a large area. The yellow patch may consist of a single plant that includes hundreds of trunks but a single root system.

sist of many different kinds of cells arranged into distinct tissues and organs.

As we look more deeply into the importance of size and complexity, keep in mind that large, complex organisms are not necessarily better or evolutionarily more successful than small or simple organisms. Consider that *Yersinia pestis*, the simple, single-celled bacterium that causes black plague, managed to kill up to half the population of large, complex humans in medieval Europe.

■ Life on Earth spans a tremendous range of size and complexity. However, neither size nor complexity reflects how successful an organism is.

Surface Area and Volume: Why Size Matters

We begin our consideration of the biological importance of size by looking at what happens when the size of an imaginary, cubical creature changes. This example will lead us to the biologically important conclusion about how organisms function: Large organisms should have more trouble than small organisms exchanging materials with their environment.

Changes in size affect various body measurements differently

To understand in a very simple, but important, way how size affects the biology of organisms, let's consider two versions of an imaginary, cubical creature, which

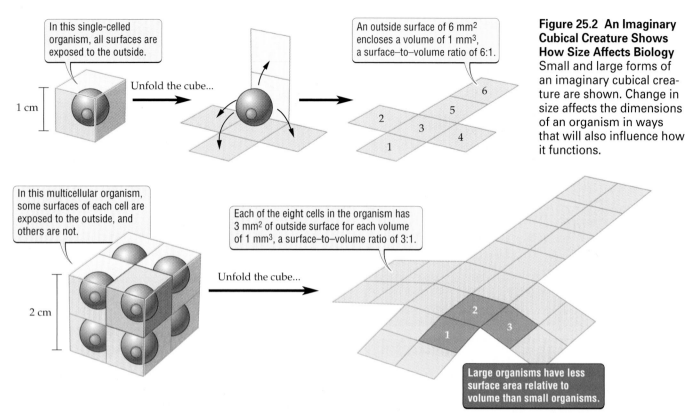

In this single-celled organism, all surfaces are exposed to the outside.

Unfold the cube...

An outside surface of 6 mm² encloses a volume of 1 mm³, a surface–to–volume ratio of 6:1.

In this multicellular organism, some surfaces of each cell are exposed to the outside, and others are not.

Each of the eight cells in the organism has 3 mm² of outside surface for each volume of 1 mm³, a surface–to–volume ratio of 3:1.

Unfold the cube...

Large organisms have less surface area relative to volume than small organisms.

Figure 25.2 An Imaginary Cubical Creature Shows How Size Affects Biology Small and large forms of an imaginary cubical creature are shown. Change in size affects the dimensions of an organism in ways that will also influence how it functions.

has a conveniently simple geometry (Figure 25.2). We can easily express the size of the cubical creature by its length, by its volume, or by its surface area.

Imagine a small cubical creature that has a length of 1 centimeter. Its length determines, for example, the size of the hole into which it fits or how high it can reach for food. The small cubical creature has a volume of 1 cubic centimeter. Volume tells us how much tissue there is, and because this tissue consists of metabolizing cells, it also tells us about how much food, water, and oxygen or carbon dioxide the creature needs. Volume relates closely to the weight of the organism, to the point that weight and volume are often used interchangeably.

Our small cubical creature has a surface area of 6 square centimeters (six sides of 1 square centimeter each). This area represents the surface across which the cubical creature could absorb the nutrients and gases that it needs, and across which it could dispose of its wastes.

Now imagine that we discover a second species of cubical creature that has a length twice that of the smaller species. Doubling the length of each side leads to several important changes (see Figure 25.2). Not surprisingly, as we double the cubical creature's length, we also increase its volume and its surface area. The larger species has a volume, and therefore a weight, that is eight times that of the smaller species (2 cm × 2 cm × 2 cm = 8 cm³). At the same time, the surface area of the

creature increases to four times that of the smaller species. The surface area increases more than the length (length two times, surface area four times), but less than the volume (eight times).

As the length of the cubical creature doubles, its volume and surface area more than double. As a result, the relationships among length, volume, and surface area differ in small and large species. Because length, volume, and surface area all affect the biology of an organism, we should expect the small species to function differently from the large species.

The surface area–to–volume ratio decreases as size increases

Of the various relationships between length, volume, and surface area, the one that we will encounter repeatedly in the next several chapters is the **surface area–to–volume ratio**:

$$\frac{\text{Surface area}}{\text{Volume}}$$

The surface area–to–volume ratio tells us about the supply of a chemical across the surface relative to the demand for that chemical by the cells making up the volume.

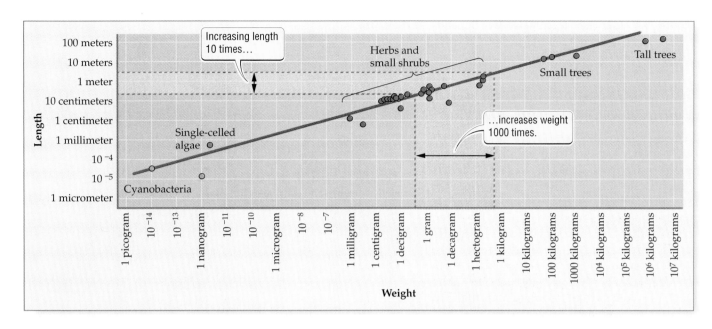

Figure 25.3 The Relationship between Length and Weight for Photosynthetic Organisms

In spite of the dramatic differences in shape among cyanobacteria, single-celled algae, and plants, they all show the same relationship between weight and length. Note especially that the relationship between the weight and length of photosynthetic organisms is the same as for the imaginary cubical creature in Figure 25.2.

In our example, the small cubical creature has twice as big a surface area–to–volume ratio (6 square centimeters of surface area per cubic centimeter of volume) as the large cubical creature (3 square centimeters of surface area per cubic centimeter of volume) (see Figure 25.2). Thus, the small cubical creature has more surface area across which it can supply each cubic centimeter of metabolizing tissue with nutrients, for example, than does the large species.

The large species must somehow compensate for the greater difficulty it faces in supplying its cells. On the other hand, the large cubical creature should have an easier time keeping valuable materials inside its body. For example, land organisms need to prevent scarce water from diffusing out of their bodies. With less surface area across which water can diffuse relative to the volume of cells in which the water is stored, the large cubical creature should lose its body water relatively more slowly than the small species. Even though they look the same, the small and large species must work differently.

Does the cubical creature model work for real organisms?

We selected a cubical creature for our discussion of the effects of size because the relationships among length, volume, and surface area in a cube are so simple. Surprisingly, the relationships hold up remarkably well for the complex shapes of real organisms. For photosynthetic organisms as different as single-celled bacteria, single-celled algae, and terrestrial plants, we find that as

length doubles, the weight, which measures the same thing as volume, increases eight times (Figure 25.3). This is the same geometric relationship that we found for the cubical creature.

> ■ The relationships among length, volume, and surface area in organisms change as size changes. Large organisms have less surface area relative to their volume than do small organisms. As a result, large organisms function differently from smaller organisms that appear similar.

Using Body Size to Understand Biology

In the chapters that follow, we will often refer to relationships between body size and a biological feature like the surface area–to–volume ratio. Such relationships are called **allometric relationships**. In most cases, allometric relationships involve the weight of the organism because weight is more easily measured than volume. In this section we introduce some examples of the different ways in which allometric relationships can help us understand how organisms function.

Figure 25.4 How Much Mammals Sleep Depends on Their Size
Cats sleep so much more than humans not because they are lazy, but because they are smaller than we are.

Allometric relationships can reveal general patterns in how organisms function

Among mammals, sleep shows some interesting patterns. Cats, for example, seem to spend their entire lives sleeping. In fact, whereas humans sleep only 8 hours, an average cat sleeps 14.5 hours each day. A broader look at patterns of sleep among mammals, however, reveals that the time devoted to sleep decreases predictably with body size (Figure 25.4). From this allometric relationship, we can conclude that large mammals spend less time each day asleep than do small mammals. In other words, cat naps reflect the small size of cats, not their laziness.

In contrast, the relationship between brain size and body size in mammals reveals that not all brains are created equal. As body size increases, brain size increases (Figure 25.5).

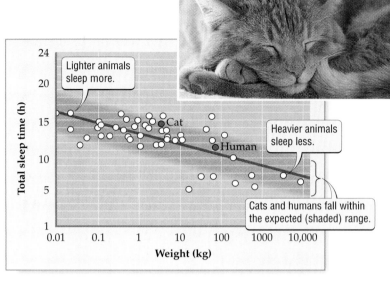

Most of the points representing individual species in Figure 25.5 cluster close to the line, indicating that most mammals have brain sizes close to those predicted for their body size. The points representing humans and dolphins, however, lie well above the line. These data suggest that for their weight, humans and dolphins have exceptionally large brains. This finding has led scientists to wonder which special selective pressures favor the evolution of large brain size in these two mammals living in very different environments.

Allometric relationships have important practical applications

The relationship between body size and function affects our everyday lives. When doctors prescribe drugs, they recommend a dosage that considers both the weight of the patient and how quickly the drug clears out of the patient's system. The drug clearance rate, it turns out, decreases with size. Large animals like humans clear drugs relatively slowly, whereas small animals clear drugs rapidly. Thus a small animal like a cat would receive a higher dose of a particular drug relative to its body weight than we would, because the cat's cells would otherwise clear the drug before it had a chance to take effect.

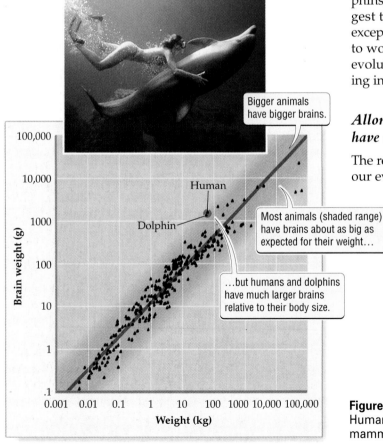

Figure 25.5 Brain Weight Increases as Body Weight Does
Humans and dolphins have exceptionally big brains for mammals of their weight.

The tale of Tusko the elephant illustrates the disastrous consequences of ignoring the allometric relationship between body size and the drug clearance rate. In a stunningly misguided effort to study an elephant behavior in which normally calm males become extremely aggressive, some psychologists decided to mimic the behavior by giving Tusko the elephant the hallucinogenic drug LSD. The researchers estimated the dose of LSD given to Tusko from laboratory experiments involving cats. A dose of 0.15 milligrams per kilogram of cat caused a definite change in behavior.

The psychologists gave Tusko what they thought to be a conservatively estimated dose: 0.1 milligrams per kilogram (or 300 milligrams LSD for a 3000-kilogram elephant). Within 2 hours, Tusko had overdosed and died. Unfortunately, these researchers had neglected to take into account the allometric relationship between size and drug clearance rates. Considering the much larger size of elephants as compared to cats, Tusko should have received no more than 0.03 milligrams per kilogram.

> ■ Allometric relationships that describe how a function changes in species of different weights are useful tools in understanding the biology of organisms.

The Evolution of Large Size

Large organisms consist of many cells. Given that large size could have evolved either by increases in cell size or by increases in the number of cells, it is interesting to consider why the latter seems to be the general approach.

We can start our search for an answer by considering single-celled organisms. Bacteria and single-celled algae both show great variation in cell size (Figure 25.6a and b).

(a)

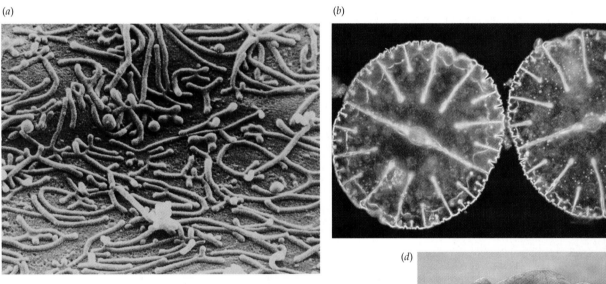

(b)

(c)

(d)

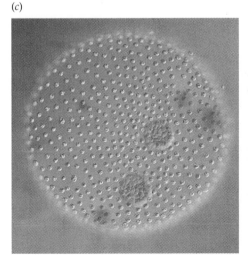

Figure 25.6 Variation in Cell Size and Number
The size of living things depends on both cell size and cell quantity. Single-celled organisms like a small bacterium (a) and a large alga (b) can vary a thousandfold in size. Alternatively, when cells remain together and specialize, organisms can reach larger effective sizes, as in the multicellular green alga *Volvox* (c). The size of mammals such as elephants (d) seems to depend largely on the number of cells that they contain, not on cell size.

The largest bacterial cell has about 100 times the volume of the smallest, and the largest single-celled alga has about 1000 times the volume of the smallest. Single-celled organisms, however, cannot simply evolve an ever-larger size. They must remain small so that their surface area–to–volume ratio is large enough to allow rapid movement of materials into and out of the cell.

When natural selection favored large size in bacteria or single-celled protists, these organisms formed collections of cells (Figure 25.6c) much as multicellular plants and animals have. In mammals, cell size does not vary much with body size, suggesting that elephants are bigger than humans entirely because elephants have many more cells than we do (Figure 25.6d). By having many small cells rather than fewer larger ones, organisms can increase in size while still accommodating the fact that the cell is the basic unit of life.

> ■ Large size evolved mainly by increases in the number, rather than in the size, of the cells making up an organism. Cell size had to remain relatively small so that cells could rapidly import and export the chemicals essential for life.

Complexity and Its Implications

In this section we briefly consider three aspects of the complexity that has arisen in most large, multicellular organisms: its evolution, advantages, and problems.

How did complexity evolve?

As Chapter 24 pointed out, plants and animals consist of relatively few cell types arranged in complex ways to form tissues and organs. A key step in the evolution of complexity seems to be the presence of duplicate structures that can specialize in function without threatening the well-being of the organism. The evolution of multicellularity, perhaps because of selective advantages associated with larger size, may have provided the raw materials for the evolution of complexity (Figure 25.7). Thus, some cells in a multicellular

organism can specialize to perform a particular function without threatening the whole, by relying on other cells to do the things they can no longer do.

On a larger scale, duplicate organs can also specialize for a particular purpose without threatening the well-being of the organism. Repeated organs within an organism can then specialize to perform particular functions. The flowers of flowering plants, for example, evolved as specialized leaves (Figure 25.8a), and the diverse legs of shrimp evolved by specialization of a basic leg design (Figure 25.8b).

The advantage of complexity is the specialization of cells

An almost universal feature of large organisms is that they have cells, tissues, or organs specialized to carry out particular functions. We have already seen, for example, that the epithelial cells lining the small intestine include two specialized epithelial cell types: secretory cells that pump salts into the gut and absorptive cells that absorb salts from the gut (see Chapter 24). Each of these epithelial cells has many carrier proteins suited to either releasing salts or absorbing salts. In contrast, a single-celled creature would have to have a number of both types of carrier proteins in its plasma membrane to carry out both functions. Presumably, a secretory cell in the gut can more rapidly discharge salt than could an unspecialized single cell.

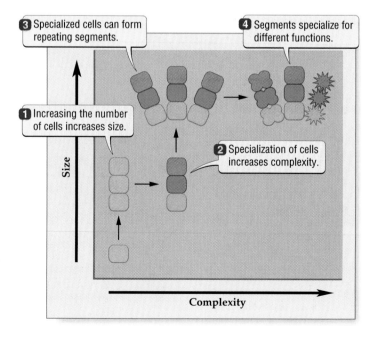

Figure 25.7 The Evolution of Complexity
This schematic diagram shows the relationship between size and complexity and a possible pathway for the evolution of complexity. The vertical size axis shows bigger organisms having an increasing number of cells; the horizontal complexity axis shows more complex organisms having an increase in kinds of cells.

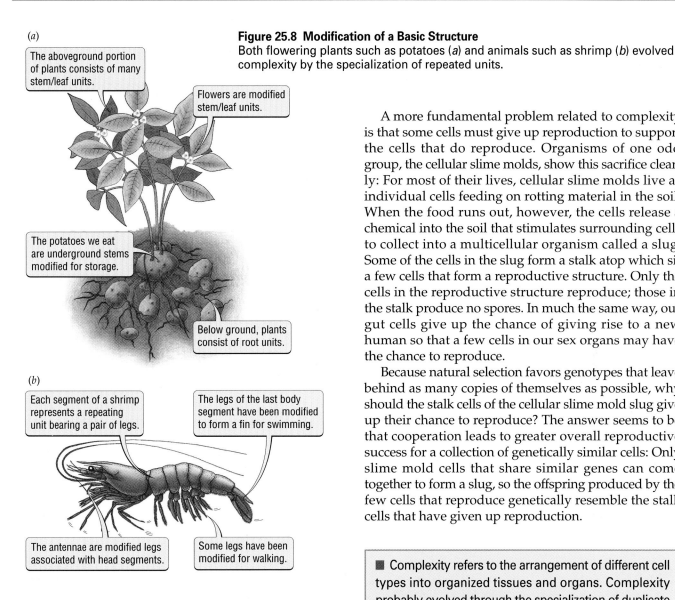

Figure 25.8 Modification of a Basic Structure
Both flowering plants such as potatoes (*a*) and animals such as shrimp (*b*) evolved complexity by the specialization of repeated units.

(a)

The aboveground portion of plants consists of many stem/leaf units.

Flowers are modified stem/leaf units.

The potatoes we eat are underground stems modified for storage.

Below ground, plants consist of root units.

(b)

Each segment of a shrimp represents a repeating unit bearing a pair of legs.

The legs of the last body segment have been modified to form a fin for swimming.

The antennae are modified legs associated with head segments.

Some legs have been modified for walking.

Complexity requires coordination among component cells

Complexity generates two closely related problems: (1) The activities of the component cells must be coordinated, and (2) some of the cells must give up the chance to reproduce. If an organism consists of more than one interdependent part, then the activity of each must be coordinated with that of the other(s). For example, plants must regulate their root growth so that they meet the demands of the aboveground parts for mineral nutrients and water. Yet roots must not grow so much that they use up the supply of sugars and proteins produced in the leaves. Complexity therefore creates a need for systems of communication and coordination among cells. The hormones that serve as chemical messengers in plants and animals and the unique nervous system of animals play this role.

A more fundamental problem related to complexity is that some cells must give up reproduction to support the cells that do reproduce. Organisms of one odd group, the cellular slime molds, show this sacrifice clearly: For most of their lives, cellular slime molds live as individual cells feeding on rotting material in the soil. When the food runs out, however, the cells release a chemical into the soil that stimulates surrounding cells to collect into a multicellular organism called a slug. Some of the cells in the slug form a stalk atop which sit a few cells that form a reproductive structure. Only the cells in the reproductive structure reproduce; those in the stalk produce no spores. In much the same way, our gut cells give up the chance of giving rise to a new human so that a few cells in our sex organs may have the chance to reproduce.

Because natural selection favors genotypes that leave behind as many copies of themselves as possible, why should the stalk cells of the cellular slime mold slug give up their chance to reproduce? The answer seems to be that cooperation leads to greater overall reproductive success for a collection of genetically similar cells: Only slime mold cells that share similar genes can come together to form a slug, so the offspring produced by the few cells that reproduce genetically resemble the stalk cells that have given up reproduction.

> ■ Complexity refers to the arrangement of different cell types into organized tissues and organs. Complexity probably evolved through the specialization of duplicate cells or organs into structures that could carry out particular functions more efficiently. Complexity carries with it the problems of coordination among cells and the sacrifice of some cells so that others can reproduce.

Highlight

The Collapse of Kong

How do allometric relationships like those discussed in this chapter apply to what we can predict would be the biology of King Kong? From Kong's approximate height of 7.2 meters and the relationship between length and mass, we arrive at an estimated mass of 13,950 kilograms. Although big, Kong was actually not astonishingly large:

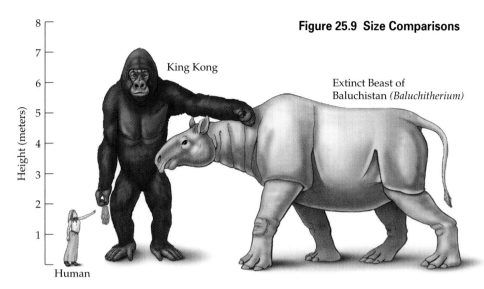

Figure 25.9 Size Comparisons

He weighed about as much as two elephants, and less than the Beast of Baluchistan (at 20,000 kilograms, the biggest mammal ever to roam the land) (Figure 25.9).

Table 25.2 makes some predictions about the biology of Kong based on his size. We should expect Kong to present various challenges in captivity: He would eat a great deal and would produce a lot of waste. Once loose, he would probably sleep little, move quickly, and cover a lot of territory in a single day. In addition, Kong should live a long time, yet produce relatively few offspring. Compared to humans, Kong should have a similarly sized heart and a small brain relative to his overall mass. Table 25.2 provides no indication of why Kong should have the foul temper that he did. We should not expect the numbers in Table 25.2 to be too precise, but they should give us a good idea of the general patterns.

In many ways, Kong represents a biologically possible creature. At least one feature of Kong's natural history, however, suggests that he could not have existed: The island on which he lived was far too small to support a population of giant gorillas. Kong alone should require around 2280 square kilometers of land to satisfy his immense appetite, but the island in the movie covered only about 15 to 20 square kilometers, as estimated from a map shown in the movie. Because Kong would not be immortal and because, as a mammal, Kong would need at least one mate to produce offspring, the existence of Kong implies the existence of several other giant gorillas. Supporting even a minimum population of Kong, Kongette, and two Konglets would require an island much larger than that depicted in the movie.

■ Using allometric relationships we can make predictions about what the biology of a creature like Kong would have been based on its size. Although within the range of possible sizes for a mammal, the island on which Kong lived was too small to support a population of gigantic apes.

Summary

An Overview of Size and Complexity

■ Life spans a tremendous range of size, from microscopic organisms to organisms that weigh thousands of times more than humans.

■ Life spans a tremendous range of complexity, from single-celled bacteria to organisms composed of millions of cells. Within complex multicellular plants and animals, each individual cell contains organelles, and groups of specialized cells are organized into tissues and organs.

■ Neither size nor complexity reflects the success of an organism.

Surface Area and Volume: Why Size Matters

■ As the size of an organism increases, surface area increases more rapidly than length, but less rapidly than volume.

25.2 Estimates about the Biology of King Kong

Trait	Value for trait (relative to weight where appropriate)	
	King Kong	Human
Metabolic rate (kcal/kg/d)	6	29
Home range size (km²)	2210	—
Organ sizes (%)		
Brain	0.1	2
Gut	16	15
Skeleton	28	13
Testes	0.01	0.07
Heart	0.4	0.5
Urine production (l/d)	77	2
Life span (yr)	149	100
Reproductive rate (offspring/yr)	0.3	1.3
Speed (km/h)	54	37
Daily sleep time (h)	3.9	8

- Volume determines how much food, water, and oxygen or carbon dioxide an organism needs. Surface area describes the amount of surface available to the organism for exchange of these materials with its environment.

- Large organisms must find ways to compensate for having less surface area relative to their volume than do small organisms. Therefore, large organisms are not simple magnifications of small organisms that appear similar.

Using Body Size to Understand Biology

- Allometric relationships are relationships between body size and a biological feature.

- Allometric relationships allow us to identify species with unique characteristics that are not simply a result of their size, and to account for the effects of size as we work with a variety of species.

The Evolution of Large Size

- Large size evolved mainly by increases in the number, rather than in the size, of the cells making up an organism. Larger organisms do not necessarily have larger cells.

- Cell size is limited by the relationship between surface area and volume. Cell size has to remain relatively small so that cells can import and export a sufficient number of molecules across the plasma membrane.

Complexity and Its Implications

- Complexity refers to the arrangement of different specialized cell types into tissues and organs.

- Complexity probably evolved through the specialization of duplicate cells or organs into structures that could more efficiently carry out particular functions.

- A cost associated with this specialization is that specialized cells are more interdependent and require communications systems that allow them to coordinate their functions. In addition, many specialized cells are not directly involved in reproduction.

Highlight: The Collapse of Kong

- Using allometric relationships, we can make predictions about the biology of the fictitious ape, King Kong, based on estimates of his size.

- Because the size of King Kong falls within the range of sizes reported for mammals, it is conceivable that an ape like King Kong could exist. The island on which King Kong lived, however, was too small to support such a large creature.

Key Terms

allometric relationship p. 389

surface area–to–volume ratio p. 388

Chapter Review

Self-Quiz

1. Large size among living organisms arises primarily from
   a. more cells.
   b. larger cells.
   c. both fewer and larger cells.
   d. increasingly complex cells.

2. Which of the following statements about the size and complexity of living things is true?
   a. Organisms fall within a narrow range of sizes.
   b. Only animals ever evolved to reach truly large sizes.
   c. Because of their simplicity and small size, single-celled organisms cannot survive in a world filled with large, complex organisms.
   d. Organisms vary greatly in both size and complexity.

3. As size increases,
   a. surface area increases more slowly than volume.
   b. length increases more rapidly than surface area.
   c. length increases more rapidly than volume.
   d. surface area increases more rapidly than volume.

4. The psychologists experimenting with Tusko the elephant could have prevented his death if they had taken into account that
   a. elephants have more cells than cats.
   b. weight affects the rate at which drugs pass through an animal's body.
   c. elephants sleep less than cats.
   d. elephants have more surface area relative to their volume or weight than cats do.

5. Cell specialization leads to
   a. increased efficiency.
   b. the need for coordination and communication between cells.
   c. both a and b.
   d. neither a nor b.

Review Questions

1. On a visit to the pet store, you encounter two different species of parrots: tiny budgies that weigh less than 25 grams and larger lovebirds that weigh five times as much. As a potential bird owner you start thinking about what it would involve to own these birds. How would the biologies of the budgies and the lovebirds differ? Think about how much they would eat, how much of a mess they would make, how much space they would need, and so on.

2. Explain why relationships between surface area and volume limit cell size.

3. Why might cells or organs need to be duplicated in a species before being able to evolve specialized functions?

25　𝕿𝖍𝖊 𝕯𝖆𝖎𝖑𝖞 𝕲𝖑𝖔𝖇𝖊

Stop Breeding Unnatural Dogs

Letter to the Editor:

Dogs have served humans for thousands of years and I feel it is time that people became more considerate of the health and well-being of our most dependable helpers and companions. Through hybridization and selective breeding we have artificially and permanently altered the genetic makeup of dogs to produce bizarre and poorly adapted varieties. Many of these varieties suffer from having surpassed natural limits on the size and shape of ancestral canine species.

A sad consequence of our quest for uniqueness is the creation of giant varieties, such as Great Danes, which struggle through life with misproportioned bodies. Great Danes are much taller than any of their ancestral canine species. These towering giants do not have physiological systems that match their large size, and as a result they often overwork their hearts, thus substantially reducing their life span compared to dogs of more natural sizes. Artificially large dogs of many breeds have shortened life spans and suffer a variety of health problems, such as abnormal structure or growth of the hips and heart disease. These problems suggest that artificial selection for size alone has produced dogs that do not possess physiological systems that match their large size. I feel that breeds of dogs should also stay within a more natural size range for their species, less than 100 pounds.

In the name of humane treatment of animals, governments should step forward and begin to right this wrong by strongly discouraging individuals from breeding giant dogs. I strongly encourage city councils to double the dog license fee for every 5 pounds by which a dog exceeds my suggested 100-pound limit.

G. S. Pointer

Evaluating "The News"

1. Early in the relationship between canines and humans, natural selection for dogs that could successfully interact with humans may have favored less aggressive and smaller dogs. More recently, artificial selection for some breeds has favored much larger dogs as well as extremely tiny dogs. What allometric relationships that influence the health and life span of dogs may have been disrupted through this artificial selection for extreme size and extreme tininess? Should those health problems or decreases in life span be of concern to people?

2. If you were a member of the city council, how would you respond to this proposal? Could you suggest any alternative solutions to the problem that this person has identified?

3. Humans have altered not only dogs, but also cats, cows, sheep, and countless crop plants—often to the detriment of the hardiness, health, or life span of the organism. Are there circumstances under which such alterations should not be allowed or are unethical?

chapter 26

Support and Shape

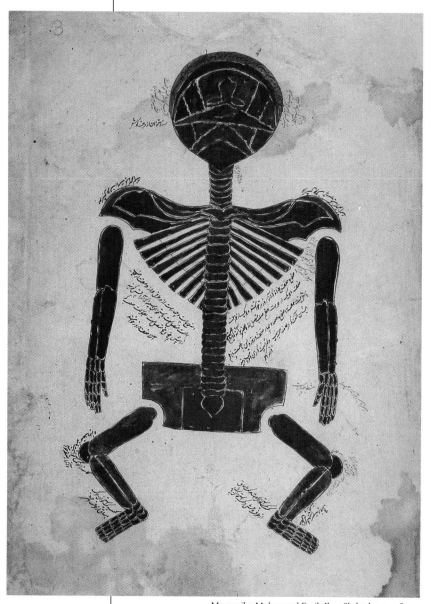

Mansur ibn Muhammad Fagih Ilyas, Skeletal system from *Five Anatomical Figures*, late fourteenth century.

Leaping Cats and Humans

Life in the city can be tough. Humans must deal with polluted air, violent crime, and speeding taxis. As our companions, cats face their own hazards in a landscape of vertical buildings and hard pavement. For example, the legendary curiosity of cats combined with the promise of better hunting grounds lures many cats through open windows or over balcony railings. Unfortunately, these adventures can turn into a quick trip that ends abruptly on the streets and sidewalks several stories down.

For humans such falls usually end in disaster: The death rate increases rapidly as people fall from ever greater heights up to the seventh story, at which point the chances of dying level off at almost 100

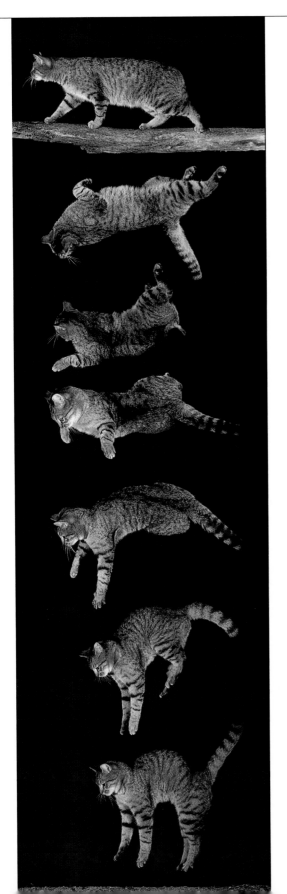

percent. In sharp contrast, information on 115 cats admitted to animal hospitals in New York City reveals two almost unbelievable facts: Although the cats fell an average distance of about five and a half stories, with one falling from the thirty-second story (this high-diving cat survived with minor injuries), only 11 of the cats died from their injuries. Even more incredibly, cats stood a better chance of surviving a fall from *above* the fifth story than from *below* the fifth story.

The secret to the survival of skydiving cats lies in the ability of their skeletons, which are mainly responsible for supporting them, to handle the tremendous forces they experience as they hit the pavement. We will explain why cats survive falls after we review some of the principles that have affected the evolution of the great diversity of support systems.

Does the Ability of Cats to Land on Their Feet Contribute to Their Proverbial Nine Lives?

Key Concepts

1. The human skeleton relies mainly on bone for support, but cartilage and hydrostats play a role as well.

2. Animals rely on specialized skeletal tissues that provide support but also accommodate movement.

3. Single-celled organisms, plants, and fungi rely for support on cell walls associated with each cell.

4. Stiff support structures resist both compression and tension to maintain shape.

5. The biologically produced materials, minerals, and hydrostats that make up support systems each offer unique advantages.

6. The support systems of terrestrial organisms must provide more support than those of aquatic organisms.

7. Joints are combinations of stiff and flexible materials that allow motion in animals.

We could not survive without a skeleton. Our skeleton gives us our shape, and our shape defines how we move, how we eat, how we breathe, how we protect ourselves, how we give birth, and, of course, how we survive falls. The human skeleton is an amazingly complex system of support that holds the body in a human shape. At the same time it allows us to change shape in a controlled way as we walk or run.

We begin this chapter by surveying the variety of support systems found in living things. Against this backdrop of diversity, we then consider the shared selective pressures that have influenced the evolution of all support systems and that have led to some surprisingly general patterns in the form and function of support systems. After this discussion we consider the unique ability of animal skeletons to provide support while at the same time allowing movement. With all this information in hand, we return to the question of why cats survive falls that are fatal to humans.

The Human Skeleton

We will use the human skeleton as a starting point in our overview of animal, and eventually all, support systems because it is most familiar to us. By understanding some basic features of the human skeleton, we have a point from which we can begin to explore some general patterns in the way support systems work.

Bone is the major support tissue

Humans have skeletons that support the body and give it shape (Figure 26.1*a*). The major component of our skeleton is **bone tissue**. Bone tissue consists of connective cells (discussed in Chapter 24) that surround them-

selves with a nonliving, mineral matrix composed largely of calcium and phosphate that we accumulate from our environment. Bone is stiff, meaning that it does not bend easily, suiting it well to providing a framework for the body. Bone supports legs and arms and forms the pelvis and skull.

Cartilage and hydrostats provide additional support

Cartilage tissue, which consists of connective cells that produce a matrix of a protein called **collagen**, supplements the support offered by bone. Cartilage supports the nose and ear lobes, and part of our rib cage, among other things. Compared to bones, the cartilage in the bridge of the nose is less stiff and bends more easily.

The final component of the human skeleton is less obvious, but occurs in the tongue and in the female's clitoris or the male's penis. Human tongues, penises, and clitorises lack bone or cartilage, yet they can become quite stiff when we wish them to. These parts of our bodies are examples of **hydrostats**, structures that stiffen when a fluid under pressure pushes against an elastic collagen matrix, as we will explain in the next section.

■ The human skeleton consists of three different types of support structures derived from specialized connective tissues: bone, cartilage, and hydrostats.

An Overview of Animal Skeletons

Animal skeletons share features that together set them apart from the support systems of other organisms and allow animals to move.

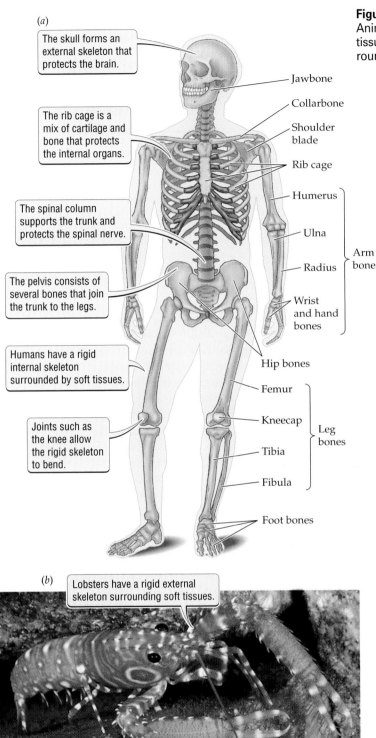

(a)

The skull forms an external skeleton that protects the brain.

The rib cage is a mix of cartilage and bone that protects the internal organs.

The spinal column supports the trunk and protects the spinal nerve.

The pelvis consists of several bones that join the trunk to the legs.

Humans have a rigid internal skeleton surrounded by soft tissues.

Joints such as the knee allow the rigid skeleton to bend.

Jawbone

Collarbone

Shoulder blade

Rib cage

Humerus

Ulna

Radius } Arm bones

Wrist and hand bones

Hip bones

Femur

Kneecap

Tibia } Leg bones

Fibula

Foot bones

(b)

Lobsters have a rigid external skeleton surrounding soft tissues.

Figure 26.1 Animal Support Systems

Animals devote specialized tissues to support their bodies. Skeletal tissues may lie inside the body, as in humans (*a*), or they may surround the body, as in lobsters (*b*).

ized tissues to support the rest of their bodies. The support tissues consist of connective cells that produce an extensive nonliving matrix around themselves. This matrix can be mineral (as with bones) or long chains of amino acids or sugars like the collagen in cartilage. The minerals that form rocklike matrices typically contain calcium, as is true of human bones and clam shells, for example. Among animals, the most widespread non-mineral substances in the matrices of support tissue are collagen and chitin. Collagen is the protein found in cartilage and many other animal support tissues, and **chitin** is a complex polysaccharide that forms the basis of support tissues in arthropods, like insects and lobsters, and many other phyla.

Hydrostats provide the main source of support to soft-bodied animals

Many familiar animals, including terrestrial and marine worms, caterpillars, and octopi, are like the human tongue in that they lack an obvious means of support (Figure 26.2*a*). These organisms depend on hydrostats, which provide support not by being made of a naturally stiff material, but through an interaction between an elastic membrane and a fluid, like water, under pressure.

Balloons illustrate how hydrostats work. When we inflate a balloon, we increase the pressure with which the air inside pushes against the elastic skin of the balloon. As we blow more air into the balloon, the pressure inside increases and the balloon becomes stiffer. Balloons highlight two important features of hydrostats: (1) As pressure increases, stiffness increases, and (2) it takes energy to create the pressure on which the hydrostat depends.

The hydrostats that support earthworms and tongues work basically like balloons. Most animals use water instead of air as the fluid, because life consists largely of water. The skin contains connective cells that typically produce a collagen or chitin matrix. Rather than increasing the pressure on their skin by increasing the amount of water in the hydrostat, animals increase pressure by contracting muscles in the skin to decrease the volume of water that it can hold. You can do this with a partially deflated balloon by pinching off some of the balloon to force the air into a small volume. Generating stiffness in a hydrostat requires muscle contraction, which requires energy.

Animals rely on specialized support tissues

Although animals have evolved a diversity of skeletal forms, all animals, including humans, rely on special-

(a)

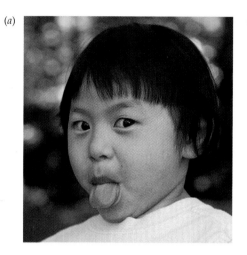

(b)

Figure 26.2 How Hydrostatic Skeletons Work
Hydrostatic skeletons give support without any rigid materials. Both the human tongue (*a*) and young plant shoots with thin cell walls (*b*) rely on hydrostatic pressure for support. When plant tissues dry out, they wilt and lose their shape.

Skeletons can lie inside or around soft tissues

Over evolutionary time, two distinct ways of arranging stiff support tissues in the animal kingdom have arisen. Some animals, including humans and most other vertebrates, have an **internal skeleton**, in which the support tissues lie inside the body surrounded by soft tissues (see Figure 26.1*a*). Other animals, such as lobsters and clams, have **external skeletons** that surround their soft tissues (see Figure 26.1*b*).

To get an idea of some of the selective advantages that might favor the evolution of internal or external skeletons, imagine you are eating a seafood dinner. It takes much less effort to pull the meat off the internal skeleton of a fish than it does to crunch your way through the external skeleton to get at the tasty meat in a lobster leg. Thus, external skeletons provide protection in addition to support. Similarly, the human skull forms a protective external skeleton that simultaneously supports and protects the brain.

At the end of your meal, the weight of the remaining lobster skeleton is greater than the weight of fish bones. This difference reflects the lower weight of internal as compared to external skeletons for the amount of support offered (Table 26.1). We will see later in this chapter that weight considerations have played an important role in the evolution of skeletons.

26.1 External Skeletons Weigh More than Internal Skeletons

Animal	Skeleton type	Percentage of body weight devoted to skeleton (%)
Clam	External shell	30
Spider	External chitin	10
Mammal	Internal bone	4
Bony fish	Internal bone	3

Joints allow animal skeletons to bend

Animals have a unique ability to move because they have muscle tissue that can contract, or shorten (see Chapter 27), and because they have cells that can change shape relatively freely compared to those of other organisms. The stiff skeletons that maintain shape, however, would make movement impossible for animals were it not for their design. Internal skeletons like those of humans or external skeletons like those of lobsters, which consist of very stiff support structures, allow motion because of **joints**, which are breaks in the skeletal system that allow the skeleton to bend in specific ways (Figure 26.3).

Skeletons without joints—for example, snail shells—severely limit movement. Later in this chapter we consider how joints combine stiff and flexible components to allow animals to bend, while at the same time giving them the stiffness needed to maintain shape.

■ All animal support systems depend on specialized tissues that provide stiffness, while at the same time accommodating the changes in shape needed for movement. The support tissues can lie within or can surround the soft tissues.

The Support Systems of Plants, Fungi, and Single-Celled Organisms

Plants, fungi, and single-celled organisms support themselves differently from animals. These organisms all rely on a **cell wall**, which animal cells lack, for support, and this dependence has important consequences for their biology.

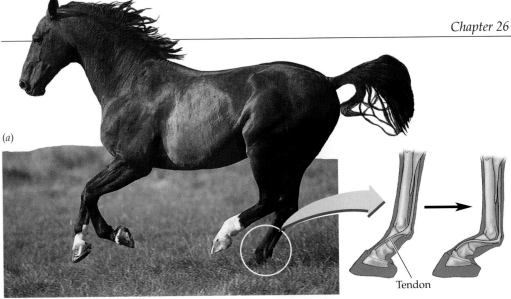

Figure 26.3 Animal Joints
The joints of horses (*a*) and grasshoppers (*b*) provide much-needed flexibility in otherwise rigid skeletons.

Tendon

(*b*)

Single-celled organisms rely on cell walls for support

Most bacterial and many protistan cells produce a cell wall that lies outside the plasma membrane. The bacterial cell wall usually contains murein, a polymer of sugar and amino acid. Protistan cell walls may consist of a complex polysaccharide called **cellulose** that is similar to the chitin that is found in arthropods, or minerals concentrated from the environment.

The cell wall either may act as a hydrostat or may be naturally stiff. When acting as a hydrostat, a single-celled organism "inflates" the cell to generate pressure by allowing water to diffuse into the cell down a concentration gradient. Creating the concentration gradient requires active transport of materials into the cell.

Multicellular plants and fungi also rely on cell walls for support

The support system of plants and fungi represents an extension of the cell wall support system of the protists from which they evolved (Figure 26.4). Each cell in a plant produces a matrix containing cellulose, and each cell in a fungus produces a matrix containing chitin. Plants and fungi rely on a combination of hydrostats and naturally rigid cell walls for stiffness.

We can see this relationship clearly in plants. The young, growing parts of plants rely on hydrostats. If allowed to dry out, these young parts wilt when a shortage of water in the plant tissues makes it impossible to keep enough water in the cells to generate much pressure (see Figure 26.2*b*). The older stems, in contrast, do not become limp even when they dry out. In the plant parts that have completed growth, the plants stiffen the cell walls by adding a complex molecule called **lignin** to the cellulose in the cell wall. Wood consists of lignified plant tissue that remains stiff even when largely dry.

The support system of plants and fungi depends on the cell walls associated with each cell, rather than with a specialized skeletal tissue as in animals; the cells of plants and fungi cannot change shape as easily as animal cells do. The inability of individual cells to change shape severely limits the ability of plants and fungi to move, as we will see in Chapter 27.

■ The support systems of plants, fungi, and single-celled bacteria and protists all rely on cell walls associated with each cell. This dependence greatly restricts the mobility of multicellular plants and fungi.

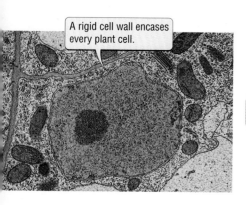

Figure 26.4 Support by Cell Walls
Plants, fungi, and many single-celled organisms have cell walls associated with individual cells that provide support.

How Do Support Systems Support Organisms?

Two physical stresses tend to change the shape of all organisms: **Compression** squeezes molecules closer together and shortens things, and **tension** pulls molecules apart and lengthens things. An organism's support system must therefore resist both compression and tension.

When a tree bends in a strong wind, tension stretches the wood on the outside of the curve in the trunk, while compression squeezes the wood on the inside of the curve (Figure 26.5*a*). How easily wood stretches and compresses in the wind determines how much the tree bends. Whereas flexible wood stretches and compresses easily, and therefore bends a lot (Figure 26.5*b*), stiff wood stretches and compresses little and bends little (Figure 26.5*c*).

If the tension generated by bending exceeds the **tensile strength** of the wood, which is the maximum tension that it can handle before molecules in the trunk pull apart, a crack will form on the outer edge of the curve in the trunk. Such a crack is bad for the tree; if it passes all the way through the trunk, the tree breaks (see Figure 26.5*c*). The ability of a support system to resist compression and tension depends both on the materials of which it is made and on the arrangement of these materials.

■ Support systems must resist changes in shape from tension, which pulls molecules apart, and from compression, which squeezes them together.

The Materials That Make Up Skeletons

We have encountered three kinds of support structures: those that depend on naturally stiff, carbon-containing macromolecules produced by the organism, those that depend on naturally stiff minerals accumulated from the environment, and those that depend on hydrostats. In this section we briefly compare their strengths and weaknesses as support materials (Table 26.2).

Carbon–containing macromolecules are light and resist tension well

Cellulose, murein, collagen, and chitin are the most widespread biologically produced support materials. These macromolecules weigh little and resist tension exceptionally well. The cellulose in wood, for example, gives it a tensile strength similar to steel cable of similar weight. The natural stiffness of these biologically produced molecules generally comes from **cross-links**, chemical connections that form between adjacent long molecules. The cross-links allow the molecules to resist compression more effectively. We can see this if we look at a caterpillar: The stiff case protecting the head consists of cross-linked chitin, whereas a flexible skin made of chitin lacking cross-links covers the rest of the body. Similarly, lignin in plants stiffens cell walls by increasing their resistance to compression.

Minerals accumulated from the environment provide stiffness

Organisms in all kingdoms use minerals accumulated from the environment in their support structures. As already mentioned, the

26.2 *Properties of Materials Used in Support Systems*			
	Material		
Property	**Biochemical**	**Mineral**	**Hydrostat**
Stiffness	Moderately stiff, if cross-linked	Stiffest	Variable, but not very stiff
Resistance to forces	Resists tension	Resists compression	Resists neither tension nor compression
Weight per volume	Light	Heavy	Light

mineral most generally incorporated is calcium, but grasses among plants, diatoms among protists, and sponges among animals, for example, may accumulate other minerals, such as the silica of which glass is made. These minerals are often abundant and easily accumulated from the environment. Mineral support systems resist compression very well and are naturally stiff.

Mineral support systems suffer from two drawbacks, however. They weigh up to twice as much as a similar volume of biologically produced material, and they tend to have low tensile strengths, causing them to break under tension. In response to selective pressures favoring light-weight support systems, organisms with mineral-based support systems tend to evolve skeletons that contain as little of the mineral as is needed to safely support them.

In response to selection for support systems that resist breakage, virtually all organisms with mineral-based support systems mix some tension-resistant biologically produced materials with the minerals. Thus the mineral matrix produced by human bone cells includes tension-resistant collagen.

Hydrostats offer relatively low stiffness but great flexibility

Even the most high-pressure hydrostats never attain the stiffness of most of the support systems based on biologically produced or mineral materials. Thus organisms such as nonwoody plants that rely on hydrostats cannot maintain their shape as well when under compression or tension. On the other hand, hydrostats offer the advantages of light weight and of a control over stiffness that is not possible with either biologically produced or mineral support systems.

Animals with hydrostats control the pressure, and hence their stiffness, by contracting muscle tissues, and other organisms do so by regulating the movement of water into or out of their cells. As a result, organisms can change the stiffness of hydrostats, allowing them to change their shape. A little experimenting with your tongue will convince you that you can make it quite stiff or quite limp.

> ■ Biologically produced support materials provide light weight and high tensile strength but relatively little stiffness. Mineral support materials offer stiffness and resistance to compression but are relatively heavy. Hydrostats provide changeable but relatively low stiffness and light weight.

(a)

(b) The ability to resist compression and tension…

(c) …determines how trees respond to the stress of hurricane-force winds.

Figure 26.5 Compression, Tension, and Tensile Strength in Support Systems
Successful skeletons resist breaking in the face of tension and compression. (*a*) When wind blows against a tree, it causes both compression and tension in the tree trunk. (*b*) Trees such as palm trees that resist these forces well can survive hurricane-force winds. (*c*) Trees such as pines break under the same conditions.

Figure 26.6 Support in Water and on Land
Much of the weight of aquatic organisms like marine algae
(*a*) and whales (*b*) is supported by the water in which they
swim. Land organisms like cows (*c*) and trees (*d*) require
stronger skeletons to resist the pull of gravity.

The Evolution of Skeletons in Water and on Land

Life in water and life on land place very different
demands on the support systems of organisms. Just as
we float in water, water supports aquatic organisms, so
they lead almost weightless existences. On land, however, the support system of an organism must support
its full weight against the pull of gravity.

Accordingly, aquatic organisms tend to have
support systems that are less stiff than
those of terrestrial organisms (Figure
26.6). Hydrostatic support systems,
which generally provide less stiffness than mineral or biologically
produced support systems, occur
much more often in aquatic habitats
than on land. The low-pressure hydro-

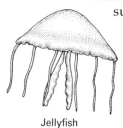

Jellyfish

stat that provides shape to a jellyfish in water is useless
on land: When a jellyfish washes ashore it collapses into
a shapeless blob on the beach. Even aquatic animals that
rely on mineral skeletons, such as fish, often have a
more delicate skeleton than terrestrial organisms of similar weight have.

Terrestrial organisms need skeletons that are stiff to
support their weight, but the support tissues themselves
add weight to the organism. As the weight of an organism increases, the weight of its support tissues increases
even more rapidly. For example, the shrew, a tiny mammal that weighs only about 8 grams, has a skeleton that
makes up 4 percent of its body weight, whereas a 7000-
kilogram elephant has a skeleton that makes up 24 percent of its body weight (Figure 26.7). This difference is
visible in the much stouter appearance of equivalent
bones in large elephants as compared to smaller cats
(Figure 26.8*a*).

This relationship sets up a vicious circle: Elephants
need heavier skeletons in part to support their heavier
skeletons. Furthermore, elephants devote less weight to
other tissues. It appears that elephants have cut corners

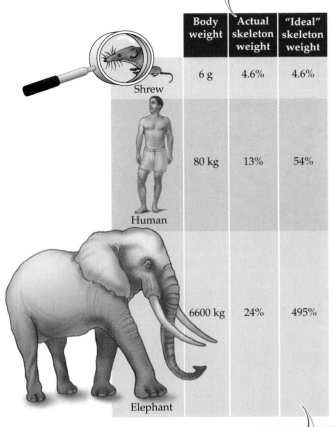

With increasing size, the proportion of a mammal's body weight devoted to skeleton also increases.

	Body weight	Actual skeleton weight	"Ideal" skeleton weight
Shrew	6 g	4.6%	4.6%
Human	80 kg	13%	54%
Elephant	6600 kg	24%	495%

To maintain an "ideal" skeleton, equally strong relative to weight as that of a shrew, humans and elephants would have to have bodies made of impossible amounts of bone.

Figure 26.7 Skeleton Weight Reflects Body Weight
Large organisms must devote more of their weight to skeletal materials than small organisms. To have skeletons proportionately as strong as those of shrews, elephants would have to devote more than 100 percent of their actual body weight to bone weight—which, of course, is impossible.

to keep their skeletons from being even heavier than they already are. Considering the properties of bone, we can estimate that for an elephant skeleton to be as strong, relative to the weight of the body, as the

Figure 26.8 Heavy Organisms Require Large-Diameter Support Structures
(*a*) Heavy elephants have bones that are larger in diameter relative to their legs than light cats. (*b*) Note the similar pattern in the thin stalk of the small mushroom as compared to the large one.

shrew's skeleton, the elephant's skeleton alone should weigh more than its entire body does in reality (see Figure 26.7). Because an elephant's skeleton actually weighs much less, it is much less strong relative to its body weight than is a shrew's skeleton.

■ Whereas aquatic organisms need relatively little support, a terrestrial organism needs a stiff skeleton that can support the organism's entire weight against the pull of gravity. Large terrestrial organisms have relatively weaker support systems than small ones because of the excessive weight of support materials.

The Human Knee

We return now to joints, using the human knee as a model. Although our knees differ in details of construction from other joints in our bodies and from joints in other animals, they make a good general model for how flexible and stiff materials combine to allow controlled motion. The human knee consists of bone and cartilage that resist compressive forces that may reach 10 times the body weight during exercise. Various arrangements of collagen surround the knee to keep the bones from pulling or twisting apart (Figure 26.9).

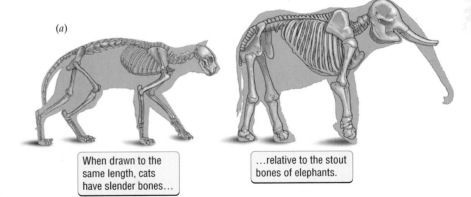

(*a*)

When drawn to the same length, cats have slender bones…

…relative to the stout bones of elephants.

(*b*)

Delicate horsehair mushrooms have more slender stalks…

…when depicted at the same size as large fly agaric mushrooms.

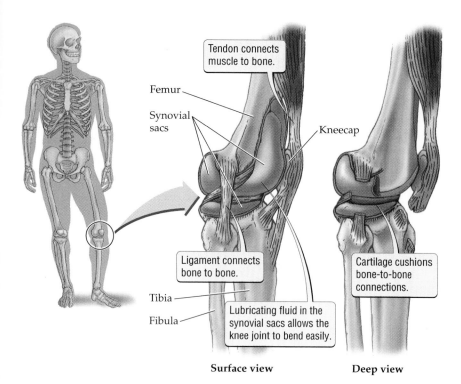

Tendon connects muscle to bone.

Femur

Synovial sacs

Kneecap

Ligament connects bone to bone.

Cartilage cushions bone-to-bone connections.

Tibia

Fibula

Lubricating fluid in the synovial sacs allows the knee joint to bend easily.

Surface view   **Deep view**

Figure 26.9 The Human Knee
The human knee is a good example of how animals balance rigidity and flexibility. The deep view in this diagram indicates what surgeons might see as they operate.

The bones in the knee joint support body weight and define how the leg bends. The end of the upper leg bone, the femur, rides in a pair of grooves at the end of the tibia, which is the larger of the two lower leg bones. This arrangement allows the leg to swing forward and back like a hinge, but not side to side, thereby providing both motion and stability. The knee must allow for the slight twisting of the tibia relative to the femur that takes place when we walk.

Ligaments are strips of collagen that connect bone to bone. Two pairs of ligaments, one on the front and back of the knee and a second on the sides of the knee, connect the femur to lower leg bones and prevent sliding along the joint surface when the knee bends, keeping the femur and the lower leg bones in their proper place. Because the collagen of the ligaments can stretch slightly, it allows some bending and twisting motion in the knee. **Tendons**, which connect muscle to bone, are another form of collagen that helps hold the knee together by connecting upper leg muscles to the bones of the lower leg.

Wherever two moving parts rub against each other, as in a joint, wear can destroy bone, and friction can waste energy. Layers of cartilage in the knee cushion the points at which the femur meets the tibia. The cartilage also works particularly well with the lubricating **synovial fluid** that reduces friction. The femur and

tibia slide past each other more easily than a skate slides over ice.

When all these components—bone, cartilage, ligaments, tendons, and synovial fluids—work together, we have a joint that can control and withstand motion over decades. When one of these components fails, however, we face potentially serious medical problems (Table 26.3).

In various kinds of arthritis, the cartilage and bone in the joint are damaged, often making bending difficult and painful. Tears in the cartilage and ligaments of the knee are common and disabling sports injuries (Figure 26.10). Resulting most often when a person rapidly speeds up or slows down while turning, torn cartilage can lead to increased wear on the knee joint. Unless repaired surgically, the remaining cartilage and underlying bone may become permanently damaged, leading to arthritis.

Ligaments tear when stretched to more than one and a half times their normal length by a severe blow to the front or sides of the knee or by severe twisting of the lower leg relative to the femur. Torn ligaments cannot hold the femur in place relative to the tibia and fibula, leading to movement in the knee joint that may make walking impossible. Like torn cartilage, torn ligaments require surgery to repair.

■ The human knee illustrates many of the basic features of the joints that allow animals to move while still maintaining their shape. Flexible materials around the joint help control the movement of the stiff skeletal components relative to one another.

26.3 *Common Ailments of the Knee Joint*				
Problem	**Bone**	**Cartilage**	**Ligaments**	**Synovial fluid**
Osteoarthritis	�	▚		
Rheumatoid arthritis	▚	▚		▚
Torn cartilage		▚		
Torn ligament			▚	
Bursitis				▚
Note: Shaded boxes indicate the tissue or body part affected by each disease.				

Figure 26.10 Damage to the Knee Joint Can Severely Restrict Mobility

Highlight

Landing Cats and Humans

Three explanations have been proposed for why cats can survive falls that are fatal to humans (Figure 26.11):

First, cat skeletons, relative to body weight, are stronger than human skeletons. Both cats and humans have strong internal skeletons made of collagen-reinforced bone; however, recall that the weight of bone increases more rapidly than the weight that it can support. The conflict between the advantages of a stronger skeleton and the disadvantages of a heavier skeleton has led to an evolutionary compromise: Humans have sacrificed skeletal strength to reduce weight. Whereas we must pay for the benefit of a tolerably light skeleton by looking carefully before we leap, smaller and more strongly constructed cats can get away with less caution.

Second, cats have an exceptional ability to soften their landing. Perhaps because falls are a normal part of life for tree-climbing creatures like cats, natural selection strongly favors cats that instinctively rotate their bodies as they fall so that they land feet first. Humans, in contrast, land arms or feet first. By landing as they do, cats spread the force of impact over four relatively stronger limbs, whereas humans concentrate a larger force over only two, weaker limbs.

Third, cats may relax a bit during the fall and prepare for a crash landing by bending their legs so that the ligaments and tendons associated with the joints can act as springs. We humans, quite understandably, tend not to relax during a free fall. By using their legs as springs, cats further reduce the force of impact by spreading it over a slightly greater period of time. Tense humans suffer the full force virtually instantaneously. Although the difference between slowing down from a fall to a dead stop in 0.002 seconds instead of 0.001 seconds may seem trivial, doubling the time to decelerate nonetheless halves the force of impact. For similar reasons, a padded dashboard in a car or a bicycle helmet can dramatically reduce the injuries suffered in a collision. The decline in deaths when cats fall from heights greater than the fifth story (see Figure 26.11) may reflect the increased time that cats have to prepare properly for impact.

Because cats climb, and really do let their curiosity have a good shot at killing them, it would be tempting to conclude that the ability of cats to survive falls from high places represents a marvelous adaptation to their way of being. In part, the amazing ability of cats to survive falls seems to be an adaptation to life in the trees. Equally important, however, may be the relatively greater strength of cat leg bones, which is a side consequence of their size rather than an adaptation.

> ■ Cats survive more falls than humans probably because their skeletons are relatively stronger than those of humans. In addition, cats may benefit from their ability to prepare for landing by relaxing, bending their joints, and landing on all four feet.

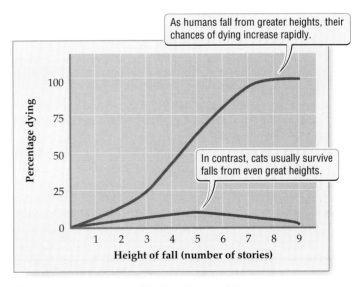

Figure 26.11 Survival of Falling Cats and Humans
Statistics collected by doctors and veterinarians show that cats survive falls from high places much better than humans do. Note both that cats are much less likely to die from a fall than are humans, and that the death rate for cats actually decreases for falls from heights greater than the fifth story.

Summary

The Human Skeleton

- Bone is the major support tissue in humans.
- Cartilage and hydrostats provide additional support.

An Overview of Animal Skeletons

- All animal support systems depend on specialized tissues that provide stiffness. Skeletons simultaneously accommodate the changes in shape needed for movement by including joints.
- Soft-bodied animals are supported mainly by hydrostats.
- Support tissues that are external to the soft tissues offer increased protection.
- Support tissues that are inside the soft tissues weigh less than external skeletons.

The Support Systems of Plants, Fungi, and Single-Celled Organisms

- The support systems of single-celled bacteria and protists and of multicellular plants and fungi rely on cell walls that surround the cells.
- Because each cell has a cell wall associated with it, this type of support system greatly restricts the mobility of multicellular plants and fungi.

How Do Support Systems Support Organisms?

- Support systems must resist changes in shape from tension, which pulls molecules apart, and compression, which squeezes them together.
- If the tensile strength of a substance is exceeded, it will begin to pull apart and break.

The Materials That Make Up Skeletons

- Biologically produced support materials, such as cellulose, murein, collagen, and chitin, provide light weight and high tensile strength but relatively little stiffness.
- Mineral support materials offer stiffness and resistance to compression, but they are relatively heavy and do not offer much tensile strength.
- Hydrostats provide changeable but relatively low stiffness and light weight.

The Evolution of Skeletons in Water and on Land

- Because water supports much of their weight, aquatic organisms can rely on hydrostats or relatively delicate mineral skeletons for support.
- Terrestrial organisms need stiff skeletons that can support the organism's entire weight against the pull of gravity.
- Large terrestrial organisms have relatively weaker support systems than small ones because of the excessive weight of support materials.

The Human Knee

- The human knee illustrates many of the basic features of the joints that allow animals to move while still maintaining their shape.
- Flexible ligaments and tendons around the joint help control the movement of the two bones—the femur and the tibia—relative to one another.
- To work efficiently and to withstand a lifetime of movement, the motion of joints requires friction-reducing components such as cartilage and synovial fluid.

Highlight: Landing Cats and Humans

- Cats survive more falls than humans probably because their skeletons are relatively heavier than those of humans.
- Cats may benefit from their apparent ability to prepare for landing by relaxing, bending their joints, and landing on all four feet.

Key Terms

bone tissue p. 400	hydrostat p. 400
cartilage tissue p. 400	internal skeleton p. 402
cell wall p. 402	joint p. 402
cellulose p. 403	ligament p. 408
chitin p. 401	lignin p. 403
collagen p. 400	synovial fluid p. 408
compression p. 404	tendon p. 408
cross-link p. 404	tensile strength p. 404
external skeleton p. 402	tension p. 404

Chapter Review

Self-Quiz

1. An important advantage of hydrostats is that they
   a. allow the organism to vary its stiffness.
   b. resist compression better than any other kind of skeleton.
   c. resist tension better than any other kind of skeleton.
   d. all of the above.

2. Collagen and chitin are
   a. biologically produced materials.
   b. connective tissues.
   c. hydrostats.
   d. minerals.

3. Which of the following is true of support in both plants and fungi?
   a. They rely on external skeletons.
   b. They rely on skeletons made of minerals.
   c. Support comes from cell walls.
   d. Their inadequate support systems force them to live in aquatic environments.

4. Large mammals have bones that are relatively heavier than the bones of small mammals because
   a. their limbs are longer.

b. their weight is greater.
c. they eat more.
d. their muscles are stronger.

5. Ligaments
   a. connect muscles to other muscles.
   b. connect muscles to bones.
   c. connect bones to other bones.
   d. all of the above

Review Questions

1. Recall that multicellular plants do not add lignin to the cell walls of growing plant parts. Why might this be?

2. Why do most organisms combine support structures that contain matrices of biologically produced molecules with structures based on mineral matrices?

3. Under what circumstances could using a hydrostatic support structure be advantageous?

---

26   The Daily Globe

Knee Injury Fuels Debate

BOSTON, MA. Less than 2 weeks after the U.S. victory in the Women's World Cup, soccer star Sarah Domore, one of the most popular and visible players on the U.S. team, tore her anterior cruciate ligament (ACL) in an exhibition game against up-and-coming college players. The injury, though not career-threatening, will require surgery and is expected to sideline Domore for at least 6 months.

Given Domore's immense popularity, some people wonder if the ACL injury will serve as a warning—or even a deterrent—to young women interested in soccer. Female athletes are more susceptible than males to ACL injuries: in some studies, females suffered such injuries at rates four to five times higher than men. The injuries are especially common in sports that require many sudden stops and turns, such as soccer, basketball, and volleyball, but why women are at higher risk for ACL tears is not clear.

Physical differences may account for some of the increased risk for women. For instance, the wider pelvis of women results in a different joint angle, and monthly changes in hormones appear to affect the flexibility of connective tissue. However, weight lifting and training routines that result in relatively stronger hamstrings, along with different jumping techniques may also help protect them from some of these injuries.

Although many athletic trainers and orthopedic specialists suggest that education and exercise programs can reduce the risk of injury to acceptable levels, not everyone agrees. Some family doctors and parents are trying to urge schools not to support soccer and other women's sports that increase the risk of injury, such as basketball and volleyball, but instead provide better funding for less risky sports such as softball and track and field. However, given the success of the U.S. women's soccer team and professional leagues such as the WNBA (Women's National Basketball Association), they may face an uphill battle.

Evaluating "The News"

1. Think about other familiar sports in which players are sometimes injured. What types of tissues seem to be more injury prone, and why do you think this is so?

2. Do schools that support sports with taxpayer dollars have any responsibilities to support only the safest sports? Should schools that support women's soccer and other teams be involved in education campaigns about how to minimize ACL injuries, or does that responsibility fall on someone else? If so, whose responsibility is it?

3. If you had a daughter, how would you decide which sport(s) you would encourage her to play?

chapter 27

Movement

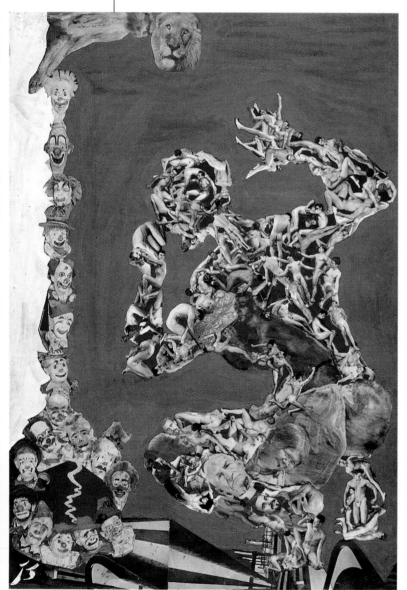

Jess, *The Mouse's Tale*, 1951.

Speeding Sperm and Racing Runners

The question of how fast we can run has motivated competitions since the beginning of organized athletics. Florence Griffith Joyner has sprinted 100 meters in a world record of 10.49 seconds. At the other extreme of distance, world record holder Tegla Laroup has run the 42.2-kilometer-long marathon in 2 hours, 20 minutes, and 47 seconds. The excitement of a sprinter like Griffith Joyner comes from seeing her reach maximum possible speeds over short distances (10.2 meters/second). The thrill with long-distance runners like Laroup is the ability to sustain a lower speed (5.0 meters/second) over much longer distances.

We stage races for the entertainment they provide, but remarkably similar tests of speed and endurance in the biological realm serve a more serious purpose. Mammalian sperm swim as quickly as possible toward unfertilized eggs moving down the fallopian tubes of the female reproductive system. From a starting field of millions of sperm released in a single ejaculation of a male, the first to reach the egg will probably be the one that succeeds in fertilizing it.

Researchers have found that sperm behave much like track stars. Sperm sprint down the short reproductive tract of female mice at 0.2

can travel. Even though the muscles that propel runners and the tiny tail-like flagella that propel sperm seem very different, we will see in this chapter that these muscles face similar challenges.

Moving Toward the Finish Line
These sperm exhibit behavior similar to this athlete.

millimeter/second. In the much longer reproductive tract of human females, however, sperm pace themselves at about half this speed as they swim the sperm equivalent of a marathon.

Comparisons between runners and sperm that travel short and long distances illustrate an important point: Although athletes and sperm seem to have little in common, similar factors determine how fast and how far they

1. The interaction of actin and myosin filaments causes muscle fibers to contract.

2. Muscles must work in pairs because an individual muscle can only contract.

3. Muscles increase strength by increasing their cross-sectional area, and they can increase speed by lengthening or including more fast muscle fibers.

4. Animals can move by walking, flying, or swimming.

5. Drag resists the forward motion of an organism.

6. Single-celled prokaryotes and eukaryotes propel themselves by hairlike structures call cilia or flagella.

7. Plants and fungi rely on wind and water currents or animals to move their pollen, seeds, and spores.

The ability of an organism to move through its environment provides it with an almost endless list of potential advantages (Figure 27.1). Consider how different our lives would be if we could not move: We could not explore our surroundings, we could not go to the refrigerator to forage for food, we could not search for mates, and we could not escape an underheated room for a warmer one, for example.

Although movement undoubtedly benefits organisms in all kingdoms of life, we saw in Chapter 26 that animals, more than any other group, have the ability to move through their world. The unique combination of support tissues that are separate from other tissues and muscle tissues that can contract makes it possible for animals to use the potential of movement more effectively than the members of any other kingdom can.

In this chapter we focus on how animals move. We begin by discussing the form and function of the muscle tissue that serves as a biological motor. We then consider how natural selection has altered the arrangement of muscle tissue, the interaction between muscle and support tissue, and the shape of organisms to allow animals to move in various ways. We close by looking at how bacteria and protists move and at how plants and fungi have overcome some of the limitations that come with their poor ability to propel themselves.

(a)

Figure 27.1 Movement Is Essential for Animals and Plants
Movement plays an important part in the lives of all organisms. (a) Cheetahs run down their antelope prey. (b) Bacteria can swim toward and away from chemical cues by using unique, hairlike flagella. (c) Plant tendrils seek things to which they can cling.

(b)

(c)

Animal Muscles as Biological Motors

The muscle tissue unique to animals plays a leading role in their ability to move. As we will see, muscle tissue is specialized to do one crucial thing: contract. Before turning to the many ways in which animals convert simple contractions into useful motion, we need to understand muscle structure and how muscles contract.

The structure of muscle reflects its function

We are all familiar with muscle tissue because when we eat fish or chicken or beef, we are eating muscles. Most of the features that relate to the role of muscle as a biological motor, however, remain hidden to the unaided eye. Each muscle—for example, the biceps of the upper arm (Figure 27.2)—consists of many basic muscle units, called **muscle fibers**. The muscle fibers in most muscles arise from the fusion of many muscle cells, so each fiber contains many nuclei. Each fiber, in turn, is packed with smaller **myofibrils** that contain structures, called **sarcomeres**, that do the work of contraction.

Sarcomeres are visible as bands when seen through a microscope (see Figure 27.2). At extremely high magnification, we can see each sarcomere clearly. It extends between two **Z discs**, visible as dark lines. Each sarcomere is based on a very specific arrangement of the two

Figure 27.2 The Fine Structure of the Biceps Muscle
Muscle contraction ultimately depends on the sliding of myosin filaments along actin filaments to shorten the length of the muscle.

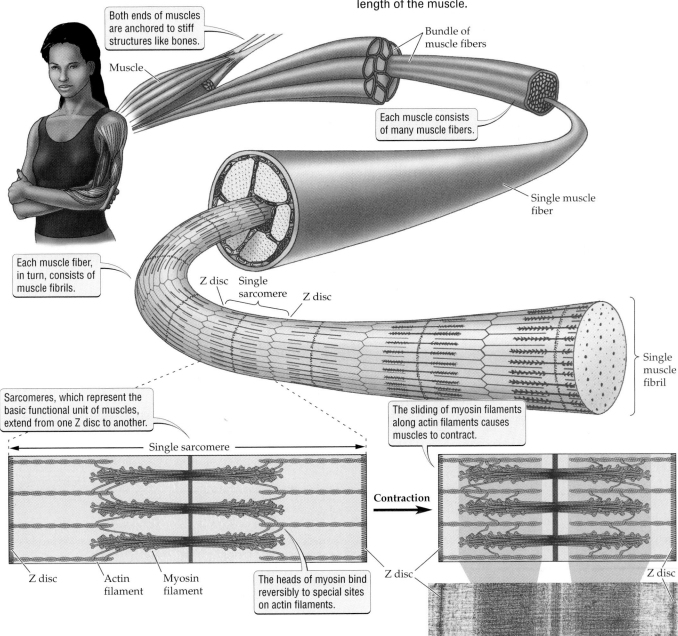

Both ends of muscles are anchored to stiff structures like bones.

Muscle

Bundle of muscle fibers

Each muscle consists of many muscle fibers.

Single muscle fiber

Each muscle fiber, in turn, consists of muscle fibrils.

Z disc   Single sarcomere   Z disc

Single muscle fibril

Sarcomeres, which represent the basic functional unit of muscles, extend from one Z disc to another.

The sliding of myosin filaments along actin filaments causes muscles to contract.

Single sarcomere

Contraction

Z disc   Actin filament   Myosin filament

The heads of myosin bind reversibly to special sites on actin filaments.

Z disc

Z disc

proteins, **actin** and **myosin**, that allow muscles to contract. Z discs are the anchor point for actin filaments, which consist of two actin molecules. The actin filaments extend toward the center of each sarcomere from the Z discs. Between the actin filaments lie myosin filaments, each of which consists of many myosin molecules.

As we describe in the discussion that follows, myosin filaments can, in the presence of the energy storage molecule ATP (see Chapter 8), pull the two Z discs closer together, causing muscle contraction. Muscles bulge when they contract because all the actin and myosin filaments must squeeze into a shorter length of muscle.

Actin and myosin interact to cause muscle contraction

When we lift weights, actin and myosin slide past one another in the biceps muscle of the arm. If we could watch the biceps at high magnification as we lifted the weight, we would see the two Z discs that define each sarcomere (see Figure 27.2) pull closer together. An even closer look would reveal that each myosin molecule ends in a protruding head that can bind to specific sites on adjacent actin filaments.

As we lift the weight, the myosin heads in the muscle repeatedly attach to a binding site on the actin filament, change shape to pull the myosin filament closer to the Z disc, and release the binding site before reattaching to a binding site closer to the Z discs (see Figure 27.2). As the

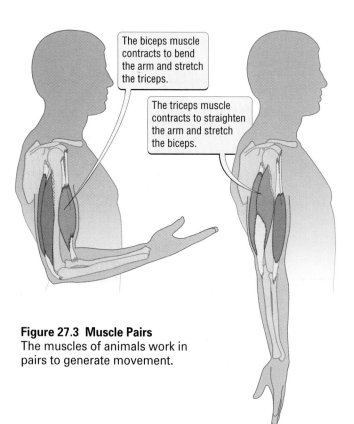

Figure 27.3 Muscle Pairs
The muscles of animals work in pairs to generate movement.

The biceps muscle contracts to bend the arm and stretch the triceps.

The triceps muscle contracts to straighten the arm and stretch the biceps.

myosin filaments "walk" their way from binding site to binding site on the actin filaments, they pull the Z discs closer together and contract the muscle. Each step of each myosin head requires the energy released by the conversion of one ATP to one ADP molecule (see Chapter 7).

Muscles must work in pairs

Because the myosin heads can only pull the Z discs together, a contracted muscle cannot stretch itself out again. Thus, muscles are arranged in pairs so that the contracting muscle stretches the other, relaxed muscle back to its starting position after a contraction. For example, in our upper arms we have triceps muscles that lie on the lower side and that contract to stretch the biceps (Figure 27.3). If we think of other muscles in our bodies, we find this pattern of paired muscles repeated.

> ■ Muscle fibers consist of sarcomeres that contract through the action of actin and myosin. Myosin molecules "walk" along the actin filaments to pull the Z discs that define a sarcomere closer together. Muscles can only contract; a contracted muscle requires the action of the muscle to which it is paired to stretch it out again.

How Animals Convert Muscles into Motion

As we have seen, muscle is a remarkable tissue that produces contraction. Within the animal kingdom, mechanisms have evolved that convert simple contractions into the wide range of motions of which most animals are capable. In this section we consider how animals vary the strength and speed with which their muscles contract.

Animals increase strength by increasing the amount of muscle tissue

All animals have essentially identical muscle tissue that contains identical actin and myosin proteins arranged in much the same way. It should not surprise us, therefore, that pieces of muscle tissue *containing similar numbers of muscle fibers* can lift about the same weight regardless of whether the tissue comes from an elephant or an earthworm.

Differences in strength between species and between muscles within an individual depend almost entirely on the cross-sectional area of the muscles. The greater the cross-sectional area, the more muscle fibers can lie side by side and contract simultaneously. For example, elephants are stronger than earthworms because elephants have muscles with bigger cross-sectional areas.

Exercise can change muscle strength as well. With weight training, muscles initially increase in strength without increasing in cross-sectional area. The number of muscle fibers in the muscles remains the same, but a greater proportion of them contract at any one time. Thus the initial increase in strength comes about because muscle fibers are being used more efficiently. With continued weight training, the number of muscle fibers in the muscle increases, leading to the greater muscle bulk for which bodybuilders strive (Figure 27.4).

The speed of muscle contraction depends on the length and type of muscle

Just as different animals have muscles that can generate different forces, they also have muscles that contract at different speeds. As with strength, these differences in speed arise not because the muscle tissues differ in any fundamental way, but because they differ in arrangement. Differences in contraction speed depend on a muscle's length and response to nerve signals.

One way to increase the speed with which a muscle contracts is to increase its length. When fully contracted, muscles shorten to about 60 percent of their resting length. Thus a 10-centimeter-long muscle in the arm shortens by 4 centimeters (to 6 centimeters) when fully contracted. In the same time, a 30-centimeter-long muscle in the thigh, by contrast, shortens by 12 centimeters (to 18 centimeters) when fully contracted. The longer leg muscle contracts three times the distance in the same amount of time as the shorter arm muscle.

Many groups of animals also have two classes of muscles that differ in how the muscle fibers respond to the signal to contract. Muscle filaments that contract quickly, like those that generate the explosive power of a runner sprinting from the starting blocks, are called **fast muscle fibers**. Fast muscle fibers contract completely in an all-or-nothing response to signals from nerves. In contrast, slowly contracting **slow muscle fibers** respond to signals from nerves by contracting in small steps (Table 27.1). Slow muscle fibers are used most often for activities sustained over long periods of time, as during the slower pace of a marathon.

Whereas fast muscles generally sacrifice force for speed, slow mus-

Figure 27.4 Exercise Can Increase Muscle Size
Weight lifting increases muscle strength by stimulating the production of additional muscle fibers, leading to an increase in muscle bulk.

cles sacrifice speed for power. Fast muscles contract so rapidly that they quickly use up the available oxygen and have to rely on anaerobic metabolism, which takes place when oxygen runs out (see Chapter 7). Slow muscle fibers, on the other hand, rely on aerobic metabolism. To ensure a steady oxygen supply, slow muscle fibers

27.1	**Comparison of Fast and Slow Muscle Fibers**	
	Type of muscle fiber	
Characteristic	**Fast**	**Slow**
Speed of contraction	Fast	Slow
Force of contraction	Weak	Powerful
Length of contraction	Brief	Sustained
Response of sarcomeres	Either none or complete contraction	Partial contraction possible
Source of ATP	Anaerobic metabolism	Aerobic metabolism
Human example	The quadriceps muscle in the thigh	Gluteus maximus muscle in the buttocks

contain a pigment called **myoglobin** that receives and holds oxygen until it is needed by the muscle.

We can usually visually distinguish slow muscle tissue by its darker color, which reflects its myoglobin content. Thus, whereas the white meat of a turkey consists mostly of fast muscle fibers used to power flight, the dark meat of the legs consists mostly of slow muscle fibers that must contract for long periods with enough force to hold up the body of the bird.

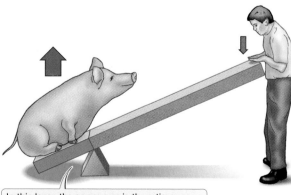

(a)

Work—lifting an object—occurs at the end of the load arm.

Our muscles apply force on the power arm.

Power arm

Load arm

The lever pivots on the fulcrum.

Because the power arm is twice as long as the load arm, we can use this lever to lift an object twice as heavy as the force we apply.

Although the power and load arms are differently arranged, this lever system is equivalent to the one above because the power arm is still twice as long as the load arm.

In this lever, the power arm is three times as long as the load arm; it allows us to lift an object three times as heavy as the force we apply.

Muscles and skeletons work together to control strength and speed of movement

The speed or strength of the movement that results from a muscle contraction also depends on the arrangement of muscles and skeletons to form **lever** systems. We have all used mechanical levers of some sort. All lever systems consists of a **fulcrum** that serves as a pivot point for stiff lever arms (Figure 27.5). We apply a force to the **power arm**, which moves the **load arm** that does the work we want done.

In our mechanical devices, we tend to use levers to increase the force that we generate from a given muscle contraction. For example, we use the claw of a hammer to pull nails that are too firmly stuck in wood to extract with our bare hands. In such cases the load arm is shorter than the power arm. Thus, the tip of the load arm moves a much shorter distance than the power arm, and the force applied to the power arm is "concentrated" over the short distance that the load arm moves. If the load arm is longer than the power arm, the force is reduced and the speed increased at the tip of the load arm as compared to that of the power arm.

In animals, the power arm is almost always shorter than the load arm. Thus the force generated at the tip of the load arm is usually less than that generated by the muscle. The arrangement of the legs of mammals with different lifestyles illustrates how natural selection can act on lever systems to maximize either strength and endurance or speed (Figure 27.6).

(b)

Figure 27.5 Lever Systems
Lever systems allow us to use our muscles more effectively. (a) Force is applied to the power arm, causing the lever to pivot about the fulcrum and do work at the load arm. (b) The lever system formed by the human arm working with a hammer allows us to pull stubborn nails using our biceps muscle.

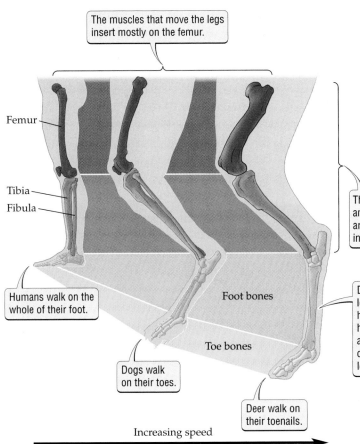

The muscles that move the legs insert mostly on the femur.

Femur

Tibia
Fibula

Humans walk on the whole of their foot.

Foot bones

Dogs walk on their toes.

Toe bones

Deer walk on their toenails.

The lengths of the femur and the lower leg bones are proportionally similar in the three animals.

Dogs and deer have longer load arms than humans because they have longer foot bones, and the foot bones contribute more to leg length.

Increasing speed →

Figure 27.6 Adaptations of Lever Systems in Mammalian Legs Human legs lend themselves less well to high-speed running than do the legs of most other animals. All of our foot and toe bones touch the ground when we run, meaning that only the upper (red) and lower (orange) leg bones contribute to leg length. With increased leg length, the length of the load arm relative to the power arm increases, decreasing the proportion of the force generated by muscle contraction that is applied where the leg touches the ground, but increasing the speed of the leg tip relative to the speed of muscle contraction.

Locomotion in Animals

As animals move through their environment, they convert the contractions of their muscles into **thrust** that propels them forward. However, forces collectively called **drag** always resist the animals' motion.

Animals can move by running, flying, or swimming

Among animals we find three main modes of locomotion: running, swimming, and flying (Figure 27.7). Animals spend energy to generate thrust with their muscles. Running animals, like humans, spend this energy to push off against the ground with their legs. Flying and swimming animals move through a fluid environment that requires a different way of generating thrust. Flying animals have evolved wings and swimming animals have evolved fins that provide broad surfaces with which to push off against air and water, respectively. These various modes of locomotion have different energetic costs, which we consider in the Box on page 421.

Drag resists forward motion

"Drag" is a blanket term that includes two very different forces that resist motion: pressure drag and friction drag.

To understand **pressure drag**, imagine riding a bicycle: A breeze begins to blow on your face as you set off. The breeze makes you work harder, as anyone who has pedaled a bicycle into a head wind can appreciate. What you feel as wind is air colliding with your face to create a zone of high pressure in front of you, and your hair streams back into a zone of low pressure behind you. High pressure in front combined with low pressure behind creates the pressure drag that pushes you backward while you do your best to drive yourself forward.

In humans, only the upper and lower leg bones contribute to leg length. In contrast, some of the bones that form part of the human foot add to leg length in dogs, and all the foot bones contribute to making a longer leg in speedy runners like deer. Increasing leg length increases the length of the load arm, which means that a given muscle contraction moves the foot a greater distance in the same amount of time. The human leg, with its relatively short load arm, works best for steady walking over long distances, dogs are suited to running over moderate distances, and deer with long legs can run very rapidly over short distances.

■ A muscle increases in strength with an increasing number of muscle fibers packed into its cross section. The length of a muscle and the response of its muscle fibers to nerve signals can enhance the speed of contraction. Lever systems that arise out of the interaction of muscle and support systems can also influence both the strength and speed with which organisms move in response to muscle contraction.

Figure 27.7 Modes of Locomotion in Animals
(*a*) Running animals generate thrust by pushing against the ground with their legs. (*b*) Swimming organisms push off against the surrounding water with paddlelike fins or cilia. (*c*) Flying animals such as this hover fly use the broad surfaces of their wings to push off against air.

(*a*)

(*b*)

(*c*)

The faster you go, the stronger the breeze feels because the pressure difference from front to back increases. The resistance offered by pressure drag increases very quickly as speed increases; thus fast-moving animals like fish and birds tend to suffer the most.

Friction drag, on the other hand, occurs when air molecules stick to you and to each other as you rush by. In other words, as you pedal down the road, you move not only yourself but extra baggage in the form of the air molecules you carry along. This friction drag can become a serious burden when moving through a **viscous** (syrupy), heavy fluid like water. If you drive your bicycle through a deep puddle, you can feel the much greater friction drag offered by relatively syrupy water as compared to that offered by air. Friction drag increases most significantly as the total surface area of the

Figure 27.8 Streamlining Reduces Pressure Drag
(*a*) Aquatic animals have converged evolutionarily on a streamlined shape as a way of minimizing the drag they experience. (*b*) Birds (such as this gannet) have also evolved a streamlined shape to reduce the pressure drag they experience during rapid flight.

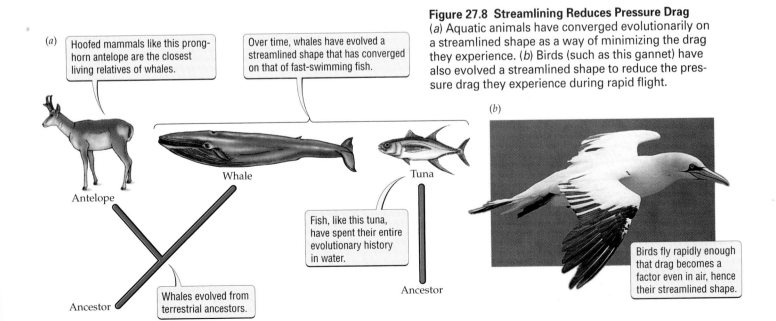

(*a*) Hoofed mammals like this pronghorn antelope are the closest living relatives of whales.

Over time, whales have evolved a streamlined shape that has converged on that of fast-swimming fish.

Antelope

Whale

Tuna

Fish, like this tuna, have spent their entire evolutionary history in water.

Whales evolved from terrestrial ancestors.

Ancestor

Ancestor

(*b*)

Birds fly rapidly enough that drag becomes a factor even in air, hence their streamlined shape.

The Scientific Process

Running, Swimming, and Flying: Which Is Most Efficient?

Which one of the three basic modes of locomotion—running, swimming, or flying—provides the most efficient way of traveling? To find out, scientists calculated the total energy animals require to move a fixed distance, and included a diverse range of species that use each mode (for example, the flying animals studied included mosquitoes, pigeons, and bats). The results, shown in the figure, are clear: no matter what an animal weighs, it is more efficient to swim (blue line) than to fly (green line), and more efficient to fly than to run (red line).

Although drag should pose a greater energy drain for swimmers than for flyers, swimmers appear to keep the effects of drag under control by moving slowly enough that neither pressure drag nor friction drag becomes a real problem. Fliers also need to spend a lot of energy simply keeping themselves aloft. Runners need not keep themselves in the air, of course, and do not need to spend much energy combating drag (no land mammals show the distinctive-

ly streamlined shape of birds or aquatic animals). When humans walk or run, however, we waste about 50 percent of the energy we put into locomotion. (As shown in the figure, a 100 kg human expends more energy running than bicycling; the wheels reduce drag, and up-and-down movement is not necessary.)

Note that the figure also shows that all three modes are more efficient for heavier animals than lighter ones. Why is this so? As you will recall from Chapter 25, larger animals have relatively less surface area than smaller ones (pages 387–389), which makes them much less susceptible to drag.

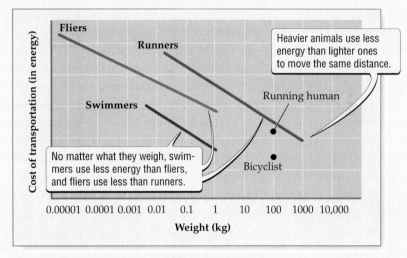

The Relative Cost of Different Modes of Locomotion
The vertical axis measures energy cost, so lower values indicate greater efficiency. Swimming is always "cheaper" than flying, and flying is cheaper than running.

organism increases and as the viscosity of the fluid increases. Friction drag is of greatest importance to organisms with a large surface area and those that live in water. Small swimming animals must spend a lot of their energy overcoming friction drag.

Natural selection favors organisms that minimize drag

Because organisms spend energy to generate thrust, drag represents a waste of valuable resources. If an organism reduces drag, then it frees up energy for other, more productive uses. Because pressure drag and friction drag have different causes, it should come as no surprise that minimizing each one requires different solutions.

The most effective way of minimizing pressure drag seems to be the streamlined shape that has evolved

independently in many fast-moving fishes, whales, and birds (Figure 27.8). In spite of the very different evolutionary histories of birds, mammals, and fish, aquatic members of these groups have all converged evolutionarily on the same streamlined shape to reduce pressure drag. A streamlined shape makes it easy for water or air to flow smoothly around the organism. Organisms reduce friction drag by having nonstick surfaces. The smooth, slimy surface that makes many algae and aquatic animals so unappealing to touch effectively reduces friction drag.

> ■ Drag resists forward motion. Pressure drag increases as the speed of an organism increases. Friction drag increases as the organism's surface area or the viscosity of the fluid increases.

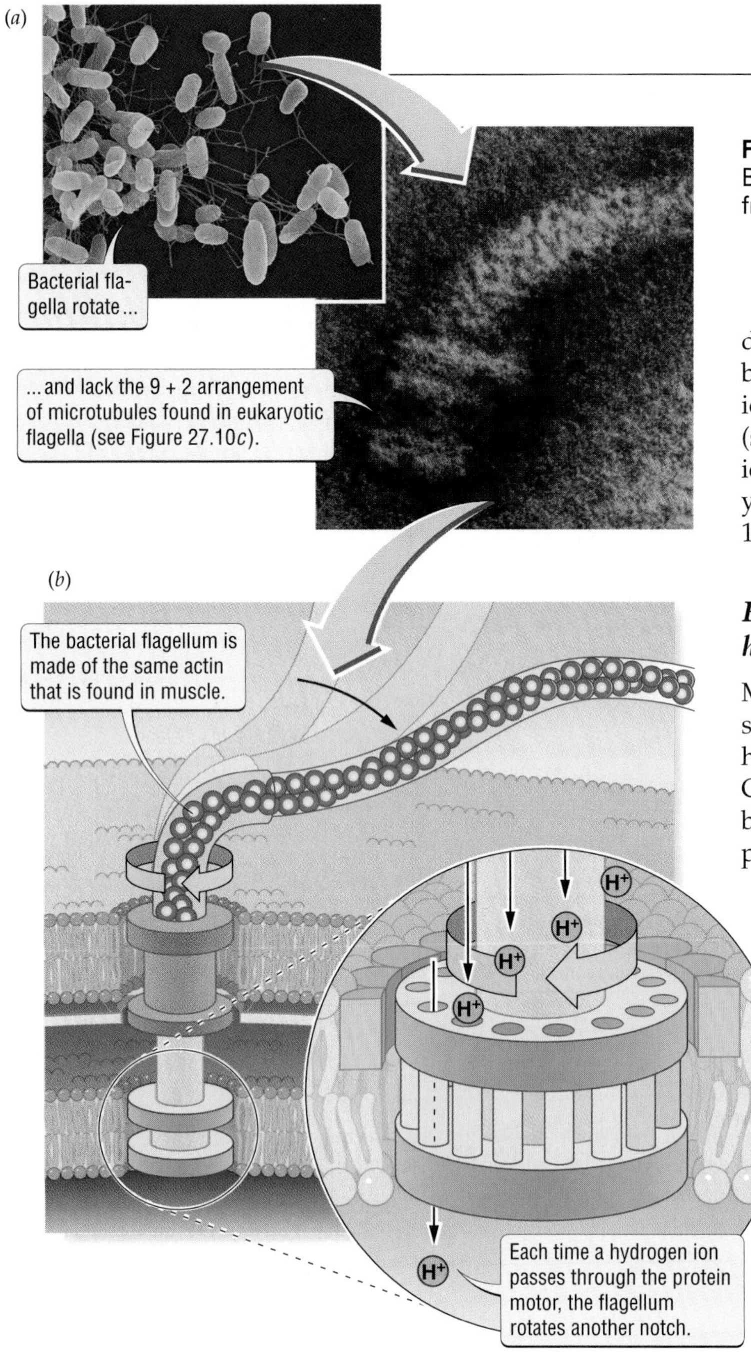

(a)

Bacterial flagella rotate...

...and lack the 9 + 2 arrangement of microtubules found in eukaryotic flagella (see Figure 27.10c).

(b)

The bacterial flagellum is made of the same actin that is found in muscle.

H^+

Each time a hydrogen ion passes through the protein motor, the flagellum rotates another notch.

Figure 27.9 Bacterial Flagella
Bacteria have flagella that differ both in form and function from eukaryotic flagella.

The motor rotates in response to hydrogen ions that diffuse into the bacterium from the environment. The bacteria use active carrier proteins to pump hydrogen ions out of the cell to create the hydrogen ion gradient (see Chapter 24). It may take as many as 100 hydrogen ions diffusing into the cell to rotate the flagellum once, yet these bacterial propellers may rotate as rapidly as 1000 times in a second.

Eukaryotes other than animals have specialized structures to help them move

Many single-celled protists and the sperm produced by some plants and all animals propel themselves using hairlike structures called **cilia** or **eukaryotic flagella**. Cilia move much as human arms do when we swim the breaststroke, and eukaryotic flagella bend in a wavelike pattern (Figure 27.10a and b).

The bending of the cilia and flagella depends on the sliding of one member of a pair of tiny tubules relative to the other in a manner similar to the sliding of actin and myosin filaments in muscle (Figure 27.10c and d). The protein that makes up the arms extending from the microtubules, dynein, shares with myosin the ability to walk up an adjacent filament and the ability to catalyze the breakdown of one ATP molecule to release the energy needed to break the cross-link. As in human muscle tissue, calcium coordinates the sliding of adjacent filaments.

■ Bacteria propel themselves by a unique rotary motor connected to a bacterial flagellum. Protists rely on eukaryotic flagella and cilia based on tubules that slide relative to one another much as actin and myosin in muscles do.

How Bacteria and Protists Move

Many bacteria and protists are quite skillful at moving through their environment. Although these organisms use various different means of propelling themselves, the dominant means are hairlike flagella and cilia.

Many bacteria have rotary motors

Although bacteria use a diversity of interesting means to propel themselves, the most widespread are **bacterial flagella**. A unique rotary motor rotates the stiff, corkscrewlike flagella of bacteria like propellers. This rotary motor is a remarkable structure (Figure 27.9).

How Plants and Fungi Get from Here to There

As discussed in Chapter 26, multicellular plants and fungi have rigid cell walls that prevent their tissues from changing shape in the way needed for motion. Nonetheless, plants and fungi need to explore their environment—for example, to find nutrients and mates. In some cases, the growth by a few millimeters in a day of plant

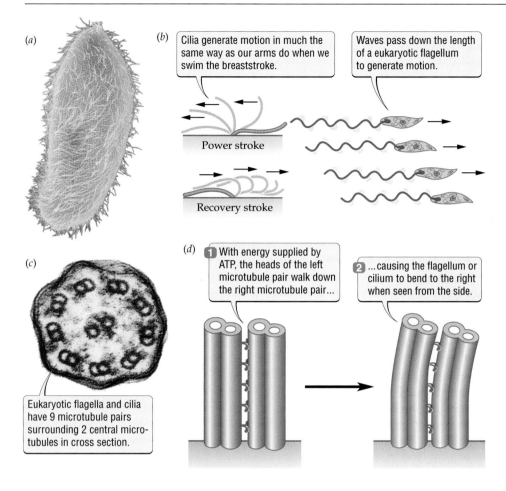

(a)

(b) Cilia generate motion in much the same way as our arms do when we swim the breaststroke.

Waves pass down the length of a eukaryotic flagellum to generate motion.

Power stroke

Recovery stroke

(c) Eukaryotic flagella and cilia have 9 microtubule pairs surrounding 2 central microtubules in cross section.

(d) 1 With energy supplied by ATP, the heads of the left microtubule pair walk down the right microtubule pair...

2 ...causing the flagellum or cilium to bend to the right when seen from the side.

Figure 27.10 Cilia and Eukaryotic Flagella
Many eukaryotic organisms, especially single-celled ones, use cilia or flagella to generate thrust (a). Although cilia look like flagella, they generate thrust by different kinds of bending motions (b). The cilia and flagella of eukaryotic organisms all share characteristic arrangement of paired microtubules (c). These microtubules slide relative to one another, much as myosin slides along actin, causing the cilium to bend (d).

The flowers that characterize flowering plants evolved to attract insects that could carry pollen grains containing plant sperm from one plant to another. The edible fruits produced by many flowering plants serve as bait to attract seed-dispersing birds and mammals. Plants and fungi also regularly use wind or water currents to disperse pollen grains, seeds, or spores over long distances. Most plants that colonize isolated islands have done so by spreading their seeds on the wind.

roots or of the hyphae through which fungi absorb nutrients keeps them always in contact with nutrients.

For plants and fungi, finding distant mates and scattering to colonize new habitats pose more difficult challenges than they do for animals (Figure 27.11). Most flowering plants and many fungi have solved this problem by taking advantage of animals, wind, or water currents.

■ Plants and fungi compensate for their inability to propel themselves by relying on animals, wind, or water currents to scatter their pollen, seeds, and spores.

(a)

(b)

Figure 27.11 How Plants and Fungi Get Around
To move a distance, plants (a) and fungi (b) must hitch a ride with animals, wind, or water currents.

Highlight

Racing Runners and Speeding Sperm Revisited

Sprinters and marathon runners, and their sperm counterparts, face a similar challenge: how best to allocate energy resources for generating thrust while overcoming drag. Both runners and sperm rely on motors based on molecules that slide past each other to generate the power needed for movement. Runners have at their disposal a body in which the muscles and rigid bones have combined to form intricate lever systems to control how the power generated by the muscles translates into motion. In addition, through training, athletes can influence the force that their muscles can generate. Sperm have only one chance to reach an egg, and they must rely on a much simpler flagellum to propel themselves.

Runners and sperm face very different problems when it comes to minimizing drag. Humans are relatively large and fast and therefore live in a universe dominated by pressure drag rather than by friction drag. Even sprinters, however, barely reach speeds at which pressure drag becomes significant as they move through air. Instead, the major way in which runners waste energy is the unnecessary up-and-down movement of the body, something that proper running technique can minimize. Small, slow-swimming sperm inhabit a watery universe dominated by friction drag that is greater than that in aquatic environments because the fluids in the reproductive tract have a greater viscosity than water. For sperm, pressure drag represents a minor problem compared to friction drag.

In spite of the functional and morphological gulf that separates runners and sperm, the differences between them diminish when viewed on the common scale of the energetic cost of locomotion. Like sprinters, sperm that move short distances can afford to consume a lot of energy quickly to attain as high a speed as possible. On the other hand, sperm that swim long distances to meet up with the egg must, like marathon runners, conserve their energy so that it lasts the whole trip.

> ■ Both humans and sperm trade off speed versus endurance. In both cases, high speed cannot be maintained indefinitely due to the cost of locomotion.

Summary

Animal Muscles as Biological Motors

■ The muscle tissue unique to animals gives them an ability to move that is not found in other multicellular organisms.

■ Muscle fibers are made up of sarcomeres, which contain actin and myosin.

■ Sarcomeres contract when myosin heads "walk" along actin filaments by attaching to successive sites along the filaments; this action pulls the Z discs anchored at the ends of the actin filaments closer together.

■ Muscles always work in pairs because individual muscles can only contract.

How Animals Convert Muscles to Motion

■ Muscles increase in strength when the number of muscle fibers within them, and therefore their cross-sectional area, increases.

■ Long muscles contract faster than short muscles.

■ Fast muscle fibers differ from slow muscle fibers in the way they respond to nerve impulses and the type of metabolism they rely on.

■ Muscles and skeletons work together as lever systems in which the relative lengths of the load and power arms help determine the force generated by muscle contraction.

Locomotion in Animals

■ Animals move by running, flying, or swimming.

■ Forward motion is resisted by two types of drag: Pressure drag increases with increased speed; friction drag increases with increased surface area or increased fluid viscosity.

■ Streamlined shapes reduce pressure drag by changing the flow of air or water around an organism.

■ Smooth surfaces reduce friction drag by allowing a viscous fluid to slip off an organism.

How Bacteria and Protists Move

■ Bacteria move by using rotary motors created by bacterial flagella that rotate like a propeller.

■ Protists move by using cilia or eukaryotic flagella, which bend. This bending results from a pair of sliding tubules, similar to the relative motion of actin and myosin in animal muscles.

How Plants and Fungi Get from Here to There

■ Plants and fungi compensate for their inability to move by relying on animals, wind, or water currents to disperse their pollen, seeds, and spores.

Highlight: Racing Runners and Speeding Sperm Revisited

■ Both runners and sperm either travel quickly or far, but not both.

■ Running consumes a lot of energy in generating an up-and-down motion, and swimming sperm spend a lot of energy overcoming drag. These energetic costs force the trade-off between speed and distance.

Key Terms

actin p. 416
bacterial flagellum p. 422
cilium p. 422
drag p. 419
eukaryotic flagellum p. 422
fast muscle fiber p. 417
friction drag p. 420
fulcrum p. 418
lever p. 418
load arm p. 418
muscle fiber p. 415

myofibril p. 415
myoglobin p. 418
myosin p. 416
power arm p. 418
pressure drag p. 419
sarcomere p. 415
slow muscle fiber p. 417
thrust p. 419
viscous p. 420
Z disc p. 415

Chapter Review

Self-Quiz

1. Which of the following is true about animal muscles?
   a. The muscle tissue in strong animals has different kinds of filaments in it than the muscle tissue of weak animals.
   b. Animals muscles can only contract, or shorten.
   c. Muscle contraction requires no energy.
   d. Both a and b are true.

2. From which of the following are grasshoppers most likely to get their great jumping power?
   a. leg muscles with a large cross-sectional area
   b. leg muscles that contain extra nuclei
   c. short leg muscles
   d. leg muscles that contain actin instead of myosin

3. To increase power from a lever system, you should
   a. make the power arm thicker.
   b. shorten the load arm.
   c. lengthen the load arm.
   d. make the load arm thicker.

4. Swimming is energetically cheaper than flying because
   a. water helps support an organism.
   b. water lubricates forward motion.
   c. there is no pressure drag in water.
   d. there is more friction drag in water.

5. The inner workings of eukaryotic flagella are most similar to
   a. bacterial flagella.
   b. plant growth
   c. lever systems.
   d. animal muscles.

Review Questions

1. Describe how myosin and actin interact to contract a sarcomere.

2. How can animals affect the speed and strength of muscle contraction if they all share essentially similar muscle tissue?

3. Consider the following pairs of organisms: (a) a walking elephant and a swimming whale, (b) a fast-swimming penguin and a fast-flying falcon (about the same size as a penguin). In each pair, which organism would you expect to face more resistance from pressure drag, and why? Similarly, which organism will face more friction drag?

4. How do fungi manage to spread so quickly in our refrigerators, where they spoil food, when they lack muscles?

27

The Daily Globe

Bike Paths Are the Answer

To the Editor:

I am writing in response to the Letter to the Editor that appeared in your newspaper last week condemning the expenditure of our taxes on maintaining bicycle paths in our city. The writer argued that tax money would be much better spent on roads used by everyone rather than on "luxuries" used by "a relative few."

I find this view not only narrow-minded but wrong.

Our city should be proud of having an excellent system of over 150 km of bike paths. These paths are more than just scenic byways along our rivers and through our parks. They provide an important commuter route connecting many of our neighbourhoods with the downtown core.

People constantly complain about the parking problems downtown, and the pollution problems that have come with the dramatic increase in car traffic around the city over the past several years.

Bicycles provide an excellent solution to these problems and more. Bicycles don't require parking places and cause no pollution, yet provide a fast and energy-free way of getting around. They make no noise, and they don't damage the environment.

Perhaps we should spend more, not less, valuable tax money on extending our bicycle paths to even more parts of our city.

Janice C. Shum

Evaluating "The News"

1. The author of this letter claims that bicycles provide an energy-free mode of transportation. Biologically, why is this claim not reasonable?

2. In what ways does the source of energy used to fuel the bicycle rider (food) also take an environmental toll?

3. Imagine that you are a member of the city council faced with letters like this and like the earlier letter favoring roads for cars over paths-for bicycles. How would you evaluate the relative costs and benefits to the environment of roads versus bicycle paths?

28

Nutrition

Rene Magritte, *L'Ile aux Tresors*, 1942.

The Rise of Agriculture and Malnutrition

*B*y all measures, the transition from a hunting and gathering way of life to an agricultural one represents a milestone in the development of civilization. Early hunter-gatherer cultures probably ate equal proportions of animal and plant foods. In contrast, agricultural societies depend mostly on plants, especially cereals such as corn and wheat, and tubers such as potatoes and cas-

sava. Cereals and tubers are rich in energy, store well when wild food is scarce, and can support more people per area than can animal-based foods. Farming thus allowed a tremendous increase in the density of human populations and created the conditions that led to cities, art, and industry.

For all its advantages, agriculture has an unexpected side effect: It can lead to poor

Main Message

Whereas producers capable of photosynthesis can obtain all their nutritional needs from the physical environment, consumers rely on organic nutrients produced by other organisms.

nutrition. Agriculture arose independently in three regions—the Middle East (10,000 years ago), southeastern Asia (8000 years ago), and Central America (5500 years ago)—from which it spread to other regions. In each case, the rise of agriculture coincided with an increase in diet-related diseases, as evidenced by the skeletons unearthed by archaeologists.

These patterns are striking at sites throughout the Americas, where agriculture traditionally depended almost completely on plant, rather than animal, crops. For example, before about AD 1000 the aboriginal populations in the Ohio River valley of the United States hunted deer and various small mammals and birds; gathered assorted fruits, berries, seeds, and nuts; and cultivated some native plants. After this time, the cultivation of corn and then beans was introduced from more southerly populations. With the introduction of agriculture, wild game dwindled to only a minor source of food in agricultural populations.

The introduction of agriculture coincided with an increase in the size and density of farming communities, indicating that the new technology increased the population capacity of

the Ohio River valley. Evidence from skeletal remains, however, suggests that, compared to the hunter-gatherers that came before them, the farmers were shorter, faced frequent food shortages, suffered nutrient deficiencies, and had rotten teeth.

In this chapter we consider how organisms obtain the nutrients they need for growth and maintenance. After introducing the basics of nutrition, we consider why an improved ability to provide food led, paradoxically, to malnutrition among early Americans.

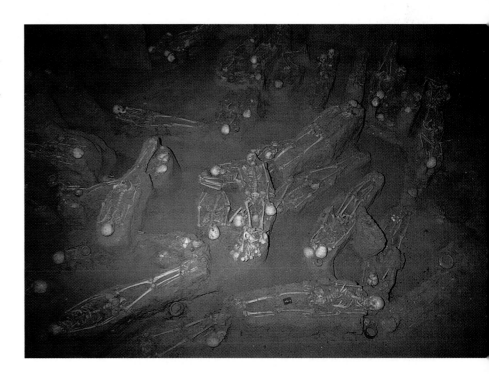

Skeletons in an Ohio River Valley Burial Mound

Key Concepts

1. Carbon, hydrogen, oxygen, nitrogen, phosphorus, and sulfur are the basic components of all chemicals produced by living organisms.

2. Animals generally process food in a fixed sequence (breaking it down first mechanically and then chemically) in a tubelike gut.

3. Plants capture energy from sunlight by photosynthesis in their leaves, and they absorb mineral nutrients from the soil using their roots.

4. Animals either seek out their food or wait for it to come to them. Plants and fungi grow toward and into their nutrient sources.

5. Organisms generally break down food into molecules that they can absorb.

6. The structures and functions of animal digestive systems reflect the animals' diets.

Some humans may live to eat, but all living things must eat to live. The absorption of nutrients from the environment is a basic function of all living things. Nutrients provide the energy and raw materials needed for all the other functions essential to life.

In this chapter we survey the different ways in which living things obtain nutrients. After quickly reviewing the nutrients basic to life, we will make the important distinction between producers and consumers. Producers photosynthesize to get the energy needed to produce larger carbon-containing chemicals from simpler substances available in their physical environment. Consumers depend on large carbon-containing substances formed by other organisms.

We will keep this important distinction in mind as we consider (1) the chemical form in which producers and consumers absorb nutrients, (2) the structures they use to absorb nutrients, (3) how they find nutrients, and (4) how they convert food into nutrients that they can absorb. We will conclude by considering some of the adaptations of consumers and producers for dealing with different sources of nutrition.

Elements of Nutrition: A Review

Life depends on various chemical elements and on energy. Carbon, hydrogen, and oxygen provide the basis for life. These three elements form the framework for all biologically produced, or **organic**, molecules, such as carbohydrates, proteins, fats, and nucleic acids. In addition, as carbohydrates and fats, they store the sun's energy in a chemical form.

Some additional elements play critical roles in the lives of organisms. These elements include nitrogen, phosphorus, and sulfur, which bond covalently to the carbon–hydrogen–oxygen backbones in amino acids, nucleotides, and the lipids that make up the lipid bilay-

er of the plasma membrane. Many other elements play equally essential roles as charged ions or in association with, but not covalently bonded to, organic molecules. For example, we have already encountered sodium and chloride ions that regulate water movement in the gut (see Chapter 24).

■ All life depends on a limited number of chemical elements, most notably carbon, hydrogen, oxygen, nitrogen, phosphorus, and sulfur.

Producers and Consumers

Life has evolved two distinctly different approaches to dealing with nutrition. On the one hand, **consumers** like animals and fungi cannot photosynthesize to capture energy directly from the sun, nor can they build all the organic molecules they need from inorganic raw materials. Consumers can only respire; thus they must convert organic chemicals made by other organisms into adenosine triphosphate (ATP) (Figure 28.1). Consumers also need organic molecules as raw materials out of which they synthesize their own carbohydrates, fats, proteins, and nucleotides. Thus, consumers must consume other organisms to obtain their energy and nutrients. ATP provides the energy used by consumers to synthesize organic chemicals.

We can classify consumers according to the form in which they obtain their organic nutrients. **Herbivores**, which include many animals and fungi, eat living plant tissues. **Carnivores** are consumers, primarily animals, that eat living animals. **Detritivores** feed on the tissues of dead organisms and include many microbes, fungi, and animals.

In contrast, **producers** can get all they need from their physical surroundings. Producers, which include cyano-

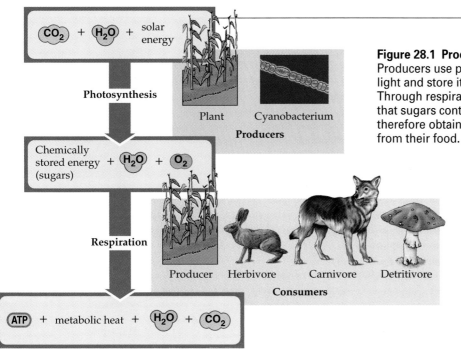

Figure 28.1 Producers and Consumers
Producers use photosynthesis to capture the energy of sunlight and store it in the chemical bonds of sugar molecules. Through respiration, producers can then release the energy that sugars contain. Consumers can only respire; they must therefore obtain organic chemicals as sources of energy from their food.

bacteria and plants, photosynthesize to convert the sun's energy into storable, chemical energy in the form of sugar molecules (see Figure 28.1). Like consumers, producers respire to release the energy stored in sugars as ATP. In producers, the ATP provides the energy needed to combine the sugars produced by photosynthesis with inorganic nutrients such as nitrates and phosphates obtained from the surrounding soil or water to synthesize the full range of organic chemicals.

■ Consumers can only respire to release the energy in organic nutrients. Producers both photosynthesize and respire.

Nutritional Requirements of Consumers and Producers

A close look at the nutritional labels on packages of human and plant foods reveals much about the differences in how producers and consumers obtain nutrients.

Consumers rely on organic nutrients for both chemical building blocks and energy

The nutritional labels on the sides of cereal boxes list the organic carbohydrates, fats, proteins, and vitamins that come from the grass seeds from which cereals are made (Figure 28.2a). Only the ionic nutrients in cereal sometimes come in an inorganic form (for example, salt, which provides sodium and chlorine).

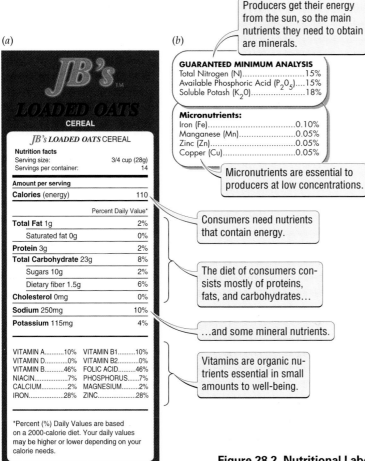

Producers get their energy from the sun, so the main nutrients they need to obtain are minerals.

Micronutrients are essential to producers at low concentrations.

Consumers need nutrients that contain energy.

The diet of consumers consists mostly of proteins, fats, and carbohydrates…

…and some mineral nutrients.

Vitamins are organic nutrients essential in small amounts to well-being.

Figure 28.2 Nutritional Labels Can Tell Us a Great Deal
(*a*) This label from a cereal box indicates the important sources of energy and nutrients for humans. We obtain energy from carbohydrates, fats, and proteins. (*b*) This label from a house plant fertilizer shows the nutritional analysis: the percentages of the essential nutrients—nitrogen (N), phosphorus (P), and potassium (K)—that the fertilizer contains.

Figure 28.3 Essential Amino Acids in the Human Diet
Humans must obtain from their food eight of the 21 amino acids they use to build proteins. To get all eight of these from plants in our diet, we must combine different grains and vegetables.

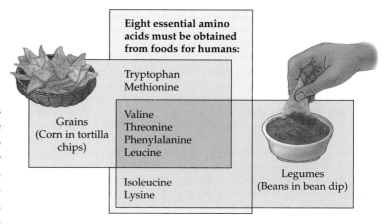

The respiration of carbohydrates and fats provides consumers like humans with most of the energy we need to survive. The organic chemicals in our food also serve as the building blocks that we reassemble into our own organic chemicals. Proteins provide as much energy per unit weight when broken down as do carbohydrates, but proteins play a much more important role as the source of amino acids used to build our own proteins. Although humans can synthesize some amino acids from sugars, others, called **essential amino acids**, we must obtain in our food (Figure 28.3).

Vitamins are a type of nutrient unique to consumers. Consumers need vitamins in only small amounts, but they are either unable to synthesize them or cannot synthesize enough to meet their own needs. Thus they must get vitamins in their food. A striking feature of vitamins is that whereas most animals have similar nutritional requirements for carbohydrates, fats, and proteins, they may differ greatly in which vitamins they need. Humans require two broad classes of vitamins: water-soluble vitamins, which help speed essential chemical reactions in the body, and fat-soluble vitamins, which play a variety of roles (Table 28.1). Lack of vitamins in the diet leads to deficiency diseases such as scurvy.

Although vitamin supplements are widely available commercially, there is no evidence for a benefit from taking more than the recommended daily intake of a vitamin: The human body quickly rids itself of excess water-soluble vitamins, and fat-soluble vitamins may build up in the body to toxic levels.

Producers rely on inorganic nutrients for chemical building blocks

The label on a typical package of plant food, or fertilizer, reveals that it provides plants with an inorganic form of three nutrients: nitrogen, phosphorus, and potassium (Figure 28.2b). These three nutrients are the ones most often in short supply in soil. Plant fertilizers provide no energy. Remember that photosynthesis lets plants store solar energy in chemical bonds as they convert the gas carbon dioxide from the air into sugars.

Organic fertilizers such as compost or manure improve plant nutrition not by providing ready-made chemical building blocks but by increasing the ability of soil to hold water and mineral nutrients. Tiny organic particles in soil tend to have negative charges, causing them to hold positively charged mineral nutrients tightly, which prevents them from being washed out of the soil. The organic nitrogen and phosphorus in manure, for example, become available to plants only after composting allows consumers such as soil bacteria and fungi to decompose the organic chemicals into mineral forms.

> ■ Consumers require carbon-based organic nutrients to provide the building blocks for their own chemistry. Producers use mineral nutrients.

Digestive Systems of Animals

Considering the diversity of form and diet among animals, their digestive systems have a remarkably consistent layout (Figure 28.4a). The human digestive system illustrates this basic pattern well (Figure 28.4b). The human gut carries food through the body from mouth

28.1	*Vitamins in the Human Diet*	
Vitamin type	**Examples**	**Main function**
Water-soluble	Thiamine, riboflavin, pyridoxine, B$_{12}$, biotin, C,[a] folates, niacin, pantothenic acid	Act with enzymes to speed metabolic reactions, or act as raw materials for chemicals that do so
Fat-soluble	Carotene, D, E, K	Produce visual pigment, calcium uptake in bone formation, protect fats from chemical breakdown, produce clotting agent in blood

[a] C is a vitamin only in primates, including humans, and a few other animals, such as guinea pigs. All other animals make the vitamin C they need.

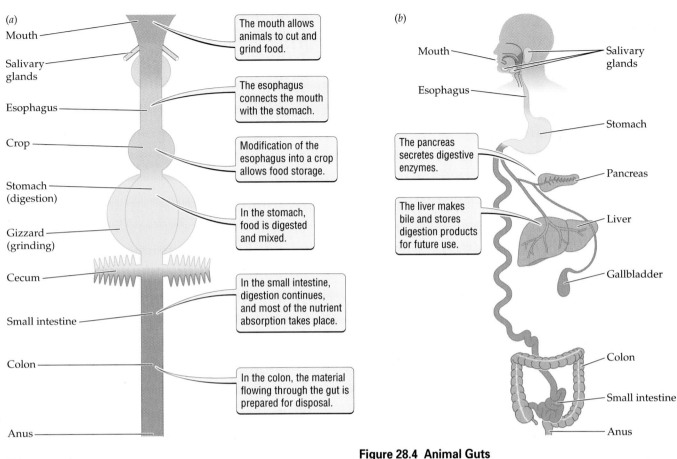

(a)

Mouth — The mouth allows animals to cut and grind food.

Salivary glands

Esophagus — The esophagus connects the mouth with the stomach.

Crop — Modification of the esophagus into a crop allows food storage.

Stomach (digestion) — In the stomach, food is digested and mixed.

Gizzard (grinding)

Cecum

Small intestine — In the small intestine, digestion continues, and most of the nutrient absorption takes place.

Colon — In the colon, the material flowing through the gut is prepared for disposal.

Anus

(b)

Mouth — Salivary glands

Esophagus

Stomach

The pancreas secretes digestive enzymes. — Pancreas

The liver makes bile and stores digestion products for future use. — Liver

Gallbladder

Colon

Small intestine

Anus

(c) **Nematode**

Mouth Intestine Anus

Earthworm Crop Intestine Anus

Mouth Esophagus Gizzard
Pharynx

Snail Intestine
Radula
Mouth Liver Stomach Anus

Cockroach Crop Gizzard
Esophagus
Anus
Rectum
Mandibles Salivary glands Intestine

Figure 28.4 Animal Guts
(*a*) The guts of all animals are organized in a remarkably consistent way. (*b*) The human digestive system follows this pattern, and even the guts of animals with specialized diets (*c*) show variations on this theme.

to anus. This one-way flow means that tissues at various points along the gut can specialize for a particular function to form a sort of food-processing assembly line.

Certain functions logically come before others along the assembly line. We mechanically break large pieces of food into small ones by chewing before swallowing. We **digest**, or break down the food chemically, mostly after chewing and before absorbing the nutrients. We absorb nutrients before preparing for disposal the waste that cannot be digested.

We can describe a generalized tubular animal gut in which food enters through a **mouth** that manipulates and physically breaks down food (see Figure 28.4*a*). Food moves through an **esophagus** that transports food to a **stomach**, which dissolves solid food into a watery fluid and where much of the chemical digestion process takes place. A **small intestine** continues the digestion of food and absorbs the nutrients. A **colon** prepares undigested material and waste for disposal through the **anus**.

■ Most animals have a tubelike gut through which food flows from a mouth through an esophagus to the stomach and intestine. Waste forms in the colon and leaves the body through the anus.

How Plants Obtain Energy and Nutrients

As producers, plants must perform two distinctly different nutritional functions: converting solar energy into a chemical form through photosynthesis and absorbing mineral nutrients from the soil. Most plants carry out photosynthesis in their leaves (Figure 28.5). Leaves are positioned on branches so that they are exposed as fully to the sun as possible. Leaves generally have a flat, thin shape that is suited to intercepting as much light as possible. Most photosynthesis takes place in the **palisade parenchyma** cells that lie mostly in the sunlit, upper half of the leaf, whereas the **spongy parenchyma** cells, specialized for absorbing carbon dioxide, generally lie mostly in the lower half of the leaf.

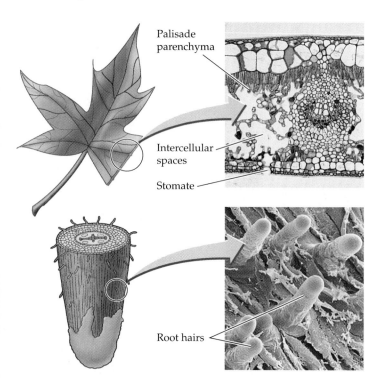

Figure 28.5 Plant Leaves and Roots
The leaf structure facilitates gas exchange and photosynthesis, with intercellular spaces (the spaces in the lower half of the leaf) and stomates (gaps in the lower leaf surface) near the lower surface and elongated cells of the palisade parenchyma nearest the sunlit surface. Underground root tips are covered with root hairs (green projections) that provide abundant surface area for nutrient absorption.

Plant roots absorb inorganic nutrients dissolved in water in the soil (see Figure 28.5). Specialized epidermal cells form fine **root hairs** that cover the root tips, providing a large surface area across which most nutrient absorption takes place. The inorganic nutrients on which plants depend generally require less processing before absorption than do the organic nutrients on which consumers depend. Thus, whereas animals must process nutrients extensively before they can be absorbed by the small intestine, plants can simply absorb nutrients.

■ Plants use thin, flat leaves to capture energy from the sun for photosynthesis, and underground roots to absorb inorganic nutrients from the soil.

Foraging for Food

The first step in nutrition, finding food, poses a difficult challenge for all organisms. How organisms forage for food depends largely on how well they and their food can change location through movement or growth.

Animals move to their food or wait for their food to come to them

Active foragers such as humans seek out food. By moving in search of their food, active foragers can feed on food that does not move (for example, herbivorous cattle and buffalo seek out patches of grass) or on food that moves actively (for example, ospreys dive into water to catch fish) (Figure 28.6*a* and *b*).

Some animals, however, wait for their food to come to them. For example, web-spinning spiders, which choose not to move in search of prey, and pitcher plants, which cannot move in search of food, must wait for food to come to them. Such creatures are called **sit-and-wait foragers**. They spend little energy on locomotion, and they depend on food that can move on its own or that is carried by air or water currents. Because sit-and-wait foragers often wait long times between meals, they must tolerate long periods without food. Many sit-and-wait foragers build elaborate traps that increase the chances that food will find them (Figure 28.6*c* and *d*).

Plants and fungi grow into and throughout their food

Although plants are producers and fungi are consumers, these two types of organisms face similar problems when it comes to finding food. Though they cannot change location by moving rapidly, they can change location by growing. The roots of plants grow through the soil that

Figure 28.6 How Organisms Forage for Food
Animals can forage actively by seeking out their food (*a* and *b*), or they can sit and wait for the food to come to them (*c* and *d*).

contains inorganic nutrients, and the rootlike **hyphae** of fungi grow into and throughout the living or dead tissues on which they feed (Figure 28.7 *a* and *b*).

The long roots and hyphae can forage over longer distances, and those with greater surface area can absorb nutrients more easily. The narrow tubes that make up roots and hyphae are the most effective way of devoting relatively little tissue to produce a lot of length and surface area. Root and hyphal systems grow in a strikingly similar pattern that allows them to distribute their absorptive surface to extract as much of the nutrients in their food source as possible. Some plants and fungi can concentrate growth specifically in places where nutrients are most abundant.

Figure 28.7 Plants and Fungi Forage in Similar Ways
The inability of both plants (producers) and fungi (consumers) to change location by rapid movement forces them to grow toward, into, and throughout their nutrient sources. (*a*) Plants absorb mineral nutrients using roots that grow through the soil. (*b*) The hyphae of fungi grow through their nutrient source. (*c*) Plant roots and fungal hyphae grow in similar ways.

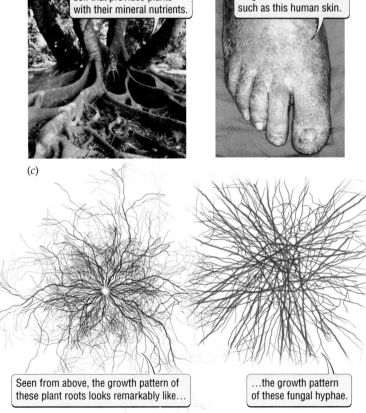

■ Animals forage for food either by actively seeking it or by waiting for it to arrive. Plants and fungi grow into and throughout their nutrient sources.

Digestion and Absorption of Nutrients

Consumers, and to a lesser extent producers, cannot absorb most nutrients in the form in which they occur in their food. Absorption requires that the nutrients cross a surface, like the the plasma membrane of a bacterium or the epithelium of the human gut, that serves in large part to separate the worlds inside and outside of an organism (see Chapter 24). Not surprisingly, the absorptive surface selectively lets some materials pass, while excluding others. Nutrients in food, therefore, must be converted into a form that can enter the organism.

Organisms must digest nutrients to absorb them

After we eat a bowl of cereal, our bodies have much to do before we can absorb and put to use the nutrients in the cereal. Flakes of cereal cannot pass from the liquid stream of food in the small intestine into the bloodstream without additional processing. Less obviously, even the organic chemicals that make up the cereal flakes are often too large and complex to make the crossing (Figure 28.8).

The human gut can absorb carbohydrates only as simple sugars, but much of the carbohydrates in cereal comes in the form of starches and fiber, which are enormous molecules consisting of many sugar units bound together. Similarly, the human gut can absorb only individual amino acids or chains of two or three units, but the amino acids in cereal come packaged as complex proteins. Fats rep-

resent a special challenge. Not only do they come as molecules too large to cross the gut wall, but they dissolve poorly in the watery mass of food moving through the gut, making absorption difficult.

Consumers other than animals, whether single-celled bacteria or multicellular fungi, must similarly break down the complex organic chemicals they encounter in their food before absorbing them. Even the roots of plants, which rely on simpler inorganic nutrients, can absorb only a small fraction of the nutrients in the soil without further processing.

Animals can break down food mechanically to speed digestion

Animals use muscles in combination with hard surfaces to mechanically grind large food particles into smaller ones. Mechanically reducing large food items into small ones provides two important benefits: (1) Large food

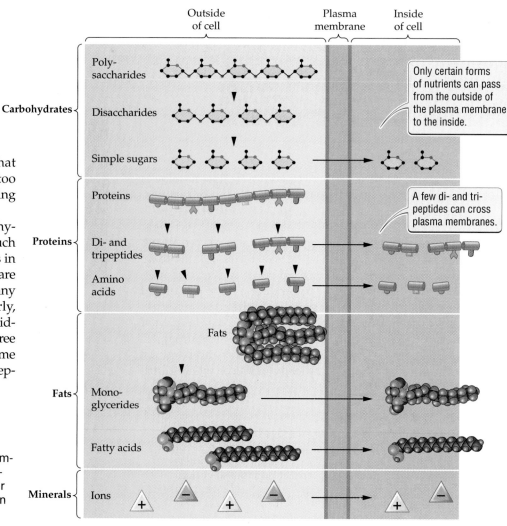

Figure 28.8 The Absorbable Forms of Nutrients
Most consumers can absorb only the simple breakdown products of the carbohydrates, proteins, and fats they encounter in their environment. Many minerals can cross membranes in their ionic forms.

items become small enough to enter the digestive system. (2) Relative to their volume, small food particles expose more of their surface area to chemical digestion, greatly speeding their breakdown compared to that of large food particles.

Like humans, many animals use hard teeth or jaws in their mouths to crush, rip, tear, or grind large food particles into small ones (see Figure 28.4*a*). Other animals, including birds and earthworms, have parts of their guts modified into muscular **gizzards**, which grind food against rocks or sand collected from the environment.

Organisms digest complex molecules into simpler ones that they can absorb

All organisms can chemically digest some complex molecules into a form that can cross the body wall. Recall that **enzymes** are proteins that speed the chemical reactions that break specific kinds of chemical bonds. As such, enzymes play a central role in converting carbohydrates, fats, and proteins into an absorbable form.

Humans produce enzymes that digest most of the nutrients in our cereal: Our saliva contains amylases that break down starches, and our pancreas releases into the small intestine disaccharidases that break two-unit sugars (disaccharides) into single sugar units, proteases that break proteins into their constituent amino acids, and

lipases that break fats into fatty acids and monoglycerides. In addition to enzymes, the stomach produces strong acids that contribute to the chemical breakdown of proteins. Plants perform a similar function by releasing various weak acids into the soil to release inorganic nutrients.

Because fats dissolve poorly in water, they remain in large clumps called **micelles** that are too large to pass from the small intestine into the blood. The large micelles must be broken down and made to dissolve more easily in the watery contents of the gut so that they can be chemically digested and absorbed. The human liver produces a substance called **bile**, which acts like dish detergent to put a water-soluble coating on small fat droplets. Fat-soluble vitamins are handled in the same way to increase their absorption. Some soil microbes produce chemicals that have a similar effect on insoluble forms of mineral nutrients, making them more available to plant roots.

When an organism lacks the enzyme needed to break down nutrients into absorbable units, those nutrients are not accessible. For example, consider the enzyme lactase, which breaks the disaccharide lactose into two simple sugars (Figure 28.9). Human infants, like those of all other mammals, need lactase when breast-feeding because lactose is the sugar in milk. Until humans began dairy farming, however, adults rarely encountered lactose. Accordingly, most humans stop producing lactase as they mature.

Only individuals of European, northern Indian, or Arabian or northern African descent frequently retain the ability to produce lactase as adults. The ability of adults in these populations to digest milk probably arose by chance mutation, but it opened up the possibility of using animal milk as a rich source of protein. In adults of other ethnic groups, bacteria living in the gut break down the undigested lactose, and the byproducts of this bacterial digestion lead to diarrhea and painful gas. Bacterially altered milk products, like cheese and yogurt, are generally digestible because the bacteria used in their production convert much of the lactose into a form we can digest.

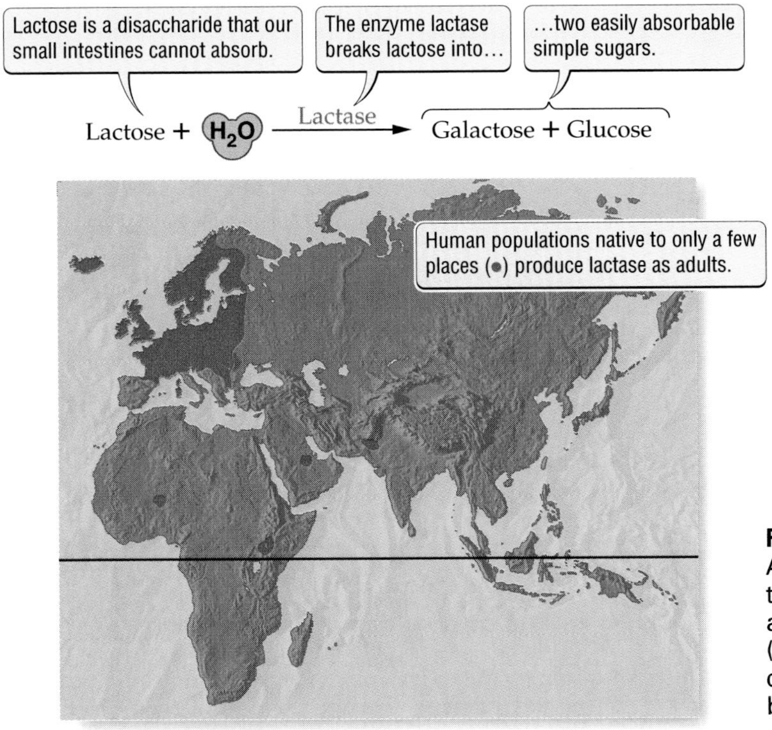

Lactose is a disaccharide that our small intestines cannot absorb.

The enzyme lactase breaks lactose into…

…two easily absorbable simple sugars.

$$\text{Lactose} + \text{H}_2\text{O} \xrightarrow{\text{Lactase}} \text{Galactose} + \text{Glucose}$$

Human populations native to only a few places (●) produce lactase as adults.

Figure 28.9 Lactase and Milk in the Human Diet
After being weaned, most people do not produce lactase, the enzyme that digests lactose, the major sugar in milk. In areas where virtually no one can digest lactose as an adult (green areas on the map), most milk products, for example, cheese and yogurt, are treated with bacteria or fungi that break down the lactose.

Absorption depends on the surface area across which nutrients can move

After converting nutrients into an absorbable form, an organism must selectively pass the ones it needs into its body. Organisms rely heavily on active transport involving carrier proteins, rather than on passive mechanisms, to move nutrients into their bodies. The end products of digestion are still too large to diffuse across the plasma membrane (see Chapter 24). In addition, the concentration of nutrients in food is often below that of the nutrients in the organism. So diffusion will not work. Plants and fungi, for example, typically grow through very dilute nutrient sources, and they often move nutrients into a body in which the nutrients are a thousand times more concentrated than in the environment.

How rapidly an organism can take up nutrients depends on the surface area available for absorption. Because nutrient demand depends mostly on the volume of the body, tiny single-celled organisms, with their high surface area–to–volume ratios need little modification to absorb enough nutrients. Multicellular plants, fungi, and animals, however, often need specialized absorptive surfaces that increase their surface area–to–volume ratio to provide an adequate supply of nutrients to their metabolizing cells.

Plant roots, roots hairs, fungal hyphae, and the small intestines of animal guts are all organs specialized for nutrient uptake. The shape of each organ maximizes the surface area–to–volume ratio. Each of these organs consists of long, narrow tubes that are branched or folded to maximize surface area while minimizing volume (see Figures 28.4 and 28.5).

The tiny root hairs formed by the epidermal cells near the tips of plant roots function specifically to increase the absorptive surface. In addition, many plants greatly increase the surface area of their roots by forming mutually beneficial associations with fungi. These associations, known as **mycorrhizae**, provide the fungi with direct access to organic nutrients made by the plants, and the threadlike hyphae of the mycorrhizal fungi extending from the root greatly increase the volume of soil from which the plants can pull inorganic nutrients (Figure 28.10).

> ■ Most animals first process food by mechanically breaking it into small pieces. Chemical digestion in all organisms converts large molecules into forms that the organism can absorb. Absorptive surfaces usually maximize surface area.

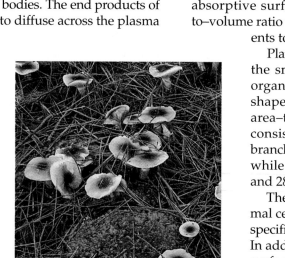

Root cells

Some mycorrhizal fungi grow around the root cells of their associated plants.

Other mycorrhizal fungi grow inside the root cells of their associated plants.

The hyphae of both kinds of mycorrhizae increase the surface area available for nutrient absorption.

Fungi

Figure 28.10 Mycorrhizal Associations
Many common forest mushrooms live in a hidden association with the trees around them. The fungal hyphae may surround the roots of the plants or may live within the plant roots. (In the bottom photo, the large purple areas inside the cells are fungi.) Mycorrhizal interactions between fungi and plant roots greatly increase the surface area for mineral absorption by the plant, and the fungi may obtain proteins and carbohydrates from the plant in return.

Adaptations of the Animal Digestive System to Different Diets

Although based on the same basic plan, the digestive systems of animals that eat different foods differ in ways that reflect the quality of the diet. In this section we illustrate some of the evolutionary modifications of the basic animal digestive system by comparing two similarly sized mammals: a dog and a sheep. As carnivores, dogs eat easily digestible and protein-rich meat. Sheep, in contrast, are herbivores that eat plant tissues low in protein and difficult to digest.

Several parts of the digestive system of dogs and sheep reflect the differences in their diet (Figure 28.11). Dogs have bladelike teeth suited for slicing meat into chunks small enough to swallow. Sheep have broad teeth that can grind tough plant tissues into small pieces. The simple stomachs of dogs produce acids to partially break down meat. The sheep's stomach has evolved into a complex, four-chambered structure that contains a thriving population of bacteria, fungi, and protists. As outlined in the Box on page 438, these microbes digest the cellulose, making the nutrients that it contains available to the sheep.

The dog has a relatively short small intestine that provides only a small surface area for digestion. The sheep, in contrast, has an extraordinarily long small intestine, which provides a huge surface area over which it can absorb the nutrients in its nutrient-poor diet. The elaborate gut of the sheep is what allows it to survive on a diet that could not support a dog.

■ All elements of animal digestive systems, from the mouth to absorptive surfaces, are designed to effectively process the specific type of food eaten by an organism.

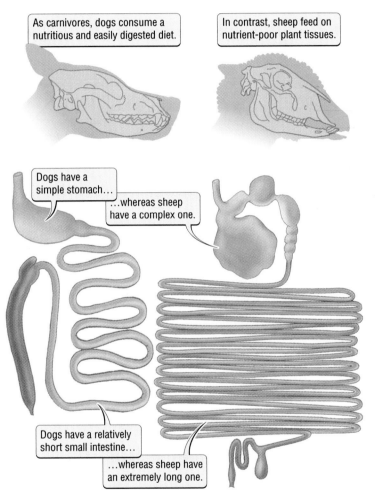

Figure 28.11 Herbivores and Carnivores Compared
Although of similar body length, carnivorous dogs and herbivorous sheep have remarkably different digestive tracts that reflect their different diets. Sheep skulls bear broad teeth suited to grinding tough, fibrous grass; dog skulls have slicing teeth suited to cutting meat into bite-sized pieces. Dogs have relatively simple stomachs and short small intestines that reflect the easily digestible nature of their food. Sheep, however, have evolved four-chambered stomachs and long small intestines that allow them to digest the difficult-to-digest grass in their diet and then to absorb most of the nutrients that are released.

Highlight

Agriculture and Malnutrition Revisited

The puzzling effect of agriculture—the simultaneous rise in malnutrition and increase in the population—stems from an increasing reliance on plants as a source of nutrients in agricultural societies. The high carbohydrate content of cereals like the corn that formed the staple food of agricultural populations in the Americas makes them rich in energy. The use of such foods greatly increased the ability of land to support larger populations compared to the protein-rich, but energy-poor, diets of hunter-gatherer societies. In addition, the easily digestible starches provided a potential energy source for infants too young to chew meat. Thus crops such as corn allowed mothers to wean infants at an earlier age than is possible among hunter-gatherers. This ability, in turn, decreased the minimum time between pregnancies and increased the population growth rate.

Although rich in energy, corn is poor in several essential nutrients. The overall protein content of corn is well below that of animal tissues, and corn proteins contain

The Scientific Process

The Case of the Missing Enzyme

Animals that feed on land plants face a serious problem: They cannot digest much of the food they eat. Much of the carbon in land plants is locked up in the complex carbohydrate **cellulose**, which forms the plant skeleton. Whereas some bacteria, single-celled eukaryotes, and fungi can produce cellulase, the enzyme that breaks down cellulose into simple sugars, most animals cannot. How herbivores handle cellulose matters a great deal because the animals that are most important in agriculture (for example, cattle, sheep, and goats) and many important pest insects (for example, leaf-cutter ants) thrive on plant tissues rich in cellulose (see the figure).

All animals that feed on food rich in cellulose enter into a mutualistic relationship with the microbes that can produce cellulase. All these animals depend on fungi, protists, or bacteria to digest the cellulose for them. The microbes, for their part, use the animals to provide them with a well-maintained home and to gather food for them (tiny microbes cannot forage widely for food).

The complex, four-chambered stomach of the cow, for example, is really an elaborate fermentation chamber that provides ideal conditions for a complex community of cellulase-producing fungi, protists, and bacteria. Leaf-cutter ants, in contrast, maintain their fungal cellulose-digesting part-

ners outside their bodies: The ants bring carefully selected leaf fragments to an underground fungus "garden" that they tend with great care.

These animal–microbe teams represent combinations of mobile foraging units (the animals) and digestive units (the fungi, protists, and bacteria) that make it possible to process the tremendous quantities of cellulose that plants produce annually.

Cows and Leaf-Cutter Ants
Cows and leaf-cutter ants rely on microbes to digest the cellulose in the plant tissues on which they subsist. The cows and ants mechanically break down the food, and microbes that produce the enzyme cellulase then digest the cellulose.

relatively little of two essential amino acids (lysine and tryptophan) that humans cannot make on their own.

Protein deficiency most strongly affects young children, in whom it stunts growth. Only by eating 0.45 kilogram (1 pound) of corn, or almost 0.8 kilogram (2 pounds) of tortillas, a day could a growing child get enough protein. Not only would this amount of food be a lot for a child to eat, but it would provide many more calories than needed. In addition, the form of iron and the vitamin niacin found in corn are not easily digested by humans.

A shortage of niacin and tryptophan leads to the diet deficiency disease **pellagra**, the symptoms of which include skin irritations, diarrhea, and dementia (Figure 28.12). A shortage of absorbable iron leads to anemia because of inadequate production of red blood cells. As a final insult, carbohydrate-rich diets cause dental decay and cavities.

Cultural approaches can overcome some of the nutritional deficiencies of early agricultural diets. The addition of lysine-rich beans to the diet improved the quality of protein in a corn-based diet (see Figure 28.3). Furthermore, aboriginal cultures in Central and North America commonly treat corn with a chemical called lye to form the corn meal used in making tortillas and to produce hominy. The treatment of corn with lye makes niacin and iron more accessible, thereby reducing the incidence of both pellagra and anemia. In these sorts of ways, ancient

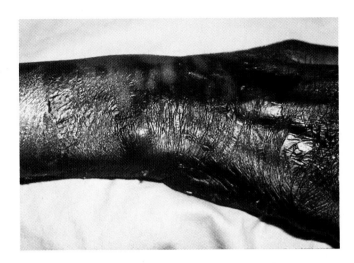

Figure 28.12 The Effect of Corn as a Staple in the Human Diet
Pellagra results from a deficiency in the vitamin niacin that often occurs among people who depend heavily on corn for their diet.

and contemporary agricultural societies have managed to reap the energetic benefits of plant-based diets, while avoiding some of the nutritional pitfalls.

■ The development of agricultural societies may have been associated with malnutrition because some crop plants lack important nutrients or contain them in a form that cannot be digested. Later modifications to agricultural diets and food preparation techniques helped overcome these difficulties.

Summary

Elements of Nutrition: A Review

■ Life depends on the elements carbon, hydrogen, and oxygen, which form the basis of all organic molecules.

■ Various other elements, especially nitrogen, phosphorus, and sulfur, play critical roles in the function of certain types of organic molecules or in cellular processes necessary for life.

Producers and Consumers

■ Consumers—herbivores, carnivores, and detritivores—can only respire, and thus must eat other organisms to acquire organic nutrients.

■ Producers can photosynthesize to capture the sun's energy in sugar molecules, which they can later process through cellular respiration.

Nutritional Requirements of Consumers and Producers

■ Consumers rely on organic nutrients for both chemical building blocks and energy; respiration releases the energy in carbohydrates and fats, and proteins are used as a source for amino acids.

■ Producers rely on inorganic nutrients for chemical building blocks only and obtain their energy from the sun.

Digestive Systems of Animals

■ Most animals have tubelike guts through which food flows.

■ Food enters through the mouth and is transported through the esophagus to the stomach, where chemicals aid in digestion.

■ After the stomach, the food is transported to the intestine, where nutrients are further digested and absorbed.

■ Transfer to the colon allows waste material to be prepared for disposal through the anus.

How Plants Obtain Energy and Nutrients

■ Plant leaves capture solar energy via photosynthesis.

■ Plant roots absorb inorganic nutrients from the soil.

Foraging for Food

■ Animals can be either active foragers that move in search of food, or sit-and-wait foragers that depend on the movement of food to them.

■ Plant roots and fungal hyphae grow into and throughout their sources of nutrients.

Digestion and Absorption of Nutrients

■ Most nutrients obtained from food must be processed before they can be absorbed.

■ In animals, mechanical breakdown allows food to enter the digestive system and increases its surface area.

■ Chemical digestion breaks large, complex molecules into simpler ones that can be absorbed.

■ Most nutrients are actively transported into the body; in multicellular organisms, specialized absorptive tissues increase the surface area available for such transport.

Adaptations of the Animal Digestive System to Different Diets

■ All elements of animal digestive systems, from the mouth to absorptive surfaces, are designed to effectively process the specific type of food eaten by an organism.

Highlight: Agriculture and Malnutrition Revisited

■ The development of agricultural societies may have been associated with malnutrition because some crop plants lack important nutrients or contain them in a form that cannot be digested.

■ Later modifications to agricultural diets and food preparation techniques helped overcome these difficulties.

Key Terms

active forager p. 432

anus p. 431

bile p. 435

carnivore p. 428

cellulose p. 438

colon p. 431

consumer p. 428

detritivore p. 428

digestion p. 431

enzyme p. 435

esophagus p. 431

essential amino acid p. 430

gizzard p. 435

herbivore p. 428

hypha p. 433

micelle p. 435

mouth p. 431

mycorrhiza p. 436

organic p. 428

palisade parenchyma p. 432

pellagra p. 438

producer p. 428

root hair p. 432

sit-and-wait forager p. 432

small intestine p. 431

spongy parenchyma p. 432

stomach p. 431

vitamin p. 430

Chapter Review

Self Quiz

1. All organic molecules
   a. occur only in producers.
   b. always contain water.
   c. are easily absorbed by consumers.
   d. always contain the elements carbon, hydrogen, and nitrogen.

2. Organisms that can both respire and photosynthesize are
   a. very rare.
   b. producers.
   c. consumers.
   d. fungi.

3. Which of the following statement is true of the biology of fungi that live as decomposers?
   a. They feed on living human skin.
   b. They absorb carbon-based nutrients through hyphae.
   c. They use leaves to photosynthesize and exchange gas.
   d. all of the above

4. Digestion is necessary because
   a. nutrients generally occur in a form that cannot be absorbed by an organism.
   b. nutrients must be metabolized before they can be used.
   c. this allows organisms to chew food more easily.
   d. both b and c

5. The guts of herbivores like kangaroos
   a. process food in an assembly-line fashion, much as we do.
   b. have relatively short small intestines.
   c. allow them to break down the large amounts of protein in the animals on which they feed.
   d. none of the above

Review Questions

1. The pitcher plant illustrated in Figure 28.6 supplements the mineral nutrients that it absorbs from the soil with nutrients obtained by trapping and digesting insects in its pitcherlike leaves. Can you think of reasons why the pitcher plant might have evolved this way of feeding? How would you expect the nutritional biology of pitcher plants to compare to the nutritional biology of other plants?

2. Our digestive system is intermediate between that of strict carnivores and strict herbivores. How might you expect our digestive system differ if we had fed exclusively on plant tissues throughout our evolutionary history?

3. Most of us have been told that we should eat a balanced diet. Using vocabulary and examples from this chapter, explain some reasons for this suggestion.

The Daily Globe

OPINION

Help Protect Our Coastal Waters

The Gulf Coast Conservation Society asks that you join us in a fight to preserve our coastal environment—using nothing more than our household budgets as a weapon. Spend a little more to buy produce grown without the use of chemical fertilizers.

Like seacoasts in many parts of the world, along our coastline today an eerie scene is appearing: large patches of once-productive sea floor, and the water above them, are lifeless for many months of the year. These "dead zones," which sometimes cover tens of thousands of square miles, appear to be caused largely by agricultural fertilizers, which run into the ocean through river systems that drain agricultural lands.

Agricultural fields are often treated with up to 100 pounds of nitrogen per acre, which allows for significant increases in crop production. However, when excess chemical fertilizer runs off fields into fresh or salt water, it has a similar growth-encouraging effect on algae, which ultimately can lead to dead zones. Fertilized algae grow and reproduce much more quickly than most of the animals that eat them, and their populations "bloom" in nitrogen-enriched water. Since consumers can't keep up with their population growth, the algae die, and bacteria populations explode as they scavenge on dead algae. The intense decomposing activity of the bacteria is what causes the dead zones; the bacteria use up all the available oxygen in the water as they feed on the algae.

Unfortunately, reducing nitrogen runoff from agricultural lands will not come cheaply. One potential solution involves reducing the amount of fertilizer applied and monitoring fields to use nitrogen only in those portions of each field that need application. Changes to other agricultural practices, including plowing and irrigation techniques, would require investment in new equipment for many farms. Setting aside large buffer zones around fields to allow for absorption of more nitrogen before it reaches rivers would be beneficial, but it would take valuable land out of cultivation.

In order to save our waters, we must be willing to help pay the bill for these solutions. Please join our campaign to inform agricultural producers that we are willing to pay a little more for "ocean-friendly" food grown using organic fertilizer, so they can continue to profit from their livelihood without damaging our aquatic resources.

Evaluating "The News"

1. Dead zones are an example of what some might call too much of a good thing—the nutrient nitrogen. What sorts of economic impacts could the dead zones of coastal waters have?

2. Having benefited in your lifetime from the ability of farmers to produce food cheaply by adding excessive nutrients into the environment, how much more would you be willing to pay for vegetables now to protect the environment? Would you be willing to pay even if most other people were still buying the cheap, environmentally unsafe vegetables?

3. In these so-called dead zones, animals that need oxygen die, but bacteria that do not require oxygen thrive. Crabs and shrimp, for example, are dying, but they are being replaced by many more bacteria, so do conservationists really need to be worried about the changes caused by excess nutrients in the water? Why should bacteria be viewed as any less important organisms than crabs or shrimp?

chapter 29

Gas Exchange

Andy Warhol, *Superman*, 1960.

In High Places

When hiking high in the Rocky Mountains, the Alps, or the Andes, you may find yourself short of breath. Shortness of breath is one of the first signs of the low oxygen content of air at high altitudes. The amount of oxygen in each liter of air that we inhale decreases as we move to higher altitudes, and it is cut in half every 5500 meters up. Physiological changes in our body resulting from low levels of oxygen start to appear at about 3000 meters, and they become more severe as we climb higher. From shortness of breath, symptoms progress to headaches, fatigue, and loss of appetite, and in severe cases of what is called mountain sickness, to nausea, heart palpitations, and hallucinations. At 7100 meters, where each breath of air contains only one-third as much oxygen as at sea level, most people lapse into unconsciousness.

Although the consequences of reduced oxygen supply at high elevations can be severe, humans can adapt to these conditions. The highest per-

442

manent human habitations lie at 5700 meters in the Andes Mountains of South America. Two Austrian mountain climbers have even made it to the top of Mount Everest (8848 meters) without the use of oxygen supplies. Other animals and plants also function at high altitudes; the record is held by an unfortunate vulture that met its end in an encounter with a transport plane flying over Africa at 11,300 meters.

In this chapter we consider how organisms move the gases they need, such as oxygen, into and out of their bodies. Once we understand the basic principles of gas exchange, we can return to consider how humans and other organisms can adjust how they exchange gases to survive in environments where the essential gases are in short supply.

High Altitude, Low Oxygen
The African vulture holds the high altitude record; residents of the Andes have adapted to low oxygen levels.

Key Concepts

1. Oxygen and carbon dioxide enter and leave organisms exclusively by passive diffusion.

2. Land- and water-dwelling organisms face very different challenges in exchanging gases.

3. Specialized gas exchange surfaces provide a lot of surface area in a small volume.

4. Whereas plants and insects transport gases directly to metabolizing cells, many animals transport gases in their blood.

5. Unavoidable water loss across the gas exchange membrane is one of the most serious problems faced by land-dwelling organisms.

More than 95 percent of the weight of living things consists of four chemical elements: carbon, hydrogen, oxygen, and nitrogen. Of these elements, organisms regularly take in carbon, oxygen, and, in a few important cases, nitrogen as gases. Almost all organisms exchange carbon dioxide and oxygen with their environment. Carbon dioxide provides the carbon from which photosynthesizing producers build the chemicals basic to life. Both consumers and photosynthesizers use oxygen when they break down carbon-based chemicals to release energy (see Chapter 8). Organisms of one very special group, the nitrogen-fixing bacteria, play an essential role in biological systems by taking up and converting abundant gaseous nitrogen, which most organisms cannot use, into biologically available forms (see the Box on page 445).

At first glance, corn plants and humans differ greatly when it comes to gas exchange. As producers, corn plants absorb from their environment carbon dioxide, which they use to make carbohydrates, and they give off oxygen as a waste gas. As consumers, humans absorb oxygen from the air we breathe and use it to release energy from the foods we eat, giving off the waste gas carbon dioxide. The leaves of the corn plants, which absorb carbon dioxide and release oxygen, seem to have little in common with our lungs, which absorb oxygen and release carbon dioxide. Nonetheless, both corn plants and humans have common gas exchange problems to solve.

In many important ways humans resemble corn plants in how we exchange gases with our environment (Figure 29.1). Both corn plants and humans process a lot of air to get the gases they need: Corn plants process about 2500 liters of air to get enough carbon dioxide to produce one-fourth teaspoon of sugar, and when we exercise intensively, humans need the oxygen in 3000 liters of air each hour.

Unlike single-celled organisms, multicellular corn plants and humans must transport oxygen and carbon dioxide to or from the individual cells where they are used. Unlike single-celled organisms and some multicellular organisms, corn plants and humans have specialized certain parts of their bodies for gas exchange. The gas exchange surface of corn lies inside the leaf and consists of a tremendous surface area provided by cells of the spongy parenchyma layer (see Figure 29.1).

Like the sponges that give it its name, the spongy parenchyma is riddled with a continuous system of holes that form the **intercellular air space**. The gas exchange surface inside our lungs covers almost 100 square meters, an area equivalent to the floor space in a large classroom. Unlike aquatic organisms, corn plants and humans have their specialized gas exchange surface on the inside rather than the outside of their bodies. As we will see, internal structures reduce water loss during gas exchange.

In this chapter we focus on how oxygen and carbon dioxide move into and out of organisms and on why organisms from evolutionary origins as different as those of corn plants and humans should have evolved functionally similar structures. To provide some basic background, we begin by describing how gases enter and leave organisms and how the availability of carbon dioxide and oxygen differs in aquatic and terrestrial habitats. We then turn to some basic biological issues: (1) What adaptations increase the rate at which organisms can absorb gases? (2) How do plants and animals transport gases to and from the cells where they are used? Throughout the chapter we refer to the different adaptations to gas exchange on land and in water.

Oxygen and Carbon Dioxide Enter Organisms by Diffusion

Given that organisms need to absorb lots of oxygen and carbon dioxide, it comes as a bit of a surprise that these gases move into and out of all living things solely by the passive process of diffusion (see Chapter 24). No known

Nitrogen, Nitrogen, Everywhere . . .

Like sailors stranded at sea, surrounded by water but unable to drink it because of its salt content, plants and animals struggle to obtain nitrogen for making amino acids while bathed by air or water that contains vast quantities of unusable nitrogen. Only a few organisms can use N_2, the gaseous form of nitrogen that makes up 78 percent of Earth's atmosphere. Nitrogen-fixing bacteria can convert N_2 into forms of nitrogen that other organisms can use.

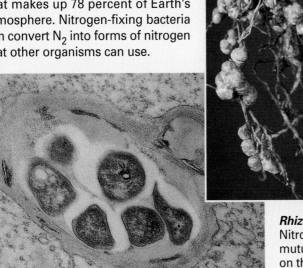

Some of these remarkable bacteria live in the soil, but many form close relationships with plants. The most famous of these are nitrogen-fixing bacteria in the genus *Rhizobium*, which live in the roots of familiar garden vegetables such as beans and peas (see the figure). *Rhizobium* take carbohydrates from the plant, and the plants benefit from the nitrogen fixed by the bacteria.

Farmers and gardeners will immediately recognize how valuable a service these bacteria perform. Every year they spend tens of millions of dollars on chemical fertilizers that provide plants with enough nitrogen to allow vigorous growth. Nitrogen-fixing bacteria, on the other hand, make nitrogen at no cost to us.

Rhizobium
Nitrogen-fixing bacteria in the genus *Rhizobium* form close, mutually beneficial relationships with peas and beans. The lumps on the roots in the photo above are nodules that contain large numbers of *Rhizobium* (the five large dark areas shown in the electron micrograph at left).

---

membrane proteins actively carry oxygen and carbon dioxide across plasma membranes. Because the exchange of oxygen and carbon dioxide depends on diffusion, three basic rules apply to gas exchange in all organisms:

Rule 1: The higher the concentration of gas in the environment compared to inside the organism, the faster the gas will diffuse in

Two biologically important points follow from the diffusion of gas down concentration gradients. Organisms that live in habitats where a gas occurs in high concentration can maintain a higher rate of diffusion of that gas into their bodies than can organisms living in habitats with low concentrations of the gas. In addition, the concentration of a gas inside the metabolizing cell must

be lower than in the environment from which it is absorbed.

Rule 2: The more surface area available for diffusion, the faster the gas will diffuse

We will focus on two biological consequences of the relationship between surface area and diffusion rate. First, single-celled and other small organisms should have fewer problems with gas exchange than large organisms because small organisms have high surface area–to–volume ratios (as discussed in Chapter 25). Second, large, multicellular organisms have evolved specialized gas exchange surfaces that maximize the area of the exchange surface, such as our lungs, the spongy parenchyma of plants, and the gills of aquatic animals (see Figure 29.1).

Rule 3: The smaller the distance over which the gas must diffuse, the faster the gas will diffuse

Oxygen and carbon dioxide cannot diffuse over long distances fast enough to meet the metabolic needs of large organisms. Thus, multicellular organisms such as plants and animals must either have a shape that brings all their cells within a few millimeters of the air or water around them or have a way of transporting gases inside their bodies. Plants have flat, thin leaves in which all cells lie near a gas exchange surface, and humans have a circulatory system that carries oxygen from our lungs to our metabolizing cells.

For example, consider how we humans absorb oxygen compared to how a trout does. We use about 2500 times more oxygen per hour than the trout, both because we are much bigger and because we use energy much more quickly. To meet our oxygen demands, we breathe about 3 kilograms of air in one hour. The trout, however, can meet its demand for oxygen only by passing 150 kilograms of heavy, oxygen-poor water over its gas exchange surfaces in the same amount of time.

Trout

> ■ Oxygen and carbon dioxide pass into and out of organisms exclusively by diffusion. Large concentration gradients, large surface areas, and short distances all help speed diffusion.

Oxygen and Carbon Dioxide in the Environment

The concentrations of oxygen and carbon dioxide in air and water set upper limits on how large the concentration gradients of these gases between the air and the plants can be. Therefore, we turn our attention to the distribution of these gases in the physical environment.

The problems of gas exchange on land are different from those in water

The circumstances under which aquatic organisms exchange gases with their environment differ from those in which terrestrial organisms exchange gases with their environment (Figure 29.2). Oxygen occurs in much higher concentrations in air than in water. Carbon dioxide, on the other hand, occurs in similar concentrations in both aquatic and terrestrial habitats.

In addition to differences in the abundance of oxygen and carbon dioxide, air and water differ physically in ways that affect the uptake of gases. For example, oxygen and carbon dioxide diffuse thousands of times more slowly in water than in air, so that in still water organisms can easily use up the oxygen or carbon dioxide supply. The much higher density of water compared to air (a liter of water weighs 1000 times more than a liter of air) also has profound implications for gas exchange.

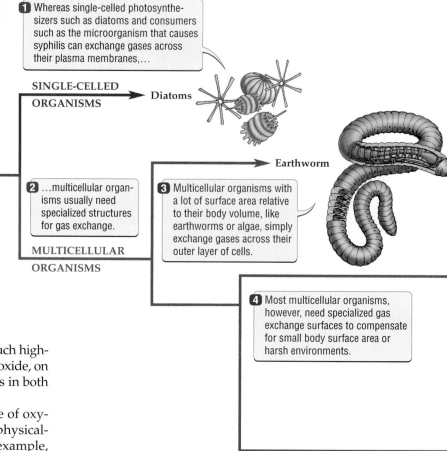

1 Whereas single-celled photosynthesizers such as diatoms and consumers such as the microorganism that causes syphilis can exchange gases across their plasma membranes,...

SINGLE-CELLED ORGANISMS   Diatoms

2 ...multicellular organisms usually need specialized structures for gas exchange.

MULTICELLULAR ORGANISMS

3 Multicellular organisms with a lot of surface area relative to their body volume, like earthworms or algae, simply exchange gases across their outer layer of cells.

Earthworm

4 Most multicellular organisms, however, need specialized gas exchange surfaces to compensate for small body surface area or harsh environments.

Figure 29.1 Gas Exchange Structures
The structures involved in gas exchange show clear patterns depending on the form of the organism and its environment.

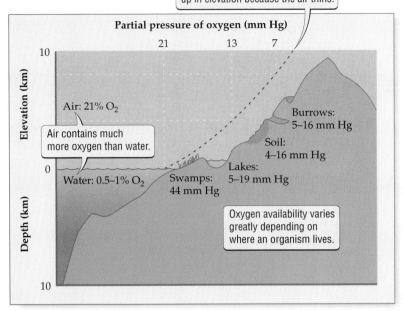

Figure 29.2 Availability of Gases in the Environment
Gas availability differs greatly among habitats. The pressure exerted by oxygen (indicated in mm Hg—the height in millimeters to which it can lift a column of heavy liquid mercury) indicates its availability: the greater the pressure, the greater the availability.

If we breathed water instead of air, our lungs would have to move about 350,000 kilograms of water in one hour. People drown not because water contains no oxygen, but because they cannot possibly pump enough heavy oxygen through their lungs fast enough.

Oxygen diffuses into animals more rapidly than carbon dioxide diffuses into plants

In absorbing oxygen, consumers such as animals deal with a much more abundant resource than carbon dioxide–absorbing, photosynthesizing plants do. Oxygen has a concentration 700 times that of carbon dioxide in air, and a concentration almost 25 times greater than carbon dioxide in water (see Figure 29.2). Rule 1 of the previous section tells us that oxygen should diffuse into our bodies many times faster than carbon dioxide diffuses into the crop plants on which we depend (Figure 29.3). In fact, many plants photosynthesize more rapidly if supplied with air that is artificially enriched with carbon dioxide to levels of as high as 20 milligrams of carbon dioxide per liter of air, a fact that farmers take advantage of when they pump carbon dioxide into greenhouses filled with crop plants.

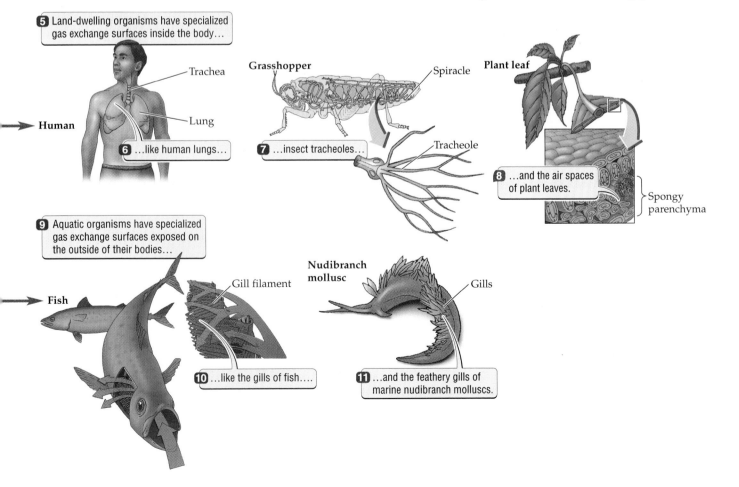

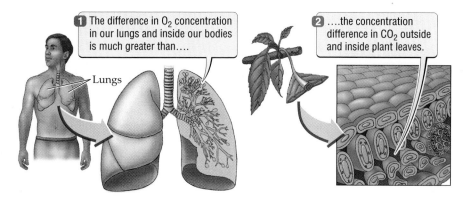

1 The difference in O₂ concentration in our lungs and inside our bodies is much greater than....

Lungs

2 ....the concentration difference in CO₂ outside and inside plant leaves.

Figure 29.3 Gas Exchange in Plants and Animals
There is a smaller difference in carbon dioxide concentration between plant leaves and the surrounding air compared to the difference in oxygen concentration between our bodies and the surrounding air. Thus, carbon dioxide diffuses into plants much more slowly than oxygen diffuses into the human body.

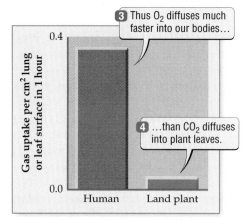

3 Thus O₂ diffuses much faster into our bodies...

4 ...than CO₂ diffuses into plant leaves.

Gas uptake per cm² lung or leaf surface in 1 hour

0.4

0.0

Human   Land plant

■ Air is richer in oxygen than water is, though carbon dioxide occurs in both environments in similar concentrations. Oxygen diffuses into animals faster than carbon dioxide diffuses into plants because environmental concentrations of oxygen are greater than those of carbon dioxide.

Gas Exchange Structures

The part of the body surface that is used for gas exchange must provide enough surface area to supply the organism's needs as rapidly as possible.

Small organisms often lack specialized gas exchange structures

As with all functions that require a large surface area–to–volume ratio, single-celled creatures can obtain oxygen and carbon dioxide with little modification (see Figure 29.1). Most bacteria and protists, whether aquatic or terrestrial and whether producers or consumers, simply exchange oxygen and carbon dioxide across their plasma membranes. Because of the higher oxygen concentrations in air compared to water, land-dwelling or-

ganisms can reach a greater size than can aquatic organisms before needing specialized gas exchange surfaces. The 9-centimeter-long red-backed salamander, one of the most common salamanders in eastern North America, absorbs all the oxygen it needs across its skin; it has no lungs.

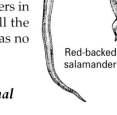

Red-backed salamander

Gas exchange surfaces are external in aquatic organisms, internal in terrestrial organisms

Organisms with bodies more than a few millimeters in diameter usually need a specialized gas exchange surface to get enough oxygen or carbon dioxide to support their needs. Aquatic animals have **gills**, which provide a large surface area for gas exchange on the outside (see Figure 29.1). Gills have evolved several times among aquatic animals, but in all cases they follow Rule 2 in providing a large surface area in a small space. The gills of fish, for example, consist of finely folded sheets of thin epithelial tissue.

In contrast, terrestrial organisms have internal gas exchange surfaces. These surfaces include the **lungs** of humans and other land-dwelling vertebrates. Our lungs are elastic sacs that allow us to pump air into and out of the body. In our lungs, gases are exchanged in tiny sacs called **alveoli** that lie at the ends of a multitude of tiny, tubular branches (Figure 29.4). Although individually tiny, the combined surface area of the 150 million alveoli in a typical human lung provides an exchange surface 90 times the surface area of our skin.

Insects have evolved a distinctly different sort of internal gas exchange surface in their **tracheoles** (see Figure 29.1). The tracheoles are a set of tiny tubes that carry oxygen from openings called **spiracles** on the outside of the insect to all the respiring cells inside.

In plants, the tubelike intercellular air spaces that run through the spongy parenchyma allow carbon dioxide

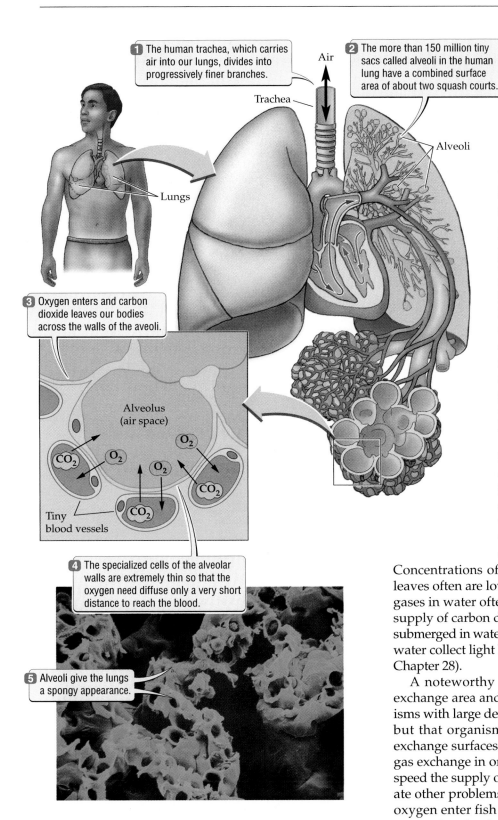

1 The human trachea, which carries air into our lungs, divides into progressively finer branches.

2 The more than 150 million tiny sacs called alveoli in the human lung have a combined surface area of about two squash courts.

Air

Trachea

Alveoli

Lungs

3 Oxygen enters and carbon dioxide leaves our bodies across the walls of the aveoli.

Alveolus (air space)

CO_2 O_2 O_2

O_2

CO_2

CO_2

Tiny blood vessels

4 The specialized cells of the alveolar walls are extremely thin so that the oxygen need diffuse only a very short distance to reach the blood.

5 Alveoli give the lungs a spongy appearance.

Figure 29.4 Human Lungs
The structure of our lungs speeds the diffusion of oxygen and carbon dioxide into and out of our bodies by providing a large surface area for exchange.

The surface area provided by the gas exchange surface matches the needs of the organism

Evidence from both animals and plants suggests that natural selection has led to gas exchange surface areas that match the oxygen or carbon dioxide demands of the organism. The lungs of active mammals, such as dogs, provide a large alveolar area compared to those of less active mammals, such as cows. In fish, the amount of gill surface area that supplies oxygen to each gram of tissue increases dramatically in fish species that have high activity levels (Figure 29.5).

The leaves of the water milfoil plant have a very different form depending on whether they grow in the water or in the air (Figure 29.6). The feathery shape of the underwater leaves provides more surface area for exchange, which allows them to absorb carbon dioxide even at low concentrations. Concentrations of carbon dioxide around submerged leaves often are low because the slow diffusion rates of gases in water often cause the plant to use up the local supply of carbon dioxide. In contrast to leaves growing submerged in water, the sheetlike leaves produced above water collect light better than the feather shapes do (see Chapter 28).

A noteworthy feature of the match between gas exchange area and demand is not so much that organisms with large demands have large exchange surfaces, but that organisms with small demands have small exchange surfaces. Although an excess surface area for gas exchange in organisms with small demand should speed the supply of gases to their tissues, it can also create other problems. The same gill surface area that lets oxygen enter fish can also, for example, contribute to generating the drag forces that resist forward motion (see Chapter 27). As we'll see later in this chapter, gas exchange surfaces in land organisms also allow precious water to escape.

to diffuse from openings called **stomata** (singular stoma) on the outside surface of the leaf to the individual photosynthesizing cells (see Figure 29.1).

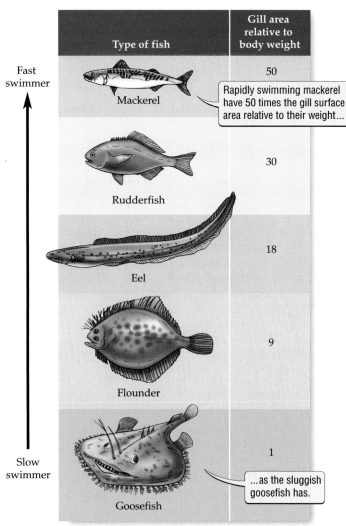

Figure 29.5 Surface Area of Fish Gills
Active fish have more gill surface area relative to body weight than do sluggish, bottom-dwelling fish. This difference suggests that natural selection favors gas exchange surface areas that meet the gas demands of the organism without being so extensive as to cause the loss of valuable nutrients by diffusion.

Figure 29.6 Adaptations of Leaves to Carbon Dioxide Availability
The different shapes of water milfoil leaves above and below the water's surface reflect the different selective pressures faced by the plant in terrestrial and aquatic habitats.

■ Single-celled and some small multicellular organisms need no specialized gas exchange surfaces. Aquatic animals use external gills to take up oxygen. Many terrestrial animals rely on internal gas exchange surfaces called lungs. Terrestrial insects have a unique internal tracheole system, which closely resembles the intercellular air spaces of plant leaves. The surface area provided by the gas exchange surface matches the needs of the organism.

How Plants and Animals Transport Gases to Metabolizing Cells

After oxygen or carbon dioxide enters the body of a multicellular plant or animal, the gas must still reach the respiring or photosynthesizing cells. As mentioned in Chapter 25, diffusion allows for the rapid movement of materials over very short distances. It works too slowly to adequately supply most metabolizing cells at distances much more than 1 millimeter. Thus, plants and animals must transport the gases they need to their individual cells.

Plants and insects bring the gas close to their cells before absorbing it

Plants and insects have evolved very similar approaches to supplying their cells with gases: The two groups both bring the atmosphere to within a fraction of a millimeter of their cells through a system of air passages. In effect, plants and insects transport the gases to the cells before they absorb them.

Oxygen and carbon dioxide move through the intercellular spaces of leaves and the tracheoles of insects by diffusion, which works only because gases diffuse rapidly in air. Even so, diffusion through the intercellular spaces and tracheoles is slow enough that it requires leaves to have a flat shape and insects to be small. If an insect were ever to achieve the large size depicted in some monster movies, it would pose no threat to anyone: It would simply drop dead from lack of oxygen.

Humans and many other animals absorb oxygen before transporting it

Humans absorb oxygen across the surfaces of our alveoli. After absorption we, like many other animals, use our circulatory system for internal transport (described in Chapter 30) to move oxygen rapidly throughout our bodies.

Unfortunately, blood is poorly suited to the transport of oxygen. The liquid portion of the blood, the **plasma**, cannot absorb enough oxygen to meet the demands of most organisms. The plasma of human blood has an oxygen capacity of only about 3 milligrams per liter, or roughly one-third the value for fresh water (see Figure 29.2)—much too low to meet our high demand for oxygen.

Oxygen-binding pigments carried in the plasma greatly increase its capacity to carry oxygen. Oxygen-binding pigments appear to have evolved independently in many different groups of animals, but they all share a similar basic design: a complex protein associated with a **heme** group that usually contains iron and that binds reversibly to oxygen. We call these molecules pigments because they have distinct colors that change depending on whether they are bound to oxygen.

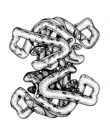

Hemoglobin molecule

Human blood contains **hemoglobin**, an oxygen-binding pigment found in many animal species that changes color from blue when it is not bound to oxygen to red when it is. Our hemoglobin does not float freely in the plasma, but is packaged inside red blood cells. Each hemoglobin molecule can carry up to four oxygen molecules, thus greatly increasing the oxygen-carrying capacity of our blood

from the 3 milligrams per liter in plasma alone to 270 milligrams per liter (Figure 29.7).

Hemoglobin is essential to our ability to supply our cells with oxygen. **Deficiency anemia** results when we do not get enough iron in our diet. Without iron, the human body cannot make enough hemoglobin; thus the oxygen-carrying capacity of blood is reduced, leading to chronic fatigue.

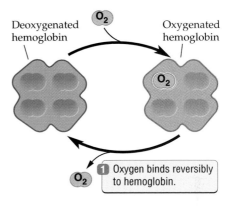

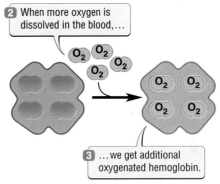

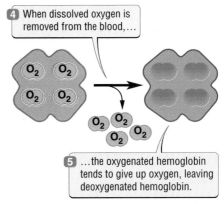

Figure 29.7 Hemoglobin Can Bind to and Release Oxygen Hemoglobin is a complex protein molecule that allows our blood to carry much more oxygen than it could otherwise. Hemoglobin picks up or releases oxygen molecules depending on the amount of oxygen dissolved in the blood around it.

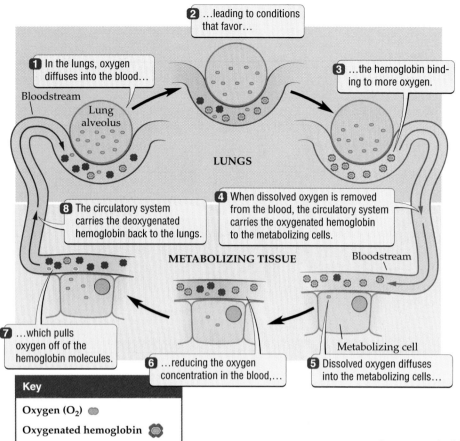

1 In the lungs, oxygen diffuses into the blood...

2 ...leading to conditions that favor...

3 ...the hemoglobin binding to more oxygen.

Bloodstream

Lung alveolus

LUNGS

8 The circulatory system carries the deoxygenated hemoglobin back to the lungs.

4 When dissolved oxygen is removed from the blood, the circulatory system carries the oxygenated hemoglobin to the metabolizing cells.

METABOLIZING TISSUE

Bloodstream

7 ...which pulls oxygen off of the hemoglobin molecules.

6 ...reducing the oxygen concentration in the blood,...

Metabolizing cell

5 Dissolved oxygen diffuses into the metabolizing cells...

Key

Oxygen (O_2)

Oxygenated hemoglobin

Deoxygenated hemoglobin

Figure 29.8 How Hemoglobin Picks Up and Delivers Oxygen
Hemoglobin picks up oxygen in the lungs and releases it as it passes through respiring tissues. Most of the oxygen in our blood at any time is bound to hemoglobin rather than dissolved in the blood fluid itself.

The transport of oxygen in the blood of organisms with oxygen-binding pigment involves several steps (Figure 29.8). When oxygen enters the blood of an organism at the exchange surface, it first diffuses into and dissolves in the plasma. The oxygen-binding pigment picks up the oxygen from the plasma. Because many animals, including humans, keep the oxygen-binding pigment in red blood cells, this process involves the diffusion of oxygen across the plasma membrane of the red blood cell to reach the binding pigment.

In the body away from the lungs, this sequence is reversed. At the site of respiration, oxygen is released from the oxygen-binding pigment into the plasma. Oxygen diffuses through the plasma and across the plasma membrane into the respiring cell. At any time, most of the oxygen in the blood binds to oxygen-binding pigments, with only a small fraction dissolved in the plasma.

■ Plants and insects allow gases to diffuse directly to each metabolizing cell. Many animals, in contrast, transport oxygen in their blood. Oxygen-binding pigments such as hemoglobin greatly increase the amount of oxygen that blood can carry.

Gas Exchange on Land

Terrestrial organisms must cope with the problems associated with the water loss that inevitably accompanies gas exchange on land. Any surface that allows the diffusion of oxygen and carbon dioxide also allows the diffusion of water out of the organism. Thus, for terrestrial organisms, gas exchange means water loss. The problem of water loss during gas exchange probably was one of the major obstacles to the evolution of terrestrial life.

Terrestrial organisms represent an evolutionary compromise between the conflicting demands of water conservation and gas exchange. Most of the surface of terrestrial creatures is waterproof and therefore useless as a gas exchange surface. Gas exchange surfaces of virtually all terrestrial organisms lie inside rather than outside the body. By moving the gas exchange surface inside their bodies, terrestrial organisms gain much more control over what crosses the membrane.

For example, kangaroo rats live in a dry desert environment and must be particularly careful to conserve water. Dry air is warmed in the kangaroo rat's lungs, where it picks up water that diffuses into the lungs from the body (Figure 29.9). The kangaroo rat recovers half of this water when the air passes through its cool nasal passages. Much as moisture condenses on the outside of a glass of ice water on a hot, humid day, moisture in the exhaled air condenses out in the cool nasal passages and is reabsorbed.

■ Surfaces that allow the diffusion of oxygen and carbon dioxide necessarily allow the diffusion of water. Gas exchange is a major source of water loss for terrestrial organisms.

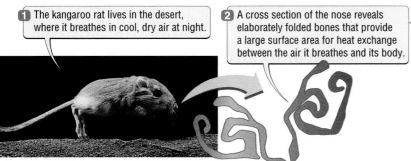

1 The kangaroo rat lives in the desert, where it breathes in cool, dry air at night.

2 A cross section of the nose reveals elaborately folded bones that provide a large surface area for heat exchange between the air it breathes and its body.

Figure 29.9 Water Conservation in Kangaroo Rats Desert-dwelling kangaroo rats recover most of the body water that is taken up by the dry air they inhale.

3 The rat inhales one liter of air, which contains little water.

4 As the inhaled air passes over the moist surfaces of the rat's nose, trachea, and lungs, the air draws 44 ml of water from the rat's body.

Inhale

Exhale

34 ml H_2O recovered in nose

9.5 ml H_2O lost to air

5 As it exhales, the rat recovers 34 ml of water when the water vapor condenses in its cool nasal passages.

6 Overall, the rat loses only 9.5 ml of body water for each liter of air that it breathes out.

0   10   20   30   40   50

Water content of breath (ml H_2O per l breath)

creased breathing rate accompanied by an increased heart rate. In essence, we respond as if we were running at lower elevations. The increased breathing rate maintains as steep an oxygen gradient into the body as possible, and a rapidly pumping heart quickly carries this blood to the metabolizing cells. As athletes and mountaineers know, the human body can adjust to the low-oxygen environment if we give it time. One recommendation is to allow one day of adjustment time for each 600 meters we wish to go above 2000 meters.

With time spent at high elevation, the human body adjusts by increasing the hemoglobin concentration in the blood and by *decreasing* the attraction of the hemoglobin to oxygen. The former increases the oxygen capacity of our blood, and the latter allows us to unload the blood more quickly at the site of metabolism. Note that the latter characteristic is just the opposite of the property of blood of evolutionarily adapted humans, thus illustrating an interesting difference between a short-term, physiological response and a long-term, evolutionary response.

■ Peoples native to high altitudes have evolved gas exchange mechanisms such as large lungs and high hemoglobin concentrations. Individuals not used to high altitudes can compensate for the low O_2 availability by breathing more rapidly, increasing blood flow, and, over time, increasing their hemoglobin concentration.

Highlight

Gas Exchange in the Mountains

People who live at high altitudes, where the oxygen concentration is low, have several genetically based features that allow them to live in their environment. Human populations that live permanently at high elevations tend to have large lungs. In addition, the hemoglobin concentration in the blood is higher, often lending a red cast to the skin of highland populations. The hemoglobin itself picks up oxygen more readily from the plasma, which allows the blood to leave the lungs 100 percent saturated, despite the smaller differences in oxygen concentration between the air and the blood. In contrast, when individuals who are native to lowland areas visit high elevations, the hemoglobin in the blood leaving their lungs is typically not fully oxygenated.

Without prior exposure to low-oxygen environments, our initial response to the thin air we would encounter in the mountains would be an emergency response: an in-

Summary

Oxygen and Carbon Dioxide Enter Organisms by Diffusion

■ Steeper concentration gradients cause gases moving from the environment to an organism to diffuse more quickly.

■ With larger surface areas available for diffusion, gases diffuse faster.

■ As the distance across which a gas must diffuse decreases, the gas diffuses faster.

Oxygen and Carbon Dioxide in the Environment

■ The oxygen concentration is higher in air than in water.

■ The concentrations of carbon dioxide are similar in air and water.

■ Because oxygen is more abundant than carbon dioxide, in both air and water, oxygen diffuses more quickly into animals than carbon dioxide does into plants.

Gas Exchange Structures

■ Small organisms, because of their large surface area–to–volume ratios, generally do not need elaborate gas exchange structures.

■ Aquatic animals more than a few millimeters long rely on external gills for gas exchange.

■ Terrestrial organisms rely on internal structures for gas exchange, such as lungs (in vertebrates), tracheoles (in insects), and intercellular air spaces (in plants).

■ The relative surface areas of gas exchange structures in different organisms match the relative metabolic demands of the organisms.

How Plants and Animals Transport Gases to Metabolizing Cells

■ Because gases diffuse slowly across long distances, multicellular organisms must transport the gases from their specialized gas exchange structures to other tissues in their bodies.

■ Plants and insects transport gases to their cells directly through an extensive system of air passages.

■ Many animals absorb the gases they take up from the air in their gas exchange structures into their blood, which then transports the gases through their bodies.

■ The blood's ability to transport oxygen is improved by specialized oxygen-binding pigments, such as the hemoglobin in human red blood cells.

Gas Exchange on Land

■ The gas exchange structures of terrestrial organisms can allow significant water loss.

■ Because water loss is a serious problem for terrestrial organisms, they have evolved internal, water-conserving systems.

Highlight: Gas Exchange in the Mountains

■ When a person first encounters the oxygen-poor air of high elevations, the body's immediate response is to increase the heart and breathing rates.

■ If the same individual stays at high elevation for a few days, the body adjusts to low oxygen concentration by increasing hemoglobin content and reducing its attraction to oxygen.

■ Human populations that have adapted to high elevations through natural selection have hemoglobin with a higher attraction to oxygen, as well as larger lungs and increased hemoglobin concentration.

Key Terms

alveolus p. 448	lung p. 448
deficiency anemia p. 451	oxygen-binding pigment p. 451
gill p. 448	plasma p. 451
heme p. 451	spiracle p. 448
hemoglobin p. 451	stoma p. 449
intercellular air space p. 444	tracheole p. 448

Chapter Review

Self-Quiz

1. The rate at which O_2 moves from the alveoli in our lungs into our blood
   a. depends on the difference in O_2 concentration between the alveoli and the blood.
   b. depends on the color of the alveoli.
   c. depends on the availability of energy to transport gases across the membrane.
   d. none of the above

2. The spiracles of insects function similarly to the
   a. lips of humans.
   b. mouths of fish.
   c. air filters of cars.
   d. stomata of plants.

3. The reason why single-celled organisms do not use specialized gas exchange structures is that their bodies
   a. are not large enough to accommodate them.
   b. have a large surface area–to–volume ratio.
   c. are not complex enough to need them.
   d. have a slow metabolism.

4. How is oxygen transported through the human body?
   a. as a dissolved gas in the plasma of the blood
   b. bound to hemoglobin molecules in red blood cells
   c. through intercellular air spaces
   d. by diffusion

5. The concentration of carbon dioxide in the air decreases with altitude in much the same way that oxygen does. Which problem would plants living high in the mountains face?
   a. maintaining a lower carbon dioxide concentration inside the plant than in the air
   b. preventing the loss of water while exchanging oxygen and carbon dioxide with the air
   c. maintaining a high enough concentration of carbon dioxide in their intercellular air spaces to allow successful diffusion of carbon dioxide into the photosynthesizing cells in the leaf
   d. all of the above

Review Questions

1. Would you expect a lungless terrestrial vertebrate, like the red-backed salamander mentioned in this chapter, to live in a relatively humid or a dry environment? Why?

2. Our oxygen demand changes dramatically depending on how active we are. Can you think of changes that take place in how your body works as you begin to exercise?

Think of how these changes help to provide more oxygen to the metabolizing cells in an active body.

3. What advantage might air-breathing aquatic mammals like dolphins have over hypothetical aquatic mammals that depend on gills for gas exchange?

---

29   𝕿𝖍𝖊 𝕭𝖆𝖎𝖑𝖞 𝕲𝖑𝖔𝖇𝖊

A WEEKLY GARDENING COLUMN

Ask The Garden Doctor

Dear Garden Doctor:
I keep reading about how human activity is causing the carbon dioxide levels in our atmosphere to rise, and it seems like the media is presenting this as a bad thing. But, don't plants use carbon dioxide to grow? Shouldn't this increase actually help our gardens and agricultural fields?
Signed, *Confused in Omaha*

Dear Confused:
Don't worry; you're not the only one who's having a tough time sorting this one out. The net effects of elevated carbon dioxide levels are still being debated by scientists. As you stated, carbon dioxide levels are rising, and plants do use carbon dioxide in the process of photosynthesis. And in fact, in some situations, plants can increase their photosynthetic rates when exposed to elevated levels of the gas. But, it doesn't happen automatically; photosynthetic rates can also be limited by the availability of other items like nitrogen, phosphorus, water, and sunlight.

Even if we could help our plants maintain higher rates of photosynthesis in the presence of extra carbon dioxide by applying expensive fertilization and irrigation, it is not clear that we would get the kind of additional growth we would be hoping for. With most agricultural crops, farmers are interested only in the production of seeds or fruit of some sort (such as wheat, corn, or apples), and many home gardeners care most about flowers, but in many studies of the combined effects of elevated carbon dioxide levels and fertilizer, the extra growth goes into leaves, stems, or even roots. Of course, in some situations, bigger leaves, stems, and roots might help us get what we're

after, so it's hard to generalize about whether this would be good or not.

The bottom line here is that nobody really knows exactly what the agricultural effects of the increasing carbon dioxide concentration in our atmosphere will be. And in this discussion, we are keeping things simple—no increase in temperature, no change in total precipitation, no change in the frequency of drought or floods. All of these additional impacts are mentioned, with varying levels of certainty, as potential effects of the accumulation of gases like carbon dioxide in our atmosphere. Given the multitude of factors interacting in this complex global phenomenon, it's no wonder that the scientific community sometimes seems a little confused when trying to predict what the impact on our crops and gardens will be.

The Garden Doctor

Evaluating "The News"

1. If we don't know the actual effect, good or bad, on crops and garden plants from increased carbon dioxide, how worried should farmers or home gardeners be?

2. If the effects might be good, should we encourage increases in carbon dioxide, until we see that the effects are harmful to plants?

3. What should you do when it appears that even the scientists are

confused, as The Garden Doctor asserts? How can our society respond to issues when we are faced with conflicting results or an incomplete understanding of a natural phenomenon?

chapter

30

Internal Transport

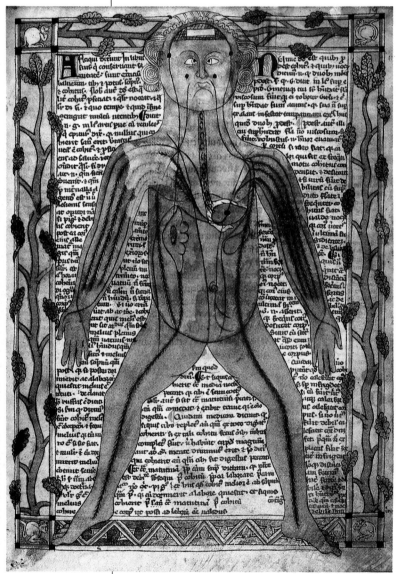

Circulatory System, late thirteenth–century manuscript.

High Blood Pressure and Life on the African Savanna

The human heart generates pressure that drives blood through the blood vessels of the circulatory system. As part of a routine checkup, doctors measure blood pressure to assess the health of the circulatory system. A blood pressure of about 120/80 millimeters of mercury (mm Hg) at rest usually indicates a heart that is working normally. The two numbers refer to the maximum and minimum, respectively, of blood pressure as the blood pulses through our blood vessels. The unit of measurement used, millimeters of mercury, refers to the height to which a pressure can lift a column of the heavy liquid mercury.

Blood pressures higher than 120/80 indicate an overworked heart. In resting adults, blood pressures of 160/90 or greater warn of potential problems. About 20 percent of North American adults have high blood pressure: Their hearts work too hard to push blood through their bodies, thus increasing their risk of heart disease.

Scientists have turned to the African savanna in their efforts to understand how the

456

Most multicellular organisms need internal transport systems to move materials among their various tissues.

human body might respond to high blood pressure. Here giraffes move gracefully across sweeping grasslands, munching on flat-topped acacia trees. Though apparently far removed from the poor diets and stressful lifestyles often blamed for high blood pressure in humans, giraffes live with blood pressures of 260/160. Why does the lifestyle of a giraffe lead to blood pressures that would threaten human lives?

The stately acacia trees on which the giraffes browse face even greater challenges. They must somehow lift sap, the plant's equivalent of blood, a height of 15 meters or more above the soil surface. Lifting sap to such heights requires an astonishing pressure of 8000 mm Hg, enough to burst a human heart.

At the end of this chapter we consider how and why giraffes and acacia trees generate the pressures needed to move blood or sap through their bodies.

Giraffes and Acacias on the East African Savanna

1. Diffusion is too slow to transport material inside multicellular organisms.

2. The human circulatory system, like that of most other animals, depends on a muscular heart that contracts and relaxes to circulate blood through the blood vessels.

3. Blood vessels must both transport blood and allow exchange of substances between the blood and the surrounding tissue.

4. Hearts move blood by creating pressure differences within the circulatory system.

5. In the phloem tissue, osmotic pressure moves the products of photosynthesis throughout plants.

6. The xylem tissue of land plants distributes water and mineral nutrients using the pulling power generated by the evaporation of water from leaf surfaces.

Multicellular organisms must move materials from certain body parts to others. Nutrients and gases must move from sites of absorption to sites of use. Coordinated activities like growth and the regulation of water and temperature also depend on the transfer of information and materials within the body.

In this chapter we consider how multicellular animals and plants transport things inside their bodies. Internal transport systems balance the demands of moving materials and exchanging these materials with the tissues they supply. As animals evolved from their single-celled ancestors, they developed internal transport systems based on muscular hearts that circulate blood through their bodies. We will see that the evolution of internal transport systems in plants has followed a different path. Plants transport nutrients and water in noncirculating sap that flows through their vascular tissues in response to diffusion and evaporation.

to a root tip, gases and dissolved materials can diffuse within a fraction of a second (Figure 30.1). However, even for sizes that we would consider small—for example, a leaf thickness or an ant's body—it takes several seconds to many minutes for materials to diffuse.

At the scale of very large organisms, like the 5.5-meter-tall giraffe and 15-meter-tall acacia tree mentioned at the beginning of the chapter, transport by diffusion would take years. Thus, organisms more than a few cells thick cannot rely on diffusion to transport materials internally. Instead, they need an internal transport system to move materials between cells in their bodies.

> ■ Diffusion works too slowly for nutrients and gases to be transported effectively in multicellular animals and plants.

Who Needs Internal Transport?

In discussing nutrition and gas exchange (see Chapters 28 and 29), we focused on the role of diffusion as a way of moving the materials essential to life. Over short distances (less than 1 millimeter), such as across plasma membranes, within a cell, or through the soil water next

Figure 30.1 The Relationship between Body Size and Diffusion Time
As the body size increases, the time that it would take to transport materials from one end of the organism to the other increases greatly. The times reported represent the time that it takes oxygen to diffuse various distances through water, the main component of living things.

Example	Cell	Small insect	Mouse	Human	Giraffe	Acacia tree
Distance	0.005 mm	1 mm	6 cm	2 m	5.5 m	15 m
Time needed for oxygen diffusion	0.0002 seconds	7 seconds	7 hours	10 months	6 years	48 years

The Human Circulatory System

How do large, multicellular organisms move materials inside their bodies? The human circulatory system—a muscular heart that circulates blood through a complex series of loops formed by blood vessels—has many of the basic elements of all the different circulatory systems found in the animal kingdom (Figure 30.2).

Blood flows to all parts of the human body through blood vessels

Tubular blood vessels form a series of loops through which blood moves from the heart to metabolizing cells and then back to the heart. In a typical adult, 5 to 6 liters of blood, carrying a rich mixture of dissolved minerals, gases, nutrients, and the products of our metabolism, circulates continuously through the blood vessels. To overcome the limited distance over which diffusion works well, nearly all of our body cells lie within 0.03 millimeter (less than one-third the thickness of this page) of a blood vessel with which they can exchange materials. For this reason blood oozes from a tiny cut made anywhere on the body.

Clearly, carrying blood so close to all the billions of cells in the body requires an extensive system of blood vessels. If laid end to end, the approximately 17 billion blood vessels in the human body would stretch an incredible 20,000 kilometers, or from Seattle to London and back again with about 1000 kilometers to spare.

There are three kinds of blood vessels in the human body: (1) Thick-walled **arteries** carry blood from the heart for distribution to the body, (2) tiny **capillaries**, less than 0.01 millimeter in diameter, allow the exchange of material between the blood and the surrounding tissue, and (3) thin-walled **veins** collect blood from the body for delivery to the heart.

The capillaries connect arteries to veins to make a **closed circulatory system** (Figure 30.3*a*). A piece of muscle tissue the size of a pencil tip may contain more than 1000 capillaries. Taken together, the capillaries in our bodies provide a combined surface area equal to the floor space in three large houses.

The human heart generates pressure that propels blood through the vessels

Contraction of the muscular heart pumps blood through our blood vessels. The human heart has four chambers: a left atrium and ventricle and a right atrium and ventricle (Figure 30.4). Each side of the heart acts as a separate pump: The larger left side pumps blood through the loops of blood vessels that serve our metabolizing tissues, and the smaller right side pumps blood through a loop that carries blood to and from the gas exchange surfaces in our lungs.

Contraction of the smaller, thinner-walled **atrium** pumps blood into the corresponding **ventricle**, which has a thicker, more muscular wall and pumps blood out of the heart and through the blood vessels. One-way valves that separate the atrium from the ventricle and the ventricle from

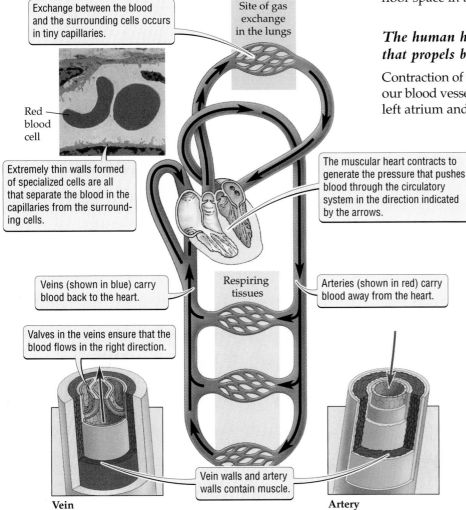

Exchange between the blood and the surrounding cells occurs in tiny capillaries.

Site of gas exchange in the lungs

Red blood cell

Extremely thin walls formed of specialized cells are all that separate the blood in the capillaries from the surrounding cells.

The muscular heart contracts to generate the pressure that pushes blood through the circulatory system in the direction indicated by the arrows.

Veins (shown in blue) carry blood back to the heart.

Respiring tissues

Arteries (shown in red) carry blood away from the heart.

Valves in the veins ensure that the blood flows in the right direction.

Vein walls and artery walls contain muscle.

Vein

Artery

Figure 30.2 The Human Circulatory System
The human circulatory system, like those of all other mammals, depends on a heart to push blood through loops formed by blood vessels.

(a) Closed circulatory systems

(b) Open circulatory systems

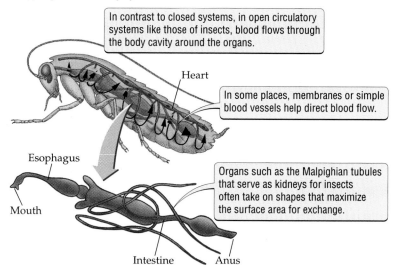

In closed circulatory systems like our own, blood flows through organs in arteries, capillaries, and veins.

Kidneys

Capillaries

The capillaries are blood vessels specialized to provide a lot of exchange surface.

Artery

Vein

In contrast to closed systems, in open circulatory systems like those of insects, blood flows through the body cavity around the organs.

Heart

In some places, membranes or simple blood vessels help direct blood flow.

Esophagus

Mouth

Organs such as the Malpighian tubules that serve as kidneys for insects often take on shapes that maximize the surface area for exchange.

Intestine   Anus

Figure 30.3 Closed and Open Circulatory Systems
Animals have two main types of circulatory systems: In closed circulatory systems such as the human kidney (a), blood flows through organs; in open circulatory systems (b), blood flows around organs.

the artery leading out of it ensure that the blood flows through each side of the heart in one direction only. Heard through a stethoscope, the *lub-dub* sound of the heartbeat is actually the sound of these valves snapping shut. The *lub* sound is the closing of the valve separating the atrium from the ventricle, and the *dub* sound is the closing of the valve separating the ventricle from the artery.

Our hearts beat, we hope, without fail about 75 times each minute, which amounts to 3 billion beats in a 70-

Figure 30.4 The Human Heart
One-way valves in the heart direct the flow of blood in response to contraction of the heart muscle and make the familiar *lub-dub* sound of our heartbeat. Arrows indicate the direction of blood flow through the heart. RA = right atrium; RV = right ventricle; LA = left atrium; LV = left ventricle.

year lifetime. At rest, our hearts pump all 5 to 6 liters of our blood each minute—the equivalent of 7000 liters of blood through the circulatory system each day. Each contraction of the heart generates a sharp increase (spike) of pressure that propels blood (Figure 30.5).

The pressure measured in a doctor's office reflects the pressure change in the arteries leading to the body from the left ventricle: For a blood pressure reading of 120/80, for example, contraction of the left ventricle generates 120 mm Hg of pressure, followed by a drop to 80 mm Hg when the ventricle relaxes and refills. Similar measurements made in the arteries leading from the smaller right ventricle into the short loop running through our lungs would reveal lower pressures, changing between only 8 and 25 mm Hg.

By the time the blood enters the capillaries, the pressure changes little and has dropped to 35 mm Hg. When the blood leaves the capillaries, the pressure is only 10 mm Hg. The drop in pressure as the blood progresses through the circulatory system reflects the effect of fric-

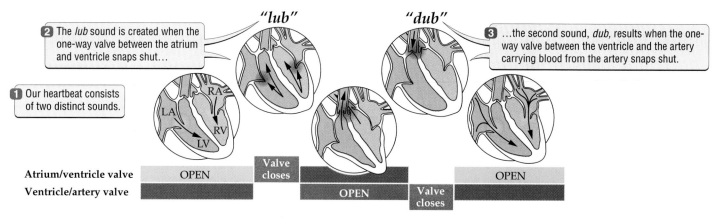

"lub"

"dub"

2 The *lub* sound is created when the one-way valve between the atrium and ventricle snaps shut...

3 ...the second sound, *dub*, results when the one-way valve between the ventricle and the artery carrying blood from the artery snaps shut.

1 Our heartbeat consists of two distinct sounds.

RA
LA
RV
LV

| Atrium/ventricle valve | OPEN | Valve closes | | | |
| Ventricle/artery valve | | | OPEN | Valve closes | OPEN |

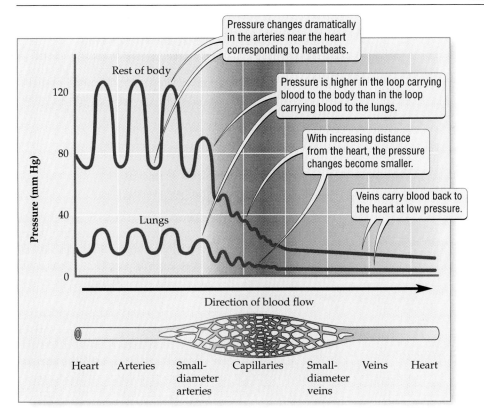

Pressure changes dramatically in the arteries near the heart corresponding to heartbeats.

Pressure is higher in the loop carrying blood to the body than in the loop carrying blood to the lungs.

With increasing distance from the heart, the pressure changes become smaller.

Veins carry blood back to the heart at low pressure.

Figure 30.5 Pressure Changes in the Human Circulatory System
Both the pressure and the changes in pressure in human blood vessels decrease as the blood moves through the circulatory system.

tion between blood flow and the blood vessel walls, which results in friction drag (see Chapter 27).

The human circulatory system responds to the changing needs of the body

When we leave the comfort of our couches and drag ourselves out for a bicycle ride, the nutrient and, in par-

ticular, the oxygen demands of our muscles increase greatly. Our blood vessels and hearts respond to this increased demand with a coordinated set of changes in heart rate, blood flow, and blood distribution (Figure 30.6).

For example, (1) the number of times the heart beats each minute doubles, thus increasing the blood pressure in the arteries to about 180 mm Hg. (2) The increased blood pressure stretches the blood vessels, allowing them to carry more blood. (3) The blood vessels supplying the muscles involved in exercise carry more blood, whereas those supplying tissues not involved in exercise, like those of the digestive system, carry less blood. As a result of these changes, the supply of blood to the active muscles may almost triple.

> ■ The human circulatory system consists of a muscular heart that contracts to push blood through loops consisting of arteries, capillaries, and veins. The arteries and veins transport blood to and from the millions of tiny capillaries, which serve as the site of exchange with the metabolizing tissues.

The Blood Vessels of Animals

Although blood vessels direct the flow of blood to specific portions of an animal's body, they also pick up and deliver the materials that they carry. The features of a blood vessel that allow it to exchange materials with surrounding cells, like thin walls and large surface area–to–volume ratios, suit it poorly for transporting blood with a minimum of resistance, which requires a small surface area–to–volume ratio to minimize friction drag.

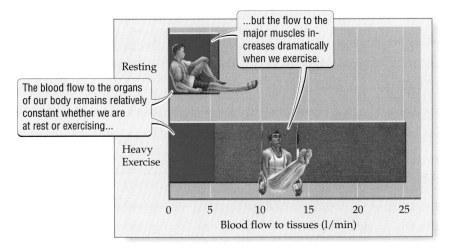

...but the flow to the major muscles increases dramatically when we exercise.

The blood flow to the organs of our body remains relatively constant whether we are at rest or exercising...

Figure 30.6 The Response of the Human Circulatory System to Exercise
The human circulatory system responds to increased activity by increasing the flow of blood to the active muscles (measured in liters/minute).

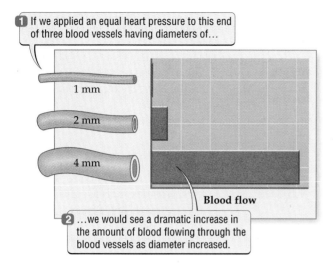

1 If we applied an equal heart pressure to this end of three blood vessels having diameters of...

1 mm

2 mm

4 mm

Blood flow

2 ...we would see a dramatic increase in the amount of blood flowing through the blood vessels as diameter increased.

Figure 30.7 Blood Vessel Size and Friction Drag
As blood vessel diameter increases, the amount of blood that it can transport increases dramatically. With the same pressure from the heart, blood flows through a 4-mm-diameter vessel 256 times as much as through a 1-mm-diameter vessel.

The large size and complex walls of arteries and veins suit them to blood transport

Blood vessels specialized for transport, like our arteries and veins, have relatively large diameters and muscular, elastic walls. Decreasing blood vessel diameter strongly increases the surface area relative to the volume of blood carried, which greatly increases the friction drag that resists blood flow. For example, decreasing the diameter of a blood vessel by half increases the resistance that blood faces by 40 times and forces the heart to work 40 times harder to maintain the same blood flow (Figure 30.7). To reduce the amount of work the heart must do to transport blood, the vessels that carry blood the farthest have the largest diameter, and the small-diameter capillaries carry the blood over only short distances.

The thick, complex walls of the transport vessels equip them to regulate blood distribution. Layers of muscle tissue in the artery walls can contract to regulate blood flow. This contraction allows animals to control the flow of blood to different parts of the body much as we can control the flow of water in our homes by opening and closing faucets. The elasticity of the artery walls allows them to stretch when heart contraction increases blood pressure.

When we feel our pulse in an artery, we actually feel the artery walls bulge in response to a surge of pressure generated by the contraction of the heart. The ability of the blood vessel walls to stretch reduces the pressure changes as distance from the heart increases, thus protecting the smallest and most delicate blood vessels, the capillaries, from damaging pressure changes (see Figure 30.5). Unfortunately, these thick elastic walls make it difficult to exchange the blood contents with the surrounding tissues.

The small diameters and thin walls of capillaries suit them to exchange

Human circulatory systems, like those of vertebrates generally, exchange materials across the tiny capillaries that connect the arteries with the veins. Such closed circulatory systems give the animal great control over where the blood flows and where exchange takes place (see Figure 30.3a).

In comparison to the arteries and veins, capillaries have extremely thin, porous walls formed exclusively of epithelial cells and across which materials diffuse easily. The small diameter of capillaries increases the surface area for exchange relative to the volume of blood passing through them. However, because of the dramatic increase in friction drag that accompanies a decrease in diameter, capillaries strongly resist blood flow. Thus, organisms with capillaries must have strong hearts that generate enough pressure to push blood through their capillaries.

Insects and some mollusks, like snails and clams, have taken a different approach from ours in dealing with the exchange function of the circulatory system. These creatures have **open circulatory systems**, in which the blood vessels leading from the heart direct blood into the spaces between the organs (see Figure 30.3b). Blood makes its way back to the heart by flowing through these spaces. Animals with open circulatory systems have relatively little control over how the blood circulates; thus blood contents are distributed less efficiently than in closed circulatory systems. The organs of animals with open circulatory systems facilitate exchange by having the organs bathed in blood shaped so as to create a lot of exchange area relative to volume.

Snail

■ The arteries and veins that carry blood to and from the capillaries have muscular walls and large diameters. These features allow them to control where blood flows and to carry blood over relatively long distances with a minimum of resistance from friction drag. Capillaries have thin walls and large surface area–to–volume ratios that facilitate exchange with the surrounding tissues.

Animal Hearts

Like human hearts, the hearts of other animals rely on muscles to pump blood through the circulatory system. The work that hearts do must both lift the blood against the pull of gravity and overcome the friction drag generated as blood flows past blood vessel walls.

Animal hearts use muscles to generate one-way flow

Hearts come in several forms, all of which use the muscle tissue that is unique to animals to generate pressure and one-way valves to direct flow. Human hearts are an example of the most widespread type, the chambered heart (Figure 30.8a). All chambered hearts consist of at least two chambers—an atrium and a ventricle—separated from each other and the blood vessels by one-way valves. Working alone, the ventricle could fill to about 75 percent of its capacity. The smaller atrium provides the force needed to fill the ventricle completely, thus increasing the amount of blood moved with each heartbeat.

Some animals also use hearts without chambers. A simple type of pump found in some animals (such as humans and sharks) consists of a blood vessel that contains one-way valves and is surrounded by muscle tissue (Figure 30.8b). Contraction of the surrounding muscle tissue during normal activity squeezes the vessel, forcing body fluid out of the squeezed portion. The one-way valves ensure that squeezing the pipe forces the body fluid to move in only one direction. In some animals, this sort of pump usually supplements a chambered heart. For example, the veins in our legs work in this way to ease the task of lifting the blood back up to the heart.

Another kind of pump used by some animals, like insects, consists of a muscular blood vessel, in which the body fluid is pushed ahead of a wave of muscular contraction, much as food is pushed down the esophagus (see Chapter 28) (Figure 30.8c).

Animal hearts overcome friction drag and lift blood

The pressure generated by the heart accomplishes two things: (1) It overcomes the drag that blood encounters as it flows past the walls of the vessels, and (2) it lifts blood against the pull of gravity. As already mentioned, anything that decreases the diameter of the blood vessels dramatically increases the pressure that the heart must generate. Blockages of blood vessels can also increase resistance, and therefore the pressure needed to maintain an adequate blood flow to the tissues.

Figure 30.8 The Hearts of Animals
Animals have three different types of hearts: chambered hearts (a), blood vessels surrounded by muscle tissue (b), and peristaltic pumps (c).

For example, diets high in saturated fats encourage the development of fatty deposits called **plaques** in our blood vessels, a condition called **atherosclerosis**. As a result, the heart must work harder to force blood through the narrowed arteries. Over many years, continued high pressure can cause changes in the heart that reduce its ability to pump blood, a cause of congestive heart fail-

(a) Chambered heart

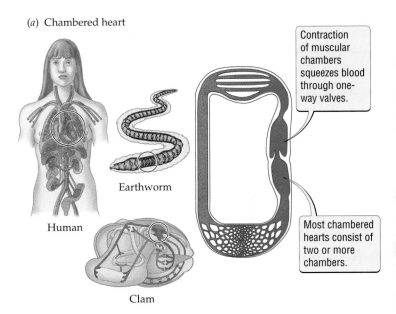

Human
Earthworm
Clam

Contraction of muscular chambers squeezes blood through one-way valves.

Most chambered hearts consist of two or more chambers.

(b) Blood vessels surrounded by muscle tissue

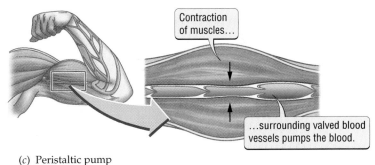

Contraction of muscles...

...surrounding valved blood vessels pumps the blood.

(c) Peristaltic pump

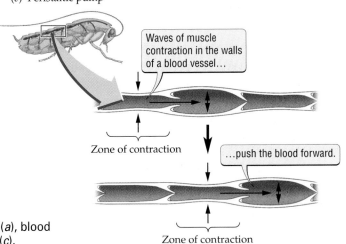

Waves of muscle contraction in the walls of a blood vessel...

Zone of contraction

...push the blood forward.

Zone of contraction

The Scientific Process

How Do You Make a Sea Serpent Faint?

The answer to this question is simple: Grab it *very carefully* by its neck and hold it with its head up for a while.

Far from being mythical creatures, sea serpents exist in the form of sea snakes that live in the waters of the Pacific Ocean (see the figure). Sea snakes are closely related to cobras, and, like cobras, they produce potent poisons to subdue their prey.

To understand why a sea snake passes out when held vertically, we need to understand the selective pressures that shaped the evolution of its circulatory system. Sea snakes spend most of their lives swimming in a horizontal position. Their hearts therefore usually do not have to lift blood much more than a centimeter. Sea snake hearts need to generate only enough pressure to overcome the friction drag their blood encounters as it flows through their blood vessels. Thus sea snakes can get by with hearts that generate pressures of between 15 and 30 mm Hg. If you hold a sea snake vertically, with its head pointing upward, its heart cannot generate enough pressure to

Sea Snakes Have Weak Hearts

pump blood to its brain. The result is an unconscious snake.

---

ure. More dramatically, if the plaques occur in one of the major blood vessels supplying the heart, they can reduce the flow of blood carrying oxygen and nutrients to heart muscles enough to kill heart muscle, causing a **heart attack**.

The heart must also provide the force needed to lift blood to the highest part of the body (see the Box above). Of the 7000 liters of blood that the human heart pumps through the circulatory system each day, roughly 1000 liters go to the brain. Because a liter of blood weighs about 1 kilogram, each day we must do the work necessary to lift 1000 kilograms of blood the 30- to 40-centimeter distance from our hearts to our heads. The extinct dinosaur *Barosaurus*, which had to lift blood the length of its extraordinary 10-meter-long neck, would have needed a heart that could generate a pressure of 646 mm Hg simply to support the weight of this column of blood.

> ■ Muscular animal hearts generate the pressure needed to lift blood against the pull of gravity and to overcome the friction drag that resists the flow of blood past blood vessel walls.

Internal Transport in Plants: A Different Approach

Although the accomplishments of the human circulatory system or that of organisms like *Barosaurus* may seem impressive, they pale in comparison to what trees must accomplish. Against the pull of gravity, trees lift soil water and dissolved nutrients from the roots to leaves 20 meters or more above the ground.

The internal-transport problem faced by plants is immense: To supply its leaves with water and nutrients a 20-meter-tall tree must generate a pressure difference between its roots and its uppermost leaves of at least 3500 mm Hg, or about 20 times the pressure generated by the human heart when we exercise. Now consider the world's tallest living tree, a 114-meter-tall giant sequoia growing in California: It must generate a pressure difference of 16,500 mm Hg to supply its uppermost leaves with water and nutrients.

The internal transport system that plants use to move materials over long distances and to great heights bears little resemblance to the circulatory system of animals. The blood equivalent of plants, called **sap**, moves through the interior of cylindrical cells that form the equivalent of blood vessels in plants. In contrast to animal blood, sap does not circulate, so we refer to the internal transport system of plants as a vascular system rather than a circulatory system. Perhaps the most puzzling aspect of the plant vascular system lies in its lack of an obvious heart equivalent, even though we know that plants must have a way of generating pressure differences greater than those achieved by animals.

> ■ Plants must generate great pressures to lift sap from the soil to the heights commonly attained by their uppermost leaves.

The Structure of Plant Vascular Tissues

Plants move sap through the insides of cells, rather than through multicellular organs as blood vessels do. Single vascular cells cannot extend from the roots to the leaves in even the smallest of plants. Instead, the vascular cells lie end to end to form vascular tissue. Pores that connect the cells allow sap to flow from one cell to the next. Moreover, cell diameter limits the diameter of plant tubes.

Typically, the cells of the plant vascular system have diameters ranging from 0.02 millimeter to a maximum of 0.5 millimeter, putting them in the range of capillaries and the smallest arteries and veins. As a result, the vascular system of plants offers very high resistance to sap flow. In a pattern that parallels that of animal circulatory systems, however, exchange usually takes place in the vascular cells that have the smallest diameter. The vascular cells with the largest diameter, although tiny compared to many arteries and veins, serve mainly for transport, not exchange.

Individual plants depend on two distinctly different vascular tissues: the phloem and the xylem (Figure 30.9). **Phloem tissue** transports sap containing the dissolved products of photosynthesis; **xylem tissue** carries sap consisting of water and dissolved mineral nutrients from the roots to the leaves.

Plants must move surplus sugars produced in photosynthesizing leaves to nonphotosynthesizing plant

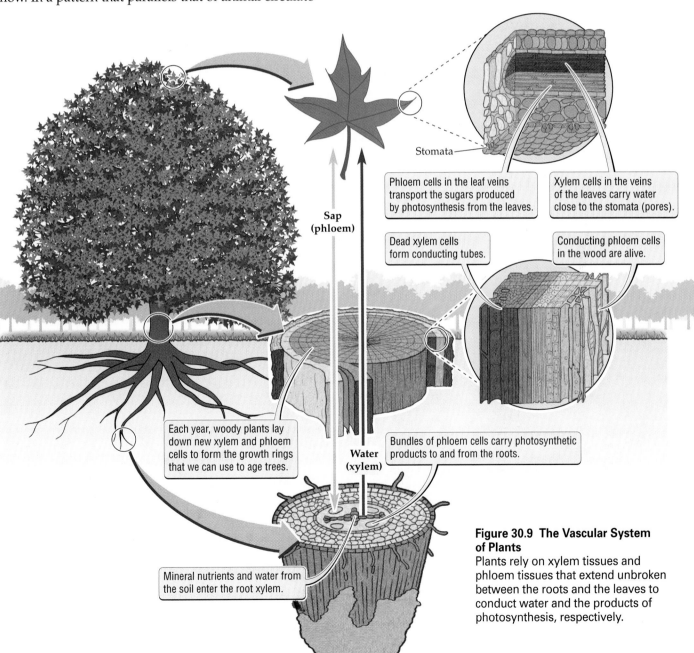

Stomata

Phloem cells in the leaf veins transport the sugars produced by photosynthesis from the leaves.

Xylem cells in the veins of the leaves carry water close to the stomata (pores).

Dead xylem cells form conducting tubes.

Conducting phloem cells in the wood are alive.

Sap (phloem)

Water (xylem)

Each year, woody plants lay down new xylem and phloem cells to form the growth rings that we can use to age trees.

Bundles of phloem cells carry photosynthetic products to and from the roots.

Mineral nutrients and water from the soil enter the root xylem.

Figure 30.9 The Vascular System of Plants
Plants rely on xylem tissues and phloem tissues that extend unbroken between the roots and the leaves to conduct water and the products of photosynthesis, respectively.

parts, such as roots or buds, that need sugar as an energy source. When dissolved in phloem sap, sugars move up and down plants through the insides of living cells that make up phloem tissue. The pores that connect phloem cells allow carbohydrates to move directly from the cytoplasm of one cell to the cytoplasm of the adjacent cell.

A major advance in the evolution of land plants was the development of a distinct vascular tissue, the xylem, through which water and mineral nutrients moved from the soil solution up to the leaves. Water and dissolved minerals move upward only through the xylem cells. Unlike the phloem cells, functional xylem cells are dead and consist of empty cylinders arranged end to end. The connections between xylem cells allow a continuous column of water to form between the roots and the leaves. As we will see, this thin column of water is critical in the movement of xylem sap. Only plants with xylem tissue have evolved a tall, upright form, suggesting the importance of xylem tissue in the lives of land plants.

■ Plants rely on two separate vascular systems for internal transport: Phloem tissue transports the products of photosynthesis throughout the plant, and xylem tissue transports water and mineral nutrients from the roots to the leaves.

How Do Plants Move Sap?

The way in which sap moves through phloem tissues differs from how it moves through xylem tissues. To generate the large pressure differences needed to move phloem and xylem sap over long distances, plants have evolved clever substitutes for the muscular hearts of animals.

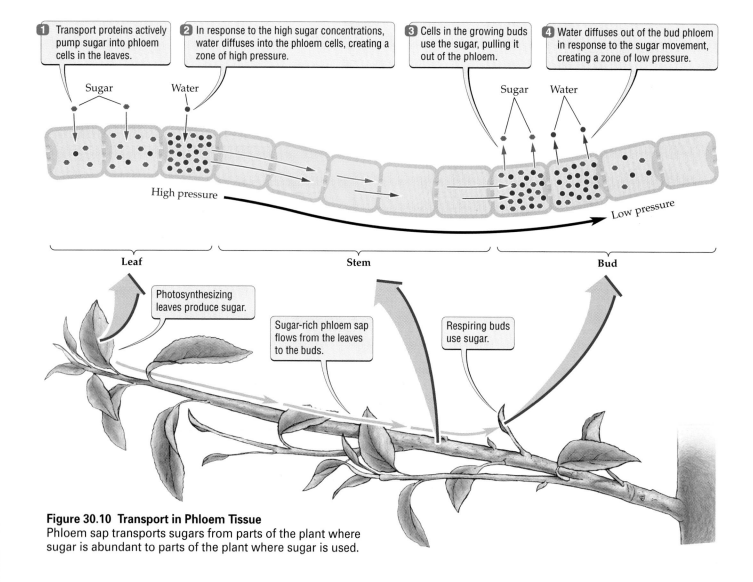

Figure 30.10 Transport in Phloem Tissue
Phloem sap transports sugars from parts of the plant where sugar is abundant to parts of the plant where sugar is used.

Phloem flow relies on osmotic pumps

Phloem sap moves by osmotically generated pressure that results when water diffuses down concentration gradients. In the leaves, specialized **companion cells** associated with the phloem use energy to actively pump the sugars produced by photosynthesis into the phloem against a concentration gradient (Figure 30.10). The active-transport proteins in the plasma membranes of companion cells can build up sugar concentrations of 10 to 30 percent in the phloem. Such high sugar concentrations mean low water concentrations in the phloem cells of the leaves, causing water to diffuse from the surrounding leaf tissue into the sugar-rich phloem cells. The accumulating water creates **osmotic pressure** as it accumulates in the phloem cells of the photosynthesizing leaf (we discuss osmotic pressure in more detail in Chapter 31).

In nonphotosynthesizing plant structures, which may be distant from leaves, cells extract sugar from nearby phloem tissue. Once the sugars have been removed from the phloem sap, the water diffuses from the now sugar-poor phloem cells into the sugar-rich surrounding cells. The diffusion of water from the phloem cells in the nonphotosynthesizing tissues leads to a region of relatively low pressure there.

The difference in pressure between the sugar-rich phloem cells in the leaves and the sugar-poor phloem cells in the nonphotosynthesizing tissues (such as roots) can easily reach 700 mm Hg, much greater than the pressure differences that the human heart generates. This pressure difference causes the phloem sap to flow from the leaves to all the tissues that consume sugar as an energy source.

Xylem flow relies on evaporation of water from leaf surfaces

The controlled evaporation of water from plant leaves creates the huge pressure differences needed to lift xylem sap to the tops of the tallest trees (Figure 30.11). To understand how simple evaporation could accomplish this feat, we must understand three points: (1) Water is pulled up to the surface of the leaves, and not pushed up by pumps in plant roots. This fact is evident when cut flowers placed in a vase continue to draw water without roots. (2) Water shows a strong tendency to be pulled into the air by evaporation. Even in humid air in which damp clothes dry slowly, a tension of more than –20,000 mm Hg draws water into the air as vapor (when a force pulls rather than pushes, pressure is negative—hence the minus sign—but it has the same effect as a push of 20,000 mm Hg). A tension of –20,000 mm Hg can lift a column of water twice as high as the tallest sequoia tree. (3) The continuous column of water established in the tubes of the xylem tissue of plants acts like a thread that resists breakage even when being pulled up by evaporation and down by gravity.

> ■ Phloem tissue relies on osmotic pressure gradients to move phloem sap. Xylem tissue relies on evaporation from the leaf surface to pull a continuous column of water up through the xylem cells. Both mechanisms generate pressure differences far beyond those generated by animal hearts.

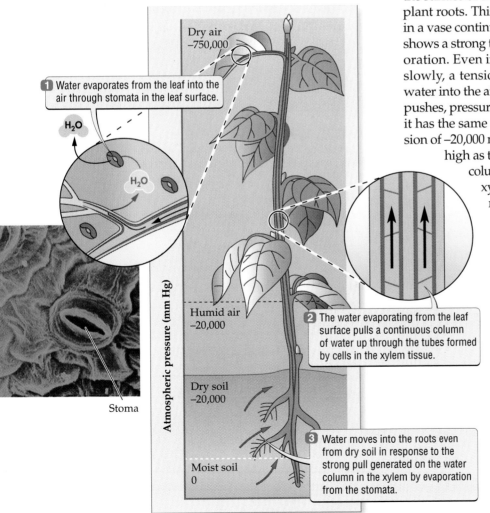

1 Water evaporates from the leaf into the air through stomata in the leaf surface.

H_2O

H_2O

Stoma

Atmospheric pressure (mm Hg)

Dry air
–750,000

Humid air
–20,000

Dry soil
–20,000

Moist soil
0

2 The water evaporating from the leaf surface pulls a continuous column of water up through the tubes formed by cells in the xylem tissue.

3 Water moves into the roots even from dry soil in response to the strong pull generated on the water column in the xylem by evaporation from the stomata.

Figure 30.11 Transport in Xylem Tissue
Evaporation from the leaf surface pulls a continuous column of water up from the soil. A more negative pressure, or greater tension, means a stronger pull.

Highlight

Acacias and Giraffes Revisited

Acacia trees and giraffes have circulatory systems that lift sap or blood to great heights. Acacias lift water from the dry soil 15 meters or more to their leaves. Although no known animal could manage this feat, the xylem tissue of the acacia easily meets this challenge. In the relatively dry air of the African savanna, evaporation can generate a pull of as much as –525,000 mm Hg, far more than the acacia needs to lift water to its leaves.

When roots cannot supply the leaves with water quickly enough, the tension generated by the dry air could break the water columns and threaten the life of the acacia. When such dangerous conditions arise during the growing season, acacias respond by closing the stomata (leaf pores) (see Chapter 29) that act as gateways to the intercellular spaces between the cells of their leaves. By closing its stomata, the acacia slows evaporation from the leaf surface greatly. However, closing stomata also cuts off the carbon dioxide supply to the leaves, preventing photosynthesis. During the 7-month dry season, therefore, closing the stomata to preserve the water columns in the xylem is not a reasonable option, so the acacias simply shed their leaves.

Giraffes have hearts that, for a mammal, can generate high blood pressures. The giraffe needs these high pressures to lift a column of blood that extends 2 meters from the heart to the brain. As an unavoidable consequence of maintaining the brain's blood supply, capillaries elsewhere in the giraffe's body experience pressures so great that they would destroy delicate human capillaries. To compensate, giraffe capillaries have thicker walls than their human equivalents. Moreover, when a giraffe bends its neck down to drink water, the blood pressure in the brain can increase from the normal 100 mm Hg to the 260 mm Hg measured in its feet.

Drinking giraffe

The capillaries in the brain lack the thicker walls of the capillaries elsewhere in the body and could easily be destroyed by such high pressures. One-way valves in the long neck veins, however, prevent a backflow of blood from the vein to the brain when the giraffe drinks, thus reducing the pressure on the capillaries of the brain. The combination of a strong heart and specialized blood vessels makes it possible for a giraffe to get a drink of water without passing out.

■ The internal transport systems of acacias and giraffes reflect the selective pressures faced by these two organisms: During periods of drought, acacias limit the water loss that accompanies xylem transport by closing stomata or shedding their leaves. Giraffes have a circulatory system suited to tolerating the high blood pressures that their hearts must generate to push water up to their brains.

Summary

Who Needs Internal Transport?

■ Diffusion can rapidly move materials across very short distances, but it is a very slow method of moving materials across more than a few cell layers.

■ Multicellular organisms need internal transport systems to supplement diffusion in order to distribute nutrients and gases.

The Human Circulatory System

■ The human heart pumps blood—which contains a mixture of dissolved minerals, gases, nutrients, and metabolic products—through closed loops of blood vessels.

■ There are three kinds of blood vessels: arteries that transport blood away from the heart, capillaries that exchange materials between the blood and adjacent tissues, and veins that carry blood back to the heart.

■ Millions of fine capillaries spread throughout our tissues so that diffusion can efficiently move materials the short distances between our cells and the nearest blood vessel.

■ The heart is composed of two separate pumps and four chambers: the right atrium and ventricle and the left atrium and ventricle.

■ The one-way flow of blood is maintained by one-way valves that separate the atria from the ventricles and the ventricles from the arteries.

The Blood Vessels of Animals

■ Transport of blood through veins and arteries requires the low friction drag associated with vessels of large diameter and low surface area–to–volume ratio.

■ With their small diameter, high surface area–to–volume ratio, and thin porous walls, capillaries maximize the exchange of materials between the blood and surrounding tissues.

■ The high surface area–to–volume ratio of capillaries greatly increases friction drag, so animals with closed circulatory systems must have strong hearts. Open circulatory systems avoid the friction drag associated with capillaries, but they also distribute materials to tissues both less precisely and more slowly.

Animal Hearts

- Chambered hearts include atria to forcefully fill ventricles, which then pump blood through vessels. Blood can also be pumped by the contraction of muscles surrounding vessels that contain one-way valves or by the contraction of muscles surrounding vessels.

- Muscular hearts of animals must create sufficient pressure to overcome friction drag associated with the flow of blood across the inner surface of blood vessels and to lift blood against the pull of gravity.

Internal Transport in Plants: A Different Approach

- The immense height of some plants produces tremendous gravitational force against which the plant must move sap from the roots to the leaves.

The Structure of Plant Vascular Tissue

- Phloem tissue, composed of living cells lined up end to end, transports sugars produced by photosynthesis both up and down the plant from photosynthesizing leaves to nonphotosynthesizing tissues.

- Xylem tissue, composed of dead cells lined up end to end, transports water and dissolved minerals up the plant from the roots to the leaves.

How Do Plants Move Sap?

- Phloem sap moves through the plant by osmotic pressure created as the diffusion of water follows the sugars that are actively pumped from photosynthesizing tissues into phloem and from phloem into nonphotosynthesizing tissues.

- Thin columns of xylem sap are pulled up the plant by evaporation of water from the leaf surfaces.

Highlight: Acacias and Giraffes Revisited

- The internal transport systems of acacias and giraffes reflect the selective pressures faced by these two organisms.

- Acacias control the evaporative water loss that accompanies xylem transport by closing stomata or shedding their leaves.

- Giraffes have a circulatory system suited to tolerating the high blood pressures that their hearts must generate to push water up to their brains.

Key Terms

artery p. 459	osmotic pressure p. 467
atherosclerosis p. 463	phloem tissue p. 465
atrium p. 459	plaque p. 463
capillary p. 459	sap p. 464
closed circulatory system p. 459	vein p. 459
companion cell p. 467	ventricle p. 459
heart attack p. 464	xylem tissue p. 465
open circulatory system p. 462	

Chapter Review

Self-Quiz

1. In which of the following ways does the circulatory system of animals differ from the vascular system of plants?
   a. The circulatory system of animals forms loops, whereas the vascular system in plants is linear.
   b. The sap of plants flows through cells, whereas the blood of animals flows through organs called blood vessels.
   c. The internal transport system of plants can lift fluids to greater heights than the internal transport system of animals.
   d. All of the above

2. Houseflies and cockroaches have an internal transport system that works at relatively low pressures and in which the blood flows around, rather than through, organs. We would call such an internal transport system
   a. an open circulatory system.
   b. a capillary.
   c. xylem tissue.
   d. a chambered pump.

3. As the diameter of a vessel increases, the friction drag that must be overcome by the heart
   a. is not affected.
   b. increases.
   c. decreases.
   d. increases and then decreases.

4. Xylem sap is
   a. pulled through the xylem by osmotic pressure.
   b. pushed through the xylem by a heart.
   c. pulled through the xylem by evaporation.
   d. pushed through the xylem by companion cells.

5. Which of the following statements about the flow of sap through phloem is *false*?
   a. Sap moves through phloem under pressure.
   b. Sap moves through phloem in one direction only: from the roots to the leaves.
   c. Phloem sap moves through the inside of living cells.
   d. Sugars move into phloem sap by active transport.

Review Questions

1. Describe and explain the variation in blood pressure throughout an individual's circulatory system during a single heartbeat.

2. After running across campus, late for a class, how would your circulatory system adjust to the lower demand for oxygen and energy once you got to the class and sat down?

3. Discuss the advantages and disadvantages of a closed circulatory system.

4. Why do you think that plants never evolved an internal transport system more like our own? Think of the two things that a plant's internal transport system must accomplish and of the limitations under which evolution by natural selection works.

The Daily Globe

Irrigation Efficiency Increases

BISMARCK, ND. Statistics provided by the North Dakota State Department of Agriculture indicate that prairie farmers are getting a greater increase in crop yields than they have in the past for every dollar spent on irrigation systems. Local farmers and officials see this as a major step forward for agriculture in this relatively rain-poor region.

In announcing the study results, North Dakota Secretary of Agriculture Bob Johanson indicated that the increase in efficiency benefited not only the farmers and consumers, but also the environment. Johanson stated: "By increasing irrigation efficiency we not only reduce the production costs of crops, but also conserve water, which is such a precious resource in our state."

A number of economically important local crops depend on irrigation, including corn and canola. Costs of installing irrigation equipment and of water rights contribute significantly to the cost of producing these crops. Recently concern has also arisen that irrigation is using up the aquifer, or underground water supply, on which most local irrigation depends. The shrinking aquifer has given rise to fears that water is being used in irrigation faster than it is being replaced by rainfall.

Canola farmer Leon Upchyk commented after the Secretary's announcement that "this is the first bit of good news that [farmers] here have had in a long time. It tells us that the investments we've made in upgrading our irrigation equipment are beginning to pay off."

Evaluating "The News"

1. Why do farmers spend money to irrigate their fields? Think of the various ways in which plants use water that have been introduced in this and preceding chapters.

2. Native plants can survive without irrigation in these same environ-ments. How do you think they manage to do this?

3. Those in favor of irrigation point out that it allows us to grow crops on otherwise less productive land, whereas those opposed to irrigation argue that we are using up valuable underground water reserves to support crops that are more easily grown elsewhere. How do you feel about the use of irrigation?

31

Maintaining the Internal Environment

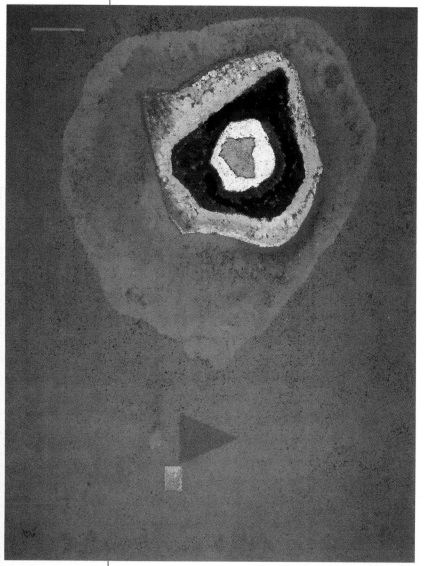

Wassily Kandinsky, *Cool Condensation*, 1930.

The Giant Guardians of the Sonoran Desert

Saguaro cacti are often used by the print and film media as the defining image of the American West. They are the largest columnlike cacti in North America. Many of the 200- to 300-year-old individuals grow to heights of 10 meters, reaching high above surrounding shrubs that often encircle their lower trunks. To see them standing against the skyline tells you you're in the Sonoran Desert of Mexico and the southwestern United States. Although their use in the media often suggests they are abundant in all western deserts, in fact saguaros occur only rarely in the other deserts of western North America, and even in the Sonoran Desert they are limited to intermediate elevations.

A close look at a saguaro reveals that it looks very different from plants in the temperate regions of North America, and even from most other plants in the Sonoran Desert. Not only are saguaros surprisingly

Main Message

Organisms use many different means to balance internal concentrations of water and solutes and to keep internal temperatures within a livable range.

tall, but they are thick, like a stack of barrels. Except at the very base, a saguaro is covered with a light green "skin" that is quite different from the bark of most plants its size. Its trunk is deeply pleated with vertical furrows; it has radiating patches of spines, especially covering its rounded top; and, like other cacti, it lacks leaves.

How have the temperature and water conditions of the Sonoran Desert selected for such a wonderfully unique plant? After considering how organisms in general respond to and manage environmental variation in temperature and water, we will return to this question and consider how the shape and physiology of the saguaro is uniquely adapted to the challenging temperature and water conditions of the Sonoran Desert.

The Saguaro Cactus Dominates the Dry, Hot Sonoran Desert

Key Concepts

1. The external environment varies greatly, but most organisms function well only over a narrow range of conditions.

2. Most organisms can regulate their internal environment within a narrow range of conditions.

3. Organisms gain and lose water by diffusion across membranes that are used to exchange gases or nutrients.

4. Organisms exchange heat with their environment by contact with it or by absorbing or radiating light. Heat generated as a by-product of respiration contributes to heat gain in some organisms.

5. The ways in which organisms gain and lose water and heat depend strongly on the environment in which they live.

On Earth, water availability and temperature vary greatly and often reach extremes that most living cells cannot tolerate. In addition, within a single habitat, temperature and water availability can change dramatically. The saguaro cactus, for example, faces swelteringly hot afternoons followed by surprisingly chilly evenings, and it withstands months of drought broken only by two brief rainy periods. To tolerate both the extremes in environment that different species encounter and the variation that a single individual may experience, organisms have evolved a tremendous diversity of ways of maintaining a relatively constant internal environment.

In this chapter, we begin by discussing the importance of regulating body temperature and water content. We then consider two closely related aspects of homeostasis: how organisms regulate the exchange of water to control both body water content and the concentration of dissolved materials in this body water. We finish by considering the importance of regulating body temperature, which we will find to be closely tied to the issue of water regulation.

Why Regulate the Internal Environment?

Many organisms function well only within a narrow range of internal environments. In humans, a losing as little as 5 percent of body weight in water (roughly the amount lost after 2 hours of continuous, vigorous activity on a warm day) can reduce our ability to do work by 30 percent or more. A loss of more than 15 percent of body weight in water generally kills mammals.

Organisms must also maintain the temperature inside their cells within a limited range. When water inside cells freezes, it expands, which can rip apart the plasma membrane (see Figure 31.2). It is for this reason that frostbitten tissue dies.

High temperatures can also interfere with cellular functions. The proteins that make up most enzymes typically start to unfold and plasma membranes are disrupted at temperatures of 45 to 50°C.

Between these lethal temperature extremes, certain temperatures allow the highest activity levels. Over the course of a day, our carefully regulated core temperature varies little: from about 36.5 to 37.5°C (Figure 31.1). Other organisms, such as the garter snake shown in Figure 31.2, can tolerate a wide range of temperatures but are much more active in a narrow range of temperatures; the snake, for example, is most active near 35°C.

> ■ Organisms can survive only when the water inside their bodies and their internal temperatures fall within a limited range of values, and they do well within an even narrower range of internal conditions.

How Organisms Gain and Lose Water

All living things constantly gain and lose water from the environment. Unless organisms can regulate how much water they gain and lose, however, they cannot maintain homeostatic control of water and solutes in the face of changing environmental conditions. Sufficient water must be maintained within and surrounding cells because it provides the ideal medium for dissolving and circulating biologically important materials and promoting biochemical reactions. The hydrostatic skeletons (see Chapter 26) that provide structural support for bacteria, fungi, plants, and many animals depend on pressure generated by water. As we will see later in the chapter, water is also important in thermal regulation. To understand how water balance can be maintained, we will first consider how water is exchanged between organisms and their environments.

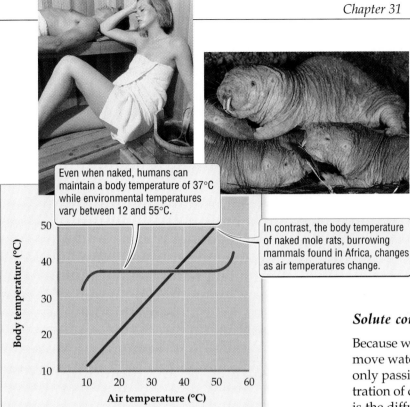

Figure 31.1 Homeostasis in a Changing Environment
Some organisms, such as humans, can maintain a constant internal environment despite changes in the external environment.

Even when naked, humans can maintain a body temperature of 37°C while environmental temperatures vary between 12 and 55°C.

In contrast, the body temperature of naked mole rats, burrowing mammals found in Africa, changes as air temperatures change.

Solute concentrations influence water movement

Because we and all other organisms lack pumps that can move water across plasma membranes, water can move only passively in response to differences in the concentration of dissolved materials (see Chapter 24). **Osmosis** is the diffusion of water across a membrane from areas of low concentration of dissolved materials to areas of high concentration. Whenever the concentration of dissolved materials (solutes) in a cell or in the organism exceeds that outside, water tends to diffuse in; alternatively, when the concentration of dissolved materials in the cell or organism is lower than that outside, water diffuses out. These osmotic movements can cause cells to shrink or expand (Figure 31.3).

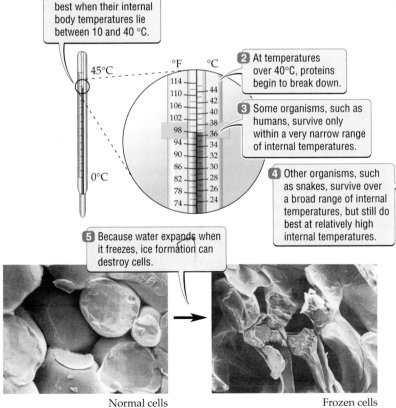

1 Most living things do best when their internal body temperatures lie between 10 and 40 °C.

2 At temperatures over 40°C, proteins begin to break down.

3 Some organisms, such as humans, survive only within a very narrow range of internal temperatures.

4 Other organisms, such as snakes, survive over a broad range of internal temperatures, but still do best at relatively high internal temperatures.

5 Because water expands when it freezes, ice formation can destroy cells.

Normal cells   Frozen cells

Figure 31.2 The Temperature of Life
Most living things have an ideal body temperature that falls between 0 and 40°C. Although most organisms can survive over a range of temperatures, they usually do best within a relatively narrow range of temperatures.

Neither gains nor loses water	Loses water	Gains water
When solute concentrations outside the cell equal concentrations inside the cell, the cell neither gains nor loses water.	When solute concentrations outside the cell exceed those inside the cell, the cell shrinks as it loses water to its environment.	When solute concentrations outside the cell are lower than those inside the cell, the cell swells as it gains water from its environment.
Plant cells (illustration)	(illustration)	(illustration)
Animal cells (image) Normal red blood cell	(image) Shrunken red blood cell	(image) Bloated red blood cell
Examples Most marine organisms	Land organisms and marine vertebrates	All freshwater organisms

Figure 31.3 Water Moves by Diffusion
Differences in the concentrations of all dissolved materials combined in water surrounding and inside cells determine how water moves into and out of cells.

(*a*) Land plants

Evaporation from leaves is the major route for water loss in land plants.

Plant roots take up water and dissolved minerals from the soil.

Water can be lost through evaporation

In terrestrial environments, water loss through evaporation presents a serious problem (Figure 31.4). Recall from Chapter 30 that water evaporates readily into the air from moist surfaces such as the intercellular spaces of plant leaves and the lungs of humans. The windier or hotter the weather, or the greater the wind speed, the greater this evaporative loss will be. Since many essential functions such as internal transport in plants (see Chapter 30) and gas exchange (see Chapter 29) necessarily involve water loss by evaporation, it is impossible for terrestrial organisms to completely eliminate it. Moreover, as we will see later, since evaporation is also an extremely effective cooling mechanism, an organism's water balance is closely related to its temperature regulation.

Metabolic processes influence water balance

Terrestrial animals typically absorb most of their water from the digestive tract. As consumers, however, animals can also gain a significant amount of water by metabolizing food: Recall from Chapter 28 that one of the by-products of the metabolism of car-

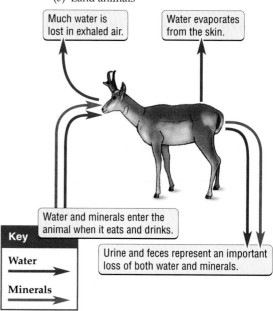

(*b*) Land animals

Much water is lost in exhaled air.

Water evaporates from the skin.

Water and minerals enter the animal when it eats and drinks.

Urine and feces represent an important loss of both water and minerals.

Key

Water →

Minerals →

Figure 31.4 How Water Enters and Leaves Organisms
Plants and animals can lose and gain water by a variety of routes.

bohydrates to release energy is water. For example, the small grain beetles that might infest a long-forgotten box of cookies can get all the water they need as metabolic water, never drinking any in liquid form.

Although it produces small, but significant, amounts of water, respiration also requires the exchange of gases with the external environment and the disposal of metabolic wastes in urine and feces, both of which are processes that result in water loss.

> ■ Water enters cells and organisms passively by diffusing from areas of low to high solute concentrations. Water can be gained through respiration and across gills and digestive systems, through gas exchange structures, and during the elimination of metabolic waste products.

How Organisms Regulate Water Balance

Organisms use many techniques to regulate the water in their tissues. The sections that follow present a few examples of the ways in which organisms regulate water and solutes.

Some organisms do not control water exchange

With the exception of fishes and mammals, single-celled organisms and multicellular animals that live in the deep ocean where water quality remains almost constant have internal water and solute concentrations that match those of the surrounding seawater. In this stable environment, organisms do not experience rapidly changing solute concentrations in the external environment, so there is little selection pressure favoring the evolution of a means of regulating their water balance.

Fish use a variety of processes to regulate water balance

Fish present an interesting example of the ability of organisms to maintain homeostatic control of water despite differences in solute concentrations between their internal and external environments. Because the

internal solute concentration of freshwater fish is higher than that of the water they live in, they tend to gain water as it diffuses in across their gills. They rid themselves of excess water by producing a very dilute urine, and they gather what salts they can by using active pumps in their gills to concentrate the scarce salts in the surrounding water into their bodies (Figure 31.5a).

How do fish live in salty seawater? Unlike the osmotic conformers already discussed, most marine fish maintain a solute concentration that differs from that of seawater. Bony marine fish probably evolved from fish adapted to fresh water, and the solute concentration in their bodies is one-third that of the surrounding ocean. In contrast to freshwater fish, they *lose* water by diffusion, especially across the surfaces of their gills, into the salt water. They avoid dehydrating in the midst of the ocean by drinking seawater and actively pumping out tremendous quantities of excess salts through their gills and as concentrated urine through specialized organs called kidneys, which we discuss below (Figure 31.5b).

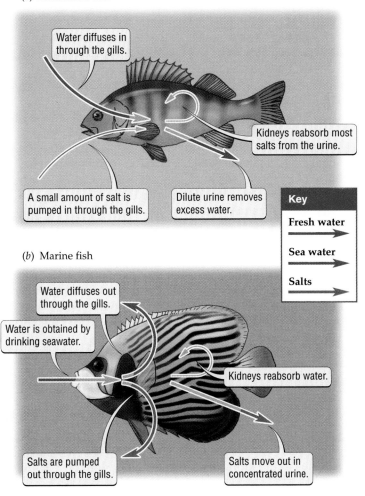

(a) Freshwater fish

Water diffuses in through the gills.

Kidneys reabsorb most salts from the urine.

A small amount of salt is pumped in through the gills.

Dilute urine removes excess water.

Key

Fresh water

Sea water

Salts

(b) Marine fish

Water diffuses out through the gills.

Water is obtained by drinking seawater.

Kidneys reabsorb water.

Salts are pumped out through the gills.

Salts move out in concentrated urine.

Figure 31.5 Osmotic Movement of Water in Freshwater and Marine Fish
Osmosis draws water into a freshwater fish (*a*) because its environment has a low relative solute concentration compared to the high relative solute concentration inside the fish. Osmosis draws water out of a marine fish (*b*) because its environment has a high relative solute concentration compared to the low relative solute concentration inside the fish.

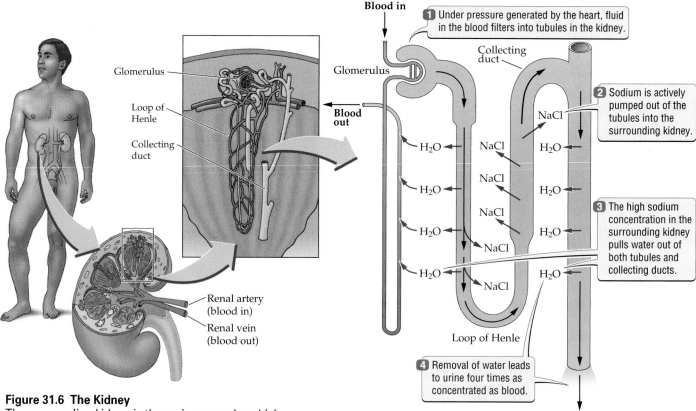

Figure 31.6 The Kidney
The mammalian kidney is the main means by which humans regulate water and solutes and dispose of waste.

Labels in figure:

Glomerulus

Loop of Henle

Collecting duct

Renal artery (blood in)

Renal vein (blood out)

Blood in

1 Under pressure generated by the heart, fluid in the blood filters into tubules in the kidney.

Collecting duct

Glomerulus

Blood out

2 Sodium is actively pumped out of the tubules into the surrounding kidney.

3 The high sodium concentration in the surrounding kidney pulls water out of both tubules and collecting ducts.

Loop of Henle

4 Removal of water leads to urine four times as concentrated as blood.

Urine

Terrestrial animals rely on specialized water-regulating structures

As plants and animals moved onto land, they faced the challenge of maintaining the water in their bodies in the face of evaporative water loss and limited water availability. Terrestrial animals, particularly those living in dry habitats such as deserts, must carefully regulate water gains and losses.

Most animals have specialized organs—kidneys—that are the major site of homeostatic regulation of solutes and water (Figure 31.6). These organs use the principle of osmosis to control the flow of filtered water back into the circulatory system. Although the designs of these organs differ greatly among species, they all rely on a two-step process: (1) Filtering the body water and a diversity of small molecules, both useful and not, through a porous membrane under pressure, and then (2) actively recovering the useful molecules that passed through the membrane.

Consider our own kidneys. In the human body, the pressure generated by the heart forces an amazing 180 liters of blood a day through the porous walls of the capillaries in the kidneys. After such filtration, only large molecules like proteins and carbohydrates remain in the blood, and the pressure-filtered water that passes into the collecting system of the kidney carries valuable sugars, amino acids, and minerals along with wastes. By active transport of solutes, however, we reabsorb almost all the water and valuable solutes back into the blood, leaving only excess solutes, wastes, and a small amount of water that is eliminated from the body in the form of urine. Thus, each day we pass out of our bodies less than 2 liters of concentrated urine, approximately 1 percent of the volume of water and solutes that the kidneys filter daily.

Terrestrial plants use roots and leaves to regulate water

Land plants face problems similar to those of land animals in reducing water loss to balance the limited amount available in the environment. Water is often tightly held by soil particles; at the same time, plants lose water during gas exchange and in supporting the upward movement of xylem sap carrying mineral nutrients.

Unlike animals, plants cannot move from one area to another to search for water and avoid dry conditions. Some plants, however can escape drought conditions by

Desert annual

dropping tissues such as leaves to reduce water loss and by greatly reducing their activity levels (for example, the acacias discussed in Chapter 30). Annual plants of the desert survive the dry season as inactive seeds that can survive for years waiting for the next rain. When the rains come, the seeds quickly absorb water and germinate, and in less than a month the desert can be blanketed with the bright colors of annuals in full bloom preparing to produce new seeds that, like those that developed into their parents, will wait through months or years of drought.

Many plants stay active even as the amount of water in the soil declines. These plants actively pump mineral nutrients into their roots, thereby encouraging the diffusion of water from the soil into the root: They adjust the solute concentration in their roots to a level sufficient to draw in water held in the soil.

In addition to encouraging the movement of water into the roots and throughout the plant, plants often conserve water by reducing evaporative water loss. Leaves and shoots bear a waterproof waxy layer that restricts most evaporative water loss to the stomata that can open and close to allow CO_2 to enter and O_2 and H_2O to leave. Although plants can close their stomata to greatly reduce water loss, they must open their stomata at some time to bring in CO_2 for photosynthesis. Many plants reduce evaporative water loss during photosynthesis by concentrating their stomata on the lower, cooler leaf surface and by covering this lower leaf surface with short hairs that reduce air movement and evaporation by holding a humid layer of air next to the surface of the leaf.

■ Many marine organisms, with the exceptions of fish and mammals, have internal concentrations of solute that matches the solute concentration of their environment. Fish maintain homeostatic control of water and solute concentrations in spite of the osmotic flow of water through the surface of their gills. Kidneys regulate water and solute concentrations in addition to disposing of waste products. Land plants escape water loss through reduced activity during drought, and they limit water loss in active tissues by limiting evaporative water loss.

How Organisms Gain and Lose Heat

In addition to regulating internal concentrations of water and solutes, many organisms must regulate temperature to survive and maintain high activity levels. A first step to understanding temperature regulation is to understand that heat enters and leaves an organism in different forms: conductive heat, radiant heat, evaporative heat, and metabolic heat (Figure 31.7).

Heat conduction requires direct contact

The exchange of **conductive heat** requires direct contact between an organism and its environment. When the rapidly vibrating molecules of a warm surface interact with the more slowly vibrating molecules of a cool surface, the molecules in the cool surface warm up (vibrate more quickly) and those in the warm surface cool down (vibrate more slowly). This transfer of heat—from a warmer object to a cooler one—by direct contact between two things is conductive heat transfer.

The rate of conductive heat transfer depends on the area of contact and the temperature difference between the two objects: If we place a whole hand on a hot surface, more heat is transferred to the skin than if we touch a warm cup of tea with the tip of one finger. The rate of conductive heat exchange is also influenced by **convection**, the circulation

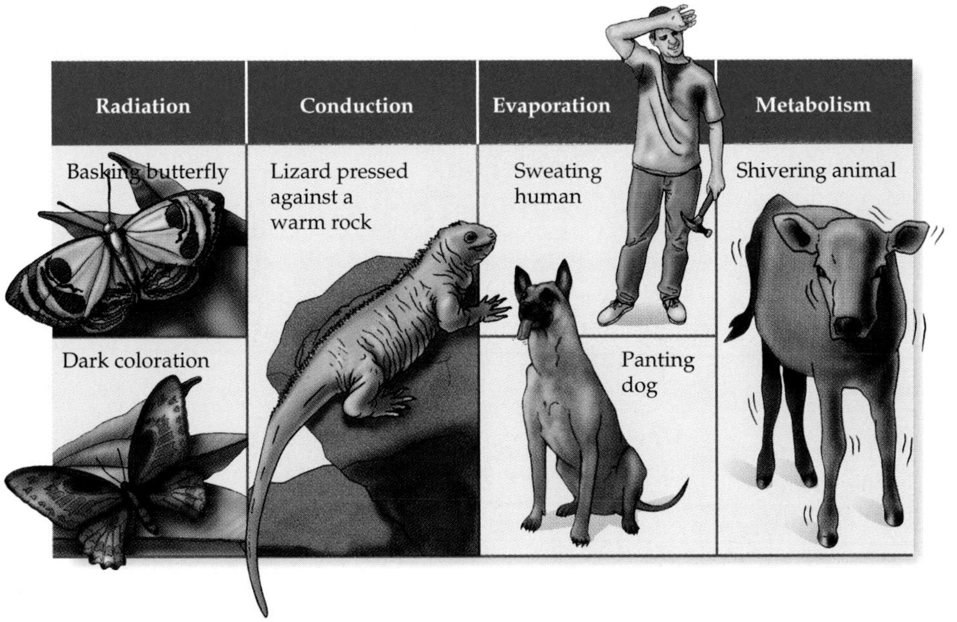

Radiation	Conduction	Evaporation	Metabolism
Basking butterfly	Lizard pressed against a warm rock	Sweating human	Shivering animal
Dark coloration		Panting dog	

Figure 31.7 How Organisms Effect Heat Exchange
Many diverse adaptations allow plants and animals to control how they lose and gain heat.

of one of the substances. For example, we may attempt to increase the transfer of heat out of an overly hot cup of tea into the air by blowing on it. The wind currents that we generate as we blow on it carry hot air away from the tea and replace it with cool air. Heat conducts rapidly from the hot tea into the cool air.

Whereas heat transfers quickly between the closely interacting molecules of solids like sand grains or rock, it transfers slowly between the loosely interacting molecules of gases like air. Biologically this difference matters a great deal because, as a liquid in which molecules interact more closely than do those of air, water conducts heat about 25 times faster than air. Consequently, conductive heat exchange is much more rapid in water than on land. Aquatic organisms exchange almost all their heat conductively because water carries heat to and from their bodies so quickly.

All organisms absorb and give off radiant heat

Radiant heat exchange involves transferring energy to and from the environment in the form of light. Light has energy that it can transfer to molecules that absorb it. When we bask in the sun, we absorb mostly visible and ultraviolet light, which causes the molecules on the surface of the skin to vibrate more vigorously.

The rules that govern radiant heat exchange differ from those for conductive heat exchange: (1) The fraction of radiant heat that an organism absorbs depends on its color, and (2) radiant heat exchange depends on the surface area presented to certain objects in the environment, particularly the sun. During cold weather, roadrunners orient their necks and backs to the sun and lift their feathers to expose the dark skin underneath that absorbs the radiant energy of the sun much more rapidly than would lightly colored skin.

Evaporation allows organisms to lose heat in hot, terrestrial environments

Whenever water evaporates from a surface, it takes with it a lot of **evaporative heat**. This phenomenon is what makes our skin feel cool after we take a dip in the ocean or when a breeze dries the sweat released from our skin glands. For water to evaporate, the interactions that hold water molecules together in a liquid form must be broken. To break enough bonds to convert 1 gram of liquid water at body temperature to 1 gram of water vapor requires 2.4 kilojoules of energy, or about six times as much energy as it takes to heat water from near freezing to near boiling. And when water evaporates from the body surface, it gets most of this energy from our body.

Perhaps the most important biological feature of evaporation is that it provides an avenue for heat loss that does not depend on the temperature of the organism relative to its environment. An organism can lose conductive or radiant heat only if the temperature of the heat source is lower than the body temperature, but no such limitation exists for evaporative heat loss. In hot environments, such as deserts, where the environmental temperature can easily exceed safe body temperatures, evaporative heat loss often makes survival possible.

In only a few organisms does metabolic heat contribute to the heat balance

Metabolic heat is an inevitable by-product of the chemical reactions essential to life. The amount of metabolic heat produced usually represents only a tiny fraction of the total heat gain of an organism. In most plants, for example, metabolic heat production amounts to less than 1 percent of the total radiant heat gain. A relatively few **endotherms** (literally "inside heat"), such as birds and mammals, generate enough metabolic heat for it to contribute in an important way to their heat gain. Thus, endotherms can maintain a high body temperature even in cold weather, when conductive heat sources are unavailable, and at night, when the radiant heat of the sun disappears. While endotherms can often maintain a constant high body temperature regardless of temperatures in the environment, that is rarely possible for organisms called **ectotherms** (literally "outside heat"), which depend on environmental heat to maintain their body temperature.

> ■ Organisms gain or lose heat through conductive movement of heat between adjacent objects, radiant heat exchange, evaporative heat loss, and metabolic heat production.

How Organisms Regulate Body Temperature

The ability of organisms to use the four basic types of heat (described in the previous section) to regulate their body temperature depends on characteristics of the organism and on the nature of the external environment.

Most aquatic organisms do not regulate body temperature

Most aquatic organisms have internal temperatures that match those of their external environment. In aquatic environments, most organisms cannot maintain a temperature that differs from that of the surrounding water. Not only does the rapid conduction of heat in water quickly carry heat to or from the organisms, but the large surface area that they need to effectively exchange gas with the water speeds heat exchange. Neither radiant heat exchange nor evaporative heat loss plays a significant role in aquatic environments.

Many terrestrial animals cannot fully regulate their internal temperatures

As animals moved onto land, they faced not only the tremendous challenges of increased evaporative water loss and reduced water availability, but also a much greater daily and seasonal variation in temperature than their ancestors had experienced in aquatic environments. Although most animals, such as insects, amphibians, and reptiles, are ectotherms, they are not completely at the mercy of the environment. Many ectotherms have behavioral and structural adaptations that enable them to avoid extremely high or low body temperatures. In addition, most can use conductive and radiant heat exchange to maintain body temperatures higher than those of their environments for at least part of each day.

Marine iguanas, which feed in the cold ocean waters surrounding the Galápagos Islands (Figure 31.8), illustrate how an ectothermic organism can exploit external heat sources to regulate body temperature. Because they lose heat quickly to the cool (20°C), surging waters of the coastal waters where they feed, they need to warm up quickly on land. Their black coloring and habit of orienting themselves perpendicular to the sun's rays maximize radiant heat gains. In addition, they gain heat conductively by pressing their bellies tightly against black volcanic rock that has been warmed by the sun, often to temperatures over 40°C.

While basking, iguanas increase their heart rate and the flow of blood to their skin to help circulate heat from the skin to the tissues inside their body. After about an hour, when their internal temperature reaches 37°C, they dive back into the water.

In the water, iguanas reduce their heart rate and the flow of blood to their skin. This reduces heat transfer from their internal tissues to their body surface. In spite of this, heat conducts rapidly from their body into the cool water, so that they can feed for only about 30 minutes before dropping body temperatures force them to return to land.

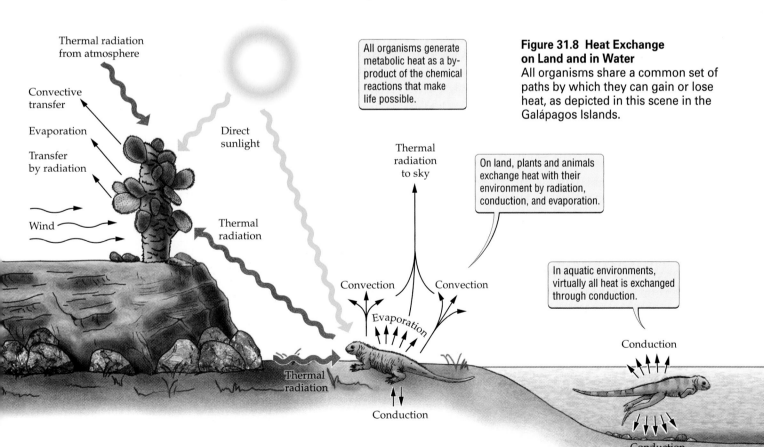

All organisms generate metabolic heat as a by-product of the chemical reactions that make life possible.

Figure 31.8 Heat Exchange on Land and in Water
All organisms share a common set of paths by which they can gain or lose heat, as depicted in this scene in the Galápagos Islands.

On land, plants and animals exchange heat with their environment by radiation, conduction, and evaporation.

In aquatic environments, virtually all heat is exchanged through conduction.

Thermal radiation from atmosphere

Convective transfer

Evaporation

Transfer by radiation

Wind

Direct sunlight

Thermal radiation

Thermal radiation to sky

Convection

Convection

Evaporation

Thermal radiation

Conduction

Conduction

Conduction

Mammals and birds use metabolic heat to regulate their temperature

As endotherms, mammals and birds can generate enough metabolic heat to maintain constant high body temperatures despite varying temperatures in their environment. The metabolic heat needed to regulate body temperature comes at a cost, however: Both at rest and during activity, the respiratory rate of endotherms is about ten times that of ectotherms. As a result, endothermic animals need to eat much more food than ectotherms do (Figure 31.9*a*).

Because endotherms rely on metabolic heat, which requires energy to generate, *conserving* body heat plays an important part in regulating body temperature. Insulating fur or feathers help to reduce the rate of conductive and radiative heat loss. The air trapped next to the skin by fur or feathers conducts heat poorly and prevents the convective movement of heat away from the body. Coats and sleeping bags work on the same principle: It is the still air that they trap rather than the materials of which they are made that keep us warm.

The insulating effect of fur and feathers to reduce conductive and radiant heat loss is a double-edged sword. When environmental temperatures are high or when increased activity generates additional metabolic heat, many endotherms must rely on evaporative heat loss to keep them cool. Panting by many birds and mammals and sweating in some mammals, such as humans and camels, are highly effective methods of cooling, but they require energy and the loss of water.

An increased reliance on metabolic heat, a reduction of radiant and conductive heat exchange, and an increased use of evaporative heat loss have produced a sophisticated system for maintaining homeostatic control of temperature that has been very successful for relatively large multicellular animals, mainly in terrestrial environments. The smallest mammals and the smallest birds are much larger than many ectotherms because as size decreases, the increase in surface area relative to volume makes it difficult to limit conductive and radiative heat loss. The rapid conduction of heat through water also makes it difficult for endotherms to limit heat loss or gain in aquatic habitats. When the influences of small size and aquatic habitats on conductive heat are combined, they present an impossible challenge to an endotherm; thus, there are no small aquatic mammals or birds.

Although endothermy depends on sophisticated temperature control in many environments, not all endotherms maintain a nearly constant temperature, just as not all ectotherms have a highly variable temperature. Although it is mainly endothermic, the roadrunner mentioned earlier can let its body temperature drop by 9°C

during the night. In the morning it suns itself to capture radiant heat to help raise its body temperature, thus reducing its overall demand for metabolic heat by approximately 40 percent. On the other hand, an ectotherm such as the crocodile can regulate its temperature remarkably well (Figure 31.9*b*). The internal temperature of large crocodiles varies by only 0.1°C between night and day as a result of their large mass and their ability to manage radiant and conductive heat exchange with basking and swimming behaviors similar to those of the marine iguana.

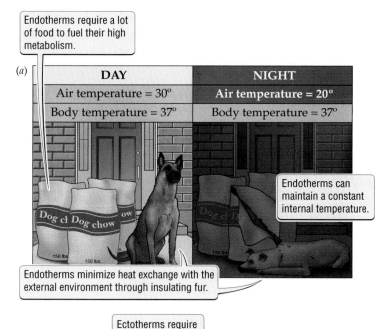

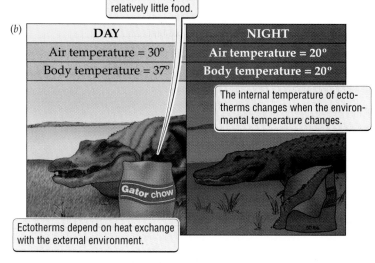

Figure 31.9 Heat Exchange in Endotherms versus Ectotherms
(*a*) An endothermic dog generates a significant amount of heat metabolically. (*b*) An ectothermic crocodile relies almost completely on environmental heat sources.

■ Small organisms and aquatic organisms tend to have body temperatures that match their environment. Many terrestrial ectothermic animals use behavioral adaptations to reduce temperature extremes. Endothermic organisms rely on metabolic heat, insulation, and evaporative cooling to regulate their body temperatures.

Highlight

The Saguaro: Master of Extremes

The saguaro (see the photo on page 473) survives brutal temperature fluctuations. It stands exposed on summer days when temperatures reach a scorching 48°C; it must also survive desert nights, during which, on the northern edge of the saguaro's range, temperatures often fall below −5°C. The saguaro may have to wait months between rains, but when the rains arrive they often come as torrential monsoons. How does the saguaro hide in broad daylight and regulate both water and temperature through the long dry spells that can last months or years?

Like other plants, saguaros depend on the sun for photosynthesis, but they belong to a select group of plants that can photosynthesize with their stomata closed. They open their stomata at night to collect CO_2 and bind the carbon into larger molecules, thereby storing CO_2. During the day while their stomata are closed, the storage molecules release CO_2 so that photosynthesis can occur. By separating photosynthesis from gas exchange, saguaros can greatly reduce water loss through their stomata.

Saguaros avoid evaporation through leaf surfaces by having no leaves. Light can shine through the thick cuticle and complex epidermis to photosynthetic tissues in the trunk. With the sunlight comes radiant heat, however, so the folded ridges on the trunk allow the light to hit the saguaro's trunk at a shallower angle, resulting in much less radiant heat absorption than on a flat surface. This feature not only protects cells from heat damage, but it also reduces the small amount of precious water that evaporates through the cuticle.

In addition to reducing the heat load, the same vertical ridges and furrows allow saguaros to take up tremendous amounts of water during the short-lived monsoons, which they can then store for months or years. Their long roots radiate out near the ground surface, so when the rain does finally pour down, they can rapidly take it up. As water is pulled into the trunk, the pleats expand, greatly increasing the plant's capacity for water storage.

The tons of water contained in a large saguaro protect it from rapid changes in temperature. However, younger individuals lack the large mass of tissue that allows large plants to slow heat exchange. Young saguaros have much denser spines at their crowns; these spines not only shade them during hot days, but also protect them from radiant heat loss during cold winter nights. The young plants are also protected from radiant heat gain (and associated water loss) during the day and radiant heat loss at night by nurse plants. The survival of young saguaros is greatly enhanced when they grow under the canopy of a desert shrub, a nurse plant —protected from temperature extremes and water stress associated with high temperature.

The ability of saguaros to regulate their temperature and water content depends on their unique size, shape, physiology, and even interactions with other organisms.

■ The ability to photosynthesize while its stomata are closed and a lack of leaves allow saguaro cacti to avoid water loss in their hot, dry desert environment. An extensive root system and a specialized trunk allow the plants to take up and store a tremendous amount of water during brief bouts of rainy weather. Saguaros avoid overheating through a trunk surface that reflects most radiant heat and a large size that prevents internal temperature fluctuations.

Summary

Why Regulate the Internal Environment?

■ Organisms can only survive if their internal environment falls within a narrow range of conditions.

■ Organisms function best under an even narrower range of conditions.

How Organisms Gain and Lose Water

■ Water diffuses across membranes in response to gradients of solute concentrations.

■ In terrestrial environments, organisms lose water through evaporation, which has a cooling effect.

■ Metabolism influences water regulation by producing metabolic water, by requiring gas exchange that often is associated with water loss, and by forcing the loss of water during the disposal of metabolic wastes.

How Organisms Regulate Water Balance

■ With the notable exceptions of fish and mammals, many marine organisms maintain internal solute concentrations that match those of the ocean around them.

- Most fish maintain homeostatic control of water concentrations in spite of the tendency of water to flow osmotically out through the gills in marine environments and in through the gills in freshwater environments.

- The kidneys of animals reduce water loss while disposing of excess salts and metabolic wastes.

- Although stomata must open for photosynthesis, plants reduce evaporative water loss through the stomata by limiting when they are open, by concentrating them on the lower leaf surfaces, and by reducing airflow across their surfaces.

How Organisms Gain and Lose Heat

- Conductive heat is transferred between touching objects from the warmer to the cooler object. The rate of conductive heat transfer is influenced by the area of contact and the difference in heat between the two objects. Conductive heat exchange happens quickly in water and slowly in air.

- The rate of radiant heat exchange depends on the color of the organism and on the surface area presented to the heat source.

- Evaporative heat loss can cool organisms to temperatures below that of the external environment.

- Metabolic heat production is important to only a few organisms, among them mammals and birds.

How Organisms Regulate Body Temperature

- Aquatic environments are associated with rapid conductive heat exchange, very little radiant heat exchange, and no evaporative heat exchange. However, because of the relatively stable water temperatures of large bodies of water, organisms living in these habitats do not experience rapid changes in body temperature.

- Ectothermic terrestrial animals use behavioral and structural adaptations to modify body temperature, mainly through conductive and radiant heat exchange.

- Endothermic mammals and birds produce large amounts of metabolic heat and retain this heat by insulating their bodies with fur and feathers.

Highlight: The Saguaro: Master of Extremes

- Saguaro cacti maintain internal water supply and temperature while living in an exceptionally hot, dry desert environment.

- The ability to photosynthesize with closed stomata and a lack of leaves reduce evaporative water loss, while a large root system and storage tissues in the stem allow saguaros to store the meager rainfall.

- Features of the trunk surface and their large size protect the saguaros from extreme changes in internal temperature.

Key Terms

conductive heat p. 479	evaporative heat p. 480
convection p. 479	metabolic heat p. 480
ectotherm p. 480	osmosis p. 475
endotherm p. 480	radiant heat p. 480

Chapter Review

Self-Quiz

1. Water moves across biological membranes
   a. because it is actively pumped.
   b. passively in response to solute concentrations.
   c. passively from high to low temperatures.
   d. actively as a result of radiant exchange.

2. Plants increase the uptake of water
   a. by increasing solute concentrations in the roots.
   b. by decreasing solute concentrations in the roots.
   c. by decreasing solute concentrations in the stomata.
   d. by increasing the number of hairs surrounding stomata.

3. Marine fish tend to
   a. gain water osmotically through gills.
   b. be osmotic conformers.
   c. lose water osmotically through gills.
   d. have excess water.

4. The primary heat source for endothermic organisms is
   a. conductive heat.
   b. radiant heat.
   c. evaporative heat.
   d. metabolic heat.

5. Fur and feathers
   a. reduce heat loss in endothermic animals.
   b. reduce convective heat transfer.
   c. reduce heat loss by trapping air, which conducts heat poorly.
   d. all of the above

Review Questions

1. How does an organism's metabolism affect its heat gains and losses? Try to think not only of the direct effects of metabolism, but also of the indirect ones.

2. Under what circumstances do organisms tend not to regulate internal water or temperature?

3. Humans are endotherms. What advantages do we have over ectotherms? What disadvantages do we face?

31

The Daily Globe

Alcohol and Sports Do Not Mix

Letter to the Editor:

I am writing to cancel my subscription to *The Daily Globe*. In last week's Sunday magazine, articles criticizing the poor examples that athletes set when they are arrested for using illegal drugs were comfortably sandwiched between pages showing healthy, active people enjoying alcohol as they enjoy sports. The subliminal message that you were sending is clear: Alcohol consumption goes hand in hand with sports.

Your readers should be informed that alcohol reduces coordination to a dangerous degree in many sports. But there are other problems when combining alcohol and exercise. You should be writing articles explaining these more subtle interactions to your readers rather than displaying advertisements that mislead your readers about the relationship between activity and alcohol.

One common and dangerous problem people face during exercise, especially in hot climates or at high elevations, is dehydration. Although drinking a cold beer may seem to quench thirst, it actually robs the body of precious water. All forms of alcohol interfere with a hormone that signals the body to reabsorb water in the kidney, and as a result water is lost. This water loss can still be felt hours later as the brain becomes dehydrated and the drinker feels hung over.

In cold-weather sports, alcohol is risky for a different reason. Although shots of schnapps or brandy can give us a seemingly warm rush on a cold winter day, they actually cause us to lose body heat at a fast rate because alcohol causes the blood vessels near the skin to expand, carrying vital heat away from the core of the body. Thus, alcohol consumption can dangerously increase the rate of heat loss in cold weather.

I encourage you to rethink your decision to accept money from advertisers that promote the dangerous combination of alcohol and sports. As long as you're taking their money, you won't be getting any more of mine.

John Singleton, Chicago, IL

Evaluating "The News"

1. Can you think of any examples of advertisements in which alcohol is depicted prominently in scenes involving exercise? What effect do you think such ads have on people?

2. Can you think of any other products advertised in a way that might distort the underlying biology?

3. If you were an owner of a magazine and you felt responsible for making a profit so that you could pay your employees, as well as for providing your readers with accurate information, what would your policy be concerning the scientific accuracy of advertisements?

chapter 32

Defense

Gilbert and George, *Deatho Knocko*, 1982.

AIDS in Africa

AIDS (acquired immunodeficiency syndrome) is caused by human immunodeficiency virus, more commonly referred to as HIV. AIDS first came to the world's attention two decades ago when doctors in the United States began to notice that members of the gay community were falling victim to a variety of previously rare diseases, including a skin cancer called Kaposi's sarcoma and an unusual kind of pneumonia, or lung infection. These diseases, it became clear, attacked the victims of AIDS when their immune systems, which are our main defense against disease, failed. The victims of AIDS invariably died.

The number of cases of this frightening new disease rapidly increased in North America and Europe, until it claimed tens of thousands of new victims each year, mostly among gay men, intravenous drug users, and people requiring frequent blood transfusions. The common bond connecting these three groups was regular contact with the body fluids of others: gay men during sex, drug users by exchanging dirty needles, and those needing blood transfusions by receiving contaminated blood.

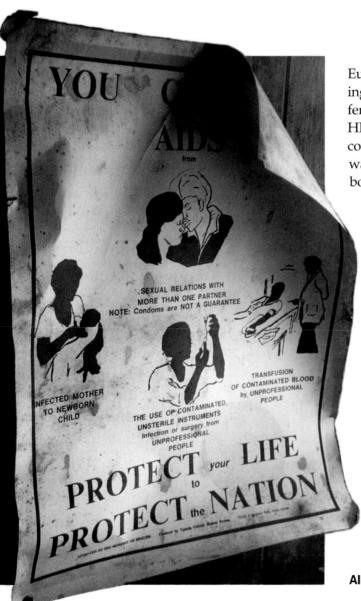

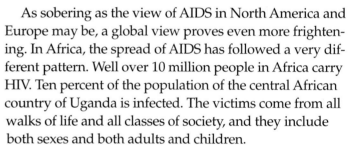
Infectious diseases, parasites, carnivo-rous predators, and herbivores are some of the greatest hazards to the well-being of organisms, making defense critical to survival and reproduction.

As sobering as the view of AIDS in North America and Europe may be, a global view proves even more frightening. In Africa, the spread of AIDS has followed a very different pattern. Well over 10 million people in Africa carry HIV. Ten percent of the population of the central African country of Uganda is infected. The victims come from all walks of life and all classes of society, and they include both sexes and both adults and children.

The spread of AIDS in Africa stems probably from two main causes. First, lack of funds forces hospitals to reuse often improperly sterilized blood transfusion equipment. Second, the social upheaval that followed the withdrawal of the European colonial powers from Africa during the 1960s created a dangerous mix of conditions: (1) Frequent use of prostitutes by men in labor camps promoted the spread of sexually transmitted diseases, which often led to open sores that allow for contact between the blood of male and female partners during sex, and (2) regular migrations of individuals and whole ethnic groups enabled the rapid spread of disease. When AIDS established itself in Africa, it spread quickly and widely. The eventual toll of AIDS in Africa is likely to be staggering.

We will return to consider how HIV knocks out the human immune system after reviewing the various ways in which living things protect themselves against invasion or attack by other organisms.

AIDS in Africa Affects Millions of Men, Women, and Children

Key Concepts

1. The outer surface of an organism forms the first line of defense against many enemies, either as a physical barrier or as the surface that makes an organism less attractive as food.

2. All organisms can distinguish themselves from other organisms, a key requirement in fending off invading disease organisms and parasites.

3. Most organisms kill invaders by using chemicals to destroy them. Animals, in addition, use mobile defensive cells that consume the invading organisms.

4. The immune systems of vertebrates can distinguish both between host cells and the cells of invading organisms (in a nonspecific defensive response) and among different species of invading organisms (in a specific defensive response that depends on unique cells called lymphocytes).

5. Living organisms avoid attacks by animals by hiding, by making themselves unpleasant to eat, distasteful, or poisonous through physical or chemical means, or by exhibiting defensive behaviors.

We often downplay the importance of diseases, parasitism, and predation. Thanks to modern medicine, we live largely sheltered from the ravages of infectious disease and in an environment stripped of predators that can kill us. This protected point of view is a by-product of life in an industrialized society. We often live into our 70s and beyond, but in the Roman Empire life expectancy ranged from 22 to 47 years and in medieval England only 20 percent of individuals saw their twentieth birthday.

Disease caused most of these early deaths. Death records from London for a particular week in 1661 show that whereas 39 people died of old age, at least 205 died of disease (Figure 32.1). The recent emergence of new infectious diseases like AIDS and the resurgence of antibiotic-resistant forms of old killers like tuberculosis (called "consumption" in Figure 32.1) (see Chapter 20) reminds us that even humans must defend themselves against other organisms.

For most organisms other than humans, not only disease and parasites, but predators as well, are major causes of death. In this chapter we introduce the ways in which organisms respond to attacks from other organisms. We first consider defenses against infectious diseases and parasites that invade organisms from within. We then turn to defenses against herbivores and carnivores that attack organisms from outside.

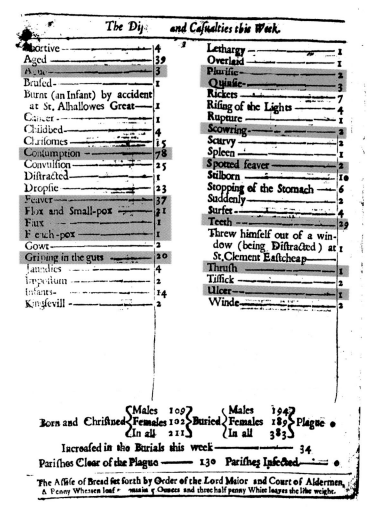

Figure 32.1 Bring Out Your Dead!
The major cause of death in seventeenth-century London was disease, not old age. This record of births and causes of death made during a week in 1661 shows that slightly more than 50 percent of the people who died succumbed to disease (shaded in orange). In addition, many of the deaths attributed to childhood or old age may well have resulted from disease. (Note that the letter *f*, when in the middle of a word, reads as the modern letter *s*.)

Defenses against Infectious Diseases and Parasites

Infectious diseases and **parasites** invade and do their damage to the organism they attack, usually called the **host**, from the inside. They often cannot survive if the host dies before they are ready to spread to new hosts. Infectious disease organisms differ from parasites mainly in that the former are usually viruses or single-celled bacteria and protists, whereas parasites are multicellular fungi, animals, or plants. Hosts defend themselves against invasion first by keeping out the invaders, then by destroying or isolating any that manage to enter.

The body wall keeps invaders out

Before invaders can cause any harm, they must enter their host. The cell walls of bacteria protect them against attack by viruses and other bacteria. The dermal tissue systems of plants defend them from attack by viruses, bacteria, and fungi. The human body wall—made up of the epithelial tissues of the skin, lungs, and gut wall—keeps out potentially dangerous microorganisms.

Many important infectious diseases and parasites of animals and plants rely on insects to penetrate the body wall of the host. For example, mosquitoes transfer the protist that causes malaria, an infectious disease that kills 1 to 2 million people annually, to humans by biting through human skin to take a blood meal. Similarly, beetles spread Dutch elm disease, which threatens to drive North American elm trees to extinction, by chewing through the protective bark to introduce into the tree's vascular tissues the fungus that causes the disease.

Elm tree

Wounds allow invaders to enter an organism

Many invaders take advantage of wounds to gain entry to their hosts. For example, the bacterium *Clostridium tetani* occurs commonly in soil and on other surfaces worldwide. Although we constantly come in contact with the bacterium, it becomes a threat only if it enters a poorly aerated break in the skin, such as a deep puncture wound. Away from oxygen, in such wounds *C. tetani* can multiply and produce a poisonous protein that causes severe and often fatal muscle spasms. Regular vaccinations protect most of us from the disease. In countries where infants are born under poor sanitary conditions, however, *C. tetani* kills more than 0.5 million babies annually by entering through the cut made at birth to the umbilical cord.

To reduce the risk posed by open wounds, plants and animals quickly seal the damage (Figure 32.2). Plants produce substances called resins and gums that help seal the wound quickly, thereby preventing entry by disease organisms. Animals with a circulatory system usually have specialized cells in their blood that quickly seal severed blood vessels. Insects seal wounds by gluing together blood cells, and mammals like us form **blood clots** consisting of sticky blood cells caught in a mesh formed from proteins that normally float in the blood. The growth of new tissues then repairs the wound more permanently.

■ The plasma membrane, cell wall, or body wall provides the first line of defense against invasion by diseases or parasites. Most multicellular organisms can quickly seal breaks in the body wall to prevent the entry of disease-causing organisms and parasites.

(*a*)

(*b*)

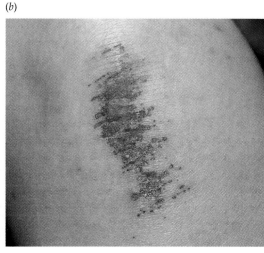

Figure 32.2 Closing Wounds Prevents Infection
(*a*) The resins produced by pines help the trees quickly seal wounds that might otherwise allow disease organisms to enter. (*b*) The scabs that seal wounds to human skin function in much the same way to keep out disease organisms.

The Vertebrate Immune Response: Consequences of a Cut

When we cut ourselves, we open a route by which disease-causing organisms can invade our bodies. Unless we can kill or isolate these invaders, they may multiply rapidly, stealing our resources and possibly, as *C. tetani* does, killing us in the process. In this section we introduce the **immune response**, a sophisticated defense against internal diseases and parasites that exists only in vertebrate animals.

The immune system defends vertebrates against invaders

In our war against invading organisms, humans, like other vertebrates, have an extremely effective immune system (Figure 32.3*a*). The human immune system consists of about 7 trillion **white blood cells** of several kinds and an assortment of defensive proteins, all of which help destroy microscopic invaders. These white blood cells and defensive proteins can move out of the vessels of the circulatory system into the fluid called **lymph** that surrounds the cells in our bodies. From there they collect in a second network of tubes, the **lymphatic ducts**, which carry them back to the blood vessels of the circulatory system. At various points along the lymphatic ducts lie **lymph nodes**, pockets containing huge numbers of white blood cells that trap bacteria and viruses roaming the body.

Although the vertebrate immune system is unique, it illustrates many of the basic problems faced by the internal defense systems of all organisms. For this reason and because of the relevance to humans, our description of the human immune system will provide a starting point for a more general survey of how organisms defend themselves against invaders.

We will describe the events that follow a finger cut in two different scenarios: (1) a cut that allows the tetanus-causing bacterium, *Clostridium tetani*, to enter

the body for the first time, and (2) a cut that leads to our second exposure to *C. tetani*, either because we survived an earlier infection or because we received a tetanus vaccination.

Scenario 1: The body defends against a new invader

The first sign that the body has joined battle against invaders is **inflammation**, a characteristic swelling and reddening of the area around the cut. This initial reaction is followed by an immune response that proceeds in stages:

1. Damaged tissues release chemicals that stimulate operation of the body's clotting substances. The invading *C. tetani* and its poison cannot easily penetrate the clot, which slows the spread of both by isolating the infected tissue.

2. A series of blood proteins known as **complement** (so called because they complement, or work

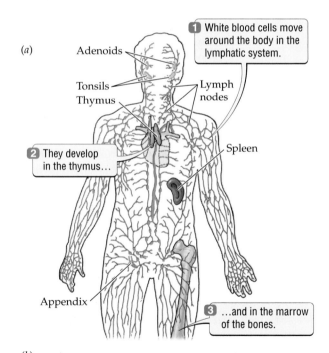

(a)

Adenoids

Tonsils

Thymus

Lymph nodes

Spleen

Appendix

1 White blood cells move around the body in the lymphatic system.

2 They develop in the thymus…

3 …and in the marrow of the bones.

Figure 32.3 The Vertebrate Immune System
(*a*) The various white blood cells in the human circulatory system form in the marrow of the bones and in the thymus. They move through a specialized system of ducts collectively called the lymphatic system (shown in green), and they accumulate in various organs that defend the various points at which invaders can enter the body. The structures that are labeled in this figure house large concentrations of white blood cells and are often associated with places at which disease organisms could easily enter the body. (*b*) The blood contains several different kinds of blood cells that enable the body to combat diseases and parasites.

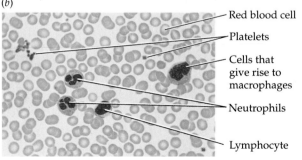

(b)

Red blood cell

Platelets

Cells that give rise to macrophages

Neutrophils

Lymphocyte

together with, white blood cells) circulate to the wound, where they bind to and kill invading cells by destroying the plasma membranes of the invaders. In addition to killing invading cells, complement helps distinguish between invading cells and our body's own cells because they remain bound only to the plasma membrane of invaders. Without distinguishing its own cells from foreign cells, no organism can defend itself effectively against invaders. The complement protein that marks invading cells allows them to be identified and destroyed by white blood cells as described in the steps that follow.

3. The clotting proteins mentioned earlier activate a type of white blood cell that is permanently stationed just below the surface of the skin. Activation converts these white blood cells (Figure 32.3*b*) into large **macrophage** (*macro*, "big"; *phage*, "eater") cells that start to prowl the wounded tissue much as predatory protists might. Although initially few in number, these macrophages bind to and then **phagocytose** (surround and digest) *C. tetani* or damaged host cells. Complement proteins that mark the surfaces of invading cells help bind the macrophages in preparation for phagocytosis.

4. Complement draws reinforcements to the wound in the form of phagocytic white blood cells of a second, smaller type, the **neutrophils** (see Figure 32.3*b*), which circulate in the blood. Individual neutrophils destroy fewer invaders than macrophages and cannot surround invaders bigger than bacteria, but they compensate in numbers. Within a few hours of injury, the neutrophil concentration may soar from 4000 to 25,000 per milliliter of blood.

5. During the next several days, additional macrophages make their way to the wound from a population circulating in the blood. Although macrophages accumulate much more slowly than neutrophils, they eventually become the dominant white blood cell at the wound. Together the macrophages and neutrophils make up a **nonspecific response** to invaders in that they attack anything other than healthy host cells. They do not distinguish between different kinds of invading organisms.

6. White blood cells of a third type, the **lymphocytes** (see Figure 32.3*b*), play a subtle but ultimately critical role in the immune response. Lymphocytes enable vertebrates to mount a **specific response** to particular species of invading organisms. This specific response sets the vertebrate immune system apart from all other internal defense systems.

Whereas complement simply distinguishes our own cells from those that are not ours, lymphocytes that bind to *C. tetani* antigens are much more specific: They distinguish *C. tetani* from other invaders.

How can lymphocytes tell *C. tetani* from a cold virus or the bacterium that causes strep throat? Each species of disease organism that invades our body carries **antigens** on its surface, molecules that have a shape that characterizes that invading species, in much the same way that fingerprints characterize individual people. At the same time, individual lymphocytes in the human body differ from each other in that differently shaped proteins are embedded in their plasma membranes (Figure 32.4). Only a few of the lymphocytes bear surface

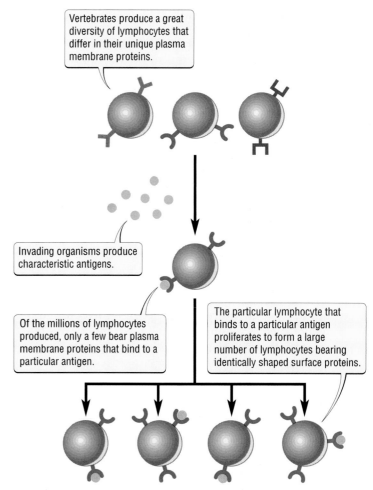

Figure 32.4 How the Human Body Recognizes Specific Invaders Lymphocytes that bear plasma membrane proteins capable of binding to molecules that characterize invading organisms multiply rapidly to give rise to identical lymphocytes, all of which bind specifically to that invading organism.

proteins that match the shape of and can bind to *C. tetani* antigens. When exposed to *C. tetani*, these few matching lymphocytes rapidly multiply and spread through the body. As we will see, a body equipped with large numbers of lymphocytes that recognize a particular invader can much more effectively fight off a second exposure to that disease.

Attacks by viruses provoke similar responses, except that the macrophages and neutrophils most often respond not to the viruses themselves, but to human cells infected with the virus. Bits of the protein coat abandoned on the plasma membrane by the virus when it penetrates a cell mark such infected human cells. In addition, cells attacked by viruses release a chemical called **interferon** that attaches to the plasma membranes of nearby cells and interferes (hence their name) with infection of these cells by the virus.

The success of the human defense against the first invasion depends on how quickly numerous macrophages and neutrophils arrive at the wound compared to how quickly *C. tetani* multiplies. Because Steps 1 through 4 of the immune response take place relatively slowly, and because bacteria like *C. tetani* multiply rapidly, we sometimes lose the race, become ill, and die.

Scenario 2: Our defenses against a second exposure to the same invader are much more effective

The vertebrate defense system differs from that of other organisms by its faster and more dramatic response to disease organisms when it encounters them a second time (Figure 32.5). For example, after our first exposure to the poison produced by *Clostridium tetani*, we can survive an exposure more than 100,000 times as severe as the one that would have killed us the first time around; thus we are essentially immune to tetanus. The key to our ability to learn from experience with a particular invader such as tetanus lies with the lymphocytes that recognize the tetanus antigen (mentioned earlier).

The much more effective defense to a second exposure to *C. tetani* comes about because the memory lymphocytes that recognized the *C. tetani* antigens have remained in the body since our initial exposure to the bacterium. Upon second exposure to the *C. tetani* antigens, the *C. tetani*–specific memory lymphocytes rapidly multiply, forming attack lymphocytes. The vast number of *C. tetani*–specific lymphocytes dramatically speed up and increase the effectiveness of all steps of the immune response.

Among the lymphocytes, we can recognize three functionally different types: (1) **B lymphocytes** produce proteins called **antibodies** that circulate freely in the

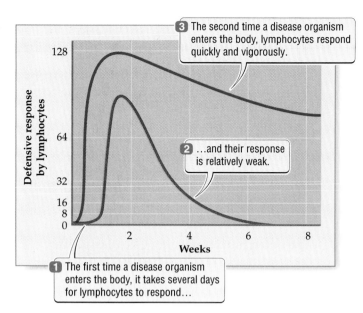

Figure 32.5 Initial versus Later Immune Responses Lymphocytes in the human immune system wages war against a particular disease organism more quickly and more vigorously in second and later attacks than in the original encounter.

blood (Figure 32.6). Antibodies are the freely soluble versions of the antigen-specific plasma membrane proteins associated with the lymphocyte proliferation. The soluble antibodies specifically target the invading organism—in this case *C. tetani*. When antibodies bind to an invading cell, they enhance the effect of the complement system already described, and they increase the ease with which macrophages and neutrophils can bind to and destroy the cell. They can also bind to and neutralize toxins, such as that produced by *C. tetani*, or to viruses, which then cannot penetrate cells. (2) **Helper T lymphocytes** allow macrophages to more rapidly bind to and consume invaders, and they stimulate their own reproduction and the production of B lymphocytes and killer T lymphocytes. (3) **Killer T lymphocytes** aid in the destruction of any of our own cells that are damaged or infected by viruses, thus aiding the macrophages and neutrophils.

Vaccines stimulate the immune system to provide protection

Preventive medicine relies heavily on **vaccines**, such as tetanus vaccines, to prime the immune system so that it is ready to attack in the event of a potentially dangerous exposure to disease organisms. This injection of a killed or harmless form of a bacterium or virus gives the immune system a chance to produce many mature lym-

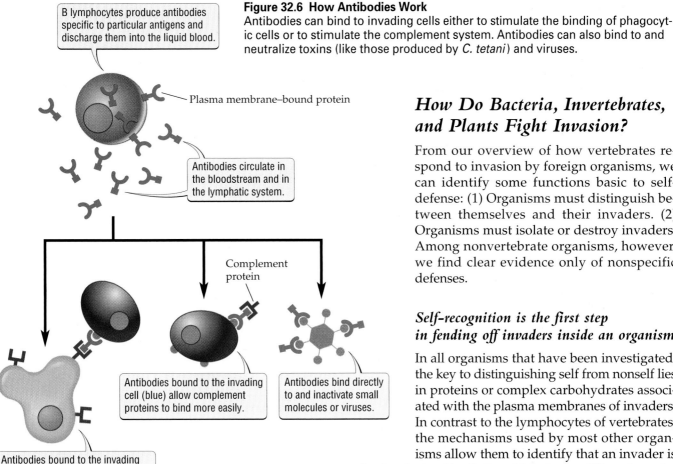

Figure 32.6 How Antibodies Work
Antibodies can bind to invading cells either to stimulate the binding of phagocytic cells or to stimulate the complement system. Antibodies can also bind to and neutralize toxins (like those produced by *C. tetani*) and viruses.

B lymphocytes produce antibodies specific to particular antigens and discharge them into the liquid blood.

Plasma membrane–bound protein

Antibodies circulate in the bloodstream and in the lymphatic system.

Complement protein

Antibodies bound to the invading cell (blue) allow complement proteins to bind more easily.

Antibodies bind directly to and inactivate small molecules or viruses.

Antibodies bound to the invading cell (blue) make it easier for phagocytic white blood cells to bind.

How Do Bacteria, Invertebrates, and Plants Fight Invasion?

From our overview of how vertebrates respond to invasion by foreign organisms, we can identify some functions basic to self-defense: (1) Organisms must distinguish between themselves and their invaders. (2) Organisms must isolate or destroy invaders. Among nonvertebrate organisms, however, we find clear evidence only of nonspecific defenses.

Self-recognition is the first step in fending off invaders inside an organism

In all organisms that have been investigated, the key to distinguishing self from nonself lies in proteins or complex carbohydrates associated with the plasma membranes of invaders. In contrast to the lymphocytes of vertebrates, the mechanisms used by most other organisms allow them to identify that an invader is foreign, but not to distinguish between different species of invaders.

Plants rely on the chemicals produced uniquely by disease organisms to signal an invasion. For example, fungi, many of which are important plant parasites, have cell walls made of chitin and other complex sugars that are not found in plant cell walls. The presence of these substances in a plant signals that it is under attack by fungi. Nonvertebrate animals recognize invaders much as vertebrates do: through a combination of marking foreign objects with proteins (as complement and antibodies did in vertebrates) and the direct binding of defensive cells to the invaders.

Hosts limit the amount of damage caused by invaders

Single-celled bacteria and protists can tolerate little damage from invaders, which most often are viruses. They cannot isolate attacking viruses and therefore must either avoid detection by viruses or kill them. Bacteria can hide by changing their plasma membrane proteins so that a virus cannot attach and initiate infection. Alternatively, bacteria may fend off a viral attack by

phocytes that recognize the antigens produced by an invader before we encounter the disease.

■ When exposed to a particular disease, vertebrates first use nonspecific white blood cells called macrophages and neutrophils to destroy the invaders. Upon their second exposure, however, the immune system permits vertebrates to mount a much more effective and specific defense. Memory lymphocytes, which recognize particular disease organisms, multiply after the first exposure and can mount an extremely rapid response to any later invasion by the disease. Three different types of lymphocytes—B lymphocytes (which produce antibodies), helper T lymphocytes, and killer T lymphocytes—assist in the specific response of the immune system. Vaccines prime the immune system so that it is ready to attack a potentially dangerous organism when it invades.

destroying the genetic material injected by the virus or by producing proteins, similar to the interferon produced by vertebrates, that stop replication of the viral genes.

Plants lack the cells specialized for defense that are found in most animals; they rely instead on a chemical defense against invasion. Plants typically respond to invasion with a **hypersensitive response** that isolates and kills the cells on which the invader depends.

In response to chemical signals that indicate invasion, plants seal off an area around the infection by reinforcing their cell walls. Then, infected plant cells release chemicals that trigger a genetically controlled "suicide program," which may include releasing enzymes that digest the cell walls or plasma membranes of invaders. The result is a patch of dead tissue that isolates or kills the invading organism (Figure 32.7). Recent evidence suggests that some plants acquire immunity to certain viral invaders after an initial attack, providing a response similar to that provided by the vertebrate immune system.

> ■ The defense of invertebrate animals resembles the response of vertebrates to their first encounter with an invader. All animals use specialized, mobile, phagocytic cells similar to vertebrate macrophages and neutrophils as a cornerstone of their internal defense. Nonvertebrates rely on nonspecific defenses.

Defenses against Carnivores and Herbivores

The response of organisms to attacks by carnivores and herbivores differs from the response to invasion by disease organisms and parasites. Because carnivores and herbivores attack their food from the outside, the victim can easily distinguish its own tissues from those of its attacker. Carnivores and herbivores inflict damage on their victims much more quickly than do disease organisms and parasites. Thus defenses against carnivores and herbivores must, as much as possible, prevent attack altogether rather than simply limit the amount of damage done.

In this section we survey four types of responses to attack by carnivores and herbivores that have evolved: avoiding discovery, physical structures that make the organism an unpleasant meal, chemical defenses that make an organism distasteful or poisonous, and behavioral defenses used by animals to reduce the likelihood of capture.

The circle of dead tissue isolates the diseased part of the leaf from its surroundings, dooming the organism that caused the response.

Figure 32.7 The Hypersensitive Response of Plants
Lesions on plant leaves result from the hypersensitive response of a plant to an invading organism.

Organisms hide from enemies to avoid discovery

Reducing the chance of discovery is one of the most effective defenses. Many organisms hide by disguising themselves to blend in with their surroundings or to look like something that cannot be eaten (Figure 32.8). The coloration or shape of many organisms lets them blend into the habitat in which they live. Ground-dwelling birds blend into the leaf litter of their habitat, and insects closely mimic the sticks, leaves, and flowers on which they live. The stripes and spots of many animals break up their shape so that they match the speckling effect of sun and shade. Even plants can take on shapes that allow them to blend into the background against which they live.

Some organisms hide by scattering themselves so widely that it becomes difficult for predators to find enough food. For example, if an insect species is rare enough, predatory birds, which often focus on the most abundant prey available, may ignore them. Many weedy plants (for example, crabgrass in our lawns or the ragweed that may invade untended gardens) normally grow in isolated patches, where they remain for only a few years before colonizing other patches. This lifestyle may help them avoid herbivores. Many of our crop plants have arisen from weedy species that may have depended on hiding to avoid their herbivores. By growing them in large fields that contain no other species, we may have made it impossible for them to hide, thus making them much more vulnerable to discovery by herbivores, leading to the current pest problems.

In other cases, organisms look like something they aren't (see Figure 32.8). For example, caterpillars of the tiger swallowtail butterfly resemble the head of a snake. Many insects have large eyespots on their wings, which make them look much more threatening than they really are. The hognose snake, a harmless species, vibrates its tail against dried leaves to make a sound like that produced by poisonous rattlesnakes. A common ploy among insects

The striped pattern of tabby cats makes them hard to see in shady places.

The harmless king snake (left) looks quite a bit like the poisonous coral snake.

The tiger swallowtail caterpillar fools predators by imitating a snake.

Figure 32.8 Looks Can Be Deceiving
Many animals avoid being attacked by blending into their background or by mimicking something dangerous.

that are active during the day is to mimic the bright coloration of species that taste bad. An interesting example is the relatively tasty viceroy butterfly, which, as we'll see shortly, resembles the poisonous monarch butterfly closely enough to fool even keen-eyed birds (see Figure 32.10).

Physical defenses make it hard for enemies to eat an organism

Armor that is tough enough to resist the jaws of the attacker is a common way in which organisms modify their surface to protect themselves (Figure 32.9). Familiar examples among animals include the many different external skeletons that simultaneously provide support and protection (see Chapter 26)—for example, the shells of turtles and clams. Some animals substitute a constructed or naturally occurring shelter for armor. The burrows of prairie dogs, the empty snail shells inhabited by hermit crabs, and the shelters built by many caterpillar species all reduce the vulnerability of these animals to predators.

Hermit crab in snail shell

Many marine algae produce a mineral-based skeleton much like those produced by coral animals, and many land plants have cell walls reinforced with so much indigestible lignin and cellulose that herbivores have trouble chewing through their leaves. Among plants especially, reinforcing the support tissues not only makes the ani-

Figure 32.9 Physical Defenses
Organisms can make themselves difficult to eat by protecting themselves with armor or spines.

The hard shell of turtles protects them against predators.

Sharp cactus spines discourage many herbivores.

mals work harder to chew up their leaves, but it has the added bonus of reducing the nutritional quality, which in plant tissues is already poor (see Chapter 28).

Some organisms have surfaces that make them dangerous to attack (see Figure 32.9). Slow-moving porcupines have fearsome spines. Though less imposing, the long hairs on many butterfly and moth larvae can make it much more difficult for birds to swallow them. A walk through the desert or through a patch of stinging nettles reveals painfully that plants, too, come equipped with an impressive arsenal of sharp weapons. Even the innocent looking hairs on the surfaces of many leaves can create serious problems for many of the small herbivores trying to feed on plant tissues.

Plants rely heavily on chemical defenses

More than any other type of organism, plants protect themselves chemically against their attackers. Plants use many of the same chemicals that offer them protection from diseases and parasites to protect themselves against herbivores. Defensive chemicals often are not part of the basic carbohydrate, protein, and fat metabolism of an organism, and they therefore play a role outside the basic biochemistry of the organism. Tens of thousands of these **secondary chemicals** have been identified in plants; milkweeds provide a well-known example (Figure 32.10).

Many defensive chemicals work as poisons that disrupt the metabolism of any attacker that eats the organism. Nicotine, the notoriously addictive chemical in cigarette smoke, protects tobacco plants extremely well against herbivore attacks. A few drops of pure nicotine will kill a human, and it is available as a potent insecticide. Similarly, the ornamental foxglove commonly grown in flower gardens produces poisons that upset heart function in vertebrates. However, this poison also provides important medical benefits: In 1785 the British physician and botanist William Withering reported that the foxglove poison, used in controlled doses, could strengthen the heartbeat of patients with weak hearts. Today, foxglove still remains the source of digitalis, a commercially important heart medicine.

Foxglove

Not all organisms produce their own poisons. Monarch butterflies defend themselves with chemicals closely related to those found in foxglove (see Figure 32.10). Monarch butterflies cannot make these heart poisons themselves. Instead, they obtain the poisons from their food plant, the common milkweed, and store them in their own tissues. As we'll discuss in Chapter 42, animals protected by poisons commonly adopt a warning behavior or bright coloration to advertise that they make a dangerous meal—hence the bright orange and black of monarch butterflies.

Behavioral defenses provide animals with additional ways to protect themselves

Some defenses that animals mount against attackers are behavioral (Figure 32.11). Animals can physically fight off attackers. Even some animals that we would not consider dangerous can seriously hurt or even kill predators, so predators avoid them in favor of weaker individuals. An adult moose, which feeds on plants, can kill attacking wolves, for example. Animals can also avoid predators by outrunning their predators. Although the predatory cheetahs are the fastest land mammals, many other mammalian speedsters are herbivores.

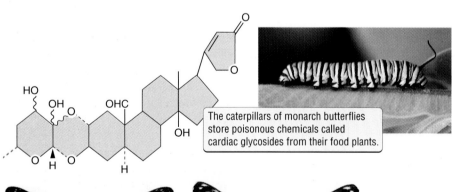

The caterpillars of monarch butterflies store poisonous chemicals called cardiac glycosides from their food plants.

The adult monarch butterflies advertise their toxicity by a bright orange and black pattern.

To protect themselves against attack, the tastier adults of the viceroy butterfly mimic the warning colors of the monarch butterfly.

Figure 32.10 The Relationship between Milkweeds and Monarch and Viceroy Butterflies
The milkweed plant produces chemicals called cardiac glycosides that disrupt heart function in vertebrates and have a foul taste. Monarch butterfly caterpillars store the cardiac glycosides from the milkweeds they eat to protect themselves against their own attackers. Viceroy butterflies mimic the appearance of the distasteful monarch butterflies, which leads potential predators of the viceroy to avoid them.

(a)

(b)

Figure 32.11 Behavioral Defenses
Animals use various behavioral defenses to avoid falling victim to predators. (a) These impalas are fleeing an approaching group of lions. (b) Groups of animals, like these flamingos, always contain a few individuals on the lookout for approaching predators.

Life as part of a group also provides many benefits when it comes to defense. A group can be more watchful while feeding than an individual can (see Figure 32.11b). For example, in a group of 20 animals, each animal need watch for approaching predators only 3 minutes of each hour to provide full coverage. A solitary individual must either eat less or make itself more vulnerable to predators. In addition, simply by being in a group, an animal has a chance to place other bodies between it and a predator. When a flock of birds or a herd of mammals scatters at the first sign of a predator, the resulting chaos may confuse the predator enough to allow the prey to escape. In addition, groups of organisms can fend off attacks from predators more effectively than individuals can: A single honeybee would stand little chance against a human, but an angry hive can drive us off.

> ■ Organisms fend off carnivores and herbivores by hiding from them, by being physically or chemically unpleasant or dangerous to eat, or via protective behaviors.

Highlight

Why Is AIDS So Deadly?

Why is AIDS such a deadly disease? As we mentioned at the beginning of this chapter, the virus that causes AIDS, HIV, kills its host not directly but by weakening the immune system. HIV attacks two components of the human immune system: the macrophages, which devour infected cells, and helper T lymphocytes, which stimulate the activity of antibody-producing B lymphocytes and killer T lymphocytes.

Evidence suggests that initially the human immune system does a good job of keeping HIV at bay. When activated, killer T lymphocytes recognize and destroy helper T lymphocytes infected by the virus. By destroying infected helper T lymphocytes, the killer T lymphocytes greatly slow the rate at which HIV reproduces and spreads, and they allow replacement of any helper T lymphocytes lost to the virus. Most AIDS victims go through a period of up to 15 years between infection by HIV and onset of the symptoms of AIDS, during which the immune system holds its ground.

Eventually, however, the virus wins, and various diseases that take advantage of a weakened immune system kill the AIDS victim. There is increasing evidence that the HIV population inside an infected individual evolves quickly following infection of a host. The antigens on the protein coat of HIV change rapidly over time. Thus, lymphocytes that fit the HIV antigen types that dominated at the time of infection may not recognize the most common antigens in the HIV population in the body a few years later. The lymphocytes seem to have an increasingly difficult time identifying HIV as the infection progresses, leading to an increasingly less effective immune response.

Eventually HIV destroys helper T lymphocytes more quickly than they can be replaced. The downfall of the helper T lymphocytes leads to a decrease in the effectiveness of the B lymphocytes and the killer T lymphocytes. Ultimately the victim's immune system collapses, and infection by normally harmless diseases leads to death.

> ■ HIV gradually weakens the immune system by infecting the macrophages and helper T cells that play a central role in helping us fend off disease.

Summary

Defenses against Infectious Diseases and Parasites

- ■ The body wall—either the cell wall of a single-celled organism or the internal and external surface tissues of a multicellular organism—helps keep diseases and parasites out of hosts.

■ Because a body wall that has been broken by a wound can allow invaders to enter an organism, plants and animals quickly seal wounds.

The Vertebrate Immune Response: Consequences of a Cut

■ The immune system of vertebrates, which uses specialized proteins, white blood cells, and the ducts and nodes of the lymphatic system, protects against microscopic invaders.

■ The human immune system reacts to a cut first with inflammation, which physically blocks the passage of invaders into the body. Complement proteins then mark any invading cells for destruction, and two kinds of specialized white blood cells, macrophages and neutrophils, in a nonspecific response, digest the invaders and any of the host's tissue that has been damaged.

■ White blood cells known as lymphocytes can distinguish among different invaders, so they enable a specific response to particular species of invading organisms.

■ If we encounter the same disease organism later, memory lymphocytes help the body mount a faster and more powerful attack on it.

■ Cells attacked by viruses release interferon, which helps prevent nearby cells from becoming infected.

■ There are three functionally different types of lymphocytes: (1) B lymphocytes produce antibodies, (2) helper T lymphocytes make macrophages more vigorous and stimulate the production of B lymphocytes and killer T lymphocytes, and (3) killer T lymphocytes help macrophages and neutrophils destroy damaged or infected cells.

■ Vaccines prime the immune system so that it is ready to attack certain potentially dangerous organisms if they invade the body.

How Do Bacteria, Invertebrates, and Plants Fight Invasion?

■ All organisms achieve the critical step of distinguishing self from nonself by recognizing proteins or carbohydrates associated with the plasma membranes of invaders.

■ Bacteria can limit the damage done by viruses by hiding from them or, if that fails, by destroying them.

■ Plants use a chemical defense system to isolate damaged tissue and destroy damaged and invading cells within that tissue.

■ Invertebrate animals depend on specialized, mobile cells to destroy invading cells in a nonspecific response to invasion.

Defenses against Carnivores and Herbivores

■ Organisms reduce the likelihood of being eaten by either avoiding detection or protecting themselves through physical, chemical, or behavioral means.

■ Some organisms avoid detection by having camouflaging colors and shapes or widely scattered populations.

■ Physical defenses include external skeletons (which act as armor) and sharp spines or hairs.

■ Plants produce secondary chemicals to poison their attackers or to fight them off.

Highlight: Why Is AIDS So Deadly?

■ HIV gradually weakens the immune system by destroying macrophages and helper T lymphocytes.

Key Terms

antibody p. 492
antigen p. 491
B lymphocyte p. 492
blood clot p. 489
complement p. 491
helper T lymphocyte p. 492
host p. 489
hypersensitive response p. 494
immune response p. 490
infectious disease p. 489
inflammation p. 490
interferon p. 492
killer T lymphocyte p. 492
lymph p. 490
lymph node p. 490
lymphatic duct p. 490
lymphocyte p. 491
macrophage p. 491
neutrophil p. 491
nonspecific response p. 491
parasite p. 489
phagocytosis p. 491
secondary chemical p. 496
specific response p. 491
vaccine p. 492
white blood cell p. 490

Self-Quiz

1. A physical barrier to keep disease-causing organisms out of the human body is provided by
   a. clothing.
   b. lymphocytes.
   c. blood clots on wounds.
   d. vaccines.

2. Which of the following statements about antibodies is true?
   a. They are produced by helper T lymphocytes.
   b. They resemble the plasma membrane–bound proteins that help lymphocytes bind to antigens.
   c. They play a critical role in the hypersensitive response that allows plants to isolate invading disease organisms and parasites.
   d. All of the above

3. What role do macrophages play in vertebrate immune systems?
   a. They help form blood clots.
   b. They mark invading cells.
   c. They remember previous infections.
   d. They digest and destroy invaders.

4. A significant difference between the defensive systems of vertebrates and other organisms is that only vertebrates
   a. have specialized cells that destroy invading cells.
   b. isolate infected tissues with a hypersensitive response.
   c. have a well-developed and stronger response to later infections by the same invader.
   d. only vertebrates can recognize invading cells.

5. Plants may reduce attack by herbivores by
   a. producing secondary chemicals.
   b. arming themselves with hairs.
   c. "hiding" among more numerous plants of another species.
   d. all of the above

Review Questions

1. Why is self-recognition important to all organisms?

2. Explain how vaccines help protect us from disease. Include in your answer a description of how the three types of lymphocytes function.

3. Compare and contrast how a single-celled bacterium defends itself from threats with how a single cell within a multicellular organism is defended from threats.

32

The Daily Globe

Medical Milestone Missed

Letter to the Editor:
June 30, 1999 should have been a day for worldwide celebration. It was the date set for the destruction of the World Health Organization's last remaining stocks of the smallpox virus, one of the world's most virulent diseases, but political and military maneuvering have postponed that date indefinitely. Instead it was a day of disappointment fueled by fear.

Smallpox killed as many as 5 million people each year as recently as 1960, but thanks to an aggressive global vaccination campaign, the wild virus was exterminated by 1977 with only laboratory stocks remaining. These stocks were used for research and vaccine production, but their eventual destruction was expected because of the elimination of the wild virus.

This May, the World Health Assembly delayed the date of the stocks' destruction until at least 2002. Proponents of preserving the stocks, who were pleasantly surprised by the lack of public outcry when the delay was announced, are concerned about intelligence reports suggesting that smallpox virus may have been acquired by rogue military powers, most notably North Korea. Proponents want to continue research into vaccines and drugs to cure the disease in case the virus is ever used as a biological weapon.

However, we as health professionals want it to be known that using laboratory stocks for renewed vaccination production may be unrealistic: It is estimated that it could take up to three years of intense effort to produce enough vaccine to have an effect on a widespread epidemic, and the resulting vaccine would be useless for people with already weakened immune systems, such as those infected with the HIV virus. Proponents are also suggesting that laboratory stocks of smallpox virus be used for research into drugs to better combat the disease; however, these calls never address what we have continually stated—that stand-ins for the smallpox virus, such as the very similar monkeypox and camelpox, might provide scientists with the tools they need for such research.

The militarization of this medical opportunity is eerily reminiscent of the nuclear arms race: a lot of speculation about what weapons other nations might develop and what they might do with them in the future is driving our decisions. A medically sound policy is the proactive one: Get rid of all known stocks of the virus.

National Medical Professionals Association

Evaluating "The News"

1. Why do you think there has been relatively little public reaction to the World Health Organization's decision not to destroy the viral stocks? Do you think that if you had known someone who had died of the disease that you might feel more strongly about its destruction?

2. Do you agree with the National Medical Professionals Association's proposal to try to remove the threat directly instead of preparing for something that might happen in the future?

3. The question of whether or not to destroy the virus will be revisited in 2002. Should it be a medical or a political decision? Who should have the final say in making the decision, and why? Scientists and medical experts? Politicians and military advisers? The general public? Religious or civic organizations?

33

Chemical Messengers

Fernand Leger, *Mona Lisa with Keys*, 1930.

Black Widow Spiders, Poison Arrows, and Cigarettes

Although most of the chemicals produced by life play an essential role in keeping organisms alive, many plants and animals also produce chemicals that are designed to kill. Some predatory animals, including spiders and snakes, produce poisons to help in catching prey. The poison often disables the would-be meal, allowing the predator to subdue a larger or faster prey animal. Black widow spiders produce a poison that disables prey by causing their skeletal muscles to contract uncontrollably. With the muscles of its prey rendered useless, the black widow can dine at its leisure.

In contrast, several species of South American frogs, called poison arrow frogs, release paralyzing substances through their brightly colored skin to defend them against predators. The poisons produced by these frogs have the opposite effect of black widow poisons: Rather than causing violent contractions, the frog poisons prevent muscle contraction. As a result, any animal that makes the mistake of eating such a frog is quickly paralyzed. Poison arrow frogs earned their name because South American peoples traditionally have used the frog poisons to coat

Main Message

Multicellular organisms use two types of chemical messengers—hormones and neuro-transmitters—to coordinate the functions necessary for survival.

the tips of their arrows so that they can more easily kill the animals they hunt.

Like poison arrow frogs, plants often produce poisons to protect themselves. One of the most familiar of these plant poisons is nicotine, which tobacco plants produce as a defense against herbivores. Humans use nicotine as a pesticide to kill insect pests. Nicotine is also the addictive chemical in cigarettes and other tobacco products. Like the poisons produced by black widow spiders, nicotine overstimulates muscles, causing a shakiness that interferes with control and coordination. To compound the problem, a lit cigarette also produces carbon monoxide, which deprives muscles of oxygen needed for the increased activity stimulated by nicotine.

What makes these substances so dangerous? The poisons produced by black widow spiders, poison arrow frogs, and tobacco plants all act by interfering with chemical messengers that animals use to coordinate the activities of their various organs.

A Collection of Dangerous Items
(*a*) The black widow spider, with its characteristic hourglass marking, can disable its prey with a single venom-filled bite. (*b*) As a defense, poison arrow frogs use bright colors to warn potential predators of the toxins they release on the surface of their skin. (*c*) Cigarettes contain a wide variety of toxic substances.

Key Concepts

1. Hormones allow distant cells within an organism to communicate and coordinate activities, and neurotransmitters do the same for adjacent nerve cells.

2. The internal transport system carries hormones from sites of production to sites of action.

3. Hormones help plants control growth and development to overcome the limitations imposed by their inability to move, cell structure, and external environment.

4. Animal hormones coordinate functions other than the rapid contraction of muscles, including the storage and distribution of essential materials, reproductive functions, and development.

5. Neurotransmitters, the chemical messengers produced by the nervous system in animals, usually act on adjacent nerve cells.

A: Knock knock.

B: Who's there?

A: Tank.

B: Tank who?

A: You're welcome!

On the surface it's a simple joke. The humor in a single joke, however, can accomplish many goals: to establish friendship, to relieve a tense moment, to insult someone, or even simply to provoke a smile. Similarly, multicellular organisms use chemical messengers to provoke a rich variety of responses.

The success of multicellular plants and animals depends on the coordinated functioning of specialized tissues and organs. Just as a group of people cannot effectively complete a task without constantly letting each other know what they are up to, the many cells within a multicellular plant or animal cannot work together to allow the individual to function effectively without communicating with each other. Unlike people, cells cannot send spoken or e-mail messages to one another. Instead they rely on chemical messages to keep one part of the organism aware of what the other is doing.

In this chapter we discuss the two types of chemical messengers: hormones and neurotransmitters. In an introductory overview we compare how hormones and neurotransmitters work. We then build on this general perspective by introducing examples that illustrate how both plants and animals use hormones to coordinate their many essential functions. This look at hormones is followed by a discussion of the way in which animals use neurotransmitters in sending signals through their nervous systems, a topic that we consider further in Chapter 34. We close by returning to consider how the

poisons produced by black widow spiders, poison arrow frogs, and tobacco plants interfere with the chemical messengers produced by animals.

An Overview of Chemical Messengers

Before discussing specific examples of how chemical messengers enable multicellular organisms to coordinate their many functions, we must understand the essentials of how a chemical can carry a message from one cell to another. In this section we compare hormones and neurotransmitters to emphasize some of the basic similarities between these two chemical messengers, as well as some of the important differences.

Chemical messengers produce a response in target cells

Both **hormones** and **neurotransmitters** are chemicals released by one cell that act on another cell, called the **target cell** (Figure 33.1). All neurotransmitters and many hormones bind to specific receptor proteins on the outer surface of the target cells. Only a very small amount of a hormone or a neurotransmitter is needed to trigger a response in target cells.

Hormones often move through the internal transport system to act on distant cells

Hormones often act on target cells located at some distance from the cells where they were produced. To reduce the travel time, most hormones travel through the internal transport system of the organism—the circulatory system of animals or the vascular system of plants (see Chapter 30). Nonetheless, hormones travel only as fast as the internal fluids move, and often sec-

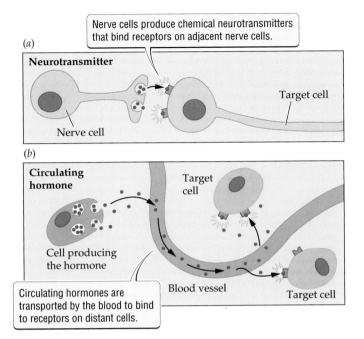

(a)

Nerve cells produce chemical neurotransmitters that bind receptors on adjacent nerve cells.

Neurotransmitter

Target cell

Nerve cell

(b)

Circulating hormone

Target cell

Cell producing the hormone

Circulating hormones are transported by the blood to bind to receptors on distant cells.

Blood vessel

Target cell

Figure 33.1 Hormones and Neurotransmitters Enable Cells to Communicate with Each Other
All organisms have internal communication systems, which coordinate the organisms' diverse activities. Common mechanisms for neurotransmitters (*a*) and hormones (*b*) require a chemical messenger that interacts with a target cell.

phobic (water-fearing) steroid molecule. The hydrophilic hormones bind to receptor proteins in the plasma membrane, as already described. The steroid hormones, however, can pass relatively easily through the phospholipid bilayer and tend to act inside the cell.

An interesting feature of many hormones is that a single hormone may produce different responses in different target cells. For example, testosterone, a steroid hormone that influences the development of many features associated with maleness in humans, stimulates the production of sperm by target cells in the testes (see Chapter 36), stimulates growth in cells throughout the body, controls the development of the male sex organs in developing babies, stimulates the production of facial hair in target cells in the skin of the face, and causes behavioral changes through interactions with target cells in the brain. As we will see later in this chapter, the fact that a single hormone has multiple effects reduces the number of hormones needed to coordinate the essential functions of an organism.

onds to hours are required for hormones to move from the producing cells to the target cells. As a result, hormones tend to coordinate processes that take place on a timescale of seconds to months (Figure 33.2). To survive their relatively long journeys, most hormones resist being broken down as they move through the internal transport system.

We find two broad classes of hormones in plants and animals: those based on a protein or other hydrophilic (water-loving) molecule and those based on a hydro-

Neurotransmitters act on adjacent nerve cells in animals

Neurotransmitters occur only in animals, where they play a critical role in transmitting nerve signals from one nerve cell to the next (see Chapter 34). In animals, nerves coordinate the contraction of muscles. Because muscles contract in fractions of a second, nerves and neurotransmitters must transmit signals more rapidly than is possible for hormones (see Figure 33.2).

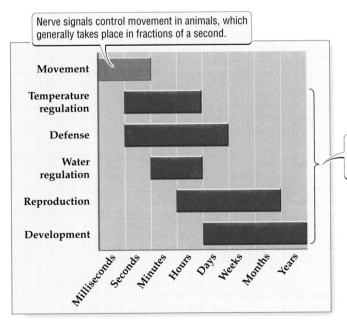

Nerve signals control movement in animals, which generally takes place in fractions of a second.

Movement

Temperature regulation

Defense

Water regulation

Reproduction

Development

Milliseconds Seconds Minutes Hours Days Weeks Months Years

Hormonal signals tend to control events that take place over a long period of time.

Figure 33.2 The Timescale over Which Chemical Messengers Work
The activities that organisms must coordinate work on timescales that range from fractions of a second to many years. Relatively slow-moving hormones (blue bars) tend to coordinate longer-term activities than the rapidly moving nerve signals controlled by neurotransmitters (red bar).

Neurotransmitters carry a signal from a stimulated nerve cell across an extremely narrow gap, called a **synapse**, to an adjacent nerve cell. Synapses are typically less than a millionth of a millimeter wide. Because of the tiny distances involved, neurotransmitters can cross a synapse virtually instantaneously by diffusion alone (see Chapter 29).

In contrast to hormones, neurotransmitters tend to be very short-lived chemicals that break down rapidly once they bind to proteins on the target nerve cell. This rapid breakdown is essential to their function because they must be removed rapidly from the target cell to avoid overstimulating it.

> ■ Chemical messengers are produced by one cell and act on another, target cell. Hormones travel from producing cells to distant target cells by means of the internal transport system of the plant or animal. Neurotransmitters occur only in animals, where they diffuse rapidly across narrow gaps, or synapses, between adjacent nerve cells.

A Survey of Plant Hormones

Plants face special challenges. Not surprisingly, plant hormones and the functions they control differ from those in animals. For example, plants cannot run from a predator or move to new or more favorable environments to find water or food, as animals do. Instead, they often cope with unfavorable or harmful environmental conditions through growth responses coordinated by hormones. Flowering plants rely on a small number of hormones to coordinate growth: Auxins, cytokinins, and gibberellins promote growth and the formation of different organs, and abscisic acid and ethylene inhibit or slow growth and the formation of organs (Table 33.1).

A plant's first task is getting out of the seed

Seeds contain embryonic plants. The embryo waits until environmental conditions, like temperature and water availability, can adequately support its growth. When enough water moves into a seed, the embryo releases hormones called gibberellins. Gibberellins stimulate the production of enzymes that break down the stored nutrients in the seed. These nutrients sustain the embryo until it has emerged from the seed and pushed through the ground, and can produce leaves to begin photosynthesis (Figure 33.3).

Plants grow in two ways and in two directions

Once established, a seedling depends on two types of growth: cell division and cell enlargement. Each is useful under different conditions. Under some environmental conditions, such as shade, plants are more sensitive to gibberellins, which, in addition to their function in seeds, promote cell elongation and stem lengthening.

Auxins, plant hormones produced in the tips of growing branches, also stimulate cell elongation by loosening the rigid cell wall fibers, which allows the cells to elongate. Auxins promote upward growth, which allows the plant to reach more sun, by suppressing the development of side branches, a phenomenon called apical dominance (see Figure 37.10). Thus, when shade limits the plant's ability to photosynthesize, auxins promote rapid upward growth. For this reason, plants that do not get enough sun become long and spindly.

Figure 33.3 Plant Hormones Control Seed Development
Plant hormones called gibberellins play an important part in the early development of plants within seeds.

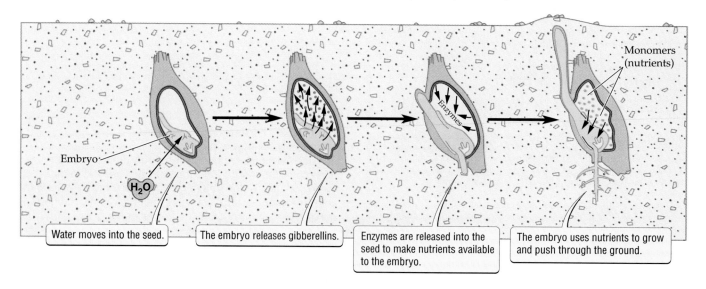

| Water moves into the seed. | The embryo releases gibberellins. | Enzymes are released into the seed to make nutrients available to the embryo. | The embryo uses nutrients to grow and push through the ground. |

33.1 *Plant Hormones and the Functions They Regulate*

Hormones	Major functions
Auxins	Promote cell elongation
	Promote growth at the top of the plant that suppresses branching (apical dominance)
	Promote vascular tissue development
	Regulate the response to light
Gibberellins	Release enzymes to feed the plant embryo
	Stimulate stem elongation
	Work with auxins to stimulate fruit development
Cytokinins	Stimulate root growth (cell division)
	Delay leaf senescence
	Promote branching (counteracts the effects of auxin in apical dominance)
	With auxins, establish balance between leaf and root mass
Abscisic acid	Promotes leaf and plant senescence
	Inhibits growth (and activity of auxins and cytokinins)
	Promotes nutrient storage in seeds
	Promotes water conservation by closing pores
Ethylene	Stimulates fruit ripening
	Promotes senescence
	Signals injury site
	Activates defense mechanisms

The other way plants grow is by cell division, which adds new cells. Cytokinins have the largest role in plant cell division. Roots produce cytokinins, stimulating their own growth at the tips. As the roots grow larger, they can support a larger plant. Cytokinins produced by the roots also travel up the plant, where they interact with auxins to control cell division in the leaves and branches.

Auxins, which normally suppress branching as already described, spread down from the growing tips of a plant. Buds along parts of the stem that are far enough from the auxin-producing tips come under the influence of cytokinins and begin to develop. Thus, cytokinins promote the growth of side branches near the base of a stem, and auxins promote rapid lengthening near the tip of the stem.

Hormones can delay or promote fruit development

Fruit ripening and seed development are necessary steps in the reproduction of flowering plants. Auxins can stimulate fruit initiation in some plants, such as strawberries. In other plants, cytokinins initiate fruit development. In general, auxins, cytokinins, and gibberellins stimulate fruit growth but tend to slow fruit ripening. In the case of gibberellins, which are abundant when the embryo is developing, the delay may ensure that healthy embryos are in place in their seeds before the fruit drops from the plant.

Two additional hormones, abscisic acid (ABA) and ethylene, promote fruit ripening and detachment from the plant. The ability of ethylene to speed the ripening of fruit has been used by fruit growers. Unripe fruit is much easier to ship and suffers fewer bruises and blemishes. Fruit can be harvested when it is immature, transported, and then placed in containers of ethylene to stimulate ripening. Some plant breeders are experimenting with genetically altering the genes responsible for natural ethylene production so that they can prevent fruit from ripening and spoiling before it gets to market.

Hormonally controlled dormancy allows plants to weather tough times

Under stressful conditions some plants shut down energy-requiring activities, entering a state called dormancy. For example, the tulip or crocus bulbs that we plant in our gardens in the fall are not seeds, but plants that have become dormant to avoid winter conditions. During these tough times, ethylene and ABA slow or stop growth so that the plant can conserve energy until conditions improve.

Ethylene and ABA also regulate **senescence**, an orderly process in which a plant systematically shuts down selected tissues and allows them to die. In fall, for example, the leaves of many nonevergreen trees senesce. As the leaves undergo a hormonally controlled death, valuable components and nutrients in the leaf are dismantled and stored in other parts of the plant. Leaf senescence is often a normal part of growth that allows a plant to respond to and survive environmental stress (Figure 33.4).

■ Plant hormones balance growth and reproductive processes with strategies for coping with stressful conditions. Auxins, cytokinins, and gibberellins generally contribute to growth and development. Abscisic acid and ethylene usually suppress growth and other energy-requiring activities.

(a)

(b)

**Figure 33.4 Hormones Enable Plants to Survive
Stressful Seasons**
(a) In a sequence of events controlled by plant hormones,
every autumn the leaves of sugar maples first change to a
brilliant orange and gold, then fall off. (b) Ocotillo plants
use senescence triggered by abscisic acid to survive long
periods of heat and drought in deserts.

Animal Hormones Regulate the Internal Environment

In animals, as in plants, hormones control many activities. Hormones can be produced in solitary cells, in groups of cells within a specialized tissue, or in an organ, called a **gland**. These hormone-producing tissues are distributed throughout the animal and make up the **endocrine system**, which works closely with the nervous system (Figure 33.5). The interaction between these systems gives organisms rapid control by the nervous system over long-term processes driven by hormones. The regulation of nutrient availability in animals provides good examples of how hormones act in animals.

Hormones regulate glucose levels

The sugar glucose may be one of the most important nutrients for animals. Glucose is the fuel for glycolysis and cellular respiration, which produce ATP (see Chapters 7 and 8). As consumers, humans and other animals must obtain glucose in their diet. After glucose is absorbed in the small intestine, it must be distributed and any excess stored until needed. The amount of glucose in the human body is managed by three hormones: insulin, glucagon, and epinephrine. In humans, both insulin and glucagon are produced and released by the pancreas.

After a meal, when the blood contains more glucose than the body can use immediately, the release of insulin into the blood stimulates the movement of glucose out of the blood and into storage tissues. To do this, insulin stimulates the binding together of individual glucose molecules into long chains to form a storage carbohydrate called glycogen. Glycogen is stored primarily in the liver and muscles.

Although insulin is most often associated with glucose storage, it also promotes the storage of fatty acids from digested fats and amino acids from digested proteins. The activity of insulin ensures that the animal does not waste excess nutrients, eliminating them from its body.

Although we always need some glucose in our blood, too much can be harmful. In a condition called diabetes mellitus, insulin cannot move enough glucose from the blood into storage cells. Diabetes results when the pancreas produces too little insulin or the insulin molecule cannot bind to proteins on the surface of the target cells. Over time, the high blood sugar may damage small blood vessels, leading to circulation problems and impaired healing of injured tissues. Damage to the small vessels in the eyes can cause vision problems (in severe cases, blindness). Most people who have diabetes can monitor and manage their own glucose levels with a combination of following dietary restrictions and taking additional insulin.

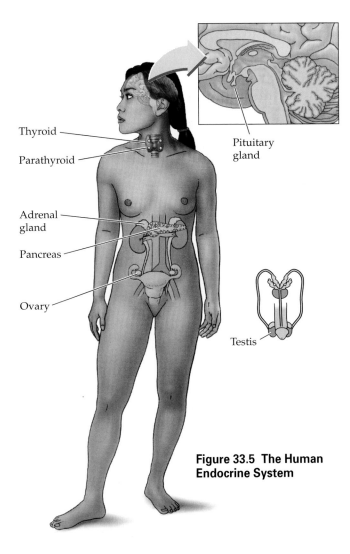

Figure 33.5 **The Human Endocrine System**

Counteracting insulin activity, the hormone glucagon is released by the pancreas when the amount of glucose in the blood falls to low levels. Glucagon increases the amount of glucose in the blood by stimulating the breakdown of liver-stored glycogen into glucose. Imagine stringing together paper clips representing glu-

cose to form a chain representing glycogen. When you need a paper clip, all you have to do is unhook one, leaving the rest of the chain intact. This is just how glucagon releases glucose from glycogen.

A third hormone, epinephrine, also affects blood glucose levels. Epinephrine functions in the nervous system as a neurotransmitter, but it also travels through the bloodstream to distant targets, as a hormone does. In response to signals from the brain during periods of stress or danger, the adrenal gland releases epinephrine and sends it to glycogen-storing cells in the liver. Like glucagon, epinephrine stimulates the breakdown of glycogen in these storage cells to release glucose. In this way, glucose becomes available to fuel a rapid escape.

Hormones regulate the concentration and distribution of calcium

In addition to regulating nutrients such as glucose, hormones regulate minerals in the body fluid. One important solute, calcium, is the major building block in the skeletal structures of vertebrates and animals that possess shells, such as clams. Furthermore, ionic calcium (Ca^{2+}) is essential in nerve cell function and muscle contraction. Finally, calcium plays a vital role as a chemical messenger within cells.

The hormones calcitonin and parathyroid hormone have opposing functions in regulating calcium concentrations in humans. The thyroid gland releases calcitonin when the amount of calcium in the blood is too high. Calcitonin removes excess calcium from the blood by promoting its storage in bones (Figure 33.6). In addition, calcitonin stimulates the kidneys to unload excess calcium into urine.

When there is too little calcium in the blood, parathyroid hormone released by the parathyroid gland stimulates the release of calcium from bones and prevents the kidneys from releasing calcium into the urine. It also stimulates special cells that dissolve some of the calcium stored in bones, releasing it back into the blood. In this way, calcitonin and parathyroid hormone maintain proper levels of blood calcium, while storing the excess and strengthening bones.

■ Hormones play a critical role in controlling levels of essential solutes such as glucose and calcium.

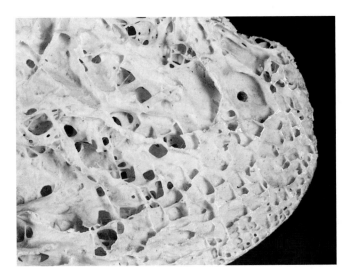

Figure 33.6 **Hormones Regulate Calcium Storage in Bones**
Bones are an important storage site for excess calcium. In times of low levels of blood calcium, parathyroid hormone stimulates the release of some of the bone's calcium salts to make calcium ions available for other processes.

Hormones in Human Reproduction

Hormones control reproduction in all animals. Hormones influence human sexual development and regulate human sexual function, pregnancy, and birth. In addition, they affect the distribution of fatty tissues, location of hair development, and bone and cartilage structure that make males and females appear outwardly different. From before birth until death, hormones have a major influence in the development and differentiation of human sexes.

Hormones play a role even before birth

Hormones influence human sexual development even before birth. If the fetus possesses a male (Y) chromosome, the male sexual organs, the testes, begin to form 4 to 6 weeks after fertilization of the egg. In the absence of a male chromosome, ovaries develop. By the seventh week, each of these developing organs produces steroid sex hormones that enter a target cell and bind proteins in the cytoplasm or the nucleus of the cell. After binding, the complex consisting of hormone and protein then activates genes that begin the process of sexual development.

Male sex hormones are called **androgens** as a group, and of these, testosterone has the major role in male sexual development. In females, in which **estrogens** and progesterone are the sex hormones, one estrogen, called estradiol, has the major influence. Despite some of the dramatic differences in their effects, male and female steroid hormones share a remarkably similar chemical structure (Figure 33.7). Individuals of both sexes contain all the sex hormones; the difference is that in males the androgens dominate, whereas in females the estrogens and progesterone dominate.

Puberty requires additional hormones

In the first 10 to 12 years of life, sex hormones circulate, but only in low concentration. During puberty, the final stage of sexual maturation, hormone levels change dramatically. A special region of the brain, the **hypothalamus**, stimulates the involvement of new hormones. The hypothalamus releases a hormone that activates two other hormones that belong to a group called **gonadotropins**: luteinizing hormone (LH) and follicle-stimulating hormone (FSH). LH and FSH are produced in a gland in the brain called the **pituitary gland**, and they stimulate the development of gametes (sex cells) in both males and females.

In addition, LH and FSH help maintain functioning sex organs and glands. In males, LH and FSH stimulate the production of sperm. At sexual maturity, sperm are

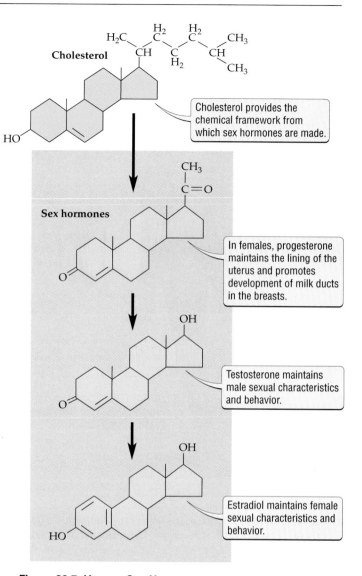

Cholesterol provides the chemical framework from which sex hormones are made.

Sex hormones

In females, progesterone maintains the lining of the uterus and promotes development of milk ducts in the breasts.

Testosterone maintains male sexual characteristics and behavior.

Estradiol maintains female sexual characteristics and behavior.

Figure 33.7 Human Sex Hormones
Closely related hormones control many aspects of sexual development and differentiation in humans.

produced continuously if a constant amount of testosterone is available. Females are born with ovaries that already contain immature egg cells. These egg cells resume development at puberty, when LH and FSH are released and stimulate the cyclical maturation and release of eggs from the ovary in a 28-day **menstrual cycle**.

Mature females have fluctuating levels of hormones

Unlike males, females do not produce mature gametes continuously, so sex hormone levels fluctuate during the menstrual cycle. Hormones that control the major events in this cycle are summarized in Figure 33.8.

At the beginning of the menstrual cycle, the levels of estrogens rise steadily during a 2-week period while a single egg completes its development in the ovary (Figure 33.8*b*). During egg development, the rising levels of estrogens also stimulate the development of a thickened lining in the uterus, the muscular organ that will receive the fertilized egg and support it during pregnancy. When the egg is fully developed, a massive and sudden release of LH and FSH from the pituitary gland stimulates ovulation (Figure 33.8*a* and *b*), the process by which the ovary ejects a mature egg and starts it on its journey to the uterus.

The spike in LH level stimulates the release of the hormone progesterone. Immediately after ovulation, LH, FSH, and estrogen levels drop suddenly, and progesterone takes over, stimulating preparation of the uterine lining for pregnancy (Figure 33.8*c* and *d*). As part of this

preparation, the lining of the uterus thickens steadily after ovulation. Progesterone maintains this thickening and promotes the development of new blood vessels, which will support an embryo through its development. If the egg is not fertilized, no embryo develops and progesterone levels drop roughly 12 days after ovulation.

Without the high levels of supporting progesterone, the new blood vessels and thickened lining of the uterus do not last long. They detach from the uterine wall, and strong uterine muscles contract to expel the unneeded tissues, releasing about 2 to 6 tablespoons of blood and uterine membranes over a period of several days. After about 7 days, rising levels of estrogens slow down the bleeding and discharge and stimulate regrowth of the uterine lining. The rising levels of estrogens also starts the maturation of a new egg as the cycle begins again.

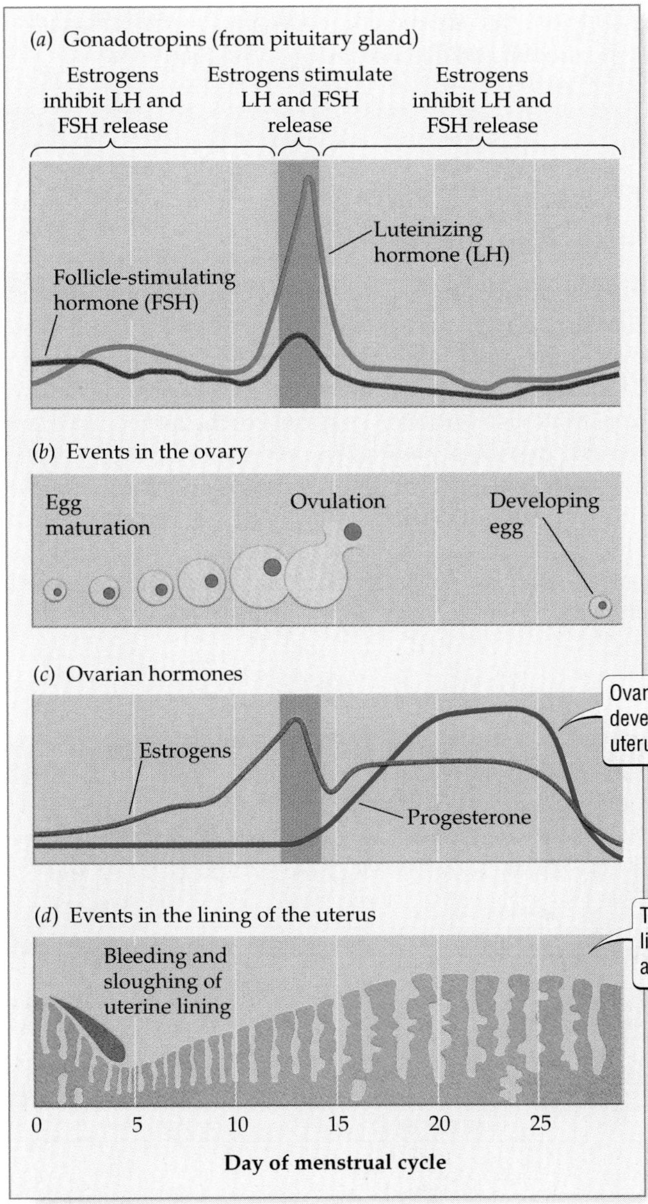

▣ Sex hormones influence sexual development and the timing of reproductive processes throughout a human's life. In addition, they facilitate sperm and egg production and optimize conditions necessary for new life.

Hormones in Metamorphosis and Color Changes

As they grow, most animals undergo dramatic changes in their body structures and the pigmentation in their skin, scales, or fur. Many insects have an external skeleton that must be shed in a process of molting before they can grow; then they must produce a new, larger skeleton. In addition, frogs and some insects have phases in which their entire body structure—inside and out—changes completely, in a process called **metamorphosis**. In other cases, the changes are minor, affecting only surface coloring. All these changes are controlled by hormones.

Molting insect with shed skeleton

(*a*) Gonadotropins (from pituitary gland)

Estrogens inhibit LH and FSH release

Estrogens stimulate LH and FSH release

Estrogens inhibit LH and FSH release

Luteinizing hormone (LH)

Follicle-stimulating hormone (FSH)

(*b*) Events in the ovary

Egg maturation

Ovulation

Developing egg

(*c*) Ovarian hormones

Estrogens

Progesterone

Ovarian hormones stimulate development of the lining of the uterus in preparation for pregnancy.

(*d*) Events in the lining of the uterus

The development of the uterine lining is controlled by estrogens and progesterone.

Bleeding and sloughing of uterine lining

0   5   10   15   20   25

Day of menstrual cycle

Figure 33.8 Hormonal Control of the Menstrual Cycle in Humans
The human menstrual cycle is closely controlled by the interaction of several hormones on a 28-day cycle. The hormones that stimulate ovulation are primarily LH and FSH; the hormones that control the onset of a menstrual period are estrogen and progesterone.

Figure 33.9 The Metamorphosis of Frogs
During their development, frogs undergo a transformation from a fishlike aquatic tadpole to the familiar terrestrial adult form.

Frogs undergo a dramatic metamorphosis from aquatic tadpoles to terrestrial adults (Figure 33.9). Tadpoles have gills that bring in oxygen only when water is flowing over them. They cannot use air. Hormones from the thyroid gland, thyroxines, drive the changes that cause tadpoles to develop an entirely modified gas exchange system and other necessary physical changes as these animals change from aquatic animals to animals that breathe air.

Figure 33.10 Hormones Allow Arctic Foxes to Blend into Their Background
Hormones trigger seasonal changes in fur color from brown in summer to white in winter, which allow arctic foxes to blend into their background throughout the year.

Color changes in animals may not be as dramatic as the physical modifications of animal metamorphosis, but they are nonetheless critical to some animals (Figure 33.10). Usually these changes are protective, designed to blend the animal visually with its surroundings. For example, some tadpoles have a daily cycle of lightening and darkening that make the animal difficult to see regardless of the time of day. In other animals color changes represent an outward display of aggression, and in others colors change to protect outer tissues from the effects of harsh sunlight. The most common hormone that controls these changes is melanocyte-stimulating hormone (MSH). MSH stimulates the production of dark pigments that can be distributed strategically in the organism.

> ■ Hormones control the changes in appearance and structure that are required for many functions, including growth and defense, in many animals.

Chemical Signaling with Neurotransmitters

Neurotransmitters are the special chemical messengers sent by the nervous system that burst out of small vesicles in the ends of nerve cells, or neurons (see Figure 34.6). The end of the nerve cell that releases the neurotransmitter is positioned so closely to the target nerve cell that the two are separated by only the narrow gap of the synapse. When a stimulated nerve cell releases the neurotransmitter, it diffuses rapidly across the synapse to bind to proteins on the adjacent nerve cell that forms the target. After binding, electrical activity in the target cell changes, often generating an electrical signal. This process will be explored in Chapter 34.

At first glance, synapses may seem like a hurdle or a limitation that the neurotransmitter must overcome. Actually, synapses represent an important site of regulation. Many synapses involve more than two nerve cells, which means that two or more stimulated nerve cells can interact to enhance or inhibit a signal in the remaining nerve cells.

Even though most neurotransmitters act at synapses, some can travel through the circulatory system to reach distant targets. These chemical messengers from nerve cells blur the lines that distinguish neurotransmitters and hormones, and some neurotransmitters behave so much like hormones that they are called neurohormones.

The type of receptor that interacts with the neurotransmitter determines the response that follows. The diversity of receptor types on the target cell determines which events can follow neurotransmitter binding, increasing the spectrum of possible communication. Thus, one neurotransmitter can bind several different receptor types and trigger different responses.

Acetylcholine is the major neurotransmitter responsible for signaling skeletal muscle contraction, yet in heart muscle it signals relaxation. Typically, the effects of a neurotransmitter in the brain and spinal cord are different from its effects in the rest of the body because the brain and spinal cord possess receptors that produce different responses from the target cells. In particular, the functions of many neurotransmitters in the brain are not well understood. Table 33.2 summarizes what we do know about the functions of the various neurotransmitters found in humans and emphasizes that each neurotransmitter plays a distinct role. In many cases, however, their precise roles have not been confirmed.

Ironically, neurotransmitters provide more information when they malfunction or lie at the center of physical problems. Much understanding of the brain has come from studying diseases like Parkinson's disease. In Parkinson's disease, the production or release of the neurotransmitter dopamine in the brain is inadequate. Because dopamine is produced in an area of the brain that controls muscle function, patients who suffer from Parkinson's disease have difficulty controlling muscular activity. Their symptoms include rigid movements, stiffness, and shaking. Looking at what goes wrong often provides valuable information about the normal function of these important chemical messengers.

> ■ Neurotransmitters participate in synaptic signaling that depends on chemical and electrical components to send messages around the body very rapidly. Their effects depend greatly on the receptors that bind them.

Highlight

Why Poisons Kill

We return now to find out what might explain the toxic activity of the black widow spider poison, poison arrow frog secretions, and the nicotine in tobacco. What's the connection?

All these poisons are neurotoxins, so called because they directly interfere with neurotransmitter function. The secret of their danger lies in their target. All three target the activity of one neurotransmitter: acetylcholine. Acetylcholine is the messenger that instructs skeletal muscles to contract. Normal movement, then, depends on its uninterrupted function.

The venom of the black widow spider causes painful muscle cramps because it stimulates massive release of acetylcholine, the main neurotransmitter initiating muscle contraction. Nicotine in cigarettes also interferes with neurotransmitter activity in normal muscle function. By mimicking the action of acetylcholine and binding to its receptors, nicotine creates a prolonged period of stimulation in the muscles, sometimes causing them to appear shaky or jumpy. Finally, toxins in poison arrow frogs block acetylcholine by binding to the receptor so that the neurotransmitter no longer has access to it. Unable to contract, the muscles become paralyzed.

Many of these and other drugs, plant compounds, animal poisons, and environmental toxins are dangerous because they mimic or disable chemical messengers that are vital to normal function.

33.2 *Common Neurotransmitters and Their Major Functions*	
Neurotransmitter	**Major functions**
Acetylcholine	Skeletal (voluntary) muscle control
Dopamine	Motor activities
GABA (gamma-aminobutyric acid)	Assisting motor coordination by inhibiting counterproductive or unneeded neurons
Epinephrine	Maintaining state of alertness, readiness
Serotonin	Temperature regulation, sensory transmission, sleep
Melatonin	Day–night cycles, such as sleep regulation
Enkephalins and endorphins	Inhibiting transmission of signals from pain receptors, processing (perception) of pain
Substance P	Regulating transmission of signals from pain receptors

■ Drugs, poisons, and a wide variety of other substances can interact at the synapse to interfere with normal neurotransmitter signaling.

Summary

An Overview of Chemical Messengers

■ Both hormones and neurotransmitters are released by one cell to act on a target cell.

■ Hormones often move through the internal transport system of an organism to act on distant cells. They tend to control the slower processes of an organism.

■ A single hormone may produce different responses in different target cells.

■ Neurotransmitters cross synapses to act on adjacent nerve cells in animals. They control the processes of an organism that must be accomplished quickly.

A Survey of Plant Hormones

■ Because plants do not have a nervous system, they use hormones to coordinate and regulate all processes throughout their bodies.

■ Plant growth is controlled by interactions between hormones such as gibberellins, auxins, and cytokinins.

■ Abscisic acid and ethylene regulate fruit ripening and the controlled senescence of tissues, including leaves and fruit.

Animal Hormones Regulate the Internal Environment

■ Together the endocrine system and the nervous system of animals coordinate and regulate processes within an organism.

■ Glucose levels are controlled by insulin, glucagon, and epinephrine, which work by regulating the amount of glucose being stored or released by tissues.

■ Calcitonin and parathyroid hormone regulate the amount of calcium circulating in the bloodstream by influencing how much is stored in bones and how much is released by the kidneys.

Hormones in Human Reproduction

■ Developing testes and ovaries in human fetuses produce hormones, including testosterone and estradiol, that influence sexual development.

■ During puberty, additional hormones called gonadotropins begin to stimulate gamete production and other reproductive processes.

■ In mature females, a monthly cycling of hormones (including estrogens, progesterone, and the gonadotropins) regulates egg production and changes in the uterus that enable it to support a developing embryo if fertilization occurs.

Hormones in Metamorphosis and Color Changes

■ Thyroxine hormones regulate the metamorphosis of amphibians from juvenile to adult forms.

■ Hormones that control the production and movement of pigments allow some animals to change color as necessary for survival.

Chemical Signaling with Neurotransmitters

■ Neurotransmitters generally carry messages from a nerve cell to a nearby cell (such as another nerve cell or a muscle fiber), but they can occasionally carry messages to more distant receptors.

■ The same neurotransmitter can have different effects when it reaches different tissues.

Highlight: Why Poisons Kill

■ Nerve poisons, or neurotoxins, interfere with the normal functioning of neurotransmitters.

■ Neurotoxins work either by mimicking neurotransmitters to overstimulate nerves or by blocking neurotransmitter receptors to prevent nerve signals from crossing synapses, leading to paralysis.

Key Terms

androgen p. 508	menstrual cycle p. 508
endocrine system p. 506	metamorphosis p. 509
estrogen p. 508	neurotransmitter p. 502
gland p. 506	pituitary gland p. 508
gonadotropin p. 508	senescence p. 505
hormone p. 502	synapse p. 504
hypothalamus p. 508	target cell p. 502

Chapter Review

Self-Quiz

1. Neurotransmitters generally can function more quickly than hormones because neurotransmitters
   a. are smaller molecules.
   b. move over much shorter distances.
   c. travel in the blood.
   d. never have to cross cell membranes.

2. The hormones that regulate growth in plants
   a. have no other functions.
   b. all work in the same way.
   c. defend against apical dominance.
   d. interact to produce a balanced growth form.

3. Two substances that can raise glucose levels in the blood are
   a. insulin and glucagon.
   b. glucagon and acetylcholine.
   c. insulin and acetylcholine.
   d. glucagon and epinephrine.

4. Male and female sex hormones
   a. are produced by the pancreas.
   b. first appear at puberty.
   c. have a similar structure.
   d. are produced by the testes and the uterus, respectively.

5. What do Parkinson's disease and black widow spider bites have in common?
    a. Both involve a hormonal imbalance.
    b. Both involve the function of neurotransmitters.
    c. Both reduce the production of insulin.
    d. Both result in a decrease in dopamine production.

Review Questions

1. Explain how hormones interact to control the growth of plants.

2. Describe the monthly cycling of hormones in adult females.

3. In what ways are neurotransmitters similar to hormones? In what ways are the two groups different?

---

33   **The Daily Globe**

Hormone-Treated Beef Is Safe

Dear Sir:

The European Union's (EU) stubborn refusal to listen to the international scientific community will continue to cost American beef producers and processors nearly $500 million annually. In 1989, the EU imposed a ban on importing our American beef because of unfounded safety concerns over residual hormones in meat, and now they are refusing to abide by the decision of two international organizations, the World Trade Organization and a joint committee organized by the World Health Organization and the United Nations, who concluded that the EU's ban was unfounded.

By giving our cattle sex hormones such as estradiol, progesterone, and testosterone, we are able to produce more flavorful, tender meat at a lower cost than by normal methods. We are not creating a hazardous food; just ask one of the millions of consumers of American beef, 90 percent of which is produced with hormone supplementation.

The EU contends that some of these hormones could cause cancer, a claim that we do not deny. Women taking such hormones for birth control or to relieve some of the symptoms of menopause have probably heard from their doctors about the breast and uterine cancer risks associated with hormone supplements. However, even without such supplements, at certain times of the month women's bodies circulate much higher concentrations of the female sex hormones than they'll ever get from eating beef with trace residues of hormones. Without even eating beef, people can ingest estrogen-like substances that occur naturally in foods such as yams, peas, dairy products, wheat germ, and soybean oil.

The EU's decision is motivated by economics, not science. The only way to fight the EU's decision is to use an economic weapon: trade sanctions that will force it to open European markets to our beef.

William Herdstrom II
Executive Vice President
Organization of American Beef
Producers

Evaluating "The News"

1. Could cultural differences between Europeans and Americans, such as the way they value scientific information or institutions, affect how they assess the health risks associated with eating agricultural products that have been grown with hormone supplements?

2. Do you find Mr. Herdstrom's argument convincing? Why or why not? Who else would you want to hear from besides a spokesman for an association of beef producers?

3. Does the argument that women have, at some times, higher levels of the hormone in their bodies than can be gotten from a meal of beef mean that beef is safe to eat? Why or why not? If different risk factors for cancer can add up to increasing risk, how safe is hormone-laced beef, even for women? What does this mean for men eating beef?

34

Nervous and Sensory Systems

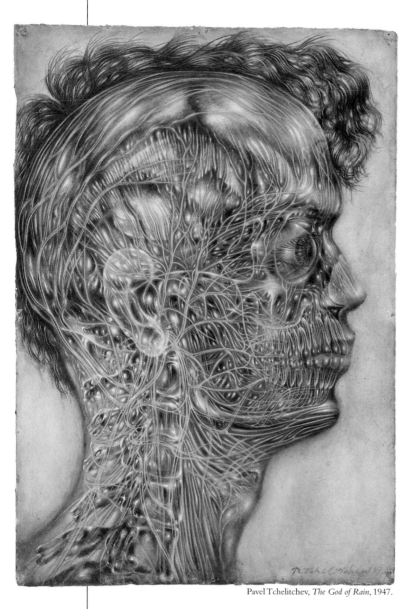

Pavel Tchelitchev, *The God of Rain*, 1947.

Phantom Pain: Missing Limbs That Really Hurt

How can a limb still itch or cause pain long after it's gone? Many people who have lost extremities to trauma or surgical amputation have persistent pain that seems to come from the missing limb. The pain is not imagined, and it isn't simply a matter of remembering previous pain.

For some, the problem is mild and can be as subtle as an itch. For others, though, it is a severe pain that seems to occur in the empty space but is painfully real nonetheless. Phantom pain is a well-documented problem in amputees. Sometimes painful physical sensations, such as arthritis pain, that occurred in the limb before its loss continue even after its removal. In some cases,

Main Message

The elements of nervous and sensory systems cooperate to collect, transmit, and interpret a variety of environmental stimuli in ways that enhance the survival of organisms.

time lessens the severity and the frequency of painful sensations.

In a related problem called reflex sympathetic dystrophy syndrome, pain occurs in an area of former injury, even after tissues are completely healed, free from residual damage. This painful condition is difficult to relieve. In some desperate cases in which the responsible nerves have been cut in an effort to stop the pain, the pain does not subside or even lessen. What could cause the painful sensations in reflex sympathetic dystrophy syndrome and phantom pain to persist in otherwise healthy tissues?

**Many Amputees Feel Real Pain
in Limbs That Are No Longer There**

Key Concepts

1. To find food, locate mates, and avoid harm, organisms must process and integrate information gathered from their environment.

2. Receptors in sense organs gather environmental signals and pass them on to the nervous system.

3. The nervous system converts environmental signals into electrical signals that can travel quickly to processing areas.

4. In most animals, the brain decodes and interprets sensory information.

As we saw in Chapter 33, multicellular organisms must coordinate *internal* body activities to function smoothly, and they accomplish this using a variety of internal chemical messengers. This is only half of the story, however. Organisms also must interpret and respond to their *external* environment.

The external world is a tricky place to navigate. Organisms must find nutrients, escape enemies, reproduce, and perform all the other functions necessary to their survival. In this chapter we examine the kinds of external information that come to organisms, the sensory organs that receive this information, and how the nervous system processes information so that an organism may respond appropriately.

We will first concentrate on light detection, discussing a range of light receptor complexity. Then we will provide a few brief examples of how organisms, especially humans, sense gravity, sound, and chemical stimuli.

Despite their amazing variety, all sensory systems must start with means of detecting a stimulus. These are called sensory receptors, which are cells or organs in multicellular organisms, specialized to detect environmental stimuli.

> ■ Organisms have a wide variety of sensory receptors to collect environmental stimuli such as light, gravity, and chemicals.

Signal Detection: Sensory Systems

Organisms use many different structures to detect stimuli in the environment. **Stimuli** are the external signals or information an organism receives, and there are several types. Some signals, such as sound, vibration, and gravity, are mechanical. Mechanical signals make a physical impact on the organism. Other signals are chemicals—in food, air, or water. Furthermore, many organisms can detect light. All these signals interact with **sensory receptors** in organisms.

The organization of sensory reception varies widely among organisms. Some single-celled organisms like bacteria and protists can sense and move away from harmful chemicals without a central nervous system. Bacteria, protists, fungi, plants, and some animals that do not possess central nervous systems can detect and respond to light by using pigment molecules that change chemically within the organism. In addition to light, plants can respond to gravity, humidity, and the chemical composition of the soil. Many insects that lack a central nervous system or brain can nevertheless smell, hear, taste, and gather information about touch.

In short, there are many more ways of gathering information about the environment than we can describe here.

Detecting Light Stimuli

In animals, sensory receptors that respond to light are eyes. The types of light sensors vary from a simple group of light-gathering pigments to the more complex, multipart eyes of insects or the image-forming eyes of vertebrates.

The simplest eyes consist of a few light-absorbing pigment molecules. These molecules change chemically when exposed to light, allowing them to initiate a signal. Some types of bacteria detect and respond to light using a version of chlorophyll. Barnacles are small marine animals attached to shore rocks that feed on microorganisms in the tides that wash over them. They have three small eyes that detect light intensities but do not focus an image. Shadows that fall over the barnacle trigger a defensive reaction, causing the barnacle to close up like a clam.

When we think of eyes, we usually think of our own set: structures that allow us to focus images and make adjustments for the available light intensity. Human eyes work as a camera does (Figure 34.1). Light travels through an opening, called the pupil, which can be adjusted to let in more or less light as needed. The light passes through a lens, which bends and focuses images

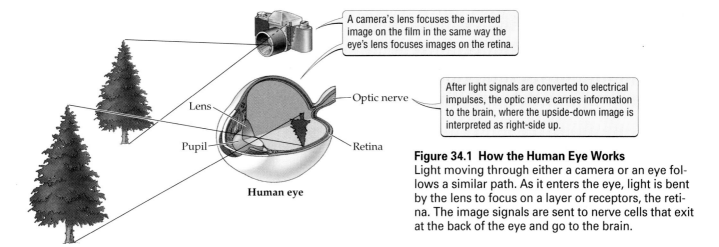

A camera's lens focuses the inverted image on the film in the same way the eye's lens focuses images on the retina.

Lens

Pupil

Human eye

Optic nerve

Retina

After light signals are converted to electrical impulses, the optic nerve carries information to the brain, where the upside-down image is interpreted as right-side up.

Figure 34.1 How the Human Eye Works
Light moving through either a camera or an eye follows a similar path. As it enters the eye, light is bent by the lens to focus on a layer of receptors, the retina. The image signals are sent to nerve cells that exit at the back of the eye and go to the brain.

onto a surface rich in light-sensitive cells called the **retina**, where the image is turned upside down.

Within the retina, two receptor cell types, rods and cones, capture different types of light. Rods gather low-intensity light but do not gather information about color. Cones need more light but contain pigments that allow them to detect red, green, or blue light. Normally, humans are born with all three types of cones, but occasionally a missing gene prevents one type from forming, causing color blindness, an inability to see one or more colors.

Rods and cones throughout the retina change the light signals into chemical signals and pass them to neighboring nerve cells. In the nerve cells, the signal changes into an electrically encoded signal. This change must occur before any sensory information can be sent to the brain. The electrically encoded visual signals, or impulses, are sent to the brain by a bundle of nerve cells that form the optic nerve (see Figure 34.1). The visual cortex in the brain decodes and interprets the language of electrical signals so that our view of the world is meaningful.

Humans are not the only animals that have complicated eye structures or color-sensitive receptors. Several types of insects that are active during the day, as well as some fish, amphibians, birds, and other primates, have color vision. Furthermore, bees can detect ultraviolet light, a type of light that is invisible to humans.

Like other animals that have a camera type of eye, insects use pigments to detect light and lenses to focus, but insect eyes contain many individual units (Figure 34.2). Each individual unit has its own lens, focusing a small part of the visual field onto the receptor surface, making a fragmented image that fits together like pieces in a puzzle. Depending on the arrangement of lens and receptors, these units can maximize the amount of light that reaches the receptors of nocturnal insects. Alterna-

tive arrangements for insects that are active during the day are not as good in low light, but they provide a clearer image in daylight.

This variety of light detection mechanisms coupled with processing in the nervous system provides a unique view of the world for each organism—one that provides the visual information that each organism needs to survive in its particular habitat.

Some organisms other than animals also have ways of sensing light. Many of these light-sensing mechanisms are simple, such as the light-sensitive chemicals in unspecialized cells that allow plants to distinguish light from dark. Others, such as the eyes used by a fungus called *Coprinus*, are specialized organs complete with lenses.

Figure 34.2 The Compound Eyes of Insects
Insects such as flies have many lens units in their eyes. The arrangement of the units can enhance the night vision of some insects by focusing more light onto the underlying sensory surface of each unit.

> ■ Eye structures that range from simple light-absorbing pigments to complex, specialized, multicellular structures receive light stimuli and convert them to electrical signals (impulses) for transmission to nerves. In humans, light moves through the eye similarly to how light moves through a camera.

Detecting Mechanical Stimuli

Gravity, sound, and physical contact are some of the mechanical stimuli sensed by a class of sensory cells called **mechanoreceptors**. Most organisms can sense gravity and orient themselves relative to its force. Both plants and animals, for example, must know which way is up. Special cells called statocysts provide gravitational information. Statocysts contain dense mineral substances (made of calcium) that fall through the cytoplasm to the "bottom" of the cell. Organisms experience their position relative to gravity when these minerals fall onto sensory structures within the cells.

The human ear (Figure 34.3) is a multipurpose mechanoreceptor. It detects two types of mechanical stimuli: sound and changes of speed. Changes of speed are detected when fluid flows past sensory cells located in three looplike tubes called the **semicircular canals**. Flexible extensions in the receptor cells, called **hair cells**, bend when fluid moves past them. These mechanical signals are converted to electrical signals for the nerve cells that will send them to the brain for processing. In the brain, signals from the semicircular canals are integrated with information gathered from the eyes. This combining of signals from the eyes and ears is the reason that a movie scene filmed by a camera spinning around can begin to make us feel dizzy, even though our chairs remain stationary.

Sound is also detected by hair cells in the inner ear. To reach the inner ear, sound waves are guided through the outer ear and vibrate a delicate membrane called the **eardrum**, which converts pressure changes in the air that we perceive as sound into physical movement (see Figure 34.3). This vibration jiggles three tiny bones—the body's tiniest—on the other side of the membrane in the middle ear. The last bone in line is attached to a snail-like structure called the **cochlea** in the inner ear. Patches of hair cells respond to a unique frequency depending on their position along the cochlea and generate the electrical signals that transmit the information to nerve cells. The encoded signals travel rapidly through the nervous system, where they are interpreted by the sound-processing centers of the brain.

Physical contact is another category of mechanical stimulus. Many types of contact can be detected by organisms. In humans, the skin is richly laced with a diverse population of receptors (Figure 34.4). Some are nerve cells with special layers or modifications that look like onions, some are bare receiving ends of nerve cells, and others are not derived from nerve cells at all. Each type of receptor responds to a specific stimulus: light touch, heavy touch, pressure, or pain.

In addition to mechanoreceptors, the skin has temperature receptors that respond to changes in temperature. In general, skin sensory receptors are distributed more densely in areas where detailed information on touch is needed. Detailed sensations from the different receptor types are combined to allow us to "picture" something with our fingers, even with our eyes closed.

> ■ In humans, mechanoreceptors in ears, statocysts, and skin allow organisms to sense sound and changes of speed, gravity, and a range of touch-based stimuli, respectively.

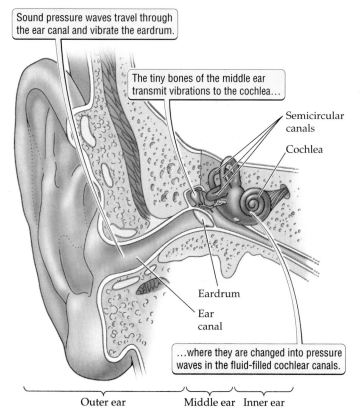

Sound pressure waves travel through the ear canal and vibrate the eardrum.

The tiny bones of the middle ear transmit vibrations to the cochlea...

Semicircular canals

Cochlea

Eardrum

Ear canal

...where they are changed into pressure waves in the fluid-filled cochlear canals.

Outer ear   Middle ear   Inner ear

Figure 34.3 The Human Ear
Structures in the ear respond to pressure waves. The tiny bones of the middle ear, called ossicles, transmit and amplify the sound waves for the cochlea. In the cochlea, patches of receptor cells distinguish different sound frequencies. The semicircular canals, angled in three different planes, detect the smallest amount of movement in any direction to help us keep our balance.

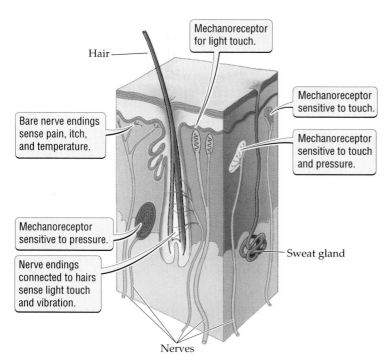

Figure 34.4 Receptors in the Skin
The skin has many different types of receptors that provide detailed information about touch and temperature.

Detecting Chemical Stimuli

Eating and reproducing are two of the most basic functions of organisms. Chemical sense organs allow some organisms to follow a trail of molecules to find food sources or other individuals (see Chapters 35 and 36). A chemical substance that binds to a receptor in a sense organ can activate chemical sensors. Different molecules bind to different receptors and are detected by the organism as different smells and tastes.

In addition to hunting for food by scent, some animals depend on chemical signals to identify each other. Experimental evidence suggests that a mother bat returning to a cave—where more than 1 million other bats may be roosting—can pick out her young by scent. Some animals use **pheromones**—airborne chemicals— to attract a mate or mark territory. The antennae of the male *Bombyx* moth, the adult form of the silkworm, may have as many as 50,000 receptors for the female pheromone. In stark contrast, the female *Bombyx* has no receptors for its own pheromone.

Smells are not the only type of chemical sensation. Humans also have chemical receptors that are distributed over the tongue surface and back of the throat that interact with four basic chemical types. **Taste buds** are small clusters of receptor cells that interact with chemicals passing through the mouth (Figure 34.5). Each taste bud identifies one of the four chemical taste character-

istics: sweet, salty, sour, or bitter. Although each type of taste bud is distributed throughout the surface of tongue, certain regions of the tongue have dense populations of one type.

Like all other sensory signals, chemical signals gathered by chemical sensors are converted to an electrical signal for travel to the brain. Ultimately, the brain decodes and interprets the combined chemical signals that allow us to experience far more than the four main chemical tastes.

Even though some people derive great pleasure from eating, taste is not merely for our entertainment. It is no accident that two of the most basic taste sensations, salt and sweet, provide information about two nutrients that are critical to our survival: salts and sugars. The other two tastes, bitter and sour, provide information about substances that are often harmful. Thus, although there are plenty of exceptions, generally foods taste good to us when they contain valuable substances and repel us when potentially harmful.

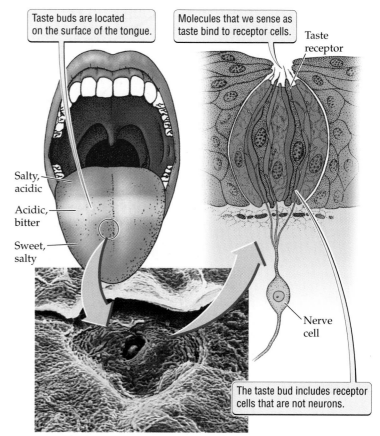

A taste bud

Figure 34.5 Taste Buds
Taste buds are groups of sensory and supporting cells that detect features of chemicals that enter the mouth and throat.

In addition to taste buds, chemical receptors in the nose provide sensory information that contributes to a rich sense of taste. Taste, then, is a sensation that depends on the integration of several types of stimuli and requires further neural processing in the brain.

> ■ Organisms detect environmental chemicals as scents and tastes to gain information about other individual organisms or potential food sources.

Signal Transmission: Action Potentials

How does an organism make sense of all the incoming signals? In most animals, the receptor cells cannot understand the signals they receive. Their job is simply to collect and pass along the information to the nervous system. It is the nervous system that sorts and interprets all the information, finally stimulating the organism to respond appropriately. In this section we look more closely at the cells that are common to many animal nervous systems: neurons and glial cells.

The **neuron**, or nerve cell, is specialized to receive information and send it long distances to the central processing areas. We have already seen some of the rich variety of sensory receptors that change the signal from mechanical, light, or physical stimulus into an electrical signal that can be transmitted by the nervous system.

Transmission of an electrical signal requires a chemical messenger to carry the signal across the synapses between nerve cells (see Chapter 33). Together, the chemical messenger and the electrical current transmit the signal to the central nervous system (see Figure 34.8), which does most of the processing. In vertebrates, the central nervous system consists of the spinal cord and the brain, which sort and interpret all incoming signals, and ultimately direct responses, which are often executed by the muscles.

Neurons transmit signals in one direction only

Several structural and functional features of neurons suit them to the task of transmitting information through the nervous system. First, signals can move through a neuron in one direction only. They start at the branched receiving ends called **dendrites** and move down the long, thin part of the cell, called the **axon** (Figure 34.6). Axons can reach lengths of 1 meter in humans and are designed to carry the electrical signals over these long distances to the next neuron, quickly and efficiently.

Since the nervous system is constantly bombarded with signals from the sensory receptors, it cannot possibly send all the information to the brain. As incoming signals arrive, they cause tiny electrical changes that accumulate in the axon. When the electrical changes in the cell are sufficiently beyond the normal resting conditions, changes in the plasma membrane of the neuron trigger a shift in the concentration of sodium and potassium ions inside and outside the cell.

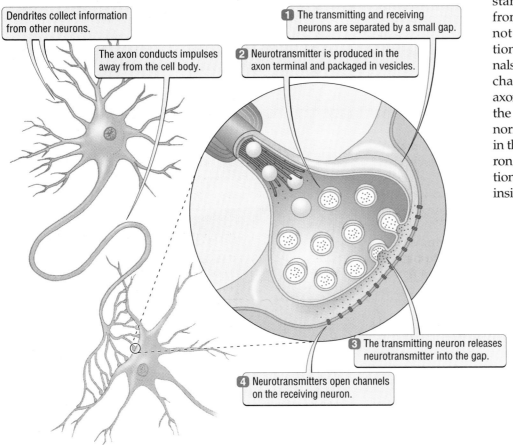

Dendrites collect information from other neurons.

The axon conducts impulses away from the cell body.

1 The transmitting and receiving neurons are separated by a small gap.

2 Neurotransmitter is produced in the axon terminal and packaged in vesicles.

3 The transmitting neuron releases neurotransmitter into the gap.

4 Neurotransmitters open channels on the receiving neuron.

Figure 34.6 Neurons and Synaptic Signaling
Neurons rapidly transmit signals in the form of action potentials. Where neurons meet, the electrical action potentials are converted into chemical signals in the form of neurotransmitters, which diffuse across a small gap (the synapse) between the neurons.

This flow of charged particles across the axon membrane generates the electrical impulse called an **action potential** that propagates down the axon to the tip. At the end of the axon, the action potential causes the release of neurotransmitter to carry the outgoing signal into the **synapse** (see Figure 34.6), a very small gap between the cells (see Chapter 33).

The synapse is a gap across which neurotransmitters from the tip of an axon diffuse to either trigger or suppress an action potential in the adjacent neuron. The signal slows down at the synapse because chemical diffusion is slower than electrical impulses. As the signal moves through the nervous system, it travels rapidly when action potentials carry it down an axon and slows down at each synapse when it is carried by neurotransmitters. Fortunately, the electrical phase of signal transmission carries the signal over the greatest distances, allowing information to travel quickly through a nervous system.

In vertebrates, insulated neurons speed up signal transmission

In vertebrates, neurons do not do the job of transmitting signals alone. Several different types of **glial cells** support and assist neurons. Glial cells perform many important functions for neurons, such as breaking down and removing neuron wastes, providing structural support systems for neurons, filtering materials that pass into the nervous system, and electrically insulating groups of neurons.

Vertebrates have an important nervous system enhancement. One important type of glial cells greatly speeds signal transmission. They wrap around the axon multiple times, covering it with several layers of membrane called **myelin**. Myelin, which contains phospholipids and cholesterol, is an effective insulator that greatly speeds up the transmission of the electrical signals along the axon.

Myelin develops in segments along the axon, and the spaces between the areas of myelination are called **nodes of Ranvier**. As the electrical signal begins to lose strength traveling down a long axon, it is reinvigorated at these nodes. Myelin prevents the current from leaking out of the insulated segments, so the action potential appears to jump from node to node.

Neurons that have very long axons or depend on very fast signal transmission are more heavily myelinated. The largest axons in humans are about 4 to 20 micrometers in diameter and bring information about balance, as well as precision touch sensations from the skin, or they send commands to skeletal muscles. The larger the diameter of an axon and the more myelinated it is, the faster the action potential can travel.

Myelin deterioration or improper formation can lead to serious problems in humans. Several diseases, collectively called demyelinating diseases, such as multiple sclerosis and Guillain-Barré syndrome, can disable and eventually kill their victims. When myelin begins to break down, vision, speech, balance, and general motor coordination are all impaired.

> ■ Neurons convert chemical signals to electrical signals that travel as action potentials to the spinal cord and brain. Myelin surrounds the axon in vertebrates, enhancing signal speed and preserving signal strength.

Signal Processing: The Human Central Nervous System

The important jobs of interpreting coded information from sensory structures, bringing together information from the internal and external environments, and controlling the activities of the other systems in the organism fall to the brain and the spinal cord, which together make up the central nervous system. This system is incredibly powerful, but we have yet to understand how much of it functions, especially in the more complicated areas of intellect, learning, memory, and emotion.

So far in this chapter we have looked at stimuli detection and transmission in animals, but unless the information can be decoded, it is not very useful. Sensory information can be processed in various routes or pathways, from very simple circuits, involving only a few neurons, to highly complicated pathways involving many more neurons in different areas of the brain and central nervous system. As a general rule, the more complicated the required response is, the more processing it will demand. In this section we explore some of the brain basics: where signals are processed and how processing affects perception.

The reflex arc is an example of simple processing

Some situations call for immediate action. If your finger touches a hot stove, you don't want to spend time sending information to be filtered by the brain, determining the degree of heat and the number of receptors in the skin that have been stimulated. You need to move your finger off the burner. Fast.

When immediate action would help an organism avoid injury, **reflex arcs** provide a means for immediate withdrawal from dangerous stimuli. The advantage of reflex arcs is that they process the rapid, protective

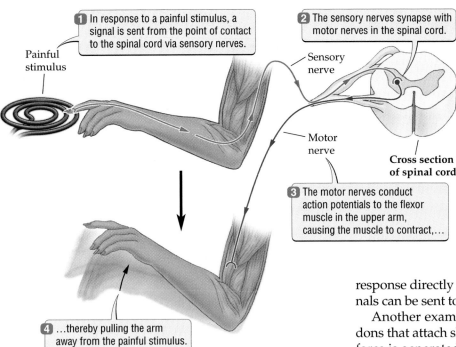

1 In response to a painful stimulus, a signal is sent from the point of contact to the spinal cord via sensory nerves.

Painful stimulus

Sensory nerve

2 The sensory nerves synapse with motor nerves in the spinal cord.

Motor nerve

Cross section of spinal cord

3 The motor nerves conduct action potentials to the flexor muscle in the upper arm, causing the muscle to contract,…

4 …thereby pulling the arm away from the painful stimulus.

Figure 34.7 Reflex Arcs
The events in a reflex arc are simplified in this figure. The entire signaling sequence—detection, processing, and response—occurs without the signal ever entering the brain. Additional events (not shown) also occur during a reflex to prevent opposing muscle groups from contracting.

response directly in the spinal cord, a lot faster than signals can be sent to the brain and processed (Figure 34.7).

Another example of a reflex pathway occurs in tendons that attach skeletal muscle to the bone. If too much force is generated by a muscle, receptors in the tendon collagen fibers activate a reflex arc to protect against tearing the muscle tissue or the tendon. Signals gener-

The Scientific Process

Spying on the Brain

Because of the unavoidable difficulties in studying activity in living brains, much remains to be discovered about function in this amazing organ. Several strategies currently provide information about normal function or damaged areas in the brain without requiring the skull or brain to be cut surgically.

Although X-ray technology has been around for nearly 100 years, it provides information only about bony or other dense structures. Soft tissue in the brain and other areas of the body will not appear on X-ray film. Hospitals currently employ two scanning techniques—magnetic resonance imaging (MRI) and computerized tomography (CT)—to reveal areas of damage in the softer tissues of the brain.

These techniques offer the benefit of showing soft tissue, but the pictures they produce are static snapshots that provide only structural information about the brain. In a hospital setting, these techniques help identify damaged areas within the skull.

In another scanning technique, called positron-emission tomography (PET), a dynamic moving image is taken of a functioning brain. PET pinpoints and ranks areas of glucose consumption in the brain, revealing how much activity is occurring in each structure during particular behaviors.

The imaging technologies described here have unveiled many activities of the brain that were once mysteries, giving us a slightly clearer picture. When it comes to a comprehensive understanding of the brain, however, there is still plenty of mystery to unravel.

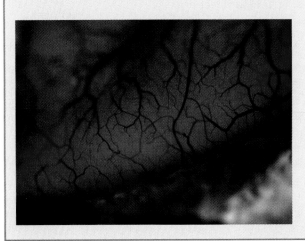

The Brain at Work
This image of a living ferret brain, generated by a process called intrinsic signal optical imaging, shows which areas are responsive to inputs from the left eye (blue) and the right eye (red).

ated in the tendon travel to the spinal cord, where they signal neurons that immediately inhibit further muscle contraction, potentially preventing injury or damage. If we pick up something too heavy, this type of reflex prompts us to drop the object.

The brain handles more complex processing

Reflexes are considered programmed responses to a particular stimulus, but they can be overridden by alternative instructions from the brain. Consider the tendon reflex just described. What if we were saving a person from falling over the edge of a cliff? The tendon reflex might normally cause us to drop the person. In this situation, however, the release of the hormone epinephrine could override the reflex and stimulate more forceful muscle contractions to help us hold onto the person.

Higher processing in the nervous system allows all incoming information to be sorted and prioritized. A single response can result from many levels of processing in the spinal cord and brain.

Organization of the human spinal cord

The spinal cord—about the thickness of a finger—is an organized collection of nerve cell bodies and axons. Areas of dense nerve cell bodies, which contain the cell's nucleus and organelles, appear darker and are called gray matter; the areas composed primarily of axons appear lighter and are called white matter (Figure 34.8).

As we have seen, depending on the origin of the stimulus, action potentials in the spinal cord can directly stimulate skeletal muscles through a reflex arc (see Figure 34.7). For further interpretation, the signals travel through the spinal cord to reach the brain (Figure 34.8). Commands from the brain must also travel back through the spinal cord to reach organs, muscles, and glands.

Organization of the human brain

The type of stimulus determines which path an action potential will take. Ultimately most information ends up in the brain, but where? One of the major relay areas is the **thalamus**, a small central switching area in the brain (see Figure 34.8). In fact, the thalamus is a critical filter, saving the brain from processing all the stimuli that constantly bombard sensory receptors. The thalamus determines which signals require an action without awareness and which signals need to be sent to the conscious perception centers in the **cerebral cortex**. For example, the act of swatting away a fly might require only a simple reflex the first time. But after the tenth fly, the information might go to the cerebral cortex for more careful or strategic contemplation.

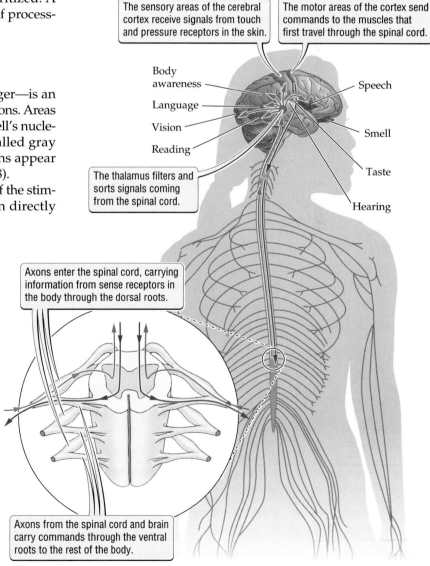

Figure 34.8 Organization of the Central Nervous System
Sensory information (shown in red) travels through the dorsal roots to the spinal cord, where it may eventually be sent to the brain for further processing. Commands from the brain (shown in blue) leave the spinal cord through the ventral roots. In the body these commands may stimulate or inhibit organs, muscles, or glands.

The human brain has several areas that handle different kinds of information (see the Box on page 522 for techniques used to study brain structures). The largest, most visible part of the brain is the **cerebrum**, which consists of two hemispheres, joined by an axon bundle that allows information to flow between them. If signals manage to pass through the thalamus, they usually arrive in an area of the cortex that corresponds to the area of the body that received the original stimulus (Figure 34.9).

Signals in the cortex are part of human conscious awareness. The cerebral cortex, the outer layer of the brain, which is rich in gray matter, looks heavily wrinkled. The folds that give the brain this wrinkled appearance increase its surface area to almost 2.2 square meters, much greater than the surface area of the skull.

The brain relies heavily on sensory information from the external environment to control the internal environment. Tight communication with the endocrine system also allows this internal regulation. In one area of the brain, neurons from the **hypothalamus** directly stimulate

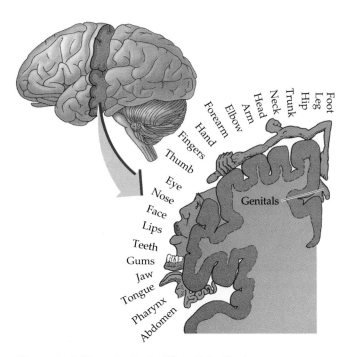

Figure 34.9 How the Brain "Sees" the Body
Different parts of the cerebral cortex are devoted to sensory information received from different parts of the body. The amount of brain tissue devoted to this sensory information does not necessarily correspond with the size of the body part. The body parts shown above look distorted because they are drawn to indicate the amount of brain area devoted to sensory information they send the brain. For example, you can see that more of the brain is devoted to receiving sensory information from the foot, a relatively small body part, than from the leg, a much larger body part.

the release of hormones from an important endocrine gland, the **pituitary gland**, coordinating many important long-term processes involved in growth, salt and water balance, and reproductive function (see Chapter 33).

> ■ The human central nervous system integrates and coordinates sensory information to allow appropriate responses to external stimuli. The central nervous system consists of the spinal cord and the brain. Reflex arcs allow the body to bypass the central nervous system in situations requiring very rapid responses.

Highlight

Perception and Pain

When we encounter light, sound, smells, or any other kind of stimulus, we receive one type of signal, but we *perceive* something completely different in each case. Our eyes do not perceive the world upside down, even though the images arrive at the retina this way, and we perceive more than the three colors our cone cells detect.

Our perception of a richly colored landscape depends on both signal transduction and decoding in the brain. Similarly, all our sensory perceptions occur after the original signals have been electrically coded, filtered by the thalamus, and sent to areas in the cerebral cortex responsible for awareness. To complicate our understanding of perception, signals may be shared among several areas of the cortex for further interpretation.

Pain is a special case of perception. The perception of pain is protective, stimulating an organism to stop behavior that provokes tissue damage or encourages it to continue. The sole function of certain receptors and neurons appears to be to notify the brain of painful stimuli, especially in the skin. In addition, the role of some neurotransmitters appears to be limited to communicating pain messages. Pain would be much easier to manage if it were a simple matter of knocking out the receptors or pain transmitters. However, the perception of pain seems to be a complicated result of processing in the brain.

Conditions such as reflex sympathetic dystrophy syndrome and phantom pain are convincing reminders of the complexity of processing pain signals. The fact that reflex sympathetic dystrophy syndrome is not helped by cutting the nerves provides an additional clue that sensory receptors are not solely responsible. The pain of reflex sympathetic dystrophy syndrome is processed in higher centers of the brain about which we know little.

Phantom limb pain is another example of the brain's role in perception. Pain pathways that once gave the brain useful information about the status of a limb appear not to reorganize after that limb is removed. Until these pathways are modified, a particular part of the brain still has neurons reporting a problem. Because the brain cannot filter out the false information, the painful sensations persist.

Modern advances in prosthetic technology (the technology of artificial limbs) take advantage of electrical activity in the remaining muscles, as well as electrical signals coming from brain itself. These tiny electrical currents can be harnessed to drive some of the prosthetic devices that help amputees reclaim some of the lost function once handled by a missing structure. These devices are being developed to manage tasks that require both strength and delicate coordination. For some amputees, the neural pathways adjust over time and the phantom sensations diminish. For others who must endure it, phantom pain, as well as reflex sympathetic dystrophy syndrome, can be a lifelong reminder of the complexities of the sensory and nervous systems.

> ■ Human understanding of the world comes from the brain's decoding and interpretation of incoming information, as in pain perception.

Summary

Signal Detection: Sensory Systems

- Organisms detect environmental stimuli—visual, mechanical, or chemical—with specialized receptors.

Detecting Light Stimuli

- Eyes use light-absorbing pigments to collect light stimuli.
- Some eyes use lenses to focus light onto a sensory surface.
- The human eye uses the rod and cone cells of the retina to detect light of different wavelengths.

Detecting Mechanical Stimuli

- Mechanoreceptors allow organisms to detect mechanical stimuli such as gravity, sound, and physical contact.
- In the human ear, sound waves produce vibrations in the eardrum that are carried by a series of structures to hair cells that trigger a nervous signal.
- Mechanoreceptors and temperature receptors in the skin allow us to detect temperature, pain, touch, and pressure.

Detecting Chemical Stimuli

- The sensations of smell and taste are produced in part by physical contact between specific molecules and receptors in the body.

Signal Transmission: Action Potentials

- Neurons and glial cells work together to carry signals from sensory receptors to the central nervous system and back again.
- Neurons change chemical signals into electrical signals, which travel along the cell's axon until they reach the synapse with another cell.
- Glial cells provide many support and maintenance functions to neurons, and they insulate neurons to speed up the transmission of electrical signals.

Signal Processing: The Human Central Nervous System

- When immediate action is required, reflex arcs bypass the brain.
- The brain handles signals that require more complex processing.
- Signals are sent to the brain, and commands from the brain are carried back to the body, through the spinal cord.
- The thalamus filters incoming signals and sends those that need further processing to the appropriate part of the brain.
- Specific areas of the cerebral cortex are involved in the conscious processing of information.

Highlight: Perception and Pain

- The perception of pain is a complex process involving higher centers of the brain.
- Even after an injury has healed or a limb has been amputated, a person can perceive pain because of continued stimulation of neural pathways.

Key Terms

action potential p. 521	**neuron** p. 520
axon p. 520	**node of Ranvier** p. 521
cerebral cortex p. 523	**pheromone** p. 519
cerebrum p. 524	**pituitary gland** p. 524
cochlea p. 518	**reflex arc** p. 521
dendrite p. 520	**retina** p. 517
eardrum p. 518	**semicircular canals** p. 518
glial cell p. 521	**sensory receptor** p. 516
hair cell p. 518	**stimulus** p. 516
hypothalamus p. 524	**synapse** p. 521
mechanoreceptor p. 518	**taste bud** p. 519
myelin p. 521	**thalamus** p. 523

Chapter Review

Self-Quiz

1. The ability to respond to light stimuli requires
   a. receptor pigments.
   b. a lens.
   c. temperature-sensitive receptors.
   d. the temporal lobe.

2. Human perception depends primarily on the function of the
   a. cerebral cortex.
   b. pituitary gland.
   c. gray matter.
   d. thalamus.

3. An electrical signal traveling along an axon
   a. diffuses rapidly in both directions.
   b. moves more rapidly than a neurotransmitter or hormone.
   c. is slowed by myelin.
   d. is carried through the blood.

4. Perception of changes of speed involves the
   a. cochlea.
   b. retina.
   c. cerebral cortex.
   d. all of the above

5. Action potentials travel more quickly
   a. through narrower axons.
   b. through more heavily myelinated axons.
   c. when moving toward the brain.
   d. when carrying a signal from a light-sensitive receptor.

Review Questions

1. Contrast the advantages and disadvantages of a reflex arc response in comparison to a response processed by the cerebral cortex.

2. Describe the movement of a signal from the dendrite of one neuron to the dendrite of another neuron.

3. Describe the pathway of energy and information from light approaching the eye to vision, a complex type of perception.

34 𝔘𝔥𝔢 𝔇𝔞𝔦𝔩𝔶 𝔊𝔩𝔬𝔟𝔢

A Whale of a Problem

WOODS HOLE, MA. For years, many species of whales were threatened by overhunting, but international cooperation has reduced that threat significantly. Now marine biologists are warning that humans may be responsible for a much more subtle problem: noise pollution that could harm many marine mammals, especially whales and dolphins.

Although humans live in a world dominated by visual perception, other species sense a very different world. In the watery depths, highly social species like humpback whales and bottlenose dolphins rely on their hearing to communicate with other individuals, coordinating cooperative activities as diverse as escaping predators

and reproducing. Some species, such as the beluga whale of arctic waters, also rely on their hearing to locate and catch their prey. The potential effects of noise pollution could be both immediate and long-term to such species: immediate if a single event disrupts their behavior temporarily, and long-term if exposure to dangerous levels of noise damages their hearing irreversibly.

One of the main sources of noise pollution in the ocean is ship engines; ironically, some people have observed that the immense popularity of whale-watching boat tours, which allow people to come very close to whales, may add to the problem. Another source of noise is military activity. For in-

stance, a test of a powerful sonar system for detecting submarines was blamed for the death of four humpback whales in 1997. Conservationists found temporary relief for the whales in 1998 when they won their lawsuit demanding a halt to the sonar testing while the effects of the sonar system were being investigated.

"Given the wide-ranging, migratory populations of many marine mammals, any successful attempt to reduce this new threat will require international cooperation, public education, and a commitment to increasing our understanding of life in the sea," said Dr. Olivia Peach, scientist for Marine Mammals Research Institute.

Evaluating "The News"

1. Can you think of any other examples of situations that might be described as sensory pollution, in which factors could affect the ability of a human or other animal to detect sensory stimuli?

2. Is sensory pollution more or less dangerous than pollution that could cause a specific disease?

3. What are the positive and negative effects on the environment of ecotourism businesses such as whale watching, swimming with dol-

phins, jeep rides through beautiful but delicate desert habitats—activities that allow people to have an "up close and personal" experience with threatened or rare organisms? Given your assessment, are they valuable enough to continue or should more of them be banned?

chapter 35

Behavior

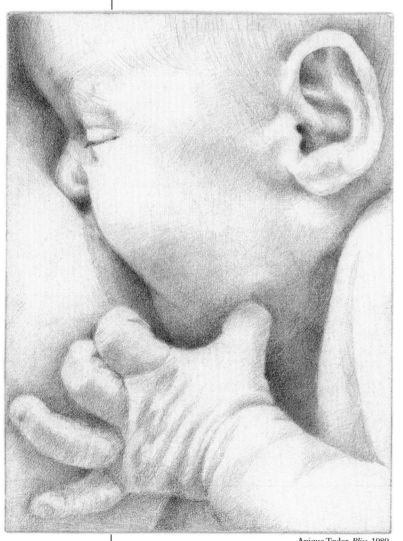

Anique Taylor, *Bliss*, 1989.

Bright Butterflies

*I*magine you are walking through the forest. On your walk you encounter a large butterfly with iridescent blue wings rimmed with yellow spots. It flies slowly, and every now and then it lands on a plant. Clearly, this is a butterfly on a mission. Curious about what the mission might be, you follow the butterfly. You discover that the butterfly usually does no more than briefly land on the plant. Every now and then, however, it lays a cluster of three to five orange eggs on the bottom of a leaf. Knowing what you do about herbivores like this butterfly, you realize that the female must be searching for plants on which its young can feed when they hatch. The food plants of this particular butterfly have small, rounded leaves and seem to grow abundantly on the forest floor.

Confident that you've figured things out, you follow another of the pretty blue butterflies. As with the first butterfly, this butterfly alights on many plants before finding one she considers a suitable place to lay her eggs. As you follow the second butterfly, you notice two things. First, when she finally lays eggs, she does so on a plant completely different from the plant selected by the first butterfly—

Through their behavior, organisms extend their ability to respond rapidly and flexibly to environmental changes.

an inconspicuous plant with long, narrow leaves. Second, most of the other plants on which she lands also bear long, narrow leaves.

Perhaps the two outwardly identical butterflies belong to different species that survive on different foods. Following a third blue butterfly, however, shows this not to be the case. The third female initially behaves like the first one—landing mostly on plants with rounded leaves and laying clusters of eggs on some of these. At one point, however, she happens to land on one of the narrow-leaved plants favored by the second female . . . and promptly lays some eggs. Thereafter, the third female switches from seeking broad-leaved plants to seeking narrow-leaved plants.

The preference for broad- or narrow-leaved plants by this butterfly, the pipe vine swallowtail butterfly of the southern United States, is a response of individuals to their environment as they search for their food plants, two different species of pipe vine. Although not a physical feature like wings for flying and eyes for seeing, the way in which pipe vine swallowtails select plants makes their survival possible. At the end of this chapter, we will revisit plant selection by pipe vine swallowtails to learn why these butterflies do what they do.

The Pipe Vine Swallowtail Lays Its Eggs Only on Small Pipe Vine Plants

1. Behavior results from interactions among the senses, coordinating mechanisms, and some means of generating motion.

2. Animals rely more heavily on behavior than do members of other kingdoms.

3. Behavior can be fixed from birth or learned based on the organism's experience.

4. Both fixed and learned behaviors can evolve.

5. Communication behavior allows two individuals to coordinate their activities.

6. Behavioral interactions within groups of organisms allow them to defend themselves and gain access to resources that are not available to individuals.

The passage at the beginning of this chapter describes the plant-finding behavior of the pipe vine swallowtail butterfly. A **behavior** is a movement made by one organism in response to another organism or in response to the physical environment. This broad definition encompasses behaviors that we recognize, such as plant finding by butterflies and the courtship behavior of humans, as well as more subtle behaviors. In this chapter we will see that behavior plays a central role in coordinating the various functions essential to survival. Behavior allows organisms, especially animals, to respond quickly and flexibly to an ever changing environment. Moreover, behavior allows organisms to coordinate their own activities with those of other organisms.

We begin this chapter by looking at the behavioral interaction between a mother and her nursing child as a way of emphasizing the relationship between behavior and the hormonal and nervous systems described in Chapters 33 and 34. We then make the point that although members of all kingdoms of life use behavior, animals rely on behavior most heavily. The subsequent discussion of behavior focuses on how it works in animals: (1) the differences between fixed and learned behaviors, (2) the genetic basis of behavior, (3) the role of communication in behavioral interactions between animals, and (4) the social behavior of animals living in groups.

Breast-Feeding in Humans as an Example of Behavior

Behavior is a crucial part of human life. To understand the function and importance of behavior we will look at the first behavioral interaction in human lives: that between newborns and their mothers during breast-feeding (Figure 35.1).

When a mother brings her breast to the mouth of her child for the first time, the child searches for the nipple with its mouth in an attempt to get milk. This behavior, called the **rooting reflex,** is stimulated by a touch against the cheek or mouth of an infant. The light touch against the infant's cheek triggers a signal carried to its brain by its nervous system. The infant's brain, in turn, coordinates nursing behavior by stimulating contractions in its mouth muscles.

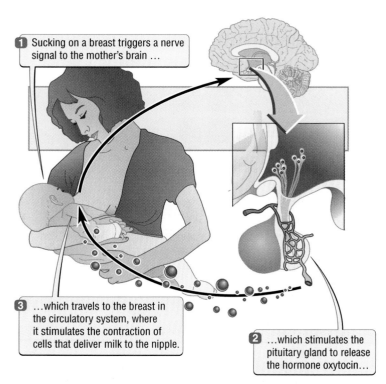

1 Sucking on a breast triggers a nerve signal to the mother's brain …

2 …which stimulates the pituitary gland to release the hormone oxytocin…

3 …which travels to the breast in the circulatory system, where it stimulates the contraction of cells that deliver milk to the nipple.

Figure 35.1 The Rooting and Milk-Letdown Reflexes of Humans
Nursing by an infant stimulates the contraction of cells in the breast to release milk into the nipple, in a behavior coordinated by both the nervous system and hormones.

Nursing stimulates a less obvious behavioral response in the mother: the **milk-letdown reflex**. Although you have probably forgotten, for the first 30 to 60 seconds that you nursed, you received no milk. The reason for the delay is that milk storage cells in a mother's breasts release milk only in response to nursing. Pressure-sensitive cells around the nipple send a nerve signal to the mother's brain, which triggers the release of a hormone called oxytocin from the pituitary gland at the base of the brain (see Figure 35.1). Oxytocin circulates in the blood to target cells in both breasts. Contraction of these cells forces milk out of storage cells and into tubes leading to the nipples. Over weeks, as her breast-feeding experience increases, a mother's milk-letdown reflex changes in an intriguing way: Instead of releasing milk in response to nursing, she releases milk in response to the hungry cries indicating that her baby needs to feed.

Behaviors give organisms the flexibility to respond rapidly to a changing environment. During breast-feeding, the rooting and milk-letdown reflexes ensure that a baby tries to feed only when a breast is available and that a mother releases milk only when her hungry baby is nearby.

The baby's mouth and the mother's breast are structures that allow babies to feed, and the behavioral interaction between mother and infant allows effective use of the mouth and breast. The switch from the touch of nursing lips to the baby's cry as the stimulus that triggers the milk-letdown reflex emphasizes the flexibility of behavioral responses. In both behaviors, the behavior depends on sensory information and on the hormones and nerve signals that coordinate functions.

> ■ The behavioral interaction between a mother and child during breast-feeding coordinates the feeding behavior of the child with milk release by the mother. Sensory stimulation that triggers nerve signals and hormone release control the behavior.

Who Behaves?

Although we typically think of behavior as something animals do, other organisms behave, too. As we have seen in the preceding chapters, members of other kingdoms can gather information about their environment and have the ability to move in response. In this section we present two examples of behavior among organisms other than animals, before focusing on animal behavior for the remainder of the chapter.

Bacteria can move toward food

Bacteria can move toward or away from an object in their environment. Such bacterial **orientation behavior** depends on simple mechanisms that emphasize the role of coordinating systems in generating behavior. Bacteria orient toward food by following two simple rules: (1) If the concentration of chemicals diffusing from the food source increases, bacteria continue in the same direction, and (2) if the concentration of chemicals diffusing from the food source decreases, bacteria turn (Figure 35.2).

Bacteria move in a straight line by rotating their flagella in one direction, and they turn by rotating the

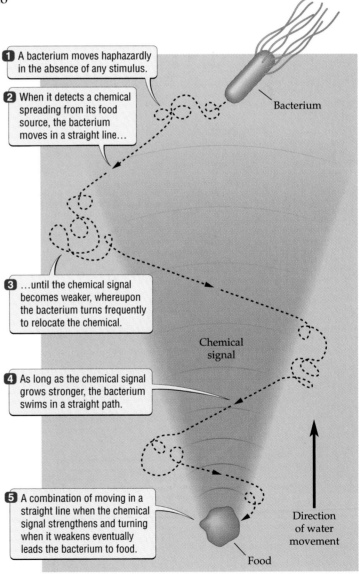

1 A bacterium moves haphazardly in the absence of any stimulus.

2 When it detects a chemical spreading from its food source, the bacterium moves in a straight line…

3 …until the chemical signal becomes weaker, whereupon the bacterium turns frequently to relocate the chemical.

4 As long as the chemical signal grows stronger, the bacterium swims in a straight path.

5 A combination of moving in a straight line when the chemical signal strengthens and turning when it weakens eventually leads the bacterium to food.

Bacterium

Chemical signal

Direction of water movement

Food

Figure 35.2 Orientation Behavior of Bacteria
Bacteria can move toward food by combining very simple behaviors.

same flagella in the opposite direction. In practical terms, then, bacteria move to food by rotating their flagella in one direction in response to increasing chemical concentrations and in the reverse direction in response to decreasing chemical concentrations. The requirements for coordinating orientation behavior are simple: (1) a stimulus (the chemical signal), (2) an ability to compare the strength of a signal at two different points (increasing or decreasing), and (3) a rule about how to respond to a change in signal strength (rotate flagella clockwise or counterclockwise).

Plants can open and close their flowers in response to temperature

If you watch flowers closely, you will notice that many of them open and close according to the time of day or the weather. These movements generally cause flowers to open when they are most likely to be visited by the pollinating animals that transfer the pollen between plants.

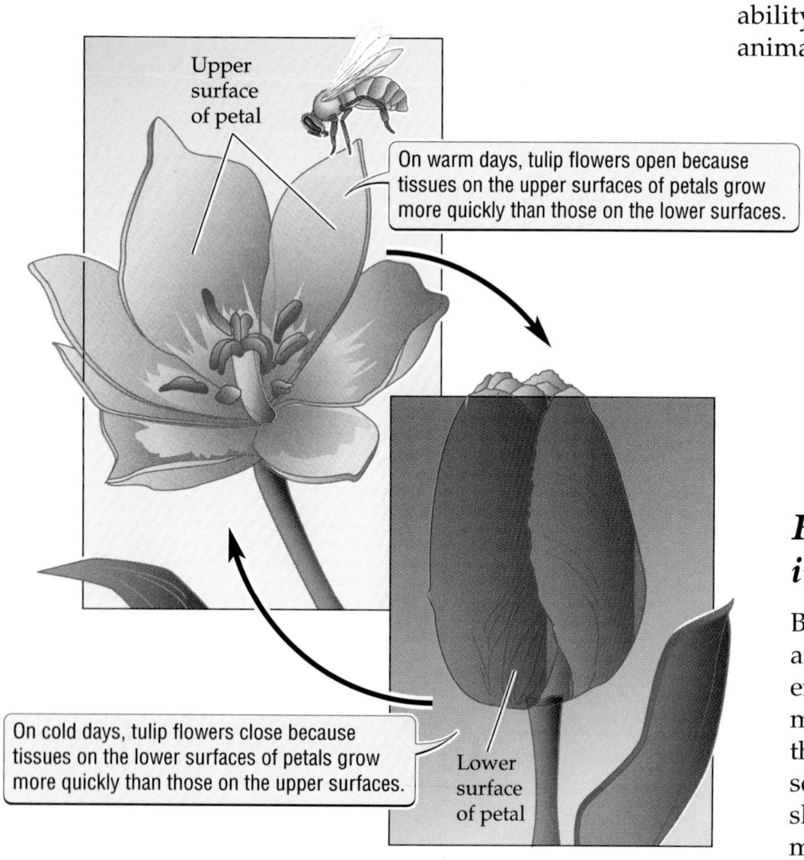

On warm days, tulip flowers open because tissues on the upper surfaces of petals grow more quickly than those on the lower surfaces.

On cold days, tulip flowers close because tissues on the lower surfaces of petals grow more quickly than those on the upper surfaces.

Upper surface of petal

Lower surface of petal

Figure 35.3 How Tulips Open and Close Their Flowers
The growth rates of tissues on the upper and lower surfaces of tulip petals respond differently to temperature, causing the flower to open in warm weather and close in cool weather.

Tulip flowers open in warm temperatures when insects are active and close when temperatures cool and insects stop visiting (Figure 35.3). When the temperature increases, the tissues on the *upper* side of the petals briefly increase their growth rate, while the growth rate of tissues on the lower surface remains constant. The lengthening of the upper petal surface relative to the lower one causes the flower to open. When the temperature drops, the tissues on the *lower* side of the petal briefly increase their growth rate relative to those on the upper side. With falling temperatures, therefore, the flower closes. Change in temperature is the stimulus, and different growth responses of tissues on the upper and lower surfaces of the petal coordinate opening and closing.

Animals rely heavily on behavior

The nervous and muscular tissues of animals suit them particularly well to coordinating sensory information with movement to generate behavior. Nerves can more rapidly relay signals than hormones can, and the collection of nerve cells into a brain gives animals a unique ability to combine sensory information. Muscles give animals an unparalleled ability to move. Not surprisingly, therefore, behavior plays a more prominent role in animals than it does in the members of any other kingdom. For this reason, the rest of this chapter focuses on animal behavior.

> ■ Animals rely more heavily on behavior than do members of other kingdoms. Nonetheless, simple behaviors play an important part in the life of members of all kingdoms.

Fixed and Learned Behaviors in Animals

Behavior patterns either can be fixed features of an animal or can develop through the experiences of the animal. For instance, males of most bird species sing a distinctive song that they use to court females. Males of some bird species know their courtship songs from birth, but others must learn them.

For example, male cowbird chicks that are raised in captivity and thus never hear an adult male cowbird give a flawless

Cowbird

White-crowned
sparrow

rendition of their species' courtship song when mature. In contrast, male white-crowned sparrows reared in captivity fail miserably in their attempts to sing a song attractive to females. To learn their courtship song, white-crowned sparrow males must hear other males of their species sing.

Animals display fixed behaviors in response to simple stimuli

In **fixed behaviors**, stimulus A leads predictably to response B. Some examples of such behaviors are familiar. For instance, newborn babies display the rooting reflex and most first-time mothers display the milk-letdown reflex without having learned how. An abandoned barn kitten that we rescue somehow knows to cover its droppings when first introduced to a litter box. Butterfly caterpillars know which food plants to eat and which to avoid from the moment they hatch.

Fixed behaviors allow organisms to behave appropriately when they have no chance to learn by experience. For example, a baby that had to learn to nurse might not learn in time to keep from starving. The cowbirds we mentioned cannot learn their courtship song from their fathers because female cowbirds have the odd habit of laying their eggs in the nests of other bird species. As a result, cowbird chicks grow up in the care of adoptive parents of a different species that sing different courtship songs.

Simple stimuli often trigger fixed behaviors. In a famous series of experiments, Nobel prize–winning scientist Niko Tinbergen showed that herring gull chicks aim their begging behavior at a conspicuous red dot on their parents' bills. Oddly colored or shaped models of herring gull heads trigger begging behavior just as effectively as lifelike models do, as long as they feature the red dot (Figure 35.4a).

Similarly, the sight of the red patches on the wings of male red-winged blackbirds stimulates neighboring male red-winged blackbirds to defend their nesting territories vigorously. It is the red of the wing patch rather than the overall appearance of a male red-winged blackbird that stimulates this behavior (Figure 35.4b). A person can cause a frenzy of territorial defense by marching into a marsh full of nesting red-winged blackbirds wearing a red shirt (but don't do this, or you will disrupt the birds' nesting!).

The stimuli that trigger fixed behaviors may do so only when conditions are right. The condition of an individual may affect whether a stimulus triggers a fixed behavior. The rooting reflex of newborn babies is a fixed behavior, but the baby exhibits it

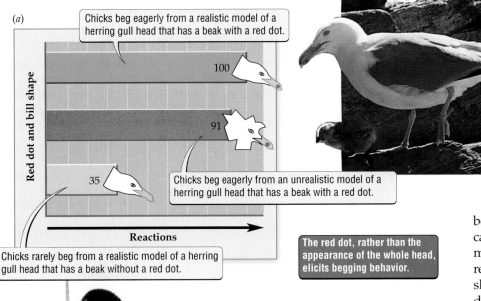

(a)

Red dot and bill shape

Chicks beg eagerly from a realistic model of a herring gull head that has a beak with a red dot.

100

91

35

Chicks beg eagerly from an unrealistic model of a herring gull head that has a beak with a red dot.

Reactions

Chicks rarely beg from a realistic model of a herring gull head that has a beak without a red dot.

The red dot, rather than the appearance of the whole head, elicits begging behavior.

(b)

The red patches on the wings of male red-winged blackbirds play a key role in territorial behavior.

Figure 35.4 Simple Stimuli Trigger Fixed Behaviors
(a) A herring gull chick will beg as eagerly at an unrealistic head model that bears the critical red dot (the trigger for begging) as they will at the real thing. (b) Red-winged blackbirds flash red wing patches at each other during territorial displays.

only when hungry. Similarly, the response of an organism to a stimulus may depend on the timing. During the breeding season, male red-winged blackbirds respond aggressively to red, but outside of the breeding season they do not. The greater level of the sex hormone testosterone in the male birds during breeding season causes the males to respond to the red wing patches.

Learned behaviors provide animals with flexible responses

In **learned behaviors**, the response to a stimulus depends on past experiences. For instance, we can train a dog to sit in response to the command "sit" or to a whistle. A dog trained to the verbal command sits in response to "sit"; a dog trained to the whistle also sits, but in response to a different stimulus. The dogs learn to respond appropriately to their different environments.

Much of our own behavior is learned rather than fixed. Mothers learn to release milk in response to a cry rather than to nursing, and students learn to answer questions properly on biology exams in response to information gained from lectures and textbooks. Most animals modify their behavior in light of their previous experiences.

The feeding behavior of rats illustrates the advantages of learned behavior. One aspect of the biology of rats that has made them so successful is that they thrive on the ever changing variety of stuff that humans discard. However, we provide rats with many potentially poisonous food items. Because of the diversity of their diet, rats cannot have a set of fixed rules about what to eat and what to avoid. Instead rats learn to avoid poisonous foods.

When they encounter a food for the first time, rats sample only an amount small enough not to kill them if the food proves toxic. If the food sample sickens them, they avoid that food in the future. If the food causes no harm, they add it to their learned list of good things to eat. By learning what to avoid based on their experiences, rats can cope with new and unexpected kinds of food, something they could not do with fixed rules about what to avoid and what to eat.

> ■ Animals display fixed behavior in response to a particular stimulus with no previous experience. Fixed behaviors allow animals to function when there is no opportunity to learn from other members of the species or when the risks associated with making a mistake are too great. Learned behaviors allow organisms to incorporate experience into their responses, and to deal with unpredictable situations.

The Evolution of Animal Behavior

Although behaviors may lack the concreteness of structural traits and although learned behaviors, in particular, depend strongly on environmental factors, behaviors have a genetic basis. For these reasons natural selection can act on behaviors just as it does on physical features.

Genes control fixed behaviors

We can easily imagine the genetic control of fixed behavior. One of the clearest examples of the direct genetic control of fixed behavior involves the nest-cleaning behavior of honeybees. Honeybees sometimes fall victim to a contagious bacterial disease that kills honeybee larvae (immatures) in the hive. The hives of some genetic varieties, or genotypes (see Chapter 12), of honeybees rarely suffer from the disease. These "resistant" genotypes become infected but reduce the spread of the bacteria by quickly removing infected larvae from the hive. Nest cleaning involves two behaviors, each controlled by a different gene: (1) cutting open chambers containing larvae killed by the bacterium, and (2) removing the dead bodies from the hive to prevent spread of the disease. "Susceptible" genotypes of honeybees neither cut open cells nor dispose of infected larvae. Genetic crosses such as those described in Chapter 12 reveal that the nest-cleaning behavior of honeybees is under simple genetic control (Figure 35.5).

More typically, studies indicate that genes influence behavior, but that interactions between genetics and environment ultimately determine the behavioral patterns shown by the individual. Careful studies of identical and fraternal twins in humans reveal that **schizophrenia**—a mental illness in which a person hallucinates, feels persecuted, and behaves strangely—has a strong genetic component.

Identical twins arise from a single fertilized egg that splits into two genetically identical individuals. Fraternal twins result when two sperm simultaneously fertilize two different eggs, leading to twins that are no more similar genetically than are typical brothers and sisters. If genes control schizophrenia, then we would expect both genetically identical twins to be schizophrenic or both to be healthy. Pairs of fraternal twins should less frequently both be schizophrenic because they differ genetically. These are indeed the results we find (Figure 35.6).

At the same time, however, if one member of a pair of identical twins has schizophrenia, the other may be completely healthy even though he or she carries all the same genes. This observation indicates that environmental influences determine whether a person with the genes for schizophrenia actually develops the disease.

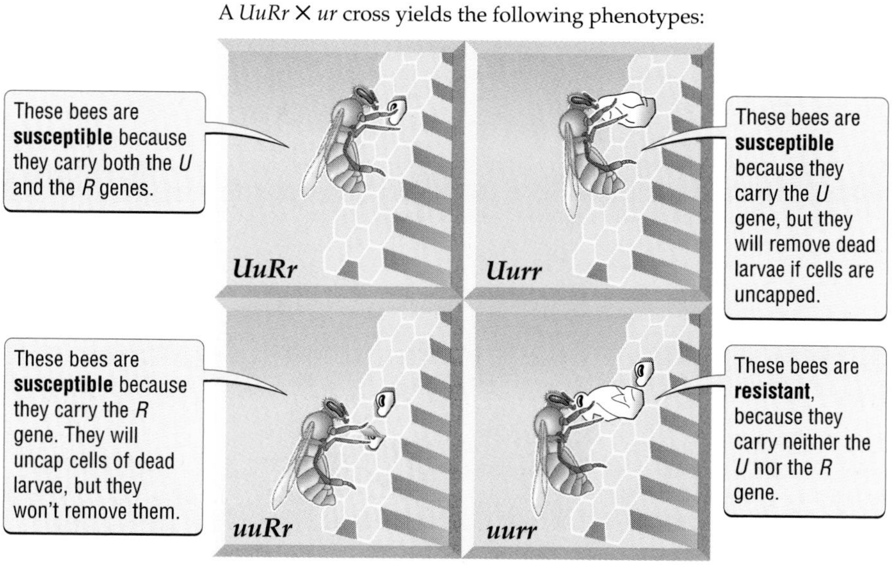

A *UuRr* × *ur* cross yields the following phenotypes:

These bees are **susceptible** because they carry both the *U* and the *R* genes.

UuRr

These bees are **susceptible** because they carry the *U* gene, but they will remove dead larvae if cells are uncapped.

Uurr

These bees are **susceptible** because they carry the *R* gene. They will uncap cells of dead larvae, but they won't remove them.

uuRr

These bees are **resistant**, because they carry neither the *U* nor the *R* gene.

uurr

Figure 35.5 Genes Control the Nest-Cleaning Behavior of Honeybees
This figure shows the results of a cross between a *UuRr* female bee and a *ur* male bee (males are haploid, hence the two genes instead of four). Any worker bee that carries either the *U* gene or the *R* gene will not work to keep the hive free of disease: Bees with the *U* gene do not uncap cells containing dead young (larvae), while bees with the *R* gene do not remove the dead young.

Genes also influence learned behaviors

Learned behaviors, by definition, have a strong environmental influence: What animals learn depends on what they experience during their lifetimes. Nonetheless, differences in what, how much, and when different species can learn reflects genetic differences in learning capacity between species.

For example, rats learn to negotiate mazes in the laboratory at different rates. By mating fast learners with other fast learners, researchers can produce offspring that also learn to negotiate the maze quickly. Mating slow learners with other slow learners produces rats that have a difficult time learning to find their way through the maze.

Species of birds, such as the Clark's nutcracker, that excel at remembering where they have stored food have brains that differ structurally, and presumably genetically, from those of closely related species that have average talent at finding stored food. Researchers have even identified genes that seem to control the ability to learn the songs that are characteristic of a bird species.

Clark's nutcracker burying a seed

■ Both fixed and learned behaviors have a genetic component and can therefore evolve.

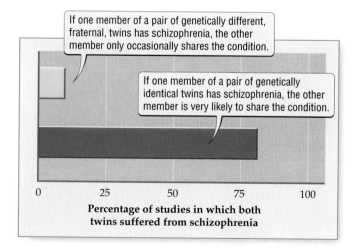

If one member of a pair of genetically different, fraternal, twins has schizophrenia, the other member only occasionally shares the condition.

If one member of a pair of genetically identical twins has schizophrenia, the other member is very likely to share the condition.

0   25   50   75   100
Percentage of studies in which both twins suffered from schizophrenia

Figure 35.6 The Genetics of Schizophrenia
Schizophrenia is a genetically based mental disorder in humans. However, environment clearly affects the expression of the genes causing schizophrenia, because even among genetically identical twins there are many cases in which only one twin develops the condition.

Communication Allows Behavioral Interactions between Animals

Communication behaviors allow one individual to exchange information with others, thereby making it possible for the animal to coordinate its activities with those of other individuals. By communicating with one another, groups of animals can do things that no individual could do on its own, such as fend off larger predators or build the pyramids.

Communication is the production of signals by one animal that stimulate a response in another

The communication behavior of animals varies widely in its complexity. At the simple end of the spectrum, the release of a chemical, called a **pheromone**, by one individual informs others of its identity and location. Probably the most widespread means of communication, pheromones occur not only in animals, but in bacteria, fungi, protists, and plants.

At the complex end of the spectrum lie human **languages**, consisting of thousands of words that represent

Figure 35.7 Honeybees Communicate by Dancing in Their Hive
Bees use a dance language in their hive to communicate complex information about the position of distant sources of food to other worker bees.

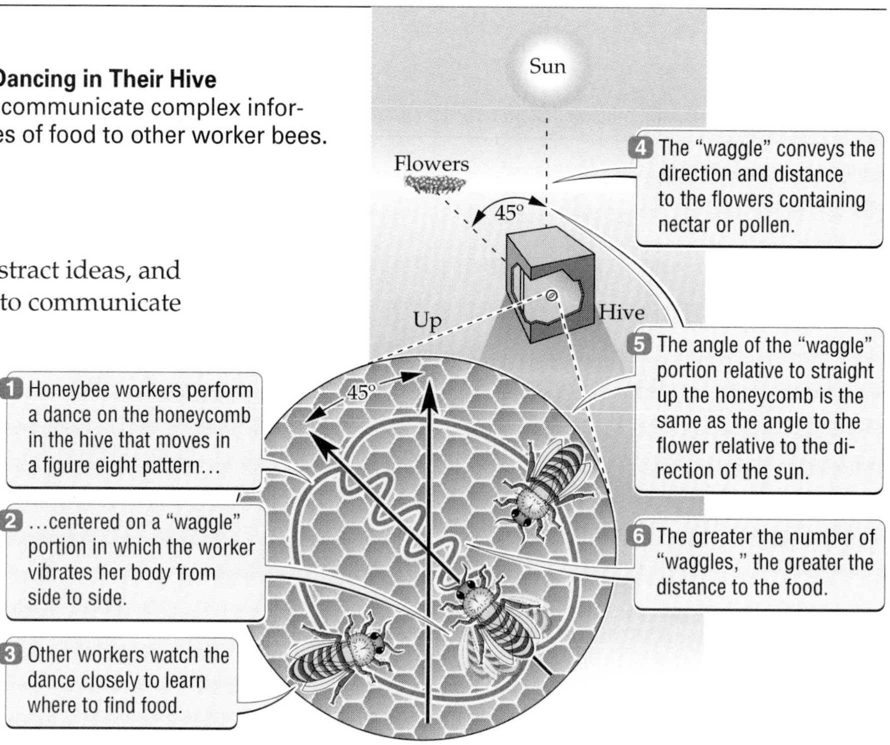

Sun

Flowers

45°

Up   Hive

4 The "waggle" conveys the direction and distance to the flowers containing nectar or pollen.

everything from objects to actions to abstract ideas, and the dance language used by honeybees to communicate the location of food to other members of the hive (Figure 35.7). The means of communication include just about any type of signal that other animals can sense: sound, visual signals, odors, electrical pulses, touch, and tastes.

1 Honeybee workers perform a dance on the honeycomb in the hive that moves in a figure eight pattern...

2 ...centered on a "waggle" portion in which the worker vibrates her body from side to side.

3 Other workers watch the dance closely to learn where to find food.

45°

5 The angle of the "waggle" portion relative to straight up the honeycomb is the same as the angle to the flower relative to the direction of the sun.

6 The greater the number of "waggles," the greater the distance to the food.

Pattern of waggle dance

Why do animals communicate?

Animals most often communicate to identify themselves, to avoid conflict, and to coordinate activities. Self-identification is probably the most common function of communication. An animal uses a variety of signals to inform other members of the species of its sex, its physical condition, and its location. When a dog marks a fire hydrant or a post with its scent, the scent communicates the dog's species, sex, breeding condition, health, and status to other dogs. The mating songs and behaviors of male birds, frogs, and crickets tell the females they court about their species, their location, and their potential quality as mates (Figure 35.8a).

(a)

(b)

Figure 35.8 Why Animals Communicate   (c)
(a) A singing male frog advertises to listening females its species, its location, and its quality as a mate. (b) A male mountain goat communicates its quality as a mate through a display that reduces the risk of injury from its short, sharp horns. (c) Humans communicate to coordinate their activities.

A second important function of communication in animals is to avoid potentially harmful conflicts. Physical conflict over food or mates can lead to serious injury of both the winner and the loser. To reduce the risk of injury in such encounters, many species communicate their fighting ability through ritual displays. When male red-winged blackbirds flash red wing patches at each other, for example, they signal their quality as fighters to other males. This allows the poorer fighter to back down without engaging in a potentially dangerous fight. By communicating, animals can resolve contests over territory without actually coming to blows.

Sheep, antelope, and their relatives bear horns that vary greatly in shape—from stout, straight weapons capable of doing great damage to impressive but ineffective weapons. The animals with the most lethal horns usually engage in displays that communicate their suitability as mates rather than in actual fight, whereas the animals with less lethal horns more often engage in physical combat (Figure 35.8b).

Animals that live in groups communicate to coordinate their behaviors to accomplish a shared task. Humans do this when they work as a team (Figure 35.8c). Wolf packs and lion prides communicate when they hunt for food. The dance language of bees (see Figure 35.7) allows the members of the hive to coordinate their foraging activity.

Language may be a uniquely human trait

We rely heavily on spoken, written, or sign languages for communication. Much of our human identity depends on our ability to express complex and abstract ideas through language. There is no strong evidence of language among other animals.

Many birds and mammals can produce a variety of sounds, each of which conveys a particular message, but these sounds are not assembled into ideas. Research on the ability of chimpanzees to communicate indicates that they can string together symbols provided by humans to express abstract ideas, although there is no indication that chimpanzees have such sophisticated communication in nature.

> ■ Communication allows animals to coordinate their activities with those of other animals.

Social Behavior in Animals

Many animals besides humans live in closely interacting groups. The behavioral interactions among members of a group offer them advantages that are not available to solitary organisms. In this section we look at some of the advantages of group living, and we consider some of the reasons why individuals might make sacrifices for the sake of the group as a whole.

Group living offers many benefits

The many benefits of group living can more than compensate for the increased competition for resources that results. Groups can find some foods better than individuals can. Groups can stir up prey more effectively than individuals can, or unsuccessful hunters can watch while others in their group find food. By working together, members of a group can use foods that are not available to individuals. The hunting of large mammals by groups of cooperating wolves or humans are excellent examples of such cooperative behavior (Figure 35.9a).

By living in groups rather than singly, animals can better avoid predators in two ways: (1) Groups of animals contain more eyes, ears, and noses, making early detection of an approaching predator more likely, and (2) the individuals on the edge of a group are more likely to be attacked than those on the inside, offering the lucky ones in the middle some protection. For example, the individuals on the inside of a herd of gazelles benefit because a lion is most likely to kill an individual at the edge of the group. Moreover, all group members benefit because at least one is likely to warn of approaching lions before the lions are close enough to attack (see Figure 35.9b).

In some cases, living in a group gives animals access to a scarce resource. Nonbreeding members of groups of scrub jays, social birds that live in Florida and in the deserts of the southwestern United States, appear to benefit because belonging to a group puts them on a sort of "waiting list" to inherit the territory from its current owner. The chance that a solitary scrub jay will find and hold control over a high-quality breeding territory is so small that it pays to put off breeding for several years while on this waiting list (see Figure 35.9c).

Individuals living in groups often act in ways that benefit other group members more than themselves

Animals that live in groups often do things that help other members of their group survive or reproduce, while decreasing their own chances of doing so. This **altruistic** behavior seemed for many years to contradict Darwin's idea that only traits that improve individual reproductive success spread through the population.

The scrub jays mentioned in the previous section provide a good example of altruistic behavior: Nonbreeding

Figure 35.9 The Benefits of Group Life (*a*) As a group, wolves can hunt larger prey than they can as individuals. (*b*) A pair of Thomson's gazelles watches for predators. (*c*) Family groups of scrub jays work together to feed young, defend nests, and hold onto valuable breeding territories.

(*a*)

(*b*)

(*c*)

members of a group help the breeding pair raise their young. This behavior improves the success of the breeding pair, while presumably placing the nonbreeding individuals at risk as they forage for food and defend the territory against intruders.

In general, social groups consist of closely related individuals. For example, a pride of lions centers around a closely related group of females. Similarly, among scrub jays, the helpers most often turn out to be offspring of the breeding pair. In these cases, the individuals that benefit from an altruistic act carry many of the same genes as do the altruist. In this way, altruistic behavior often helps the spread of many of the same genes that the altruist carries.

Do social insects represent the ultimate in group living?

Social insects such as ants, bees, and termites include some of the most successful species on Earth. Although they evolved independently of one another, these three groups share some remarkable traits.

Each species lives in large colonies run by individuals belonging to distinct groups that serve distinct functions within the colony. The closely related workers work together to build complex nests (Figure 35.10), forage widely for food, and defend the colony effectively against predators in ways that individuals never could. The queen spends her life producing massive numbers of eggs while being tended carefully by the workers. The workers in a hive are offspring of the queen and therefore are closely related. The workers represent the extreme in altruism in that they give up reproduction to help their mother raise massive numbers of young.

■ The social behavior of animals living in groups allows them to feed on otherwise unavailable food types, to avoid predators more effectively, and to defend scarce resources. Social behavior often involves altruism, in which an individual reduces its chances of reproducing by aiding other members of its group.

Figure 35.10 Cooperation Creates Termite Mounds
The giant termite mound on which this cheetah sits was built through the cooperative effort of millions of closely related and reproductively inactive termite workers. Such workers make up most of the population of each colony.

on plants only when there is a good chance the plant is a pipe vine.

The next time you see a butterfly, remember some of the remarkable and sophisticated behaviors that these animals use to find quickly what they seek.

> ■ The ability of pipe vine swallowtails to change their searching behavior depending on the relative abundance of broad- and narrow-leaved plants reduces the time needed to find plants on which to lay eggs.

Highlight

Why Pipe Vine Swallowtails Learn

We don't usually think of insects as intelligent animals. The switching of the pipe vine swallowtail's preference for broad- or narrow-leaved pipe vines, however, is a remarkably complex mechanism that allows female butterflies to respond rapidly to changes in food plant availability. These butterflies can search effectively for only one leaf shape at a time. Just as we have much more success in looking for something once we know what the object of our search looks like, butterflies that search exclusively for either broad- or narrow-leaved plants find pipe vines faster than do individuals looking for both types of plants at once.

The relative abundance of broad- and narrow-leaved pipe vines changes from place to place within the forest. A female butterfly looking for narrow-leaved plants where broad-leaved plants predominate finds pipe vines on which to lay eggs at a much lower rate than a female searching for the predominant pipe vine species. The ability of a female to change the leaf shape she seeks to match the changing forest floor over which she flies allows her to lay more eggs in less time spent searching.

The very fact that female pipe vine swallowtails focus on leaf shape in their search reflects a second kind of learning. The true measure of the quality of a plant as a place to lay eggs depends on its species. Females can tell for sure whether a plant is a pipe vine only by landing on and tasting the leaves. Because landing takes time, females learn to associate leaf shape with their food plant species, and thus can identify plant species without landing. In this way, females take the time to land

Summary

Breast-Feeding in Humans as an Example of Behavior

- ■ Sensory stimulation initiates responses, including nerve signals and the release of hormones, that stimulate muscle contractions, which allow breast-feeding to occur.

- ■ The behavioral interaction between infant and mother coordinates the responses of both so that the child tries to feed only when milk is available and the mother releases milk only when her hungry baby is nearby.

Who Behaves?

- ■ Bacteria exhibit orientation behavior that allows them to move toward food or other elements in their environment.

- ■ Plants can respond to stimuli from their environment, such as by opening and closing flowers in response to temperature.

- ■ The nervous and muscular tissues of animals allow them to rely heavily on behavior.

Fixed and Learned Behaviors in Animals

- ■ Animals display fixed behaviors in response to simple stimuli, which can enhance an organism's survival when there is no time to learn how to respond.

- ■ Learned behaviors, in which the response depends on past experience, provide animals with flexible responses.

The Evolution of Animal Behavior

- ■ The ability of genes to control both fixed and learned behaviors allows behaviors to evolve.

- ■ Many behaviors, both fixed and learned, can be influenced by the environment.

■ Learning is one example of how environmental influences can affect behaviors under genetic control, but the capacity to learn may also be under genetic control.

Communication Allows Behavioral Interactions between Animals

■ Communication is the production of signals by one animal to stimulate a response in another.

■ Animals communicate to identify themselves, to avoid conflict, and to coordinate activities.

■ Language may be a uniquely human trait.

Social Behavior in Animals

■ Living in groups can enable animals to avoid predation and to gain access to resources, including food and breeding territories, more effectively.

■ Altruistic behavior can evolve within groups of closely related animals.

Highlight: Why Pipe Vine Swallowtails Learn

■ The ability of pipe vine swallowtails to change their searching behavior depending on the relative abundance of broad- and narrow-leaved plants reduces the time needed to find sites for eggs.

Key Terms

altruism p. 537	milk-letdown reflex p. 531
behavior p. 530	orientation behavior p. 531
fixed behavior p. 533	pheromone p. 535
language p. 535	rooting reflex p. 530
learned behavior p. 534	schizophrenia p. 534

Chapter Review

Self-Quiz

1. The rooting reflex
   a. occurs in response to milk production.
   b. is an example of a fixed behavior.
   c. occurs in response to a visual stimulus.
   d. is an example of a learned behavior.

2. Behavior
   a. occurs only in animals.
   b. always involves communication.
   c. allows organisms to evolve.
   d. allows organisms to respond quickly to changes in their environment.

3. Learning
   a. occurs only in humans.
   b. overrides all genetic control of behavior.
   c. depends on an organism's past experience.
   d. requires communication.

4. Which of the following is not an example of altruistic behavior?
   a. A butterfly lays an egg.
   b. A scrub jay chases a snake away from another jay's nest.
   c. A sterile worker ant defends the nursery of its anthill from invading ants.
   d. A lioness regurgitates food for her sister's cubs.

5. Animals that live in groups
   a. have no fixed behaviors.
   b. usually rely on language for communication.
   c. can use behavior to compensate for increased competition.
   d. have no genetically controlled behaviors.

Review Questions

1. What tissues and organs allow animals to use behavior so extensively, and why?

2. Explain how learning is influenced by both genetic and environmental factors.

3. What sorts of fixed behaviors might be advantageous to animals that live in a group, and why?

4. Describe the three ways in which communication can benefit organisms, including examples not given in this chapter.

Protesters Pick on New Breed Restrictions

CINCINNATI, OH. In response to several recent attacks by dogs, the city council today passed a new regulation banning the ownership of pit bulls and Rottweilers within the city limits.

In the past 18 months, at least six people have been seriously injured in attacks by dogs within the city limits, and Joshua Corona, a 4-year-old boy, was mauled to death by a neighbor's Rottweiler. Four of the nonlethal injuries were caused by pit bulls or pit bull crosses, leading to the concern over this dog breed.

Outside city hall, about 20 dog enthusiasts gathered to protest the new regulation. John Sykes, a Rottweiler breeder and trainer, complained that the regulation unfairly targets the breed rather than the behavior. "Sure, Rottweilers are big, scary-looking dogs. But when raised in a loving environment and trained by a conscientious owner, they can become wonderful family pets. The problem is not the dog breed; it's irresponsible owners who encourage ferocious behavior and don't keep their dogs under control."

Leading her pit bull "Chucky" on a short leash, 20-year-old Melissa Whitsen held a sign that said, "A dog is this woman's best friend." She bought Chucky 3 years ago, when she first moved out on her own. "As a woman, living alone, I wanted both the companionship and security that a dog like this could give me." She credits Chucky with scaring away an intruder just three nights ago.

Across town, Joshua's father, Michael Corona, applauded the council's decision. He explained that his son had been playing in his own backyard when the attack occurred, and he stated, "These dogs have been bred to be guard dogs, which makes them ferocious and strong. They have no business in residential areas where kids and families are trying to lead normal lives."

After the protesters left, the city council defended its decision in a short statement prepared for the press, noting that the regulation was modeled on similar laws passed by cities in neighboring states.

Evaluating "The News"

1. Consider the wild ancestors of domestic dogs. Do you think the behaviors we now consider fierce in domestic dogs would ever have developed without human intervention? Why or why not?

2. Explain how you think dog attacks of the sort described in this article could most effectively be reduced. In other words, if you lived in this city, would you have supported the council's decision or another alternative, and why?

3. How do you think these sorts of regulations can (or should) deal with the complicating factor of mixed-breed dogs?

chapter *36*

Reproduction

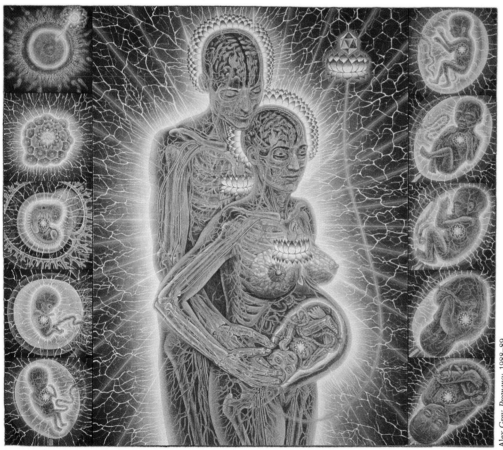

Alex Grey, *Pregnancy*, 1988–89.

Gender–Bending Plants

We generally consider maleness or femaleness an easily identified characteristic of an individual. Most humans live their lives as either a sperm-producing male or an egg-producing female. For many organisms, however, human notions of gender have little relevance.

Jack-in-the-pulpits are plants whose name comes from the unique shape of their flowers: A modified leaf shaped like a church pulpit surrounds a thick standing object, which is actually a spike of densely crowded, tiny flowers. These plants grow in moist forests throughout eastern North America. A

Reproduction, or the making of offspring, is the central event in an organism's life.

close look at jack-in-the-pulpits reveals that some are male, producing only powdery, yellow pollen, and the rest are females that eventually produce a cluster of bright scarlet berries.

Among jack-in-the-pulpits, however, all is not as simple as it seems. This year's male may have been last year's female, or vice versa. The gender of a jack-in-the-pulpit depends on the amount of energy the plant manages to store in its roots rather than on the presence or absence of genes on the Y chromosome, as it does in humans. Jack-in-the-pulpits switch from being male when small to being female when large. Individuals that capture a lot of energy this year stand a good chance of producing female flowers next year.

Females produce energetically expensive seeds or eggs. Large size reflects more resources available to make this heavy investment in offspring. Jack-in-the-pulpits that store away only a little energy typically produce male flowers in the next year. Because pollen production requires less of an investment, even small plants with small reserves can produce at least some pollen.

A Change in Size Means a Change in Gender
Jack-in-the-pulpits switch from male to female or from female to male depending on their size.

Key Concepts

1. Often in reproduction the genes of two individuals are combined.

2. Prokaryotes reproduce asexually, and they only sporadically combine genes from two individuals, by a variety of means.

3. Eukaryotes can reproduce both sexually and asexually. In sexual reproduction, the male and the female each contribute half of the genes to each offspring.

4. Both animals and plants improve their reproductive success by moving close to a potential mate before releasing eggs or sperm and by selecting high-quality mates.

5. Plants and animals can select mates carefully to increase the likelihood that the genes contributed by their mates will produce high-quality offspring.

6. Both sexual and asexual reproduction offer important advantages.

*I*f life has a purpose, it is to make more of itself. Thus **reproduction**, the making of offspring, characterizes life. Like most other organisms, humans **reproduce sexually**, meaning that reproduction involves a mixing of the genes from two parents. The genetic rearrangements that take place during sexual reproduction maintain the genetic variation needed for evolution by natural selection. Many organisms, however, **reproduce asexually**; that is, a single parent produces offspring that are genetically identical copies of itself.

Is sexual reproduction better than asexual reproduction? The answer seems to depend on the environment in which an organism lives (Figure 36.1). Sexual reproduction leads to offspring that differ genetically from their parents. As a result, sexual reproduction may work best when offspring must survive in a habitat that differs from the one in which their parents had succeeded. Thus, the species that stand to benefit most from sexual reproduction (1) have young that move away from their parents to live in slightly different habitats or (2) have long-lived individuals that inhabit an environment that changes from the time of their parents to the time when their own offspring develop. In contrast, the mixing up of genes that characterizes sexual reproduction may be a disadvantage for species in which the offspring develop under conditions identical to the one in which their parents thrived.

Because an organism's evolutionary success hinges on its reproductive success, it devotes much of its time and energy to ensuring that it produces as many successful offspring as possible. In a very true sense, all the functions and structures described so far in Unit 5 allow the organism to accumulate the resources it needs to reproduce.

The strawberry plants produced asexually on runners develop close to the plants that produced them.

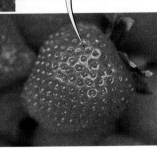

The sexually produced seeds that dot the outside of a strawberry may be carried by animals to distant habitats and may not sprout for many years.

Figure 36.1 Sexual and Asexual Reproduction in Strawberries Asexually produced offspring usually end up living close to their mother. Sexually produced offspring often move away from their mother.

We begin this overview of reproduction by comparing and contrasting reproduction and its relationship to genetic recombination in bacteria and in eukaryotic organisms. For the rest of the chapter, we focus on reproduction in animals and flowering plants, the two most familiar groups of eukaryotes. We describe reproduction in animals and plants to introduce the reproductive structures, to give an idea of the frequency of sexual compared to asexual reproduction in these groups, and to show the different evolutionary directions in which animals and plants have taken the basic pattern of eukaryotic sexual reproduction.

We then look at how animals and plants deal with two important problems associated with eukaryotic sexual reproduction: (1) How do the sperm produced by males manage to reach the eggs produced by females? (2) How do individuals select high-quality mates? Finally, we consider why sexual reproduction evolved.

How Bacteria and Eukaryotes Reproduce and Mix Genes

The relationship between reproduction and the mixing of genes in prokaryotes such as bacteria differs greatly from that relationship in eukaryotes such as animals and plants. Three outstanding features distinguish the mixing of the genes of two prokaryotic individuals from that of eukaryotes: (1) Whereas in eukaryotes genes mix during reproduction, in prokaryotes the two processes occur separately. (2) Whereas two eukaryotic parents contribute genes equally to their offspring, mating bacteria contribute genes unequally to their offspring. (3) Whereas the combination of genetic material from two eukaryotic parents is highly predictable, the transfer of genes in prokaryotes is often much more variable.

Prokaryotes transfer genes in one direction in a process separate from reproduction

Prokaryotes such as bacteria reproduce when one cell divides asexually into two independent cells (Figure 36.2*a*). Reproduction in prokaryotes does not involve the combining of genes from two individuals.

(*a*)

Figure 36.2 Reproduction and Gene Transfer in Prokaryotes
(*a*) Bacteria reproduce by dividing asexually into two independent cells. (*b*) In transformation, one of several gene transfer mechanisms in bacteria, DNA fragments are transferred in one direction only: from a donor individual into the genome of a recipient.
In some bacteria that undergo transformation, only single-stranded DNA can enter the recipient, as shown here. In other transforming bacteria, double-stranded DNA can enter the recipient.

(*b*)

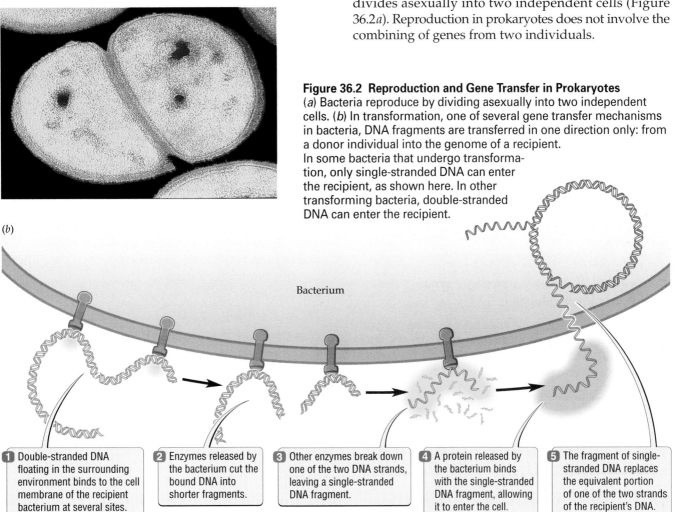

Bacterium

1 Double-stranded DNA floating in the surrounding environment binds to the cell membrane of the recipient bacterium at several sites.

2 Enzymes released by the bacterium cut the bound DNA into shorter fragments.

3 Other enzymes break down one of the two DNA strands, leaving a single-stranded DNA fragment.

4 A protein released by the bacterium binds with the single-stranded DNA fragment, allowing it to enter the cell.

5 The fragment of single-stranded DNA replaces the equivalent portion of one of the two strands of the recipient's DNA.

Prokaryotes transfer genes in two general ways. In **transformation**, bacterial individuals, called **recipients**, take up fragments of double-stranded DNA released into the environment by other bacteria, called **donors** (Figure 36.2b). Donor bacteria release their DNA into the environment when they die from natural causes or in response to chemicals produced by other bacteria. Transformation almost certainly evolved as a way of incorporating new genes into bacterial DNA.

Only bacterial cells that have produced the proper binding proteins in their plasma membranes can take up the DNA released into the environment. Once inside the recipient, a single strand of the donor DNA fragment becomes incorporated into the recipient DNA. Thus transformation produces a single, genetically unique bacterium from two parent bacteria, clearly showing the distinction between gene transfer and reproduction in prokaryotes. The relative contribution of the two parent bacteria to the transformed individual depends on how much donor DNA is incorporated into the recipient DNA.

A second means of gene transfer involves plasmids (small, circular pieces of DNA) or viruses inside a host bacterium. While being distributed to other bacteria, plasmids and viruses may accidentally transfer some of the host's DNA along with their own. Usually when viruses and plasmids transfer to a new host bacterium, only viral and plasmid DNA enters the new host. Occasionally, however, some bacterial DNA is transferred in the process. Although almost certainly not a mechanism that evolved to exchange bacterial DNA, these accidental transfers of bacterial DNA by viruses and plasmids can play an important role in the transfer of genetic information between bacteria. Biotechnologists use viruses and plasmids as tools to insert potentially useful genes into bacteria, or even into animals and plants.

Both eukaryotic parents contribute genes during sexual reproduction

In eukaryotes, the combining of genes from two parents to form a genetically unique offspring always takes place during sexual reproduction. Eukaryotic species that reproduce sexually produce individuals with distinct genders: males and females. **Males** produce specialized sex cells called sperm. **Sperm** can move, and they contain little beyond chromosomes and the cellular machinery needed for them to move. **Females** produce a second type of sex cell, called eggs. **Eggs** are typically much larger than sperm and generally do not move. Eukaryotic organisms produce sperm and eggs through meiosis (see Chapter 10), in which diploid cells

(containing two copies of every chromosome) are converted into haploid cells (containing only one member of each pair of chromosomes). Thus sperm and eggs are haploid, containing only half of the genetic information carried by the diploid individual that produced them.

Haploid sperm and eggs fuse to form a diploid cell during **fertilization**. Eggs of many eukaryotes release chemical pheromones that allow sperm to find them. Flowering plants seem to physically guide the sperm to the egg, as we will describe later in the chapter. Chemical interactions between the sperm and the egg allow the sperm to enter the egg. Once inside the egg, the sperm's nucleus separates from the rest of the sperm and fuses with the egg's nucleus to form a genetically unique, diploid, fertilized egg. One member of each pair of chromosomes in the fertilized egg comes from the male and one from the female. The organelles and other cellular machinery of the fertilized egg come almost entirely from the female, however.

■ Prokaryotes transfer genes in one direction only: from a donor to a recipient. In prokaryotes genes are transferred in a process separate from reproduction. In eukaryotes the mixing of genes always occurs during sexual reproduction. Sexual reproduction involves a combining of genetic material, of which half comes from the female and half from the male.

Animal Reproduction

Having shown some dramatic differences between prokaryotic and eukaryotic reproduction, we now turn to the different ways of reproducing that have evolved within the eukaryotes. First, to set the stage for a more general discussion of asexual and sexual reproduction and of male and female genders in animals, we briefly describe human reproduction.

Human reproduction illustrates many basic features of sexual reproduction in animals

Sexual reproduction in humans provides a good picture of sexual reproduction in animals. As we all know, humans come in two genders: male and female. Males produce haploid sperm (containing 23 chromosomes) in specialized organs called **testes** (singular: testis). Human sperm can swim through fluid by means of a flagellum (see Chapter 27). Females produce haploid eggs (containing 23 chromosomes) by meiosis in their **ovaries**.

During intercourse, the male **penis** directs almost 300 million tiny sperm into the **vagina** of the female (Figure

36.4). The female usually releases one egg from her ovaries into the **oviduct** each month. The egg moves down the oviduct, and during intercourse massive numbers of sperm swim up the oviduct in response to a pheromone released by the ovary. The female aids the progress of the swimming sperm by a combination of muscular contractions of the **uterus** and the release of chemicals that stimulate sperm to swim. However, only a few hundred sperm reach the egg at the site of fertilization in the oviduct. Only one of these sperm successfully fuses with the egg because changes in the egg shortly after fusion prevent entry by other sperm. Fusion of the nucleus of the egg with that of the sperm produces a diploid, fertilized egg containing 23 pairs of chromosomes that come equally from both parents.

Humans, like most organisms, only fertilize eggs following a sequence of activities that increase the chances that a sperm will encounter an egg and result in a fertilized egg that develops into a new individual. During courtship, we spend an amazing amount of energy, time, and effort on finding and impressing a potential mate (see Chapter 35). Careful mate selection is important to avoid some of the hazards involved in sexual reproduction (see the Box on page 549), and it is also important in ensuring the genetic quality of the mate. Once we manage to win someone's heart, we consummate the relationship by having sexual intercourse, during which the male releases sperm directly into the female's body. As a fertilized egg develops into an adult, it receives a great deal of care, first inside then outside its mother's body. This parental care greatly increases the chances that it will survive to become a successful adult (see Chapter 37).

Many animals reproduce asexually

Many animal species reproduce asexually, including some members of all animal phyla. Asexually reproducing animals have lifestyles as different as those of oceangoing jellyfish and desert-dwelling lizards. The offspring that result from asexual reproduction contain exactly the same genes as their parent.

Although some animals rely exclusively on asexual reproduction, most asexually reproducing species switch between sexual and asexual reproduction depending on environmental conditions. For example, aphids, small insects that sometimes infest our greenhouses or garden plants, usually reproduce asexually. The crowds of aphids that we see sucking sap from plants in the summer are all genetically identical copies of the females that originally colonized the plant. Aphids rely on sexual reproduction, however, when environmental conditions drive them from their summer food plant (Figure 36.3).

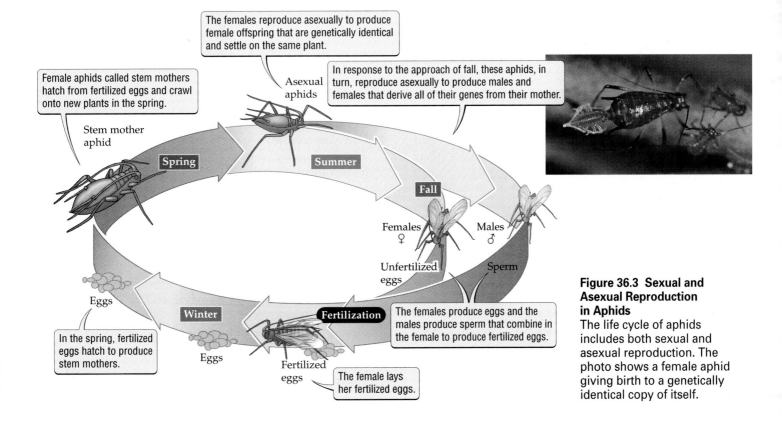

The females reproduce asexually to produce female offspring that are genetically identical and settle on the same plant.

Female aphids called stem mothers hatch from fertilized eggs and crawl onto new plants in the spring.

In response to the approach of fall, these aphids, in turn, reproduce asexually to produce males and females that derive all of their genes from their mother.

Asexual aphids

Stem mother aphid

Spring

Summer

Fall

Females ♀

Males ♂

Unfertilized eggs

Sperm

Eggs

Winter

Fertilization

In the spring, fertilized eggs hatch to produce stem mothers.

Eggs

Fertilized eggs

The females produce eggs and the males produce sperm that combine in the female to produce fertilized eggs.

The female lays her fertilized eggs.

Figure 36.3 Sexual and Asexual Reproduction in Aphids
The life cycle of aphids includes both sexual and asexual reproduction. The photo shows a female aphid giving birth to a genetically identical copy of itself.

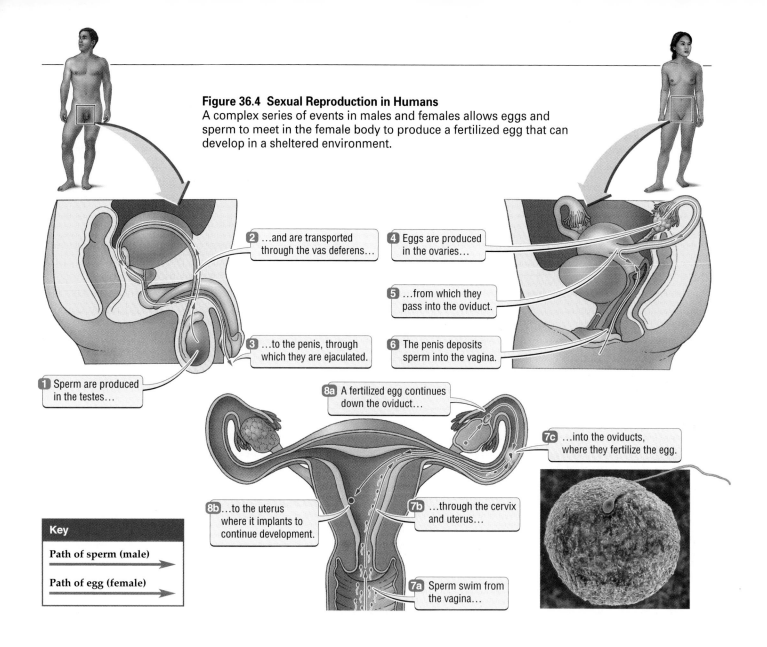

Figure 36.4 Sexual Reproduction in Humans
A complex series of events in males and females allows eggs and sperm to meet in the female body to produce a fertilized egg that can develop in a sheltered environment.

2 …and are transported through the vas deferens…

4 Eggs are produced in the ovaries…

5 …from which they pass into the oviduct.

3 …to the penis, through which they are ejaculated.

6 The penis deposits sperm into the vagina.

1 Sperm are produced in the testes…

8a A fertilized egg continues down the oviduct…

7c …into the oviducts, where they fertilize the egg.

8b …to the uterus where it implants to continue development.

7b …through the cervix and uterus…

Key

Path of sperm (male)

Path of egg (female)

7a Sperm swim from the vagina…

Because humans reproduce sexually, we may find the idea of asexual reproduction, which eliminates the need for two sexes, strange. Keep in mind, however, that our ability to manipulate mammalian reproduction has advanced to the point that scientists have asexually produced Dolly, a sheep cloned from and genetically identical to her mother (see page 561). The day when humans can reproduce asexually with the help of medical technology cannot be far off.

Animals may be male, female, or both

Gender in animals is more variable than our human perspective might lead us to realize. As already noted, humans may be sperm-producing males or egg-producing females. The majority of animal phyla, however, include many species made up of individuals that produce both functional testes and ovaries and are therefore both female and male. We call such individuals **hermaphrodites**.

Hermaphrodites live in a wide variety of habitats and even include members of our own phylum (Chordata). Probably the most familiar hermaphrodite is the common earthworm (Figure 36.5). Mature earthworms simultaneously have functional testes and ovaries, but many other hermaphrodites, including many fish species, change gender depending on their size (as the jack-in-the-pulpit plants described at the beginning of this chapter do).

A common misconception about hermaphrodites is that they fertilize their own eggs. Most hermaphrodites that have functional testes and ovaries at the same time must still mate with another individual. Nonetheless, they have an advantage over animals that are male only or female only in that they can mate with every individual they encounter.

Sexually Transmitted Diseases: The Dark Side of Sex

AIDS, gonorrhea, syphilis, chlamydia, and herpes are some of the best known human sexually transmitted diseases (STDs). As the name implies, these diseases are transmitted during sexual contact between two individuals. Some of the most common, serious infectious disease organisms among humans are STDs: In the United States alone, 12 million people contract an STD each year, half of the population contracts at least one STD by the age of 35, and the annual cost related to STDs (other than AIDS) is over $10 billion.

The success of any STD, from the perspective of the disease organism, lies in its use of sexual contact as a way of spreading from one host to another. The most obvious way to reduce the risk of sexually transmitted disease is to abstain from sex. Avoiding sex, however, prevents an individual from contributing genes to the next generation. Thus, STDs have found the perfect way of spreading, since sexual intercourse is essential to our reproductive success.

Efforts to fight the spread of STDs focus on three approaches: (1) reducing the number of sexual partners that an individual has; (2) promoting safe sex involving the use of condoms and other barriers to eliminate the exchange of bodily fluids between sex partners; and (3) rapid detection and treatment of the disease following infection.

Sexually transmitted diseases may affect over half the United States population during their lifetimes.

Figure 36.5 The Familiar Earthworm Is a Hermaphrodite
Each earthworm produces both eggs and sperm. For this reason, all earthworms of the same species are potential mates. In contrast, in animals that have separate males and females, only members of the opposite gender are potential mates.

■ Among animals, males produce haploid sperm in their testes and females produce haploid eggs in their ovaries. Hermaphrodites produce both functional testes and ovaries. Many animals reproduce asexually instead of, or more commonly in addition to, reproducing sexually.

Plant Reproduction

Plants share with animals most of the basic features of their reproductive biology: (1) A sperm fertilizes an egg to form a new individual containing pairs of chromosomes that come equally from the male and female parents. (2) Asexual reproduction is, if anything, even more common than in animals. (3) An individual plant may be exclusively male, exclusively female, or, as a rule rather than an exception, hermaphroditic.

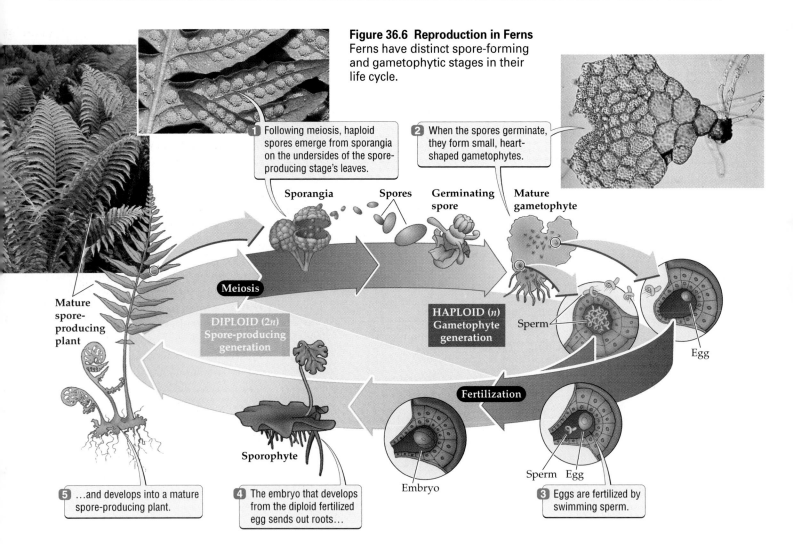

Figure 36.6 Reproduction in Ferns
Ferns have distinct spore-forming and gametophytic stages in their life cycle.

1 Following meiosis, haploid spores emerge from sporangia on the undersides of the spore-producing stage's leaves.

2 When the spores germinate, they form small, heart-shaped gametophytes.

Sporangia   Spores   Germinating spore   Mature gametophyte

Meiosis

DIPLOID (2*n*) Spore-producing generation

HAPLOID (*n*) Gametophyte generation

Sperm

Egg

Mature spore-producing plant

Fertilization

Sporophyte

Sperm Egg

Embryo

5 ...and develops into a mature spore-producing plant.

4 The embryo that develops from the diploid fertilized egg sends out roots...

3 Eggs are fertilized by swimming sperm.

Plant evolution, however, has generated some unique variations on the basic eukaryotic life cycle. One important difference between plant and animal reproduction is that, whereas animals produce sperm and eggs in specialized organs (the testes and ovaries), plants produce sperm and eggs in specialized individuals called male and female **gametophytes**. Some gametophytes produce sperm; others produce eggs.

A second, unique aspect of plant reproduction arises because plants, unlike most animals, cannot walk, fly, or swim to find a mate. Flowering plants, the group with which we are most familiar and on which we focus in this section, have overcome this problem by relying heavily on animals and wind to locate mates for reproduction.

Plants produce sperm and eggs in separate individuals

Because ferns illustrate the process so well, we begin our discussion of sexual reproduction in plants by describ-

ing the process in ferns. The ferns that we most commonly see (Figure 36.6) consist of typical diploid eukaryotic cells containing paired chromosomes. On the undersides of their leaves, ferns produce specialized structures that release dustlike haploid (*n*) **spores**, each of which contain just one member of each chromosome pair.

Although spores may seem like the equivalent of sperm or eggs, they differ in one critical way: Unlike sperm or eggs, spores do not fuse with other spores to form diploid embryos. Instead, the spores released by a fern sprout to form small, independent, photosynthesizing plants consisting of cells that contain unpaired chromosomes.

These small, haploid plants are the gametophytes, and they produce swimming sperm and large, stationary eggs. The fern sperm swim through water to fuse with an egg. As in animals, this fertilization results in a genetically unique, diploid (2*n*) fertilized egg. The fertilized egg develops into a spore-producing fern to start the cycle again.

Flowering plants follow the basic plant model in their sexual reproduction

For those of us who have grown flowering plants in our homes or gardens, the reproductive cycle described for ferns seems to have little in common with flower and seed production. A closer look, however, reveals a similar reproductive cycle in flowering plants (Figure 36.7). The plants that we think of as apple trees, for example, consist of diploid cells, just like the spore-producing stage of ferns. The flowers of the apple tree that represent the reproductive structures produce neither sperm nor eggs. Instead, like the spore-producing structures on fern leaves, flowers produce haploid spores that do not fuse with other spores.

Unlike ferns, however, flowering plants do not release these spores. Instead, the spores remain in the flower and develop into simple nonphotosynthesizing gameto-

(a)

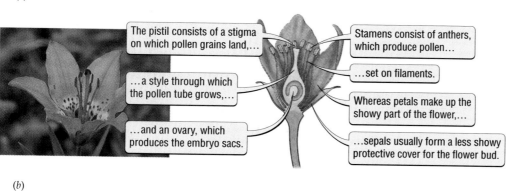

The pistil consists of a stigma on which pollen grains land,…

…a style through which the pollen tube grows,…

…and an ovary, which produces the embryo sacs.

Stamens consist of anthers, which produce pollen…

…set on filaments.

Whereas petals make up the showy part of the flower,…

…sepals usually form a less showy protective cover for the flower bud.

Figure 36.7 Reproduction in Flowering Plants
Flowering plants have gametophytic stages that cannot live as separate plants and instead develop as an embryo sac (female) and pollen grain (male) dependent on and embedded in the tissue of the spore-forming stage.

(b)

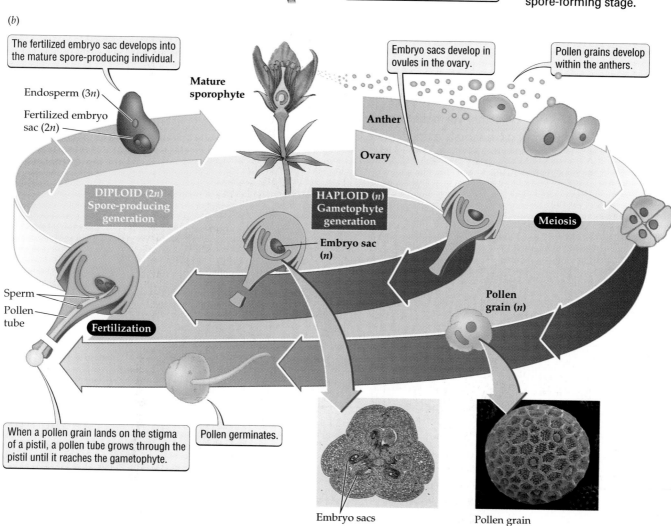

The fertilized embryo sac develops into the mature spore-producing individual.

Endosperm (3*n*)

Fertilized embryo sac (2*n*)

Mature sporophyte

Embryo sacs develop in ovules in the ovary.

Pollen grains develop within the anthers.

Anther

Ovary

DIPLOID (2*n*) Spore-producing generation

HAPLOID (*n*) Gametophyte generation

Meiosis

Embryo sac (*n*)

Sperm

Pollen tube

Pollen grain (*n*)

Fertilization

When a pollen grain lands on the stigma of a pistil, a pollen tube grows through the pistil until it reaches the gametophyte.

Pollen germinates.

Embryo sacs

Pollen grain

phytes that live as a sort of parasite on the flower tissues while they produce the sperm and eggs. **Pollen grains**, which are released from the **stamens**, are the male gametophytes. The female gametophytes, called **embryo sacs**, remain embedded in the ovary tissue of their spore-producing parent.

Fertilization of the eggs produced by the embryo sac requires that the pollen grain somehow move from the stamens of its parent plant to another flower. It also requires that the pollen grains release the sperm in such a way that they can penetrate the flower tissues surrounding the embryo sac. The shape of the mature pollen grain allows it to ride wind or water currents, or to stick to the body of a mobile animal, so that it can move to other flowers, as we will describe a little later.

When a pollen grain reaches a flower of its species, it catches on a sticky surface in the flower, called the **stigma**, and produces a **pollen tube.** The pollen tube grows through the tissue of the style to the embryo sac. Two sperm cells move through the pollen tube and enter the embryo sac. Inside the embryo sac one of the sperm cells fertilizes the egg cell to form an embryo, and the other fertilizes a second cell to form the **endosperm**, a nutrient storage tissue that makes up most of a seed.

> ■ Plants produce haploid sperm and haploid eggs in separate haploid individuals called gametophytes. In flowering plants the male gametophyte (pollen grain), which produces sperm, and the female gametophyte (embryo sac), which produces eggs, live in the tissues of the diploid plant that produced them.

Ensuring That Sperm Meet Eggs

For eukaryotic organisms, fertilization of an egg by a sperm is the last step in a complex sequence of events leading up to fertilization. Finding a mate allows the male to release its sperm close to eggs, reducing the distance that the short-lived sperm must travel before reaching their goal. Careful mate selection increases the chances that the individual chosen for mating is of sufficiently high quality that the resulting offspring stand a good chance of surviving.

Decreasing the distance between sperm and egg increases fertilization success

Males and females must move close to each other before releasing sperm or eggs, or the chances of fertilization fall to near zero. Under ideal conditions, human sperm can swim about 10 meters in a day. In the enclosed space of the female reproductive tract this rate of travel is more than adequate, but if released into the ocean the chances of the sperm reaching an egg would become very small.

In addition, as sperm and eggs move away from their sources, their concentration rapidly decreases with increasing distance. To fertilize human eggs in test tubes

Figure 36.8 Finding a Mate
Animals use diverse signals to find and evaluate potential mates. (*a*) A male sharp-tailed grouse attracts females with a dance display. (*b*) Fireflies signal mates after dark using flashs of green light. (*c*) Female moths attract males by releasing a species-specific chemical odor (a pheromone) from brushlike structures.

(*a*)

(*b*)

(*c*)

during in vitro fertilization (described later in this chapter), 60,000 sperm are needed to give a reasonable chance of fertilization. Clearly it is important that males and females either seek one another out or grow in sufficiently dense populations that sperm and egg need not travel far.

Mobile animals actively seek out mates

Finding a potential mate may seem less challenging than the other component of mating behavior—choosing a mate—because we live in towns and cities and need not travel far to find a member of the opposite gender. For many organisms, however, encountering a potential mate is difficult: The chance that a slowly moving animal, which includes most small animals (see Chapter 27), or a species whose members live widely scattered will bump into a potential mate is relatively small.

Animals can find other individuals of their species by using signals of various sorts (Figure 36.8). The most common signals are sounds or chemicals, both of which can travel great distances. The chirps and croaks produced by male birds, crickets, and frogs all guide females to eager mates. Sex pheromones (characteristic chemical scents) released by female moths can draw in males from distances of many kilometers. Some animals, such as fireflies and deep-sea fish, produce flashes of light to draw in mates through their dark environments.

Plants and immobile animals need other means of sperm transport

Although stationary, plants and the marine animal species that cannot move as adults must nonetheless transport their sperm close enough to the egg that the sperm have a chance of encountering the egg. Immobile animals and plants commonly rely on wind or water currents to bring the sperm to the eggs. In the case of plants, the sperm are contained within pollen.

Allergy sufferers know that in the spring and fall the air is filled with pollen grains carrying plant sperm. Sperm and eggs released in the water go where the currents scatter them. Typically organisms such as corals in a coral reef or grasses in a meadow that rely on water and wind to transport their eggs or sperm must live close together to give sperm a reasonable chance of encountering an egg.

The reproductive organs we call flowers that characterize flowering plants originally arose as a way of overcoming the limitations of water and wind as pollen dispersers. Flowers attract and manipulate the movements of mobile insects, birds, and even mammals so that they carry the sperm-containing pollen from the stamens to the stigmas on other individuals of the same species (Figure 36.9).

Plants bribe the pollinating animals to visit their flowers with foods such as sugary nectar or protein-rich pollen. Species-specific combinations of flower color, shape, and smell ensure that the pollinator carries the pollen that it picks up to another flower of the same species. If you follow a honeybee as it moves through a field, for example, you will see that it visits only a few of the available flower types. But with each visit to a flower, it deposits and picks up pollen caught in the hairs covering its body.

Insects such as honeybees carry the pollen that dusts their bodies from flower to flower as they search for food.

Birds play an important role as pollinators as they search for nectar in flowers.

The distinctive colors, shapes, and smells of flowers induce pollinators to selectively visit one or a few flower types.

Figure 36.9 Pollinators Provide Stationary Flowering Plants with a Way of Getting Their Sperm to Other Flowers
The spectacular colors, shapes, and smells of flowers, in combination with rewards such as nectar, direct pollinating animals to visit other flowers of the same species, where they transfer pollen in the process.

(a) Male land animals typically transfer sperm directly into the female to avoid exposing it to the environment.

(b)

Aquatic animals like these salmon may release vast quantities of sperm and eggs into the water, where fertilization takes place.

(c)

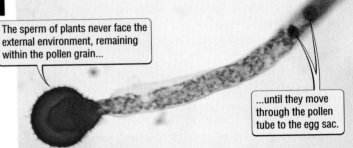

The sperm of plants never face the external environment, remaining within the pollen grain...

...until they move through the pollen tube to the egg sac.

Figure 36.10 Sperm Transfer in Animals and Plants

How close together the sperm and eggs must be released depends on the environment

The environment strongly limits the length of time that sperm and unfertilized eggs can survive outside the organism (Figure 36.10). If released onto land, neither eggs nor sperm survive more than a few minutes, and the lack of water makes it impossible for the sperm to move. In ocean water, however, sperm can survive because the environment resembles body fluids. Thus many aquatic organisms, such as squid, simply release sperm into the water surrounding a female, whereas land-dwelling organisms almost always release their sperm directly into the female.

Squid

Copulation (see Figure 36.10*a*) is the means by which animal males release sperm directly into females (think of the human sexual act). Flowering plants release their sperm only after the pollen grain encasing the sperm lands on the stigma of a flower.

■ Animals and plants move the sperm closer to the egg to increase the likelihood that the sperm will fertilize the egg. Mating behavior brings mobile male and female animals close together, but plants and immobile animals rely on pollinating animals or water or wind currents to bring pollen grains closer to embryo sacs.

Choosing a Mate

Humans try to select their mates carefully. Only after considerable thought do most of us decide with whom we want to produce children. A few species, including humans, select mates for their ability to help raise offspring. Of much more general concern, however, is the fact that sexual reproduction involves mixing genetic information with that of another individual. If an organism selects its mate poorly, it risks combining its own genes with inferior ones to produce poor offspring.

Males and females select mates differently

Because males and females produce different kinds of sex cells, they choose their mates differently. Whereas each sperm costs males relatively little in the way of resources, each egg takes up more of the female's resources. This simple observation has two important consequences: (1) Males can produce more sperm than females can produce eggs, and (2) a female stands to lose more if one of her eggs results in poor-quality offspring than does the male that fertilized that egg. As a result, females choose mates carefully to ensure that each offspring has as great a chance of success as possible, and males tend to mate with as many females as possible.

Females try to select high-quality males

Attention to mate selection has typically focused on the elaborate contests among males for the chance to mate with females (Figure 36.11), such as those described in Chapter 35. Scientists who study mating behavior have

(a)

The traits that males use to convince females of their suitability as mates can reach extremes, as in the fantastic tail displays that peacocks use to attract peahens.

Figure 36.11 Some Females Judge Males by Their Displays
Careful selection of mates by females may increase the success of offspring.

(b)

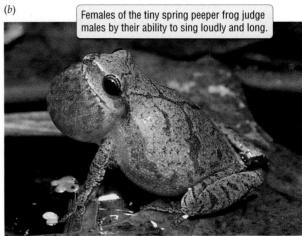

Females of the tiny spring peeper frog judge males by their ability to sing loudly and long.

become increasingly aware that females play an active role in mating. Females actively accept some and reject other males. Because females often invest so much in their offspring, they stand to lose a great deal if they mate with a poor-quality male. The competitive mating behaviors in which males engage often indicate the quality of their genes. The offspring of the winners often prove more successful than the offspring of the losers.

In plants, too, the outcomes of competitions determine the fathers of seeds: When pollen grains land on the stigma, they release sperm cells into pollen tubes that grow through flower tissues to the egg sacs. The faster a pollen tube grows, the more likely its sperm will find an unfertilized egg sac. Several recent studies have shown that seeds that result from pollen that produces rapidly growing pollen tubes produce more vigorous seedlings than do seeds that result from pollen that produces slowly growing pollen tubes.

■ Selection of a high-quality mate increases the chances that the union of genetic information from both parents will produce successful offspring.

Highlight

Technology and Human Reproduction

On the surface the events surrounding human reproduction seem to have turned us into remarkably efficient breeding machines. After all, more than 6 billion human beings blanket Earth, and millions more are added each year.

In spite of the rapid growth of the human population, humans have a hard time becoming pregnant. In healthy couples actively trying to have a baby, the chances of pregnancy in any given month are only about one in four. Moreover, about one in 12 couples are infertile and fail to become pregnant during one year of trying. Evidence suggests that the proportion of infertile couples is gradually increasing.

Recent advances in reproductive technology have made it possible for infertile couples to have children. Since the first successful fertilization of a human egg outside the body in 1978, in vitro fertilization (fertilization in glass, commonly referred to as IVF) has advanced to the point where doctors can overcome many infertility-related problems.

In IVF, hormone injections stimulate the female's ovaries to mature up to 20 eggs instead of the usual one in a particular month. Doctors collect these eggs surgically just before they are released into the oviducts. Some of the eggs collected in this way are then placed in a small tube with a sample of about 60,000 sperm from the male. Under these conditions, 70 to 80 percent of the eggs typically become fertilized. Two days later, if all goes well, two to three healthy-looking embryos are selected and placed directly into the uterus of the female. IVF leads to pregnancy in about 50 percent of attempts.

Although reproductive technologies such as IVF make it possible for some infertile couples to have children, they also raise some difficult issues. For example, during IVF doctors usually collect more eggs than they fertilize, and they usually fertilize more eggs than they place in the uterus. What do we do with the extra eggs and the extra embryos?

Typically both excess unfertilized embryos and eggs are frozen to preserve them in case the couple would like to try IVF again, but in most cases this approach simply delays the difficult decision about what to do with the excess eggs and embryos. In addition, difficult legal and moral questions arise if the couple divorces or one member of the couple dies. For instance, who has control over the eggs and the embryos? In many cases, laws have yet to catch up with the possibilities raised by these technologies.

> ■ New reproductive technologies like in vitro fertilization help couples overcome fertility problems, but raise many moral and legal questions that have yet to be answered adequately.

Summary

How Bacteria and Eukaryotes Reproduce and Mix Genes

- Prokaryotes reproduce asexually when one parent cell divides to produce two cells, each identical to the original cell.

- Prokaryotes transfer genetic information from one individual to another through transformation and sometimes as a plasmid or virus moves from one host cell to another.

- In eukaryotes, genetic material is exchanged during reproduction. Meiosis distributes half the genetic information into each sperm in males and each egg in females; fertilization combines the genetic information from the sperm and egg.

Animal Reproduction

- Human males release many millions of sperm from their penis into the female's vagina. A few hundred of these sperm manage to reach the egg, and only one fertilizes it.

- Human females produce approximately one egg each month that moves from an ovary into the oviduct, where it may be fertilized.

- Unlike humans, which rely solely on sexual reproduction, some animals reproduce only asexually and others can reproduce either asexually or sexually.

- Many animals are hermaphrodites, which produce both eggs and sperm and are able to reproduce with any other individual of their species.

Plant Reproduction

- Diploid plants undergo meiosis to produce spores, which grow into haploid gametophytes. These gametophytes then produce either sperm or eggs.

- Whereas fern gametophytes develop independently of the diploid plant, flowering-plant gametophytes remain attached to the diploid plant as they develop.

- In flowering plants the male gametophyte, pollen, is carried to the stigma of another flower. The sperm move out of the pollen and through a pollen tube to the female gametophyte, the embryo sac, where one sperm fertilizes the egg to form the embryo and a second sperm fertilizes another cell to form endosperm.

Ensuring That Sperm Meet Eggs

- To succeed at fertilization, sperm and eggs must be near each other.

- Animals use calls, pheromones, and even flashes of light as signals to search for and move toward potential mates.

- Immobile marine animals that use water currents and plants that use wind to carry sperm to eggs often must live near many individuals of the same species. Some flowering plants use animals to carry pollen to the stigmas of other flowers of the same species.

- In the ocean, many organisms rely on external fertilization; on land, most organisms fertilize the egg inside the female parent, where the egg is protected from drying out.

Choosing a Mate

- Females choose mates more carefully than males because eggs require a significant investment of resources and are in limited supply.

- Competition between males and mate choice by females each plays a part in determining the identity of the male parent and therefore the quality of the set of genes contributed by the male.

Highlight: Technology and Human Reproduction

- Many couples are infertile, and the frequency of infertility seems to be increasing.

- In vitro fertilization helps many otherwise infertile couples produce offspring.

- In vitro fertilization often produces extra embryos and unfertilized eggs that present society with challenging new ethical dilemmas.

Key Terms

asexual reproduction p. 544

copulation p. 554

donor p. 546

egg p. 546

embryo sac p. 552

endosperm p. 552

female p. 546

fertilization p. 547

gametophyte p. 550

hermaphrodite p. 549

male p. 546

ovary p. 548

oviduct p. 548

penis p. 548

pollen grain p. 552

pollen tube p. 552

recipient p. 546

reproduction p. 544

sexual reproduction p. 544

sperm p. 546

spore p. 550

stamen p. 552

stigma p. 552

testis p. 548

transformation p. 546

uterus p. 548

vagina p. 548

Chapter Review

Self-Quiz

1. Which of the following organisms do *not* depend on reproduction to exchange genetic information?
   a. animals
   b. plants
   c. bacteria
   d. eukaryotic cells

2. Human eggs are produced in the
   a. oviduct.
   b. testes.
   c. uterus.
   d. ovary.

3. Like animals, plants also produce
   a. many more sperm than eggs.
   b. a few more sperm than eggs.
   c. equal numbers of sperm and eggs.
   d. fewer sperm than eggs.

4. Asexual reproduction produces offspring that are
   a. genetically identical to their parents.
   b. genetically identical to their siblings.
   c. none of the above
   d. both a and b

5. In many species, mate choice is
   a. more important to females than to males.
   b. more important to males than to females.
   c. equally important to males and females.
   d. important only during asexual reproduction.

Review Questions

1. Contrast the production of eggs and sperm in humans with the production of eggs and sperm in flowering plants.

2. What is it about the reproduction of many flowering plants that allows them to live in more diverse communities, with many different plant species, in comparison to wind-pollinated plants, which tend to live among many individuals of their own species?

3. How might the structure of the female part of a flower be modified to increase the intensity of the competition between male pollen tubes as they race toward the embryo sac, and how might this increased competition benefit the female parent?

4. How does habitat affect the reproduction of plants and animals?

Disappearing Bugs Threaten Wildflowers and Crops

ITHACA, NY. Researchers studying bees and other animals say a critical link in the life cycle of many flowering plants is being broken. The flowering plants that brighten our gardens and the crops that are the foundation of our agricultural production depend on pollinators—bees, butterflies, and even birds and bats—to carry pollen from plant to plant. As our wild lands have been replaced by agricultural and urban areas, the number of native insects that work as pollinators has dropped dramatically.

Nearly half of the pollination of agricultural plants has been taken over by managed and wild European honeybees. These honeybees also pollinate some wildflowers, although scientists say they don't yet know what percentage of wild-flowers depends on honeybees. Unfortunately, the number of managed honeybees in the United States has declined by more than 25 percent since the early 1990s. Researchers estimate that this loss of honeybees has resulted in 5.7 billion dollars worth of damage to agricultural production around the world each year.

Although there are many reasons for the decline in honeybees and other insect pollinators, including introduced mites, disease, introduced competitors such as the Africanized honeybee, and bad weather, Dr. Pauline Nation of the University of West Idaho points to two reasons for the dramatic drop in pollinators that every person can try to prevent: habitat destruction and the misuse of pesticides. The Environmental Protection Agency now requires some pesticides to have warning labels that recommend against application near blooming crops or weeds, but these warnings have not reduced the effects on pollinators. Dr. Nation noted that pesticides continue to be used inappropriately.

Burt Umbleby of the Ithaca Metro Gardening Club suggests that local gardeners may be able to help by growing plants that provide nectar for insects and may help reverse the pollinator loss. He added that many gardening books now list plants that are particularly beneficial to pollinators.

"We're part of the problem," said Mr. Umbleby, "and now gardeners need to become part of the solution."

Evaluating "The News"

1. In home gardens and commercial farms, reproduction—the successful pollination of flowers—is the key to successful harvests for most plants. If pesticides are useful to farmers for protecting plants so they can yield a good harvest, what are farmers to do? How do you weigh the detriment to pollinators against the protection of plants?

2. Is there any way to incorporate aesthetic values associated with flowers into economic arguments about the costs and benefits of pesticide use?

3. If legislators are going to attempt to promote conservation to bring back plants that require pollinators, should conservation legislation focus on saving insects and other animal pollinators, plant species, specific environments, or interactions? How would you conserve an interaction?

37

Development

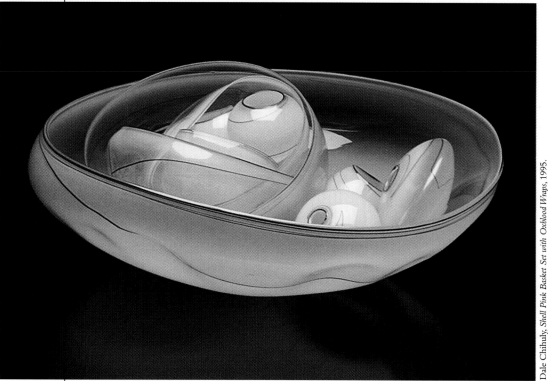

Dale Chihuly, *Shell Pink Basket Set with Oxblood Wraps*, 1995.

Cloning Carrots and Lambs

In groundbreaking research that sounds more like an adventure in cooking than like science, in 1958 F. C. Steward and his coworkers managed to clone a single, mature plant cell to give rise to an entire new plant. They placed tiny pieces of phloem tissue taken from a carrot root in a large container filled with coconut milk and stirred gently. The phloem cells divided as they circulated through the coconut milk to

form lumps of callus, a type of poorly orga- nized tissue that typically forms in plants after wounding.

Coconut milk, which is actually the unique, liquid endosperm, or nutritive tis- sue, in coconut seeds, provided the nutri- ents needed to support the dividing cells. The constant movement of the callus in the stirred coconut milk broke off individual callus cells. These individual cells divided,

Main Message

The coordinated action of genes directs the development of the fertilized egg into a whole organism with specialized cells, tissues, and organs.

and instead of forming another mass of disorganized callus tissue, they differentiated to form a recognizable root and a mass of cells that eventually formed the stems and leaves of a new carrot plant.

To understand just how amazing this event was, imagine putting a few cells from your little finger into a big vat full of stirred mother's milk . . . and returning a few weeks later to find embryonic copies of yourself floating around in there.

In 1996, almost 40 years after the work of Steward and his colleagues, a group of scientists working in Scotland used a single cell taken from the udder of a sheep to produce a now famous lamb called Dolly. Cloning a sheep proved considerably more complicated than cloning a carrot. The scientists first had to remove the haploid nucleus from an unfertilized sheep egg. They then inserted into the egg another nucleus taken from an udder cell. The egg now had a diploid nucleus containing two copies of every chromosome, just as a fertilized egg would. The egg was then implanted into another female sheep, just as is done in human in vitro fertilization. A perfectly healthy lamb, Dolly, was the result.

The research into cloning plants and animals raises interesting questions about how complex, multicellular organisms develop from the single cell of a fertilized egg. If we can grow whole carrots and sheep from a single cell, what prevents a phloem cell or an udder cell from doing so in nature? Conversely, what determines that a cell becomes a phloem cell rather than a xylem cell, or an udder cell rather than a brain cell? Why is it easier to clone carrots than to clone sheep? As we will see in this chapter, these questions go right to the heart of how organisms control their development, and they apply equally to all multicellular organisms.

A Clone and Her Offspring
Dolly the sheep (left) is the first mammal to have been cloned from a mature cell. She has produced a lamb, Bonnie (right), by normal reproduction.

Key Concepts

1. The various organs that make up most animals develop in predictable ways from three distinct layers of cells. These cell layers, called germ layers, appear early in development.

2. All plants differentiate three tissue types that are distinct from the tissues of animals early in development. The mature plant forms as a combination of cell growth and cell death arranges the three tissue types into recognizable plant parts.

3. During development the potential form and function of cells become more specialized as progressively more genes in the cell are prevented from activation.

4. Chemicals called morphogens play a central role in coordinating how the genes in each cell are activated and inactivated.

5. Developmental changes have played an important part in generating evolutionary changes in plants and animals.

As described in Chapter 36, we start life as a fertilized egg containing equal genetic contributions from our mother and father. This single cell multiplies and develops into a complex human consisting of many different cell types organized into tissues and organs that interact in a coordinated way.

Our transformation from a fertilized egg into what we are today involved much more than an increase in number of cells. It required a precisely controlled differentiation of one cell into trillions of cells, each specialized for specific functions. During our development these cells appeared not only in a carefully defined sequence, but also in carefully defined sites in our bodies.

In Unit 3 we learned that DNA encodes the blueprint for the body's design. However, because each cell in the body contains exactly the same DNA, any differences in cell structure or function must stem from differences in how the body interprets its DNA blueprint as it develops.

In this chapter we investigate how a developing, multicellular organism converts the information contained in its DNA into many different and interacting tissues. To provide a foundation for our discussion, we first contrast patterns of development in plants and animals. We will see that in both groups, development of different cell and tissue types depends on a closely regulated sequence of activating and inactivating genes. Finally, we explore the evolutionary role of changes in the timing of developmental events.

Animal Development

For all their outward differences in size, shape, and complexity, animals share remarkably similar developmental patterns. We will use human development as an example to illustrate important steps in differentiation.

Early animal development is characterized by rapid cell division

During the first stages of development of most animals, the fertilized egg divides quickly into a many-celled, hollow sphere called the **blastula**. Never again in its life will an animal's cells divide as rapidly as they do right after fertilization. In part, the rapidness of cell division in the early embryo results because, unlike later cell division, it involves mainly subdivision of the egg rather than cell growth. Cells are increasing in number while decreasing in size. For example, after many cell divisions, a frog embryo containing a thousand cells is still about the same size as the original fertilized egg (Figure 37.1).

A few features of the development of mammals like humans are unique and noteworthy. Although mammalian cells divide without growth early in development, they divide more slowly than in other animals. In the week that it takes the human egg to travel from the site of fertilization in the oviduct to the uterus (see Chapter 36), it divides only four times, forming the 16-cell stage. A second difference is that the product of early cell divisions, the blastula, is not just a simple, hollow sphere, but one that contains two distinct cell types.

37.1	**The Three Germ Layers and Their Fates**
Germ layer	**Corresponding adult structure**
Endoderm	Gut, liver, lungs
Mesoderm	Skeleton of limbs and body, muscles, reproductive structures, kidneys, blood, blood vessels, heart, inner layer of skin
Ectoderm	Skull, nerves and brain, outer skin, teeth

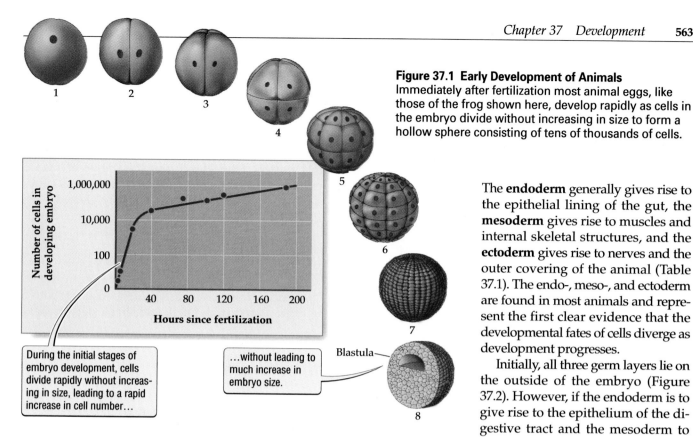

Figure 37.1 Early Development of Animals
Immediately after fertilization most animal eggs, like those of the frog shown here, develop rapidly as cells in the embryo divide without increasing in size to form a hollow sphere consisting of tens of thousands of cells.

During the initial stages of embryo development, cells divide rapidly without increasing in size, leading to a rapid increase in cell number…

…without leading to much increase in embryo size.

The **endoderm** generally gives rise to the epithelial lining of the gut, the **mesoderm** gives rise to muscles and internal skeletal structures, and the **ectoderm** gives rise to nerves and the outer covering of the animal (Table 37.1). The endo-, meso-, and ectoderm are found in most animals and represent the first clear evidence that the developmental fates of cells diverge as development progresses.

Initially, all three germ layers lie on the outside of the embryo (Figure 37.2). However, if the endoderm is to give rise to the epithelium of the digestive tract and the mesoderm to muscles and skeletal tissue, these germ

The outer layer of the blastula will become the developing embryo's portion of the **placenta**, and an inner set of cells clustered at one end of the blastula will become the actual embryo. The uniquely mammalian placenta is a structure in the uterus that permits transfer of nutrients and wastes between the mother and the developing embryo. This first identifiable distinction between cell types in mammalian development occurs at the 32-cell stage, and this differentiation must take place before the embryo can embed itself in the uterine wall.

The second stage of animal development produces three germ layers

During the second stage of typical animal development, two critical events take place. First, depending on their position within the blastula, cells in the embryo form three **germ layers**—endoderm, mesoderm, and ectoderm—that follow three distinct developmental paths to give rise to different tissues in the mature animal.

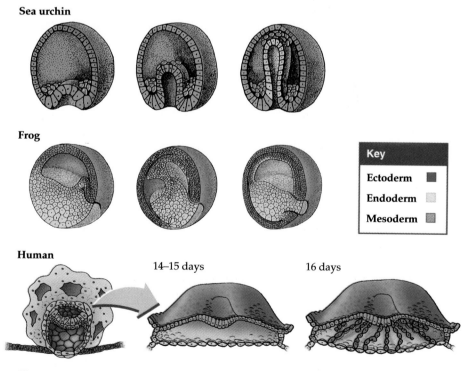

Figure 37.2 Animals Produce Three Germ Layers with Developmentally Distinct Fates
In spite of the difference in patterns of early cell division, all animals have recognizable endoderm, mesoderm, and ectoderm cells once the stage of rapid cell division ends.

layers must move to more appropriate positions inside the embryo. This remarkable rearrangement takes place during **gastrulation**, a complex sequence of events in which the germ layers move relative to one another so that the endoderm lies inside the embryo and is surrounded by mesoderm. As a result, the ectoderm forms the entire outer layer of the embryo.

In humans, the formation of the germ layers and their rearrangement during gastrulation take place during the 2 weeks following implantation of the embryo in the uterus wall.

Cell proliferation, differentiation, and rearrangement generate the mature body

After gastrulation, animals undergo a period of rapid differentiation of the germ layers into the various organs needed to make up a functioning individual. For example, just 3 weeks into our own 40-week-long development (Figure 37.3), humans have an identifiable heart. By the eighth week of development, the head is clearly identifiable in the 2.5-cm-long human embryo, the endoderm has differentiated into an identifiable liver, the mesoderm has differentiated into red blood cells and

kidneys, and the beginnings of all the organs found in an adult are present. By 12 weeks, the fingers and toes of the 8-cm-long embryo have nails, external genitalia are recognizably male or female, and the gut has developed from the endoderm to the point that it can absorb sugar.

Genetic or environmental problems that disrupt the normal sequence of events during this period of intense differentiation can lead to severe developmental problems. In humans, most miscarriages take place during the critical first third of pregnancy because any developmental problems in the embryo are most likely to be expressed at this time.

To illustrate just how sensitive early development can be to even brief disruptions, consider the impact of thalidomide, a drug sometimes prescribed as a sedative in the 1950s. Thalidomide interferes with the development

Figure 37.3 Human Development
Rapid differentiation of organs marks the early stages of human development, whereas rapid weight gain marks the later stages. The three trimesters, or three-month stages, commonly used to divide up human pregnancy correspond roughly to important developmental events. Improvements in medical technology now allow even some babies born in the middle of the second trimester to survive.

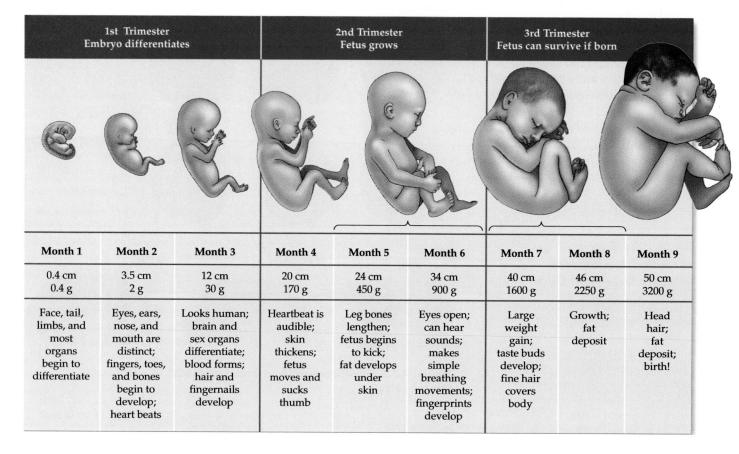

1st Trimester Embryo differentiates			2nd Trimester Fetus grows			3rd Trimester Fetus can survive if born		
Month 1	**Month 2**	**Month 3**	**Month 4**	**Month 5**	**Month 6**	**Month 7**	**Month 8**	**Month 9**
0.4 cm 0.4 g	3.5 cm 2 g	12 cm 30 g	20 cm 170 g	24 cm 450 g	34 cm 900 g	40 cm 1600 g	46 cm 2250 g	50 cm 3200 g
Face, tail, limbs, and most organs begin to differentiate	Eyes, ears, nose, and mouth are distinct; fingers, toes, and bones begin to develop; heart beats	Looks human; brain and sex organs differentiate; blood forms; hair and fingernails develop	Heartbeat is audible; skin thickens; fetus moves and sucks thumb	Leg bones lengthen; fetus begins to kick; fat develops under skin	Eyes open; can hear sounds; makes simple breathing movements; fingerprints develop	Large weight gain; taste buds develop; fine hair covers body	Growth; fat deposit	Head hair; fat deposit; birth!

Figure 37.4 The Effects of Thalidomide
The limb deformities of infants whose mothers took thalidomide as a mild sedative during the 1950s resulted when the drug interfered with limb development. Even a single use during the narrow window of time between the twenty-fourth and thirty-fifth days of pregnancy interferes with a critical phase of bone elongation, leading to deformed, flipperlike appendages.

ment to give us a basis for comparison to animal development. At the same time we examine the mechanisms that control development.

Plants begin development inside seeds

The fertilized plant egg begins development surrounded by a protective seed in conifers and flowering plants. As is the case in animal development, distinctive structures develop during the first stages of plant development. The first divisions of the embryo result in a mass of cells, all of which divide. Soon, however, two important events take place within the developing embryo.

The cells of the embryo differentiate into three distinct types: (1) **protoderm** (*proto*, "first"; *derm*, "skin"), which develops from the outer layer of embryo cells, and (2) **ground meristem** and (3) **procambium**, both of which develop from the inner cells of the embryo (Figure 37.5). We can trace all the tissues in a mature plant back to one of these three embryonic tissues.

of limb bones. If taken by a pregnant woman between 24 and 35 days after fertilization, when the embryo normally develops its arms and legs, the drug can lead to devastating limb deformities (Figure 37.4).

■ Animal development has several identifiable stages. After fertilization, cells divide repeatedly, without increasing in size, to form a blastula. Then cells differentiate into three germ layers, each with a distinct developmental fate. Gastrulation moves the cells in each germ layer into appropriate positions. After gastrulation, animal organs differentiate relatively quickly.

Plant Development

Like animals, plants begin life as a fertilized egg. This single cell gives rise to a diversity of cell types carefully organized into the various tissues that make up a mature plant. In this section we survey the patterns of plant develop-

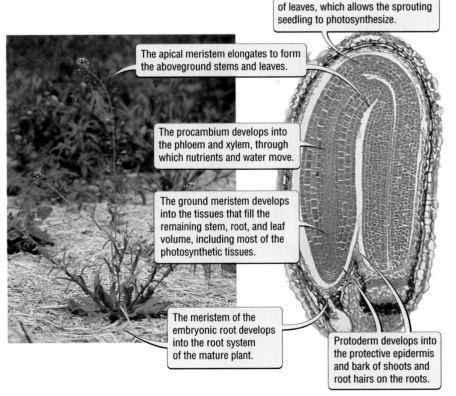

Seed leaves develop into the first pair of leaves, which allows the sprouting seedling to photosynthesize.

The apical meristem elongates to form the aboveground stems and leaves.

The procambium develops into the phloem and xylem, through which nutrients and water move.

The ground meristem develops into the tissues that fill the remaining stem, root, and leaf volume, including most of the photosynthetic tissues.

The meristem of the embryonic root develops into the root system of the mature plant.

Protoderm develops into the protective epidermis and bark of shoots and root hairs on the roots.

Figure 37.5 The Development of Plant Tissues
Plant embryos produce three distinct tissue types—protoderm, ground meristem, and procambium—each of which gives rise to a predictable set of tissues in the mature plant.

The second important event is the restriction of cell division to zones called **apical meristems** that lie at the two tips of the elongated embryo (see Figure 37.5). One of these apical meristems gives rise to the root system of the plant; the other forms the shoot system, including the stems, leaves, and reproductive structures.

Plants that produce seeds show some special twists to development

Within the seeds of conifers and flowering plants, early development includes a few features not found in other plants. The protoderm, ground meristem, and procambium develop into the various tissues that make up leaves, called **seed leaves** (see Figure 37.5), within the seed. The seed leaves surround the apical meristem, which develops into the stems, leaves, and flowers that make up the aboveground shoot system. In addition, special nutritive tissues form around the developing embryo (Figure 37.6).

In conifers these nutritive tissues form from tissues of the female gametophyte that produced the egg. In flowering plants the nutritive tissues form from the development of the **endosperm** tissue that resulted from

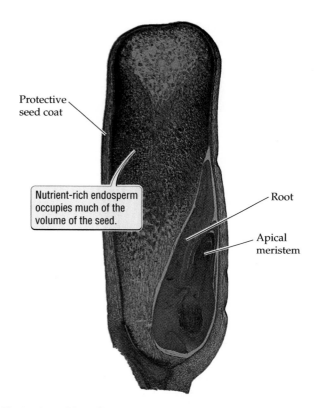

Figure 37.6 Plant Seeds
Plant seeds contain developing embryos and nutritive tissues, encased in a protective seed coat.

Labels on figure:
Protective seed coat
Nutrient-rich endosperm occupies much of the volume of the seed.
Root
Apical meristem

the fertilization of the polar nuclei by a second sperm cell (see Chapter 36). With many flowering plants, such as corn and coconuts, endosperm fills much of the seed volume. Finally, the seeds develop a hard external coat that protects them.

The endosperm and seed leaves produced early in the development of seed-producing plants allow them to enter a dormant stage in which development stops. For most plants, the transition between an embryo developing within a seed or spore and life as an independent, photosynthesizing plant is the most vulnerable time of life. The well-stocked, well-protected, and well-developed embryos of conifers and flowering plants can lie dormant within the seed and resume development and growth only when conditions in their harsh land environment are favorable. In contrast, plants like mosses and ferns, cannot interrupt development at this vulnerable stage and must continue developing regardless of the condition of their environment.

The procambium, ground meristem, and protoderm differentiate to form plant organs

As in animals, each of the three basic tissues in the plant embryo gives rise to specific tissues in the mature plant. The protoderm gives rise to the outer covering of plants. The procambium gives rise to the tissues that make up the internal transport system of plants, which consists of water-conducting xylem and nutrient-conducting phloem (see Chapter 30). The ground meristem gives rise to the photosynthetic tissues that make up the bulk of leaves and to the nonphotosynthetic tissue that lies between the xylem and phloem in the stems and roots.

Also as with animals, the way in which these three basic tissues differentiate produces the remarkable diversity of form and function in plants. Plant cells have a rigid cell wall that prevents them from moving during development as animal cells can. Instead, the shape of plant structures arises from patterns in cell division, cell enlargement, and cell death.

Plant development depends heavily on the production of identical modules

A universal feature of plant development is its reliance on repeated units, or **modules**, of a relatively small number of different structures. The aboveground portion of a plant consists of stem/leaf units repeated many times (Figure 37.7). Each stem/leaf unit includes meristem tissue that can differentiate into any of the tissues in a stem or a leaf.

When we look at a plant branch from above, we can see that each stem/leaf unit is rotated at a constant angle

(a)

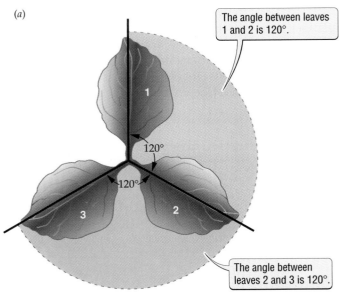

The angle between leaves 1 and 2 is 120°.

120°

120°

The angle between leaves 2 and 3 is 120°.

(b)

Figure 37.7 Modular Development in Plants
(a) A view down the shoot of a plant shows the spiral arrangement of repeated stem/leaf units. (b) Each stem/leaf unit is rotated by a constant angle from the one below it, giving rise to complex spiral patterns.

from the one beneath it to generate a predictable spiral arrangement (see Figure 37.7). Thus plants appear to control the spatial arrangement of cells and tissues very precisely during growth and development.

Because of their modular growth, plants often continue to develop new tissues throughout their lives. Modular development gives plants tremendous flexibility in their growth form: For example, repeatedly mowed weeds in a lawn can nonetheless produce seeds from the undamaged modules missed by the mower's blades. On the other hand, nonmodular animals, such as humans, reach a distinct developmental end point.

■ In early plant development, three developmentally distinct cell types form: protoderm, ground meristem, and procambium. Nutritive tissues in the seeds of conifers and flowering plants allow the embryos to remain dormant until conditions are favorable for germination and growth. Plants develop as a series of repeated modules. Modular growth requires that plants have developmentally flexible cells, arranged in meristems, associated with each module.

What Controls Development?

The preceding descriptions of animal and plant development are simplified overviews of how the diversity of form and function develops in multicellular organisms. It should be clear from these descriptions that development of a fertilized egg into a mature organism requires precise control over the fates of the cells as they divide. The different cells that are arranged into tissues in an organism differ not because they contain different genetic information, but because they express the genetic information that they contain differently.

One of the major advances in biology in the past few decades has been a rapid growth in our understanding of how the expression of the genetic information in cells is controlled during development. We will focus on two important aspects of this regulation: the activation and inactivation of genes within cells during development and the factors that control which genes are activated or inactivated.

Genes that are activated and inactivated during development determine cell fate

The reading of the DNA blueprint in each cell proceeds in two stages: transcription, in which the DNA message is converted into RNA in the nucleus, and translation, in which the RNA message crosses into the cytoplasm and directs protein production (see Chapter 15). If transcription is prevented or promoted, or translation is altered, the identical genetic information in each cell may be expressed differently.

For example, humans and other mammals produce different kinds of oxygen-binding hemoglobin pigments (see Chapter 29) at different stages of development: embryonic hemoglobin (0 to 8 weeks), fetal hemoglobin (8 to 12 weeks), and then progressively greater proportions of adult hemoglobin (12 weeks and beyond). The embryonic and fetal hemoglobins have a greater tendency to bind to oxygen than adult hemoglobin does, allowing the developing baby to pull oxygen from its mother's blood by way of the placenta. The genes that code for the embryonic, fetal, and adult globin proteins are associated with many DNA sequences that, when bound to regulatory proteins (such as the morphogens that we will describe shortly), allow transcription of the DNA. At certain stages during development specific cells activate transcription of the appropriate globin genes, while the other globin genes remain inactive.

In other cases, cells transcribe a particular gene, but the processing of the resulting messenger RNA differs in cells with different fates. An example is the gene in sea urchins that controls the production of the protein actin, one of the molecules in muscle tissue. Actin is produced specifically in the ectoderm of sea urchins early in development. Although ectoderm, mesoderm, and endoderm cells all transcribe the actin gene into messenger RNA (mRNA), only in the ectoderm cells is the transcribed RNA converted into a piece of mRNA that can move out of the nucleus into the cytoplasm of the cell where the young sea urchin produces the actin.

Another way in which organisms can change gene expression during development involves conversion of the transcribed RNA into two different products by being cut up in different ways to make different messenger RNAs. For example, male and female fruit flies form different mRNAs from a single gene that they both express. The resulting proteins, which determine gender in the developing flies, play an important part in controlling the sexual differences in these organisms.

Cell destiny becomes increasingly narrowly defined during development

As organisms develop, the range of potential fates of cells in their bodies becomes more and more limited. For example, the single cell of the fertilized egg has the potential to become any cell type in the mature organism. But by the time cells have differentiated into endo-, meso-, and ectoderm in animals or into protoderm, ground tissue, and procambium in plants, each cell faces a more restricted future.

In the embryo the potential fates of a cell depend on the particular genes in its DNA that can still be activated. In animals, the future of a cell usually becomes fixed during development. In plants, the future seems to depend somewhat on a continued interaction with surrounding tissues. Even leaf cells specialized for photosynthesis or stem cells specialized for transporting water often retain the ability to produce all the cell types in the plant.

As development proceeds in both animals and plants, the range narrows for the cell types that can be differentiated. This growing limitation creates a problem: For multicellular organisms to reproduce, they must retain some cells that have unrestricted futures and can produce all the structures needed in their offspring.

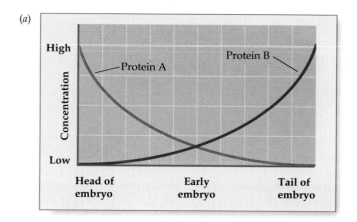

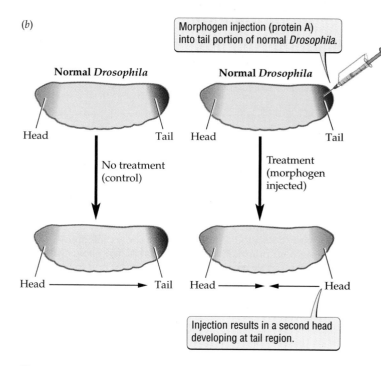

Figure 37.8 Morphogens Often Control Development
(*a*) Patterns of increasing or decreasing morphogen concentration provide information about the positions of cells relative to the head and tail ends of developing fruit fly embryos. (*b*) Experimental manipulation of the morphogen gradients can dramatically disrupt development. By injecting protein A, which identifies the head end of the embryo, into the tail end, cells near the tail are "fooled" into developing into head tissues.

In plants this problem is solved because reproductive structures develop from cells in the undifferentiated meristems that lie at the tip of each branch and root (see Figure 37.5). Because their pattern of growth is generally more defined than that of plants, animals overcome this problem by setting aside a few cells to form the **germ line** early in development. These unique cells, which eventually end up in the egg-producing ovaries or sperm-producing testes, remain unspecialized and separate from the other cells.

Specific molecules activate or inactivate genes in the developing organism

Genes generally switch on or off in response to the presence of chemicals, usually proteins, called **morphogens**. One method by which morphogens can work is to spread outward from a cell, decreasing in concentration with increasing distance from the source. Cells with the proper receptors in their plasma membrane can interact with the morphogens and activate or inactivate genes. In this way, genes can be expressed at specific locations in the developing organism.

The role of morphogens in controlling development has been studied extensively in fruit flies. In the embryonic fruit fly we find that patterns in the concentration of several protein morphogens provide developing cells with information about their position relative to the head and tail (Figure 37.8a). By experimentally introducing these morphogens into a developing embryo, researchers can disrupt the morphogen patterns and, as a result, the developmental patterns. If a morphogen that identifies the head end is injected into the tail end of the embryo, a two-headed embryo can be created (Figure 37.8b).

Morphogens act on a family of homeotic genes (see Chapter 16), called *Hox* genes, that identify the various segments in the body of the fruit fly. The *Hox* genes appear to regulate the genes that control the form of a particular segment. *Hox* genes and morphogenic proteins that resemble those in fruit flies play a major role in directing head-to-tail differences in the development of tissues in all animals, including humans.

Morphogens can also simply act on adjacent tissues to provide information on the identity of adjacent tissue. In birds, for example, the interaction between mesoderm- and ectoderm-derived tissues in the skin determines the form of the skin (Figure 37.9). Chicken ectoderm from the wing develops into wing feathers when placed next to mesoderm from the wing, but into scales and claws when placed next to mesoderm from the foot. Chemical interactions between the mesoderm and ectoderm tissue determine which genes in the ectoderm are activated.

Hormones (chemical messengers that move through the internal transport system of organisms) also help direct development. For example, sex hormones influence the development of testes and ovaries in animals. Hormones play a particularly important role in the ongoing development of plant tissues. Auxins are plant hormones produced in the apical meristem of plant shoots. Auxin affects the differentiation of plant tissue into phloem and xylem: At low concentrations it leads to phloem only; at higher concentrations it leads to both phloem and xylem. In addition, auxin produced by the apical meristem stops the development of other meristems (Figure 37.10). A decrease in the auxin concentration that reaches the other meristems, either as a result of removal of the apical meristem or because of long distances separating the two, allows the other meristems to develop.

Environment can influence development

The developmental patterns of many organisms depend not only on genetically programmed events, but also on environmental influences. In Chapter 36 we saw, for

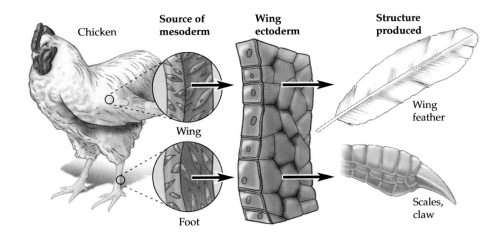

Figure 37.9 Contact between Cells Can Influence Development
Cells in the ectoderm-derived layers of chicken skin develop differently depending on the source (wing or foot) of the mesoderm-derived layers of chicken skin that they contact. When placed next to ectoderm from the wing, mesoderm from the wing stimulates the production of feathers typical of wings. When placed next to the same wing ectoderm, mesoderm from the foot stimulates the production of scales and claws typical of feet.

(a)

Auxin, a plant hormone produced in the apical meristem, moves down the stem to stop the development of other buds lower down on the plant.

Removal of the apex by pruning or by insect damage blocks the transport of auxin to the lower buds.

As a result, one or more of the lower buds starts to develop, producing a new plant apex that stops development of the lower buds.

(b)

This was the apex of the plant before it was removed by pruning.

Figure 37.10 Hormonal Control of Plant Development
The plant hormone auxin, produced in the apical meristem of plants, stops the development of other meristems, leading to a phenomenon known as apical dominance (a). Removal of the apex allows the other meristems to develop (b). This is why we prune plants: Pruning leads to bushier, fuller growth.

example, that the sex of a jack-in-the-pulpit plant depends on its ability to accumulate resources. Many plants trigger the development of flowers in response to the length of days relative to nights. Some plants, like the poinsettias that brighten up our homes in the middle of winter, flower only when days become progressively shorter relative to nights. Others, like the spinach in our vegetable gardens, flower only when days become progressively longer relative to nights.

Animals, too, change developmental patterns in response to environmental influences. Gender in many turtles and in alligators depends on the temperature of the nest in which the eggs develop: Turtle eggs from cool nests tend to develop into males, and alligator eggs from cool nests tend to develop into females. Many aquatic invertebrates develop into a spiny or bristly form when exposed to chemicals released by nearby predators in their environment. The spines and bristles make it harder for the predators to eat their prey. Aphids are common, small insects that show several environ-

mental influences on their development: Scarcity of food and high densities of other aphids nearby stimulate the development of winged forms, and temperature and day length regulate whether newly formed individuals reproduce sexually or asexually.

■ Cells and tissues become specialized as a result of the regulation of gene expression within cells, not of genetic differences between cells. Specialization limits the ability of a cell to produce other cell types, but unspecialized cells (meristems in plants and germ lines in animals) are required for reproduction. Morphogens help coordinate development by influencing the expression of homeotic genes, which influence the expression of other genes. Other factors that influence development include hormones and the environment.

Development and Evolution

Changes in development that seem small can cause dramatic changes in the form and function of the mature organisms. Because developmental processes have such a strong effect, rather simple genetic changes can lead to complex changes in shape and function. As a result,

Figure 37.11 Development and the Evolution of Flower Shape
Simple changes in the genes that control flower development may be all that separate radially symmetrical flowers such as daisies from bilaterally symmetrical flowers such as snapdragons.

Radially symmetrical flowers, like those of daisies, look similar regardless of which side is up.

A single gene expressed only in the upper part of the developing flower slows growth there, creating bilateral symmetry.

Bilaterally symmetrical flowers, like those of snapdragons, have a left side that is the mirror image of the right side, and a definite top and bottom.

developmental changes have played an important role in the evolution of life. In this section we discuss three examples of the role of development in evolutionary change.

Changes in flower shape may arise from changes in one gene

Shape plays an important role in helping pollinating animals identify flowers. We can distinguish two broad groups of flowers: radially symmetrical (for example, daisy and dandelion flowers) and bilaterally symmetrical (such as snapdragon flowers) (Figure 37.11).

Experiments indicate that in snapdragons at least, a single gene that is expressed only in the uppermost portions of the flower meristem creates the bilateral symmetry. Where expressed, this gene leads to reduced petal growth and to a reduction in the number of nearby flower parts. Mutant snapdragons that have an inactive form of this gene produce radially symmetrical flowers with six petals instead of the usual five. Pollinators that visit normal, bilaterally symmetrical snapdragon flowers might not recognize the radially symmetrical snapdragon flowers as worth visiting. Thus a simple genet-

ic change leads to a far-reaching change in flower characteristics that could affect the plant's reproduction.

Changes in the expression of one gene could dramatically change the shape of a chicken's foot

Chicken feet have four distinct toes (Figure 37.12). Chick embryos, however, have webbed feet more like those of an adult duck than those of an adult chicken. The dis-

Figure 37.12 Why Chickens Don't Have Ducks' Feet
Adult chickens have feet with separate toes, but early in development a webbing connects their toes that is similar to the webbing of adult ducks' feet. In chickens, the webbing disappears during development, when a gene turns on in the cells of the embryonic webbing that triggers the deaths of these cells.

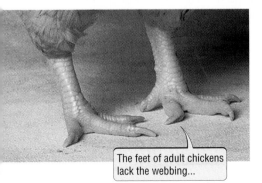

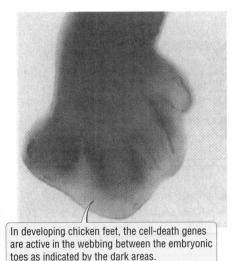

The feet of adult chickens lack the webbing...

...that stretches between the toes of adult ducks, allowing them to swim.

In developing chicken feet, the cell-death genes are active in the webbing between the embryonic toes as indicated by the dark areas.

appearance of the webbing between the embryos' toes seems to depend on a single gene. This particular gene, when active in a cell, causes that cell to die. During development, this gene turns on specifically in the cells in the webbing and not in the toes (see Figure 37.12). A chick embryo that had a mutant form of this gene that failed to turn on in the webbing might well grow up to have ducklike, rather than chickenlike, feet. We can see how a minor genetic change—a mutation in a single gene—could lead to a dramatic morphological change, the development of a completely different kind of foot.

Humans and chimpanzees differ in the timing of developmental events

Chimpanzees are among our closest relatives. Outwardly, adult humans look little like adult chimpanzees. Adult chimpanzees have jaws that extend farther than human jaws, and their skulls are much less domed than human skulls (Figure 37.13). The heads of baby chimpanzees, on the other hand, are shaped much like baby human skulls,

and newborn chimpanzees share with adult humans a relatively flat face and domed skull. During human development, the shape of the skull changes little, but during chimpanzee development, skull shape changes greatly (see Figure 37.13).

The difference in skull shape between adult humans and chimpanzees may simply reflect differences in the timing of the development of the bodies relative to the rate of sexual maturation that defines adulthood. Evidence suggests that, compared to chimpanzees, the rate at which the human skull develops has slowed relative to the rate at which the reproductive organs develop. In other words, humans gain sexual maturity while still in a youth's body. Here again, the effects of developmental changes are pronounced and complex, but the causes may be relatively simple.

> ■ Small developmental changes can have profound evolutionary effects. A single gene regulating development can dramatically alter the form or function of an organism.

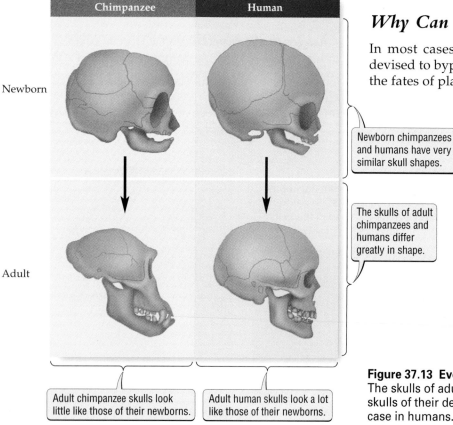

Newborn chimpanzees and humans have very similar skull shapes.

The skulls of adult chimpanzees and humans differ greatly in shape.

Adult chimpanzee skulls look little like those of their newborns.

Adult human skulls look a lot like those of their newborns.

Figure 37.13 Evolution of the Human Skull
The skulls of adult chimpanzees differ in shape from the skulls of their developing young relatively more than is the case in humans.

Highlight

Why Can We Clone Carrots and Sheep?

In most cases, cloning is a way that humans have devised to bypass the factors that normally determine the fates of plant or animal cells during development. For example, when the phloem cells of a carrot are isolated, they no longer receive positional information from neighboring cells and they no longer come under the influence of hormones released elsewhere in the plant. Similarly, the sheep's udder cell, once inserted into a sheep's egg, is cut off from all the morphogens that normally control its fate.

However, cloning sheep required more elaborate methods than those needed to clone carrots and many other plants. Success in cloning mammals has up to now required the insertion of a diploid nucleus into an egg cytoplasm. The egg cytoplasm seems to provide the conditions necessary to restore some developmental flexibility to mammalian nuclei taken from specialized udder cells of mature organisms. To clone the sheep described at the beginning of the chapter, the Scottish researchers had to deprive the udder cells of nutrients for 5 days for the cells to recover some developmental flexibility.

In general, plants commit their cells less firmly to a specific fate during development. The modular growth form of plants means that cells capable of developing into any plant tissue are scattered all over the plant. In animals, in contrast, such cells are isolated in germ lines. In fact, we take advantage of this developmental flexibility to clone houseplants. If you place African violet leaves in moist soil, for example, they will readily take root to produce a new African violet plant.

> ■ Cloning mammals takes extensive manipulation, and to date it has been successful only when a nucleus from the mature cell to be cloned is placed into an unfertilized egg. Plants, with their many meristems and less restrictive cell fates, can be cloned more easily.

Summary

Animal Development

- Early development of animal embryos involves mainly cell division, with little, if any, cell growth or specialization.

- After the period of rapid cell division, the three germ layers (endo-, meso-, and ectoderm) form.

- During gastrulation, the germ layers migrate into their appropriate positions within the embryo.

- After gastrulation, the germ layers differentiate into specialized organs.

Plant Development

- After a period of rapid cell division, plant embryos form three basic tissues: protoderm, ground meristem, and procambium.

- The development of nutritive tissues around the embryos of conifers and flowering plants allows them to remain dormant until conditions favor germination and growth.

- The rigid cell walls of plant cells prevents cell migration during development. Plants take shape by controlling patterns of cell division, cell enlargement, and cell death.

- The apical meristem tissue in each repeated plant module enables plants to continue to grow throughout their lives, even after a significant portion of the plant has been damaged.

What Controls Development?

- Cell and tissue specialization is a result of the regulation of gene expression within cells, not of genetic differences between cells.

- Gene expression can be regulated by control of transcription of DNA to make RNA, by control of the "editing" or cutting up of mRNA, or by regulation of transcription of RNA to form proteins.

- As cells become more specialized during development, their ability to produce other cell types becomes more limited. The need for reproduction, however, requires that some cell types—meristems in plants and germ lines in animals—retain the ability to produce cells that are not locked into a specialized type.

- Patterns in morphogen concentrations inform cells of their positions within the developing embryo and influence the expression of homeotic genes, which in turn influence the expression of many genes that influence structure and function.

- In addition to morphogens, hormones influence patterns of development.

- The internal systems that regulate development respond to internal environmental cues such as nutrient status and external environmental cues such as temperature, light regime, or even the chemicals produced by potential predators and competitors.

Development and Evolution

- Changes in the expression of a small number of genes that influence development can have profound effects on the resulting organism.

- Changes in the rate of growth of one structure in relation to another can dramatically influence structure and function.

- Changes in the pattern of programmed cell death during development can influence structure and function.

Highlight: Why Can We Clone Carrots and Sheep?

- The fates of mature mammal cells are so limited that, to date, they can only be cloned by placing the nucleus into an unfertilized egg.

- The fates of mature plant cells appear to be much less restricted than those of mature animal cells, allowing them to be cloned much more easily.

Key Terms

apical meristem p. 566

blastula p. 562

ectoderm p. 563

endoderm p. 563

endosperm p. 566

gastrulation p. 564

germ layer p. 563

germ line p. 569

ground meristem p. 565

mesoderm p. 563

module p. 566

morphogen p. 569

placenta p. 563

procambium p. 565

protoderm p. 565

seed leaf p. 566

Chapter Review

Self-Quiz

1. Which of the following events characterizes the early stages of human development?
   a. the appearance of distinct endoderm, mesoderm, and ectoderm cells
   b. a complex rearrangement of cells that leaves the ectoderm to form the outer layer of the developing embryo
   c. rapid differentiation of the various organs
   d. all of the above

2. Even after the top half of a plant has been grazed, it can continue to grow because
   a. gastrulation has not occurred.
   b. cell division accelerates.
   c. developmental genes have not been lost.
   d. ungrazed modules have undamaged meristems.

3. Development can be regulated by control of which of the following processes?
   a. translation
   b. transcription
   c. editing of messenger RNA
   d. all of the above

4. Which of the following differences between plant and animal cells may allow gastrulation in animals and prevent gastrulation in plants?
   a. the presence of cell walls in plants and the absence of cell walls in animals
   b. the absence of cell walls in plants and the presence of cell walls in animals
   c. the difference in plasma membranes between plants and animals
   d. hormonal differences between plants and animals

5. The development of form and function is influenced by
   a. programmed cell death.
   b. the rate of cell growth.
   c. both a and b
   d. none of the above

Review Questions

1. Why do you think that the fates of cells become increasingly narrowly defined as development progresses? What problems might arise if all cells in a mature plant or animal could become any other kind of cell?

2. What type of information does a cell need to appropriately regulate its growth and specialization? And how does a cell obtain this information?

3. Can you think of reasons why many biologists believe that developmental changes have played a particularly important role in the evolution of plants and animals?

The Daily Globe

Deformed Frogs May Indicate Environmental Damage

MINNEAPOLIS, MN. When the second grade class from St. Ingmar Elementary School visited a local marsh they made a disturbing discovery: they found frogs with extra legs and eyes.

Students in Ms. Pedersen's second grade class found the frogs while looking for tadpoles in a marsh in this small farming community north of Minneapolis. Ms. Pedersen noted that her class found many frogs with deformities of all sorts during their trip.

Local activists have become particularly concerned since the marsh is part of the drainage that feeds the local reservoir and provides the town's drinking water.

The St. Ingmar students appear to have learned what scientists at major universities have been discovering over the past decade. According to Dr. Rick Kikker of Lutheran Community College, si-

milar deformities have been found in frogs across the United States and Canada. "In some places 96% of the frogs collected suffer from at least some degree of deformity," he noted.

Scientists emphasize that the cause of the deformities remains a mystery. Two prime suspects in the search for a culprit are chemical pollutants and increasing ultraviolet radiation. Peter Lundgren, a fisheries biologist at the Walleye Lake Fish Hatchery, explained that some chemical pollutants mimic the action of homones, which act like chemical messengers inside the bodies of animals. He explained that these chemicals may disrupt events normally controlled by the animals' own hormones. He pointed out that other researchers have found that exposing frog tadpoles or fish to high levels of ultraviolet light causes similar

deformities. Lundgren notes that ultraviolet radiation has been increasing as a result of damage to Earth's ozone layer.

Walter Cleaver, a student at Stanton University in California, has a very different explanation. He has found that worms living inside the frogs can cause deformities in the laboratory.

Harold Simpson, a spokesman for the United Paper Company, which has a factory on the edge of the St. Ingmar marsh, points out that his company uses the latest in pollution control technology and that researchers at United Paper favor the worm theory.

The mayor of St. Ingmar agreed, saying at a recent press conference, "There is no scientific evidence that our local water, our drinking water, is contaminated or at any such risk."

Evaluating "The News"

1. Do you think that a single chemical or increased levels of ultraviolet radiation could cause deformities such as those found in the frogs? Explain why or why not.
2. Many of the people of St. Ingmar are now buying bottled water,

which can be quite expensive. Should the people of St. Ingmar be concerned about their water quality or not?
3. How would you respond to a request for funding from researchers interested in studying

the causes of deformities in frogs in the St. Ingmar marsh? Are animals valid indicators of threats to human well-being? Should we spend public money to study economically unimportant species?

38 *Human Form and Function*

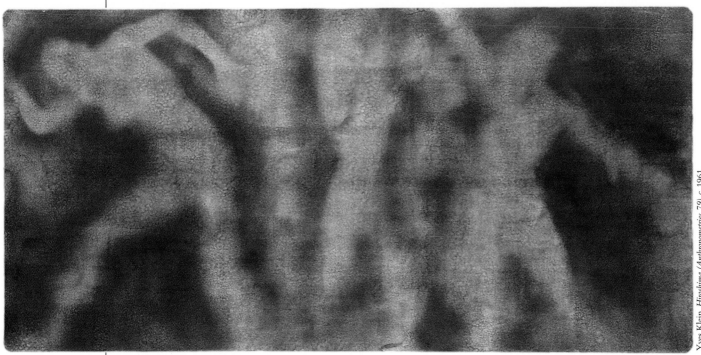

Yves Klein, *Hiroshima (Anthropometries 79)*, c. 1961.

Lucy's Footprints

Long before our species came into being, three individuals no more than a meter and a half tall walked upright across a desolate African landscape. The recent eruption of a nearby volcano had covered the landscape with a layer of fine gray ash, smothering all ground vegetation. The small group hurried almost due north across

the lifeless plain, probably on a desperate search for refuge from the devastation that surrounded them. At one point the group stopped to look back in the direction from which it had come. Quickly they turned north again to resume their flight to safety.

Amazingly we *know* that something very like this scene took place between 3.4 and 3.8

Main Message

The way in which the modern human body functions reflects a compromise among the many conflicting selective pressures that humans have faced during their evolutionary history.

million years ago near what is now Laetoli in the East African nation of Tanzania. In 1978, a team of scientists led by Mary Leakey stumbled across the fossilized footprints that tell this tale. The three creatures that unknowingly recorded a fragment of their lives in the rocks of Laetoli were among our ancestors. They belonged to an important species in human evolutionary history called *Australopithecus afarensis*, first made famous by the discovery of a nearly complete female skeleton nicknamed Lucy.

In a way that no bones can, these footprints illuminate our evolutionary history. They reveal something of how our ancestors behaved, how they moved, and even the predicaments they encountered. The footprints go beyond the information that fossilized skeletons provide about appearance, giving us a rare glimpse of how our ancestors lived.

The discovery of the Laetoli footprints takes on even more significance when we consider that *A. afarensis* is one of the first members of our lineage to have many distinctively "human" characteristics. Lucy's remains suggested that *A. afarensis* stood upright, a characteristic feature of our own biolo-

gy. Lucy's remains and the footprints tell us that although her brain had yet to evolve to the large size expected for a member of our own genus, *Homo*, she walked much as we do. In addition, although no evidence exists that *Australopithecus* made or used the stone tools such as those produced by later *Homo*, most researchers believe that *Australopithecus* used simple wooden tools. By understanding Lucy, we gain an important insight into our own origins.

Australopithecus afarensis
Provides Insight into Human Evolution

Key Concepts

1. Upright, bipedal walking is a distinctively human trait that may have made possible the use of tools by freeing our hands, but also forced adjustments in our backbone and in our reproductive biology.

2. The increase in brain size during human evolution

involves mostly the part of the brain used to make sense of and respond appropriately to sensory information, especially that coming from the hands.

3. Human reproductive biology reflects the need to accommodate babies born with large brains.

Throughout Unit 5, we have surveyed the various structures and functions that allow organisms to deal with the challenges of staying alive. We have dealt with each functional system separately, focusing on how different organisms accomplish the same essential function. Largely lost in this approach is an appreciation of how the various functions interact, both during the lifetime of an individual and over the evolutionary history of a lineage. In this chapter we draw on ideas presented in many of the preceding chapters to focus on a single, familiar species—humans—and consider how the various parts of the human body evolved and work together as a functioning whole.

Because we cannot hope to cover everything about human form and function in a single chapter, we will concentrate on three interrelated aspects of human biology: (1) Humans walk on two legs. (2) Humans have very large brains relative to body size. (3) Birth in humans is unusually complex. The first two of these topics deal directly with features that set humans apart from other mammals. Our way of walking and the intellectual capacities provided by the human brain have played a central role in our evolutionary success and in defining human beings. The third topic, reproduction, depends on all functions of the body and, as we will see, has been shaped by the way we walk and by our ability to think.

Each of these three aspects of our biology—locomotion, coordination, and reproduction—has been dramatically influenced by changes in the other. We will see that the human body, far from being a perfect machine, represents a remarkable set of compromises.

A Brief Review of Human Evolution

Since we already discussed human evolution in detail in Chapter 23, we will review just a few points relevant to our discussion of human form and function. Humans belong to the species *Homo sapiens* (literally, "thinking man"), in a group of mostly tree-dwelling mammals called the primates. Three trends distinguish primate evolution: (1) the evolution of opposable thumbs and big toes, which allowed primates to climb more effectively; (2) the evolu-

tion of eyes positioned on the front of the face, which made possible the excellent depth perception needed for movement from tree branch to tree branch; and (3) the evolution of large brains, the reason for which remains a mystery.

The first humanlike primate, or hominid, appears in the fossil record about 4.4 million years ago. These early African hominids lived mainly on the ground and walked **bipedally**—that is, on two legs. The first member of the genus *Homo* appeared in Africa about 2.5 million years ago. Members of this genus characteristically used tools and may have used language. A rapid increase in brain size marks the evolutionary history of *Homo*. Our own species, *H. sapiens*, probably evolved in Africa within the past 200,000 years. Since their first appearance, humans have spread over the face of Earth and increased in number to 6 billion individuals.

> ■ Humans have evolved relatively recently from primates, a group of tree-climbing mammals characterized by opposable thumbs, stereoscopic vision, and brains that are large relative to body size.

Walking on Two Legs

Walking upright on two legs sets humans apart from our primate ancestors, as well as from all other four-limbed mammals. An upright posture freed our hands for carrying and manipulating things. In this way, our mode of locomotion probably created the opportunity for the evolution of tool use. At the same time, however, converting the skeleton of our ancestors that walked on all fours into a skeleton suited for life on two legs has created some problems, including, as we will see, backaches and difficult births.

Life in an increasingly treeless environment may have favored an upright posture

If we trace human history back far enough, we find tree-dwelling ancestors that used both hands and feet to

(a)   (b)

Figure 38.1 Early Human Ancestors
Ramapithecus (a) was an ancestral ape alive at the time of transition to a drier climate on Earth. It may have used an upright posture in much the same way that modern grassland-dwelling baboons (b) do.

tion to life in trees and by creating the selective pressure that favored limbs suited to an upright posture.

Our most likely ancestors during this period of climatic change were members of, or at least resembled, the extinct genus *Ramapithecus*. *Ramapithecus* stood just over 1 meter tall and weighed about 20 kg (Figure 38.1). These small apes probably used an upright posture much as modern baboons that live on African grasslands do: to see over the grass tops when watching for predators and to make themselves more threatening to predators (by appearing larger because of their increased height) when caught far from the safety of the scattered trees.

grasp branches. Life in the trees favored the use of all four hands and feet to grasp branches.

During the period when walking on two legs is thought to have first arisen, roughly 12 million years ago, Earth's climate became increasingly dry. The drier conditions allowed drought-tolerant grasses to spread into formerly forested areas. Life in these grassy habitats may have set the stage for walking on two legs by weakening the selective pressures that favored adapta-

A nonclimbing life is associated with the evolution of hands that manipulate

The transition to life in a grassland environment also led to an increasing dependence on grass seeds for food. The ability to carefully collect the small grass seeds required a more precise grip than was possible with a hand suited to life in the trees.

Our tree-dwelling ancestors had hands and feet suited for grasping tree branches (Figure 38.2). In tree-dwelling apes, both the big toes of the feet and the thumbs of the hand are opposable; that is, they can

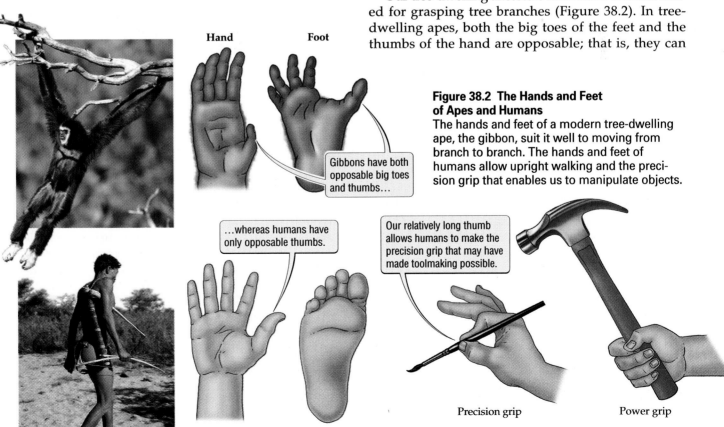

Hand   **Foot**

Gibbons have both opposable big toes and thumbs...

...whereas humans have only opposable thumbs.

Our relatively long thumb allows humans to make the precision grip that may have made toolmaking possible.

Figure 38.2 The Hands and Feet of Apes and Humans
The hands and feet of a modern tree-dwelling ape, the gibbon, suit it well to moving from branch to branch. The hands and feet of humans allow upright walking and the precision grip that enables us to manipulate objects.

Precision grip   Power grip

move independently of the other toes or fingers to grasp things. Compare what we can do using our **opposable thumb** with the range of motion of our big toe, which is not opposable. To keep from getting in the way when moving from branch to branch, however, the opposable big thumbs of tree-dwelling apes are much shorter than the human opposable thumb.

The opposable thumb suited for life in trees restricts hands to grasping objects relatively powerfully but imprecisely (see Figure 38.2). The success of apes living in grasslands no longer depended as much on their ability to grasp tree branches. As a result, selective pressures favoring hand shapes suited to other functions could more readily influence the evolution of hands. The less the hands were needed for locomotion, the more they could specialize for precise manipulation. Having well-developed precision, in turn, probably set the stage for our sophisticated tool use.

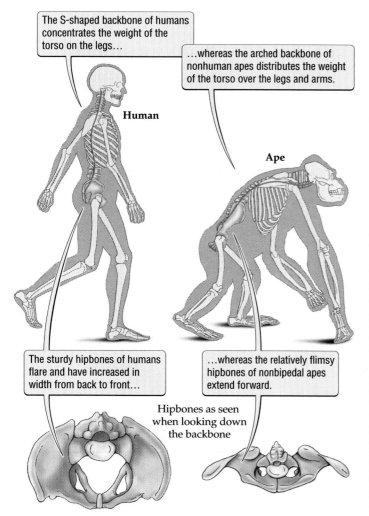

Figure 38.3 Effects of Bipedalism on the Human Skeleton Human skeletons differ from those of apes that walk on all fours in ways that reflect the different demands of supporting an upright posture.

Upright posture required some dramatic changes in the human skeleton

A fully upright posture had evolved by Lucy's time. The skeleton of *Australopithecus afarensis* shows a thorough modification of the typical mammalian skeleton, which suits life on four legs, to one that suits life on just two. The conversion of the skeleton of a mammal that moves on all four legs to one that allows bipedal walking required many adjustments.

Two changes in particular are striking. The backbone of mammals that walk on all fours, like cats, dogs, and gorillas, forms an arch that rests on two sets of pillars, the fore- and hind legs (Figure 38.3). Arches appear commonly in bridges and buildings because they convert the downward pull of gravity on the mass of a structure into a squeezing, compressing force that stone and concrete can resist very well. Similarly, the arch formed by the backbones of four-legged animals allows them to support their weight by converting it into a compressive force that the stonelike mineral matrix of our bones can resist (see Chapter 26) (Figure 38.4*a*).

The arched backbone allows bone to easily support the weight of the guts that hang suspended from it in a sling formed from a broad sheet of ligament. If this backbone were the same in modern humans, the upright posture would rotate the arch to form a C shape that would collapse easily under the weight of our guts (Figure 38.4*b*). To accommodate the changed orientation, our backbone takes on an S shape shortly after birth that allows it to act more like a pillar. In essence, the weight of our upper body and all its organs is supported by a pole (Figure 38.4*c*).

The pressure of body weight is distributed evenly over the arched backbone of gorillas, but in humans it is concentrated on the lower back. This difference is evidenced in the relatively uniform size of gorilla vertebrae, in comparison to the vertebrae of humans, which in the lower back are massive compared to those in the shoulder area, where they support less weight.

The second conspicuous difference between the human skeleton and that of animals that walk on four legs lies in the hipbones and their associated muscles (see Figure 38.3). The four legs that support most mammals provide a naturally stable base, like a table. In contrast, humans remain upright when we stand or walk only through the balancing action of the muscles of our hips. Most of the hip muscles that in a gorilla serve to straighten its legs and drive it forward have in humans taken on primarily a stabilizing role.

Think of what happens when we walk, for example: Every time we plant our foot, the upper body tends to continue moving forward. Without the stabilizing contraction of the biggest muscle in the body, the gluteus

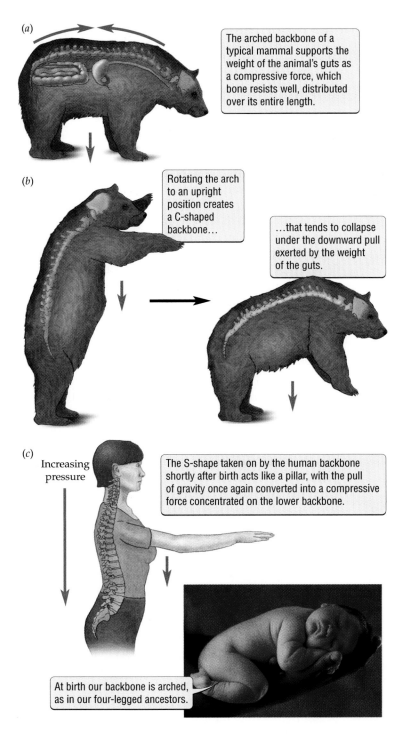

(a) The arched backbone of a typical mammal supports the weight of the animal's guts as a compressive force, which bone resists well, distributed over its entire length.

(b) Rotating the arch to an upright position creates a C-shaped backbone...

...that tends to collapse under the downward pull exerted by the weight of the guts.

(c) Increasing pressure

The S-shape taken on by the human backbone shortly after birth acts like a pillar, with the pull of gravity once again converted into a compressive force concentrated on the lower backbone.

At birth our backbone is arched, as in our four-legged ancestors.

Figure 38.4 The Design of Backbones
The human backbone is a modified version of the backbone of four-legged mammals. (*a*) Mammals that walk on four legs have a backbone that forms a curved arch that rests on four pillars, the fore- and hind legs. Weight pulling down on an arch creates the compressive forces that bone is so good at resisting. (*b*) When a four-legged mammal stands in an upright position, the arch of the backbone no longer supports the animal's weight, and tends to fold in half under the downward pull of the animal's weight. (*c*) The human backbone has lost the arched shape of most mammalian backbones; instead, it acts as a pillar that resists the compressive forces created by the downward pull of our guts. Human infants less than a few months old still have an arched backbone.

the side of its torso, human hipbones flare outward to form a dishlike shape. Human hipbones extend beyond the joint that connects the upper leg bone, or femur, with the hip to place the stabilizing muscles in a better position to keep the body upright.

The skeletal changes required for an upright posture create several problems

Tinkering with a skeleton designed for life on four legs to convert it into one suited for life on two legs has created its share of problems. We will briefly consider two of these problems here. Because we concentrate all our weight on the joint between the backbone and the leg bones at the hips, the hipbones come under tremendous stress. Coincidentally, it is here that we suffer most of our skeletal aches and pains. Lower back problems plague many people.

In addition, modifying the shape of the hipbones to accommodate an upright posture has affected the shape of the **pelvic aperture**, the opening through the hipbones that serves as the birth canal. In humans, the pelvic aperture has become relatively smaller than it is in primates that walk on all fours (see Figure 38.7). This decrease in size makes childbirth more difficult, as we will see later in the chapter.

maximus muscle that makes up most of the buttocks, we would flop forward at the waist with each step. In the gorilla, the gluteus maximus is a minor muscle that helps straighten the leg.

The shape of the human hipbone differs from the gorilla's hipbone to accommodate the stabilizing muscles that attach to it (Figure 38.5; see also Figure 38.3). Whereas the gorilla has bladelike hipbones that extend forward along

■ The evolutionary changes involved in the transition from walking on four legs to bipedalism created some problems for modern humans. A major consequence of bipedalism was the freeing up of hands that made it possible for humans to make and use tools.

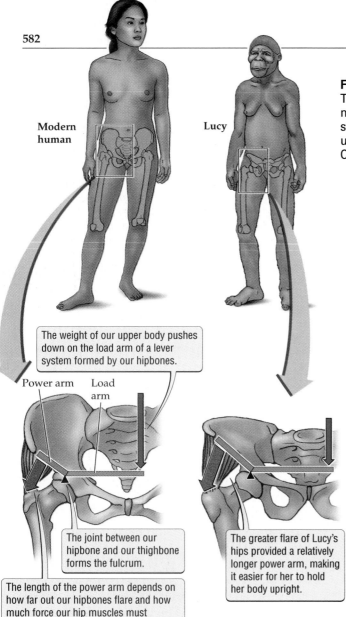

Figure 38.5 Hominid Hips Accommodate an Upright Posture
The outward flare of bipedal human hips helps provide the hip muscles with leverage in supporting the body. Lucy had more strongly flared hips and therefore could probably maintain an upright posture more easily than modern humans can. (See Chapter 27 for a review of lever systems.)

Modern human

Lucy

The weight of our upper body pushes down on the load arm of a lever system formed by our hipbones.

Power arm   Load arm

The joint between our hipbone and our thighbone forms the fulcrum.

The length of the power arm depends on how far out our hipbones flare and how much force our hip muscles must generate to offset the weight of our torso.

The greater flare of Lucy's hips provided a relatively longer power arm, making it easier for her to hold her body upright.

Humans Have Big Brains

Although the human brain is not the largest in the animal kingdom, an honor that belongs to the 6-kilogram brain of the 130,000-kilogram blue whale, it is the largest in relation to body size. In this section we explore the connections between the structure of our unique brain and two equally unique characteristics of being human: making tools and communicating through spoken language.

How and why has the large human brain evolved?

As the size of the mammalian brain changes, so does the size of the brain's various subdivisions. However, the various parts of the brain do not increase in size at the same rate as overall brain size increases. As a result, an increase in brain size causes a change in the relative importance of

the various subdivisions of the brain. For example, the size of the mammalian **neocortex**, the part of the brain that allows mammals to make sense of and respond appropriately to sensory information, increases in size more rapidly with increasing brain size than does the part of the brain (the medulla) devoted to regulating basic body functions, such as breathing and digesting (Figure 38.6). Thus, simply because our brain is big, our neocortex makes up a much larger proportion of our brain than it would in a smaller-brained organism.

Big brains evolved rapidly and much later than bipedalism

The large size that characterizes the human brain has arisen very recently in our evolutionary history. Four million years ago *Australopithecus* had a brain volume of about 420 milliliters. Two to 2.5 million years later, the first member of our own genus, *Homo habilis*, had a brain that was about 200 milliliters bigger. In the past 1.5 to 2 million years, however, our brain size has more than doubled, to the current volume of about 1400 milliliters. Human brain size began its rapid increase in size roughly 2 million years after clear evidence of upright posture, and more or less in parallel with the increasing use of stone tools. From the preceding discussion, we would conclude that much of this recent, rapid increase in our overall brain size resulted from an increase in the size of the neocortex.

The reason why human brain size has increased so rapidly during our recent evolutionary history remains a mystery. Among mammals, the only features of the biology of an organism that consistently seem to go hand in hand with larger-than-expected brain size are life in trees and feeding on foods that are both scarce and patchily distributed. Both of these characteristics require the handling of complex spatial information. These two explanations seem unlikely to account for the dramatic increase in the size of the human brain over its recent evolutionary history. Early, bipedal humans neither spent much time climbing trees nor fed on food types that were any more scarce or patchily distributed than those of other organisms.

Instead, once brains reached the size of *Homo erectus* brains, a small increase in overall brain size may have led to a relatively large increase in the size of the neocortex, which plays so important a role in how we perceive and respond to our world. The neocortex may

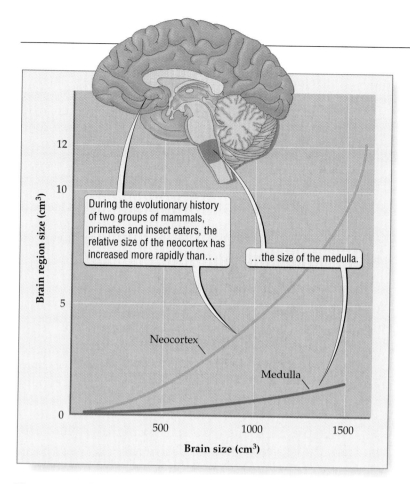

Figure 38.6   The Effects of Having a Bigger Brain
As brain size increases in primates and insect-eating mammals, the neocortex, devoted to responding appropriately to sensory information, increases in size faster than the medulla. In the brain of a modern human, therefore, the neocortex is larger relative to the medulla than it was in the brain of *Australopithecus*.

have proven especially useful to humans who lived and hunted in social groups, and who were beginning to develop language and technology.

In what would appear to be a logical extension of this trend in human evolution, we have come to equate big brains with greater intelligence. Thus, our culture is filled with images of highly intelligent earthlings and aliens with big heads that presumably contain large brains. Recent studies relating brain size to scores on intelligence tests support this stereotype, although they also show that test scores can vary widely for a given brain size.

Despite the associated increase in intelligence, it is unlikely that the human race will continue to evolve ever larger brains. As suggested earlier in this unit, the support of our large brains consumes a lot of energy (see Chapter 30). At rest the human brain, which makes up 2 percent of our body weight, consumes 15 percent of the oxygen we take in. In addition, as we will see later in this chapter, the human brain is already big enough to complicate birth.

Large parts of the human brain are devoted to hands and to speech

In comparing the organization of the modern human brain to that of our closest living relatives, the chimpanzees, we find notable differences related to our extensive use of our hands and to language. More of the brain processes sensory inputs from the hands and controls their movements in humans than in chimpanzees. This feature of the human brain almost certainly followed from the evolution of a precision grip in our first upright ancestors. In combination with the depth perception skills that we inherited from our primate ancestors, our ability to make fine movements with our fingers may have contributed to our unique toolmaking abilities.

In addition, whereas a large portion of the large neocortex of humans is devoted to understanding and producing speech, only a small portion of the chimpanzee's neocortex is devoted to these functions. The left- and right-hand sides of our brains have specialized to handle different aspects of using our hands and of language. Chimpanzees process information related to hands and to sound production and detection similarly on both sides of their brain.

> ■ Modern humans have big brains relative to their body size. Much of this increase in brain size has come relatively late in our evolutionary history and has involved an increase in the size of the neocortex, the portion of the brain that allows us to make sense of and respond appropriately to sensory information.

Human Birth: A Success Story?

For all our extraordinary success at filling Earth with people, human birth is remarkably complex and hazardous. In this section we consider some of the general features of human reproduction. Then we turn to some of the problems arising from the conflicting evolutionary pressures for upright posture and increased brain size.

Human babies depend on their parents for an unusually long time

For a species that has managed to cover practically every corner of the world with an incredible population of 6 billion, our reproductive biology proves remarkably unremarkable. We are not prolific breeders like the orchids or puffball fungi that can pump out hundreds of thousands of offspring in a lifetime. Instead, we produce one baby at a time, usually at intervals of 2 years or more. Over our lifetimes, only very exceptional individuals produce more than 20 offspring. Nor are our babies particularly well prepared to face the world at birth. Newly hatched cockroaches can feed themselves from hatching, and the tiny seedlings of giant sequoia trees carry all they need to get started in life in a small seed. A newborn human can do little more for itself than make a great deal of noise and mess.

What sets humans apart is the tremendous investment we make in each of our young, both in time and in resources. Compared to other animals, we humans depend on our parents for an unusually long time. For 9 months we live sheltered in our mother's uterus. For a year or more we obtain all our nutrients by nursing at her breast. For an additional 10 or more years we depend on our parents to provide food and, more importantly, information and skills. Our remarkable ability to learn means that we can benefit from long periods of training.

The prolonged physical and intellectual development of humans takes place within a complex social structure. This social structure traditionally revolved around an extended family that included parents, aunts, uncles, cousins, grandparents, and so on. Once a child becomes independent of its mother's breast, all these individuals can help feed, protect, and train the child. These social exchanges with other humans almost certainly contributed to the development of the remarkably complex societies we live in today.

therefore dictates the ability of the baby to pass through the pelvic aperture. Whereas the skull of a baby chimpanzee passes easily through the pelvic aperture of its mother during birth, a human infant must make some elaborate maneuvers to fit its large skull through the pelvic aperture of its mother (Figures 38.7 and 38.8).

To emerge from the uterus, we first had to bend our necks and then rotate our heads as we passed through the hipbones (see Figure 38.8). At birth, the diameter of a human baby's skull actually exceeds the maximum diameter of the pelvic aperture. Thus, in addition to the baby's bending and turning, both the baby's skull and the mother's birth canal must change shape for the baby to squeeze through.

This incredibly complex and tight fit is possible only because of modifications to the reproductive biology of humans. The first of these was a change in the arrangement of the hipbones of females to create a rounder opening than that found in the hipbones of the smaller-brained *Australopithecus* (see Figure 38.7). This modification required a widening of the hips from front to back. In addition, around the time of birth the ovaries of expectant mothers release a hormone called relaxin, which enables the ligaments that hold together the hipbones to stretch a bit. Without this stretching, the baby's head could not fit through the pelvic aperture. A side effect of relaxin production, however, is that the expectant mother often suffers from joint pains because relaxin affects the ligaments that hold together bones in other joints of the body as well.

Another feature of human reproduction that sets our species apart is that human babies are born at an earli-

The mechanics of human birth are complex

One of the riskiest stages in our lives and those of our mothers occurs at birth. Before the development of sterile medical procedures, the baby, the mother, or both commonly died during childbirth.

The difficulty of birth in humans reflects the large size of the skull and the brain that the skull protects. During birth humans, like all other mammals, must exit the mother's body through a pelvic aperture that is defined by rigid hipbones. The skull is a baby's largest rigid structure, and it

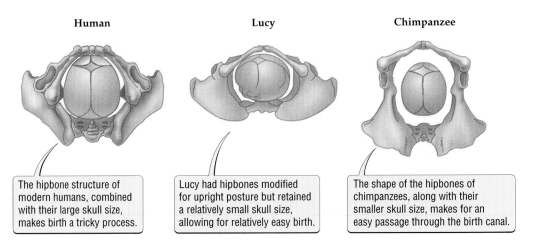

Human	Lucy	Chimpanzee
The hipbone structure of modern humans, combined with their large skull size, makes birth a tricky process.	Lucy had hipbones modified for upright posture but retained a relatively small skull size, allowing for relatively easy birth.	The shape of the hipbones of chimpanzees, along with their smaller skull size, makes for an easy passage through the birth canal.

Figure 38.7 The Pelvic Apertures of Apes and Hominids
Modern humans have difficult births because of our unusually large skull size and because the changes in the structure of the human hip to accommodate an upright posture affected the shape of the birth canal. The human ligaments outlined in red loosen to allow the pelvis to enlarge during birth.

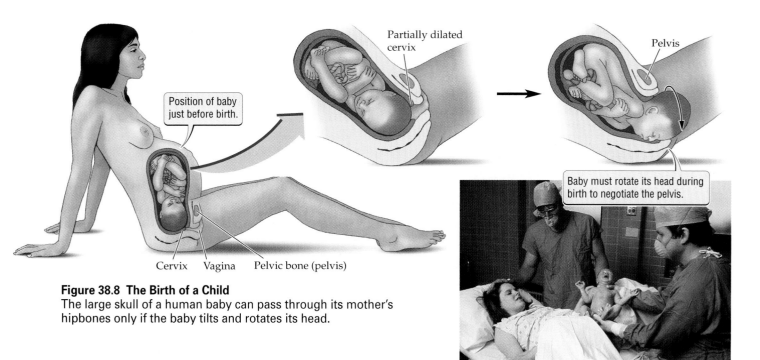

Figure 38.8 The Birth of a Child
The large skull of a human baby can pass through its mother's hipbones only if the baby tilts and rotates its head.

er stage in development than other primates are (Table 38.1). This difference is apparent in the fact that the faces of adult humans are much more like those of their young than are the faces of chimpanzees. If we extended our development in the uterus to the same stage as other primates, however, the skull would be too big to pass through the birth canal. Our necessarily "premature" birth explains our exceptional helplessness at birth, and the resulting long dependence on our parents may contribute to the strong social bonds that link human babies with their families.

■ Humans depend on their parents for an unusually long time. Human birth is difficult and risky because the skull of an infant fits through the pelvic aperture only with difficulty. These problems stem largely from the reduced size of the pelvic aperture that accompanied the evolution of bipedalism and the increased size of the skull that accompanied the evolution of large brain size.

38.1 A Comparison of the Development of Humans and Chimpanzees at Birth		
Trait	**Humans**	**Chimpanzees**
Percentage of adult brain size at birth	23	40
Bone development at birth	Incomplete	Complete
Time to complete set of baby teeth	6–24 months	3–12 months

Highlight

Evolution Doesn't Always Make Us Better

We like to picture ourselves perched atop the evolutionary tree, an example of nature's steady progress toward perfection. From Lucy, however, we can learn a valuable and humbling lesson. An analysis of the structure of Lucy's hipbones suggests strongly that the long extinct *Australopithecus afarensis* was better suited to upright posture 4 million years ago than we are today. Lucy's hipbones flare out more dramatically than do our own (see Figure 38.5). This increased breadth of Lucy's hipbones placed the stabilizing muscles farther out from the joint connecting the hipbone to the femur than are our own. From this position, Lucy's muscles could hold her upright with relatively little effort.

Our own hipbones appear to have become less well suited to an upright posture as a compromise designed to accommodate modern human babies. The large skull size of modern human babies at birth compared to the skull size of *Australopithecus afarensis* newborns requires a differently shaped passage through the pelvic aperture (see Figure 38.7). Unconstrained by excessively brainy babies, *Australo-*

pithecus afarensis hipbones could come closer to the ideal for upright posture.

We often mistakenly think of evolution as a linear process that creates a progressively better product. The Nobel prize–winning biochemist François Jacob, however, has pointed out that natural selection can work only on available variation: It does what it can with the materials at hand. The products of evolution are therefore often filled with compromises and solutions that fall short of ideal. As this chapter has shown, our own evolutionary progress illustrates this point wonderfully.

> ■ Evolution necessarily involves compromises between different systems of the body. The skeletal system of our ancestors was better adapted to an upright posture than ours is. We have compromised that system to accommodate a larger brain.

A Brief Review of Human Evolution

■ Humans belong to a group of mammals known as primates.

■ Distinguishing characteristics of primates include opposable thumbs, which allow for grasping; forward-facing eyes, which enables excellent depth perception; and large brains.

■ Our species, *Homo sapiens*, probably evolved in Africa within the last 200,000 years.

Walking on Two Legs

■ An increasingly dry climate and a shift from forests to grassland environments may have selected for bipedalism and for hands better able to manipulate objects with precision.

■ Full upright posture required a change from a C-shaped spine to an S-shaped spine, as well as modification of the hipbones and muscles to support the body.

■ These skeletal changes have resulted in problems: Lower back pain is caused by the tremendous physical stress placed on the hip area, and some childbirth complications are caused by the reduced pelvic aperture associated with upright posture.

Humans Have Big Brains

■ As brain size increases, some parts of the brain, such as the neocortex (involved in the processing of and response to sensory input) increase in size more quickly than others.

■ After the evolution of upright posture, brain size in members of the genus *Homo* began to increase more rapidly.

■ The large neocortex of humans may have made possible the development of more complex social behaviors and tool use.

■ Human brains dedicate relatively large areas to processing information concerned with using our hands and with speaking.

Human Birth: A Success Story?

■ Humans depend on their parents much longer than other animals do.

■ Birth in modern humans is difficult because of both the relatively large size of the baby's skull and the relatively small pelvic aperture of the mother.

■ Human babies are born less completely developed than are the young of other primates. As a result, after birth, human growth and development require a tremendous amount of parental care, as well as social interactions with an extended community.

Highlight: Evolution Doesn't Always Make Us Better

■ Evolution necessarily involves compromises between different systems of the body.

■ The skeletal system of our ancestors was better adapted to an upright posture than ours is. We have compromised that system to accommodate a larger brain.

Key Terms

bipedalism p. 578   opposable thumb p. 580

neocortex p. 582   pelvic aperture p. 581

Chapter Review

Self-Quiz

1. The earliest members of the genus *Homo*
   a. walked on four limbs.
   b. had C-shaped spines.
   c. walked bipedally.
   d. had a relatively large neocortex, compared to *Homo sapiens*.

2. Opposable thumbs probably evolved in primates initially to enable
   a. climbing.
   b. grass seed collecting.
   c. tool use.
   d. better precision in gripping.

3. The different shape of human hipbones as compared to those of gorillas reflects
   a. the need to allow for the birth of a baby with a bigger skull.
   b. the greater height of humans as compared to gorillas.
   c. the greater weight of humans as compared to gorillas.
   d. the need to allow attachment of large muscles that help stabilize the upright body.

4. As brain size increases,
   a. all parts of the brain increase at the same rate.
   b. the part of the brain associated with processing and responding to information increases more quickly than other parts.
   c. seed eating increases.
   d. skull size decreases.

5. Human babies are born relatively less developed than chimpanzee babies because
   a. 9 months of pregnancy is the maximum an organism can endure.
   b. chimpanzees have a shorter life span than humans.

c. a longer pregnancy would allow the baby's skull to grow so large that it could not pass through the pelvic aperture of the mother.

d. chimpanzees devote less of their brain to speech.

Review Questions

1. Explain how the ability of humans to reproduce represents a compromise between walking on two legs and having a big brain.

2. Compare and contrast the evolution of larger brains in early primates with the rapid evolution of large brains in the genus *Homo*.

3. How did changes in the way hands function interact with both the development of upright posture and the development of the modern human brain?

---

38   𝕿𝖍𝖊 𝕭𝖆𝖎𝖑𝖞 𝕲𝖑𝖔𝖇𝖊

Do Changing Birth Practices Threaten Babies?

WASHINGTON, DC. Almost a decade ago, a U.S. governmental health initiative declared war on the cesarean section (C-section), an operation that is used to avoid the risks of birth through the vaginal canal by allowing the baby to be delivered through a surgical incision in the mother's abdomen. The initiative arose out of concern over skyrocketing C-section rates in the United States: Whereas only 5 percent of births were cesarean deliveries in 1965, 25 percent were C-sections in 1988. In contrast, in some European countries, cesarean rates are closer to 10 percent. Thus, the initiative set the goal of reducing the rates to 15 percent by the year 2000, and seems to have worked. By 1998, cesarean rates had dropped to about 21 percent.

Some doctors fear that avoiding C-sections may put the health of some babies and mothers at risk. They cite increasing rates of certain types of birth injuries in the 1990s as evidence for their concern.

Obstetrician Julius Della Roma cautions that instead of blindly forging ahead toward a magical number, researchers should be trying to understand why high C-section rates exist. Recent studies suggest that the fear of malpractice suits may cause doctors to end difficult labors and resort more quickly to C-sections. In addition, the increasing use of hormones to induce labor and of painkillers to ease labor has been blamed with ultimately making vaginal deliveries less likely to succeed, forcing doctors to perform emergency C-sections. Even fetal monitors, devices

that allow doctors and mothers to track the condition of a baby constantly during labor, have been blamed for unnecessary surgeries.

Changes in our society may play a role as well. Many women are delaying having families until they are older, or they are having babies while simultaneously pursuing a professional career, and Dr. Ellen P. Doral of the Children's Hospital notes that these women are more likely to request a cesarean because of their lifestyle: "Simply telling doctors to put down the knife to meet some arbitrarily defined standard is not going to help mothers or their babies," said Dr. Doral. "Understanding what makes for the safest delivery for any particular woman, whether it's a C-section or vaginal delivery, will."

Evaluating "The News"

1. If you were a doctor and a woman asked for a cesarean in order to avoid the rigors of labor, even though there were no medical indications that her delivery would have a high risk for complications, how would you respond? Should nonmedical concerns be allowed to influence a mother's decision about whether to deliver her baby surgically?

2. Do you think the comparisons to cesarean rates a few decades ago or in other countries is a valid way to determine whether current rates in your country are too high? Why or why not?

3. Some people have theorized that the high rate of cesareans is due to better nutrition and prenatal care

for American mothers. The result is big babies—too big to fit comfortably through a vaginal canal that evolved in the days of smaller babies. How could you test whether such a theory were valid? If it were, what should mothers and doctors do?

Kit Williams, *The Death of Spring*, 1980.

Interactions with the Environment

39

The Biosphere

Dale Chihuly, *Sky Blue Soft Cylinder with Golden Yellow Lip Wrap*, 1993.

El Niño

*T*hroughout much of the world something was seriously amiss. Off the western coast of South America, the fish on which both seabirds and local fishermen depended were gone. As a result, thousands of seabirds starved to death. Farther north, along the coast of California, underwater "forests" dominated by long strands of the brown algae, kelp, were destroyed or heavily damaged by storms. Still farther north, the Canadian and Alaskan coasts experienced high sea levels and unusual amounts of rainfall.

Global patterns of air and water circulation can cause events in one part of the world to affect biological communities all over the world.

On the other side of the ocean throughout the entire western Pacific, the sea level dropped, killing huge numbers of animals that live in coral reefs. Reefs were also hit hard in the eastern Pacific, where 50 to 98 percent of the coral of all species died. Lands separated by thousands of kilometers experienced unusual and violent swings in the weather: Tropical rainforests in Borneo were ravaged by catastrophic drought and fire, and deserts in Peru were flooded by torrential rains. The dramatic changes in the weather were associated with crop failures and disease outbreaks: Corn yields dropped drastically in southern Africa, Australian wheat fields were destroyed, and cholera ravaged parts of South America.

This list of disasters sounds like the beginning of a doomsday science fiction movie. But these events—and many others like them—actually happened in 1982 to 1983, and similar events occurred in 1997 to 1998. These catastrophes were not caused by alien invaders, or even by humans. They were set in motion by natural changes in wind and water currents, changes known collectively as El Niño events.

As we will see in this chapter, El Niño events are just one example of how wind and water currents help shape Earth's climate and thus have a huge impact on all life.

A Rainforest Burns in Brazil during the 1997–98 El Niño.

Key Concepts

1. Ecologists study interactions between organisms and their environment. All ecological interactions occur in the biosphere, which consists of all the living organisms on Earth together with the environments in which they live.

2. Climate has a large effect on the biosphere. Climate is determined by incoming solar radiation, global movements of air and water, and major features of Earth's surface.

3. The biosphere can be divided into major terrestrial and aquatic life zones called biomes.

4. Terrestrial biomes cover large geographic regions and are usually named for the dominant plants that live there. The location of terrestrial biomes is determined by climate, primarily temperature and precipitation, and by the actions of humans.

5. Aquatic biomes cover more than 70 percent of Earth's surface and are usually characterized by physical conditions of the environment, such as temperature and salt content. Aquatic biomes are heavily influenced by their surrounding terrestrial biomes, by climate, and by the actions of humans.

There are millions of known species on Earth, and many millions more are yet to be discovered. Each of these organisms lives in a characteristic place, or **habitat**. Collectively, these organisms and their habitats make up the **biosphere**, which is all living organisms on Earth together with the environments in which they live. The biosphere is very complex. It includes grasslands, deserts, tropical rainforests, streams, and lakes, to name just a few of its many parts. It also includes the bottoms of our feet and the bacteria that grow there, as well as deep-sea vents on the bottom of the ocean and the bacteria that grow there.

In this chapter we discuss the climate and its importance for the biosphere. We also describe terrestrial and aquatic biomes, the major parts into which the biosphere can be divided. First, however, we provide an overview of **ecology**, the scientific study of interactions between organisms and their environment.

Ecology: Studying Interactions between Organisms and Their Environment

All organisms interact with their environment. These interactions go both ways: Organisms affect their environment, and the environment affects organisms. Ecologists study interactions in nature at several different levels, including individual organisms, groups of individuals of a species (populations), groups of different species (communities), and **ecosystems**, a term that refers to the different species in an area plus the environment in which they live. Collectively, all ecosystems on Earth make up the biosphere.

Ranging as it does from the study of individual organisms to the study of the biosphere, ecology is a broad and complex subject. It is also an important area of applied biology. As we will learn in this unit, humans are changing the biosphere, often in ways that are not intended or that have unexpected consequences. A major goal of ecology is to document and understand the consequences of human actions for life on Earth.

For example, chlorofluorocarbons used in refrigerants and in aerosol sprays have created a hole in the ozone layer, a potentially dangerous side effect that people did not want or expect. The ozone hole is potentially dangerous because ozone prevents much of the sun's ultraviolet light from reaching Earth, thus protecting organisms from DNA mutations caused by exposure to ultraviolet light (see Figure 14.9). Ecologists seek to understand both the formation of the ozone hole and its consequences for life on Earth.

In the chapters of this unit, our study of ecology will take us from individual organisms, to populations, to interactions among organisms, to communities, to ecosystems, and finally, to global change caused by humans. But all ecological interactions, no matter at what level they occur, take place in the biosphere. Thus, by beginning our study of ecology with an overview of the biosphere, we will be able to use the material in this chapter to help us understand the different levels (individuals to ecosystems) at which ecology can be studied.

■ Ecologists study interactions between organisms and their environment at different levels, ranging from individuals to ecosystems. A major goal of ecology is to document and understand the consequences for life on Earth of human actions that are changing the biosphere.

Climate

Weather refers to the temperature, precipitation, wind speed, humidity, cloud cover, and other physical conditions of the lower atmosphere at a specific place over a short period of time. The weather changes quickly and is hard to predict. But **climate**, the prevailing weather conditions experienced in an area over relatively long periods of time (30 years or more), is predictable and has a great impact on ecological interactions. In fact, organisms in the biosphere are more strongly influenced by climate than by any other feature of the environment.

In this section we consider factors that shape climate: the amount of incoming solar radiation, the movement of air and water, and major features of Earth's surface.

Incoming solar radiation shapes climate

Tropical areas near Earth's equator are much warmer than polar regions. The reason for this difference in temperature is that sunlight strikes Earth directly at the equator but at a slanted angle near the North and South Poles (Figure 39.1).

The difference in the angle at which rays from the sun strike Earth in polar and equatorial regions has two main effects. First, rays from the sun travel a greater distance through the atmosphere in polar regions than they do in the Tropics (roughly 0 to 30 degrees latitude). Because the atmosphere reflects, absorbs, and scatters sunlight, this increased distance that the rays travel means that less solar energy reaches Earth's surface in polar regions than near the equator. Second, for a given amount of solar energy that reaches the surface of Earth, that energy is spread over a larger area in polar regions than in the Tropics (see Figure 39.1). Together, these two effects cause polar regions to receive only 40 percent of the solar radiation that reaches the Tropics, making them much colder than the Tropics.

The amount of solar radiation received by regions outside the Tropics varies greatly during the year, giving rise to the seasons. The large seasonal variation in solar radiation occurs because Earth is tilted 23.5 degrees on its axis. As Earth revolves around the sun, the Northern Hemisphere is tilted toward the sun in June (and hence receives more energy) and away from the sun in December (and hence receives less energy). Careful examination of Figure 39.2 will show why

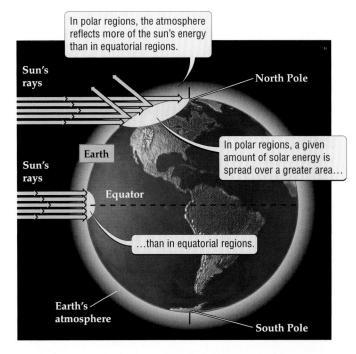

Figure 39.1 Solar Radiation Strongly Influences Climate
Sunlight strikes Earth less directly in polar regions than in tropical regions. For this reason, polar regions receive less energy from the sun than do tropical regions, thus making polar regions colder than tropical regions.

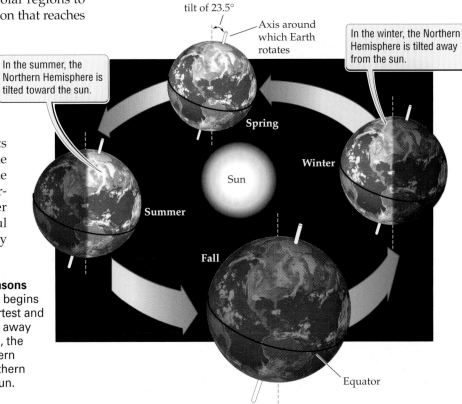

Figure 39.2 The Tilt of Earth Causes the Seasons
In the Northern Hemisphere, winter officially begins on December 21, when days are at their shortest and the Northern Hemisphere is maximally tilted away from the sun. The reverse occurs on June 21, the first day of summer, when days in the Northern Hemisphere are at their longest and the Northern Hemisphere is maximally tilted toward the sun.

it is winter in the Southern Hemisphere (for example, Chile) when it is summer in the Northern Hemisphere (for example, Canada).

The movement of air and water shapes climate

Near the equator, intense sunlight heats moist air, causing the air to rise from the surface of Earth. The air cools as it rises, which causes rain (see Figure 39.3); rain falls as the air cools because cold air holds less water than does warm air. Usually cold air sinks. In this case, however, the cool air cannot sink immediately because of the warm air that is rising beneath it. Instead, the cool air moves to the

north and south, tending to sink back to Earth at about 30 degrees latitude. The cool air warms as it descends, allowing it to hold more water. As the air flows back toward the equator, it absorbs available water (see Figure 39.3). By the time it reaches the equator, the air is once more hot and moist, so it rises to begin the cycle again.

Earth has four **giant convection cells** in which moist, warm air rises and dry, cool air sinks (Figure 39.3). The four convection cells are located in tropical and polar regions, where they generate relatively consistent wind patterns. In temperate regions (roughly 30 to 60 degrees latitude), winds are more variable and there are no stable convection cells. The variable winds form when cool, dry air from polar regions collides with warm, moist air that moves toward the poles from the Tropics.

The winds produced by the four giant convection cells do not move straight north or straight south (Figure 39.4). Instead, the rotation of Earth causes winds to curve as they travel near the surface of Earth. For example, winds that travel toward the equator curve to the west. When winds curve to the west, they blow from the east; hence such winds are called easterlies ("from the east"). Similarly, winds that travel toward the poles blow from the west and are called westerlies. In southern Canada and in much of the United States, winds blow mostly from the west; thus storms in these regions usually move from west to east.

The rotation of Earth, differences in water temperature, and the directions of prevailing winds all contribute to the formation of ocean currents. In the Northern Hemisphere, ocean currents tend to run clockwise between the continents; in the Southern Hemisphere, they tend to run counterclockwise (Figure 39.5).

Ocean currents carry a huge amount of water and can have a large influence on regional climates. For example, the Gulf Stream moves 25 times the water carried by all the world's rivers combined. Without the warming effect of the water carried by this current, countries such as Great Britain and Norway would have a subarctic to arctic climate. Overall, the Gulf Stream causes cities in Europe to be much warmer than cities of similar latitude in North America, as illustrated by Rome versus Boston, Paris versus Montréal, and Stockholm versus Fort-Chimo (a town of 1400 people in Québec, Canada).

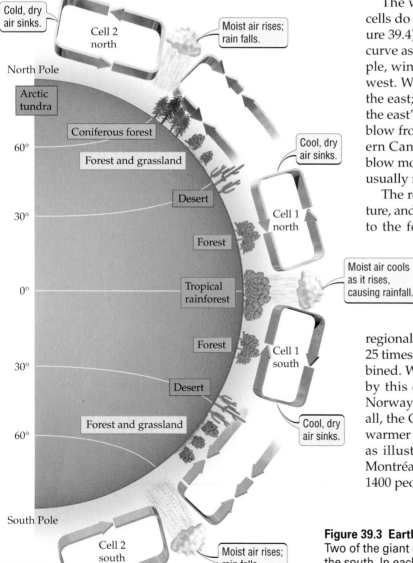

Figure 39.3 Earth Has Four Giant Convection Cells
Two of the giant convection cells are in the north, and two are in the south. In each of the four convection cells, relatively warm, moist air rises, cools, and then releases moisture as rain or snow. The cool, dry air then sinks to Earth and flows back toward the region where the warm air is rising.

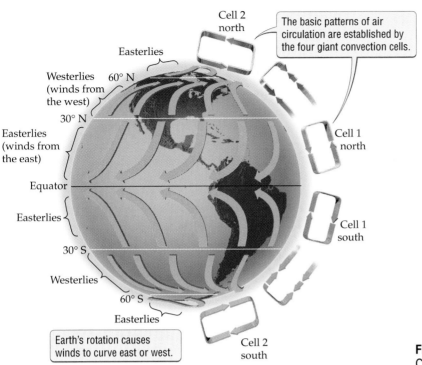

Cell 2 north

The basic patterns of air circulation are established by the four giant convection cells.

Easterlies

Westerlies (winds from the west) 60° N

Cell 1 north

30° N

Easterlies (winds from the east)

Equator

Cell 1 south

Easterlies

30° S

Westerlies

60° S

Easterlies

Earth's rotation causes winds to curve east or west.

Cell 2 south

Figure 39.4 Global Patterns of Air Circulation
The four giant convection cells determine the basic pattern of air circulation on Earth. Earth's rotation then causes winds to curve to the east or to the west. The direction the winds curve depends on location, but for any given geographic region on Earth, the winds usually blow from a consistent direction.

Figure 39.5 The World's Major Ocean Currents
Cold ocean currents are colored blue; warm ocean currents are colored red.

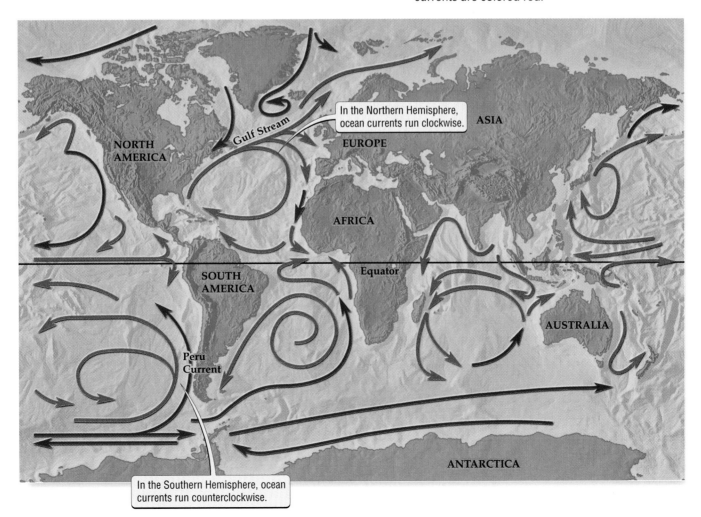

In the Northern Hemisphere, ocean currents run clockwise.

ASIA

EUROPE

NORTH AMERICA

Gulf Stream

AFRICA

SOUTH AMERICA

Equator

AUSTRALIA

Peru Current

ANTARCTICA

In the Southern Hemisphere, ocean currents run counterclockwise.

Major features of Earth's surface shape climate

Heat is absorbed and released more slowly from large bodies of water than from land. Because they retain heat comparatively well, oceans and large lakes moderate the climate of surrounding lands. Mountains also can have a large effect on a region's climate. For example, mountains often cause a **rain shadow** in which little precipitation falls on the side of the mountain that faces away from prevailing winds (Figure 39.6). In the Sierra Nevada range of North America, five times as much precipitation falls on the west side of the mountains (which faces toward winds that blow in from the ocean) as on the east side of the mountains. Mountain ranges in northern Mexico, South America, Asia, and Europe also create rain shadows.

Formation of deserts

The formation of deserts illustrates how the climate of a region can be affected by a combination of incoming solar radiation, air and water currents, and major features of Earth's surface.

Many of the world's largest deserts are located at about 30 degrees latitude (see Figure 39.7). Deserts are found at these latitudes for two reasons. First, the dry air that descends at these locations brings little precipitation (see Figure 39.3). Second, these latitudes are relatively close to the equator, so they are hot. Of course, hot regions with little precipitation are well suited for the formation of deserts.

Other deserts, especially those in the temperate zone, are caused mainly or in part from rain shadows created by mountains. Rain shadows helped form the Mojave Desert and the Great Basin (a desert region) in North America, as well as the Gobi Desert of Asia. Finally, some of the driest deserts in the world, such as the Atacama (South America) and Namib (southern Africa) Deserts, occur in places where winds blowing from the sea pass over cold ocean currents. The air cools as the winds pass over the cold ocean currents, causing rain to fall before the winds reach the shore. When the cool, dry winds reach the shore, they warm again, bringing little rain and causing extremely arid conditions; the Atacama Desert, for example, receives only 0.6 centimeters of precipitation per year.

Figure 39.6 Rain Shadow Effect
The side of a high mountain that faces the prevailing winds (the windward side) receives more precipitation than the side of the mountain that faces away from the prevailing winds (the leeward side). The leeward side is thus said to be in a rain shadow.

> ■ The climate of a region is shaped by three major factors: the amount of incoming solar radiation, global patterns of movement of air and water, and major features of Earth's surface, such as mountains. Climate has a great impact on the organisms in the biosphere.

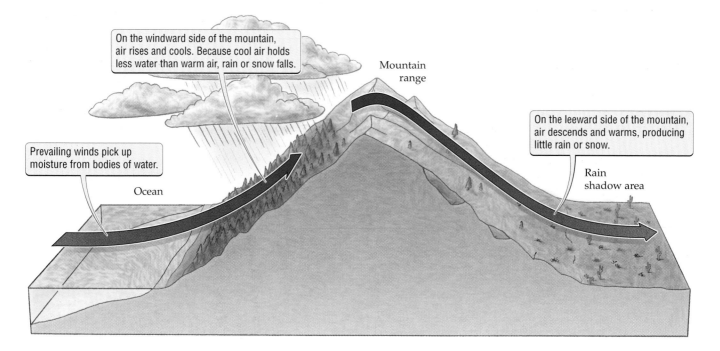

On the windward side of the mountain, air rises and cools. Because cool air holds less water than warm air, rain or snow falls.

Mountain range

On the leeward side of the mountain, air descends and warms, producing little rain or snow.

Prevailing winds pick up moisture from bodies of water.

Ocean

Rain shadow area

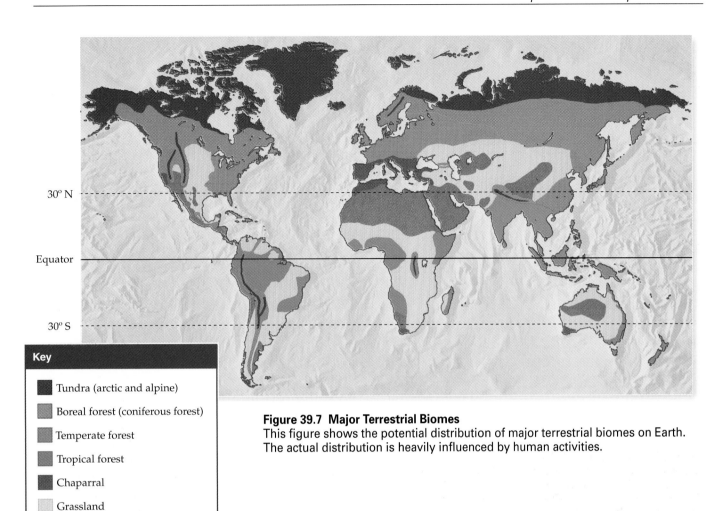

Key

- Tundra (arctic and alpine)
- Boreal forest (coniferous forest)
- Temperate forest
- Tropical forest
- Chaparral
- Grassland
- Desert

Figure 39.7 Major Terrestrial Biomes
This figure shows the potential distribution of major terrestrial biomes on Earth. The actual distribution is heavily influenced by human activities.

Terrestrial Biomes

Now that we've discussed factors that influence climate, let's turn to a description of components of the biosphere. The biosphere can be divided into major terrestrial and aquatic life zones called **biomes**. Terrestrial biomes, such as grasslands and tropical rainforests, cover large geographic areas and are usually named after the dominant vegetation of the region.

There are seven major terrestrial biomes: tropical forest, temperate forest, grassland, chaparral, desert, boreal forest, and tundra (including the tops of high mountains) (Figure 39.7). Figure 39.8 illustrates each of these biomes with a representative photograph. Keep in mind, however, that maps and photographs of a biome can give the misleading impression that the entire biome is the same. For example, to produce a worldwide map of major terrestrial biomes like the one shown in Figure

39.7, regions that actually are very different are lumped together into a single biome. If this were not done, the map would be so complicated that it would not be worth much.

Consider grasslands, for example. The grassland biome includes arid grasslands (characterized by short, drought-resistant grasses) in southwest North America, tallgrass prairie (characterized by tall grasses and many wildflowers) in north central North America, and savanna (grasslands dotted with occasional trees) in Africa. What is true of grasslands is true of all biomes: Both the species found in the biome and the ecological conditions of the biome can vary greatly from place to place.

The location of terrestrial biomes is determined by climate and human actions

Climate is the single most important factor controlling the potential or natural location of terrestrial biomes. The climate of a region, most importantly the temperature and the amount and timing of precipitation, allows some species to thrive and prevents other species from living there. Overall, the impact of temperature and

(a) Tropical forest

(b) Temperate forest

(d) Chaparral

(e) Desert

(f) Boreal forest

Figure 39.8 Terrestrial Biomes

(a) Tropical forests form in warm, rainy regions and are dominated by a rich diversity of trees, vines, and shrubs. (b) Temperate forests are dominated by trees and shrubs that grow in regions with cold winters and moist, warm summers. (c) Grasslands are common throughout the world and are dominated by grasses and many different types of wildflowers. Grasslands often occur in relatively dry regions with cold winters and hot summers. (d) Chaparral is characterized by shrubs and small, nonwoody plants that grow in regions with mild summers and winters, and by low to moderate amounts of precipitation. (e) Deserts form in regions with low precipitation, usually 25 centimeters per year or less. (f) Boreal forests are dominated by coniferous trees that grow in northern or high-altitude regions with cold, dry winters and mild, humid summers. (g) Tundra is found at high latitudes and high elevations and is dominated by low-growing shrubs and nonwoody plants that tolerate extreme cold.

(c) Grassland

(g) Tundra

moisture on species causes particular biomes to be found under a consistent set of conditions (Figure 39.9).

Climate can exclude species from a region directly or indirectly. Species that cannot tolerate the climate of the region are excluded directly. Species that can tolerate the climate but that are outperformed by other organisms that are better adapted to the climate are excluded indirectly.

Although climate places limits on where biomes *could* be found, the actual location of biomes in the world today is very strongly influenced by people. Humans have converted portions of the natural biomes in North America to urban (housing and industry) and agricultural areas, a situation common in many other parts of the world as well. We will return to the impact of humans on natural biomes when we discuss global change in Chapter 45.

> ■ There are seven major terrestrial biomes. Climate is the most important factor controlling the potential or natural location of terrestrial biomes; climate can exclude a species from an area directly or indirectly. The actual location of terrestrial biomes is heavily influenced by people.

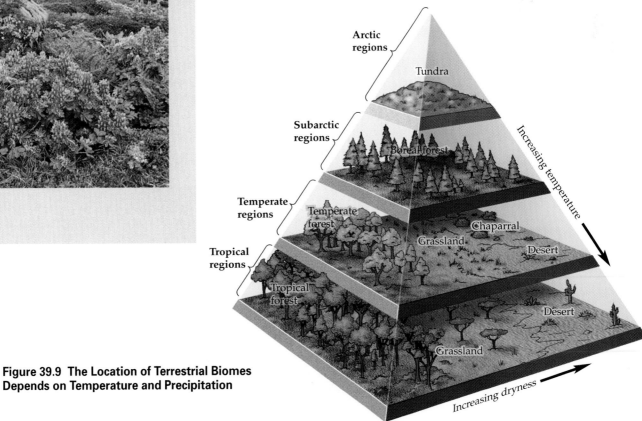

Figure 39.9 The Location of Terrestrial Biomes Depends on Temperature and Precipitation

Aquatic Biomes

Life evolved in water billions of years ago, and aquatic habitats make up 70 percent of the biosphere. There are eight major aquatic biomes: river, lake, wetland, estuary, intertidal zone, coral reef, ocean, and benthic zone (Figure 39.10). Unlike terrestrial biomes, aquatic biomes are usually characterized by physical conditions of the environment, such as salt content, water temperature, water depth, and the speed of water flow.

As with terrestrial biomes, the photographs in Figure 39.10 capture only a small portion of the diversity of aquatic biomes. Lake biomes, for example, include bodies of water that range in size from small ponds to very large lakes. In addition, very different species can be found in two areas of an aquatic biome. For example,

(a) River

(b) Lake

(c) Wetland

(d) Estuary

Figure 39.10 Aquatic Biomes

(*a*) Rivers are relatively narrow bodies of fresh water that move continuously in a single direction. (*b*) Lakes are standing bodies of fresh water of variable size, ranging from a few square meters to thousands of square kilometers. (*c*) Wetlands are characterized by shallow waters that flow slowly over lands that border rivers, lakes, or ocean waters. (*d*) Estuaries are tidal habitats where rivers flow into the ocean. (*e*) Intertidal zones are found in coastal areas where the tides rise and fall on a daily basis, periodically submerging a portion of the shore. (*f*) Coral reefs form in warm, shallow waters located in the Tropics and are named after the corals on which many of the reef's other organisms depend. (*g*) Oceans cover the majority of Earth's surface. They include a shallow layer (100 to 200 meters deep) in which photosynthesis can occur, and deep ocean waters that little light can penetrate. (*h*) Benthic zones are home to a wide variety of organisms that live on bottom surfaces below all other aquatic biomes.

(*e*) Intertidal zone

(*f*) Coral reef

(*g*) Ocean

A deep-ocean-water fish

(*h*) Benthic zone

algae may thrive in a lake that contains large amounts of nitrogen and phosphorus. The remains of dead algae are consumed by bacteria. As the bacteria grow and reproduce, they may use so much of the oxygen dissolved in the water that fish in the lake die. In contrast, a lake with little nitrogen and phosphorus may have few algae, few bacteria, high oxygen levels, and many species of fish. Thus, although both areas are lakes, the species that live there differ tremendously.

Aquatic biomes are influenced by terrestrial biomes, climate, and human actions

Aquatic biomes, especially lakes, rivers, wetlands, and coastal portions of marine biomes, are heavily influenced by the terrestrial biomes that they border or through which their water flows. For example, high and low points of the land determine the location of lakes and the speed and direction of water flow. In addition, when water drains from a terrestrial biome to an aquatic biome, it brings with it dissolved nutrients (such as nitrogen or phosphorus) that were part of the terrestrial biome. Because nutrients are available only in low amounts in many aquatic biomes, nutrients imported from the surrounding terrestrial biome can have a large effect. For example, the addition of nutrients from a farm field to a lake can cause algae to thrive and fish to die (as described in the preceding section).

Aquatic biomes also are strongly influenced by climate. In temperate climates, for example, seasonal changes in temperature cause the oxygen-rich water near the top of the lake to sink in the fall and the spring, bringing oxygen to the bottom of the lake. In tropical climates, seasonal differences in temperature are not great enough to cause a similar mixing of water from the top and bottom of lakes. This lack of mixing causes the deep waters of tropical lakes to have low oxygen levels and relatively few forms of life.

Climate also has very important effects in the open ocean. For example, climate helps determine the temperature, sea level, and salt content of the world's oceans. As we saw at the beginning of this chapter, the physical conditions of the ocean have dramatic effects on the organisms that live in the ocean; thus the climate has a powerful impact on marine life.

Finally, as with terrestrial biomes, the actions of humans strongly affect aquatic biomes. Portions of some aquatic biomes, such as wetlands and estuaries, are often destroyed to allow for development projects. Rivers, wetlands, lakes, and coastal marine biomes are negatively affected by pollution in most parts of the world. Aquatic biomes also suffer when humans destroy or modify the terrestrial biomes in which they are situated.

For example, when forests are cleared for timber or to make room for agriculture, the rate of soil erosion can

Figure 39.11 Effects of Erosion on a River
This river is clogged by sediments that washed into it as a result of erosion.

increase dramatically because trees are no longer there to hold the soil in place (Figure 39.11). Increased erosion can cause streams and rivers to be clogged with silt, which harms or kills invertebrates, fish, and many other species.

> ■ There are eight major aquatic biomes. Usually characterized by physical conditions of the environment, aquatic biomes are strongly influenced by their surrounding terrestrial biomes, by climate, and by human actions.

Highlight

One World

Spanish conquistadores are said to have cooled their water flasks in the unusually cold waters found off the coast of Peru. Cold waters are brought to this region from the south by the Peru Current (see Figure 39.5). The low sea surface temperatures of this region are also caused by cold, nutrient-rich waters that rise from the ocean depths to the surface. These waters provide food and suitable temperatures for enormous numbers of plankton, which can support a rich Peruvian fishery. Before it crashed in 1972 (during an El Niño event), the Peruvian anchovy fishery was the largest fishery in the world; its 1970 catch for this single species of fish equaled almost 20 percent of the total world harvest of fish.

Every year, the Peruvian fishing season ends when a current of warm water, known by local fisherman as **El Niño** (literally, "the child," so named because it often appears close to Christmas), pushes south along the coast of Ecuador. Usually, the southward flow of El Niño has a temporary and local effect. But once every 2 to 10 years, El Niño triggers floods, fires, torrential

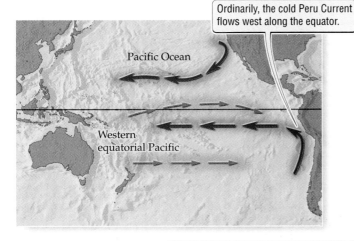

Ordinarily, the cold Peru Current flows west along the equator.

Pacific Ocean

Western equatorial Pacific

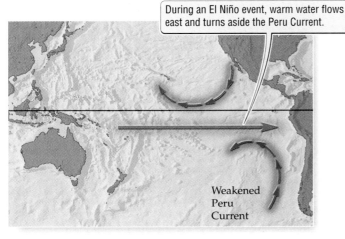

During an El Niño event, warm water flows east and turns aside the Peru Current.

Weakened Peru Current

Figure 39.12 El Niño Events
During an El Niño event, winds from the west push warm surface water from the western Pacific to the eastern Pacific. The resulting changes in sea surface temperatures help cause additional changes in ocean currents, wind currents, sea levels, and patterns of precipitation throughout much of Earth. Cold ocean currents are colored blue; warm ocean currents are colored red.

rains, droughts, disease, crop failures and more, as was described at the beginning of this chapter.

Strong El Niño events change the flow of the cold Peru Current and push warm water from the western Pacific to the east (Figure 39.12). The arrival of warm water from the western Pacific changes the usual pattern of sea surface temperatures, leading to changes in ocean currents, atmospheric pressure systems, wind systems, sea levels, and patterns of rainfall over much of the world. These and other drastic changes in the weather then cause the tremendous disruptions to natural systems described at the beginning of the chapter and shown in Figure 39.13.

Interesting in and of themselves, El Niño events also illustrate an important general point: Global patterns of air and water circulation can cause events that occur in one part of the world to affect biological communities all over the world. As we will see in the upcoming chapters, it can be difficult to understand the ecological inter-

actions that occur within a local area. In many cases, however, what ecologists (and policy makers) really need to understand are both local interactions and the impact of distant events. Such an understanding requires a detailed understanding of the biosphere, a monumental, complex task. To anyone who argues that there are no major scientific frontiers left to explore, the biosphere looms large as proof that they are wrong.

■ Once every 2 to 10 years the southward flow of a warm water current known as El Niño changes the flow of the Peru Current, triggering a series of events that alter weather patterns all over the world. The changes in weather brought about by El Niño have a large impact on the organisms of the biosphere, providing a vivid example of how events in one part of the world can have far-reaching consequences.

Summary

Ecology: Studying Interactions between Organisms and Their Environment

■ Ecologists study interactions between organisms and their environment at different levels, ranging from individuals to ecosystems.

■ A major goal of ecology is to document and understand the consequences for life on Earth of human actions that are changing the biosphere.

Climate

■ Climate depends on incoming solar radiation. Tropical regions are much warmer than polar regions because sunlight strikes Earth directly at the equator but at a slanted angle near the North and South Poles.

■ Climate is strongly influenced by four giant convection cells that generate relatively consistent wind patterns throughout much of Earth.

■ Ocean currents carry an enormous amount of water and can have a large impact on regional climates.

■ Regional climates are greatly affected by major features of Earth's surface. For example, mountains can create rain shadows, thereby contributing to the formation of deserts.

■ Climate has a great impact on the organisms in the biosphere.

Terrestrial Biomes

■ There are seven major terrestrial biomes: tropical forest, temperate forest, grassland, chaparral, desert, boreal forest, and tundra.

■ Climate is the most important factor controlling the potential location of terrestrial biomes; climate can exclude a species from an area directly or indirectly.

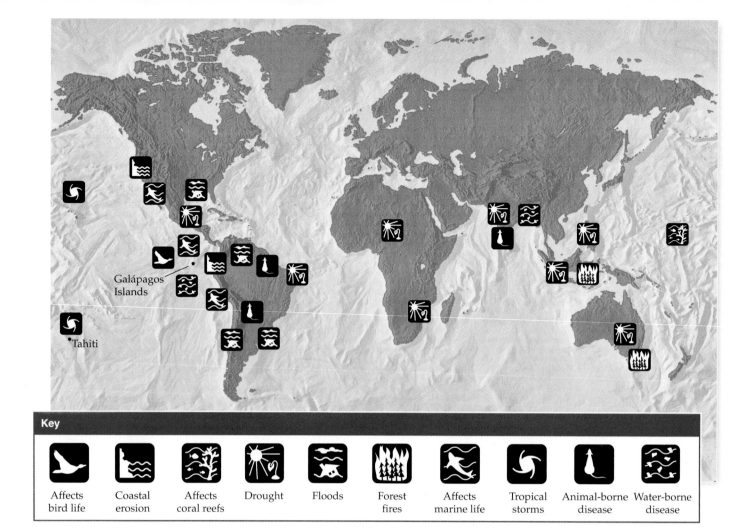

Key

| Affects bird life | Coastal erosion | Affects coral reefs | Drought | Floods | Forest fires | Affects marine life | Tropical storms | Animal-borne disease | Water-borne disease |

Figure 39.13 El Niño Events Have Large and Varied Effects throughout Earth

- The location of terrestrial biomes is heavily influenced by people. Humans have destroyed large portions of natural biomes so that the land could be used for agriculture, housing, or industry.

Aquatic Biomes

- There are eight major aquatic biomes: river, lake, wetland, estuary, intertidal zone, coral reef, ocean, and benthic zone.

- Aquatic biomes are usually characterized by physical conditions of the environment.

- Aquatic biomes are strongly influenced by their surrounding terrestrial biomes, by climate, and by human actions.

Highlight: One World

- El Niño events illustrate how global patterns of air and water circulation can cause events that occur in one part of the world to have far-reaching consequences.

Key Terms

biome p. 597	**El Niño** p. 602
biosphere p. 592	**giant convection cell** p. 594
climate p. 593	**habitat** p. 592
ecology p. 592	**rain shadow** p. 596
ecosystem p. 592	**weather** p. 592

Chapter Review

Self-Quiz

1. At which of the following levels do ecologists *not* study nature?
   a. ecosystem
   b. individual
   c. organelle
   d. population

2. Which of the following can contribute to the formation of a desert?
   a. winds that pass over warm ocean currents
   b. cool, dry air that sinks toward Earth at about 30 degrees latitude
   c. rain shadows
   d. both b and c

3. London, England, and Winnipeg, Canada, are located at a similar latitude, yet London is much warmer than Winnipeg. Why?
   a. The Gulf Stream warms Europe.
   b. The Peru Current keeps Canada cold.
   c. Easterlies occur in London, Westerlies in Winnipeg.
   d. There is a rain shadow effect in Winnipeg.

4. The biosphere consists of
   a. all organisms on Earth only.
   b. only the environments in which organisms live.
   c. all organisms on Earth and the environments in which they live.
   d. none of the above

5. What aspect(s) of climate most strongly influence the locations of terrestrial biomes?
   a. rain shadows
   b. temperature and precipitation
   c. only temperature
   d. only precipitation

Review Questions

1. Discuss the difference between potential and actual locations of biomes. What factors control the potential and actual locations of biomes?

2. Explain in your own words how global patterns of air and water circulation can cause local events to have far-reaching ecological consequences. Give an example that shows how local ecological interactions can be altered by distant events.

3. Explain how climate can exclude species from a region, in both terrestrial and aquatic biomes.

---

39 — *The Daily Globe*

Climate Change and El Niños: What Will the Future Bring?

LOS ANGELES, CA. El Niños are bad now, as recent drought and fires in Indonesia, floods in South America, and massive die-offs of fish populations around the world readily attest. But now scientists have begun predicting that as the climate heats up, El Niños may become even worse.

Though El Niños begin regionally as a current of warm ocean water that moves south along the Pacific coast of South America, they leave a huge global "foot-print." And it looks like the globe will leave a footprint of its own on El Niño: According to Dr. Brent Holigan, climatologist at The University of Las Vegas, increasing numbers of new studies have confirmed that the climate is warming, and as the climate warms, El Niños will happen more often and will be even stronger than the ones we have now.

For example, Dr. Marcia Gonzales, an ecologist with the Nature Defense Fund, argues that when humans warm the climate by burning fossil fuels, chances increase that a strong El Niño event will occur. Recent El Niño events bear these predictions out: the El Niño event of 1982 to 1983 was called the strongest of the twentieth century, but just 15 years (and three El Niños) later, the El Niño event of 1997 to 1998 was even stronger.

"If we don't act quickly to halt global warming," said Dr. Gonzales, "I hate to think what the future will bring."

Evaluating "The News"

1. As we will describe in Chapter 45, most scientists think that global warming is occurring and is caused in large part by the burning of fossil fuels. Why might the burning of fossil fuels in one part of the world (say, the United States) be expected to cause global temperatures to increase?

2. Nations are often viewed as independant islands, each able to decide what should be done within their borders. Is this a reasonable view? How can human society prevent events that occur in one part of the world from harming biological communities in other parts of the world?

3. If the link between global warming and an increase in the severity of El Niños is real, do people in countries such as Indonesia, which experiences the worst effects of El Niños, have the right to demand that people in countries such as the United States, which contributes the most to global warming, take action to stop global warming?

40

Why Organisms Live Where They Do

George Stubbs, *Zebra*, 1763.

Starlings and Mynas

Watch a large flock of birds as they land to rest for the night. It is an impressive sight: The sky darkens and churns with the motion of thousands of birds, and the sound of birdcalls fills the night air.

If you live in North America, you may well have seen European starlings as they settled for the night. Although originally from Europe, starlings now live on all of the world's continents except South America and Antarctica. Starlings feed and sleep in flocks.

Organisms live where they do because of where they evolved, continental drift, the presence of suitable habitat, and dispersal.

They are considered pests in North America because they are bold and aggressive, they have displaced native bird species such as bluebirds, and they cause extensive damage to wheat and some fruit crops.

Starlings did not migrate from Europe to North America on their own. In fact, the circumstances that surround their arrival are rather curious. Starlings plague North American farmers today because they were mentioned in the works of Shakespeare and a New York City man named Eugene Schieffelin sought to introduce to North America all the birds referred to by the Bard. Shakespeare had no idea of the trouble he was going to make when he innocently mentioned starlings. In April 1890, Mr. Schieffelin organized the release of a flock of 80 birds in Central Park, New York City. The following year, another flock of 80 birds was released. It took the starlings about 10 years to establish themselves in the New York City area. Since 1900 they have swept across the entire North American continent.

The crested myna is a close relative of the starling, and like the starling it was introduced to North America in the 1890s. But there the similarity ends. Mynas were introduced to Vancouver, British Columbia, in 1897. The Vancouver population of mynas reached a peak of about 20,000 birds in the late 1920s, declining since then to 5000 to 6000 birds. Despite early reports of mynas located as far south as Portland, Oregon, they are now found only in Vancouver.

Thus, roughly 100 years after both species were released, millions of starlings occur throughout the North American continent, but mynas are few in number and live only where they were originally released. Why have these two closely related species had such different fates?

Starlings Landing to Rest for the Night

1. The area over which we can find a species is referred to as its distribution or range.

2. Species are not distributed evenly throughout the environment, in part because the environment varies greatly from place to place and in part for various biological reasons.

3. The distribution of a species is determined by one or more of the following: history (where the species evolved

and continental drift), the presence of suitable habitat, and dispersal.

4. The distribution of species changes naturally with time as a result of migration, climate change, and loss of suitable habitat. Humans are causing large changes in the distribution of species, mainly by destroying or degrading suitable habitat.

Ecology is a young science, having begun about 100 years ago. Throughout the brief history of this science, ecologists have tried to understand why organisms live where they do. This information is of fundamental scientific interest because it enriches our knowledge of the world around us.

Knowledge of why organisms live where they do also helps us solve practical problems. For example, to control a pest species we need to understand what features of the environment cause the pest to thrive and what features cause it not to thrive. With this knowledge we can then modify the environment so that it is less suitable for the pest species. On a much broader scale, human activities are causing the decline and extinction of species throughout Earth. We need to understand why species live where they do so that we can reduce

our impact on other species while still allowing for sustainable forms of economic development.

Life is not distributed evenly on Earth: Any sampled area, even a continent, contains only a small fraction of all species on Earth. In this chapter we will explore some of the reasons why an organism is found in one place but not another, and we will discuss how such knowledge can be used in practical applications.

The Distribution of Species: Where Organisms Live

The area over which we can find a species is its **distribution** or **range**. European starlings, for example, range throughout North America, whereas crested mynas are restricted to Vancouver, British Columbia (Figure 40.1).

Like the myna, many species have a small geographic range. Some species of plants live on only a single hillside in tropical forests, and the Organ Mountain primrose is found in only 11 canyons of the Organ Mountains, a small mountain chain in southern New Mexico. Other species are common throughout one or a few continents. Relatively few spe-

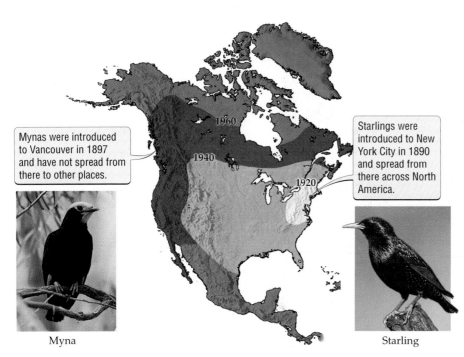

Mynas were introduced to Vancouver in 1897 and have not spread from there to other places.

Starlings were introduced to New York City in 1890 and spread from there across North America.

1960

1940

1920

Myna

Starling

Figure 40.1 The Ranges of Two Introduced Species
Mynas and starlings are closely related birds that were introduced to North America at roughly the same time. Since their introduction, mynas have not spread but starlings have spread across North America. The dates printed on the map show the edge of the starling's range at different points in time.

cies live on all or most of the world's continents; starlings, humans, and Norway rats are notable exceptions.

The place in which a species lives is called its **habitat**. The habitat of a species has a characteristic set of environmental conditions. These conditions may include both physical (for example, the amount of sunlight) and biological (for example, the specific species present) components of the environment. Consider the forest and meadow shown in Figure 40.2. Some species of plants, such as starflower, thrive only in the relatively cool, low-light environment found on the forest floor. Other plant species, such as New England aster, are found only in more open areas. Still other plants, like young sugar maple trees and the wildflower rough-leaved goldenrod, are found in both forest and open habitat.

The distribution of a species includes the area it occupies during all its life stages. We know so little about many organisms that we do not understand their distribution very well. This lack of knowledge is particularly true for insects, fungi, and plants that are hard to find or have life stages that are hard to study. We may know, for example, under what conditions the adult organism lives, yet have no idea about where or how other life stages live. The distribution of animals that migrate also can be hard to discover because it may include geographic regions separated by thousands of kilometers. Ignorance about the distribution of a species can hold even for large or striking organisms, such as the monarch butterfly (see the Box on page 610).

> ■ Some species have a small geographic range; others are found throughout one or a few continents. Relatively few species are found on most or all the world's continents. The habitat of a species has a characteristic set of environmental conditions, which include both physical and biological components of the environment.

Species Have Patchy Distributions

The natural world is patchy; that is, physical conditions, resources, and organisms vary greatly from place to place. Focusing first on organisms, relatively few species in nature are distributed in an even, or **regular**, fashion (Figure 40.3). For example, trees in a forest may either grow in an unpredictable, or **random**, fashion, or they may be grouped together, or **clumped**.

The patchy distribution of organisms is very clear if you take a bird's-eye view of the world. Viewed from above, even landscapes without farm fields are composed of patches: Here there may be a patch of aspen trees, there a meadow with pine seedlings. Trees are not unique in this regard: Over large areas, all species have a patchy distribution.

Why do organisms have patchy distributions? First, the underlying physical conditions often differ greatly from one location to the next. Factors such as the amount of nitrogen in the soil (which plants need to grow), the temperature, and the moisture available to organisms often vary greatly, even over distances as short as a few centimeters or meters. Variation in the physical environment can affect what organisms live there; often the patchy distribution of organisms matches the patchy distribution of underlying physical conditions. For example, the shrub *Clematis fremontii* lives only on dry, rocky outcrops in Missouri.

The shrub
Clematis fremontii

Starflower

New England aster

Figure 40.2 A Boundary between Forest and Meadow

Nearly regular   Random   Clumped

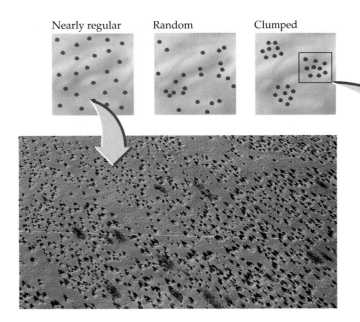

Figure 40.3 Distribution of Organisms
Although some organisms, such as creosote bush in the Mojave Desert, have a nearly regular distribution, most organisms have either a clumped or a random distribution.

The Scientific Process

Monarch Mysteries

Individual monarch butterflies are beautiful, and it is a breathtaking sight when thousands of them gather in eastern North America during their annual migration. Monarch caterpillars feed almost exclusively on milkweed plants, and in the spring, large numbers of monarchs migrate to the Gulf coast states of eastern North America, lay eggs on milkweed, and die. The offspring of these butterflies migrate north, and several more generations of butterflies are born and spend their lives in the summer breeding range. In the autumn, a final generation of butterflies migrates south.

The monarch migration gave rise to two great mysteries. First, although millions of butterflies migrate south from eastern North America each year, for a long time no one knew where they went. Beginning in 1857, scientists tried to answer this question. It took almost 120 years. Finally, in January of 1975, biologists discovered that the monarchs that migrate from eastern North America spend the winter in a few spots in a mountainous area west of Mexico City.

In the spring, monarch butterflies migrate north from Mexico to breed throughout much of eastern North America. Because monarch caterpillars depend on milkweed plants for food, the northern limit of the monarch's summer breeding range (shown in light blue) closely matches the northern limit of milkweeds.

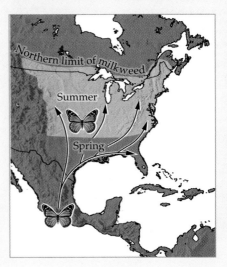

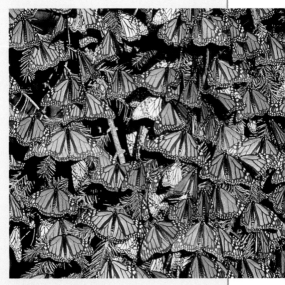

Overwintering monarchs

The second mystery has yet to be solved: How do the monarchs know where to go? Remember that as the monarchs migrate north, the adult butterflies lay their eggs on milkweed, and they die shortly thereafter. No monarch makes the entire journey. How the final generation can find its way back to the mountains of Mexico remains an unanswered and fascinating question.

Few parts of the state have such outcrops; within regions that do, the outcrops are found in clumps and hence so are the shrubs.

Organisms also can have patchy distributions for biological reasons that are not related directly to the physical environment. For example, the vast majority of seeds produced by plants land close to the mother plant, thus causing plants of a given species to clump together. Animals that feed on plants can have a patchy distribution for various reasons, the simplest of which is that their host plants have a patchy distribution.

> ■ Over large geographic areas, all organisms have patchy distributions. The patchy distribution of an organism often matches the patchy distribution of an underlying physical resource. Organisms also can have patchy distributions for biological reasons.

> Polar bears live where land meets the Arctic Ocean, which is shown in yellow. They hunt on ice packs far out into the sea.

Figure 40.4 The Range of Polar Bears

Factors That Determine the Distribution of Species

The distribution of a species is determined mainly by one or more of the following: history, the presence of suitable habitat, and dispersal.

History: Evolution and continental drift determine distribution

Past events can have a profound effect on where organisms live today. Why, for example, are polar bears found in the Arctic (Figure 40.4) but not in Antarctica? Polar bears hunt on ice packs and eat seals, both of which abound in Antarctica. In part, the answer to our question lies in an accident of history: Polar bears evolved in the Arctic, and the tropical regions in between appear to have prevented them from reaching Antarctica. As the polar bear example shows, organisms do not necessarily live everywhere they are well suited to live.

In the early part of the twentieth century, the German scientist Alfred Wegener made the seemingly far-fetched suggestion that the continents move over time (see Figure 22.6). At first, geologists hotly disagreed, but in the 1960s Wegener's suggestion was shown to be correct. Some biologists were early supporters of Wegener's theory because it helped them explain the curious distribution of many groups of organisms.

For example, the southern beech tree is found today in Australia, New Zealand, and the southern parts of South America (Figure 40.5). The southern beech has heavy seeds and is not able to spread long distances across ocean waters, so how did it come to live where it

does? The answer is simple: It evolved more than 65 million years ago, when South America, New Zealand, and Australia were connected to each other. The movement of the continents during the past 50 million years brought the southern beech to its current locations.

The presence of suitable habitat determines distribution

Good and poor places exist for all species, and the distribution of species is limited by the presence of suitable habitat. This concept sounds simple, but what makes for "suitable habitat" can be complex. Let's consider how the physical environment, biotic (living) environment, interaction between the physical and the biotic environments, and disturbance influence what is suitable habitat for a species.

1. *Physical environment.* The growth of organisms is limited by essential environmental factors. For example, some plants, like the New England aster of Figure 40.2, cannot get enough energy from photosynthesis in the shade of a forest; such species are found only in open areas. Likewise, cold-intolerant plants (Figure 40.6) and animals cannot live in regions of excessive cold.

2. *Biotic environment.* Organisms interact with and often require the presence of other organisms. An animal that feeds on only one type of plant cannot

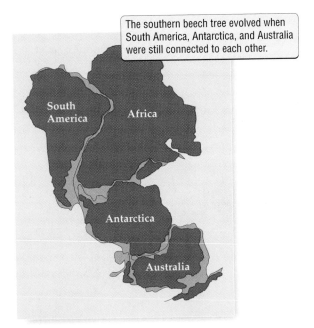

The southern beech tree evolved when South America, Antarctica, and Australia were still connected to each other.

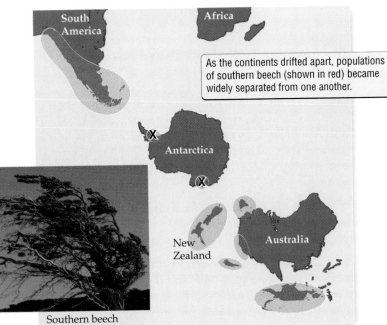

As the continents drifted apart, populations of southern beech (shown in red) became widely separated from one another.

Southern beech

Figure 40.5 How Did It Get There?
The modern distribution of the southern beech tree resulted when once-connected regions in South America, Australia, and New Zealand were separated by continental drift. Fossil pollen of the southern beech has been found at the two sites in Antarctica marked **X**.

Figure 40.6 It Can't Take the Cold
The distribution of saguaro cacti (shown in tan) in Arizona matches the region where the temperature never remains below freezing (0°C) for more than 36 hours.

live where that plant is absent, and a plant that requires a specific insect to pollinate its flowers cannot reproduce without that insect. In general, species that depend completely on one or a few other species for growth, reproduction, or survival cannot live where the species on which they depend are absent. Organisms also can be excluded from an area by predators or by competitors, either of which can greatly reduce the survival or reproduction of a species.

3. *Interaction between the physical and the biotic environments.* The physical and biotic environments often act together to determine the distribution of organisms. For example, the barnacle *Balanus balanoides* cannot survive when summer air temperatures are above 25°C, and it cannot reproduce if winter air temperatures do not remain below 10°C for 20 days or more. On the Pacific coast of North America, temperatures are such that the barnacle should be found 1600 kilometers farther south than it currently is (Figure 40.7). *B. balanoides* is absent from the region shown in purple in Figure 40.7 because competition from other species of barnacles prevents it from living in what would otherwise be suitable habitat. To the north, as temperatures become increasingly low, a point is reached where *B. balanoides* outcompetes the other barnacles and maintains healthy populations. Thus, the physical and biotic environments interact to determine where this barnacle is found.

In the blue-colored region, temperatures occasionally remain below freezing for more than 36 hours, preventing saguaro cacti from living there.

Figure 40.7 Competition Limits a Species' Range
The barnacle *Balanus balanoides* attaches to rocks of intertidal zones, as shown in this photograph taken along the coast of Alaska. *B. balanoides* can tolerate the temperatures found in the region shaded purple on the map at right, but competition with other barnacles prevents it from living there. It is found only in the region shaded red.

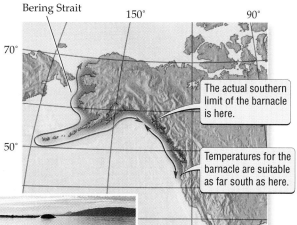

The actual southern limit of the barnacle is here.

Temperatures for the barnacle are suitable as far south as here.

Many species are not found in areas of suitable habitat at least in part because they cannot or have not yet dispersed to that habitat. Although polar bears are known to travel more than 1000 kilometers in a year, they have never dispersed from where they evolved, the Arctic, to Antarctica, a region in which they would probably thrive. Similarly, plant species often are not found in regions in which they can grow, presumably because they were not able to reach those areas by seed dispersal.

The failure of species to disperse to areas of suitable habitat is especially clear on islands. The Hawaiian Islands, for example, have only one native species of mammal, a bat, which was able to fly to the islands. No other mammals were able to disperse to Hawaii on their own, although cats, pigs, wild dogs, rats, goats, mongooses, and other mammals now thrive in Hawaii following their introduction to the islands by humans.

Mongoose

■ The distribution of organisms is strongly influenced by history, including continental drift and where the organisms evolved. Dispersal and the presence of suitable habitat also have a great impact on where organisms live. The presence of suitable habitat is determined by the physical environment, biotic environment, interaction between the physical and the biotic environments, and disturbance.

4. *Disturbance.* The distribution of some organisms depends on regular forms of **disturbance**, which is defined as an event that kills or damages individuals to create an opportunity for other individuals to become established. For example, many species persist in an area only if there are periodic fires. If humans prevent fires, such species are replaced by species that are not as tolerant of fire but that are superior competitors in the absence of fire. Floods, windstorms, droughts, and other changes from the usual conditions of a local environment are other forms of disturbance that can exclude some species but give others an advantage.

Dispersal determines distribution

Dispersal occurs when individuals travel long distances away from other members of their species. Dispersal may be active, as when organisms walk, fly, or swim. Dispersal can also be passive, as when organisms are transported by water or wind.

Changes in the Distribution of Species

The distribution of species can change over time. If a species migrates on its own or is introduced by humans to new habitat, it may spread through the new territory, thereby expanding its geographic range. As discussed earlier, for example, starlings spread throughout North America after they were introduced to New York City (see Figure 40.1). The distribution of a species also may change in response to relatively gradual climate change or to more rapid changes caused by humans, such as habitat loss.

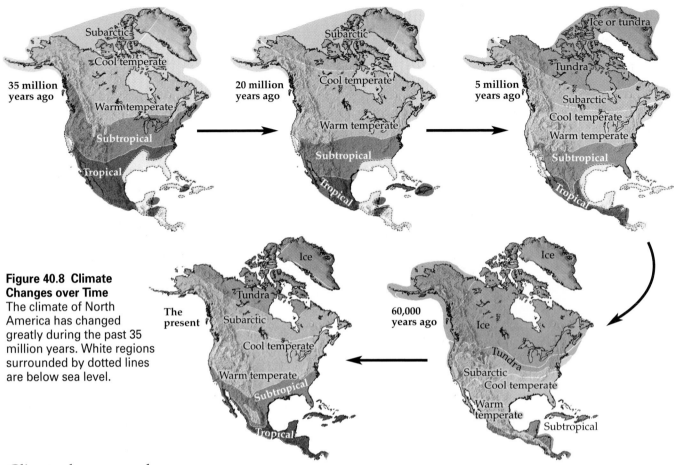

Figure 40.8 Climate Changes over Time
The climate of North America has changed greatly during the past 35 million years. White regions surrounded by dotted lines are below sea level.

Climate change may change the distribution of species

A region's climate sets limits on the species that can live there (see Chapter 39). The climate of Earth, however, is not constant. What we experience today as "normal" climate actually is much warmer than what has been typical for the past 400,000 years. Over even longer periods of time, the climate of North America has changed drastically (Figure 40.8). As the climate changed, the plant and animal species that live in North America also changed: For example, 35 million years ago fossil evidence indicates that the deserts of southwestern North America were covered with tropical forests.

Loss of habitat may reduce the distribution of species

Humans are now destroying or degrading the habitat of wild species at a rate that far exceeds what occurs naturally (Figure 40.9; also see Chapter 45). For any species, as its habitat is lost, its geographic range is reduced. Species that are adapted to living in only one habitat type are hit particularly hard; such species can decline dramatically or go extinct when their habitat is reduced or destroyed. For example, species of desert pupfish are in danger of ex-

tinction throughout the southwestern United States because their unique habitat—salty desert ponds—is being degraded and destroyed.

Desert pupfish

> ■ The distribution of a species can increase when the species migrates on its own or is introduced by humans to a new geographic region. The distribution of a species also can change in response to gradual changes in the climate or to rapid changes caused by humans, such as the degradation or destruction of natural habitat.

Using Knowledge about the Distribution of Species

There are many practical applications of knowledge about the distribution of species. For example, if biologists know what makes the environment suitable for a rare plant or for a species of economic importance, they can use that information to ensure that the environment remains suitable for the species of interest. Similarly, knowing why

1620

Old-growth forest is shown in red.

1920

Each dot represents 25,000 acres of old-growth forest.

Figure 40.9 The Destruction of Natural Habitat
From 1620 to 1920, vast regions of original (old-growth) forest in the United States were cut down to make room for agriculture, housing, and industry.

pest species live where they do can guide efforts to control them, thereby saving considerable time and money.

As humans alter natural habitats throughout Earth, it is becoming increasingly urgent that we understand why organisms live where they do. For example, to understand how many acres of old-growth forest are required for populations of spotted owls to survive in the northwestern United States, biologists must know what makes the environment suitable for the owl and how far the owl can disperse. Thus, since the presence of suitable habitat and dispersal are key factors in determining the distribution of spotted owls, biologists must know why the owl lives where it does if they are to improve its chance for survival.

Knowledge about why organisms live where they do has far-reaching consequences. In the case of the spotted owl, such knowledge has influenced political decisions about how much logging can be allowed in the remaining stands of the old-growth forest that owls need for their survival. In general, if we are to make accurate predictions about the impact of humans on natural environments—thus enabling us to make intelligent decisions about what actions we should and should not take—we must understand why species live where they do.

■ Knowledge about why organisms live where they do can be used to protect species of value to humans, including rare species and species of economic importance; such knowledge also can be used to control pest species. To understand the impact of humans on the natural environment, we must understand why organisms live where they do.

Highlight

The Fate of Introduced Species

Some of the most spectacular stories in ecology concern the introduction of species. In some cases, introduced species have been used with great success to control pest outbreaks (see Chapter 41). Unfortunately, other species introductions have caused great damage, as with the starlings that were introduced to North America. However, despite the striking nature of some species introductions, most attempts to introduce species fail, typically for unknown reasons.

The nearly simultaneous introduction to North America of two closely related birds, crested mynas and starlings, provides a unique glimpse into factors that influence the fate of species after they are introduced. We will use two levels of explanation—what ecologists call proximate and ultimate causes—to examine why starlings spread across North America while mynas remained in Vancouver, British Columbia.

Proximate causes address how an organism responds to its immediate environment. There are several proximate reasons why mynas have not dispersed from Vancouver. For example, in experiments conducted in the Vancouver area, myna females laid fewer eggs than starling females did. Once the eggs were laid, adult mynas incubated their eggs less frequently than starlings did. As a result, temperatures were colder in myna nests than in starling nests, and many of the myna eggs died. For the eggs that did hatch, myna chicks ate less and grew more slowly than starling chicks. Finally, adult mynas suffered more from cold during the winter than starlings did.

Overall, a clear proximate explanation emerges: Mynas reproduce less well, use food energy less efficiently, grow more slowly, and are less tolerant of cold than starlings. For these reasons, they have not spread from Vancouver to the rest of North America.

Ultimate causes address the historical and evolutionary reasons why organisms respond as they do to their immediate environment. In the case of mynas and starlings, the evolutionary past of these two birds helps us understand their performance in North America. Mynas are native to the subtropics of Southeast Asia. As such, they evolved in and are adapted to a much warmer and more stable environment than that found in most of North America. Their adaptation to warm, stable environments kept mynas from spreading from their Vancouver point of introduction. In contrast, starlings evolved in Europe, a region with a much colder climate than that of Southeast Asia. Hence, when starlings arrived in North America, they were well equipped to deal with cold winters and variable environmental conditions.

As is often true in ecological studies, a combination of proximate and ultimate causes provides the best explanation for the dramatically different fates of mynas and starlings.

> ■ Proximate explanations address how an organism responds to its environment. Ultimate explanations focus on historical and evolutionary reasons for why organisms respond as they do to their environment.

Summary

The Distribution of Species: Where Organisms Live

■ Some species have a small geographic range; others are found throughout one or several continents.

■ Few species are found on most or all of the world's continents.

■ The habitat of a species has a characteristic set of environmental conditions, which include both physical and biological components of the environment.

Species Have Patchy Distributions

■ Over large geographic areas, all organisms have patchy distributions.

■ The patchy distribution of an organism often matches the patchy distribution of an underlying physical resource.

■ Organisms can have patchy distributions for biological reasons.

Factors That Determine the Distribution of Species

■ The distribution of organisms is strongly influenced by history, including continental drift and where the organisms evolved.

■ Dispersal and the presence of suitable habitat have a great impact on where organisms live.

■ The presence of suitable habitat is determined by the physical environment, biotic environment, interaction between the physical and the biotic environments, and disturbance.

Changes in the Distribution of Species

■ The distribution of a species can increase when the species migrates on its own or is introduced by humans to a new geographic region.

■ The distribution of a species can change in response to gradual changes in the climate or to rapid changes caused by humans, such as the degradation or destruction of natural habitat.

Using Knowledge about the Distribution of Species

■ Knowledge about why organisms live where they do can be used to protect rare species and species of economic importance, as well as to control pest species.

■ To understand the impact of humans on the natural environment, we must understand why organisms live where they do.

Highlight: The Fate of Introduced Species

■ Proximate explanations address how an organism responds to its environment.

■ Ultimate explanations focus on historical and evolutionary reasons for why organisms respond as they do to their environment.

Key Terms

clumped distribution p. 609	proximate cause p. 615
dispersal p. 613	random distribution p. 609
distribution p. 608	range p. 608
disturbance p. 613	regular distribution p. 609
habitat p. 609	ultimate cause p. 616

Chapter Review

Self-Quiz

1. The distribution of saguaro cacti shows how _____ can limit the distribution of an organism.
   a. temperature
   b. predation
   c. sunlight
   d. water

2. Plant seeds usually do not travel far. The short distances that seeds travel tend to cause offspring to have a _____ distribution.
   a. random
   b. nearly regular
   c. clumped
   d. proximate

3. If you were to go back in time 100 million years and then briefly examine what Knoxville, Tennessee, was like every 10 million years, what would you find?
   a. Climate and organisms changed little over time.
   b. The climate changed, but the organisms did not.
   c. The organisms changed, but the climate did not.
   d. Both climate and organisms changed over time.

4. Why don't polar bears live in Antarctica?
   a. There is no food for them in Antarctica.
   b. They used to live in Antarctica, but they went extinct.
   c. There are not enough ice packs in Antarctica.
   d. They evolved in the Arctic and have not dispersed to Antarctica.

5. What type of explanation addresses historical and evolutionary reasons why organisms live where they do?
   a. proximate
   b. ultimate
   c. environmental
   d. none of the above

Review Questions

1. Describe factors that make habitat suitable for a species.

2. Explain why species are not distributed evenly throughout the environment.

3. Where a species evolved can influence its distribution. Do you think that *when* a species evolved might also be important? Relate your answer to the impact of continental drift and dispersal on the distribution of species.

4. How would you determine whether dispersal, the presence of suitable habitat, or history was most important in limiting the distribution of a species?

40   **The Daily Globe**

Some Like It Salty

EL TASCO, TX. This spring, like every spring, the banks of the Rio Grande are lined with an attractive tree bearing sprays of pink flowers. But while this tree might look attractive, beneath those pretty flowers lies a dangerous invader, one that is slowly poisoning the other plant species of its river habitat.

The dangerous beauty is salt cedar, a tree introduced to the southwestern United States for use as an ornamental tree in gardens. The tree escaped from cultivation and spread rapidly along rivers and streams. Salt cedar grows so well in the wild that it is crowding out trees native to our riverbanks, such as willows and cottonwoods. How does the invader do it? In part, it uses a secret weapon: Salt cedar likes salty soil, and as it grows it discharges salt from its roots, degrading the habitat for other tree species.

"Salt cedar is very hard to kill," says biologist Glenn Mack with the Texas Department of Game and Fish and one of the researchers working hard to get rid of the beautiful invader, "but we finally may have figured out a way to give the native trees an edge."

After much experimentation, the biologists learned that if they cut or burned the salt cedar, then released waters from the dam that is located upstream, the extra water created the flood conditions that allowed seeds of the native tree species to germinate and grow. By hitting the salt cedar hard and improving the habitat for native tree species, the biologists hope that the native tree species will regain lost ground. That would be good news for the native trees, as well as for the many bird species that prefer cottonwoods and willows to that deadly invader, the salt cedar.

Evaluating "The News"

1. Humans have altered many habitats in ways similar to how the introduction and spread of salt cedar altered the river and stream habitats of the southwestern United States. As with salt cedar, many of these changes seem subtle, until they are closely examined. If humans alter the environment in ways that only a professional biologist would realize, do such changes matter?

2. It takes considerable effort and money to try to control an introduced species. At the time of purchase, should people who buy such species be taxed to provide funds to pay for efforts to control them? Do you think nonnative species should be banned for use as ornamental plants or pets?

3. Do native species have a right to be protected from introduced species? Do people have an obligation to protect native species from introduced species? Many species have moved around the world from their place of origin, including humans. How do people fit into your conception of the rights of native versus nonnative species?

chapter *41*

Growth of Populations

Rosamund Purcell, *European Moles*, 1992.

Easter Island

Imagine standing at the edge of a cliff on Easter Island, looking into the long-abandoned quarry of Rano Raraku. Scattered about the grassy slopes of the quarry lie hundreds of huge, eerie statues carved from stone many years ago. The scene is beautiful, yet also ghostly and disturbing. Some of the statues stand upright but unfinished; they look almost as if the artists dropped their tools in midstroke. Others are complete but lie fallen at odd angles. As you leave the quarry, you see hundreds more statues scattered along the coast of Easter Island. Who carved these statues? Why were so many left unfinished? What happened to the people who made them?

The mystery of these ancient statues deepens when we consider where Easter Island is and what the island looks like today. Extremely isolated, the island is a small, barren grassland with little water and little potential for agriculture. How could such a remote and forbidding place support a civilization capable of carving, moving, and maintaining these enormous stone statues?

618

No population can continue to increase in size for an indefinite period of time.

The answers to the questions posed here provide a scary lesson for humans today. Easter Island was not always a barren grassland; at one time most of the island was covered by forest. According to archaeological evidence, no humans lived on the island until about AD 400. At that time about 50 Polynesians arrived in large canoes, bringing with them crops and animals with which to support

themselves. These people developed a well-organized society capable of sophisticated feats, such as moving 15- to 20-ton stone statues long distances without the aid of wheels (they probably rolled the statues on logs).

By the year 1500, the population had grown to about 7000 people. By this time, however, virtually all trees on the island had been cut down to clear land for agriculture and to provide the logs used to roll the statues from one place to another. The cutting of trees and other forms of environmental destruction caused increased soil erosion and decreased crop production, leading to mass starvation.

With no large trees remaining on the island, the people could not build canoes to escape the ever-worsening conditions. The society collapsed and sank into warfare, cannibalism, and living in caves for protection. The population crashed, and even 400 years later (1900) there were only 2000 people on the island, less than one-third the number that had lived there in 1500.

What caused the events on Easter Island? In essence, the number of people and the patterns of resource use increased above the level that the land could support. And that's the scary part: Many scientists now think that Earth's human population is at or near the level the planet can support. Is the story of Easter Island a sneak preview of what will happen to the whole human race?

Easter Island
Easter Island is small (166 square kilometers) and extremely isolated, located 3700 kilometers west of the coast of Chile and 4200 kilometers east of the nearest major population center in Polynesia (Tahiti). Three of the abandoned stone statues are shown above.

Easter Island ⊙

Key Concepts

1. A population is a group of interacting individuals of a single species located within a particular area.

2. Populations increase in size when birth and immigration rates exceed death and emigration rates, and they decrease when the reverse is true.

3. A population that increases by a constant proportion from one generation to the next exhibits exponential growth; such a population has a J-shaped growth curve.

4. Eventually, the growth of populations is limited by factors such as food shortages, predators, and disease.

5. The world's human population is increasing exponentially. Rapid human population growth cannot continue indefinitely; either we will limit our growth, or the environment will do it for us.

Ecology is the study of interactions between organisms and their environment. Two important questions ecologists ask are, Where do organisms live (and why)? And how many organisms live there (and why)? Efforts to understand the distribution and abundance of organisms provide insight to the natural world and are essential for the solution of real-world problems, such as the protection of rare species or the control of pest species.

We described the distribution of organisms in Chapter 40. In this chapter we focus on how abundant organisms are, emphasizing the growth of populations over time. We begin by defining what populations are; then we describe examples of how populations grow over time. Finally, we consider limits to growth that are faced by all populations, including the human population.

What Are Populations?

A **population** is a group of interacting individuals of a single species located within a particular area. This definition sounds simple to use, and in some cases it is. For example, the human population on Easter Island consists of all the people that live on the island.

Ecologists usually describe the number of individuals in a population by the **population size** (the total number of individuals in the population) or by the **population density** (the number of individuals per unit area). To return to the Easter Island example, in the year 1500 the population size was 7000 and the population density was 42 people per square kilometer (7000/166 = 42, where 166 square kilometers is the area of the island), a higher density than the 30 people per square kilometer that lived in the United States in 1999. This is an easy example; islands have well-defined boundaries, and human individuals are easy to count. But often it is difficult to determine the size or density of a population.

For example, suppose a farmer wants to know whether the aphid population that damages his or her crops is increasing or decreasing (Figure 41.1). Aphids are small and hard to count. More importantly, it is not obvious how to define the aphid population. Should only the aphids in the farmer's field be counted? What about the aphids in the next field over?

In general, what constitutes a population often is not as clear-cut as in the Easter Island example. Overall, the area appropriate for a particular population depends on the questions of interest and aspects of the biology of the organism, such as how far the organism moves.

> ■ A population is a group of interacting individuals of a single species located within a particular area. What constitutes an appropriate area depends on the questions of interest and the biology of the organism under study.

Changes in Population Size

All populations change in size over time—sometimes increasing, sometimes decreasing. In one year, abundant rainfall and plant growth may cause mouse populations to increase; in the next, mouse populations may decrease dramatically. Such changes in population size can have important consequences for people. For example, an increase in the number of deer mice, carriers of the deadly Hantavirus, is thought to have been responsible for the 1993 outbreak of Hantavirus in the southwestern United States.

Deer mouse

Whether a population increases or decreases in size depends on the number of births and deaths in the population, as well as on the number of individuals that

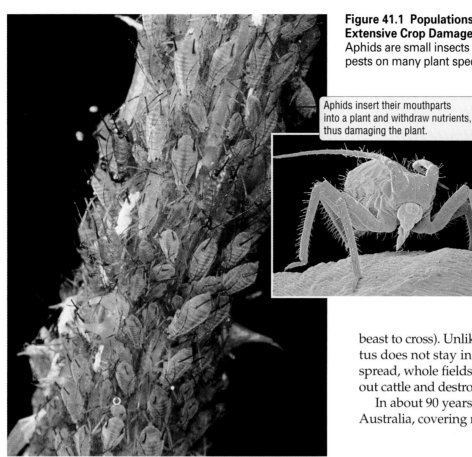

Figure 41.1 Populations of Aphids Can Cause Extensive Crop Damage
Aphids are small insects with sucking mouthparts. They are pests on many plant species, such as the rose stem shown here.

Aphids insert their mouthparts into a plant and withdraw nutrients, thus damaging the plant.

proportion from one generation to the next (Figure 41.2).

It is not uncommon for populations to increase exponentially, at least initially. Consider the following tale of woe: In 1839, a rancher in Australia imported a species of *Opuntia* (prickly pear) cactus and used it as "living fence" (a thick wall of this cactus is nearly impossible for human or beast to cross). Unlike a real fence, however, *Opuntia* cactus does not stay in one place; it spreads. As the cactus spread, whole fields were turned into "fence," crowding out cattle and destroying good rangeland.

In about 90 years, *Opuntia* cacti spread across eastern Australia, covering more than 243,000 square kilometers

immigrate to (enter) or emigrate from (leave) the population. A population increases in size whenever births and immigration are greater than deaths and emigration. The environment plays a key role in the increase or decrease of a population because birth, death, immigration, and emigration rates all may depend on the environment.

In some cases, populations increase or decrease rapidly in size, as discussed in the next section.

Exponential growth

Many organisms produce vast numbers of young, and if even a small fraction of the young survive, the population appears to grow without limit. **Exponential growth** is an important type of rapid population growth that occurs when a population increases by a constant

Figure 41.2 Exponential Growth
In this hypothetical population, each individual produces two offspring, so the population increases by a constant proportion each generation (it doubles). The number of individuals added to the population size increases each generation, causing the J-shaped curve that is characteristic of exponential growth.

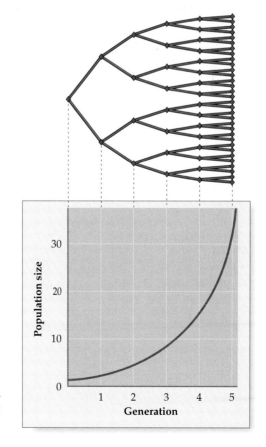

and causing great economic damage. All attempts at control failed until 1925, when scientists introduced a moth appropriately named *Cactoblastis*. This moth killed billions of cacti and successfully brought the cactus under control (Figure 41.3). Overall, the number of *Opuntia* individuals increased exponentially at first, then declined even more rapidly after introduction of the moth. Exponential growth has been found for introduced species other than *Opuntia*, as well as for species that expanded naturally to new territories.

Cactoblastis cactorum moth

(a)

(b)

Figure 41.3 Control of Introduced Cacti in Australia
(a) A dense stand of *Opuntia* cactus 2 months before the release of the moth *Cactoblastis*. (b) The same stand 3 years later, after the moth had killed the cacti by feeding on its growing tips.

When a population grows exponentially, the proportional increase is constant, but the numerical increase becomes larger each generation. For example, in terms of its proportional increase, the population in Figure 41.2 doubles in every time period. With respect to its numerical increase, however, the population increased by 1 individual between generations 1 and 2, but by 16 individuals between generations 4 and 5.

The time it takes a population to double in size—the **doubling time**—can be used as a measure of how fast the population grows. It is nice when our bank accounts double rapidly, but when populations grow exponentially in nature, problems eventually result. What kinds of problems do you think would arise in an exponentially growing population?

■ All populations change in size over time. Populations increase when birth and immigration rates are greater than death and emigration rates, and they decrease when the reverse is true. Exponential growth occurs when a population increases by a constant proportion from one generation to the next. Populations may initially increase at an exponential rate when organisms are introduced or migrate to a new area.

Populations Cannot Increase without Limit

Two giant puffball mushrooms can produce up to 7000 billion offspring (Figure 41.4). If each offspring survived and reproduced, the descendants of two giant puffballs would weigh more than Earth in just two generations. Humans and prickly pear cacti have much longer doubling times than giant puffballs, but given enough time they would also produce an astonishing number of descendants. Obviously, however, Earth is not covered with giant puffballs or even with humans. Giant puffballs illustrate an important general point: No population can increase in size indefinitely. Limits exist.

The reasons why populations cannot continue to increase are simple. For example, imagine that a few bacteria are placed in a closed jar that contains a source of food. The bacteria absorb the food and then divide, and their offspring do the same. The population of bacteria grows exponentially, and in short order there are billions of bacteria. Eventually, however, the food runs out and metabolic wastes build up. All the bacteria die.

This example may seem extreme because it involves a closed system: No new food is added, and the bacteria and the metabolic wastes cannot go anywhere. In many respects, though, the real world is similar to a closed system: Space and nutrients, for example, exist in limited amounts. In the *Opuntia* example of the previous section, even if humans had not introduced the moth *Cactoblastis*, the cactus population could not have sustained exponential growth indefinitely. Eventually the growth of the cactus population would have been limited by an environmental factor, such as a lack of suitable habitat.

Some populations grow according to an S-shaped curve. Such populations grow rapidly at first, but then stabilize at the maximum population size that can be supported indefinitely by their environment, a concept which is known as the **carrying capacity** (Figure 41.5). The growth rate of the population decreases as the population size nears the carrying capacity because features of the environment, such as food or water, begin to be in short supply. At the carrying capacity, the population growth rate is zero.

For example, in the 1930s the Russian ecologist G. F. Gause found that laboratory populations of *Paramecium* increased to a stable, maximum population size (see Figure 41.5). In these experiments, Gause added new nutrients at a steady rate and removed the old solution at a steady rate. At first, the *Paramecium* population increased rapidly in size. But as the population continued to increase, the paramecia used nutrients so rapidly that

Giant puffball mushrooms can produce 7000 billion offspring in a single generation. Large-sized giant puffballs weigh 40 to 50 kilograms each (a small one is shown here).

Figure 41.4 Will They Overrun Earth?

food began to be in short supply, slowing the growth of the population. Eventually the birth and death rates in Gause's experiments equaled each other. When birth and death rates are balanced, the population stops growing and stabilizes at the carrying capacity. Unlike a natural system, there was no immigration or emigration in Gause's experiments. In natural systems, populations reach and remain in balance when

birth + immigration = death + emigration

for extended periods of time.

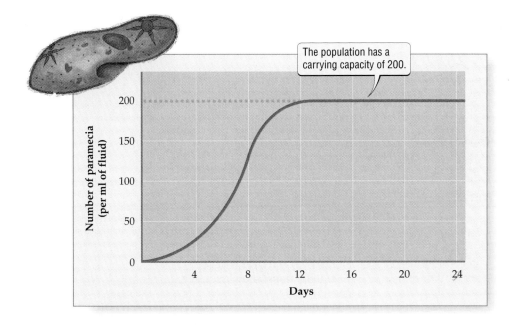

The population has a carrying capacity of 200.

Figure 41.5 Carrying Capacity
A population of the single-celled protozoan *Paramecium caudatum* increases rapidly at first, then stabilizes at the maximum population size that can be supported indefinitely by its environment, that is, the carrying capacity.

■ Because the environment has a limited amount of space and resources, no population can continue to increase in size indefinitely. Some populations increase rapidly at first, then stabilize at the carrying capacity.

Factors That Limit Population Growth

Natural populations experience limits to population growth (Figure 41.6), just as the laboratory populations of bacteria and *Paramecium* described in the preceding section did. The growth of natural populations can be held in check by factors such as food shortages, disease, predators, habitat deterioration, weather, and natural disturbances. Let's consider how such factors limit the growth of natural populations.

Any area has a limited amount of food and other essential resources. Thus, as the number of individuals in the population increases, there are fewer resources per individual. Individuals produce fewer offspring when they have fewer resources, causing the growth of the population to decrease. Similarly, when there are large numbers of individuals in a population, disease spreads more rapidly and predators may pose a greater risk (since many predators prefer to hunt abundant sources of food). Overall, when there are more individuals in a population, birth rates may drop or death rates may increase, and either effect may limit the growth of the population.

Large populations also can cause habitat deterioration. For example, if a population exceeds the carrying capacity of its environment, the population may decrease rapidly and the carrying capacity may be lowered (Figure 41.7; see also pages 618–619 for a human example). A lowered carrying capacity means that the habitat can no longer support the number of individuals it once could.

In the examples given so far in this section, factors that limited the growth of populations acted more strongly as the number of individuals in the population increased. When the number of individuals increases, the density also increases (since density is the number of individuals in a population divided by the area of the population). Factors that limit the growth of populations more strongly as the density of the population increases are said to be **density-dependent**.

In other cases, populations are held in check by factors like weather that are not affected by the density of the population; such factors are said to be **density-independent**. Density-independent factors often prevent

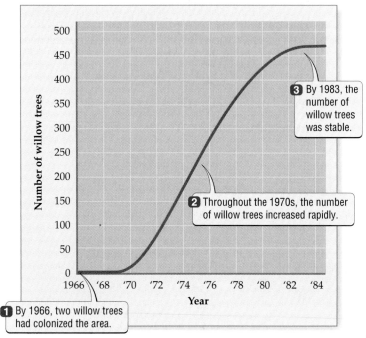

Figure 41.6 An S-Shaped Curve in a Natural Population
Rabbits heavily grazed young willow trees, preventing willows from colonizing a site in Australia. Rabbits were removed in 1954. Willows colonized the area by 1966 and increased rapidly in number, and the population then leveled off at about 475 trees.

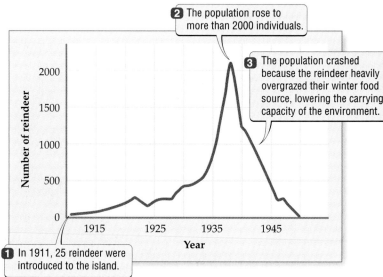

Figure 41.7 Boom and Bust
A reindeer population increased rapidly after it was introduced in 1911 to Saint Paul Island off the coast of Alaska, then crashed. In 1950, only eight reindeer remained.

populations from reaching high densities in the first place. For example, year-to-year variation in weather may cause conditions to be suitable for rapid population growth only occasionally. Poor weather conditions may reduce the growth of a population directly (by freezing the eggs of an insect, for example) or indirectly (by decreasing the number of plants available as food to that insect). Natural disturbances like fire and floods also limit the growth of populations in a density-independent way.

> ■ The growth of populations can be limited by food shortages, disease, predators, habitat deterioration, weather, natural disturbances, and other factors. Density-dependent factors limit the growth of populations more strongly when the density of the population is high. Density-independent factors also can limit the growth of populations.

gether in a tightly linked cycle (Figure 41.8). Populations of two species can cycle together when at least one of the species is very strongly influenced by the other species. In Figure 41.8, for example, the lynx depends on the hare for food, so lynx populations increase when hare populations increase.

There are relatively few examples from nature in which the abundance of two species shows regular cycles like those of the hare and lynx. However, the populations of most species do rise and fall over time—just not as regularly as in Figure 41.8. Irregular fluctuations are far more common in nature than is the smooth rise to a stable, maximum population size shown in Figure 41.6.

Finally, different populations of the same species may have different patterns of growth. Understanding these differences can provide critical information on how best to manage endangered or economically important species. For example, data on birth rate and territory size of the rare spotted owl were used to predict how the growth of spotted owl populations would be affected by the number and location of patches of the bird's

Patterns of Population Growth

Under the best of conditions, the population size of any species increases rapidly. An initial period of rapid population growth is found in both J-shaped curves (see Figure 41.2) and S-shaped curves (see Figure 41.6). In a J-shaped curve, rapid population growth may continue until resources are overused, causing the population size to drop dramatically (see Figure 41.7). In an S-shaped curve, the rate of population growth slows as the population size nears the carrying capacity. Predators, disease, and many other factors may then keep the population near the carrying capacity for long periods of time.

As discussed earlier, populations change in size over time, increasing at some times and decreasing at others. Even populations that have an S-shaped growth curve do not remain indefinitely at a single, stable population size; instead they fluctuate slightly over time but remain close to carrying capacity.

In some cases, the population sizes of two species change to-

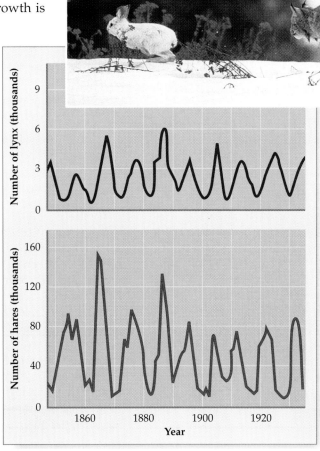

Figure 41.8 Population Cycles
Lynx eat snowshoe hares, and populations for both species cycle. The lynx depends on the hare for food, so the number of lynx is strongly influenced by the number of hares. Experiments indicate that hare populations are limited by their food supply and by their lynx predators. (Numbers of lynx and hares were determined from the number of furs sold by trappers to the Hudson's Bay Company, Canada.)

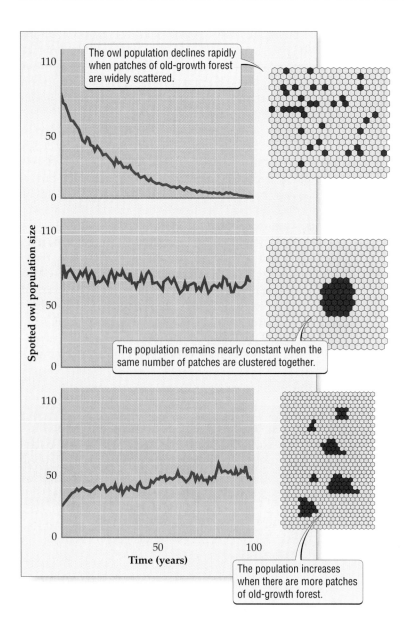

The owl population declines rapidly when patches of old-growth forest are widely scattered.

Spotted owl population size

The population remains nearly constant when the same number of patches are clustered together.

Time (years)

The population increases when there are more patches of old-growth forest.

Figure 41.9 Same Species, Different Outcomes
Populations of the endangered spotted owl are predicted to show different patterns of growth over time, depending on the arrangement and amount of their preferred habitat, old-growth forest.

Highlight

Human Population Growth

The human population is growing at a spectacular rate (Figure 41.10). It took more than 100,000 years for the human population to reach 1 billion people, but now our population increases by a billion people every 11 to 12 years. The impact of humans on the planet has increased even faster than our population size. For example, from 1860 to 1991 the human population increased fourfold, but our energy consumption increased 93-fold.

The global human population passed the 6 billion mark in 1999. At present, the world's population increases by 80 million to 90 million people each year, roughly 10,000 people per hour. The rapid increase of the human population is all the more sobering when the following facts are considered:

- More than 1.3 billion people live in absolute poverty.

- Two billion people lack basic health care or safe drinking water.

- More than 2 billion people have no sanitation services.

- Each year 15 million people, mostly children, die from hunger or hunger-related problems.

preferred habitat, old-growth forest (Figure 41.9). Predictions like those in Figure 41.9 help forest managers decide where and how much (if any) forest can be cut without harming the owl.

- Populations can exhibit various different patterns of growth over time, including J-shaped curves, S-shaped curves, tightly linked cycles, and irregular fluctuations. Understanding why different populations have different patterns of growth can provide critical information on how best to manage endangered or economically important species.

By the year 2025, the global population is projected to have grown from the present 6 billion to 8.3 billion. Even if birth rates dropped immediately from the current 2.9 children per female to a level that ultimately would allow the human population to replace itself but not increase (about 2 children per female), the human population would continue to grow for at least another 60 years. The population will continue to increase long after birth rates drop because a huge number of existing children have not yet had children of their own.

How did the human population increase so rapidly, apparently escaping earlier limits to population growth? First, as our ancestors emigrated from Africa they encountered and prospered in many kinds of new habi-

Figure 41.10 Rapid Growth of the Human Population

As on Easter Island, many problems facing humans today relate to increased population growth and environmental deterioration. More people means more environmental deterioration, which in turn makes it harder to feed the people we already have. Already much of Africa depends on imported food to prevent starvation, and urban areas in California persist only because of water imported from other states. In addition, like Easter Island, our economy and culture are not based on the sustainable use of resources; **sustainable** refers to an action or process that can continue indefinitely without using up resources or causing serious damage to the environment. Let's examine implications of the nonsustainable use of resources for one aspect of society, agriculture.

In the past, humans met the demand of feeding more people by increasing the amount of land under cultivation (primarily by irrigation) and by producing higher yields per unit area of farmland. However, the best farmland is already taken, and the land that is easiest to irrigate is already irrigated. Yields of major grains like wheat are no longer increasing faster than the human population size: Per person, the worldwide yield of grains has remained roughly constant since about 1990. In addition, increases in food production have environmental costs, such as deforestation, soil erosion, loss of biodiversity, depletion of groundwater, and pollution from pesticides, fertilizers, and energy use. As we degrade the environment, it becomes harder to produce the food needed to feed the billions of people on our planet.

Will humans limit the growth and impact of our global population, or will the environment do it for us? To limit the growth and impact of the human population, we must address the interrelated issues of population growth, poverty, unequal use of resources, environmental decay, and sustainable development. It is especially important for people who live in North America, Japan, and Europe to address such issues because people in these regions consume far more than their share of Earth's resources. For example, given patterns of resource use in India and the United States, it would take more than 11 *billion* people living in India to consume the same amount of copper and petroleum that just 273 million Americans used in 1999.

tats. Few other species can thrive in places as different as grasslands, coastal environments, tropical forests, deserts, and arctic regions.

Second, humans greatly increased the carrying capacity of the places in which they lived. For example, the development of agriculture allowed more people to be fed per unit area of land. Similarly, in the last 300 years death rates have dropped as a result of various factors, including improvements in medicine, sanitation, and the transportation of food. However, birth rates did not drop at the same time; hence the population grew. Finally, the heavy use of oil and other energy sources and nitrogen fertilizers in the twentieth century led to great increases in crop yields per unit area of farmland.

Viewed broadly, human inventiveness and technological development have allowed us to sidestep limits to population growth for some time. Ultimately, however, like all other populations, the human population cannot continue to increase without limit.

What does the future hold?

We began this chapter with a description of the rise and fall of the civilization on Easter Island. At least initially, the people who colonized Easter Island maintained a culturally rich and densely populated society. But their society did not persist. The people of Easter Island temporarily increased the carrying capacity of the island by cutting down the forests to create farm fields. Ultimately, however, cutting down the trees led to environmental deterioration, starvation, and the collapse of the civilization.

The hope for the future—for your future and the future of your children—lies in realistically assessing the problems we face, and then committing ourselves to take bold actions to address those problems. In the end, it is up to all of us to help ensure that humankind does not repeat on a grand scale the tragic lessons of Easter Island.

■ The global human population is growing very rapidly, increasing by 1 billion people every 11 to 12 years. The current rapid rate of human population growth cannot continue indefinitely, since it is based on a nonsustainable use of resources. To prevent the environment from limiting the growth of the human population, humans must act to address the interrelated problems of population growth, poverty, unequal use of resources, environmental decay, and sustainable development.

Summary

What Are Populations?

■ A population is a group of interacting individuals of a single species located within a particular area.

■ What constitutes an appropriate area depends on the questions of interest and the biology of the organism under study.

Changes in Population Size

■ All populations change in size over time.

■ Populations increase when birth and immigration rates are greater than death and emigration rates, and they decrease when the reverse is true.

■ A population grows exponentially when it increases by a constant proportion from one generation to the next.

■ Populations may initially increase at an exponential rate when organisms are introduced or migrate to a new area.

Populations Cannot Increase without Limit

■ Because the environment has a limited amount of space and resources, no population can continue to increase in size for an indefinite period of time.

■ Some populations increase rapidly at first, then stabilize at the carrying capacity, the maximum population size that can be supported by their environment.

Factors That Limit Population Growth

■ The growth of populations can be limited by food shortages, disease, predators, habitat deterioration, weather, natural disturbances, and other factors.

■ Density-dependent factors like food shortages and disease limit the growth of populations more strongly when the density of the population is high.

■ Density-independent factors like weather and natural disturbances also can limit the growth of populations.

Patterns of Population Growth

■ Populations can exhibit various different patterns of growth over time, including J-shaped curves, S-shaped curves, tightly linked cycles, and irregular fluctuations.

■ Understanding why different populations have different patterns of growth can provide critical information on how best to manage endangered or economically important species.

Highlight: Human Population Growth

■ The global human population is growing very rapidly, increasing by 1 billion people every 11 to 12 years.

■ The current rapid rate of human population growth cannot continue indefinitely, since it is based on a nonsustainable use of resources.

■ To prevent the environment from limiting the growth of the human population, humans must act to address the interrelated problems of population growth, poverty, unequal use of resources, environmental decay, and sustainable development.

Key Terms

carrying capacity p. 623

density-dependent p. 624

density-independent p. 624

doubling time p. 622

exponential growth p. 621

population p. 620

population density p. 620

population size p. 620

sustainable p. 627

Chapter Review

Self-Quiz

1. A population of plants has a density of 12 plants per square meter and covers an area of 100 square meters. What is the population size?
   a. 120
   b. 1200
   c. 12
   d. 0.12

2. A population that is growing exponentially increases
   a. by the same number of individuals each generation.
   b. by a constant proportion each generation.
   c. some years and decreases other years.
   d. none of the above

3. In a population that has an S-shaped growth curve, after an initial period of rapid increase, the number of individuals
   a. continues to increase.
   b. drops rapidly.
   c. remains near the carrying capacity.
   d. cycles regularly.

4. The growth of populations can be limited by
   a. natural disturbances.
   b. weather.
   c. food shortages.
   d. all of the above

5. Factors that limit the growth of populations more strongly at high densities are
   a. density-dependent.
   b. density-independent.
   c. exponential factors.
   d. sustainable.

Review Questions

1. Assume a population grows exponentially, increasing by a constant proportion of 1.5 per year. Thus, if the population initially contains 100 individuals, it will contain 150 individuals in the next year. Graph the number of individuals in the population versus time for the next five years, starting with 150 individuals in the population.

2. (a) What factors prevent populations from continuing to increase without limit? (b) Why is it common for species that enter a new region to grow exponentially for a period of time?

3. Describe the difference between density-dependent and density-independent factors. Give two examples of each.

4. Different populations of a species can have different patterns of population growth. Explain why an understanding of the cause of these different patterns of population growth can help managers protect rare species or control pest species.

5. List five specific actions that you can take to limit the growth or impact of the human population.

41

The Daily Globe

New Hope to Fight Outbreaks of Deadly Disease

WASHINGTON, D.C. The huge floods of 1998, brought in by the strongest El Niño of the twentieth century, caused a large amount of direct damage and created ideal conditions for disease outbreaks. Waterborne diseases like cholera and mosquito-spread diseases like malaria, dengue fever, and Rift Valley fever hit hard, killing thousands of people.

Now scientists at the Walter Reed Army Institute of Research, Washington, D.C., and the Centers for Disease Control (CDC), Atlanta, Georgia, say they have learned enough about the ecology of the deadly Rift Valley fever, a disease

that affects both cattle and people, to successfully predict outbreaks of the disease in Kenya. Like other studies in the young, growing field of ecological epidemiology, this newly published research provides a beacon of hope: We may be able to predict disease outbreaks months in advance, giving time for public health officials to take steps to prevent them from occurring.

Scientists were able to predict past outbreaks of Rift Valley fever using historical data on the disease, previously collected data on sea surface temperatures, and satellite images showing the amount of

plant growth—all of which allowed them to predict when mosquito populations would skyrocket. In the future, such advance warning, which could provide as much as five months' lead time, would give to health officials the notice they would need to control mosquito populations, thus limiting the spread of the disease.

As Dr. Roy West of the CDC put it, "Mosquitoes spread the disease from cattle to people, so what we needed to be able to do was predict when mosquito populations would rise to dangerous levels. This new approach lets us do just that."

Evaluating "The News"

1. So far, the new method for predicting the outbreak of Rift Valley fever has been used with historical data on sea temperatures, plant growth, and disease outbreaks. What should scientists do next?

2. As the world's human population grows, why might disease outbreaks be increasingly common?

3. The control of diseases like Rift Valley fever relies on knowledge about the ecology of populations

and the impact of air and water currents on the biosphere. Is it more important to use public funds to deepen our understanding of basic ecology or to seek ways to predict and control specific diseases?

Joan Miró, *The Tilled Field*, July 1923–Winter 1924.

chapter 42

Interactions among Organisms

Lumbering Mantises and Gruesome Parasites

Praying mantises have been seen to walk to the edge of a river, throw themselves in, and drown shortly thereafter. If they are rescued from the water, the mantises immediately throw themselves back in. What causes them to do this?

This bizarre behavior appears to be driven not by the mantises themselves, but by a worm. Less than a minute after the praying mantis lands in the water, a worm emerges from the mantis's anus. This worm attacks and infects terrestrial insects, such as praying mantises, but it also depends on an aquatic host for part of its life cycle. The

630

worm has performed a neat trick: It has evolved the ability to cause its insect host to jump into the river, a behavior that kills the insect but increases the chance that the worm will eventually reach its aquatic host.

Moving from the bizarre to the gruesome, examine the photographs below. The fungus that killed the ant first grew throughout the entire body of the ant, dissolving portions of its body and using it for food. Finally, the fungus sprouted reproductive structures (indicated by arrows), thereby enabling it to spread and attack other ants. Fungi attack many other species, including the corn plant shown. The human example illustrates an important point: Hundreds of millions of people are disabled every

year by protozoan, fungal, and animal (for example, worm) attackers, each of which has unique and harmful effects on the human body.

Parasites, such as the worms that plague praying mantises and the fungi that riddle the bodies of ants, are organisms that live in or on other organisms (their hosts). They obtain nutrients from their hosts, often causing harm but not immediate death. The effects of parasites on their hosts illustrate one important type of interaction among organisms: a relationship in which one species benefits and the other is harmed. In this chapter we discover how ecological interactions like parasitism help determine the distribution and abundance of organisms.

Parasitic Relationships
(*a*) An ant was killed by a fungus. (*b*) A person infected with a protozoan that attacks the skin. (*c*) An ear of corn destroyed by the fungus known as corn smut.

(*a*)

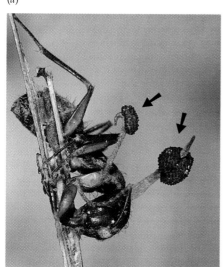

(*b*)

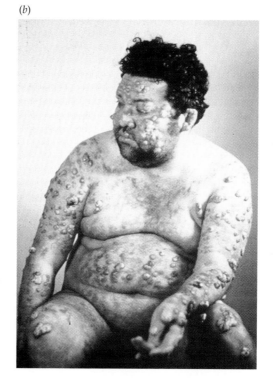

(*c*)

Key Concepts

1. Two species can interact to benefit both species (mutualism). Mutualisms evolve when the benefits of the interactions outweigh the costs for both species.

2. In consumer–victim interactions, such as parasitism, one species in the interaction benefits (the consumer) while the other is harmed (the victim). Victims have evolved elaborate ways to defend themselves against consumers.

3. In competition, two species that share resources have a negative impact on each other. Competition can result in

the evolution of increased structural differences between species.

4. Mutualism, consumer–victim interactions, and competition are common in nature and help determine where organisms live and how abundant they are.

5. Interactions among organisms have a large effect on communities and ecosystems.

Ecologists study interactions between organisms and their environment. An organism's environment includes the other organisms that live there. Thus, the subject of this chapter—interactions among organisms—is central to the very definition of ecology.

We have already seen how important interactions among organisms are. For example, the climate sets broad limits on where organisms live, but as we learned in Chapter 40, interactions among organisms can prevent a species from living everywhere that it otherwise could (see Figure 40.7). Similarly, in Chapter 41 we saw how the moth *Cactoblastis*, which feeds on the cactus *Opuntia*, caused *Opuntia* populations to crash in Australia. Overall, interactions among organisms influence all levels at which ecology can be studied, including individuals, populations, communities, and ecosystems.

The millions of species on Earth can interact in many different ways. In this chapter we classify interactions among organisms by whether the interaction is beneficial (+) or harmful (–) to each of the interacting species. In addition, we focus on only the three most important kinds of ecological interactions: +/+ interactions, in which both species benefit (mutualism); +/− interactions, in which one species benefits and the other is harmed (consumer–victim interactions); and −/− interactions, in which both species are harmed (competition). As we will see, each of these three types of interactions among organisms is very common, and each plays a key role in determining where organisms live and how abundant they are.

Mutualism

A **mutualism** is an interaction between two species in which both species benefit (a +/+ interaction). Mutualism is common and important to life on Earth: Many species receive benefit from and provide benefit to other

species—benefits that have the important effect of increasing the survival and reproduction of the interacting species.

Types of mutualisms

There are many kinds of mutualisms, including behavioral mutualisms, pollinator mutualisms, farming mutualisms (see the Highlight on page 639), seed dispersal mutualisms, and others. Here we elaborate on two common types: behavioral mutualisms and pollinator mutualisms.

Mutualisms in which each of the two species has altered its behavior to benefit the other species are called **behavioral mutualisms**. The relationship between shrimps and goby fish is a good example of a behavioral mutualism (Figure 42.1). Shrimps in the genus *Alpheus* live in an environment that has plenty of food but little shelter. They dig burrows in which to hide, but they see poorly and so are vulnerable to predators when they leave the burrows to feed. These shrimps have formed a fascinating relationship with goby fish: The gobies act as "seeing-eye" fish, warning the shrimps of danger when the shrimps venture out of their burrows to eat. In return, the shrimps share their burrows with the goby fish, thus providing the fish with a safe haven.

In **pollinator mutualisms**, an animal like a honeybee transfers male reproductive cells (pollen) from one plant to the female reproductive organs of another plant of the same species. These animals are known as pollinators, and without them the plants could not reproduce. To ensure that pollinators come to them, plants offer pollinators a reward, such as pollen or nectar for food. Thus, both species benefit. Pollinators are important in both natural and agricultural sys-

Honeybee

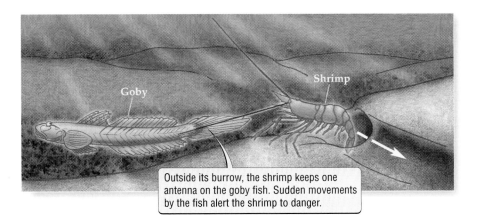

Figure 42.1 A Behavioral Mutualism
Alpheus shrimps build burrows and provide goby fish with shelter. The goby fish provide an early-warning system to the nearly blind shrimps when they leave their burrows to feed.

Outside its burrow, the shrimp keeps one antenna on the goby fish. Sudden movements by the fish alert the shrimp to danger.

tems. For example, the oranges we buy at the supermarket are available only because honeybees pollinated the flowers on orange trees, thus enabling the trees to produce their fruit.

Mutualists are in it for themselves

Although both species in a mutualism benefit from the relationship, what is good for one species may come at a cost for the other species. For example, a species may use energy or increase its exposure to predators when it acts to benefit its mutualistic partner. From an evolutionary perspective, mutualisms evolve when the benefits of the interactions outweigh the costs for both species. In general, mutualisms are not cost-free, and the interests of the two species may be in conflict.

Consider the mutualism between the yucca and the yucca moth, in which both the plant and its moth pollinator depend absolutely on each other. Female yucca moths collect pollen from yucca flowers, fly to another group of flowers, and lay their eggs on the ovary of a newly opened flower. After they have laid their eggs, the female moths climb up the flower's style and deliberately place the pollen they collected earlier onto the plant's female reproductive parts (Figure 42.2). When the moth larvae hatch, they feed on the seeds of the yucca plant. Thus, the moth both pollinates the plant and eats some of the plant's seeds.

In a cost-free situation for the plant, the moth would transport pollen but would not destroy any of the plant's seeds. In a cost-free situation for the moth, the moth would produce as many larvae as possible and thereby consume many of the plant's seeds. In actuality, an evolutionary compromise has been reached: The moths usually lay only a few eggs per ovary, and the plant tolerates the loss of a few of its seeds. Yucca plants have a defense mechanism that helps keep this compromise working: If a moth lays many eggs in one of the plant's ovaries, the plant can selectively abort that ovary, hence killing the moth's eggs.

Mutualisms are everywhere around us

Mutualism is very common. Most of the plant species that dominate forests, deserts, grasslands, and other biomes are mutualists. For example, about 80 percent of plant species have a mutualism with fungi in which plant roots obtain some nutrients from the fungi and provide other nutrients to the fungi (see the discussion of mycorrhizae in Chapter 2).

Figure 42.2 A Pollinator Mutualism
The yucca and the yucca moth are dependent on one other for survival.

As mentioned in the previous section, many animal species are involved with plants in pollinator mutualisms. Other examples of mutualisms involving animals include the spectacular reefs found in tropical oceans (Figure 42.3). These reefs are built by corals (soft-bodied animals), most of which house photosynthetic algae—their mutualistic partners—inside their bodies. Corals provide algae a home and several essential elements, like phosphorus, and algae provide corals with some of the carbohydrates they produce during photosynthesis.

Mutualisms can determine the distribution and abundance of species

A particular mutualism can influence the distribution and abundance of organisms in two ways. First, because each species in a mutualism survives and reproduces better where its partner is found, the two species strongly influence each other's distribution and abundance. Second, a mutualism can have indirect effects on the distribution and abundance of species that are not part of the mutualism. For example, coral reefs (see Figure 42.3) are home to many different plant and animal species. The corals that build the reefs depend on their mutualism with algae, and thus, indirectly, so do the many other species that live in coral reefs.

■ Common in nature, mutualisms evolve when the benefits of the interactions are greater than the costs for both partners. Mutualisms help determine the distribution and abundance of the mutualist species and of species that depend directly or indirectly on the mutualist species.

Consumer–Victim Interactions

Consumer–victim interactions are +/− interactions in which one species benefits (the consumer) and the other is harmed (the victim). There are three main types of consumers in such interactions: **predators**, which kill their prey; **parasites** (see definition on page 631); and **pathogens** (disease-causing organisms); and **herbivores**, which eat plants.

The three major types of +/− interactions are very different from one another. For example, whereas predators such as wolves kill their prey immediately, herbivores such as cows or parasites such as fleas do not. Although the three types of consumer–victim interactions have obvious and important differences, in this section we take a look at general principles that apply to all three of them.

Consumers can be a strong selective force

Many species have evolved elaborate strategies to avoid being eaten. For example, plants produce spines and poisonous chemicals as defenses against herbivores. In some cases, a plant is much more likely to produce spines when it has been grazed by herbivores than when it has not (Figure 42.4).

Many organisms have bright colors or striking patterns that warn potential predators they are heavily defended, usually by chemical means (Figure 42.5a). Such **warning coloration** can be highly effective. For example, blue jays quickly learn not to eat monarch butterflies, which are brightly colored and contain chemicals that make the birds very

Figure 42.3 The Home a Mutualism Built
Tropical reefs are home to many different species. The great diversity of life in tropical reefs depends on corals, which form the reefs in and around which many species live. The corals that build the reefs depend on a mutualism with algae.

Figure 42.4 A Cactus That Produces Spines When Grazed
On three islands off the coast of Australia, the percentage of cacti with spines is higher on the island that has cattle. Field and laboratory experiments show that grazing directly stimulates the production of spines.

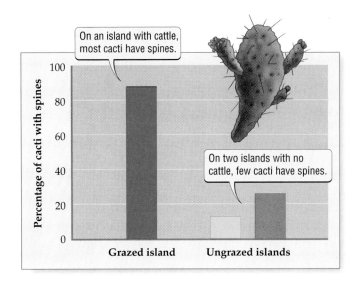

sick (Figure 42.5*b*). Other prey seek to avoid predators by being hard to find or hard to catch (see Figure 19.1). And finally, as we discussed in Chapter 32, organisms have evolved molecular defenses (immune systems) to help them fight off the ravages of disease and parasitic infection.

The many ways in which victim species protect themselves against consumers indicate that consumers often apply strong selection pressure to their victims.

Consumers can restrict the distribution and abundance of victims

The American chestnut used to be a dominant tree species across much of eastern North America. In 1900, however, a fungus that causes the disease chestnut blight was introduced to the New York City area. This

Chestnut tree

fungal disease spread rapidly, killing most of the chestnut trees in eastern North America.

Today the American chestnut survives throughout its range only in isolated patches and only as sprouts that arise from the base of otherwise dead tree trunks. The impact of chestnut blight on chestnut trees shows how a consumer (fungus) can limit

the distribution and abundance of a victim (chestnut); in this case, a formerly dominant tree species was nearly eliminated from its entire range.

Consumers can drive victims to extinction

Laboratory experiments with protozoans and with mites have shown that predators can drive prey extinct. Consumer–victim interactions can drive victim species to extinction in natural systems as well. The impact of chestnut blight on the American chestnut provides one clear example: Although the chestnut tree is not extinct throughout its entire range, it has been driven to extinction in many local populations. Similarly, cactus moths drove many populations of the *Opuntia* cactus in Australia extinct (see Chapter 41).

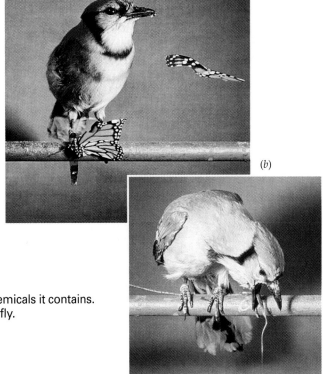

Figure 42.5 Warning Coloration
(*a*) The bright colors of this poison dart frog warn of the deadly chemicals it contains.
(*b*) An inexperienced blue jay vomits after eating a monarch butterfly.

Consumers can alter the behavior of victims

The bizarre story of praying mantises that jump to their deaths in rivers (see the beginning of the chapter) provides a dramatic example of how consumers can alter the behavior of their victims. But consumer–victim interactions can alter the behavior of victim species in more subtle ways as well.

For example, predators can be a driving force that causes animals to live or feed in groups. In some cases it is advantageous for prey to live in groups, since several prey acting together may be able to prevent predators from attacking (Figure 42.6). Large numbers of prey also may be able to provide better warning of a predator's attack. Because more birds can watch for predators, the approach of a goshawk (a predatory bird) is detected much sooner by a large flock of wood pigeons than by a single pigeon. The early detection by large flocks of wood pigeons causes the success rate of goshawk attacks to drop from nearly 80 percent when they attack single birds to less than 10 percent when they attack flocks of more than 50 birds (Figure 42.7).

> ■ Consumers such as predators, parasites and pathogens, and herbivores can be a strong selective force, leading their victims to evolve various ways to avoid being eaten. Consumers can restrict the distribution and abundance of their victims, in some cases driving them to extinction.

Competition

In **competition**, each of two interacting species has a negative impact on the other (a –/– interaction). Competition is most likely when two species share an important resource, such as food or space that is in short supply.

There are two main types of competition. In **interference competition**, one organism directly excludes another from the use of resources; for example, two species of birds may fight over the use of tree holes to use as nest sites. In **exploitation competition**, species

> Although a single musk ox may be vulnerable to predators like wolves, a group that circles is a difficult target.

Figure 42.6 Come and Get Us

compete indirectly for shared resources, each reducing the resource levels available to the other. There are many cases of exploitation competition; for example, two plant species may compete indirectly by their shared use of a resource that is in short supply, such as soil nitrogen.

> With flocks of more than 50 pigeons, goshawks caught a wood pigeon in fewer than 10% of their attacks.

Figure 42.7 Safety in Numbers
The success of goshawk attacks on wood pigeons decreases greatly when there are many pigeons in a flock.

Evidence for competition

Competition is very common and often has important effects in natural populations. For example, along the coast of Scotland the larvae of two species of barnacles, *Balanus balanoides* and *Chthamalus stellatus*, both settle on high and low portions of the shore. However, *Balanus* adults appear only on rocks on the lower portion of the shore, which are more frequently covered by water, and *Chthamalus* adults occur only on high portions of the shore, which are more frequently exposed to air.

In theory, the distribution of *Balanus* and *Chthamalus* could have been caused by competition or by the environment. In an experimental study, ecologists discovered that *Chthamalus* could perform very well on low portions of the shore, but only if *Balanus* was removed (Figure 42.8). Hence, competition with *Balanus* ordinarily prevents *Chthamalus* from living low on the shoreline. The distribution of *Balanus* on the other hand, depends mainly on physical factors: The increased heat and dryness found at higher levels of the shoreline prevent *Balanus* from surviving there.

In some cases, resources shared by two species may be so readily available that little competition occurs. For example, competition is relatively uncommon among leaf-feeding insects. The reason is simple: A huge amount of leaf material is available for the insects to eat, and usually there are too few insects to cause their food to be in short supply. Since their food remains abundant, little competition occurs.

Competition can limit the distribution and abundance of species

A great deal of field evidence shows that competition can limit the distribution and abundance of species, as in the barnacle example discussed in the previous section. A second example concerns wasps of the genus *Aphytis*. These wasps attack scale insects, which can cause serious damage to citrus trees. Female wasps lay eggs on the scale insects, and newly hatched wasp larvae pierce the scale insect's outer skeleton and then consume its body parts, killing the scale insect.

In 1948, the wasp *Aphytis lingnanensis* was released in southern California to curb the destruction of citrus trees

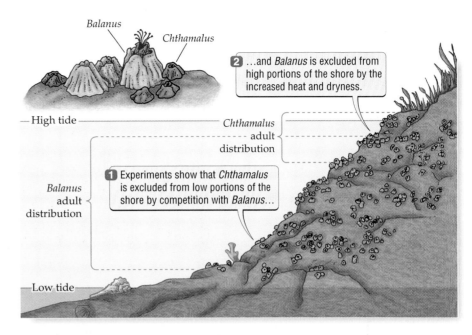

Figure 42.8 What Keeps Them Apart?
On the rocky coast of Scotland, adult *Balanus* barnacles are found only on low portions of the shore, and adult *Chthamalus* individuals are found only on high portions of the shore.

caused by scale insects. *A. lingnanensis* was released in the hope that it would provide better control of scale insects than the closely related wasp *A. chrysomphali* could. *A. lingnanensis* did prove to be a superior competitor (Figure 42.9), driving *A. chrysomphali* to extinction in most locations and, as hoped for, providing better control of scale insects. Although *A. lingnanensis* drove *A. chrysomphali* to extinction in many areas, both wasp species had a negative effect on the other because each used resources (scale insects) that otherwise could have been used by its competitor.

Competition can increase the differences between species

As Charles Darwin realized long ago, competition between species can be intense when the two species are very similar in form. For example, birds whose beaks are similar in size eat similar seeds and compete intensely, and birds whose beaks differ in size eat different seeds and compete less intensely. Intense competition between similar species may result in **character displacement**, in which the forms of competing species evolve to become more different over time. By reducing the similarity in form between species, character displacement should reduce the intensity of competition.

Some evidence for character displacement comes from observations that the forms of two species are

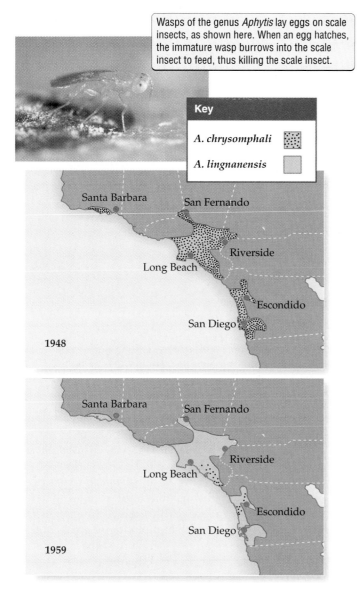

Figure 42.9 A Superior Competitor Moves in
After being introduced to southern California in 1948, the wasp *Aphytis lingnanensis* rapidly drove its competitor, *A. chrysomphali*, extinct in most locations. Both species of wasp prey on scale insects that damage citrus crops such as lemons and oranges.

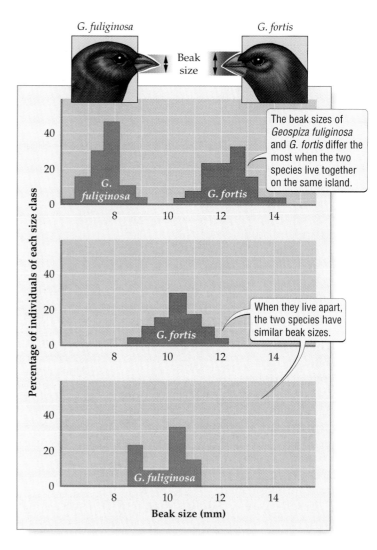

Figure 42.10 Character Displacement
In character displacement, competition for resources causes competing species to evolve to be more different in form over time. Competition between two species of ground finches, *Geospiza fuliginosa* and *G. fortis*, may be the driving force that causes the beak sizes of these birds to be more different when they live on the same island than when they live apart.

more different when they live together than when they live in isolation. Among birds, for example, beak size, and hence the preferred size of the seeds the birds eat, of two species of Galápagos finches are more different on islands that have both species than on islands that have only one of the species (Figure 42.10). Recent experiments with fish and lizards also suggest that character displacement is important in nature.

■ In interference competition, an organism directly excludes its competitor from the use of resources. In exploitation competition, species compete indirectly, each reducing the resource levels available to the other. Competition is common and has a strong effect on the distribution and abundance of species. Competition can increase structural differences between species.

Interactions among Organisms Shape Communities and Ecosystems

Throughout this chapter we have seen how interactions among organisms help determine the distribution and abundance of organisms. Interactions among organisms also have large effects on communities (populations of different species that live in an area) and ecosystems (each of which is a community of organisms plus the environment in which the organisms live).

When dry grasslands are overgrazed, for example, grasses may become less abundant and desert shrubs may become more abundant. These changes in the abundance of grasses and shrubs can change the physical environment. For example, the rate of soil erosion may increase because shrubs do not stabilize soil as well as grasses do. Ultimately, if overgrazing is severe, the ecosystem can change from a dry grassland to a desert.

Dingo

Different types of interactions among organisms are connected to each other and together can have complicated effects on natural communities. For example, dingoes are the largest predators on the Australian continent. When dingoes were excluded from a particular region in Australia, the population of their preferred prey, red kangaroos, increased dramatically (166-fold). The resulting increase in grazing by kangaroos changed the outcome of competition among plant species, causing some species to increase in abundance and others to decrease. Overall, the removal of dingoes resulted in (1) an increase in red kangaroos, (2) a decrease in plants consumed by red kangaroos, (3) changes in the competitive interactions among plants, and (4) changes in the composition of plant communities.

In two examples discussed in this section, an interaction among organisms had a ripple effect, changing the abundance of populations, the community of species that lived in an area, and even, in the case of the dry grasslands, converting one ecosystem (dry grassland) to another (desert shrubland). In general, interactions among organisms can affect all levels at which ecology can be studied: the individual organisms involved in an interaction, populations of organisms, communities of organisms, and ecosystems.

> ■ The outcome of one type of interaction among organisms can change other types of interactions among organisms. Interactions among organisms affect individuals, populations, communities, and ecosystems.

Highlight

Agriculture: Consumer–Victim Interaction or Mutualism?

Many people eat meat, and if you live in a rural area, you may have killed or witnessed the killing of animals that provide people with meat. At one level, the relationship between humans and the animals they eat is a simple consumer–victim interaction: Humans kill some domesticated animals for food. But is the relationship between humans and the food we eat really this simple?

Humans raise cattle in many parts of the world; thus there are many more cattle than there would be otherwise. In this sense, what appears to be a consumer–victim interaction also has elements of a mutualism: Individual cows produce more calves than they could on their own (a "+" for cows), and cows provide people with food (a "+" for people).

What is true for cows is true for all our agricultural crops: Although we consume the plants and animals that we raise for food, we also produce these species in far greater numbers than they could achieve without our aid. In addition, many of our crops can be viewed as species that we created by guiding their evolution from wild species. For example, corn and cows were each created by humans from a wild ancestor (see the Box in Chapter 21, page 333).

"Farming mutualisms" also can be found in the natural world. For example, some ants "herd" aphids: The ants protect the aphids from predators, and the aphids provide the ants with a carbohydrate food source. In some cases, interactions in nature share with human farming systems the blurred distinction between consumer–victim interactions and mutualisms. For example, leaf-cutter ants in South America cut portions of leaves from plants, bring the leaf material back to their nests, and cultivate fungi that grow on the leaves (mutualism). The ant larvae then eat portions of the fungi for food (consumer–victim interaction). Although they eat some of the fungi, the ants also care for the fungi and move them from place to place, thus increasing both the distribution and the abundance of the fungi.

> ■ Humans kill and consume the plants and animals we raise for food, a consumer–victim interaction. Our interactions with the species we raise for food also have elements of a mutualism: Crop plants and domesticated animals produce more offspring than they could without our help, and they provide us with food. Farming mutualisms are also found in the natural world.

Summary

Mutualism

■ Common in nature, mutualisms evolve when the benefits of the interactions are greater than the costs for both partners.

■ Mutualisms help determine the distribution and abundance of the mutualist species and of species that depend directly or indirectly on the mutualist species.

Consumer–Victim Interactions

■ Consumers such as predators, parasites and pathogens, and herbivores can be a strong selective force, leading their victims to evolve various ways to avoid being eaten.

■ Consumers can restrict the distribution and abundance of their victims, in some cases driving their victims to extinction.

Competition

■ In interference competition, an organism directly excludes its competitor from the use of resources.

■ In exploitation competition, species compete indirectly, each reducing the resource levels available to the other.

■ Competition is common and has a strong effect on the distribution and abundance of species.

■ Competition can increase structural differences between species.

Interactions among Organisms Shape Communities and Ecosystems

■ The outcome of one type of interaction among organisms can change other types of interactions among organisms.

■ Interactions among organisms affect individuals, populations, communities, and ecosystems.

Highlight: Agriculture: Consumer–Victim Interaction or Mutualism?

■ Humans kill and consume the plants and animals we raise for food, a consumer–victim interaction.

■ Our interactions with the species we raise for food have elements of a mutualism: Crop plants and domesticated animals produce more offspring than they could without our help, and they provide us with food.

■ Farming mutualisms are also found in the natural world.

Key Terms

behavioral mutualism p. 632

character displacement p. 637

competition p. 636

consumer–victim interaction p. 634

exploitation competition p. 636

herbivore p. 634

interference competition p. 636

mutualism p. 632

parasite p. 634

pathogen p. 634

pollinator mutualism p. 632

predator p. 634

warning coloration p. 634

Chapter Review

Self-Quiz

1. Which of the following statements about consumers is true?
   a. They cannot drive their victims extinct.
   b. They are not important in natural communities.
   c. They can apply strong selection pressure to their victims.
   d. They cannot alter the behavior of their victims.

2. In what type of competition do organisms directly confront each other over the use of a shared resource?
   a. interference
   b. exploitation
   c. physical
   d. unstable

3. Interactions among organisms
   a. do not influence the distribution and abundance of organisms.
   b. are rarely beneficial to both species (that is, mutualism is not common).
   c. have a strong influence on communities and ecosystems.
   d. cannot drive species to extinction.

4. Advantages received by a partner in a mutualism can include
   a. food.
   b. protection.
   c. increased reproduction.
   d. all of the above

5. The shape of a fish jaw influences what the fish can eat. Researchers found that the jaws of two fish species were more similar when they lived in separate lakes than when they lived in the same lake. The increased difference in jaw structure when the fish live in the same lake is a potential example of
   a. warning morphology.
   b. character displacement.
   c. mutualism.
   d. consumer–victim interactions.

Review Questions

1. A mutualism typically has costs for both of the species involved. Why then is mutualism so common?

2. How can a species that is an inferior competitor have a negative effect on the superior competitor?

3. Rabbits can eat many plants, but they prefer some over others. Assume that the rabbits in a grassland that has many plant species prefer to eat a species of grass that happens to be a superior competitor. If the rabbits were removed from the region, which of the following do you think would be most likely to happen? The plant community would (a) change to have fewer species, (b) change to have more species, or (c) remain largely unchanged. Explain your answer.

42   **The Daily Globe**

Human Interference Causes Imbalance

To the Editor:

I recently read a story about Bob Hardy, a self-proclaimed "nature lover" who camped illegally on a small island off the coast of New Zealand. As Bob put it, "I just wanted to see a brown kiwi, a rare flightless bird that lives on the island. I couldn't get a permit to go out there, so I rented a boat and went there myself." Bob may have meant no harm, but he brought his dog Max, and Max killed 26 of the 157 rare birds in just four nights on the island.

Bob's careless actions drove the brown kiwi a step closer to extinction. But I mention Bob and his dog to make a larger point: All over the world humans are introducing new predators or killing native predators, thereby changing the balance of nature. When humans introduce a predator, like Bob's dog Max, native species may go extinct because they have not evolved defenses to cope with the new predator. Such would have been the fate of the flightless brown kiwis if Max had remained on the island longer.

Similarly, when we remove a native predator, we may shift the balance of nature in undesirable ways. For example, here in the Midwestern United States we hunted wolves to extinction, thus removing a major predator that kept deer populations in check. With their primary predator gone, deer populations periodically increase rapidly in size. When this happens, deer can become so numerous that they overgraze their food sources, damaging many tree and shrub populations; as a result, many deer starve to death.

The point is that species are interconnected: When we introduce or remove predators, we may cause undesirable effects, like the explosion of deer populations and the resulting damage to plant populations. We must be much more careful about what we do; otherwise, thoughtless actions like those of Bob Hardy will drive species extinct and change the entire web of life.

Carol Simpson
Wisconsin Park Service

Evaluating "The News"

1. Small consumers that eat plants, such as insects, are often introduced to new regions by accident. What steps would have to be taken for people to be more careful about introducing such species? Would you be willing to have your actions restricted in order to prevent the introduction of such species?

2. Do you think people who import nonnative species to a region for use as pets or as ornamental plants should be allowed to do so? Why or why not?

3. Because wolves may kill livestock, many ranchers oppose reintroducing them into regions where the wolves were hunted to extinction. Do you think wolves should be reintroduced despite the objection of ranchers? Why or why not?

4. Are the interactions of humans with other organisms part of "nature," or are humans outside the "balance of nature" that is referred to in this letter to the editor? If humans are a part of nature, should we be careful what actions we take, since, by definition, any action we take is "natural"? Explain your answer.

43

Communities of Organisms

The Birth of an Island

Richard Ross, *Deyrolle Taxidermy, Paris, France*, 1986.

Hawaii is a volcanic island, formed over hundreds of thousands of years by eruptions of lava that gradually built the island up from the ocean floor. Finally, as molten rock poured from the volcano and broke the surface of the water, the island emerged from the ocean about 600,000 years ago. Then, the lava cooled rapidly to create the first bit of land where Hawaii now stands.

The birth of the island of Hawaii was just the most recent of many such events that have occurred during the last 70 million years. The Hawaiian Islands are a chain of volcanic islands, each formed as the Pacific continental plate moved over a stationary, weak point in Earth's mantle. Kure, the oldest island still visible, is about 15 million years old. Northwest of Kure lie the remnants of even older islands, all of which have now sunk back below the sur-

Communities change naturally and can recover rapidly from some, but not all, forms of disturbance.

face of the ocean. The oldest of these now submerged islands is approximately 70 million years old.

From an ecological and evolutionary point of view, when an island rises out of the sea it marks the beginning of a huge and exciting natural experiment. What organisms will first colonize the island? How will these organisms interact and evolve over time? Will new and unusual communities of organisms form on the island? Or will the island communities come to resemble communities on nearby mainlands?

The Hawaiian Islands are very remote, and before the arrival of humans they were colonized by relatively few species. Over time, however, the species that colonized the islands gave rise to many new species, with the result that communities of organisms on Hawaii are very different from those anywhere else on Earth. Today, many of the unique communities on Hawaii are threatened by introduced species and other forms of human impact. Is there something about island communities that makes them particularly vulnerable to human impact? On Hawaii and elsewhere, what can be done to prevent human actions from having undesirable effects on ecological communities?

Now under water, this former island was formed 70 million years ago.

Pacific Ocean

Hawaii was formed 600,000 years ago.

An Ecological Community
The Hawaiian Islands are part of a chain of volcanic islands that have risen from the sea over the past 70 million years. As newly formed islands were colonized by plants and animals and as new species evolved on the islands, ecological communities like this one were formed.

Key Concepts

1. A community is an association of populations of different species that live in the same area.

2. Food webs document the feeding relations within a community. Keystone species play a critical role in determining the type and abundance of species in a community.

3. All communities change over time. As species colonize new or disturbed habitat, they tend to replace each other in a directional and fairly orderly process called succes-

sion. Communities also change over time in response to changes in climate.

4. Communities can recover rapidly from some forms of natural and human-caused disturbance.

5. It may take thousands of years for communities to recover from other forms of human-caused disturbance, suggesting the need for restraint when we take actions that affect natural communities.

A **community** is an association of populations of different species that live in the same area. There are many different types of communities, ranging from those found in grasslands and forests to those found in the digestive tract of a cow. Most communities have many species of organisms, and as we learned in Chapter 42, the interactions among these organisms can be complex. As we'll discover in this chapter, ecologists

seek to understand how interactions among organisms influence natural communities.

Ecologists also seek to understand how human actions affect communities. At present, humans are having a great effect on many kinds of ecological communities. When we cut down tropical forests we destroy entire communities of organisms, and when we give antibiotics to a cow we alter the community of micro-

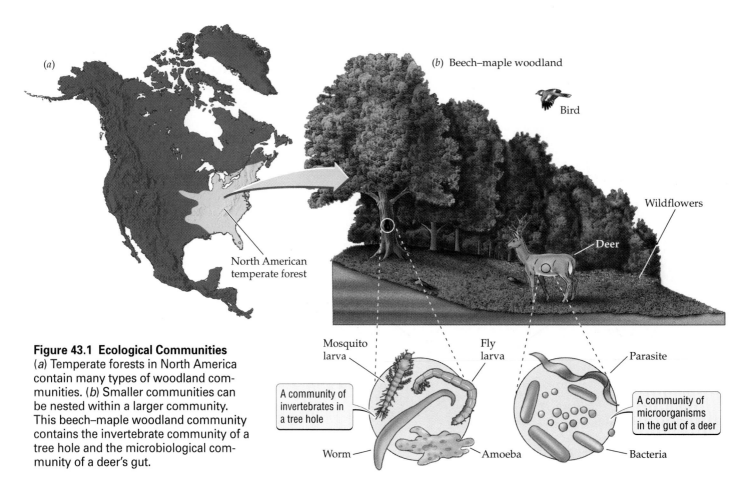

Figure 43.1 Ecological Communities (*a*) Temperate forests in North America contain many types of woodland communities. (*b*) Smaller communities can be nested within a larger community. This beech–maple woodland community contains the invertebrate community of a tree hole and the microbiological community of a deer's gut.

(*a*)

North American temperate forest

(*b*) Beech–maple woodland

Bird

Wildflowers

Deer

Mosquito larva

Fly larva

Parasite

A community of invertebrates in a tree hole

A community of microorganisms in the gut of a deer

Worm

Amoeba

Bacteria

organisms living in its digestive tract. To prevent our actions from having undesirable effects, we must understand how communities work and how they respond to both natural and human-caused disturbances.

In this chapter we describe factors that influence what species are found in a community. We pay particular attention to how communities change over time and how they respond to disturbance, including forms of disturbance caused by people. We begin by discussing the nature of ecological communities.

Nature of Communities

Communities vary greatly in size and complexity, from the community of microorganisms that inhabits a small temporary pool of water, to the community of plants that lives on the floor of a forest, to a forest community that stretches for hundreds of kilometers. Communities also can be nested within each other, as Figure 43.1 shows. In

general, just as for the study of populations, what constitutes a community depends on the organisms under study and the biological questions of interest.

Ecological communities are shaped by the individual species that live in the community, by interactions among species, and by interactions between the species and the environment. In this section we focus on how individual species and interactions among species affect communities of organisms. At the close of the chapter, we discuss how interactions between species and their environment influence communities.

Food webs consist of multiple food chains

One important aspect of a community is who eats whom. Feeding relations can be described by **food chains**, each of which is a single sequence of who eats whom in a community (see Figure 43.2). The movement of food through a community can be summarized by connecting the different food chains to each other to form a **food web**, which consists of the interconnected and overlapping food chains of a community (Figure 43.2).

Food webs and the ecological communities that they describe are based on a foundation of producers. **Producers** are organisms that use energy from an external source like the sun to produce their own food without having to eat other organisms or their remains. On

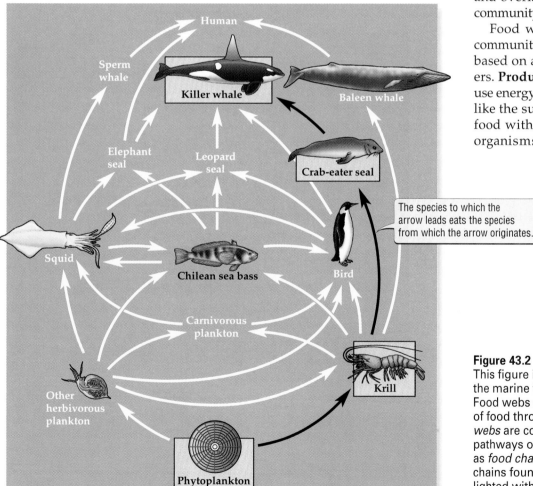

Figure 43.2 A Food Web
This figure is a simplified version of the marine food web in the Antarctic. Food webs summarize the movement of food through an ecosystem. *Food webs are composed of many specific pathways of who eats whom, known as food chains.* One of the food chains found in this food web is highlighted with black arrows and yellow boxes.

land, green plants harvest energy from the sun and are the producers. In aquatic biomes, a wide range of organisms serve as producers, including plankton (oceans), algae (intertidal zones and lakes), and bacteria (deep-sea vents).

Primary consumers are organisms such as herbivores, like cows or grasshoppers, that eat producers. **Secondary consumers** are organisms such as predators, like humans and birds, that feed on primary consumers. This process of eating organisms that ate other organisms can continue: A bird that eats a spider that ate a beetle that ate a plant is an example of a tertiary consumer.

Species interactions have profound effects on communities

Mutualism, consumer–victim interactions, and competition each influence the distribution and abundance of species found in a community, as shown by the coral reef, dingo, and barnacle examples discussed in Chapter 42. Here we focus on the effects of **keystone species**, species that relative to their own abundance have a large impact on the type and abundance of the other species in a community.

In architecture, a keystone is the central, topmost stone that keeps an arch from collapsing. As originally conceived by the ecologist Robert Paine, a keystone species serves a similar role with respect to the "architecture" of a food web: A keystone species is a predator at the top of the food web, and if it is removed, the entire food web will change drastically.

In an experiment conducted along the rocky coast of Washington State, Paine removed the sea star *Pisaster* from one site and left an adjacent, undisturbed site as a control. He found that in the absence of the sea star, of the original 18 species in the community, only mussels remained (Figure 43.3). When the sea stars were present, they ate mussels, thereby keeping the number of mussels low enough so that the mussels did not crowd out the other species of the community.

In other communities, herbivores such as sea urchins, snow geese, and elephants have been found to function as keystone species. Thus, the term "keystone species" is no longer restricted to top predators. In general, keystone species have large effects on communities because they alter the

results of interactions among organisms that determine the types and abundance of species in a community.

When humans remove a keystone species, the community changes greatly. For example, when people removed rabbits from a region in England, they unintentionally converted grasslands with many plant species to grasslands with just a few species of grasses. This change occurred because rabbits were a keystone species that had held the grasses in check. In the absence of rabbits, the grasses crowded out other plant species.

When the sea star *Pisaster* was removed from a community experimentally, the number of species dropped from 18 to 1, a mussel.

Figure 43.3 A Keystone Species
The sea star *Pisaster ochraceus* is a keystone species that feeds on mussels, thereby preventing the mussels from crowding out other species in the community.

A *Pisaster* sea star feeding on a mussel

■ Communities can be described by food webs, which summarize the interconnected food chains of who eats whom in a community. Relative to their abundance, keystone species have large effects on the food web of a community because they alter the results of how organisms interact, thus changing the types or abundance of species in a community.

Communities Change over Time

All communities change over time. For example, the number of individuals in a species often changes as the seasons change: For example, we would not find a butterfly flying in a North Dakota field in the middle of winter. Similarly, every community shows year-to-year changes in the abundance of organisms (see Chapter 41). In addition to such seasonal and yearly changes, communities show broad, directional changes in species composition over longer periods of time. In this section we consider directional changes in communities that occur over relatively long time periods.

Succession establishes new communities and replaces altered communities

How do groups of species come together to form a community? A community may begin when new habitat is created, as when a volcanic island rises out of the sea or when rock and soil is deposited by a retreating glacier. New communities also may form in regions that have been disturbed, as by a fire or hurricane. Species that arrive early in new or disturbed habitat tend to be replaced later by other species, which in turn may be replaced by still other species. Earlier species are re-placed by later species because the replacement species are better able to grow and reproduce under the environmental conditions of the area.

The process by which species are replaced over time is called **succession**. For a given location, the order in which species replace one another is fairly predictable (Figure 43.4). A sequence of species replacements sometimes ends in a **climax community**, which for a particular climate and soil type is a community whose species are not replaced by other species. But in many, perhaps most, ecological communities, disturbances such as fire or windstorms occur so frequently that communities are constantly changing in response to a previous disturbance event. Such a community does not form a climax community.

Primary succession is succession that occurs in newly created habitat, as when a glacier retreats or an island rises from the sea. In some cases of primary succession, the first species to colonize the area alters the habitat in ways that allow later species to thrive. In other cases, the early colonists hinder the establishment of later species. For example, an experimental study on primary succession in marine intertidal communities found that the first species of algae to colonize new habitat (concrete blocks) inhibited the establishment of species that ultimately replace them on the rocky coast near Santa Barbara, California. In cases like this, the early species lose their hold eventually because they are more susceptible to a particular feature of the environment, such as disturbance (for example, fire, wind, waves), grazing by herbivores, or extremes of heat or cold.

Figure 43.4 Ecological Succession
When strong winds cause moving sand dunes to form at the southern end of Lake Michigan, ecological succession often leads to a community dominated by black oak. Succession in such dunes occurs in three stages and forms black oak communities that have lasted up to 12,000 years. Under different local environmental conditions, succession in Michigan sand dunes can lead to the establishment of stable communities as different as grasslands, swamps, and sugar maple forests.

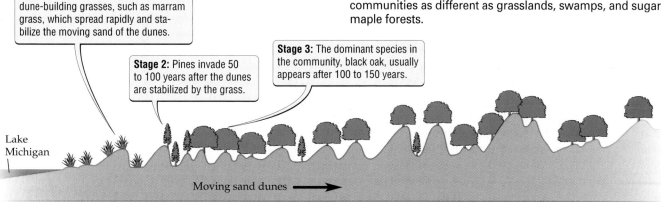

Stage 1: Bare sand is first colonized by dune-building grasses, such as marram grass, which spread rapidly and stabilize the moving sand of the dunes.

Stage 2: Pines invade 50 to 100 years after the dunes are stabilized by the grass.

Stage 3: The dominant species in the community, black oak, usually appears after 100 to 150 years.

Lake Michigan

Moving sand dunes ⟶

(a)

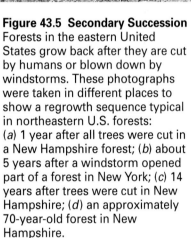

(b)

Figure 43.5 Secondary Succession
Forests in the eastern United States grow back after they are cut by humans or blown down by windstorms. These photographs were taken in different places to show a regrowth sequence typical in northeastern U.S. forests: (a) 1 year after all trees were cut in a New Hampshire forest; (b) about 5 years after a windstorm opened part of a forest in New York; (c) 14 years after trees were cut in New Hampshire; (d) an approximately 70-year-old forest in New Hampshire.

(c)

(d)

Secondary succession is the process by which communities recover from disturbance, as when a field ceases to be used for agriculture or when a forest grows back after a storm that has blown down many trees (Figure 43.5). Compared to primary succession, the soil in secondary succession is well developed and may contain seeds from species that usually occur late in the successional process. The presence of such seeds in the soil can considerably shorten the time required for the later stages of succession to be reached.

Communities change as the climate changes

Groups of species can stay together for long periods of time. For example, some plant communities in the southeastern United States and in China resemble an extensive community that once stretched across the northern parts of Asia, Europe, and North America. As the climate grew colder during the past 60 million years, plants in these communities migrated south, forming communities in Southeast Asia and southeastern North America that are similar in composition to the community from which they came.

Although groups of plants can remain together for millions of years, the community located in a particular place changes as the climate of that place changes. The climate of a given location can change over time for two reasons: global climate change and continental drift.

First, as discussed in Chapter 40, the climate of Earth changes over time. Historically, changes in the global climate have been due to relatively slow natural processes, such as the advance and retreat of glaciers, but evidence is now mounting that human activities are causing rapid changes in the global climate (see Chapter 45).

Second, the continents do not remain in one place (see Figure 22.6), and as they move their climates change. To give a dramatic example of continental drift, 1 billion years ago Queensland, Australia, which is now located at 12 degrees south latitude, was located near the North Pole. Roughly 400 million years ago, Queensland was at the equator. The species that thrive at the equator and in the Arctic are very different; thus the movement of the continents resulted in large changes in the communities of Queensland over time.

> ■ All communities change over time. Directional changes that occur over relatively long periods of time have two main causes: succession and climate change. Primary succession occurs on newly created habitat, secondary succession in communities recovering from moderate forms of disturbance. The climate where a community is located can change because of global climate change or continental drift.

Communities Can Recover from Disturbance

Ecological communities are subject to many natural forms of disturbance, such as fires, floods, and severe windstorms. Following such disturbances, secondary succession can reestablish previously existing communities. Thus, communities can recover from some forms of disturbance.

Communities have been exposed over long periods of time to natural forms of disturbance. In contrast, humans may introduce entirely new forms of disturbance, such as the dumping of hot wastewater into a river. Humans also may alter the frequency of an otherwise natural form of disturbance—for example, causing a dramatic increase or decrease in the frequency of fires or floods.

Can communities also recover from disturbance caused by humans? For some forms of human-caused disturbance, the answer is yes. Throughout the eastern United States, for example, are many places where forests were cut down and used for farmland and then, years later, the farmland was abandoned. Forests called second-growth forests have grown at these abandoned farms, often within 40 to 60 years after farming stopped.

Second-growth forests are not identical to forests that were originally present; the size and abundance of tree species are different, and fewer plant species grow beneath a second-growth forest than beneath a forest that has never been cut down. However, the forests already have recovered partially, and over the next several hundred years there will probably be fewer and fewer differences between original and second-growth forests.

In some cases, communities can also recover from pollution. Lake Washington is a large, clear lake in Seattle, Washington. As the city of Seattle grew, raw sewage was dumped into the lake. This practice declined after 1926 and was stopped by 1936. However, beginning in 1941 treated sewage was discharged into Lake Washington from newly constructed sewage treatment plants.

A major effect of discharging sewage into Lake Washington was that phosphorus in the sewage provided extra nutrients for algae. The numbers of algae soared, decreasing the clarity of the water. As the algae reproduced and then died, the resulting huge numbers of dead algae provided a food source for bacteria, whose populations also increased. Bacteria use oxygen when they consume dead algae, causing oxygen concentrations in the water to decrease. Decreased oxygen concentrations can kill invertebrates and fish.

By the early 1960s Lake Washington was highly degraded, so much so that it was referred to in the local press as Lake Stinko. From 1963 to 1968 less and less sewage was dumped into the lake, and virtually none was dumped after 1968. Once inputs of sewage were stopped, algae populations declined, oxygen concentrations increased, and Lake Washington returned to its former, clear state.

> ■ Communities can recover from some forms of natural and human-caused disturbances. Depending on the community, the time required for recovery varies from years to decades or centuries.

Humans Can Cause Long-Term Damage to Communities

Communities do not always recover from disturbances caused by humans. For example, northern Michigan once was covered with a vast stretch of white- and red-pine forest. From 1875 to 1900 nearly all of these trees were cut down, leaving only a few scattered patches of virgin forest. The loggers left behind large quantities of branches and sticks, which provided fuel for fires of great intensity. In some locations, the pine forests of northern Michigan have never recovered from the combination of fire and logging.

A combination of logging and fire has also changed large regions of South America and the Pacific islands from tropical forests to grasslands (Figure 43.6). Scientists estimate that it will take tropical forests hundreds to thousands of years to recover from such changes.

Finally, in some dry grasslands in the American Southwest, overgrazing by cattle has transformed healthy grasslands into desert shrublands (Figure 43.7). How do cattle cause such large changes? Grazing and trampling by cattle decrease the amount of grass in the community. With less grass to cover the soil and hold it in place, often the soil becomes drier and erodes more

Figure 43.6 Communities Do Not Always Recover from Disturbance
(a) A low elevation Indonesian rainforest, photographed in 1980. (b) In 1982, the rainforest shown in (a) burned. As seen in this 1985 photograph, three years after the fire, few trees remained alive. (c) A second fire occurred in 1998, killing the remaining trees and converting the former rainforest to a grassland dominated by an introduced species of grass. This photograph was taken from a site near that shown in (a) and (b).

rapidly. Desert shrubs thrive under the new soil conditions, but grass does not. Overall, the observed changes in soil characteristics can make it very difficult to reestablish grasslands, even when cattle are removed.

In the three examples discussed in this section—the pine forests of Michigan, the tropical forests of South America and the Pacific islands, and the grasslands of the American Southwest—humans have altered communities so greatly that it will take hundreds or thousands of years for the communities to recover. The long period of time

required for communities to recover from some human-caused disturbances suggests the need for restraint when we take actions that affect natural communities.

Figure 43.7 Overgrazing Can Convert Grasslands into Deserts
More than 200 years ago, large regions in the American Southwest were covered with dry grasslands. Most of these grasslands have been converted to deserts, in large part because of overgrazing.

(a)

This dry grassland is in southern New Mexico.

(b)

This nearby, former dry grassland is now a desert.

■ It can take hundreds to thousands of years for communities to recover from some forms of human-caused disturbance. It is important to consider that we can have long-lasting effects when we take actions that affect natural communities.

Disturbance, Community Change, and Human Values

Change is a part of all communities. What, if anything, is special about human-caused community change? What human values are affected by community change?

Human actions can change communities very rapidly across large geographic regions. Disturbances not caused by people also can occur rapidly and affect large geographic regions. For example, most scientists think the impact of a large asteroid contributed to the sudden extinction of many species, including the last of the dinosaurs, 65 million years ago (see Chapter 22). However, human-caused community change is unique in that we have direct control over whether or not we take the actions that cause community change. Our unique ability to control our actions brings with it the responsibility to use our power to cause community change wisely, a topic to which we will return in Chapter 45.

What human values are affected by human-caused community change? First, human actions that degrade or destroy ecological communities have ethical consequences. When humans disrupt communities, our actions kill individual organisms, alter communities that may have persisted for thousands of years, and threaten species with extinction. According to results from surveys in North America and Europe, many people find such impacts of human actions on individuals, communities, and species ethically unacceptable.

Second, human-caused community change often reduces the aesthetic value of the community. Tropical forests have unique aesthetic value to many people. When our actions cause tropical forests to be destroyed and replaced by introduced grasses (see Figure 43.6), we deprive current and future generations of experiencing the beauty of those forests.

Finally, in many cases when humans change a community, we reduce its economic value. For example, economic value is lost when we convert grasslands to deserts, in part because there is more plant material to support grazing in grasslands than in deserts. In general, when our actions cause major damage to ecological communities, we often harm our own long-term economic interests. At present, it is common for people to take actions, such as allowing livestock to overgraze

grasslands, that result in short-term economic benefit but cause long-term economic harm.

Communities change constantly in the face of both natural and human-caused disturbances. The challenge for human society is to manage Earth's changing communities while maintaining the diversity of their species and the ethical, aesthetic, and economic value that they provide to humans. What do you think must be done so that we can successfully meet this challenge? We describe one ecologist's thoughts on this question in the Box on page 653.

■ Natural and human-caused disturbances can change communities greatly. Humans can control whether or not we perform actions that cause community change. Our potential to control our actions gives us the responsibility to act in ways that do not reduce the ethical, aesthetic, and economic value that communities provide to humans.

Highlight

Community Change in the Hawaiian Islands

The Hawaiian Islands are 4000 kilometers from the nearest continent. They are the most isolated chain of islands on Earth. Because the islands are so remote, entire groups of organisms that live in most communities never reached Hawaii. For example, there are no native ants or snakes, and there is only one native mammal (a bat).

The few species that arrived at the Hawaiian Islands found themselves in an environment that lacked most of the species from their previous communities. The sparsely occupied habitat and the lack of competitor species resulted in the evolution of many new species. For example, there are many different species of Hawaiian silverswords found on the islands today, all of which arose from a single ancestor (Figure 43.8). The new silversword species evolved to have very different forms, enabling them to live in very different habitats. Because other groups of plants and animals on the islands also evolved unique new species, the Hawaiian Islands have many distinctive natural communities.

Since the arrival of people in the islands about 1500 to 2000 years ago, Hawaii's unique biological communities have been threatened by habitat destruction, overhunting, and introduced species. Of these threats, the impact of introduced species can be easy to overlook because such species often wreak their havoc quietly, behind the scenes. Introduced Argentine ants, for example, may drive native insects to extinction, but it can take years before even a trained biologist realizes what the introduced ants have done.

Islands are particularly vulnerable to the destruction caused by introduced species. Relatively few species colonize newly formed islands, and these species then evolve in isolation. Because of this isolation, species on islands may be ill equipped to cope with new predators or new competitor species that are brought from the mainland. In addition, introduced species often arrive without the predator and competitor species that held their populations in check on the mainland. Thus, on islands the potential exists for populations of introduced species to increase dramatically.

In addition to causing individual species to become extinct, introduced species can destroy entire communities. For example, most of the native plants in Hawaii are not adapted to fire. An introduced species that alters the frequency or intensity of fire can have devastating effects. Consider what happened with the introduced species beard grass.

Beard grass was introduced to Hawaii by humans and invaded the seasonally dry forests of Hawaii Volcanoes National Park in the late 1960s. Before that time, fires occurred on average every 5.3 years, and each fire burned only 0.25 hectare (1 hectare is about 2.5 acres). Since the introduction of beard grass, fires have occurred at a rate of more than one per year, and the average burn area of each fire has increased to more than 240 hectares.

Why did the introduction of beard grass increase the frequency and intensity of fire? As beard grass grows, it deposits a large amount of dry plant matter on the ground. This material catches fire easily, and the fires burn much hotter than they would in the absence of beard grass. Beard grass recovers well from large and hot fires, but the native trees and shrubs of the seasonally dry woodland do not. As a result, former woodlands have now been converted to open pastures filled with beard grass and other (even more fire-prone) introduced grasses.

The Hawaiian dry woodlands are destroyed, probably forever. Because there is no hope of restoring the native community, ecologists are now trying to construct a new community that is tolerant of fire yet contains native trees and shrubs. This is a difficult challenge, and it is uncertain whether the effort will succeed. If not, what was once woodland will remain indefinitely as open meadows filled with introduced grasses.

> ■ Island communities are very vulnerable to introduced species in part because species on islands evolved in isolation and hence may not be able to cope with introduced predators or competitors. In the Hawaiian Islands, introduced species have caused the extinction of individual species and the destruction of entire communities of organisms.

Figure 43.8 Great Diversity from a Single Ancestor
Hawaiian silverswords are a diverse group of plant species that evolved from a single ancestor and are found only on the Hawaiian Islands. Although the three silversword species shown here are closely related, they live in very different habitats and differ greatly from each other.

Peter Vitousek Examines the Human Impact on Earth

Peter Vitousek, an ecologist at Stanford University who grew up in Hawaii and who performs research in the Hawaiian Islands, has written several thought-provoking articles about the impact of humans on ecological communities. His time in Hawaii has given him firsthand knowledge of the ways humans affect the unique ecological communities found there (see the Highlight on p. 651). That firsthand knowledge has contributed to the passion with which he seeks to understand and limit the negative impact of humans on all ecological communities.

As discussed in this chapter, humans can change ecological communities in ways that take thousands of years to recover from. As we enter a new century, how can a growing human population prevent the extinction of species and maintain the ethical, aesthetic, and economic value that we get from ecological communities? In his scientific and other writings, Dr. Vitousek argues that we must take a multipronged approach.

Peter Vitousek

First, we must reduce our impact on Earth by reducing the rate of growth of our population and reducing the rate at which we use resources (see Chapters 41 and 45).

Second, we must recognize that humanity is a dominant force on Earth: No place on this planet is free from significant human impact. It is not enough to leave nature alone; our influence is too great for that. Instead, we must manage ecological communities actively.

The need to manage ecological communities implies a third need: We must devote far greater effort to understanding how ecological communities work. By understanding communities better, we will have a much better chance of managing them in ways that prevent the extinction of species while preserving the different values that ecological communities provide people. In addition, as Dr. Vitousek and others have pointed out, to understand the world better is to enjoy it more and to want to take better care of it.

Finally, Dr. Vitousek argues that to maintain ecological communities and the values they provide humans, we must all work together. Many scientists are better at figuring out what is going on and why than they are at seeing and implementing alternative ways of doing things. Thus, partnerships among scientists, economists, social scientists, governments, businesses, and private landowners will be essential as humanity strives to solve the complex problems that we will face throughout the twenty-first century.

Summary

Nature of Communities

- Communities can be described by food webs, which summarize the interconnected food chains of who eats whom in a community.

- Relative to its abundance, a keystone species has a large effect on the food web of a community.

- Keystone species have large effects because they alter the results of interactions between organisms, thus changing the types or abundance of species in a community.

Communities Change over Time

- All communities change over time.

- Directional changes that occur over relatively long periods of time have two main causes: succession and climate change.

- Primary succession occurs on newly created habitat, secondary succession in communities recovering from moderate forms of disturbance.

- The climate where a community is located can change because of global climate change or continental drift.

Communities Can Recover from Disturbance

- Communities can recover from some forms of natural and human-caused disturbances.

- Depending on the community, the time required for recovery varies from decades to centuries.

Humans Can Cause Long-Term Damage to Communities

- It can take hundreds to thousands of years for communities to recover from some forms of human-caused disturbance.

- It is important to consider that we can have long-lasting effects when we take actions that affect natural communities.

Disturbance, Community Change, and Human Values

- Natural and human-caused disturbances can change communities greatly.

- Humans can control whether or not we perform actions that cause community change.

- Our potential to control our actions gives us the responsibility to act in ways that do not reduce the ethical, aesthetic, and economic value that communities provide to humans.

Highlight: Community Change in the Hawaiian Islands

- Island communities are very vulnerable to introduced species in part because species on islands evolved in isolation and hence may not be able to cope with introduced predators or competitors.

- In the Hawaiian Islands, introduced species have caused the extinction of individual species and the destruction of entire communities of organisms.

Key Terms

climax community p. 647
community p. 644
food chain p. 645
food web p. 645
keystone species p. 646
primary consumer p. 646

primary succession p. 647
producer p. 645
secondary consumer p. 646
secondary succession p. 648
succession p. 647

Chapter Review

Self-Quiz

1. A species that has a large impact on a community relative to its abundance is called a
   a. top predator.
   b. top herbivore.
   c. keystone species.
   d. dominant species.

2. Organisms that can produce their own food from an external source of energy without having to eat other organisms are called
   a. suppliers.
   b. consumers.
   c. producers.
   d. keystone species.

3. Ecological communities
   a. cannot recover from disturbance.
   b. can recover from natural but not human-caused disturbance.
   c. can recover from all forms of disturbance.
   d. can recover from some but not all forms of natural and human-caused disturbances.

4. Which of the following was *not* caused by the introduction of beard grass to Hawaii?
   a. an increase in the growth of native trees and shrubs
   b. an increase in the frequency and intensity of fire
   c. the decline of native trees and shrubs
   d. the conversion of dry woodlands to grasslands

5. A directional process of species replacement over time in a community is called
   a. global climate change.
   b. succession.
   c. competition.
   d. community change.

Review Questions

1. Describe how each of the following factors influences ecological communities: (a) species interactions, (b) disturbance, (c) climate change, (d) continental drift.

2. What is the difference between primary and secondary succession?

3. Provide an example of how the presence or absence of a species in a community can alter a feature of the environment, such as the frequency of fire.

4. Do you think it is ethically acceptable for people to change natural communities so greatly that it takes thousands of years for the communities to recover? Why or why not?

The Daily Globe

Strange Bedfellows?
Environmentalists and Ranchers Unite

ALBUQUERQUE, NM. In the ranching country of the southwestern United States, there are bumper stickers and signs with humorous but angry messages like: "Hungry? Out of work? Eat an environmentalist," an expression of the long-standing conflict between ranchers trying to make a living and environmentalists who oppose the environmental damage caused by ranching. Now, however, the times may be beginning to change as environmentalists and ranchers have begun to join forces to save the habitat that has so long supported ranchers and is so prized by conservationists.

"I grew up hating environmentalists," said Jake Simms, a rancher in Socorro County, New Mexico. "They were the enemy. But some-

thing had to be done. My land could hardly support cattle anymore."

As Mr. Simms and many other ranchers are discovering, while their great grandparents knew a Southwest in which grass stretched for miles and grew to lengths that came above a person's knees, those days and those grasslands are gone, replaced by desert shrubs that provide little food for cattle. Regions once covered with black grama grass were heavily grazed for many years, then invaded by the desert shrub creosote bush, a plant that cattle don't like to eat. With such changes occurring throughout his ranch, Mr. Simms, worried that he would have to give up ranching, turned to an unusual source for

help: the Nature Defense Fund, a large environmental organization. Like an increasing number of ranchers, Mr. Simms found that he and environmentalists had a common goal: improving the land.

The former foes have agreed to a plan in which ranchers remove cattle from environmentally sensitive areas, graze cattle only at a level that the remaining grasslands can support, and devote a small portion of land toward a new industry, ecotourism.

"So far it seems to be working," said Ellen Deen, spokeswoman for the Nature Defense Fund. "Environmentalists and ranchers can cooperate to the benefit of both."

Evaluating "The News"

1. When the goals of environmentalists differ from those of ranchers, loggers, and others, the conflict is often portrayed as a simple choice of jobs versus the environment. If the economic activity in question is not occurring at a rate that can be sustained indefinitely, is the issue really as simple as this?

2. A major factor in our society is the economic bottom line. Such a focus leads people to maximize profits today. Do you think society should place such a strong emphasis on short-term economic gain? Why or why not?

3. Does the current generation have the responsibility to protect the environment for use by future generations? Or does the current generation have the right to use the land as they see fit, even if it harms the environment for generations to come?

44

Ecosystems

Roger Minick, *Woman with Scarf at Inspiration Point,*
Yosemite National Park (Sightseer Series), 1980.

Operation Cat Drop

*I*n 1955, the World Health Organization staged an unusual mission in Borneo, Malaysia, called Operation Cat Drop, in which cats were parachuted into local villages. The story behind this strange event began 16 years earlier, with the development of the pesticide DDT.

In 1939, Paul H. Müller discovered that DDT could be used to kill insects, a finding that earned him the 1948 Nobel prize in physiology or medicine. DDT was one of many chemical poisons developed in the 1940s to kill insects that spread human diseases. Soon DDT and other pesticides were also widely used to control insects that eat crop plants and insects that are annoying but do not threaten human health or crop production. The new poisons were an immediate success: They were cheap, easy to apply, and highly toxic to insects. Although some biologists expressed doubts about their use as early as 1945, most people viewed DDT and similar substances as miracle pesticides.

This view began to change in 1962 with the publication of Rachel Carson's *Silent Spring*. Carson questioned the widespread use of highly toxic pesticides, noting, "Future historians may well be amazed by our distorted sense of proportion. How could intelligent beings seek to control a few unwanted species by a method that contaminated the entire environment and brought the threat of disease and death even to their own kind?"

Rachel Carson's concerns stemmed in part from the toxic nature of DDT to humans and other organisms and in part from the tendency of DDT to persist in the environment for long periods of time. After DDT enters an organism, most of it remains within the body and is stored in fatty tissue. When predators eat organisms that are contaminated with DDT, the predators store in their bodies most of the DDT that was contained in each of their prey. Because organisms toward the top of a food chain have eaten many smaller organisms, they have the highest concentrations of DDT in their bodies.

To return to the parachuting cats, in 1955 as many as 90 percent of the people in some areas of Borneo suffered from malaria. To combat the disease, the World Health Organization sprayed the island with DDT to kill the mosquitoes that transmitted malaria. The disease was brought under control, but soon a new health threat emerged: Villages were overrun with great numbers of rats, and hence there was danger of an outbreak of other diseases, such as plague and typhus.

What had gone wrong? The DDT did kill the mosquitoes, but it also contaminated houseflies and other insects, which were then eaten by small lizards. Because the insects contained high levels of DDT, the DDT was even more concentrated in the lizards. When the lizards were eaten by the village cats, the cats died of DDT poisoning. With the cats out of the way, the rat population exploded, disrupting village life and bringing with them the potential for many dreadful diseases.

Although the cats delivered by Operation Cat Drop brought the rat population under control, the unexpected side effects of spraying with DDT demonstrate the need for caution when adding toxic chemicals to the environment. In addition, the movement of DDT through the food chain (insects to lizards to cats) provides an example of how natural systems recycle materials and follow a few simple rules, such as "everything goes somewhere." By studying how energy and materials flow through natural systems, we can learn much about how natural systems work and why our actions can have surprising, unintended, and sometimes devastating consequences.

A Forest in Malaysia

Key Concepts

1. An ecosystem consists of a community of organisms together with the chemical and physical environments in which the organisms live. Energy, materials, and organisms may move from one ecosystem to another.

2. Energy cannot be recycled. Energy captured by producers is lost as metabolic heat as it moves through a food chain (for example, from plant to herbivore to carnivore). Because of the steady loss of energy, more energy is available toward the bottom than toward the top of the food chain.

3. Earth has a fixed amount of nutrients, the chemical elements needed to sustain life. Nutrients are recycled between organisms and the physical environment. Without this recycling, life on Earth would cease.

4. Ecosystems provide humans with essential services, such as nutrient cycling, flood control, and the filtering of pollutants out of air and water. Our civilization depends on these and many other ecosystem services.

To survive, all organisms need energy to run their metabolism and materials to construct and maintain their bodies. With respect to energy needs, almost all life on Earth depends directly or indirectly on solar energy, the supply of which is renewed continually. Materials, on the other hand, such as the carbon, hydrogen, oxygen, and other elements of which we are made, are added to our planet as small amounts of meteoric matter from outer space. Earth, therefore, has an essentially fixed amount of materials for organisms to use. This simple fact means that for life to persist, natural systems must recycle materials.

This chapter introduces ecosystem ecology, the study of how energy and materials flow through natural systems. We begin by discussing the capture and movement of energy in ecosystems; then we consider how elements essential to life are cycled between organisms and the physical environment. At the close of the chapter, we discuss human impacts on the cycling of materials in ecosystems.

An Overview of How Ecosystems Function

An **ecosystem** consists of a community of organisms together with the chemical and physical environment in which the organisms live. As with communities, an ecosystem may be small or very large: A puddle teeming with protozoans is an ecosystem, as is the Atlantic Ocean. In fact, global patterns of air and water circulation (see Chapter 39) may be viewed as linking all the world's organisms into one giant ecosystem, the biosphere.

Ecosystems are considered open systems because energy, materials like nitrogen and water, and organisms may move from one ecosystem to another. The open nature of ecosystems forces ecosystem ecologists to measure as precisely as possible the movement of ener-

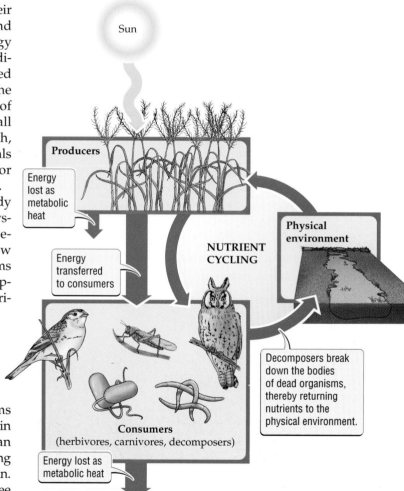

ONE-WAY FLOW OF ENERGY THROUGH THE ECOSYSTEM

Figure 44.1 How Ecosystems Work
At each step in a food chain, a portion of the energy captured by producers is lost as heat given off during cellular respiration (metabolic heat). Thus, energy flows through the ecosystem in a single direction and is not recycled. Nutrients cycle between organisms and the physical environment.

gy and materials into and out of the system. A focus on energy and materials allows very different ecosystems to be compared in a meaningful and standardized way. For example, the effect people have on the movement of energy and nutrients can be used to examine the relative impact of human actions on ecosystems as different as deserts and tropical rainforests.

In broad overview, ecosystems function as shown in Figure 44.1. A portion of the energy captured by producers is lost as metabolic heat at each step of a food chain. Thus, energy is not recycled, but exhibits a one-way flow: It enters the ecosystem from the sun and leaves the ecosystem as metabolic heat. The nutrients required for life, on the other hand, are cycled between organisms and the nonliving environment. Nutrients are absorbed by producers from the environment, cycled among consumers for varying lengths of time, and eventually returned to the environment by decomposers that break down the dead bodies of organisms.

> ■ Ecosystems are open systems in which energy and materials can move from one ecosystem to another. Energy is not recycled in ecosystems; rather, a portion of the energy captured by producers is lost at each step of a food chain. Nutrients are recycled in ecosystems; they pass from the environment to producers to consumers, then back to the environment when decomposers break down the bodies of dead organisms.

Energy Capture in Ecosystems

Most life on Earth depends directly or indirectly on the capture of solar energy by photosynthetic organisms like green plants. Energy captured by plants and other photosynthetic organisms is stored in their bodies in chemical forms, such as carbohydrates. Herbivores that eat plants, carnivores that eat herbivores or other carnivores, and decomposers that consume the remains of all dead organisms depend indirectly on solar energy originally captured by plants.

The amount of energy that organisms fix in photosynthesis, minus the amount lost as heat in cellular respiration, is called **net primary productivity**, or **NPP** (Figure 44.2). Ecologists usually measure NPP as the amount of new biomass produced by photosynthetic organisms during a specified period of time (the term "biomass" refers to the mass of organisms per unit area). Measured in this way, NPP is a rate with units of mass per unit area per unit time, such as kilograms of biomass produced in a square meter in a year. Units based on mass can be converted easily to units based on energy.

(a)

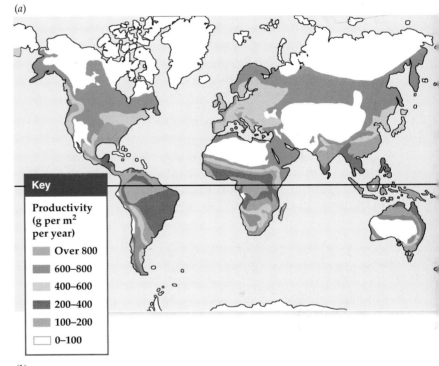

(b)

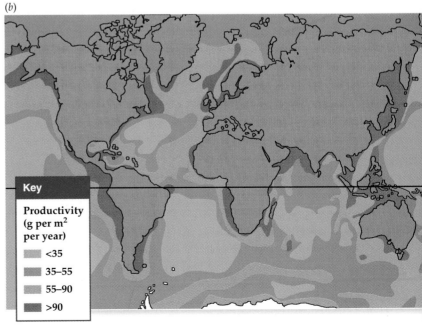

Figure 44.2 Net Primary Productivity
The amount of net primary productivity varies greatly in different parts of the world in both terrestrial (a) and marine (b) ecosystems. Net primary productivity is represented as grams per square meter of area (g per m^2 per year).

NPP is not distributed evenly across the globe. On land, NPP tends to decrease from the equator toward the poles (Figure 44.2*a*). This decrease occurs because the amount of solar radiation available to green plants, the most important producers in terrestrial systems, is highest at the equator and lowest at the poles (see Chapter 39). But there are many exceptions to the general decrease in NPP that occurs from the Tropics to the poles. For example, there are large regions of very low productivity in northern Africa, central Asia, central Australia, and the southwestern portion of North America. Each of these regions is the site of one of the world's major deserts.

The low NPP in deserts emphasizes the fact that sunlight alone is not sufficient for NPP to be high; water is also essential. In addition to water and sunlight, productivity in terrestrial ecosystems can be limited by temperature and soil nutrients. On land, the most productive ecosystems are tropical rainforests and cultivated land (Table 44.1). The least productive terrestrial ecosystems are deserts and tundra (including alpine communities).

The global pattern of NPP in marine ecosystems (Figure 44.2*b*) is very different from that on land. There is little tendency for NPP to decrease from the equator to the poles. The open ocean has low productivity and is, in essence, a marine desert. The productivity of marine ecosystems is often high in regions of the ocean that are close to land. Streams and rivers that drain from the land to the ocean contain nutrients that are in short supply in the ocean. Addition of these nutrients to the ocean stimulates the growth and reproduction of phytoplankton, the microorganism producers at the bottom of aquatic food webs, thus causing the NPP to be high where oceans border land.

High productivity also occurs in the ocean wherever there are upwellings of nutrients from sediments or deep water. Upwellings can cause high productivity even at high latitudes, where both temperatures and incoming solar radiation are low. Finally, the NPP in marine and other aquatic ecosystems can be strongly limited by light and temperature. Although the world's oceans generally have very low productivity, coral reefs and estuaries are among Earth's most productive ecosystems (see Table 44.1). Some wetlands, such as swamps and marshes, also have high productivity.

> ■ Most life on Earth depends on energy that is captured by photosynthetic organisms and then stored in their bodies in chemical form. Net primary productivity varies greatly in different parts of the world. On land, NPP tends to decrease from the equator toward the poles. In marine ecosystems, NPP is often high where the ocean borders land or where upwellings provide scarce nutrients to marine organisms.

Energy Flow through Ecosystems

There are two major ways in which organisms get energy. As described in Chapter 43, producers get their energy from nonliving sources, such as the sun. The organisms at the bottom of a food web, such as green plants, algae, and photosynthetic bacteria, are producers. **Consumers** get their energy by eating other organisms or by eating the remains of once living organisms. Consumers include herbivores, carnivores, and decomposers.

Energy pyramids

Light energy is stored by plants in chemical forms. Chemical compounds that contain energy captured by plants or other producers can be transferred from organism to organism in a food chain. However, energy cannot be recycled; once the energy in a given chemical compound is used, it is gone. To illustrate this point, we will follow the fate of energy from the sun after it strikes the surface of a grassland.

44.1 Net Primary Productivity (NPP) for Selected Ecosystems of the World

Ecosystem	Total area on Earth (10^6 km^2)	NPP (g per m^2 per year) Range	NPP (g per m^2 per year) Average
Terrestrial			
Tropical rainforest	17	1000–3500	2200
Cultivated land	14	100–3500	650
Temperate forest	12	600–2500	1240
Grassland	24	200–2000	790
Tundra and alpine community	8	10–400	140
Desert	42	0–250	40
Aquatic			
Coral reef	0.6	500–4000	2500
Swamp and marsh	2	800–3500	2000
Estuary	1.4	200–3500	1500
Lake and stream	2	100–1500	250
Ocean upwelling zone	0.4	400–1000	500
Open ocean	332	2–400	125

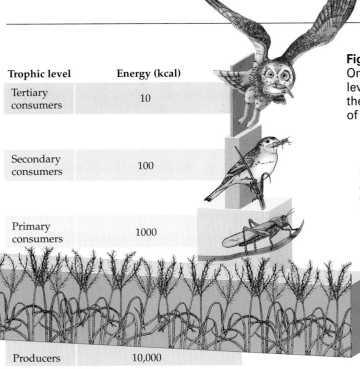

Trophic level	Energy (kcal)
Tertiary consumers	10
Secondary consumers	100
Primary consumers	1000
Producers	10,000

Figure 44.3 An Idealized Energy Pyramid

On average, roughly 10 percent of the energy at each trophic level is transferred to the next trophic level. This figure shows the energy at each trophic level for each 1,000,000 kilocalories of energy from the sun (1 kilocalorie equals 1000 calories).

taken up by the body (for example, we cannot digest the cellulose that is contained in the apple), or it is lost as metabolic heat given off as a by-product of cellular respiration.

Secondary productivity

The rate of new biomass production by consumers is called **secondary productivity**. Because consumers depend on producers for both energy and materials, secondary productivity is highest in areas with high net primary productivity. For example, tundra has a much lower NPP than grassland. For this reason fewer herbivores per unit area can be supported on tundra than on grassland.

New biomass made by plants and other producers is consumed either by herbivores or by decomposers. In some ecosystems, 80 percent of the biomass produced by plants goes directly to decomposers, such as bacteria and fungi (Figure 44.4). Eventually, since all organisms die, all biomass produced by plants, herbivores, and carnivores is consumed by decomposers. As discussed in the next section, decomposers are tremendously important to all life: They recycle nutrients, thereby ensuring that nutrients do not remain locked up as dead material once an organism dies.

A portion of the energy captured by grasses is transferred to the herbivores that eat the grasses, and then to the predators that eat the herbivores. The transfer of energy from grasses to herbivores to predators is not perfect, however. For example, once a unit of energy is used by an organism to run its metabolism, that energy is lost from the ecosystem as unrecoverable heat. Thus, energy moves through ecosystems in a single direction: As one proceeds up a food chain (for example, from grass to grasshopper to bird), portions of the energy originally captured by photosynthesis are steadily lost. For this reason the energy available to organisms in an ecosystem can be represented by a pyramid.

Each level of an energy pyramid corresponds to a step in a food chain and is called a **trophic level** (Figure 44.3). The grass–grasshopper–bird example given in the previous paragraph has three trophic levels: Grass is on the first trophic level, the grasshopper is on the second trophic level, and the bird is on the third trophic level. On average, roughly 10 percent of the energy at one trophic level is transferred to the next trophic level. The energy that is not transferred between trophic levels is either not consumed (for example, when we eat an apple, we eat only a small part of the apple tree) or not

Figure 44.4 Decomposers Are Extremely Important to Any Ecosystem

In ecosystems of all types, more than 50 percent of net primary productivity is used directly by decomposers. This example shows a forest in which 80 percent of the NPP is used directly by decomposers, and the remaining 20 percent is used by other consumers (herbivores and carnivores).

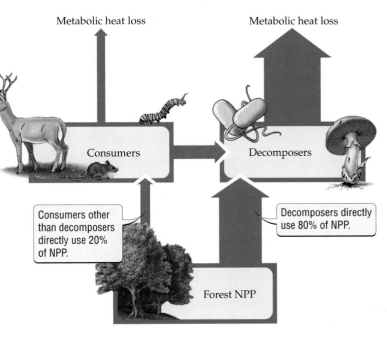

Metabolic heat loss

Metabolic heat loss

Consumers

Decomposers

Consumers other than decomposers directly use 20% of NPP.

Decomposers directly use 80% of NPP.

Forest NPP

■ Chemical compounds that contain energy captured by producers can be transferred from organism to organism in a food chain. Once an organism uses the energy in a chemical compound to run its metabolism, that energy is lost from the ecosystem. Secondary productivity, the new biomass made by consumers, is highest in areas of high net primary productivity.

Nutrient Cycles

Chemical elements like carbon, hydrogen, oxygen, and nitrogen are used by organisms to construct their bodies (Figure 44.5). Producers obtain these and other essential chemical elements from the soil, water, or air in the form of ions like nitrate (NO_3^-) or inorganic molecules like carbon dioxide (CO_2). Consumers obtain essential elements by eating producers or other consumers.

Essential elements required by producers are called **nutrients**. Such essential elements include carbon, hydrogen, oxygen, nitrogen, phosphorus, sulfur, and many others. In humans and other animals, "nutrients" refers to vitamins, minerals, essential amino acids, and essential fatty acids. In the ecosystem context of this chapter, however, we'll use the term "nutrients" to mean the essential elements required by producers.

The nutrients needed for life are cycled between organisms and the physical environment (see Figure 44.1). First, nutrients are taken up from the physical environment by producers. These nutrients are then passed either directly or indirectly through herbivores and carnivores to decomposers. **Decomposers** break down once living tissues into simple chemical components, thereby returning nutrients to the physical environment. Without decomposers, nutrients could not be returned from organisms to the physical environment. Thus without decomposers, nutrients could not be repeatedly reused, and life would cease because all essential nutrients would remain in the bodies of dead organisms.

Within the physical environment, a nutrient may pass between air, water, soil, and rock before it is captured once again by a producer. Nutrients can cycle between organisms and the physical environment rapidly (in days to months) or very slowly: It may take many millions of years before an element that was once in an organism is finally reabsorbed by a producer.

The cyclical movement of a nutrient between organisms and the physical environment is called a **nutrient cycle**. There are two main types of nutrient cycles: sedimentary and atmospheric.

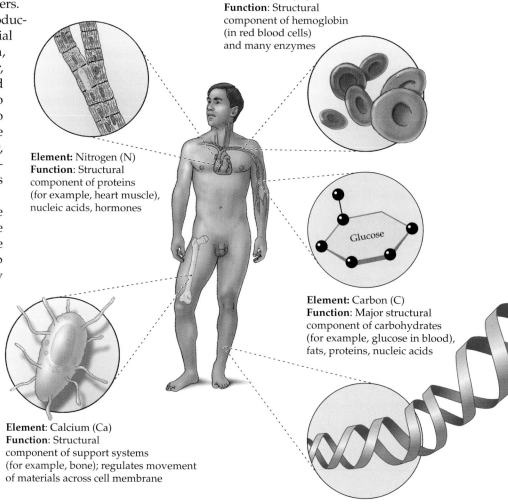

Element: Iron (Fe)
Function: Structural component of hemoglobin (in red blood cells) and many enzymes

Element: Nitrogen (N)
Function: Structural component of proteins (for example, heart muscle), nucleic acids, hormones

Glucose

Element: Carbon (C)
Function: Major structural component of carbohydrates (for example, glucose in blood), fats, proteins, nucleic acids

Element: Calcium (Ca)
Function: Structural component of support systems (for example, bone); regulates movement of materials across cell membrane

Figure 44.5 Some of the Chemical Elements Required by Organisms

Element: Phosphorus (P)
Function: Structural component of nucleic acids (for example, DNA of a skin cell), phospholipids, bone

Sedimentary nutrient cycles

A nutrient that does not enter the atmosphere easily is said to have a **sedimentary cycle**. Such nutrients first cycle within terrestrial and aquatic ecosystems for variable periods of time; then they are deposited in the ocean as sediments. Nutrients may remain in sediments, unavailable to most organisms, for hundreds of millions of years. Eventually, however, the bottom of the ocean is thrust up by geologic forces to become dry land, and once again the nutrients may be available to organisms. Sedimentary nutrients usually cycle very slowly, and therefore they are not replaced easily once they are lost from an ecosystem.

Phosphorus is an important nutrient that has a sedimentary cycle (Figure 44.6). Phosphorus is important because it often limits net primary productivity, especially in aquatic ecosystems. For example, NPP usually increases when phosphorus is added to lakes. Such an increase in productivity can have undesirable effects: When phosphorus is added to lakes in the form of sewage or water that drains from agricultural fields and contains fertilizer (agricultural runoff), the added phosphorus can cause algae and other phytoplankton populations to increase dramatically. The increase in phytoplankton populations can lead eventually to the death of aquatic plants, fish, and invertebrates (see the Lake Washington example in Chapter 43, page 649).

The phosphorus cycle begins when a phosphorus atom is released from rock by weathering. Once this atom is absorbed by a producer, it may cycle with the terrestrial ecosystem for years or centuries before it leaves the ecosystem in a stream. From there, the phosphorus atom cycles within streams, rivers, or lakes, but within a few weeks to years it enters the ocean. Within the ocean, the atom cycles between deep and surface waters for an average of 100,000 years, after which it is deposited in ocean sediments. It can remain trapped in ocean sediments for 100 million years or more. Finally, geologic forces cause the ocean floor to rise and become dry land, and the cycle begins again.

Atmospheric nutrient cycles

Nutrients such as nitrogen, carbon, and sulfur that cycle between terrestrial ecosystems, aquatic ecosystems, and the atmosphere are said to have an **atmospheric cycle**. Because such nutrients enter the atmosphere easily, their cycling in a local ecosystem, such as a lake in a remote region, may be affected by events that occur in distant parts of the globe.

Sulfur is an important nutrient that has an atmospheric cycle. There are three natural ways by which sulfur enters the atmosphere from terrestrial and aquatic ecosystems (Figure 44.7): in sea spray, as a metabolic by-product (the gas hydrogen sulfide, H_2S) in some types of bacteria, and least importantly in terms of overall amount, as a result of volcanic activity. As we discuss in the following section, human activities also cause sulfur to enter the atmosphere.

Sulfur enters ecosystems on land by the weathering of rocks and as sulfate (SO_4^{2-}) that is lost from the atmosphere. Sulfur enters the ocean as stream runoff from land and again as sulfate lost from the atmosphere. Once in the ocean, sulfur cycles within the marine ecosystem before being lost in sea spray or in ocean sediments. Compared to phosphorus, sulfur cycles through terrestrial and aquatic ecosystems quickly.

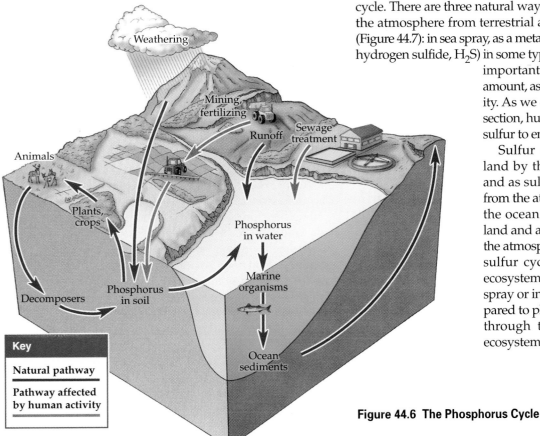

Weathering

Mining, fertilizing

Runoff

Sewage treatment

Animals

Plants, crops

Phosphorus in water

Decomposers

Phosphorus in soil

Marine organisms

Key

Natural pathway

Pathway affected by human activity

Ocean sediments

Figure 44.6 The Phosphorus Cycle

Figure 44.7 The Sulfur Cycle

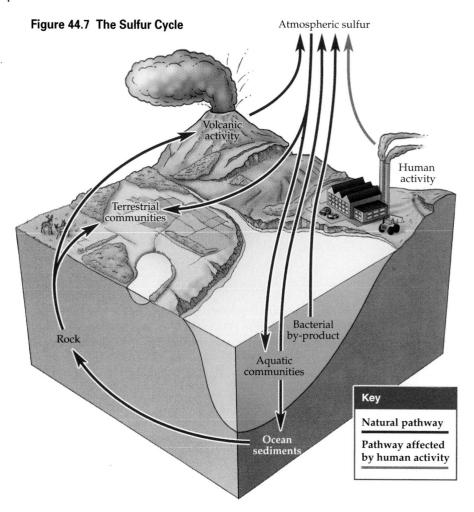

Atmospheric sulfur

Volcanic activity

Human activity

Terrestrial communities

Rock

Bacterial by-product

Aquatic communities

Ocean sediments

Key

Natural pathway

Pathway affected by human activity

■ Nutrients are cycled between organisms and the physical environment. Decomposers return nutrients from the bodies of dead organisms to the physical environment, thus allowing the cycling of nutrients, on which all life depends. Nutrients such as phosphorus that don't enter the atmosphere easily have sedimentary cycles that usually take a long time to complete. Nutrients such as carbon, nitrogen, and sulfur that enter the atmosphere easily have atmospheric cycles that occur relatively rapidly and that can transfer nutrients between distant parts of the world.

Human Activities Can Alter Nutrient Cycles

Human activities can have a major impact on nutrient cycles. For example, ecologists have shown that clear-cutting a forest, followed by spraying with herbicides to prevent regrowth, causes the forest to lose large amounts

of NO_3^- (nitrate), an important source of nitrogen for plants (Figure 44.8). On a larger geographic scale, the nitrogen and phosphorus used to fertilize a farmer's fields can be carried by streams to a lake hundreds of kilometers away. The addition of nitrogen and phosphorus to the lake may disrupt local nutrient cycles and cause some species (for example, algae) to thrive at the expense of others (for example, fish and invertebrates).

Pollutants that contain sulfur can cause acid rain

Effects are often felt across international borders when humans alter a nutrient cycle. Consider sulfur dioxide, which is released to the atmosphere when we burn fossil fuels like oil and coal. The burning of fossil fuels has altered the sulfur cycle greatly: Annual human inputs of sulfur to the atmosphere are more than one and a half times the inputs from natural sources.

Most human input of sulfur to the atmosphere is concentrated in heavily industrialized areas such as northern Europe and eastern North America. Once in the air, sulfur dioxide (SO_2) is dissolved in water and converted to sulfuric acid (H_2SO_4), which then returns to Earth in rainfall. Rainfall normally has a pH of 5.6, but sulfuric acid (and nitric acid, caused by nitrogen-containing pollutants) has caused the pH of rain to drop to values as low as 2 or 3 in the United States, Canada, Great Britain, and Scandinavia. (See Chapter 5 to review pH.) Rainfall with a low pH is called **acid rain**.

Acid rain can have devastating effects on human-made structures (Figure 44.9) and on natural ecosystems. With respect to its effects on ecosystems, acid rain has drastically reduced fish populations in thousands of Scandinavian and Canadian lakes. Much of the acid rain that falls in these lakes is caused by sulfur dioxide pollution that originates in other countries (for example, Britain, Germany, the United States).

The international nature of the acid rain problem has led nations to agree to reduce sulfur emissions. In the United States, sulfur emissions were cut by 50 percent between 1980 and 1998. Such reductions are a very positive first step, but problems resulting from acid rain will be with us for a long time: Acid rain alters soil chemistry and thus has effects on ecosystems that will last many decades after the pH in rainfall returns to normal levels.

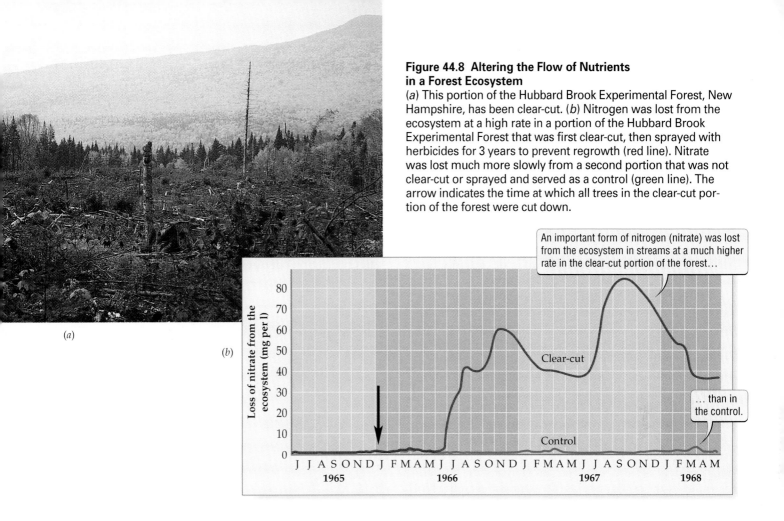

Figure 44.8 Altering the Flow of Nutrients in a Forest Ecosystem
(*a*) This portion of the Hubbard Brook Experimental Forest, New Hampshire, has been clear-cut. (*b*) Nitrogen was lost from the ecosystem at a high rate in a portion of the Hubbard Brook Experimental Forest that was first clear-cut, then sprayed with herbicides for 3 years to prevent regrowth (red line). Nitrate was lost much more slowly from a second portion that was not clear-cut or sprayed and served as a control (green line). The arrow indicates the time at which all trees in the clear-cut portion of the forest were cut down.

An important form of nitrogen (nitrate) was lost from the ecosystem in streams at a much higher rate in the clear-cut portion of the forest...

... than in the control.

Implications of nutrient cycles

The nutrients needed for life are cycled between air, water, soil, rock, and living organisms. Because many nutrients enter the atmosphere easily, human alteration of nutrient cycles can have local, regional, and global consequences. As ecologist Charles Krebs has written, humans "are all tied together to the fate of the earth. Because of the world scale of nutrient cycles, we are linked [to] communities all over the world. There are no more islands."

> ■ Human activities can alter nutrient cycles on local, regional, and global scales. In some cases, human inputs to a nutrient cycle exceed those from all natural sources combined, creating problems of an international scope, such as acid rain. Because many nutrients enter the atmosphere easily, actions in one part of the globe can affect ecological communities located in distant parts of the world.

Figure 44.9 Acid Rain Has Damaged Many Man-made Structures
This obelisk was in good condition when it was moved from Egypt to New York City in the late nineteenth century. Since then, acid rain has damaged the left side of the structure, which faces into prevailing winds, more than the right side of the structure, which faces away from the winds.

Highlight

Ecosystems Provide Essential Services

In 1996 and 1997, floods struck the western United States with a fury. Damage amounted to billions of dollars in the states of Nevada, California, Oregon, and Washington (Figure 44.10a). Most news reports about the floods and associated mud slides (Figure 44.10b) said that they were caused by unusually large amounts of rain and snowfall. Although the weather certainly had a major impact, it is only part of the story.

A huge flood does not always just happen, beyond our control. Human actions may help set the stage for flooding: By logging hillsides, diverting rivers, and building on areas where rivers typically overflow (floodplains), we damage the ability of ecosystems to respond as they normally would to heavy rainfall. Increasingly, the effects of our actions return to us in the form of terrible floods.

How do logging and other types of development alter the frequency and severity of floods? Trees hold soil in place with their roots, and a vast amount of water is removed from the soil by trees. By cutting down trees, logging increases the rate of soil erosion and the amount of water in the soil; much of the extra water in the soil leaves the ecosystem in streams (stream runoff). Higher levels of stream runoff and erosion make both mud slides (see Figure 44.10b) and floods more likely.

When we build dikes and levees and divert the flow of rivers, we seek to control rivers and to prevent floods. We do this to protect homes or industrial areas located in what were once floodplains. But by preventing rivers from overflowing into floodplains, we reduce the ability of the ecosystem to handle periods of heavy rainfall. Floodplains function as huge sponges: When streams and rivers overflow, floodplains absorb excess water and prevent even more severe floods from occurring farther downstream. By building on floodplains and attempting to control floods, we unintentionally make it much more likely that when a flood does occur, it will be a big one.

Floodplains provide us with a free service: They act as safety valves for major floods. Ecosystems provide us with many other free services (Figure 44.11). For example, wetlands and estuaries have a great capacity to filter wastes from water and convert them to harmless substances. Throughout the world, humans take advantage of this fact by using wetlands and estuaries as cost-free waste treatment plants. Unfortunately, the filtering capacity of wetlands and estuaries can be overrun by large cities, as happened in Chicago and New York in the 1970s.

Other free ecosystem services include pollination by insects (essential for many crops), removal of atmos-

pheric pollutants by forest ecosystems, maintenance of breeding grounds for shellfish and fish of commercial importance, control of the concentration of CO_2 and O_2 in the atmosphere, prevention of soil erosion by plants, screening of dangerous ultraviolet light by the atmospheric ozone layer, moderation of the climate by the ocean, and, a major focus of this chapter, nutrient cycling (essential for all life).

Too often we humans manipulate the environment for temporary advantage while neglecting to consider the full implications of our actions. Our civilization would

(a)

(b)

Figure 44.10 Flood Devastation in the Pacific Northwest
(a) Following heavy flooding, people use boats and float tubes to get from place to place in Oregon City. (b) Mud slides can kill people, contaminate stream ecosystems, and have undesirable aesthetic effects.

**Natural services provided
by ecosystems**

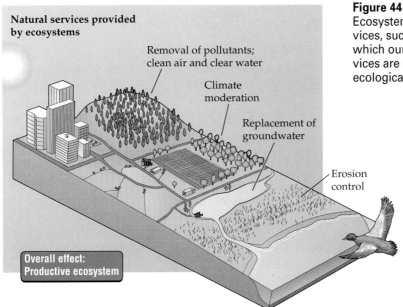

Removal of pollutants;
clean air and clear water

Climate
moderation

Replacement of
groundwater

Erosion
control

Overall effect:
Productive ecosystem

**Damage to ecosystems resulting
in loss of natural services**

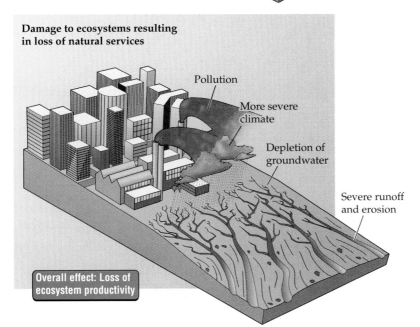

Pollution

More severe
climate

Depletion of
groundwater

Severe runoff
and erosion

Overall effect: Loss of
ecosystem productivity

collapse if not for the many free and essential services provided to us by ecosystems. We are not even remotely close to being able to duplicate with technology what ecosystems provide us for free. Thus, when we destroy or degrade ecosystems, we do so at our own peril.

■ Human society depends on free and essential ecosystem services such as nutrient cycles, flood control, and the filtering of pollutants from air and water.

Figure 44.11 Services Provided by Ecosystems
Ecosystems provide humans with free and essential services, such as nutrient cycling and erosion control, on which our civilization depends. The same ecosystem services are also essential to maintaining the productivity of ecological communities.

Summary

An Overview of How Ecosystems Function

■ Ecosystems are open systems in which energy and materials can move from one ecosystem to another.

■ Energy is not recycled in ecosystems. A portion of the energy captured by producers is lost at each step of a food chain.

■ Nutrients are recycled in ecosystems. They pass from the environment to producers to consumers, then back to the environment when decomposers break down the bodies of dead organisms.

Energy Capture in Ecosystems

■ Most life on Earth depends on energy that is captured by photosynthetic organisms and then stored in their bodies in chemical form.

■ Net primary productivity (NPP) varies greatly in different parts of the world.

■ On land, NPP tends to decrease from the equator toward the poles.

■ In marine ecosystems, NPP is often high where the ocean borders land or where upwellings provide scarce nutrients to marine organisms.

Energy Flow through Ecosystems

■ Chemical compounds that contain energy captured by producers can be transferred from organism to organism in a food chain.

■ Once an organism uses the energy in a chemical compound to run its metabolism, that energy is lost from the ecosystem.

■ Because energy is lost once it is used, the energy available to organisms in an ecosystem can be represented as an energy pyramid.

■ Secondary productivity is highest in areas of high net primary productivity.

Nutrient Cycles

■ Nutrients are cycled between organisms and the physical environment.

■ Decomposers return nutrients from the bodies of dead organisms to the physical environment, thus allowing the cycling of nutrients, on which all life depends.

■ Nutrients that do not easily enter the atmosphere have sedimentary cycles, which usually take a long time to complete.

■ Nutrients that easily enter the atmosphere have atmospheric cycles, which occur relatively rapidly and can transfer nutrients between distant parts of the world.

Human Activities Can Alter Nutrient Cycles

■ Human activities can alter nutrient cycles on local, regional, and global scales.

■ In some cases, human inputs to a nutrient cycle exceed those from all natural sources combined, creating problems of international scope, such as acid rain.

■ Because many nutrients enter the atmosphere easily, actions in one part of the globe can affect ecological communities located in distant parts of the world.

Highlight: Ecosystems Provide Essential Services

■ Human society depends on free and essential ecosystem services such as nutrient cycles, flood control, and the filtering of pollutants from air and water.

Key Terms

acid rain p. 664	nutrient p. 662
atmospheric cycle p. 663	nutrient cycle p. 662
consumer p. 660	secondary productivity p. 661
decomposer p. 662	sedimentary cycle p. 663
ecosystem p. 658	trophic level p. 661
net primary productivity (NPP) p. 659	

Chapter Review

Self-Quiz

1. The amount of energy fixed by plants in photosynthesis, minus the amount lost as heat in cellular respiration, is
   a. secondary productivity.
   b. consumption efficiency.
   c. NPP.
   d. photosynthetic efficiency.

2. The movement of nutrients between organisms and the physical environment is called
   a. nutrient cycling.
   b. ecosystem services.
   c. NPP.
   d. nutrient pyramids.

3. Free services provided to humans by ecosystems include
   a. control of atmospheric CO_2 concentration.
   b. prevention of soil erosion.
   c. filtering of pollutants from water and air.
   d. all of the above

4. Each step in a food chain is called a
   a. trophic level.
   b. consumer level.
   c. food web.
   d. producer.

5. In all ecosystems, what type of organisms consume 50 percent or more of the NPP?
   a. herbivores
   b. decomposers
   c. producers
   d. carnivores

Review Questions

1. What prevents energy from being recycled in ecosystems?

2. What is the essential role of decomposers in ecosystems?

3. Explain why human alteration of nutrient cycles can have international effects.

4. Because nutrient cycles have effects that cross international borders, would it be in the self-interest of nations to form and abide by strict, enforceable international agreements on inputs to nutrient cycles?

5. Describe key ecosystem services and discuss the extent to which human economic activity depends on such services.

44   **The Daily Globe**

New York City Buys Land to Save Water

NEW YORK, NY. This week, New York City officials announced the purchase of 1000 acres of stream-side property in the Catskills. The largest purchase yet in the city's pioneering efforts to preserve the drinking water that comes from this region, this latest acquisition has drawn attention around the country as municipalities of all sizes—from tens of thousands to hundreds of thousands—have begun looking for solutions to the decline in the quality of their drinking water.

As in other areas, for years residents of New York City took their drinking water for granted, receiving it from the Catskills region of upstate New York. There in the pristine countryside, water was kept pure by the root systems, soil microorganisms, and natural filtration processes of the forest ecosystems. But in recent years, increases in sewage, fertilizers, and pesticides caused water quality to deteriorate.

When the water no longer met Environmental Protection Agency (EPA) standards, New York City officials could have built a water treatment plant, like other cities, for a cost of 6 to 8 billion dollars, plus another 300 million dollars per year to operate. Instead, for an estimated cost of 1 to 1.5 billion dollars, the city has embarked on an ambitious but simple plan: protect the environment so that natural ecosystems can once again supply the city with clean water. The city is buying land that borders rivers in the Cat-

skills, protecting the land from development to minimize fertilizer and pesticide inputs into the water, and building sewage treatment plants for rural communities in the Catskills, thus decreasing sewage inputs.

"We're talking about a no-brainer here," said city water official Joe Marin, who has already been contacted by a number of cities hoping to follow New York's model. "For a one-time investment of less than 25 percent of the cost of building a treatment plant, we get clean water and we save huge annual expenses." In addition, the city protects the environment in a case in which what's good for the ecosystem is also good for the bottom line.

Evaluating "The News"

1. The choice in New York City was clear: The city could save money and protect the environment at the same time. In other situations, protecting the environment may not be cost-effective in a strictly economic sense. Should society protect the environment regardless of cost? How can a balance between ecological and economic factors be reached?

2. Should individuals, companies, or cities that damage the ability of the ecosystem to provide free services

(such as pure drinking water) be charged for harming the ecosystem? Why or why not?

3. The profit motive can drive people to come up with creative solutions to complex problems, including environmental problems. One solution to New York City's problems would be to form a corporation that could manage the improvement of the city's water supply. This corporation would have the right to sell an ecosystem service—in this case, the provision of water that meets

EPA standards for clean drinking water. Ownership of this right would enable the firm to raise the money needed to improve New York City's drinking water. Driven by profit-making, the firm might develop new and cost-effective ways of protecting the ecosystem's ability to provide pure drinking water. Should such profit-driven approaches be used to protect the environment? Why or why not?

45

Global Change

Oliver Burston, *Hypervoxel Embolism*, 1999.

Devastation on the High Seas

Nearly 75 percent of the surface of Earth is covered with ocean. The oceans are so deep and so vast that many scientists once thought humans could not drive marine species extinct. They reasoned that no matter how much we overhunted a species or polluted local portions of its habitat, there would always be places where the species could thrive. Now it seems that even marine species are not safe from human impact.

Consider the white abalone's tale of woe. This large marine shellfish once was common along 1200 miles of the California coast. It lives on rocky reefs in relatively deep water (25 to 65 meters or deeper). The fact that it lives in deep waters protected it for a while: White abalone is delicious to eat, but people first hunted other species of abalone that live in shallow waters and hence are easier to find. When the shallow-water species became rare, fishermen turned to the white abalone. After only 9 years of commercial fishing, the fishery collapsed. This species, which once cov-ered the seafloor with up to 10,000 individuals per hectare, is now on the verge of extinction.

And the white abalone is not alone. The barn-door skate, a large fish that once had a wide geo-graphic distribution, has also been hunted to near extinction. Overall, humans have had an enormous negative impact on fish populations worldwide. Recent studies indicate that 66 percent of the world's marine fisheries are in trouble from overfishing.

Another sign of our impact on marine ecosystems concerns what fishermen are catch-ing. In the past 45 years, the catch from fish-eries across the globe has included fewer large, carnivorous fish and more invertebrates and small fish that feed on plankton. Thus it appears we have altered the food webs of many ocean communities, not just reduced the popu-lations of individual species.

The extinction of species and other effects of humans on marine ecosystems are part of the many changes we are causing to Earth. Other examples include changes to the global sulfur cycle (see Chapter 44), effects on the location of biomes (see Chapter 39), and the extinction or decline of many species (see Chapter 4). These and other worldwide environmental changes are referred to as global change, the subject of this chapter.

White Abalone and Its California Ocean Home

Key Concepts

1. Human impacts on the world's lands and waters are thought to be the main causes of the current high rate of extinction of species.

2. Human inputs to the global nitrogen cycle now exceed those of all natural sources combined. If unchecked, changes to the nitrogen cycle may have large, negative effects on many ecosystems.

3. The concentration of carbon dioxide (CO_2) gas in the atmosphere is increasing at a dramatic rate, largely because of the burning of fossil fuels. Increased CO_2 levels

are expected to have large but hard-to-predict effects on life on Earth.

4. Increased concentrations of CO_2 and certain other gases are predicted to cause global warming, a rise in Earth's temperature. Most scientists think that global warming will occur, but its extent and consequences remain uncertain.

5. Because global change caused by humans is expected to have large, negative consequences for many species, including ourselves, people need to manage Earth's ecosystems in a sustainable fashion.

Statements by politicians, talk show hosts, and others often give the impression that worldwide change to the environment, or **global change**, is a controversial topic. Such statements cause many in the general public to think that global change may not be occurring, causing them to wonder whether anything really needs to be done about it.

The impression of controversy is unfortunate. Contrary to what some reports in the media might lead us to believe, we know with certainty that global change is occurring. For example, biological invasions have increased worldwide (see examples in Chapters 40, 41, and 43), large losses of biodiversity have occurred (see Chapter 4), and pollution has altered ecosystems throughout the world. Each of these three examples illustrates an important type of global change that we know with certainty is happening today.

Although the examples of global change just mentioned are caused by people, the biosphere has always changed over time. The continents change locations, the climate changes, and natural forms of disturbance and succession change communities (see Chapters 39 and 43). Since global change occurs naturally, even in the absence of human impact we would know with certainty that global change is happening today.

In this chapter we describe how humans have influenced global change. We begin with two types of global change that we know have occurred and that we know are caused by people: changes in land and water use, and changes in the cycling of chemicals through ecosystems. We also discuss an aspect of global change that is likely but not certain: global warming (a worldwide increase in temperature). Finally, we discuss why scientists are increasingly

convinced that humans are causing global change at a rate and intensity that is unmatched by natural patterns of change.

Land and Water Use

Humans have a major impact on the lands and waters of Earth. For example, we have drastically altered how water cycles through ecosystems. Humans now use more than half of the world's accessible fresh water, and we have altered the flow of nearly 70 percent of the world's rivers. Since water is essential to all life, our heavy use of the world's waters has many and far-reaching effects, including changing where water is found and what species can survive in a given location.

In this section we discuss our impact on the lands and waters of Earth in terms of land and water transformation. **Land transformation** refers to physical and biotic changes that humans make to the land surface of Earth. Such changes include the destruction of natural habitat to allow for urban growth, agriculture, and resource use (as when a forest is clear-cut for lumber). Land transformation also includes many human activities that alter natural habitat to a lesser degree, as when we graze cattle on grasslands. Similarly, **water transformation** refers to physical and biotic changes that humans make to the waters of our planet.

Evidence of land and water transformation

Aerial photos and satellite data, changing urban boundaries, and local instances of the destruction of natural habitats show how humanity is changing the face of Earth (Figure 45.1). Together, these and many other pos-

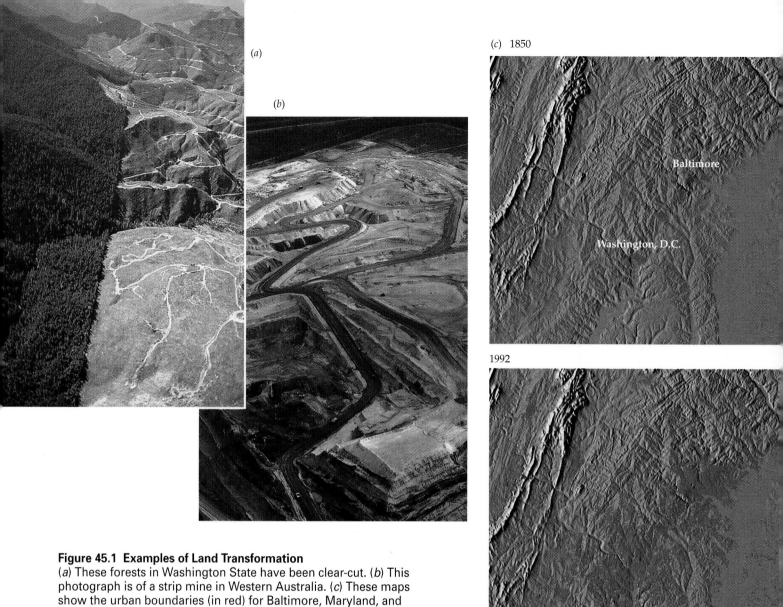

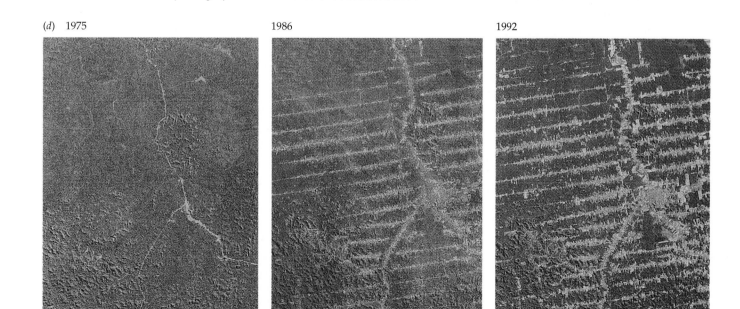

Figure 45.1 Examples of Land Transformation
(*a*) These forests in Washington State have been clear-cut. (*b*) This photograph is of a strip mine in Western Australia. (*c*) These maps show the urban boundaries (in red) for Baltimore, Maryland, and Washington, D.C., at different points in time (1850 and 1992).
(*d*) These satellite images show the destruction of tropical forests in Rondônia State, Brazil, over a 17-year period. Light-colored regions have been clear-cut. The photographs cover an area of about 5000 km².

sible examples show that land and water transformation is occurring, is caused by humans, and is global in scope.

To estimate the total amount of land transformation caused by people, the impacts of many different human activities must be summed across every acre of the world. This task may seem nearly impossible, but with the use of satellites and other new technologies, we can now measure our total impact on Earth for the first time in history. Although we are just beginning to determine the extent to which we have transformed the planet, one reasonable estimate is that humans have substantially altered one-third to one-half of the land surface of Earth. Thus, although we do not know the exact amount of land transformation, we do know that we have altered a large percentage of Earth's land surface.

In modifying the land and water for our use, we have had a great impact on many ecosystems. For example,

(*a*) Wetland distribution, circa 1780s

(*b*) Wetland distribution, circa 1980s

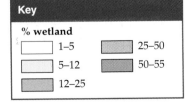

Key

% wetland

☐ 1–5	▨ 25–50
☐ 5–12	▩ 50–55
▨ 12–25	

large regions of the United States were once covered by wetlands. Many of these wetlands have been drained so that the land could be used for agriculture or other purposes. During the 200-year period beginning in the 1780s, wetlands declined in every state in the United States (Figure 45.2). Wetlands are still declining today. Other examples of human impact on ecosystems include the ongoing destruction of tropical rainforests (see Figure 45.1*d*) and the conversion of once vast grasslands in the American Midwest to croplands.

Consequences of land and water transformation

Many ecologists think that land transformation and water transformation are now—and will remain for the immediate future—the two most important components of global change. There are several reasons why the transformation of the lands and waters of Earth is so important.

First, as humans alter the land and water to produce goods and services for an increasing number of people, we use a very large share of the world's resources. For example, estimates suggest that humans now control (directly and indirectly) more than 40 percent of the world's total net primary productivity (see Chapter 44) on land.

To control such a large portion of the world's terrestrial net primary productivity, humans have reduced the amount of land and other resources available to nonhuman species, causing many species to go extinct (see Chapter 4). Water transformation has similar effects. Together, land transformation and water transformation are thought to be the main factors responsible for another major component of global change, the extinction of species.

The transformation of land and water has other effects as well. For example, when a forest is cut down, the local temperature may increase and the humidity may decrease. Such changes to the climate can make it less likely that a forest will reappear. In addition, the cutting and burning of forests contribute to current increases in global CO_2 levels (as we'll see shortly), an aspect of global change that may alter the climate worldwide. With respect to water transformation, when humans overfish or pollute the water, we may cause major changes in the abundance and type of species found in the world's aquatic ecosystems (see Chapters 43 and 44).

■ Human activities are changing the land and waters of the entire planet. Land and water transformation has caused the extinction of species and has the potential to alter local, regional, and global climate.

Figure 45.2 The Decline of Wetlands
Wetlands in the United States declined greatly from the 1780s (*a*) to the 1980s (*b*).

Changes to the Chemistry of Earth

Life on Earth depends on and is heavily influenced by chemicals in ecosystems. Net primary productivity often depends on the amount of nitrogen available to producers, and the amount of sulfuric acid in rainfall has many effects on ecological communities. The nitrogen that stimulates net primary productivity and the sulfur in acid rain are just two of many possible examples of chemicals that are cycled through ecosystems.

Humans are changing how many chemicals are cycled through ecosystems. For example, chlorofluorocarbons (CFCs) are synthetic chemical compounds used as coolants in refrigerators. Because they are synthetic, until recently there were no CFCs in the environment. CFCs are not toxic, but their use and subsequent release to the atmosphere had large and surprising negative effects: They caused an unexpected decrease in the ozone layer, as illustrated by the hole in the ozone layer over the Antarctic (Figure 45.3). Because the ozone layer shields the planet from harmful ultraviolet light, damage to it poses a serious threat to life. Fortunately, the international community responded quickly to this threat, and treaties to halt the production of CFCs are now in place.

By adding synthetic and naturally occurring chemicals to the environment, humans have altered how many chemicals cycle through ecosystems. In some cases, some of the harm caused by changes to chemical cycles has been undone (see the Box on page 677). In other cases, great challenges lie ahead for coping with human-caused changes in the cycling of chemicals through ecosystems. We consider two instances in which such challenges remain in the sections that follow: the global nitrogen cycle and the global carbon cycle.

> ■ Human activities are changing how many chemicals are cycled through ecosystems.

The Global Nitrogen Cycle

In the global nitrogen cycle (Figure 45.4), a large amount of nitrogen is in the atmosphere, where N_2 gas makes up 78 percent of the air we breathe. However, plants and most other organisms cannot use N_2 directly. Instead, the nitrogen in N_2 gas must be converted to other forms, such as nitrate (NO_3^-) or ammonium (NH_4^+). The conversion of N_2 to NH_4^+, called **nitrogen fixation**, is accomplished by several species of bacteria and, to a much lesser degree, by lightning. Once nitrogen is converted to NH_4^+, other bacteria can convert it to NO_3^-. The two forms of nitrogen that can be used by plants (NH_4^+ and NO_3^-) then cycle among plants, animals, and microorganisms. The amount of nitrogen that cycles among organisms is much smaller than the amount in the atmosphere.

Humans have changed the global nitrogen cycle greatly

The amount of nitrogen available to an ecosystem depends on how much nitrogen is converted from N_2 gas to forms that can be used by plants and other producers, like nitrate (NO_3^-) or ammonium (NH_4^+). In recent years, the amount of nitrogen converted to usable forms (fixed) by human activities has exceeded the amount fixed by all natural sources combined (Figure 45.5).

Much of the nitrogen fixation by humans is the result of the industrial fixation of nitrogen to produce fertiliz-

(*a*) 1979

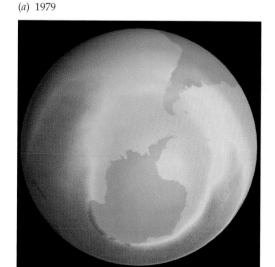

(*b*) 1998

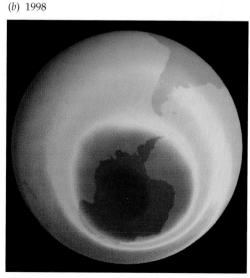

Figure 45.3 The Antarctic Ozone Hole
These satellite images show average ozone levels over Antarctica for the months of September 1979 (*a*) and September 1998 (*b*). Ozone levels declined slowly in the 1970s, then dropped dramatically in the 1980s. Thus, the September 1979 image represents near-normal conditions. In the September 1998 image, regions with the greatest ozone loss appear blue. At the time this book went to print, the largest ozone hole ever recorded occurred on September 19, 1998.

Figure 45.4 The Global Nitrogen Cycle

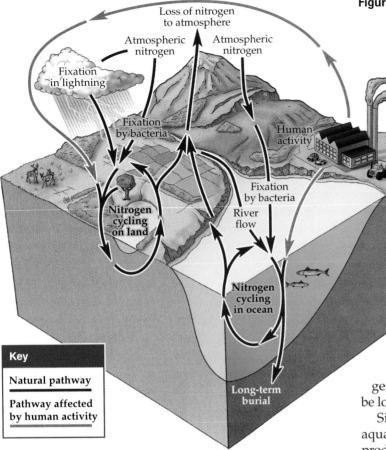

Key

Natural pathway

Pathway affected
by human activity

ers. Other major sources of nitrogen fixed by human activities include fixation by car engines and fixation by bacteria that have a mutualistic relationship with peas and other crop plants. The fact that human inputs of nitrogen to ecosystems are greater than all natural inputs tells us that our activities have greatly changed the global nitrogen cycle.

Potential effects of changing the nitrogen cycle are far-reaching. For example, when nitrogen is added to terrestrial communities, net primary productivity usually increases but the number of species usually decreases (Figure 45.6). Ecosystems in the Netherlands receive more added nitrogen than those in any other country in the world. The addition of nitrogen to grasslands in the Netherlands that historically were poor in nitrogen has caused more than 50 percent of the species to be lost from some of these communities.

Similarly, when nitrogen is added to nitrogen-poor aquatic ecosystems, such as many ocean communities, productivity increases but species are lost. In general, an increase in productivity caused by the addition of nitrogen is not necessarily better for the ecosystem.

The impact of human activities on the global nitrogen cycle also has increased the atmospheric concentration of nitrous oxide, one of the greenhouse gases that contribute to the occurrence of global warming (a topic we will address shortly). Finally, our impact on the nitrogen cycle is expected to cause an increase in acid rain and smog.

■ Nitrogen must be fixed or converted from N_2 gas before it can be used by producers. Human activities fix more nitrogen than all natural activities combined. The extra nitrogen fixed by human activities has altered the global nitrogen cycle, leading to increases in productivity that can cause the loss of species from ecosystems.

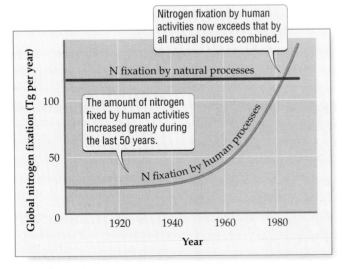

Figure 45.5 Human Impact on the Global Fixation of Nitrogen
Nitrogen is fixed naturally by bacteria and by lightning at a rate of about 130 teragrams (Tg) per year (1 Tg = 10^{12} grams, or 1.1 million tons). Human activities such as the production of fertilizer now fix more nitrogen than all natural sources combined.

The Global Carbon Cycle

There are vast amounts, or pools, of carbon in organisms, soils, the air, and especially the ocean. Carbon cycles readily among these pools (Figure 45.7). We will focus on

The Scientific Process

Back from the Brink of Extinction

Before being banned in 1972, the synthetic pesticide DDT (see Chapter 44) was used widely in the United States for about 30 years. Since this chemical does not occur naturally, its use introduced an entirely new substance to ecosystems. The cycling of DDT through ecosystems had undesirable and unanticipated effects; for example, birds with high levels of DDT in their bodies produced such fragile eggs that they could not reproduce.

The bald eagle was nearly driven extinct by DDT poisoning. By the early 1960s in the United States, there were only 400 breeding pairs of bald eagles in the lower 48 states.

The effect of DDT on bald eagles prompted the ban on the use of DDT within the United States, a ban that has allowed bald eagles and other birds, like the peregrine falcon, to bounce back from a perilous brush with extinction.

Today, there are more than 4000 breeding pairs of bald eagles in the lower 48 states. Although their numbers are still low, bald eagles represent one of many environmental success stories. When humans recognize a problem and take decisive action to fix it, as with the ban on DDT use, species, communities, and ecosystems often recover from the negative effects of human impact.

Bald eagle

---

one portion of the global carbon cycle, the concentration of the gas carbon dioxide (CO_2) in the atmosphere.

Although CO_2 makes up less than 0.04 percent of Earth's atmosphere, its importance is far greater than its low concentration might suggest. For example, CO_2 is an essential raw ingredient for photosynthesis, on which most life depends. CO_2 is also the single most important greenhouse gas (see the discussion of global warming later in the chapter). Thus, scientists took notice in the early 1960s when new measurements showed that the concentration of CO_2 in the atmosphere was rising rapidly.

Global CO_2 levels have risen dramatically

The red points in the graph of Figure 45.8 plot the concentration of CO_2 measured in the atmosphere from 1958 to the present. By measuring CO_2 concentrations in air bubbles trapped in ice for hundreds to hundreds of thousands of years, scientists have also estimated the concentration of CO_2 in both the recent and relatively distant past (open circles on Figure 45.8).

(a)

Native grasslands in Minnesota often have 20 to 30 plant species per square meter. This plot received no nitrogen and lost no species from 1982 to 1994.

(b)

Figure 45.6 Nitrogen Addition Can Decrease the Number of Species
(a) No nitrogen added.
(b) Nitrogen added.

After nitrogen was added, most of the native species disappeared, and an introduced species, European quackgrass, took over the plot. When the experiment began, this plot looked much like the one in (a).

Figure 45.7 The Global Carbon Cycle

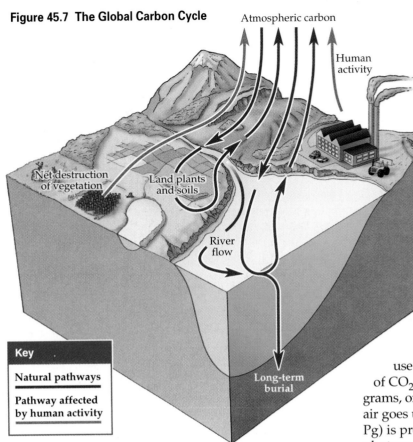

At points in time for which direct measurements from the air and estimates from ice bubbles have both been made, the two ways of measuring the concentration of CO_2 agree, giving us confidence that the ice bubble measurements are accurate. Both types of measurements show that CO_2 levels rose greatly during the past 200 years.

The recent increase in CO_2 levels is striking for two reasons. First, the increase happened quickly: The con-

centration of CO_2 increased from 280 parts per million (ppm) to 365 ppm in roughly 200 years. This rate of increase is greater than even the most sudden increase that occurred naturally during the past 420,000 years. Second, although the concentration of CO_2 has ranged from about 200 ppm to 300 ppm during the past 420,000 years, CO_2 levels are now higher than those estimated for any time during this period. Overall, global CO_2 levels have changed very rapidly in recent years and have reached concentrations that are unmatched for the last 420,000 years.

Increased CO_2 concentrations are caused by humans

Roughly 75 percent of current increases in atmospheric CO_2 levels are due to human use of fossil fuels (for example, oil and gas). The use of fossil fuels releases about 5.6 petagrams (Pg) of CO_2 to the atmosphere each year (1 Pg equals 10^{15} grams, or 1.1 billion tons), and the amount of CO_2 in the air goes up by about 3.5 Pg per year. The difference (2.1 Pg) is probably absorbed partially by plants and other photosynthetic organisms and partially by the ocean. Thus, the use of fossil fuels releases more than enough CO_2 to account for the current increases in atmospheric CO_2 concentrations. In addition, the cutting down and burning of forests is thought to cause about 25 percent of the current yearly increase in global CO_2 concentrations.

Increased CO_2 concentrations have many biological effects

An increase in the concentration of CO_2 in the air can have large effects on plants (Figure 45.9). For example, at least initially, many plants increase their rate of photosynthesis and use water more efficiently when more CO_2 is available. The increased rate of photosynthesis and the more efficient use of water causes such plants to grow more rapidly at high CO_2 levels than at low CO_2 levels.

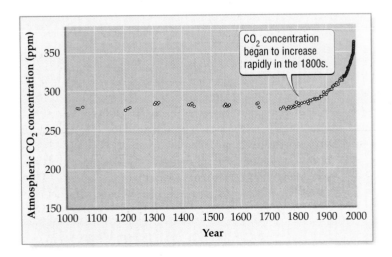

Figure 45.8 Atmospheric CO_2 Concentrations
Atmospheric CO_2 levels (measured in parts per million, or ppm) have increased greatly in the past 200 years. Solid red circles show results from direct measurements of the concentration of CO_2 in the atmosphere. Open circles indicate CO_2 levels measured from bubbles of air trapped in ice.

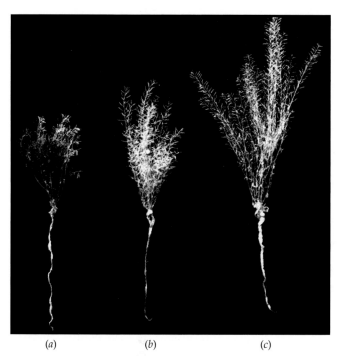

(a)   (b)   (c)

**Figure 45.9 High CO$_2$ Concentrations
Can Increase Plant Size**
This photograph shows three *Arabidopsis thaliana* plants of
the same genotype grown under different CO$_2$ concentrations. Plants grew larger as the concentration of CO$_2$ was
increased. CO$_2$ concentrations were (*a*) 200 ppm, a level similar to that found roughly 20,000 years ago; (*b*) 350 ppm, the
level found in 1988; and (*c*) 700 ppm, a predicted future level.

If CO$_2$ levels remain high, some plant species keep
growing at higher rates, but others drop their growth
rates over time. As CO$_2$ concentrations in the atmosphere
rise, species that maintain rapid growth at high CO$_2$ levels might outcompete other species in their current ecological communities or invade new communities.

Differences in how individual species respond to
high CO$_2$ levels could cause changes to entire communities. But it will be difficult (at best) to predict exactly
how communities will change under high CO$_2$ levels.
For example, an increase in CO$_2$ concentration can
cause the amount of carbon in the leaves of plants to
increase, which in turn may cause some insect pest populations to increase in number. Increases in pest populations could have very large effects on plant communities, but the exact nature of these effects is hard to
predict. In general, increased CO$_2$ levels will probably
alter many ecological communities, but often in unpredictable ways.

Increased CO$_2$ levels are also likely to cause Earth's
climate to warm. It is to this topic that we turn next.

■ The use of fossil fuels and the destruction of forests
have caused the concentration of CO$_2$ in the atmosphere to increase greatly in the past 200 years, thus
changing the global carbon cycle. Increased CO$_2$ concentrations can alter the growth of individual plants in
ways that will probably cause changes to many ecological communities.

Global Warming

Gases in the atmosphere such as carbon dioxide (CO$_2$),
methane (CH$_4$), and nitrous oxide (N$_2$O) absorb heat that
radiates from Earth's surface to space. These gases are
called **greenhouse gases** because they function much as
the walls of a greenhouse or the windows of a car do:
They let in light but trap heat. As the concentration of a
greenhouse gas in the atmosphere goes up, more heat
should be trapped, thus raising the temperature of Earth.

Global temperatures appear to be rising

CO$_2$ is the most important of the greenhouse gases
because so much of it enters the atmosphere. Scientists
have predicted that because CO$_2$ traps heat, current
increases in atmospheric CO$_2$ concentrations will cause
temperatures on Earth to rise. This aspect of global
change, known as **global warming**, has proven controversial in both the media and the political arena.

We know that CO$_2$ concentrations are increasing, but
is the global climate getting warmer? Although the four
hottest years on record have occurred since 1990, there
is so much year-to-year variation in the weather that it
can be hard to show that the climate really is getting
warmer. In 1995, however, the United Nations–spon-
sored Intergovernmental Panel on Climate Change concluded for the first time that the climate was warming
(Figure 45.10). The panel also concluded that the increase in global temperatures was most likely due to
human-caused increases in the concentration of CO$_2$
and other greenhouse gases.

Since 1995, new statistical analyses have supported
the panel's conclusion that recent rises in global temperatures represent a real trend, not just ordinary variation in the weather. In addition, results from new scientific studies suggest that recent temperature increases
may have already changed ecosystems. For example, as
temperatures have increased in Europe during the twentieth century, dozens of bird and butterfly species have
shifted their geographic ranges to the north. Similarly,
plants in northern latitudes have increased the length of
their growing season as temperatures have warmed
since 1980.

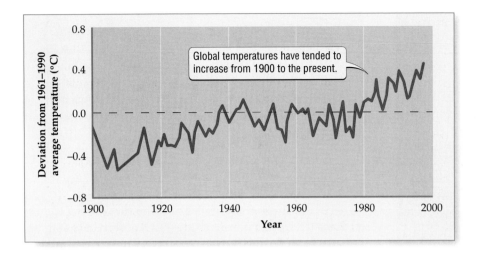

Figure 45.10 Global Temperatures Are on the Rise
Global temperatures are plotted here relative to the average temperature for the period 1961 to 1990. Portions of the curve below the dotted line represent lower-than-average temperatures; portions above the dotted line represent higher-than-average temperatures.

(a) Current climate

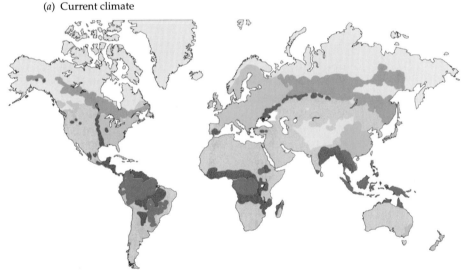

(b) Future climate (3.5°C increase in temperature)

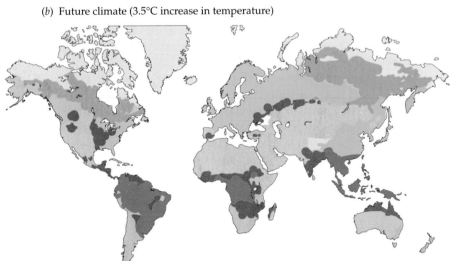

Other studies published since 1995 indicate that the warming that has occurred since 1950 has been caused largely by human activities. Overall, an increasing amount of scientific evidence suggests that global warming is already happening, is affecting ecological communities, and is caused at least in part by human activities, such as the use of fossil fuels.

What will the future bring?

Because there is no end in sight to the rise in CO_2 levels, the current trend for increased global temperatures seems likely to continue. How will increased temperatures affect life on Earth? Not surprisingly, the effects depend on how much global warming occurs.

For example, scientists predict that 100 years from now average temperatures on Earth will have risen by 1 to 3.5°C (2°C is often cited as a best estimate). Although these may not sound like large increases, an increase of 3.5°C would likely have a large impact on Earth's biomes (Figure 45.11). Many species might go extinct simply because they are unable to migrate north fast enough to keep up with the changing climate. A 3.5°C increase in global temperatures by the year 2100 would also be disastrous for the world's agricultural systems, especially since by then there probably will be about 5 billion more people to feed.

In addition, at the high end of climate change predictions (a 3.5°C increase), humanity may face a rise of 5 meters in the global sea level. Such a rise would sub-

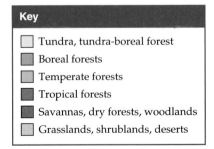

Key

☐ Tundra, tundra-boreal forest
■ Boreal forests
☐ Temperate forests
■ Tropical forests
■ Savannas, dry forests, woodlands
☐ Grasslands, shrublands, deserts

Figure 45.11 Biomes on the Move
If the climate warms by 3.5°C, the distribution of forests, grasslands, deserts, and other biomes could be altered greatly by global climate change.

merge many cities and even entire island nations. Even at the low end of global warming predictions (a 1°C increase), the effects of increased global temperatures are likely to be considerable, ranging from negative effects on agriculture to an increased occurrence of severe weather (for example, floods, hurricanes, and extremes of hot and cold).

Overall, estimates related to the timing, extent, and consequences of global warming are filled with uncertainty. This uncertainty puts us in the difficult position of choosing to act now, perhaps unnecessarily, or choosing to wait until it is too late to do anything about the problem. Efforts to curb global warming will have social costs, but delays may have far greater costs. Given such uncertainties, what do you think we should do?

> ■ The global climate has warmed during the twentieth century, apparently at least in part as a result of human activities. Although the amount of global warming that will occur in the twenty-first century is uncertain, if high-end predictions are correct, the social and economic costs will be extremely large.

Highlight

A Message of Ecology

The science of ecology has important and timely messages for humanity, such as the one we learned in Chapter 41: No population can continue to increase without limit. Although related, the message of this chapter is more complex. As one ecologist has written, "We are changing the world more rapidly than we are understanding it." In a very real sense, the world is in our hands. What we do to change it will determine our future and the future of all other species on Earth.

As we have seen in this chapter, human activities have caused global change and have had a large impact on life on Earth. Depending on the actions we take, global change has the potential to have even greater effects in the future. With these considerations in mind, we leave you with the following thoughts.

For the first time in history, satellite data and other new technologies allow us to measure our impact on the entire planet. The ability to recognize how much we are changing the planet provides a new source of hope: We can use that information to guide our efforts to reduce the human impact on life on Earth.

Figure 45.12 From Pure Selfishness to Concern for All
This figure illustrates an ethical sequence in which the individual expands his or her concern from a focus only on the self to a concern for the entire biosphere.

As scientists, we think the main message provided by knowledge of how much humans have changed the planet is that we must reduce the rate at which we alter Earth's ecosystems. This course of action is not only good for other species, it is in our own self-interest: As described in Chapter 44, our entire civilization depends on many ecosystem services. If we continue to ignore the impact of our actions on natural systems, ultimately we will harm ourselves.

To reduce our impact on natural systems, we must limit the growth of the human population, and equally importantly, we must use resources more efficiently. Simply put, we must strive to have a sustainable impact on Earth, ceasing to alter natural systems at a rate that leads to short-term gain but long-term damage.

To achieve the goal of having a sustainable impact on the planet, humans must anticipate the effects of our actions before they have disastrous consequences. No other species is capable of such forethought, but we are. Will we do it? Will we be bold enough, creative enough, intelligent enough to take responsibility for our impact on Earth? Can we shift our worldview from one that seeks to dominate nature to one that recognizes the value and intrinsic worth of other species (Figure 45.12)—indeed, of the entire biosphere?

> ■ Knowledge of the large impact of human activities on the planet suggests that humans should reduce the rate at which we alter Earth's ecosystems. To prevent human-caused global change from having a negative effect on ourselves and other species, we humans must learn to have a sustainable impact on Earth.

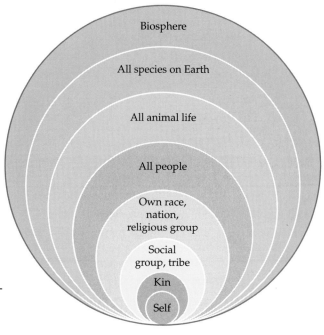

Biosphere

All species on Earth

All animal life

All people

Own race, nation, religious group

Social group, tribe

Kin

Self

Summary

Land and Water Use

- Human activities are changing the lands and waters of the entire planet.

- Land and water transformation has caused the extinction of species and has the potential to alter local, regional, and global climate.

Changes to the Chemistry of Earth

- Human activities are changing how many chemicals are cycled through ecosystems.

The Global Nitrogen Cycle

- Nitrogen must be fixed or converted from N_2 gas before it can be used by producers.

- Human activities fix more nitrogen than all natural activities combined.

- The extra nitrogen fixed by human activities has altered the global nitrogen cycle, leading to increases in productivity that can cause the loss of species from ecosystems.

The Global Carbon Cycle

- The use of fossil fuels and the destruction of forests have caused the concentration of CO_2 in the atmosphere to increase greatly in the past 200 years, thus changing the global carbon cycle.

- Increased CO_2 concentrations can alter the growth of individual plants in ways that will probably cause changes to many ecological communities.

Global Warming

- The global climate has warmed during the twentieth century, apparently at least in part as a result of human activities.

- Although the amount of global warming that will occur in the twenty-first century is uncertain, if high-end predictions are correct, the social and economic costs will be extremely large.

Highlight: A Message of Ecology

- Knowledge of the large impact of human activities on the planet suggests that humans should reduce the rate at which we alter Earth's ecosystems.

- To prevent human-caused global change from having a negative effect on ourselves and other species, we humans must learn to have a sustainable impact on Earth.

Key Terms

global change p. 672

global warming p. 679

greenhouse gas p. 679

land transformation p. 672

nitrogen fixation p. 675

water transformation p. 672

Chapter Review

Self-Quiz

1. Which of the following do most ecologists think are mainly responsible for the current high rate of extinction of species?
   a. increased CO_2 concentration
   b. global warming
   c. hunting and fishing
   d. land and water transformation

2. CO_2 absorbs some of the _____ that radiates from the surface of Earth to space.
   a. ozone
   b. heat
   c. ultraviolet light
   d. smog

3. The conversion of N_2 gas to a form of nitrogen that can be used by plants (NH_4^+) is called
   a. nitrogen fixation.
   b. nitrogen assimilation.
   c. nitrogen cycling.
   d. nitrogen uptake.

4. The concentration of CO_2 in the atmosphere is now about ___ ppm, a level that is about 30 percent higher than preindustrial levels.
   a. 165
   b. 265
   c. 365
   d. 465

5. Human-caused changes to the nitrogen cycle are expected to result in
   a. an increase in acid rain.
   b. an increase in the loss of species from ecosystems.
   c. higher concentrations of a greenhouse gas.
   d. all of the above

Review Questions

1. Summarize the major types of global change caused by humans.

2. Compare human-caused examples of global change to examples of global change not caused by people. What is different or unusual about human-caused global change?

3. What activities do you perform that are good for the global environment? What activities do you perform that harm the global environment?

4. What changes to human societies would have to be made for people to have a sustainable impact on Earth?

The Daily Globe

Society Taxes the Wrong Activities

To the Editor:

As a society, what do we want individuals and corporations to do? Have jobs, work hard, create new and profitable businesses? Or do we want people and corporations to pollute the environment, destroy natural habitat, and contribute to global warming? If the way we are taxed is any guide, the answer is pollute, destroy, and warm the planet.

We tax people when they perform activities that benefit society, such as working hard at their jobs or creating new businesses. If the efforts of people or businesses cause them to make more money, we tax them even more. In contrast, we charge few taxes to people or corporations that damage the environment. Emissions from the vehicles we drive contribute to global warming; pesticides and other chemicals we use alter how materials are cycled through ecosystems; and activities like mining or building housing developments—which we even subsidize!—destroy natural habitat.

Why does our society punish people (by taxation) for activities that benefit society? Why do we reward people (by not taxing them or by providing them subsidies) for activities that damage the environment? When individuals and corporations damage the environment, society as a whole ultimately ends up sharing the cost of fixing the damage. Is it sensible for us to reward people for actions whose results society must later pay to fix?

We are doing everything backward. We should lower the taxes we charge on income and profits from business and start to punish people and businesses that harm the environment by taxing them more heavily. How about high taxes on gasoline to help stop global warming? How about charging heavy taxes for environmentally harmful activities to pay for tomorrow's environmental repairs?

It is time to reward people who do what we say we want them to do and punish those who do the opposite.

Lisa Conracia

Evaluating "The News"

1. The author of this letter to the editor states that we punish people for socially beneficial activities and reward people for socially harmful activities. Do you think this is true? If so, why? If you disagree with the letter writer, explain why.

2. If gasoline were taxed heavily, do you think pressure from con- sumers would cause car manufacturers to respond by designing cars that got more miles per gallon? What do you think would happen if corporations were taxed heavily on the amount of pollution they released into the environment?

3. An assumption of this letter to the editor is that people are most likely to change their behavior if such changes can save them money. Do you agree with this assumption? Are there other ways that might be as effective or more effective in encouraging people to make changes that would benefit the global environment?

Suggested Readings

Unit 1 Diversity of Life

FOR FURTHER EXPLORATION

Futuyma, D. J. 1995. *Science on Trial: The Case for Evolution.* Sinauer Associates, Sunderland, MA. A leading evolutionary biologist summarizes the evidence for evolution and refutes the arguments made by creationists.

May, R. M. 1988. "How Many Species Are There on Earth?" *Science* 241: 1441–1449. Robert May, a distinguished English biologist, evaluates Terry Erwin's estimate of the number of species on Earth, which were based on fogging of canopies.

Margulis, L. and Schwartz, K. V. 1982. *Five Kingdoms: An Illustrated Guide to the Phyla of Life on Earth.* W. H. Freeman and Co., San Francisco. An instructive, detailed description of the five kingdoms and the many organisms that comprise them.

The New York Times Book of Science Literacy. Volume 2: *The Environment from Your Backyard to the Ocean Floor.* 1994. Times Books, New York. Collected articles from *The New York Times* on biodiversity and related environmental issues.

Pool, R. 1990. "Pushing the Envelope of Life." *Science* 247: 158–247. An exploration of what is known about *Archaebacteria* and their extreme lifestyles.

Takacs, D. 1997. *The Idea of Biodiversity.* Johns Hopkins University Press, Baltimore, MD. A discussion of the history of biodiversity and its meaning for biologists, written by a biologist–historian.

Wilson, E. O., et al. 1989. *Biodiversity.* National Academy Press, Washington, D.C. A discussion by a leading evolutionary biologist of the evolution of diversity.

FOR YOUR ENJOYMENT

Attenborough, D. 1995. *Private Life of Plants.* Princeton University Press, Princeton, NJ. A popular natural history about plant "behavior."

Gould, S. J. 1980. *The Panda's Thumb.* W. W. Norton and Co., New York. Essays by a leading evolutionary biologist on evolution and evolutionary relationships.

Howarth, F. G., Sohmer, S. H. and Duckworth, W. D. 1988. "Hawaiian Natural History and Conservation Efforts." *BioScience* 38: 232–238. The story of Hawaii's many dwindling native species and the efforts to save them.

Raup, D. 1991. *Extinction: Bad Genes or Bad Luck?* W. W. Norton and Co., New York. In this book for lay audiences, renowned paleontologist David Raup examines why some organisms survive mass extinctions and others do not.

Watson, J. D. 1968. *The Double Helix: A Personal Account of the Discovery of the Structure of DNA.* Atheneum, New York. A lively account of the Nobel Prize–winning discovery of the structure of DNA, written by one of its discoverers.

Wilson, E. O. 1991. "Rain Forest Canopy: The High Frontier." *National Geographic* 180: 78–107. A beautifully illustrated account of the diversity found in the highest branches of the tropical rainforest.

Unit 2 Cells: The Basic Units of Life

FOR FURTHER EXPLORATION

Atkins, P. W. 1987. *Molecules.* Scientific American Library, New York. Accurate and revealing renderings of important molecules.

Cooper, G. M. 1998. *The Cell: A Molecular Approach.* Sinauer Associates, Sunderland, MA. A concise introductory text on the molecular biology of cells.

Hall, D. O., and Rao, K. K. 1999. *Photosynthesis (Studies in Biology).* Cambridge University Press, Cambridge. An easily understandable and methodical treatment of photosynthesis.

Needham, J. (ed.) 1970. *The Chemistry of Life.* Cambridge University Press, Cambridge. A selection of thoughtful essays on the history of biochemistry

Salway, J. G., and Salway, J. D. 1994. *Metabolism at a Glance.* Blackwell Science Inc., Cambridge, MA. A concise and complete overview of metabolism with clear diagrams.

Stossel, T. 1994. "The Machinery of Cell Crawling." *Scientific American*, September. Insightful discussion of how different cytoskeletal proteins contribute to cell movement.

Stryer, L. 1995. *Biochemistry,* 4th Ed. W. H. Freeman and Co., New York. Exceptionally well written and organized introductory text on biochemistry.

FOR YOUR ENJOYMENT

Bloch, K. 1997. *Blondes in Venetian Paintings, the Nine-Banded Armadillo, and Other Essays in Biochemistry.* Yale University Press, New Haven. Colorful anecdotes by a Nobel Prize–winning scientist on the chemical basis for diverse human and biological phenomena.

Emsley, J. 1998. *Molecules at an Exhibition: Portraits of Intriguing Materials in Everyday Life.* Oxford University Press, New York. Fascinating and entertaining anecdotes about the many chemical compounds encountered in daily life.

Galston, A. W. 1992. "Photosynthesis as a Basis for Life Support in Space." *Bioscience*, July/August. Provocative discussion of how photosynthetic organisms may be used to support artificial ecosystems in space.

Goodsell, D. S. 1998. *The Machinery of Life.* Springer Verlag, New York. Beautifully illustrated tour of the molecular structure of cells.

Loewenstein, W. R. 1999. *The Touchstone of Life: Molecular Information, Cell Communication, and the Foundations of Life.* Oxford University Press, New York. A lucid discussion of how the flow of information, molecular biology, and cell structure work together in living processes.

Rasmussen, H. 1991. *Cell Communication in Health and Disease: Readings from Scientific American Magazine.* W. H. Freeman and Co., New York. A collection of articles from Scientific American in which researchers examine the complexity of cell communication and its role in various human diseases such as atherosclerosis and diabetes.

Weinberg, R. A. 1998. *Racing to the Beginning of the Road: The Search for the Origin of Cancer.* W. H. Freeman and Co., New York. The fascinating history of cancer's origins, written by one of the most prominent researchers in cancer biology. Contains revealing renderings of important molecules.

Unit 3 Genetics

FOR FURTHER EXPLORATION

"Making Gene Therapy Work." *Scientific American,* June 1997. A collection of articles outlining current progress in human gene therapy.

Mange, E. J. and A. P. Mange. 1999. *Basic Human Genetics,* 2nd Ed. Sinauer Associates, Sunderland, MA. An introduction to human genetics.

"Microchip Arrays Put DNA on the Spot." *Science,* October 16, 1998. A collection of news articles discussing the current and likely future impact of DNA chips.

Orel, V. 1996. *Gregor Mendel, the First Geneticist.* Oxford University Press, New York. A biography of the founder of the science of genetics.

Snustad, D. P., Simmons, M. J. and Jenkins, J. B. 1997. *Principles of Genetics.* John Wiley and Sons, New York. A general genetics textbook.

Subramanian, S. 1995. "The Story of Our Genes." *Time Magazine,* January 16. An overview piece on genetic screening and what our genes can tell us.

FOR YOUR ENJOYMENT

Judson, H. F. 1996. *The Eighth Day of Creation: The Makers of the Revolution in Biology.* Cold Spring Harbor Laboratory Press, Cold Spring Harbor, NY. An engaging history of molecular genetics.

Keller, E. F. 1983. *A Feeling for the Organism: The Life and Work of Barbara McClintock.* W. H. Freeman and Co., San Francisco. The story of the life and work of a Nobel Prize–winning female scientist working in an all-male world of science.

Kitcher, P. 1996. "Junior Comes Out Perfect." *New York Times Magazine,* September 29. An article on the potential for producing the perfect baby through genetic screening.

Watson, J. D. 1968. *The Double Helix.* Atheneum, New York. A personal and controversial account of the Nobel Prize–winning discovery of the structure of DNA.

Unit 4 Evolution

FOR FURTHER EXPLORATION

Axelrod, R. 1984. *The Evolution of Cooperation.* Basic Books, New York. A far-reaching book on cooperation that shows how phenomena ranging from mutualism to trench warfare can be partially explained by an evolutionary perspective.

Darwin, C. 1859. *On the Origin of Species.* J. Murray, London. (reprint edition: 1964, Harvard University Press, Cambridge, MA.) The single most important book on evolution ever published. A landmark scientific work that revolutionized biology and had a large impact on many other areas of human society. With careful reading, this text is very clear to a nonscientific audience.

Freeman, S. and Herron, J. C. 1998. *Evolutionary Analysis.* Prentice Hall, Englewood Cliffs, NJ. A general textbook on evolution.

Futuyma, D. J. 1995. *Science on Trial: The Case for Evolution.* Sinauer Associates, Sunderland, MA. A leading evolutionary biologist summarizes the evidence for evolution and refutes the arguments made by creationists.

Gould, S. J. 1989. *Wonderful Life: The Burgess Shale and the Nature of History.* W. W. Norton and Co., New York. An engaging account of the extraordinary organisms found in a bed of fossils from the Cambrian period and the implications of these organisms for the history of life on Earth.

Slatkin, M. (ed.) 1995. *Exploring Evolutionary Biology: Readings from American Scientist.* Sinauer Associates, Sunderland, MA. An interesting collection of articles on evolution, originally published in the journal *American Scientist* and intended for both scientists and the general public.

FOR YOUR ENJOYMENT

Diamond, J. 1992. *The Third Chimpanzee: The Evolution and Future of the Human Animal.* HarperCollins Publishers, New York. A fascinating account of human evolution and the future of our species by one of the leading writers on science for the general public.

Gould, S. J. 1985. *The Flamingo's Smile: Reflections in Natural History.* W. W. Norton and Co., New York. Essays by a leading evolutionary biologist on evolution and evolutionary relationships.

Johanson, D. and Shreeve, J. 1989. *Lucy's Child.* Avon Books, New York. The story of the discovery of the *Australopithecus afarensis* skeleton that became known in the popular press as Lucy.

Nesse, R. and Williams, G. 1995. *Why We Get Sick: The New Science of Darwinian Medicine.* Vintage Books, New York. A compelling argument for why both doctors and patients need to understand basic evolutionary principles.

Weiner, J. 1994. *The Beak of the Finch: A Story of Evolution in our Time*. Knopf, New York. A Pulitzer Prize–winning account of evolution in the Galápagos finches.

Wilson, E. O. 1992. *The Diversity of Life*. W. W. Norton and Co., New York. A thought-provoking description of how the diversity of life evolved and how it is being threatened by human actions.

Unit 5 Form and Function

FOR FURTHER EXPLORATION

Bonner, J. T. 1980. *The Evolution of Culture in Animals*. Princeton University Press, Princeton, NJ. An interesting treatment of the evolution of cultural behavior.

Eisenberg, A., Murkoff, H. E. and Hathaway, S. E. 1996. *What to Expect When You're Expecting*. Workman Publishing, New York. This and other books on human pregnancy give a good overview of the steps in and timing of human repro-duction and development.

Gordon, J. E. 1978. *Structures*. Pelican Books, London. Draws many parallels between the structure of plant and animal support systems and the way in which engineers assemble materials into bridges and buildings.

Owen, J. 1980. *Feeding Strategy*. University of Chicago Press, Chicago. A broad and elementary review of nutrition among living things.

Schmidt-Nielsen, K. 1972. *How Animals Work*. Cambridge University Press, London. This book presents diverse and entertaining examples of how and why animals function the way that they do.

Schmidt-Nielsen, K. 1984. *Scaling: Why Is Animal Size So Important?* Cambridge University Press, New York. A read-able but thorough overview of the effect of size on animal biology.

Vogel, S. 1988. *Life's Devices*. Princeton University Press, Princeton, NJ. A thorough overview of the mechanics of how organisms deal with their environment.

FOR YOUR ENJOYMENT

Andrews, M. 1976. *The Life that Lives on Man*. Taplinger Publishing Company, New York. A novel look at how vari-ous microbes, fungi, and animals make their living on the surfaces of our own bodies.

Bakker, R. T. 1986. *The Dinosaur Heresies*. Zebra Books, New York. A thorough presentation of the idea that dinosaurs were endotherms.

Bonner, J. T. 1980. *The Evolution of Culture in Animals*. Princeton University Press, Princeton, NJ. An interesting treatment of cultural evolution.

Gordon, J. E. 1978. *Structures*. Pelican Books, London. Draws many parallels between the structure of plant and animal support systems and the way in which engineers assemble materials into bridges and buildings.

Unit 6 Interactions with the Environment

FOR FURTHER EXPLORATION

Garrett, L. 1994. *The Coming Plague: Newly Emerging Diseases in a World Out of Balance*. Farrar, Straus and Giroux, New York. A description of how the human impact on Earth can affect the development of new human diseases.

Gates, D. M. 1993. *Climate Change and its Biological Consequences*. Sinauer Associates, Sunderland, MA. A clear description of past and present examples of climate change and how those changes affect life on Earth.

Kareiva, P. (ed.) 1998. *Exploring Ecology and its Applications: Readings from American Scientist*. Sinauer Associates, Sunderland, MA. An interesting collection of articles on ecol-ogy, originally published in the journal *American Scientist* and intended for both scientists and the general public.

Newton, L. H. and Dillingham, C. K. 1997. *Watersheds 2: Ten Cases in Environmental Ethics*. Wadsworth Publishing Company, Belmont, CA. An excellent collection of environ-mental case studies, each of which provides readers with a clear description of the important biological and ethical con-siderations that relate to the issue at hand.

Primack, R. B. 1998. *Essentials of Conservation Biology*, 2nd Ed. Sinauer Associates, Sunderland, MA. A good introduction to the fast-developing field of conservation biology, a branch of science focused on the conservation of the diversity of life on Earth.

Ricklefs, R. E. and Miller, G. L. 1999. *Ecology*, 4th Ed. W. H. Freeman and Co., New York. A general ecology textbook.

FOR YOUR ENJOYMENT

Botkin, D. B. 1996. *Our Natural History: The Lessons of Lewis and Clark*. Perigee, New York. A provocative and interesting book that uses the journeys of Lewis and Clark as a spring-board from which to discuss how ecological systems are always in a state of change and how endangered ecosystems can be saved.

Carson, R. 1962. *Silent Spring*. Houghton Mifflin Co., New York. An award-winning book that galvanized public inter-est in ecology and in the environmental movement.

Hubbell, S. 1999. *Waiting for Aphrodite: Journeys into the Time Before Bones*. Houghton Mifflin Co., Boston. A beautifully written collection of essays about the lives of animals with-out backbones, the invertebrates.

Leopold, A. 1949. *A Sand County Almanac and Sketches Here and There*. Oxford University Press, New York. A collection of essays from one of the founders of the conservation movement.

McKibben, B. 1995. *Hope, Human and Wild: True Stories of Living Lightly on the Earth*. Little, Brown and Co., Boston. Beautifully written and penetrating essays that focus on suc-cessful community efforts to preserve wilderness and reverse environmental damage.

Orr, D. W. 1994. *Earth in Mind: On Education, Environment, and the Human Prospect*. Island Press, Washington, D.C. A thought-provoking collection of essays about causes of and solutions to global environmental problems.

Table of Metric–English Conversion

Metric unit and abbreviation	Equivalent	Common conversions		

Length

Metric unit and abbreviation	Equivalent	To convert	Multiply by	To yield
nanometer (nm)	$0.000000001\ (10^{-9})$ m	inches	2.54	centimeters
micrometer (μm)	$0.000001\ (10^{-6})$ m	yards	0.91	meters
millimeter (mm)	$0.001\ (10^{-3})$ m	miles	1.61	kilometers
centimeter (cm)	$0.01\ (10^{-2})$ m			
meter (m)	—	centimeters	0.39	inches
kilometer (km)	$1000\ (10^{3})$ m	meters	1.09	yards
		kilometers	0.62	miles

Weight (mass)

nanogram (ng)	$0.000000001\ (10^{-9})$ g	ounces	28.35	grams
microgram (μg)	$0.000001\ (10^{-6})$ g	pounds	0.45	kilograms
milligram (mg)	$0.001\ (10^{-3})$ g			
gram (g)	—	grams	0.035	ounces
kilogram (kg)	$1000\ (10^{3})$ g	kilograms	2.20	pounds
metric ton (t)	$1{,}000{,}000\ (10^{6})$ g $(=10^{3})$ kg			

Volume

microliter (μl)	$0.000001\ (10^{-6})$ l	fluid ounces	29.57	milliliters
milliliter (ml)	$0.001\ (10^{-3})$ l	quarts	0.95	liters
liter (l)	—			
kiloliter (kl)	$1000\ (10^{3})$ l	milliliters	0.034	fluid ounces
		liters	1.06	quarts

Temperature

degree Celcius (°C)	—	To convert Fahrenheit (°F) to Centigrade (°C): $^{\circ}C = \frac{5}{9}(^{\circ}F - 32^{\circ})$
		To convert Centigrade (°C) to Fahrenheit (°F): $^{\circ}F = \frac{9}{5}\,^{\circ}C + 32^{\circ}$

Answers to Self-Quiz Questions

Chapter 1
1. *b*
2. *b*
3. *a*
4. *c*
5. *b*
6. *a*

Chapter 2
1. *b*
2. *a*
3. *d*
4. *c*
5. *b*
6. *a*
7. *a*

Chapter 3
1. *a*
2. *c*
3. *d*
4. *c*
5. *a*
6. *d*

Chapter 4
1. *c*
2. *c*
3. *b*
4. *c*
5. *d*
6. *b*

Chapter 5
1. *a*
2. *c*
3. *d*
4. *a*
5. *c*
6. *b*

Chapter 6
1. *d*
2. *a*
3. *c*
4. *b*
5. *a*
6. *b*

Chapter 7
1. *c*
2. *a*
3. *b*
4. *c*
5. *a*
6. *d*

Chapter 8
1. *b*
2. *d*
3. *a*
4. *a*
5. *a*
6. *c*

Chapter 9
1. *d*
2. *d*
3. *a*
4. *b*
5. *b*
6. *d*

Chapter 10
1. *b*
2. *a*
3. *d*
4. *b*
5. *d*
6. *c*

Chapter 11
1. *c*
2. *a*
3. *b*
4. *b*
5. *a*
6. *d*

Chapter 12
1. *a*
2. *c*
3. *b*
4. *d*
5. *d*

Chapter 13
1. *d*
2. *b*
3. *c*
4. *c*
5. *a*

Chapter 14
1. *c*
2. *c*
3. *b*
4. *d*
5. *a*

Chapter 15
1. *b*
2. *c*
3. *b*
4. *d*
5. *a*
6. *d*

Chapter 16
1. *d*
2. *c*
3. *b*
4. *a*
5. *d*

Chapter 17
1. *c*
2. *a*
3. *b*
4. *d*
5. *a*

Chapter 18
1. *c*
2. *a*
3. *c*
4. *a*
5. *a*
6. *a*

Chapter 19
1. *d*
2. *d*
3. *c*
4. *a*
5. *b*

Chapter 20
1. *c*
2. *b*
3. *a*
4. *c*
5. *d*
6. *a*

Chapter 21
1. *b*
2. *b*
3. *c*
4. *a*
5. *d*

Chapter 22
1. *d*
2. *d*
3. *b*
4. *a*
5. *c*

Chapter 23
1. *c*
2. *a*
3. *b*
4. *d*
5. *c*

Chapter 24
1. *a*
2. *c*
3. *a*
4. *a*
5. *d*

Chapter 25
1. *a*
2. *d*
3. *a*
4. *b*
5. *c*

Chapter 26
1. *a*
2. *a*
3. *c*
4. *b*
5. *c*

Chapter 27
1. *b*
2. *a*
3. *b*
4. *a*
5. *d*

Chapter 28
1. *d*
2. *b*
3. *b*
4. *a*
5. *a*

Chapter 29
1. *a*
2. *d*
3. *b*
4. *b*
5. *d*

Chapter 30
1. *d*
2. *a*
3. *c*
4. *c*
5. *b*

Chapter 31
1. *b*
2. *a*
3. *c*
4. *d*
5. *d*

Chapter 32
1. *c*
2. *b*
3. *d*
4. *c*
5. *d*

Chapter 33
1. *b*
2. *d*
3. *d*
4. *c*
5. *b*

Chapter 34
1. *a*
2. *a*
3. *b*
4. *d*
5. *b*

Chapter 35
1. *b*
2. *d*
3. *c*
4. *a*
5. *c*

Chapter 36
1. *c*
2. *d*
3. *a*
4. *d*
5. *a*

Chapter 37
1. *d*
2. *d*
3. *d*
4. *a*
5. *c*

Chapter 38
1. *c*
2. *a*
3. *d*
4. *b*
5. *c*

Chapter 39
1. *c*
2. *d*
3. *a*
4. *c*
5. *b*

Chapter 40
1. *a*
2. *c*
3. *d*
4. *d*
5. *b*

Chapter 41
1. *b*
2. *b*
3. *c*
4. *d*
5. *a*

Chapter 42
1. *c*
2. *a*
3. *c*
4. *d*
5. *b*

Chapter 43
1. *c*
2. *c*
3. *d*
4. *a*
5. *b*

Chapter 44
1. *c*
2. *a*
3. *d*
4. *a*
5. *b*

Chapter 45
1. *d*
2. *b*
3. *a*
4. *c*
5. *d*

Answers to Review Questions

Chapter 1

1. Dr. Larson's honeysuckle work:

   Observations: Japanese honeysuckle spreads in the wild faster than the native coral honeysuckle. Vines need supports.

   Hypothesis: Japanese honeysuckle spreads faster than the native coral honeysuckle because it is better at finding supports, especially supports far from where it grows.

   Experiment: Grow both species of honeysuckle and place supports close to and far away from the plants.

   Further **observations** during the experiment: The Japanese species grew differently around a central axis, allowing it to produce roots faster. These observations led to another **hypothesis**: The difference in rotation patterns contributes to the rapid spreading of the Japanese species.

2. From smallest to largest, the elements of the biological hierarchy are: molecules, cells, tissues, organs, organ systems, individual organisms, populations, species, communities, ecosystems, biomes, biospheres.

3. Energy flows in one direction through biological systems from sunlight to producers (photosynthesizers), where it is converted to chemical energy, then to consumers (animals, fungi), which eat the producers and emit heat energy.

Chapter 2

1. Factors that likely contributed to the success of the Monera include their rapid rate of reproduction and their ability to obtain nutrients in very diverse ways.

2. *Giardia* is of interest to biologists concerned with evolution of eukaryotes because unlike most eukaryotic cells, it contains two nuclei but no mitochondria. As a result, *Giardia* provide a look at one of many evolutionary experiments in assembling eukaryotic cells in the history of life. Slime molds are of interest because individual slime molds can exist as single-celled, then multicellular organisms, providing biologists a window into the transition that eukaryotes made from single-celled to multicellular living. Studying these two stages may provide scientists with information about the evolution of multicellular organisms.

3. Sponges are the group in which the first specialized cells appeared. However, sponges are merely loose collections of such cells. The cnidaria were among the first animals to evolve true, differentiated tissues. Flatworms were among the first animals to evolve true organs.

4. While viruses exhibit some characteristics of living organisms, they are not placed into any existing kingdom. Viruses are not placed in any one kingdom because these highly degenerate organisms are thought to have evolved many times from many kingdoms.

Chapter 3

1. Systematists compare characteristics of organisms, including structural characteristics, behavior patterns, and DNA. Systematists then can place an organism on the evolutionary tree of life with those organisms with which it has the greatest number of shared derived features.

2. A "real group" is one which contains all the descendants of a single common ancestor and only the descendants of that single common ancestor. Many systematists feel that it is important only to designate organisms into "real groups" because they are based strictly on evolutionary relationships. Others, however, feel that doing away with groups like the reptiles would be too radical a move.

3. An evolutionary tree is like a hypothesis because both involve predictions that may be supported or rejected by the results of further testing.

Chapter 4

1. Terry Erwin estimated the number of insect species in the world by fogging a rainforest tree with insecticide and counting the number of beetle species that fell to the ground. Using other estimates, he extrapolated an estimate of the total number of insect species in the world. These indirect measures, which depend on other estimates, are difficult to do because they involve making a lot of assumptions about which scientists have little information.

2. Biodiversity has fluctuated ever since life began, declining rapidly during mass extinctions followed by slow recovery periods. Five previous mass extinctions (around 440, 350, 250, 206, and 65 million years ago) were not man-made, but are thought to have been caused by natural forces such as climate changes, volcanic eruptions, changes in sea level, and atmospheric dust from an asteroid collision. Recovery periods involved millions of years. The current rapid expansion of human populations may lead to another such mass extinction.

3. As human populations grow, cities, suburbs and commercial areas expand and natural habitats are destroyed and degraded, driving out other species. Human population growth has increases pollution of water and the atmosphere, further

threatening biodiversity. Expansion of human populations has also increased the introduction of nonnative species, which is another threat to biodiversity.

Chapter 5

1. Polymers are complex molecules, often with attached functional groups. Combinations of these functional groups give polymers chemical properties not present in smaller molecules.

2. The pH of pure water should be 7. As a measurement, pH units represent the concentration of free hydrogen ions in water. In the presence of a base, the pH will be above 7, indicating that there are more hydroxyl ions than hydrogen ions; thus, the solution is basic. In the presence of an acid, the pH will be below 7, indicating more free hydrogen ions; thus, the solution is acidic. Pure water has equal amounts of hydrogen and hydroxyl ions and is thus neutral.

3. Each carbon atom has four bonding areas that can bond covalently to other carbons or to different atoms, creating large molecules containing hundreds of atoms.

4. Proteins are polymers of amino acids. Living organisms use proteins for physical structures such as hair and for enzymes to facilitate chemical reactions needed by the body.

Chapter 6

1. One function of proteins embedded in the plasma membrane is to form gateways that allow selective passage of ions and molecules in and out of the cell. This allows wastes to be removed and needed substances to enter. Another function is to recognize changes in the outside environment and to communicate with other cells so a quick response to change is possible.

2. Unlike lysosomes in animals cells, large vacuoles in plant cells can fill with water and push against the cell wall, giving it rigidity.

3. The enzyme is produced on ribosomes associated with the endoplasmic reticulum. A vesicle transports the enzyme to the Golgi apparatus, where it is packaged into a specialized vesicle called a lysosome. The lysosome is released into the cell. Since the enzyme is enclosed in a membrane or vesicle, it would not be manufactured on a free-floating ribosome.

4. Both microtubules and actin filaments can change length quickly. With microtubules, this allows movement of organelles within the cell. Movement of the cell itself occurs when actin filaments lengthen or shorten and change cell shape.

Chapter 7

1. If cells oxidized food molecules in one step, so much energy would be released that we would burst into flame. A step-by-step process allows some of the energy to be safely stored in chemical bonds and released when needed.

2. The second law of thermodynamics requires that the organization of the cell be compensated for by the transfer of disorder to the environment. Heat fulfills this requirement by increasing the disorderly and random movement of molecules in the environment.

3. Higher temperatures increase random molecular collisions while enzymes selectively bind to specific reactants and bring them closer together to facilitate a chemical reaction. This closeness lowers the energy of activation required for the reaction.

4. Specific organelles, such as mitochondria, increase the efficiency of catalysis by concentrating both the enzymes and substrates needed for a particular process. In mitochondria, free-floating enzymes are concentrated in the lumen, and other enzymes are physically arranged next to each other in the membrane. Both of these arrangements facilitate enzyme catalysis.

5. Enzymes in a detergent increase its efficiency by binding to the stains in clothing, bringing them in close contact with the detergent. The stain and detergent are the reactants.

Chapter 8

1. The transfer of electrons down an ETC produces a proton gradient in both chloroplasts and mitochondria. The protons move through a membrane channel known as the ATP synthase, which in turn catalyzes the production of ATP.

2. Sugar made in the leaves of a plant by photosynthesis is transported to the roots. Cellular respiration in the mitochondria of root cells converts this sugar to usable energy in the form of ATP.

3. O_2 is consumed when it picks up electrons and combines with protons to form water. Thus, O_2 is the last electron acceptor in the series of reactions that make up oxidative phosphorylation.

4. When protons pass through an ATP synthase channel, ADP is converted to ATP. Drugs that allow this channel to be bypassed will deter the production of ATP.

Chapter 9

1. The receptor could be located in the cytosol or the nucleus, since hydrophobic molecules, such as steroids, are able to pass through the lipid core of the plasma membrane.

2. Signal amplification is important to a cell's ability to communicate with other cells, because it activates many proteins at each step of the signal cascade, increasing the effect of the original signal and causing a faster response.

3. Phosphorylation by a kinase changes the shape of an enzyme, thus altering its activity. It activates some enzymes and inhibits others.

4. Adrenaline binds to and activates a receptor on the outside of a cell. This activates a G protein inside the cell by inducing it to exchange a guanine nucleotide (GDP) for another nucleotide (GTP) in the cytosol. This turns on an enzyme that converts ATP into another nucleotide (cAMP). The cAMP then activates a cascade of enzymes needed for the cell's response to the adrenaline.

Chapter 10

1. A horse cell undergoing mitosis would have 64 chromosomes. A horse cell undergoing meiosis II would have 32 chromosomes.

2. The mitotic spindle directs the movement of chromosomes during cell division.

3. Mitosis: Spindle microtubules from each centrosome attach to the duplicated centromere of a chromosome at the kinetochores. During anaphase, these microtubules pull the sister chromatids to opposite poles of the cell. Each sister chromatid forms a new chromosome in the daughter cells. Meiosis I: A single microtubule from opposite poles attaches to each homologous chromosome. During anaphase I, the homologous chromosomes (sister chromatids still joined) are pulled to opposite poles. This halves the number of chromosomes in each daughter cell.

4. Meiosis is important for the production of sex cells because it allows cells to maintain a constant chromosome number when fertilization occurs. If sex cells underwent mitosis instead, the gametes would contain a complete set of chromosomes and the zygote formed after fertilization would have twice the number of chromosomes as its parents. The developing embryo would not have the same karyotype as its parents and could not survive.

Chapter 11

1. p 53 prevents the passing on of harmful mutations by stopping a cell with DNA damage from dividing until the damage is repaired. If the damage is too severe, p53 kills the cell.

2. Peyton Rous filtered the extract from cancerous tumors to remove bacteria. If the chickens injected with the filtered extract developed cancer, Rous could rule out bacteria as the cause.

3. Malignancies develop over decades through an accumulation of several DNA mutations. Activation of at least one oncogene and deactivation of several tumor suppressors usually must occur first. Therefore, the less long-term exposure to chemical mutagens, the less risk of developing cancer.

4. If only one homologous chromosome is missing the *Rb* gene, some protein can still be produced and the disorder is less severe, usually involving a single tumor in one eye. A mutation in the remaining gene, however, causes multiple tumors in both eyes. If *Rb* were an oncogene, severe cancer would be produced with only one active gene. Therefore, *Rb* must be a tumor suppressor.

Chapter 12

1. The genotype of the unknown parent can be determined by crossing it with a with a pea plant that is homozygous recessive for this trait (*pp*). This parent can only provide recessive alleles, so if there are any progeny with the recessive phenotype (white flowers), the unknown parent must have been heterozygous (*Pp*) to provide the second recessive allele.

2. A lethal genetic disorder caused by a dominant allele would kill both the individuals with a homozygous dominant genotype as well as those with the heterozygous genotype. If the disorder kills people before they reach reproductive age, then the allele will rapidly be selected out of the population. A disorder caused by a lethal recessive allele would only kill individuals with the homozygous recessive genotype, so the allele could still be carried by heterozygous individuals and passed on to the next generation.

3. Probability refers to the chance that a certain event will occur. If a coin is only flipped a few times, for instance, then there is no way of knowing whether or not the results obtained were representative. However, if the coin is tossed many times, or if many pea plants are observed, then predictable patterns can be discerned.

Sample Genetics Problems

1. a. *A* and *a*

   b. *BC*, *Bc*, *bC*, and *bc*

   c. *Ac*

   d. *ABc*, *Abc*, *AbC*, *aBC*, *aBc*, *abC*, and *abc*

   e. *aBC* and *aBc*

2. a. genotype ratio: 1:1 phenotype ratio: 1:1

	A	*a*
a	*Aa*	*aa*

   b. genotype ratio: 1:0 phenotype ratio: 1:0

	B
b	*Bb*

   c. genotype ratio: 1:1 phenotype ratio: 1:1

	AB	*Ab*
ab	*AaBb*	*Aabb*

   d. genotype ratio: 1*BBCC*:1*BBCc*:2*BbCC*:2*BbCc*:1*bbCC*:1*bbCc* phenotype ratio: 6:2, reduced to 3:1

	BC	*Bc*	*bC*	*bc*
BC	*BBCC*	*BBCc*	*BbCC*	*BbCc*
bC	*BbCC*	*BbCc*	*bbCC*	*bbCc*

   e. genotype ratio:
   1*AABbCC*:2*AABbCc*:1*AABbcc*:1*AAbbCC*:2*AAbbCc*:1*AAbbcc*: 1*AaBbCC*:2*AaBbCc*:1*AaBbcc*:1*AabbCC*:2*AabbCc*:1*Aabbcc* phenotype ratio: 6:2:6:2, reduced to 3:1:3:1

	ABC	*ABc*	*AbC*	*Abc*	*aBC*	*aBc*	*abC*	*abc*
AbC	*AABbCC*	*AABbCc*	*AAbbCC*	*AAbbCc*	*AaBbCC*	*AaBbCc*	*AabbCC*	*AabbCc*
Abc	*AABbCc*	*AABbcc*	*AAbbCc*	*AAbbcc*	*AaBbCc*	*AaBbcc*	*AabbCc*	*Aabbcc*

3.

	S	*s*
S	*SS*	*Ss*
s	*Ss*	*ss*

genotype ratio: 1*SS*:2*Ss*:1*ss*
phenotype ratio: 3 healthy:1 sickle-cell anemia

Each time two *Ss* individuals have a child there is a 25% chance that the child will have sickle-cell anemia.

4. a. *NN* and *Nn* individuals are normal; *nn* individuals are diseased.

b.

	N	n
N	NN	Nn
n	Nn	nn

genotype ratio: 1*NN*:2*Nn*:1*nn* phenotype ratio: 3 normal:1 diseased

c.

	N	n
N	NN	Nn

genotype ratio: 1:1 phenotype ratio: 2 healthy:0 diseased

5. a. *DD* and *Dd* individuals are diseased; *dd* individuals are normal.

b.

	D	d
D	DD	Dd
d	Dd	dd

genotype ratio: 1*DD*:2*Dd*:1*dd* phenotype ratio: 3 diseased:1 normal

c.

	D	d
D	DD	Dd

genotype ratio: 1:1 phenotype ratio: 2 diseased:0 healthy

6. The parents are most likely *AA* and *aa*. The white parent must be *aa*. The purple parent could potentially be *AA* or *Aa*, but if it were *Aa*, we would expect about half of the offspring to be white. Therefore, if the cross yields many offspring and all are purple, it is extremely likely that the purple parent's genotype is *AA*.

7. The yellow allele is dominant. Since each parent breeds true, that means each parent is homozygous. When a homozygous recessive parent is bred to a homozygous dominant parent, the F$_1$ generation will exhibit only the dominant phenotype. Therefore, the phenotype of the F$_1$ generation, yellow, is produced by the dominant allele.

Chapter 13

1. Nonparental genotypes are formed by genetic recombination events such as crossing-over, independent assortment of chromosomes, and fertilization. Mutation can also lead to the creation of new, nonparental alleles.

2. It is unlikely that alleles that cause lethal dominant genetic disorders will be as common in the population as alleles that cause lethal recessive genetic disorders, because lethal dominant disorders affect both individuals with homozygous dominant genotypes and those with heterozygous genotypes. This would cause lethal dominant alleles to be removed from the population at a faster rate than lethal recessive alleles, which only cause the death of individuals with the homozygous recessive genotype. Since these dominant alleles are *lethal*, they will be selected out of the population more quickly, and not be "protected" by heterozygote carriers as a recessive allele would be.

3. If someone had not yet had children and found he or she carried the allele, he or she would have the option to choose not to have children. Even if the partner did not carry the allele, the fact that the disease-causing allele is dominant means that even the one allele may affect any children the couple had. Generally speaking, knowing whether a person has a gene that carries a genetic disorder may help determine whether that person should undergo certain treatments or lifestyle modifications. For instance, if parents learn that a baby has PKU, they could modify the baby's diet and avoid the disease; otherwise, the child's mental abilities would be affected. The major disadvantage of genetic tests is that often the presence of a gene indicates a *predisposition* toward a disease, rather than the actual existence of a disease. This might lead to unnecessary preventative actions, such as a woman deciding to have a healthy breast removed because she fears a predisposition to breast cancer, or an insurance company increasing a person's premiums, without knowing whether the person would have actually contracted the disease.

Sample Genetics Problems

1. a. Males inherit their X chromosome from their mothers, since their Y chromosome must come from their fathers. Their mothers do not have a Y chromosome to give them, and they must have one in order to be male.

b. No, she does not have the disorder. If she has only one copy of the recessive allele, her other X chromosome must then have a copy of the dominant allele. She is a carrier, but she does not have the disorder herself.

c. Yes, he does have the disorder. The trait is X-linked, he has only one X chromosome, and that X chromosome carries the recessive disorder-causing allele. His Y chromosome does not carry an allele for this gene, so cannot contribute to the male's phenotype relative to this trait.

d. If the female is a carrier of an X-linked recessive disorder, her genotype is $X^D X^d$, where D = the dominant allele and d = the recessive, disorder-causing allele. This means she can produce two types of gametes relative to this trait: X^D and X^d. Only the X^d gamete carries the disease-causing allele.

e. None of their children will have the disorder, since the mother will always contribute a dominant non-disorder-causing allele to each child. However, all of the female children will be carriers, since their second X chromosome comes from their father, who only has one X chromosme to contribute, and it carries the disorder-causing allele.

2. a. 50% chance of *aa* cystic fibrosis genotype

	A	a
a	Aa	aa
a	Aa	aa

b. 0% chance of *aa* cystic fibrosis genotype

	A	A
A	AA	AA
a	Aa	Aa

c. 25% chance of *aa* cystic fibrosis genotype

	A	a
A	AA	Aa
a	Aa	aa

d. 0% chance of *aa* cystic fibrosis genotype

	A	A
a	Aa	Aa
a	Aa	Aa

3. a. 50% chance of Huntington's disease genotype, *Aa*

	A	a
a	Aa	aa
a	Aa	aa

b. 100% chance of Huntington's disease genotype, *AA* or *Aa*

	A	A
A	AA	AA
a	Aa	Aa

c. 75% chance of Huntington's disease genotype, *AA* or *Aa*

	A	a
A	AA	Aa
a	Aa	aa

d. 100% chance of Huntington's disease genotype, *Aa*

	A	A
a	Aa	Aa
a	Aa	Aa

4. a. 0% chance of a child with hemophilia

	X^a	Y
X^A	X^AX^a	X^AY
X^A	X^AX^a	X^AY

b. 50% chance of a child with hemophilia

	X^a	Y
X^A	X^AX^a	X^AY
X^a	X^aX^a	X^aY

c. 25% chance of a child with hemophilia

	X^A	Y
X^A	X^AX^A	X^AY
X^a	X^AX^a	X^aY

d. 50% chance of a child with hemophilia

	X^A	Y
X^a	X^AX^a	X^aY
X^a	X^AX^a	X^aY

No, male and female children do not have the same chance of getting the disease. Male children are more likely to have hemophilia since they do not possess a second allele for this trait to mask a recessive allele that they may inherit.

5. The terms "homozygous" and "heterozygous" refer to pairs of alleles for a given gene. Since a male has only one copy of any X-linked gene, it does not make sense to use these pair-related terms.

6. The disease-causing allele is recessive and is carried by both parents, because although neither the mother nor the father expresses the trait in question, some of their children do. The disease-causing allele is located on an autosome. If it were on the X chromosome, the father would express the gene, since we have already determined that he must carry one recessive copy of the gene, and he would not have another copy of the gene to mask this recessive allele. Both individuals 1 and 2 of generation I have the genotype *Dd*.

7. The disease-causing allele is recessive. Neither individual number 1 nor number 2 in generation II exhibits the disease phenotype, yet they have a child who does. The disease-causing allele is located on the X chromosome. If individual 1 in generation II is not a carrier, as stated, then the only way he can have a child (individual 2, generation III) that expresses this recessive trait is if that child does not carry a second gene for the trait, which is true for males relative to X-linked traits. Otherwise, not being a carrier, he (individual 1, generation II) would have two copies of the dominant allele, and each of his children would express the dominant non-disease-causing phenotype no matter what allele they got from their mother. This is not what the pedigree shows for his offspring in generation III. His son inherited his Y chromosome from his father and happened to get his mother's recessive allele for the trait, which is expressed since the son does not have a second allele to mask it. The daughters inherit a dominant non-disease-causing allele from their father, so it does not matter which of their mother's alleles they inherit. They will still not show the disease phenotype, as shown by the pedigree. In addition, if individual 6 in generation II is not a carrier as stated, then she will give a dominant non-disease-causing allele to every child, male or female, so none of them would express the disease phenotype, even though their

father contributes a recessive disease-causing allele to each daughter (and no allele for this trait to his sons). This is consistent with the pedigree.

8. a. If the two genes are completely linked:

	AB	ab
aB	AaBB	aaBb

b. If the two genes are on different chromosomes:

	AB	Ab	aB	ab
aB	AaBB	AaBb	aaBB	aaBb

Chapter 14

1. The two strands of the double helix follow specific base-pairing rules. This means that when the two strands are separated, each can act as a template for a new DNA molecule. Each of the two new DNA molecules will have one original strand and one newly synthesized strand.

2. An allele is made up of a precise sequence of DNA bases. If this sequence is changed by mutation, a new allele has been produced. This new allele may no longer produce the correct gene product. These incorrectly made products can cause human genetic disorders. An example of mutation causing a genetic disorder is the defective hemoglobin molecule, which is caused by a single base mutation; this mutation is the cause of sickle-cell anemia.

3. Cancers are caused by uncontrolled cell division. If a mutation in a gene that controls cell division is not repaired due to a defective (mutated) repair protein, the cell may begin to divide rapidly and uncontrollably, thus causing cancer.

Chapter 15

1. A gene is a segment of DNA that codes for a particular molecule of protein or RNA. Although most genes code for proteins, there are genes that produce a final product of rRNA or tRNA.

2. rRNA is a component of ribosomes. Ribosomes make the bonds that link amino acids together to make proteins during the process of translation. During translation, tRNA transfers the correct amino acid to the ribosome. Each tRNA has a specific anticodon that matches with an mRNA codon that specifies the particular amino acid that should next be added to the growing protein. mRNA specifies the order of amino acids in a protein. During translation, tRNA will "read" the mRNA codons and transfer the correct amino acid onto the growing protein molecule.

3. Transcription: A DNA molecule is partially unwound so that the RNA polymerase enzyme can have access to the base pairs of a given gene. The RNA polymerase matches complimentary bases to the exposed DNA sequence, thus forming an RNA molecule. When the RNA polymerase reaches a DNA terminator sequence, transcription ends, the new RNA molecule is released, and the DNA strands bond back to each other. Translation: mRNA molecules contain the information

necessary to build proteins. A molecule of mRNA binds to a ribosome, and translation begins at the start codon nearest to where the mRNA is bound to the ribosome. A given tRNA binds to a specific mRNA codon and also to a specific amino acid. As the ribosome moves one codon at a time, different tRNAs bind to the mRNA codons and the ribosome links together the amino acids brought by the tRNAs. This is how the sequence of codons on mRNA is translated to a sequence of amino acids in a protein.

4. tRNA must be able to bind to both the mRNA codon and the amino acid to ensure that the amino acid being added to the protein chain is the correct one as coded by the mRNA codon.

5. A gene is a portion of a DNA molecule that contains information for the synthesis of a protein or RNA final product. That information is transcribed to a complimentary RNA molecule. In the case of RNAs that have protein products, the specific order of bases in the mRNA codes for the order of amino acids in the gene's protein product. The protein product is what determines a phenotype (for example, the presence of brown pigment in the iris of an eye).

Chapter 16

1. a. Although all the cells of a given organism contain the same DNA, different genes are expressed in different cells. The expression of different genes can affect cell structure.

b. Although all the cells of a given organism contain the same DNA, different genes are expressed in different cells. This allows different cells to perform various metabolic tasks.

2. Crossing-over occurs more often between genes that are located further apart. Thus, long sequences of non-coding DNA between genes should increase the frequency of recombination between the genes.

3. The tryptophan operator controls whether or not the gene that codes for tryptophan is transcribed. If tryptophan is present, a repressor protein is able to bind to the operator and prevent transcription, since the presence of tryptophan indicates that the cell does not need to waste energy by making more. If tryptophan is absent, the repressor is unable to bind to the operator, thus allowing the gene to be transcribed so that the tryptophan needed by the cell is produced. This example follows the general eukaryotic and prokaryotic pattern of gene control by regulation of transcription. Regulatory DNA sequences such as the tryptophan operator can switch genes on and off. In order to do this, the regulatory DNA must interact with regulatory proteins, such as the tryptophan repressor protein. Control of transcription determines whether or not the gene product is expressed.

4. The advantage of studying groups of genes together is that it gives a more accurate picture of how they interact in a living organism. The disadvantage is that studying several genes at once makes it harder to determine which genes are having what effect, or which combination of genes is having that effect.

Chapter 17

1. a. DNA cloning is the process of isolating a gene and making many copies of it. Two methods of cloning a gene are the construction of a DNA library and the use of the polymerase chain reaction (PCR) technique. To construct a gene library, the DNA of a given organism is digested into fragments by a restriction enzyme. These fragments are inserted into a vector that transfers the gene into bacteria. As these bacteria reproduce, they copy the inserted gene along with their own genes, thus cloning the gene. In PCR, short primers that can bond with the target DNA are mixed with that DNA. Heat is used to separate the strands of the organism's DNA so that the primer can bond to each target area. DNA polymerase then fills in a complementary strand of DNA for each of the separated strands that have a primer attached. In this way, many copies of the gene can be made; thus, it is cloned.

   b. The advantage of gene cloning is that it easier to study a gene once you have many copies of it. The cloned gene can then be sequenced or transferred to other organisms or used in various experiments.

2. To screen a DNA library for a particular gene, a probe for the gene of interest is needed. The DNA of the bacterial colonies that make up the library is tested to see if they can base-pair with the probe. If they can, they contain the gene of interest.

3. Some people argue that humans do not have the right to alter the DNA of other organisms and, hence, that all forms of genetic engineering are not ethical. However, many people think that it may be ethical to alter the DNA of all organisms, depending on the reasons for doing it. If altering the DNA of a child allowed that child to overcome a genetic disease and live a healthy life, many people would support such a use of DNA technology. If the genes were being altered merely for cosmetic reasons, many people would find such changes objectionable.

4. Many people would argue that altering the DNA of an individual suffering from a genetic disease should be legal. However, since altering DNA in any way is a serious and not always predictable matter, limits should be placed on the kinds of traits that DNA technology is used to alter. One way to draw the line between "unacceptable" and "acceptable" changes to the DNA of humans would be to allow changes to genes that cause genetic disorders but not to genes that control cosmetic features.

Chapter 18

1. Some people want genetic testing done that would reveal whether a fetus would suffer such a disease. These families may use that information to plan for the future of the child and their family. Others may use the information to terminate the pregnancy and prevent the suffering that would otherwise ensue. Some people do not want the information from such testing because they fear a bad result. Still others, particularly those who would not opt to terminate the pregnancy either way, find the information useless.

2. Some believe that testing for incurable diseases should not be allowed under any circumstances because the results steal all hope from those who discover they have the disease. The same people argue that even for those lucky enough to learn that they are free of the disease, there is a burden of guilt for having escaped what relatives must endure. Others believe that the new technology gives all people the right to know their genetic fate, good or bad. Many say such knowledge allows a person to plan more wisely for the future.

3. Some believe that people—whether parents, employers, or insurers—have the right to whatever information the new technologies afford them. Others, for religious and other ethical reasons, object to the use of such knowledge. Many believe that employers and others do not have the right to genetic information about their children, employees or customers, whatever its potential use. Others worry that the potential for the abuse of that information is too great a risk to take, despite possible gains.

Chapter 19

1. Life on Earth is characterized by adaptations (a match between organisms and their environment), by a great diversity of species, and by many examples in which very different organisms share morphological or other characteristics. Each of these aspects of life on Earth can be explained by evolution: (a) Adaptations result from natural selection, (b) the diversity of life results from speciation, which occurs when one species splits to form two or more species, and (c) shared characteristics of life are due to common descent.

2. Overwhelming evidence indicates that evolution is a fact. Support for evolution comes from five lines of evidence:

   a. The fossil record provides clear evidence of the evolution of species over time and documents the evolution of major groups of organisms from previously existing organisms.

   b. Organisms contain evidence of their evolutionary history. For example, when anatomical data alone are used to determine evolutionary relationships, scientists find that the proteins and DNA of closely related organisms are more similar than those of organisms that do not share a recent common ancestor. In this and many similar examples, the extent to which organisms share characteristics other than those used to determine evolutionary relationships is consistent with scientists' understanding of evolution.

   c. Scientists' understanding of evolution and continental drift has allowed them to correctly predict the geographic distribution of certain fossils depending on whether the organisms evolved before or after the breakup of Pangea.

   d. Scientists have gathered direct evidence of small evolutionary changes, in thousands of studies, by documenting genetic changes in populations over time.

   e. Scientists have observed the evolution of new species from previously existing species.

3. In any area of science, new pieces of information are still being added to our knowledge of the subject. The debate between scientists as to which mechanisms are most important only means that evolution is not fully understood, not that it does not occur.

4. Genetic drift has a greater effect on smaller populations. If the bee population were larger, the likelihood that all bees of

a certain type would die in one winter would be smaller, and the dramatic shift in the frequency of the *A* and *a* plant alleles would therefore also be less likely.

Chapter 20

1. Nonrandom mating occurs when an organism is more likely to mate with members of a portion of the population than with individuals chosen at random from the population at large. Gene flow is the exchange of genes between populations. Gene flow makes the genetic makeup of populations more similar to one another. Genetic drift is a process in which alleles are sampled at random over time. Genetic drift can have a variety of causes, such as chance events that cause some individuals to reproduce and other individuals not to reproduce. Natural selection is a process in which individuals with particular heritable characteristics survive and reproduce at a higher rate than other individuals.

2. Potential benefits include providing a larger population, which is therefore less susceptible to chance events and genetic drift, and providing an input of new alleles on which natural selection can operate. Drawbacks include the introduction of individuals with genotypes that are not well-matched to the local environmental conditions of the smaller population. Throughout time, some species have gone extinct locally or worldwide. This is a natural process. However, humans have increased the rate that populations and species have become extinct. If there are numerous other populations of a species, it may not be worth introducing new members to the smaller population. If the smaller population is one of the few populations of that species left, then it may be more important to introduce new individuals in an attempt to allow the population to recover and survive.

3. The genotype frequencies for the original population are:

$$AA: \frac{280}{280 + 80 + 60} = 0.67$$

$$Aa: \frac{80}{280 + 80 + 60} = 0.19$$

$$aa: \frac{60}{280 + 80 + 60} = 0.14$$

The allele frequencies for this population are:

$$p = 0.67 + \frac{0.19}{2} = 0.765 \text{ and}$$

$$q = 0.14 + \frac{.019}{2} = 0.235$$

Note that the sum of the frequencies of the 2 $2p$ allele + q allele = 1.0

Hardy–Weinberg predicts that the frequency of genotype *AA* should be

$$p^2 = (0.765)(0.765) = 0.585;$$

that the frequency of genotype *Aa* should be

$$2pq = (2)(0.765)(0.235) = 0.360;$$

and that the frequency of genotype *aa* should be

$$q^2 = (0.235)(0.235) = 0.055.$$

Note that the sum of the genotype frequencies,

$$p^2 + 2pq + q^2 = 1.0$$

These calculated genotype frequencies do not match those of the original population. This difference could be due to mutation, nonrandom mating, gene flow, a small population size, and/or natural selection.

Chapter 21

1. Classifying species by their inability to sexually reproduce with other species is a convenient definition, but there are many nonreproductive alleles that could cause these organisms to be different enough that they can be classified as separate species even though they can produce hybrids.

2. Some of the populations in Lake Victoria may have had so little contact with one another that they evolved into separate species in geographic isolation, despite the fact that they lived in the same lake. Other populations may have evolved into new species in the absence of geographic isolation.

3. Species that interbreed in nature may still be distinct species due to a host of nonreproductive alleles that may cause distinctions. For this reason, many people would argue that the rare species is separate from the common species and should remain classified as rare and endangered.

Chapter 22

1. A mass extinction may be associated with rapid environmental changes that have no relation to the conditions that favored a particular adaptation. Thus, organisms with wonderful adaptations can (and have) become extinct during mass extinction events.

2. Although speciation can happen within a single year, it often takes hundreds of thousands to millions of years to occur. Thus, it is not surprising that it usually takes one to seven million years for the number of species found in a region to rebound after a mass extinction event. The time required to recover from mass extinction events provides a powerful incentive for humans to halt the current, human-caused loss of species; otherwise, it will take millions of years for the number of species on Earth to recover.

3. Microevolution refers to the changes in allele or genotype frequencies within a population of organisms. This is fundamentally different than macroevolution, which deals with the rise and fall of entire groups of organisms, the large-scale extinction events that cause some of these changes, and the evolutionary radiations that follow extinction events. Macroevolutionary changes cannot be predicted solely from an understanding of the evolution of populations. Of the evolutionary processes studied in Unit 4, speciation occupies a "middle ground" between macroevolution and microevolution.

Chapter 23

1. Human prejudices might be decreased if individual humans looked more alike. However, a great diversity of cultural riches would be lost.Nonrandom mating, natural selection, and genetic drift all can prevent gene flow from making the genetic make-up of human populations more similar.

2. Classification systems are constructed so that we can better understand and speak about various organisms, but the separations we make between organisms are sometimes somewhat arbitrary. Although we classify ourselves in a separate genus from the three species of chimpanzees, a hypothetical observer from outer space might well classify humans and chimpanzees as members of the same genus.

3. Information about genes that cause differences between humans and chimpanzees may lead to better understanding of both species. However, many people would argue that we lack the perspective or ethical right to create "designer chimps."

Chapter 24

1. The plasma membrane serves two related functions essential to the survival of cells: 1) it selectively filters biologically important molecules entering and leaving the cell, and 2) it allows the cell to maintain an internal environment that is both suitable for life's chemistry and different from the external environment. Destroying the plasma membrane disrupts both of these functions, killing the bacterium.

2. Evolution teaches that all living things descended from a common ancestor. This idea is supported by evidence compiled from biogeography, fossil records, comparative anatomy, and molecular biology. Therefore, it comes as no surprise that different species share features.

3. There are too many examples of interactions between functions to list all of them here. The two following examples are familiar ones that illustrate the nature of such interactions: 1) When we eat, we rely on our eyes, nose, and tongue (senses) to identify food; muscles (movement) transport the food through our digestive system (nutrition). Once the food is digested and the nutrients are absorbed, the blood (internal transport) carries the nutrients to the metabolizing cells. 2) When we run, we rely on our eyes and sense of balance (senses) to keep us on our path. Muscles (movement) in our legs actually propel us. The oxygen and nutrients needed for activity are carried from our lungs (gas exchange) and guts (nutrition) to the muscles by our circulatory system (internal transport). Our nervous system helps to coordinate all of these activities (internal communication).

4. Although trees consist of relatively few cell types, these cell types are arranged into a variety of tissues that serve specific functions such as fluid conduction, support, photosynthesis, and active transport. These tissues combine in various ways into three tissue systems. Different arrangements of the three tissue systems lead to roots specialized for absorption, stems specialized for support and transport, and leaves specialized for photosynthesis.

5. A single-celled organism must be a generalist, performing all the functions needed to survive. On the other hand, cells in the multicellular human body are specialists. Although they can perform certain functions very efficiently, they must rely on other cells to meet some of their needs. This requires a speedy and coordinated system of communication and transport.

Chapter 25

1. Allometric relationships that relate the size of an organism to its form or function describe many aspects of an organism's biology. We will list just a few of the effects here; a careful search through the chapter will reveal more. The smaller budgies would have a larger surface area–to–volume ratio than the larger lovebirds. This means that all sorts of functions involving exchange, such as losing body heat and water, and absorbing nutrients, might be faster in the budgie than in the lovebird. Budgies would probably sleep longer than lovebirds. A sick budgie might require more medication relative to its body weight than would a sick lovebird. The lovebird would need more food and make more mess than a budgie, but would probably eat less than five times as much food and make less than five times as much waste. The budgie could probably get by in a smaller cage than the lovebird.

2. The size of a cell is limited by its ability to exchange with its environment a sufficient amount of material to support cell function. As cell size increases, the surface area across which exchange can take place decreases relative to the volume of the cell. Thus, cells that are too big cannot supply their needs across the relatively small surface area provided by their plasma membrane.

3. If a single cell or organ performed several functions, the species could not survive if that cell or organ became modified to do just one thing. Duplication of cells and organs is necessary before specialization can occur so that the various functions can be divided amongthem.

Chapter 26

1. Lignin resist stretching under tension. Growth of the cell would cause the surrounding cell wall to stretch. If growing cells had cell walls rich in lignin, the cell wall would interfere with cell growth.

2. Biologically produced molecules and mineral materials have complementary properties that together produce effective support structures. Lightweight, biologically produced materials generally resist breakage very well under tension, but are often not very stiff and may not resist compression well. Heavy mineral materials, on the other hand, are stiff and resist compression, though they often break relatively easily under tension. Biologically produced and mineral materials combine to form support structures, such as our own bones, which resist bending, compression, and tension remarkably well.

3. Organisms that need relatively little support (such as many marine invertebrates) can get by with a hydrostat and can take advantage of the hydrostat's light weight. Hydrostats

also provide a flexible support system without requiring joints, and they allow organisms to vary their stiffness. The hydrostat that supports earthworms takes advantage of both of these properties to allow the worm to move through the soil.

Chapter 27

1. In a microscopic view of muscle, a sarcomere is an area between two Z discs. Actin filaments extend from each Z disc toward the center of the sarcomere. Myosin filaments lie between actin filaments and have heads which, in the presence of ATP, can bind to the actin, change shape, and in the process, pull the Z discs closer together. This causes the sarcomere to shorten or contract.

2. The arrangement of stiff support structures into lever systems allows organisms to effectively increase or decrease the strength or speed of a muscle contraction. If the muscle that moves a limb is located near the joint that acts as the fulcrum in the lever system, the muscle contraction leads to relatively slow movement with more force at the end of the load arm of the lever, where the work is done. Alternatively, if the muscle is located relatively far from the joint, the muscle contraction leads to relatively rapid but weak movement at the end of the load arm. Strength is increased if many muscle fibers work side by side in a bulky muscle. Slow contraction of a muscle often allows the muscle to generate a stronger contraction, whereas rapid contractions lend speed but little strength.

3. a. The walking elephant would experience less pressure drag and less friction drag than the swimming whale. Water is both denser and more viscous than air, greatly increasing both the pressure and friction drag experienced when moving at a given speed. In addition, whales swim more rapidly than elephants walk, which will also cause them a further increase in the pressure and friction drag that whales face.

   b. The flying falcons have to deal with a lot of pressure drag, but relatively little friction drag, whereas the swimming penguins have to deal with a lot of both types of drag. Although air is less dense than water, falcons fly fast enough so that pressure drag becomes an issue, much as it does for the aquatic penguins. For this reason, both have streamlined shapes. The greater viscosity of water ensures that friction drag remains a much bigger problem for penguins than for falcons.

4. Fungi lack the muscles, cilia, and flagella that allow other organisms to move. In addition, they have rigid cell walls that restrict movement. To some extent, fungi spread as their rootlike hyphae grow through their food. Much more important, however, are the tiny spores which float long distances in even gentle air currents. Fungi spread primarily because their spores can exploit the movement of the air around them.

Chapter 28

1. There are several possible reasons why pitcher plants supplements the mineral nutrients that they obtain from the soil with nutrients absorbed from the insects that they catch,

including the following: 1) The soil in which pitcher plants grow may not provide enough nutrients for them to survive. 2) The roots with which pitcher plants absorb minerals from the soil may not provide enough surface area with which to absorb an adequate supply of mineral nutrients. 3) Pitcher plants may have evolved carnivory because this allows them access to richer nutrient sources than are available to most plants. The natural history of pitcher plants shows that Reason 1 is most likely to explain carnivory in pitcher plants. Pitcher plants, like almost all other carnivorous plants, live in extremely nutrient-poor soils.

2. If humans fed exclusively on plant leaves, our digestive systems might have evolved differently from the ones that we have in several ways: 1) Our teeth would have been better suited to grinding up the tough cell walls of plant tissues. Most herbivores have well-developed grinding teeth with broad, flat surfaces. 2) Our small intestine would have evolved to provide a greater surface area for absorption to compensate for the low nutrient content in most plant tissues. Most herbivorous mammals have very long small intestines that allow them to absorb most of the nutrients from plant tissues. 3) If we fed on very nutrient-poor plant tissues, we may have evolved a mutualistic relationship with microbes to help break down the cellulose in the cell walls of plant tissues. The microbes often live in specialized structures in the guts of animals, as is the case in the modified stomachs of sheep.

3. To survive and grow, we need to obtain essential nutrients such as vitamins, minerals, and amino acids from our food. The balance of these nutrients in our own bodies differs from those in the foods that we eat, forcing us to mix and match foods so that our intake corresponds to what we need. Vitamins are essential for our nutritional well-being, but we must obtain them from our food. Most foods contain only a few of the vitamins that we need. Similarly, not all foods contain all of the amino acids essential for our survival

Chapter 29

1. The lungless salamander would need to exchange gases through its skin without water loss. Therefore, it would need a relatively humid environment to keep its skin moist.

2. When we begin to exercise, our muscles start to contract more vigorously. As a result, the demand for oxygen, particularly in the muscles, increases. We supply this increased oxygen demand by increasing the rate at which we breathe in and out. This keeps the oxygen concentration in our alveoli high, allowing for the rapid diffusion of oxygen into our blood. An increase in our heart rate once we start exercising increases the speed with which the oxygen in the blood, bound to hemoglobin, moves to the active muscles. The rapid movement of oxygenated blood past the metabolizing muscle cells ensures that the oxygen can diffuse rapidly into the muscle cells.

3. The major advantages of obtaining oxygen from air instead of water are, first, that air contains a greater proportion of oxygen than does water, and, second, that air offers much less resistance to ventilation or to moving over the gas exchange surface. An air-breathing aquatic mammal gets the oxygen

that it needs from relatively few liters of air, as compared to the number of liters of water that a hypothetical, gill-bearing mammal would have to process to get an equivalent amount of oxygen. The animals that rely on gills, moreover, must have a large gill surface area in contact with relatively viscous water (Chapter 27), which means that gills represent a potentially great source of friction drag. The air-breathing mammal, with internal lungs, has no such problem.

Chapter 30

1. The highest blood pressure (approximately 120 mm Hg) occurs when the left ventricle contracts, pumping blood into arteries to be carried throughout the body. When this ventricle relaxes and refills, the pressure drops to about 80 mm Hg. Blood pumped from the right ventricle goes to the lungs. Since it has a shorter distance to travel, less pressure needs to be generated: 25 mm Hg drops to 8 mm during relaxation. As the blood passes into capillaries, friction drag lowers the pressure to 35 mm Hg. Pressure of blood leaving the smaller-diameter capillaries drops to 10 mm Hg. Resistance to blood flow is increased when the diameter of a vessel is decreased, slowing down the blood flow. Blood that must travel uphill to return to the heart depends on skeletal muscles to push it upward against gravity. Valves in the veins keep it from flowing backward. The consistent decrease in blood pressure throughout the human circulatory system is necessary to allow the blood to complete a circuit of the circulatory system by always flowing from a region of higher blood pressure to a region of lower blood pressure.

2. A lower demand for oxygen from your running muscles would lead to a reduced rate at which the circulatory system supplies the oxygen. After sitting down following your run, your heart rate would decrease, reducing the blood pressure generated by the ventricles. Because the steepness of the pressure gradient in the circulatory system determines how fast the blood flows, the drop in ventricular pressure leads to a decreased blood flow. In addition, local blood, and therefore oxygen, supply is adjusted by the opening and closing of the blood vessels that supply tissues. By reducing the diameter of the blood vessels feeding your running muscles, these muscles would receive less blood once you stopped running.

3. Advantages of a closed circulatory system: It allows more control over where the exchange of gases and nutrients occurs. Organs can be large because blood goes through the organs instead of flowing into spaces around them. Disadvantages of a closed circulatory system: A strong pump is needed to supply enough pressure to push blood through vessels, overcome friction drag, and lift blood against the flow of gravity. This requires energy.

4. The plant vascular system must lift sap under high tensions and pressures, and it must accomplish this without the muscle tissue that propels blood through animal circulatory systems. Natural selection can only act on available variation. Because only animals have muscle tissue, and the chances of a mutation in plants leading to a similar contractile tissue are slim, the evolution of internal transport systems in plants has followed a very different path. Plants rely on different means

of moving fluid: osmotic pressure moves the phloem sap, and tension generated by evaporation of water from the leaves' surfaces moves the xylem sap. Although the movement of phloem sap requires active transport, and therefore the expenditure of energy on the part of the plant, xylem transport is accomplished using the sun's energy. Even if a plant were to evolve a muscle-like tissue capable of pumping fluid, it is not clear that natural selection would favor the spread of this feature, given the pressures that plants would need to generate and that their current xylem transport requires little energy expenditure on the part of the plant.

Chapter 31

1. With organisms, such as endotherms, that produce large amounts of metabolic heat, this form of heat can figure prominently in their heat budgets. Thus, endotherms such as humans rely heavily on metabolic heat as a heat input. Because organisms generate metabolic heat by spending energy, they can conserve energy by using other heat sources, including conductive and radiant heat inputs, to reduce the amount of metabolic heat they must generate to maintain body temperatures. Heat loss in organisms that depend on metabolic heat can be costly. Consequently, endotherms tend to minimize the conductive, convective, and radiative heat loss to the environment with various insulating body covers, including fat, fur, and feathers.

2. Organisms do not regulate internal temperature or water when they cannot or when they obtain little benefit from doing so. Aquatic organisms generally do not regulate their body temperature. The high rate of conductive heat loss in water makes it virtually impossible to maintain a body temperature that differs much from the surrounding water. Organisms that live in a very constant environment often stand to gain little by having the machinery to regulate body temperature. Bodies of water often change in temperature little, if at all, from day to night, or even over the course of a year. Therefore, even though aquatic organisms cannot regulate body temperature, they suffer little, because their body temperature—like that of their environment—remains constant. Marine animals have solute concentrations in body water that closely match the concentration of salts in their environment. Thus, marine organisms have little need to regulate their body water.

3. Unlike ectotherms, humans do not rely on environmental heat sources to maintain their body temperature. This comes at a great energetic cost that results from the energy used to generate the metabolic heat. Thus, humans must eat much more, relative to their body weight, then ectotherms. Endothermy, however, allows us to maintain constant body temperatures and levels of activity during hot days and cool nights and during hot summers and cold winters.

Chapter 32

1. Recognizing their own tissues as different from the cells of invaders makes it possible for organisms to protect themselves against diseases that invade their body. In the absence of an ability to distinguish self from non-self, organisms

could only kill invading cells by killing many of their own cells as well.

2. A dead or weakened disease organism in the vaccine has identifying proteins, called antigens, on its surface. A few of our white blood cells have surface proteins that can recognize these antigens. B lymphocytes are stimulated to reproduce themselves rapidly and produce antibodies, which are proteins that can circulate in the blood and neutralize a specific antigen. These antibodies remain in the bloodstream and can attack quickly if the virulent form of the organism enters the body. T lymphocytes are present in two forms: helper T cells, which allow other white blood cells (neutrophils and macrophages) to attach to and engulf the antigen, and killer T cells, which consume any cells damaged by the invader. Once stimulated by the vaccine, both B and T lymphocytes will "remember" and respond faster to a second exposure to the antigen.

3. Single-celled organisms defend themselves against invaders primarily by preventing invasion. Only by keeping bacteria or viruses from entering their cell can single-celled organisms prevent the consequences of disease. In addition to defenses that make it difficult for invaders to penetrate their body, multicellular organisms can mount a number of defenses should the invaders get in. Nonspecific defenses are widespread, and allow multicellular organisms to selectively destroy foreign cells inside their body. In addition to the general defense, vertebrates have an immune system that acts as a specific defense that can distinguish between different invading cells and mount a response appropriate to each individual disease.

Chapter 33

1. Auxins suppress the growth of side branches, while cytokinins promote growth of side branches. Auxins, cytokinins, and gibberellins stimulate growth of fruit, but they slow ripening. ABA and ethylene promote ripening, but can also slow plant growth under stressful conditions, thus conserving energy.

2. A menstrual cycle begins as an egg develops in response to estrogen levels that increase over a two-week period. The high estrogen level triggers the pituitary gland to release follicle-stimulating hormone and luteinizing hormone. These two hormones act together to promote the release of an egg from the ovary. The high luteinizing hormone levels stimulate the release of another hormone, progesterone, that prepares the uterus for embryo implantation following the release of the egg. The high progesterone levels cause the pituitary gland to stop releasing follicle-stimulating hormone and luteinizing hormone. If the egg is not fertilized by a sperm, the progesterone level decreases after about 12 days. Without progesterone in the system, the uterine lining sloughs off and estrogen levels once again begin to increase.

3. Both neurotransmitters and hormones are released by one set of cells and act on a second set of target cells. Both neurotransmitters and hormones work at very low concentrations. Hormones differ from neurotransmitters in that they are found in organisms other than animals. In addition, hormones tend to travel to the target cells through an organism's internal transport system, whereas neurotransmitters diffuse minute distances across a synapse to act on a target nerve cell. Hormones generally act on a longer time scale and are broken down more slowly than neurotransmitters.

Chapter 34

1. The advantage of a reflex arc response is its immediacy in an emergency situation. Signals filtered through the brain require longer response time but are needed for more complex responses. Example: A reflex action would let you pull away from a pinprick, but further processing by the cerebral cortex would allow you to inject a needed vaccine or medication, such as insulin.

2. In response to an action potential traveling down the axon, the dendrites of a neuron release chemical neurotransmitters from their tips. The neurotransmitters are released into the gap, or synapse, separating the neuron from adjacent neurons. Because of the narrow width of the synapse, the neurotransmitter diffuses rapidly. Receptors on the dendrites of adjacent neurons bind the neurotransmitter. Binding of the neurotransmitter to the receptors elicits a response in the second neuron. In some cases, this response is a new action potential that then travels the length of the second neuron's axon. In other cases, the binding of the neurotransmitter inhibits the formation of an action potential.

3. In image-forming eyes, a lens gathers and focuses light that reflects off of an object on to light-sensitive cells, which are gathered together into a retina. The interaction between light energy and light-sensitive chemicals inside the light-sensitive cells triggers an action potential that travels down nerve cells to the brain. The brain then interprets the nerve signals coming from various parts of the retina as an image.

Chapter 35

1. Behavior in animals depends on the interaction of the nervous, muscular, and endocrine systems. Of these systems, the nervous and muscular systems are unique to animals. The nervous system allows animals to quickly integrate sensory information, and to coordinate a movement in response by stimulating contraction of the appropriate muscles.

2. Evidence indicates that the ability to learn depends both on opportunities to learn from others or from experience (environmental effects), and from a genetic predisposition to learn (genetic effects). Organisms that live in groups or spend some time growing up with their parents can learn by imitating the other individuals of their own species. Organisms that live alone from birth or hatching can learn by remembering and selectively repeating only those behaviors that allow them to better find food or shelter or mates. For organisms to learn, however, they must have a nervous system and brain that is sufficiently well developed to allow them to interpret and remember what they have experienced. The development of these structures is under genetic control. In addition, species capable of learning often tend to excel in learning the types of behaviors most important to their success. Thus, organisms that hide away food throughout their territory are extremely good at remembering where things are.

3. Fixed behaviors that might be important to group members might involve responses to stimuli about which an organism might never have a chance to learn. Thus, a group-living animal might have a fixed behavior that led it to avoid snake-like objects or to hide at the sight of a silhouette that resembles that of a predatory bird. There might not be a chance to learn the appropriate behaviors from others, because one incorrect response could easily be fatal. Behaviors associated with surviving during the first few days following birth or hatching, before the creature has a chance to learn, are fixed behaviors.

4. Communication can make group life possible, it can help it find and evaluate a mate, it can help it learn from others. Most group-living organisms spend much time remaining in contact with each other through calls of various sorts. The calls can also inform group members of danger or of the discovery of food. Courtship behavior in humans, birds, frogs, and insects often involves elaborate sequences of actions that communicate the quality of one individual to the other. Learning in humans and other primates depends strongly on the ability of parents and other family members to communicate with the young.

Chapter 36

1. Humans produce eggs and sperm in distinct tissues (the gonads) within each individual. Plants have complex life cycles that involve the production of eggs and sperm by separate individuals called gametophytes. In flowering plants, these gametophyte individuals live within the tissues of the sporophyte that produced them, which makes it look on the surface as though the sporophyte produces the eggs and sperm within tissues.

2. Flowers attract insects, birds, and mammals. These animals may accumulate pollen on their bodies when they come into contact with the flower. This pollen can be carried long distances and eventually dropped off onto a flower of the same species. Wind, however, will not carry the pollen far, so without the diverse methods of pollination the flowering plant uses, wind-pollinated plants must live in close proximity to others of their species.

3. The structure of the female parts of a flower can be modified in many ways to increase competition among male gametes. One prominent mechanism is to increase the length of the style that separates the stigma (where the pollen lands) from the ovaries (where the embryo sacs reside). Lengthening the style lengthens the race between individual pollen tubes that grow down through the style tissue, making it easier for small differences in pollen quality to translate into earlier arrival of the winning sperm. A second approach is to increase the number of pollen grains landing on the stigma at any time by encouraging animal pollination. Having many pollen grains on the stigma at once is like having a large field of runners in the race. The best of a large field of runners is likely to be better than the best in a small field of runners. All of this male–male competition benefits the female by increasing the likelihood that the sperm that fertilizes the embryo sac is of high quality.

4. Aquatic habitats allow for external fertilization of eggs. Sperm and eggs can survive in seawater for some time because seawater has a solute concentration that resembles that inside the sperm and egg. As environments become more challenging (e.g., freshwater with much lower solute concentrations than those inside the sperm and eggs, and the dry terrestrial environment), the life span of sperm and unfertilized eggs outside the organism is much reduced. This means that sperm must either be deposited outside of the organism directly onto the eggs so that fertilization can take place quickly, or fertilization must take place internally. Most terrestrial organisms rely on internal fertilization.

Chapter 37

1. The increasingly narrowly defined fates of cells as development progresses help to ensure that fewer developmental mistakes will be made. A cell that retains a completely unlimited fate could give rise to descendant cells of any of a number of functional types. Such cells, if present late in development, could lead to organs that are supposed to fulfill a particular function containing a variety of functionally unrelated cells.

2. During development, the fates of cells depend on the identity of neighboring cells, their position within the developing organism as a whole, and the stage of development. They gain this information from a variety of chemicals released by other cells in the developing organism, including hormones and morphogens. In some cases, the presence or absence of the chemicals provides the necessary information; in others it is their concentration.

3. Small genetic changes during development can lead to large changes in form or function. Differences in human and chimpanzee head shapes may result in large part from timing of events during development of the head. The loss of webs in chicken feet results from the action of a single gene during development.

Chapter 38

1. When humans became bipedal, the modified shape of the pelvic opening that serves as a birth canal became relatively smaller than in four-legged primates. However, the increase in brain size made the human baby's skull size greater than the diameter of the pelvic aperture. To allow the baby to maneuver through the opening, female hips evolved to became correspondingly wider from front to back. Also, when birth is imminent, a hormone, relaxin, causes ligaments holding hipbones together to stretch. These modifications were not needed for early bipeds because of their smaller brains.

2. Large brains in early primates may well have evolved as a way of processing the information needed to move through the three-dimensional world of the trees. Alternatively, the complex seasonal and distributional patterns of the fruits that provided food for many primates may have selected for the evolution of larger brains. In humans, the rapid evolution of brain size seems to be linked to tool use and the use of language for communication.

3. A milestone in the evolution of the human hand has been the evolution of an opposable thumb suitable for a precision grip. This hand structure was made possible by the evolution of an upright posture, which freed the hand from the constraints of having to be used in locomotion. Having hands freed for tool use may have created selective pressures that favored the evolution of larger brains: The precise manipulations made possible by human hand structure require sophisticated integration of sensory information and fine muscle movements. This integration takes place in the neocortex that makes up much of the human brain. In addition, tool use and creation also required sophisticated intelligence.

Chapter 39

1. The potential or natural location of a biome is where conditions are such that the biome could, in principle, be found, while the actual location is where the biome currently is found. The potential locations of terrestrial biomes are most strongly influenced by climate, particularly by temperature and precipitation, while the potential locations of aquatic biomes are strongly influenced by their terrestrial biomes and by climate. The actual locations of both terrestrial and aquatic biomes are strongly influenced by human actions.

2. Atmospheric wind convection cells and oceanic water currents carry the results of local events (a volcanic eruption or an oil spill) to distant areas around Earth. For example, oil spilled into an ocean current next to one continent's shore may be carried by that current and end up coating the shores of other continents. This distant event can alter local ecological interactions when shorebirds are killed when they become coated with oil from that spill. In such a case, the birds may no longer keep certain organisms (their food) under control, and they are no longer available as a food source for their predators.

3. Climate can directly exclude species from a region, as when the lack of rainfall excludes many plant and animal species from desert biomes. Climate also can exclude species from a region indirectly, as when a species that can tolerate the climate is excluded by other organisms that are better adapted to the region's climate.

Chapter 40

1. An organism needs food and water, not too many predators, shelter, and acceptable climate conditions. Both the physical and biotic environments, and the interaction between them, will have an influence on the suitability of a given habitat for a given organism. Disturbances such as fire, flood, earthquakes, windstorms, and droughts will also determine whether a given habitat is suitable for a given organism.

2. Species distribution relies heavily on habitat distribution, and habitats are not evenly distributed throughout Earth's environment. In addition, plant distribution is often clumped due to the fact that plants cannot move to disperse themselves, and their offspring can only get so far away. This factor may cause animals that feed on these plants to have a distribution matching that of the plants.

3. If a species evolved before the continents began drifting apart, it is likely to have a wider distribution since it may have existed on each (current) landmass before they separated. A species that evolved later may be more isolated, since it might not have the ability to disperse the great distances between continents. In addition, the differences in climate between the continents is greater now than when they were closer together, and while organisms that existed on the supercontinent had a long amount of geologic time to adapt to these changes, organisms that evolved later might have had to be able to tolerate a more sudden change in climate as they attempted to colonize a new continent.

4. Studying the reproductive and social habits of the organism would help determine whether the distribution was, for instance, naturally clumped (as plants or family groups), or whether the offspring travel widely but cannot survive in other habitats. The fossil record for a species would help determine whether it evolved before or after the breakup of the continents. Finally, experiments could be performed to examine the impact of other species. For example, a species could be moved to a region where it does not occur and competitor species could either be removed or left alone to see if the competitors or local environmental conditions prevented the species from living there.

Chapter 41

1. Coordinates for graph, in the notation (x coordinate, y coordinate):

(1, 150) (2, 225) (3, 337.5) (4, 506.3) (5, 759.4)

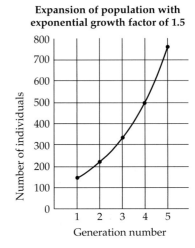

Expansion of population with exponential growth factor of 1.5

2. a. Some factors that limit population growth include available habitat, available food and water, disease, weather, natural disturbances, and predators.

   b. Species new to an area often do not have an established predator. In addition, they have not yet reached the carrying capacity of their habitat.

3. A density-dependent factor is one whose intensity increases as the density of the population increases. An example of a density-dependent factor is the fact that disease spreads more rapidly in densely populated areas. A density-inde-

pendent factor is not affected by the density of the population. Temperature is a density-independent factor. If the temperature drops below what a certain kind of plant can tolerate, it does not matter how dense that population of plants is; they will still die. Fire is also a density-independent factor.

4. If the pattern of population growth is understood, managers may be able to manipulate or protect the factors that most directly affect the population growth rate. If a population of organisms is more successful because, for example, it has adequate access to water, a manager would be sure that nearby rivers are not drained off for agricultural needs.

5. Limit reproduction to no more than one child per parent; reduce the consumption of unnecessary goods; reuse and recycle items to promote sustainable use of resources; work to develop and follow environmentally friendly policies and activities (for instance, using energy-efficient cars and light bulbs); purchase goods that have a lower impact on the environment, including organically grown clothing and food items.

Chapter 42

1. Mutualisms are common because their costs to organisms are outweighed by the benefits they provide organisms. Yucca plants may lose a few seeds to the offspring of their moth pollinators, but they still end up with more seeds than if the moth had not mediated pollination to begin with.

2. The inferior competitor is still using resources that the superior competitor needs, thus possibly limiting its distribution and/or abundance.

3. The plant community would most likely change to have fewer species. When the rabbits are removed, the grass that is no longer being eaten can assert its dominance as the superior competitor and will probably drive some of the other species in that area to extinction.

Chapter 43

1. a. Interactions among species, such as competition, mutualism, and consumer–victim interactions, have a large effect on the distribution and abundance of species found in a community. Some species, called keystone species, have a particularly strong effect in that, relative to their abundance, they have a disproportionately large impact on the type and abundance of other species in the community.

b. Disturbances such as fire occur so often in many ecological communities that the communities are constantly changing in response to a previous disturbance event. Depending on the type and severity of disturbance, a given community may or may not be able to recover from a particular disturbance event.

c. The climate is a main factor in determining what organisms can live in a given area. If the climate changes, the habitat changes, leading to changes in the success of various organisms in that area.

d. The effect of continental drift on ecological communities is largely based on the climate changes that occur as the continents move to different longitudes (or areas in a different wind or water current pattern).

2. Primary succession occurs in a newly created habitat. Secondary succession is the process by which communities recover from disturbance. The newly created habitat on which primary succession occurs can result from a "disturbance," such as the retreat of a glacier, but such disturbances are of a more catastrophic nature than those found in secondary succession.

3. The addition of beard grass to Hawaii has increased the rate and area of fires on the island. This is due to the large amount of dry matter the grass produces, which burns more easily and hotter than the native plants. In this way, one species in the community has altered the frequency of fire.

4. Change is a part of all ecological communities. However, human-caused change is unique in that we can consider the impact of our actions and we can decide whether or not to take the actions that cause community change. Whether or not a particular change is viewed as ethically acceptable will depend on the type of change, the reason for it, and the perspective of the person evaluating the change. For example, a person might find it ethically acceptable to alter a region so as to produce a long-term source of food for the growing human population, yet not ethically acceptable to take actions that result in short-term economic change but cause long-term economic loss and ecological damage.

Chapter 44

1. Energy captured by producers from an external source such as the sun is stored in the bodies of producers in chemical forms, like carbohydrates. At each step in a food chain, a portion of the energy captured by producers is lost from the ecosystem as metabolic heat. This steady loss prevents energy from being recycled.

2. Decomposers are essential because they recycle nutrients from dead organisms to the physical environment, where the nutrients may again be used by another organism.

3. Nutrients can cycle on a global level. When sulfur dioxide pollution is put into the atmosphere in one area of the world, wind convection cells move that pollution around the world where it can affect other ecosystems.

4. Internal agreements on inputs to nutrient cycles can in principle provide protection to the economic interests of nations (and they have in practice). Such agreements may also help nations realize that their actions can have far-reaching effects.

5. Human economic activity is interwoven with several key ecosystem services. Pollination is essential for the production of crop (and other) plants. Floodplains act as safety valves for major floods if we do not build on them or separate them from the body of water they help control, and wetlands and estuaries act as cost-free waste filtration systems. We rely on nutrient cycling to keep us alive. When ecosystem services such as these are damaged, human economic interests are damaged as well.

Chapter 45

1. Major types of global change caused by humans include land and water transformation, changes to the chemistry of Earth (for example, changes to nutrient cycles), increases in biological invasions, and increases in the extinction rate of species.

2. Human-caused global changes often happen at a much more rapid rate than changes due to natural causes. The speed of continental drift or natural climate change is much slower than the measurable changes humans have caused to carbon dioxide levels in the atmosphere or to the amount of nitrogen fixed. Also, humans have a choice about the global changes we cause.

3. Some examples of actions that are good for the global environment: Reduce the quantity of nonfood items purchased; reuse items until they are no longer usable; buy used items rather than getting everything new. Recycle paper, plastic, glass, and metal. Bring reusable cloth bags when shopping. Rarely use paper cups, plates, or towels. Plant trees and other native plants, especially those that help feed the native wildlife. Reduce water use by not leaving water running when brushing teeth, by adjusting the water level of washing machines to match the size of the load, and by using water-saving fixtures. Reduce fossil fuel use by choosing a gas-efficient car and by using household heating and air conditioning only as needed. Use compact fluorescent light bulbs, and turn off lights that are not in use. Support organic farmers by purchasing organically grown food. Some examples of actions that are harmful for the global environment: Use nonrecyclable products and products that are not from recycled sources. Drive alone rather than carpooling or taking public transportation. Use a gas- or electric-powered lawn mower. Make excessive purchases of nonessential items (gadgets, many sets of clothes, extra cars, vacation homes, etc.). Drive a low-mileage vehicle, such as a sport utility vehicle or truck, out of choice, not necessity.

4. For people to have a sustainable impact on Earth, we must reduce the rate of growth of the human population and we must reduce the (per person) rate of resource use. To achieve these goals, many aspects of human society would have to change. For example, our views of nature would have to change from one of a limitless source of goods and materials that can be exploited for short-term economic gain to one that accepts limits and seeks always to take only those actions that can be sustained for long periods of time. Many specific actions would follow from such a change in our view of the world, such as: an increase in recycling; the development and use of renewable sources of energy; a decrease in urban sprawl; an increase in the use of technologies with low environmental impact (such as organic farming); and a concerted effort to halt the ongoing extinction of species.

Photography Credits

SPL = Science Photo Library
SS = Science Source
NASC = National Audubon Society
   Collection

Table of Contents *Jellyfish*: © Fred Bavendam/Minden Pictures. *Frogs*: © Kitchin & Hurst/Tom Stack & Assoc. *Dividing bacterium*: © Dr. Tony Brain/ SPL/SS/Photo Researchers, Inc. *E. coli DNA*: © Dr. Gopal Murti/SPL/SS/ Photo Researchers, Inc. *Wolves*: © Art Wolfe. *Meadow*: © Barbara Gerlach/Dembinsky Photo Assoc. *Chimpanzee family*: © Erwin & Peggy Bauer/Tom Stack & Assoc.

Chapter 1 *protesters*: © Diana Walker/Time Magazine. 1.2: Don Gettinger, courtesy K. C. Larson. 1.3: Ash Kaushesh, courtesy K. C. Larson. 1.4: © Stan Osolinsky/Dembinsky Photo Assoc. 1.4 *inset*: © Gail Nachel/ Dembinsky Photo Assoc. 1.5: © Inga Spence/ Tom Stack and Assoc. 1.7*a*: © Jany Sauvanet/ NASC/Photo Researchers, Inc. 1.8*b*: © John Eastcott/YVA Momatiuk/Planet Earth Pictures. Box: © CORBIS/Roger Ressmeyer.

Chapter 2 *foraminiferan*: © Dennis Kunkel, University of Hawaii. 2.3 *chlamydia*: © Dr. Kari Lounatmaa/SPL/SS/Photo Researchers, Inc. 2.3 *archaebacteria*: © Dr. Kari Lounatmaa/SPL/SS/Photo Researchers, Inc. 2.3 *Borrelia*: © David M. Phillips/SS/Photo Researchers, Inc. 2.3 *Streptomyces*: © Dr. Jeremy Burgess/SPL/SS/Photo Researchers, Inc. 2.3 *E. coli*: © Biology Media/SS/Photo Researchers, Inc. 2.5: © Breck P. Kent/Earth Scenes. 2.6: © Krafft/Hoa-qui/Photo Researchers, Inc. 2.7: © Claudia Adams/ Dembinsky Photo Assoc. 2.8 *seaweed*: © Science VU/Visuals Unlimited. 2.8 *Paramecium*: © M. Abbey/Visuals Unlimited. 2.8 *Plasmodium*: © Omikron/SS/Photo Researchers, Inc. 2.8 *dinoflagellate*: © Dennis Kunkel, University of Hawaii. 2.8 *diatoms*: © Biophoto Assoc./SS/Photo Researchers, Inc. 2.10 *ferns*: © Greg Vaughn/Tom Stack & Assoc. 2.10 *sequoia*: © Tom Stack/Tom Stack & Assoc. 2.10 *mosses*: © Michael P. Gadomski/ NASC/Photo Researchers, Inc. 2.10 *Rafflesia*: © Compost/Visage/Peter Arnold, Inc. 2.10 *orchids*: © Kjell B. Sandred/Visuals Unlimited. 2.13*a*: © Inga Spence/Tom Stack & Assoc. 2.13*b*: © Georgette Douwma/Planet Earth Pictures. 2.14 *Pilobolus*: © Cabisco/ Visuals Unlimited. 2.14 *stinkhorns*: © Sharon

Cummings/Dembinsky Photo Assoc. 2.14 *Penicillium*: © Dennis Kunkel, University of Hawaii. 2.16: © Noble Proctor/SS/Photo Researchers, Inc. 2.17: © Dr. Edward S. Ross. 2.18: © John Shaw/Tom Stack & Assoc. 2.19 *jellyfish*: © Brian Parker/Tom Stack & Assoc. 2.19 *sponges*: © Tom Zuraw/Animals Animals. 2.19 *clam*: © Dave Fleetham/Tom Stack & Assoc. 2.19 *flatworm*: © Newman & Flowers/NASC/Photo Researchers, Inc. 2.19 *fire worm*: © Susan Blanchet/Dembinsky Photo Assoc. 2.19 *sea star*: © Copr. F. Stuart Westmorland/NASC/Photo Researchers, Inc. 2.19 *Morpho*: © G. W. Willis/Animals Animals. 2.19 *frogs*: © Brian Kenney/Planet Earth Pictures. 2.19 *fish*: © Andrew J. Martinez/NASC/Photo Researchers, Inc. 2.19 *kangaroos*: © Esther Beaton/Planet Earth Pictures. 2.19 *chimpanzees*: © Zig Leszczynski/ Animals Animals. 2.24 *worms*: © Steve Hopkin/Planet Earth Pictures. 2.24 *abdomen*: © Michael Neveux.

Chapter 3 *ice man*: © Paul Hanny/Gamma Liaison. 3.1*a Prince William and Prince Harry*: © AP/Wide World Photos. 3.1*a adults*: © Tim Graham/Corbis Sygma. 3.1*b*: James S. Miller. 3.5*a*: Mick Ellison.

Chapter 4 *golden toads*: © CORBIS/Michael & Patricia Fogden. 4.1: © Mark Moffett/ Minden Pictures. 4.5: © Greg Vaughn/Tom Stack & Assoc. 4.6*a*: David McIntyre. 4.6*b*: © Bernd Witich/Visuals Unlimited. 4.7*a*: © Jim Battles/Dembinsky Photo Assoc. 4.7*b*: © Jack Jeffrey. 4.8: Harald Pauli. 4.9: J. H. Lawton/ Centre for Population Biology. Box *canopy*: © Michael Fogden/Earth Scenes. *dirigible*: © L. Pyot/OCV-PN/Eurelios.

Chapter 5 *Mars life*: NASA. 5.9*b*: © Biophoto Assoc./SS/Photo Researchers, Inc. 5.9*c*: © CNRI/SPL/SS/Photo Researchers, Inc.

Chapter 6 *Listeria*: Justin Skoble and Daniel A. Portnoy. 6.3 *envelope*: © Dennis Kunkel, University of Hawaii. 6.3 *nucleus*: © Dr. Gopal Murti/SPL/SS/Photo Researchers, Inc. 6.4 *smooth ER*: © David M. Phillips/ Visuals Unlimited. 6.4 *rough ER*: © K. G. Murti/Visuals Unlimited. 6.6: © Dennis Kunkel, University of Hawaii. 6.7 *upper*: © David M. Phillips/Visuals Unlimited. 6.7 *lower*: © Dr. Gopal Murti/SPL/SS/Photo Researchers, Inc. 6.8: © Biophoto Assoc./ SS/Photo Researchers, Inc. 6.9: © Bill Longcore/SS/Photo Researchers, Inc. 6.10: ©

Dr. Kari Lounatmaa/SPL/SS/Photo Researchers, Inc. 6.11*a*: © K. G. Murti/ Visuals Unlimited. 6.11*b*: Dr. Mark McNiven. 6.12: Dr. Louise Cramer. Box *left*: © Nancy Kedersha/Immunogen/SPL/SS/Photo Researchers, Inc. *right*: © Dennis Kunkel, University of Hawaii.

Chapter 7 *kit*: Photo by Frank Simonetti, courtesy Dr. Michael Echols, Medical Antique Collectors (www.antiquelures.com/ medical/med.htm). Box *runner*: © M. Gratton/ Vision/SS/Photo Researchers, Inc. *mouse*: © Stephen Dalton/Animals Animals. *nematode*: © Sinclair Stammers/SPL/SS/Photo Researchers, Inc.

Chapter 8 *brain*: © Alexander Tsiaras/SS/ Photo Researchers, Inc. 8.3: © Dr. Jeremy Burgess/SPL/SS/Photo Researchers, Inc. 8.8*a*: © E. Brenckle/Explorer/SS/Photo Researchers, Inc. 8.8*b*: © Denton Taylor.

Chapter 9 *Vibrio cholerae*: © Dennis Kunkel, University of Hawaii. 9.1 *alveoli*: © Oliver Meckes/SS/Photo Researchers, Inc. 9.1 *capillary*: © SPL/SS/Photo Researchers, Inc.

Chapter 10 *blood cells*: © Andrew Syred/ SPL/SS/Photo Researchers, Inc. 10.1: © Peter Skinner/SS/Photo Researchers, Inc. 10.4*a*: © Alfred Pasieka/Peter Arnold, Inc. 10.4*b*: © Leonard Lessin/Peter Arnold, Inc. 10.5: Andrew S. Bajer, Univ. of Oregon. Box: © Andrew Paul Leonard/SS/Photo Researchers, Inc.

Chapter 11 *adenovirus*: © P. Webster/Custom Medical Stock Photo. 11.2: © K. G. Murti/ Visuals Unlimited. 11.6: © Abramson/ Servodidio.

Chapter 12 *tabby*: Steve Cousins/Visual Concepts Photography. 12.1: The Mendelianum. 12.7*a*: © Elisabeth Weiland/ NASC/Photo Researchers, Inc. 12.7*b, c*: Carol Worsham. 12.8: © NCI/Photo Researchers, Inc. 12.9: Walter Chandoha. *Reginald Punnett*: Courtesy the Master and Fellows of Gonville and Caius College, Cambridge.

Chapter 13 *chromosome 4*: Dr. Marcy McDonald. 13.10: Courtesy Dr. Pragna Patel. 13.11 *child*: Dr. George Herman Valentine. 13.11 *chromosomes*: Dr. Irene Uchida. *Thomas Morgan*: © Corbis Bettman.

Chapter 14 *DNA model*: © Ken Eward/ Biografx/SS/Photo Researchers, Inc. 14.8: © Kenneth Greer/Visuals Unlimited. 14.9 *basal cell*: © Paul Parker/SPL/SS/Photo Researchers, Inc. 14.9 *melanoma*: © Kenneth Greer/Visuals Unlimited. Box: Courtesy the Principal and Fellows of Newnham College, Cambridge.

Chapter 15 *tRNA model*: © Ken Eward/SS/ Photo Researchers, Inc. 15.7: © Ken Eward/ SS/Photo Researchers, Inc. 15.10 *left*: © Dr. Tony Brain/SPL/SS/Photo Researchers, Inc. 15.10 *right*: © Meckes/Ottawa/SS/Photo Researchers, Inc.

Chapter 16 *lamb*: Lynn James, USDA Poisonous Plant Research Laboratory. 16.5: Dr. F. Rudolf Turner. 16.10: Joseph L. DeRisi, Vishwanath R. Iyer, and Patrick O. Brown. Box: Cold Spring Harbor Laboratory Archives.

Chapter 17 *tobacco*: © Keith V. Wood/ Science VU/Visuals Unlimited. 17.5: PE Biosystems, Foster City, CA. 17.6: Huntington Potter, U. South Florida, and David Dressler, U. Oxford. 17.10: Purdue University Entomology Department. 17.11: © Ted Thai/Time Magazine.

Chapter 18 *autoradiograph*: © SIU/Visuals Unlimited. 18.2: © Abramson/Servodidio. 18.3: © Nick Kelsh/Peter Arnold, Inc. 18.4 *left*: © Custom Medical Stock Photo. 18.4 *right*: Courtesy Cooley-Dickinson Hospital, Mammography Dept. 18.6a: © Richard Hutchings/SS/Photo Researchers, Inc. 18.6b: © Biophoto Assoc./SS/Photo Researchers, Inc.

Chapter 19 *Galápagos*: © Breck P. Kent/Earth Scenes. 19.1: © Michael Fogden/Oxford Scientific Films/Animals Animals. 19.2: © M. W. F. Tweedie/NASC/Photo Researchers, Inc. 19.8: Dr. Peter Boag.

Chapter 20 *HIV model*: © Graphico/SPL/ SS/Photo Researchers, Inc. 20.1: © Darrell Gulin/Dembinsky Photo Assoc. 20.6: © Dominique Braud/Tom Stack & Assoc. 20.10: Thomas Bates Smith.

Chapter 21 *stream*: David Reznick. 21.1: Erick Greene. 21.2: © Dr. Edward S. Ross. 21.3a: © Bill Beatty/Visuals Unlimited. 21.3b: © Robert A. Lubeck/Animals Animals. 21.5a: © Joe McDonald/Visuals Unlimited. 21.5b: © Ana Laura Gonzalez/Animals Animals. 21.6: Dr. William Boecklen. 21.8: Jason Rick, Courtesy Loren Rieseberg.

Chapter 22 *landscape*: © Rod Planck/ Dembinsky Photo Assoc. *inset plants/lichens*: Dr. David G. Campbell. 22.1a, b: © Stanley M. Awramik/Biological Photo Service. 22.1c: Courtesy Niles Eldredge. 22.1d: © B. Miller/ Biological Photo Service. 22.1e: © David M. Dennis/Tom Stack & Assoc.

Chapter 23 *Neanderthal*: The Neanderthal Museum. 23.4a: © John Reader/SPL/ SS/Photo Researchers, Inc. 23.4b: © Manoj Shah/Planet Earth Pictures. 23.6c: Jean-Jacques Hublin, reprinted by permission from *Nature* 394:721 © 1998 Macmillan Magazines Ltd. 23.6d: © Tom McHugh/SS/Photo Researchers, Inc. 23.9: © David Perrett & Duncan Rowland, Univ. of St. Andrews/ SPL/SS/Photo Researchers, Inc.

Chapter 24 *lawn*: © Robert M. Ballou/Earth Scenes. 24.3: © Reuters/Herb Swanson/ Archive Photos. 24.5: © Don Fawcett/ SS/Photo Researchers, Inc. 24.8: © Bruce Brander/SS/Photo Researchers, Inc.

Chapter 25 *Kong*: © CORBIS/Bettmann. 25.1 *bacterium*: © Dennis Kunkel, University of Hawaii. 25.1 *rotifer*: © Roland Birke/Peter Arnold, Inc. 25.1 *aspen*: © Adam Jones/ Planet Earth Pictures. 25.1 *sponge*: © Marilyn Kazmers/Dembinsky Photo Assoc. 25.4: © Robert Winslow/Animals Animals. 25.5: © Jeffrey L. Rotman/Peter Arnold, Inc. 25.6a: © David Phillips/Visuals Unlimited. 25.6b: © Manfred Kage/Peter Arnold, Inc. 25.6c: © Alex Rakosy/Dembinsky Photo Assoc. 25.6d: © Stan Osolinski/Dembinsky Photo Assoc.

Chapter 26 *falling cat*: © Gerard Lacz/Peter Arnold, Inc. 26.1: © Dave Fleetham/Tom Stack & Assoc. 26.2a: Christopher Small. 26.2b: © Martyn F. Chillmaid/SPL/NASC/ Photo Researchers, Inc. 26.3a: © Lothar Lenz/OKAPIA/NASC/Photo Researchers, Inc. 26.3b: © William J. Weber/Visuals Unlimited. 26.3b *inset*: © Bill Beatty/Visuals Unlimited. 26.4: © Biophoto Assoc./ SS/Photo Researchers, Inc. 26.5b: © Carl Purcell/SS/Photo Researchers, Inc. 26.5c: © Ted Levin/Earth Scenes. 26.6a: © Flip Nicklin/Minden Pictures. 26.6b: © François Gohier/NASC/Photo Researchers, Inc. 26.6c: © Ralph Reinhold/Animals Animals. 26.6d: © Carr Clifton/Minden Pictures. 26.8 *Marasmius*: © Richard L. Carlton/NASC/ Photo Researchers, Inc. 26.8 *Amanita*: © Martin H. Chester/NASC/Photo Researchers, Inc. 26.10: © Reuters/Pierre duCharme/Archive Photos.

Chapter 27 *sprinter*: © L. S. Stepanowicz/ Visuals Unlimited. *sperm*: © Dennis Kunkel, University of Hawaii. 27.1a: © Gerard Lacz/Peter Arnold, Inc. 27.1b: © Dennis Kunkel, University of Hawaii. 27.1c: © Stephen P. Parker/NASC/Photo Researchers, Inc. 27.2: © Quest/SPL/SS/Photo Researchers, Inc. 27.4: © Larime Photographic/Dembinsky Photo Assoc. 27.5: © Dr. Blanche C. Haning/The Lamplighter. 27.7a: © Ray Wolfe/NASC/Photo Researchers, Inc. 27.7b: © Peter Scoones/Planet Earth Pictures. 27.7c: © Hans Pfletschinger/Peter Arnold, Inc. 27.8: © John Cancalosi/Peter Arnold, Inc. 27.9 *bacteria*: © Dennis Kunkel, University of Hawaii. 27.9 *flagellum*: © Julius Adler/

Visuals Unlimited. 27.10a: © Dennis Kunkel, University of Hawaii. 27.10c: © Dr. Gopal Murti/SPL/SS/Photo Researchers, Inc. 27.11a: © Hans Pfletschinger/Peter Arnold, Inc. 27.11b: © Chris Howes/Planet Earth Pictures.

Chapter 28 *skeletons*: Brian Seed. 28.5 *leaf*: © Dr. John D. Cunningham/Visuals Unlimited. 28.5 *root*: © Dennis Kunkel, University of Hawaii. 28.6a: © William J. Weber/Visuals Unlimited. 28.6b: © Tom J. Ulrich/Visuals Unlimited. 28.6c: © Carroll W. Perkins/ Animals Animals. 28.6d: © Bill Lea/ Dembinsky Photo Assoc. 28.7a: © David M. Dennis/Tom Stack & Assoc. 28.7b: © Ken Greer/Visuals Unlimited. 28.10 *upper*: © Michael P. Gadomski/Dembinsky Photo Assoc. 28.10 *lower*: © M. I. Walker/SS/Photo Researchers, Inc. 28.12: © Ken Greer/Visuals Unlimited. Box *cow*: © Mitsuaki Iwago/ Minden Pictures. *ants*: © Dr. Edward S. Ross.

Chapter 29 *vulture*: © Joe McDonald/ Animals Animals. *Andean*: © Pete Oxford/ Planet Earth Pictures. 29.4: © Manfred Kage/ Peter Arnold, Inc. 29.6: © David Sieren/ Visuals Unlimited. 29.9: © Tom McHugh/ NASC/Photo Researchers, Inc. Box *roots*: © David M. Dennis/Tom Stack & Assoc. *Rhizobium*: © John Innes Institute/SPL/ SS/Photo Researchers, Inc.

Chapter 30 *giraffes*: © Jonathan Scott/Planet Earth Pictures. 30.2: © Dennis Kunkel, University of Hawaii. 30.11: © Gerald Van Dyke/Visuals Unlimited. Box: © Fred Bavendam/Peter Arnold, Inc.

Chapter 31 *cactus*: © John Lemker/Earth Scenes. 31.1 *sauna*: © Lee Snyder/Photo Researchers, Inc. 31.1 *mole rats*: © Raymond A. Mendez. 31.2 *cells*: Roger S. Pearce and Edward N. Ashworth. 31.2 *snake*: © Leonard Lee Rue III/NASC/Photo Researchers, Inc. 31.3 *left*: © Stan Flegler/Visuals Unlimited. 31.3 *center, right*: © David M. Phillips/ Visuals Unlimited.

Chapter 32 *poster*: Nan Sinauer. 32.2a: © Inga Spence/Tom Stack & Assoc. 32.2b: © Christian Grzinek/Okapia/SS/Photo Researchers, Inc. 32.3: © Ed Reschke/Peter Arnold, Inc. 32.7: © Donald Specker/Earth Scenes. 32.8 *cat*: © Martin Rugner/Planet Earth Pictures. 32.8 *caterpillar*: © Stephen J. Krasemann/Nature Conservancy/ NASC/Photo Researchers, Inc. 32.8 *king snake*: © David M. Dennis/Tom Stack & Assoc. 32.8 *coral snake*: © Zig Leszczynski/ Animals Animals. 32.9 *turtle*: © Jeff Lepore/NASC/Photo Researchers, Inc. 32.9 *cacti*: © Dr. Edward S. Ross. 32.10 *caterpillar*: Hans Damman. 32.10 *butterflies*: Dr. Thomas Eisner. 32.11a: © S. J. Krasemann/Peter Arnold, Inc. 32.11b: © Nigel Dennis/NASC/ Photo Researchers, Inc.

Chapter 33 *spider*: © J. P. Jackson/NASC/ Photo Researchers, Inc. *frog*: © Zig Leszczynski/Animals Animals. *cigarette*: © Science VU/Visuals Unlimited. 33.4a: © Laurence Pringle/NASC/Photo Researchers, Inc. 33.4b: © C. C. Lockwood/Earth Scenes. 33.6: © Bernard, G. I. OSF/Animals Animals. 33.9: © Patrice Ceisel/Stock, Boston/PNI. 33.10 *summer fox*: © Russ Gutshall/ Dembinsky Photo Assoc. 33.10 *winter fox*: © Jim Roetzel/Dembinsky Photo Assoc.

Chapter 34 *amputee*: © Reuters/Jeff J. Mitchell/Archive Photos. 34.2: © Dennis Kunkel, University of Hawaii. 34.5: © P. Motta/Photo Researchers, Inc. Box: Dr. Leonard E. White and Dr. David Fitzpatrick, Duke University Medical Center.

Chapter 35 *swallowtail*: © S. McKeever/ NASC/Photo Researchers, Inc. 35.4a: © Jan Robert Factor/NASC/Photo Researchers, Inc. 35.4b: © Ray Richardson/Animals Animals. 35.8a: © Christopher Crowley/Tom Stack & Assoc. 35.8b: © Stan Osolinski/ Dembinsky Photo Assoc. 35.8c: © Michael J. Minardi/Peter Arnold, Inc. 35.9a: © Anup & Manoj Shah/Planet Earth Pictures. 35.9b: © Thomas Kitchin/Tom Stack & Assoc. 35.9c: © Glen Woolfenden. 35.10: © Gunter Ziesler/ Peter Arnold, Inc.

Chapter 36 *jack-in-the-pulpit*: © Nada Pecnik/ Visuals Unlimited. 36.1: © Alan & Linda Detrick/NASC/Photo Researchers, Inc. 36.1 *inset*: © Dr. Blanche C. Haning/The Lamplighter. 36.2: © CNRI/SPL/SS/Photo Researchers, Inc. 36.3: Dr. Thomas Eisner. 36.4: © Dennis Kunkel, University of Hawaii. 36.5: © Hans Pfletschinger/Peter Arnold, Inc. 36.6 *fern*: © Walter H. Hodge/Peter Arnold, Inc. 36.6 *sporangia*: © Milton Rand/ Tom Stack & Assoc. 36.6 *gametophyte*: © Biophoto Assoc./SS/Photo Researchers, Inc. 36.7 *lily*: © Ed Reschke/Peter Arnold, Inc. 36.7 *embryo sac*: © Dr. John D. Cunningham/ Visuals Unlimited. 36.7 *pollen*: © David Scharf/ Peter Arnold, Inc. 36.8a: © Carl R. Sams II/Peter Arnold, Inc. 36.8b: © Paul Stevens/ Planet Earth Pictures. 36.8c: © Kjell B. Sandved/Visuals Unlimited. 36.9a: © Hans Pfletschinger/Peter Arnold, Inc. 36.9b: © Anthony Mercieca/NASC/Photo Researchers, Inc. 36.10a: © Y. Arthus-Bertrand/Peter Arnold, Inc. 36.10b: Dr. Thomas Quinn. 36.10c: © Ed Reschke/Photo Researchers, Inc. 36.11a: © Alan G. Nelson/Animals Animals. 36.11b: © Skip Moody/Dembinsky Photo Assoc. Box: © Robert Fox/Impact Visuals/PNI.

Chapter 37 *Dolly*: Roddy Field/Roslin Institute. 37.4: © Deutsche Presse/Archive Photos. 37.5 *shepherd's purse*: © Stubben/ Visuals Unlimited. 37.5 *seed*: © Biophoto Assoc./SS/Photo Researchers, Inc. 37.6: © Dr. John D. Cunningham/Visuals Unlimited.

37.7b: © Ray Ellis/NASC/Photo Researchers, Inc. 37.7c: © Greg Gawlowski/Dembinsky Photo Assoc. 37.10: © Dr. Blanche C. Haning/ The Lamplighter. 37.11 *daisy*: © John Shaw/ Tom Stack & Assoc. 37.11 *snapdragon*: © John Sohlden/Visuals Unlimited. 37.12 *chicken feet*: © Dowling, R. OSF/Animals Animals. 37.12 *duck feet*: © William S. Paton/Planet Earth Pictures. 37.12 *developing chicken*: Lee Niswander.

Chapter 38 *Australopithecus*: The Natural History Museum, London. 38.1: © Renee Lynn/NASC/Photo Researchers, Inc. 38.2 *gibbon*: © Brian Kenny/Planet Earth Pictures. 38.2 *man*: © Dr. Edward S. Ross. 38.4: © Paul Hirata/Stock Connection/PNI. 38.8: © SIU/ SS/Photo Researchers, Inc.

Chapter 39 *fire*: © H. John Maier/Image Works/Time, Inc. 39.8a: © Doug Locke/ Dembinsky Photo Assoc. 39.8b: © Michael P. Gadomski/Earth Scenes. 39.8c: © Willard Clay/Dembinsky Photo Assoc. 39.8d: © Earl Scott/NASC/Photo Researchers, Inc. 39.8e: © Willard Clay/Dembinsky Photo Assoc. 39.8f: © Francis Lepine/Earth Scenes. 39.8g: © John Shaw/Tom Stack & Assoc. 39.10a: © Terry Donnelly/Tom Stack & Assoc. 39.10b: © Carr Clifton/Minden Pictures. 39.10c: © Neil McIntyre/Planet Earth Pictures. 39.10d: © Kevin Magee/Tom Stack & Assoc. 39.10e: © Jim Zipp/NASC/Photo Researchers, Inc. 39.10f: © Linda Pitkin/Planet Earth Pictures. 39.10g: © Chet Tussey/NASC/Photo Researchers, Inc. 39.10g *inset*: © Gregory Ochocki/NASC/Photo Researchers, Inc. 39.10h: News Office, Woods Hole Oceanographic Institution. 39.11: NASA.

Chapter 40 *starling flock*: © Edward Coleman/ Planet Earth Pictures. 40.1 *myna*: © Toni Angermayer/NASC/Photo Researchers, Inc. 40.1 *starling*: © Susan & Allan Parker ARPS/ Planet Earth Pictures. 40.2 *starflower*: © Farrell Grehan/NASC/Photo Researchers, Inc. 40.2 *meadow*: © Dr. Blanche C. Haning/ The Lamplighter. 40.2 *aster*: © Adam Jones/ Planet Earth Pictures. 40.3 *creosote bush*: © Ed Darack/Planet Earth Pictures. 40.3 *trees*: © Michael P. Gadomski/NASC/Photo Researchers, Inc. 40.4: © Thomas D. Mangelsen/Peter Arnold, Inc. 40.5: © COR- BIS/Wolfgang Kaehler. 40.6: © Carr Clifton/ Minden Pictures. 40.7: © Carr Clifton/ Minden Pictures. 40.9: © Gerry Ellis/Minden Pictures. Box: © Frans Lanting/Minden Pictures.

Chapter 41 *moai*: © George Holton/Photo Researchers, Inc. 41.1: © Dr. Jeremy Burgess/ SPL/NASC/Photo Researchers, Inc. 41.1 *inset*: © Volker Steger/SPL/SS/Photo Researchers, Inc. 41.3: Reproduced with per- mission of the Department of Natural Resources, Queensland, Australia. 41.4: © Scott Camazine/NASC/Photo Researchers,

Inc. 41.7: © Mark Newman/Tom Stack & Assoc. 41.8: © Alan G. Nelson/Dembinsky Photo Assoc. 41.11: © Fritz Prenzel/Peter Arnold, Inc.

Chapter 42 *ant*: © Gregory G. Dimijian/ NASC/Photo Researchers, Inc. *skin disease*: © Science VU/Visuals Unlimited. *smut*: © Inga Spence/Tom Stack & Assoc. 42.2: Dr. Olle Pellmyr. 42.3: © Fred Bavendam/Peter Arnold, Inc. 42.5a: © Mike Bacon/Tom Stack Assoc. 42.5b: Dr. Lincoln Brower. 42.6: © Fred Bruemmer/Peter Arnold, Inc. 42.9: J. K. Clark/ University of California, Davis.

Chapter 43 *rainforest*: © Greg Vaughn/Tom Stack & Assoc. 43.3: Dr. Robert T. Paine. 43.5: Dr. Peter Marks. 43.6: © Dr. Johann Goldammer. 43.7: Robert Gibbens, Jornada Experimental Range, USDA. 43.8: Dr. Gerald D. Carr.

Chapter 44 *forest*: © Michael Fogden/Earth Scenes. 44.8a: © Dr. John D. Cunningham/ Visuals Unlimited. 44.9: © Dr. John D. Cunningham/Visuals Unlimited. 44.10a: © Bob Galbraith/AP Photo. 44.10b: Steven Holt.

Chapter 45 *ocean*: © Carr Clifton/Minden Pictures. *abalone*: Ron McPeak. 45.1a: © D. Dancer/Peter Arnold, Inc. 45.1b: © Georg Gerster/Photo Researchers, Inc. 45.1c: USGS. 45.1d: NASA. 45.3: NASA. 45.6: Dr. David Tilman. 45.9: Joy Ward and Anne Hartley. Box: © Frans Lanting/Minden Pictures.

Illustration and Text Credits

The following illustrations are, in full or in part, *from* Purves, W. K., Orians, G. H., Heller, H. C., and Sadava, D. 1998. *Life: The Science of Biology*, Fifth Edition. Sinauer Associates, Inc., Sunderland, MA, and W. H. Freeman and Co., New York: 2.1, 2.4, 19.7, 21.7, 27.2, 28.3, 29.1, 30.8, 31.5, 31.6, 32.3, 33.1, 34.1, 34.3, 34.6, 34.9.

The following illustrations are, in full or in part, *after* Purves, W. K., Orians, G. H., Heller, H. C., and Sadava, D. 1998. *Life: The Science of Biology*, Fifth Edition. Sinauer Associates, Inc., Sunderland, MA, and W. H. Freeman and Co., New York: 1.11, 3.3, 5.8, 10.5, 14.1, 16.2, 16.6, 17.8, 19.9, 20.7, 20.10, 22.3, 23.8, 27.10, 28.4, 30.2, 30.3, 33.3, 33.8, 34.4, 34.5, 35.7, 42.8.

Chapter 3 3.5b,c: © Edward Heck, American Museum of Natural History.

Chapter 4 4.2: After data from E. O. Wilson, 1992. *The Diversity of Life.* The Belknap Press of Harvard University Press, Cambridge, MA.

Chapter 13 13.7: from *Time*; data from Dr. Victor A. McKusick, Johns Hopkins University.

Chapter 19 Display quotation from Richard C. Lewontin: As cited in Futuyma, D. 1995. *Science on Trial*, p. 161. Sinauer Associates, Inc., Sunderland, MA.

Chapter 20 20.8: after Curtis, C. F., Cook, L. M., and Wood, R. J. 1978. "Selection for and against insecticide resistance and possible methods of inhibiting the evolution of resistance in mosquitoes." *Ecological Entomology* 3: 273–287. 20.9: after Karn, M.N. and Penrose, L.S., 1951. "Birth weight and gestation time in relation to maternal age, parity, and infant survival." *Annals of Eugenics* 16: 147–164. 20.11: after Gould, J. H. and Keeton, W. T. 1996. *Biological Science*, Sixth Edition. W. W. Norton and Co., New York.

Chapter 21 21.4: after Clutton-Brock, T., Guinness, F. E. and Albon, S. D. 1983. "The costs of reproduction to red deer hinds." *Journal of Animal Ecology* 52: 367–383.

Chapter 22 22.4: © D. W. Miller. 22.9: after Hickman, C. P., Jr. and Roberts, L. S. 1994. *Biology of Animals.* William C. Brown, Dubuque, IA. 22.10: after Benton, M. J. 1995. "Diversification and extinction in the history of life." *Science* 268: 52–58.

Chapter 23 Table 23.1: after Ruff, C. B., Trinkaus, E., and Holliday, T. W. 1997. "Body mass and encephalization in *Pleistocene Homo.*" *Nature* 387: 173–176; Strickberger, M. W. 1996. *Evolution.* Jones and Bartlett, Boston; Jurmain, R., Nelson, H., and Trevathan, W. 1997. *Introduction to Physical Anthropology.* West/Wadsworth, Belmont, MA.

Chapter 25 25.3: after Niklas, K. J. 1994. *Plant Allometry: The Scaling of Form and Process.* University of Chicago Press, Chicago. 25.5: after Martin, R. D. 1981. "Relative brain size and basal metabolic rate in terrestrial vertebrates." *Nature* 293: 57–60.

Chapter 26 26.7: after data from Calder, W. A. 1984. *Size, Function, and Life History.* Harvard University Press, Cambridge, MA; Schmidt-Nielsen, K. 1984. *Scaling.* Cambridge University Press, Cambridge, England. 26.8: after Vogel, S. *Life's Devices.* Princeton University Press, Princeton, NJ. 26.11: after Diamond, J. M. 1988. "Why cats have nine lives." *Nature* 332: 586–587.

Chapter 27 Box Figure: after Tucker, V. A. 1975. "The energetic cost of moving about." *American Scientist* 63: 413–419.

Chapter 28 28.9: map data after Faltz, G. and Rotthauwe, H. W. 1977. "The human lactose polymorphism: physiology and genetics of lactose absorption and malabsorption." In Steinberg, A. G., Bearn, A. G., Motolsky, A. G., and Childs, B. (Eds.) *Progress in Medical Genetics, NS, II*, pp. 205–249. Saunders, Philadelphia.

Chapter 29 29.2: after Graham, J. B. 1990. "Ecological, evolutionary, and physical factors influencing aquatic animal respiration." *American Zoologist* 30: 137–146. 29.9: after data in Welch, W. R. 1984. "Temperature and humidity of expired air: Interspecific comparisons and significance for loss of respiratory heat and water from endotherms." *Physiological Zoology* 57: 366–375.

Chapter 31 31.1: after data in Guyton, A. C. and Hall, J. E. 1996. *Textbook of Medical Physiology*, Ninth Edition. W. B. Saunders Co., Toronto; Louw, G. 1992. *Physiological Animal Ecology.* Longman Scientific and Technical, Harlow, England.

Chapter 32 32.5: after Guyton, A. C. and Hall, J. E. 1996. *Textbook of Medical Physiology*, Ninth Edition. W. B. Saunders Co., Toronto.

Chapter 37 37.1: after Sze, L. C. 1953. "Changes in the amount of deoxyribonucleic acid in the development of *Rana pipiens.*" *Journal of Experimental Zoology* 122: 577–601.

Chapter 38 38.6: after Finlay, B. L. and Darlington, R. B. 1995. "Linked regularities in the development and evolution of mammalian brains." *Science* 268: 1578–1584.

Chapter 39 39.3: after Miller, G. T., Jr. 1996. *Living in the Environment.* Wadsworth Publishing Co., Belmont CA. 39.13: after *Our Changing Planet: The FY 1997 U.S. Global Change Research Program.* Supplement to the President's Fiscal Year 1997 Budget; NOAA Office of Global Programs.

Chapter 40 40.5: top map from Pielou, E. C. 1979. *Biogeography.* John Wiley and Sons, Inc., New York. Reprinted by permission. 40.8: after Dorf, E. 1960. "Climatic changes of the past and present." *American Scientist* 48: 341–364.

Chapter 41 41.5: after Gause, G. F. 1934. *The Struggle for Existence.* Williams and Wilkins, Baltimore, MD. 41.6: after data from Alliende, M. C. and Harper, J. L. 1989. "Demographic studies of a dioecious tree. I. Colonization, sex, and age structure of a population of *Salix cinerea.*" *Journal of Ecology* 77: 1029–1047. 41.8: after data from Scheffer, V. B. 1951. "The rise and fall of a reindeer herd." *Scientific Monthly* 73: 356–362. 41.10: after McKelvey, K., Noon, B. R., and Lamberson, R. H. 1993. "Conservation planning for species occupying fragmented landscapes: The case of the Northern Spotted Owl." In Kareiva, P. M., Kingsolver, J. G., and Huey, R. B. (Eds.) *Biotic Interactions and Global Change*, pp. 424–450. Sinauer Associates, Inc., Sunderland MA.

Chapter 42 42.4: after Myers, J. H. and Bazely, D. 1991. "Thorns, spines, prickles, and hairs: Are they stimulated by herbivory and do they deter herbivores?" In Tallamy, T. W. and Raupp, M. J. (Eds.) *Phytochemical Induction by Herbivores*, pp. 325–344. Wiley, New York. 42.7: after Kenward, R. E. 1978. "Hawks and doves: Factors affecting success and selection in goshawk attacks on wood-pigeons." *Journal of Animal Ecology* 47: 449–460. 42.9: after Ricklefs, R. E. and Miller, G. L. 2000. *Ecology*, Fourth Edition. W. H. Freeman and Co., New York; based on DeBach, P. and Sundby, R. A. 1963. "Competitive displacement between ecological homologues." *Hilgardia* 34: 105–106. 42.10: after Lack, D. 1947. *Darwin's Finches*. Cambridge University Press, Cambridge, England.

Chapter 43 43.3: after data from Paine, R. T. 1974. "Intertidal community structure: Experimental studies on the relationship between a dominant competitor and its principal predator." *Oecologia* 15: 93–120. 43.4: after Krebs, C. J. 1994. *Ecology*, Fourth Edition. HarperCollins College Publishers, New York; based on Olson, J. S. 1958. "Rates of succession and soil changes on southern Lake Michigan sand dunes." *Botanical Gazette* 119: 125–170.

Chapter 44 44.3: from Campbell, N. A. (Ed.) © 1987, 1990, 1993, 1996. *Biology*, Fourth Edition. The Benjamin/Cummings Publishing Co., Inc. Reprinted by permission. 44.8b: after Krebs, C. J. 1994. *Ecology*, Fourth Edition. HarperCollins College Publishers, New York; based on Likens, G. E., Bormann, F. H., Johnson, N. M., Fisher, D. W., and Pierce, R. S. 1970. "Effects of forest cutting and herbicide treatment on nutrient budgets in the Hubbard Brook watershed-ecosystem." *Ecological Monographs* 40: 23–47. Table 44.1: after Whittaker, R. H. 1975. *Communities and Ecosystems*. MacMillan, London.

Chapter 45 45.2: from Dahl, T. E. 1990. *Wetland losses in the United States, 1780s to 1980s*. U.S. Department of Interior, Fish and Wildlife Service, Washington, D. C. 45.5: from Vitousek, P. M. 1994. "Beyond global warming: Ecology and global change." *Ecology* 75: 1861–1876. Reprinted by permission of Ecological Society of America. 45.8: after Post, W. M., Peng, T.-H., Emanuel, W. R., King, A. W., Dale, V. H., and DeAngelis, D. L. "The global carbon cycle." *American Scientist* 78: 310–326. 45.10: data from the National Climatic Data Center, 1997. 45.11: from *IPCC Second Assessment Report: Climate Change*. 1995. A report of the Intergovernment Panel on Climate Change. IPCC, Geneva, Switzerland. 45.12: from Noss, R. F. 1992. "Issues of scale in conservation biology." In Fiedler, P. L. and Jain, S. K. (Eds.) *Conservation Biology: The Theory and Practice of Nature Conservation, Preservation, and Management*, pp. 239–250. Chapman and Hall, New York. Reprinted by permission.

Fine Art Credits

Frontispiece Alexis Rockman, *Biosphere: Hydrographer's Canyon*, 1994, 48 × 40 in. Photograph: Gorney Bravin & Lee.

Unit 1 Alexandre-Isidore Leroy de Barde, *Still Life with Exotic Birds*, 1804. Gouache and watercolor, Musée du Louvre, Paris.

Unit 2 James Barsness, *The World All Around*, 1998. Acrylic, ink, collage of paper mounted on canvas, 73.75 × 94.25 in. Private collection: Pelham Manor, New York. Photograph: George Adams Gallery, New York.

Unit 3 Sonia Delaunay, *Playing Cards (Jeu de Cartes)*, 1959. Lithograph, 37.5 × 25.5 in. © L&M Services B.V. Amsterdam 991104. © ARS/NY.

Unit 4 Philip Taaffe, *Passage II*, 1998. Mixed media on canvas, 26.5 × 36.5 in. Private collection, New York.

Unit 5 Robert Kushner, *Chrysanthemum Brocade*, 1994. Oil, glitter, and acrylic on canvas, 72 × 72 in. Private collection, courtesy DC Moore Gallery, New York City.

Unit 6 Kit Williams, *The Death of Spring*, 1980. 51 × 76 cm.

Chapter 1 Joan Miró, *Carnival of Harlequin*, 1924–25. Oil on canvas, 26 × 36.63 in. Albright-Knox Art Gallery, Buffalo, New York.

Chapter 2 Georgia O'Keeffe, *Jack-in-the-Pulpit No. IV*, 1930. Oil on canvas. National Gallery of Art, Washington, Alfred Steiglitz Collection, bequest of Georgia O'Keefe. Photograph: Richard Carafelli.

Chapter 3 Alexis Rockman, *The Bounty*, 1991. Photograph: Gorney Bravin & Lee.

Chapter 4 Martin Johnson Heade, *Study of an Orchid*, 1872. Oil on canvas, 18 × 23 in. Collection of the New York Historical Society.

Chapter 5 Matt Mullican, *Dallas Project, Cosmology Model,* panels 1-8, 1987.

Chapter 6 Terry Winters, *Direction Field*, 1996.

Chapter 7 Ben Shahn, *Helix and Crystal*, 1957. 53 × 30 in, tempera, San Diego Museum of Art. © Estate of Ben Shahn/ Licensed by VAGA, New York City.

Chapter 8 Paul Klee, *Around the Fish (Um den Fisch)*, 1926. Oil on canvas, 18.38 × 25.13 in. The Museum of Modern Art, New York, Abby Aldrich Rockefeller Fund. Photograph © 1999, The Museum of Modern Art, New York.

Chapter 9 Ross Bleckner, *Healthy Spot*, 1996. Oil/linen, 84 × 72 in. Photograph: Mary Boone Gallery, New York.

Chapter 10 Hilma af Klint, *Group 4. The Ten Greatest no. 7 Manhood (Lnr 108)*, 1907. Tempera on paper. The Hilma af Klint Foundation, Stockholm, Sweden.

Chapter 11 Eugene Von Bruenchenhein, *EVB 390 Untitled, no. 659 (GS261)*, 1957, Oil on Masonite, 24 × 24 in. Photograph: Ricco/Maresca Gallery.

Chapter 12 Max Miller, *Cause and Because*, 1997. Watercolor and gouache, 11 × 9 in.

Chapter 13 Victor Brauner, *Little Morphology (Petite Morphologie)*, 1934. Oil on canvas, 25.75 × 17.88 in. The Menil Collection, Huston. Photograph: Hickey-Robertson, Houston.

Chapter 14 Hilma af Klint, *Group 9 Series UW. Dove no. 25*, 1915. Tempera on Paper. The Hilma af Klint Foundation, Stockholm, Sweden.

Chapter 15 Ross Bleckner, *In Replication*, 1998. Oil on canvas, 84 × 72 in. Mary Boone Gallery, New York.

Chapter 16 Alfred Jensen, *Physical Optics*, 1975. Photograph: Pace Wildenstein Gallery.

Chapter 17 Arthur Tress, *Fish Tank Sonata, Corn Farmer*, 1990.

Chapter 18 Sandy Skoglund, *Babies at Paradise Pond*, 1996. Sandy Skoglund/ Superstock.

Chapter 19 Franz de Hamilton, *Concert of Birds*, c. 1682. Oil on copper, 61.8 × 77.4 cm. Staatliche Kunsthalle, Karlsruhe.

Chapter 20 Betty La Duke, *Africa: Osun's Children*, 1990. Acrylic, 72 × 68 in.

Chapter 21 Richard Ross, *Muséum National d'Histoire Naturelle, Paris, France*, 1982.

Chapter 22 William Morris, *Artifact Series #11 (Man and Beast)*, 1988. Glass, 18 × 108 × 60 in.

Chapter 23 Rosamund Purcell, *Homo sapiens, Gorilla gorilla, Ourangutan*, 1992. Leningrad, Museum of Zoology.

Chapter 24 Remedios Varos, *Planta insumisa*, 1961. Oil on masonite, 84 × 62 cm (oval).

Chapter 25 Charles Ray, *Artist With Fall '91*, 1992. Photograph: Regen Projects, Los Angeles.

Chapter 26 Mansur ibn Muhammad Fagih Ilyas, Skeletal system from *Five Anatomical Figures*. Original, late 14th century illustration from Tashrih Munsori. Opaque watercolor, 16 × 9.5 in. The British Library.

Chapter 27 Jess, *The Mouse's Tale*, 1951. Collage and gouache on paper, 47 × 32 in. Collection, San Francisco Museum of Modern Art, gift of Frederic P. Snowden.

Chapter 28 René Magritte, *Treasure Island (L'Ile aux Tresors)*, 1942. 60 × 80 cm. Musées Rouaux des Beaux-arts, Brussels. Photograph: Hersovici/Art Resource, New York. © ARS, New York.

Chapter 29 Andy Warhol, 1960. Courtesy of The Andy Warhol Foundation Inc./Art Resource, New York. © The Andy Warhol Foundation for the Visual Arts/ARS, New York.

Chapter 30 *Circulatory System*, late thirteenth-century manuscript. Pen and wash on parchment. 10.5 × 7.5 in. The Bodleian Library, Oxford.

Chapter 31 Wassily Kandinsky, *Cool Condensation*, 1930. Oil on cardboard, 15.75 × 14.56 in. © ARS/NY.

Chapter 32 Gilbert and George, *Deatho Knocko*, 1982. 165.25 × 157.5 in. Photograph: Anthony d'Offay Gallery, London.

Chapter 33 Fernand Leger, *Mona Lisa with Keys*, 1930. Oil on canvas, 35.19 × 28.38 in, Musée National Fernand Leger, Biot. © ARS/NY.

Chapter 34 Pavel Tchelitchev, *The God of Rain*, 1947. Gouache on paper, 22.63 × 15.63 in. New York State Museum, Trenton, New Jersey. Gift of Lloyd B. Westcott.

Chapter 35 Anique Taylor, *Bliss*, 1989. Pencil, 4 × 6 in.

Chapter 36 Alex Grey, *Pregnancy*, 1988–89. From *Sacred Mirrors: Visionary Art of Alex Grey*. Oil on linen, 50 × 56 in. © Alex Grey. www.alexgrey.com

Chapter 37 Dale Chihuly, *Shell Pink Basket set with Oxblood Wraps*, 1995. 9 × 22 × 22 in. Photograph: Claire Garoutte.

Chapter 38 Yves Klein, *Hiroshima (Anthropmetries 79)*, c. 1961. 139.5 × 280.5 cm. The Menil Collection, Houston. Photograph: Rick Gardner, Houston.

Chapter 39 Dale Chihuly, *Sky Blue Soft Cylinder with Golden Yellow Lip Wrap*, 1993. Glass, 17 × 21 in. Photograph: Claire Garoutte.

Chapter 40 George Stubbs, *Zebra*, 1763. Yale Center for British Art, Paul Mellon Collection.

Chapter 41 Rosamund Purcell, *European Moles*, 1992.

Chapter 42 Joan Miró, *The Tilled Field*, July 1923–Winter 1924. Solomon R. Guggenheim Museum, New York. Photograph by David Heald, © The Solomon R. Guggenheim Foundation, New York. © ARS/NY.

Chapter 43 Richard Ross, *Deyrolle Taxidermy, Paris, France*, 1986.

Chapter 44 Roger Minick, *Woman with Scarf at Inspiration Point, Yosemite National Park (Sightseer Series)*, 1980. Photograph: Jan Kesner Gallery, Los Angeles.

Chapter 45 Oliver Burston, *Hypervoxel Embolism*, 1999. © 1999 Oliver Burston/debut art.

Glossary

acid A chemical compound that can release a hydrogen ion. Compare *base* and *buffer*.

acid rain Rainfall with a low pH. Acid rain is a major environmental problem and is a consequence of the human release of sulfur dioxide and other pollutants into the atmosphere; in the atmosphere these pollutants are converted to acids that then fall back to Earth in rain or snow.

actin A protein found in muscle tissue and in bacterial flagella. In muscles, actin forms the framework bound to the Z discs of sarcomeres.

actin filament A protein fiber composed of actin monomers. Actin filaments are part of the cell's cytoskeleton.

action potential An electrical signal generated by the flow of ions across the plasma membrane of animal nerve cells. Action potentials are self-amplifying and can travel down a nerve cell in only one direction.

activation energy The small input of energy required for a chemical reaction to proceed.

active carrier protein A protein in the plasma membrane of a cell that, with the energy stored in a molecule like adenosine triphosphate (ATP), changes shape to transfer a molecule across the plasma membrane. Compare *passive carrier protein*.

active forager A forager that can move toward its food.

active movement Movement of molecules that requires energy. Active movement may occur either up or down a concentration gradient. Compare *passive movement*.

active site The specific region on the surface of an enzyme where substrate molecules bind.

adaptation A characteristic of an organism that improves the organism's performance in its environment. Adaptations help organisms accomplish important functions such as food capture, defense against predators, and reproduction. Adaptations often are complex and cause the organism to appear well designed for its environment.

adaptive evolution The process in which natural selection causes the quality of an adaptation to improve over time.

adenosine triphosphate See *ATP*.

adenylate cyclase The enzyme that catalyzes the conversion of ATP into the cyclic nucleotide cAMP. Adenylate cyclase is a component of many different signal cascades in cells.

adrenaline A hormone produced by the adrenal gland in response to stress. Adrenaline activates a G protein signal cascade in cells.

aerobic Of or referring to a metabolic process or organism that requires oxygen gas. Compare *anaerobic*.

aerobic respiration A general term used to describe a series of oxidation reactions that use oxygen to produce ATP.

allele An alternative version of a gene. Each allele has a different DNA sequence than that of all other alleles of the same gene.

allele frequency The proportion (percentage) of a particular allele in a population.

allometric relationship A relationship between size and another functional or structural property of an organism.

altruism A behavior that harms the individual performing the behavior while aiding other individuals.

alveolus (pl. alveoli) Any of the small pockets in the mammalian lung where most gas exchange takes place.

amino acid A chemical compound that has an amino group, a carboxyl group, and a variable side chain attached to a single carbon atom. Proteins are polymers of amino acids.

amniocentesis A procedure in which fluid containing fetal cells is extracted from a woman's uterus. Chromosomes from the cells are examined for genetic defects or DNA is extracted from the cells for genetic screening.

anaerobic Of or referring to a metabolic process or organism that does not require oxygen gas. Compare *aerobic*.

anaphase The stage of mitosis during which sister chromatids separate and move to opposite poles of the cell.

androgen Any of a group of steroid hormones that maintain male sexual and behavioral characteristics. Compare *estrogen*.

angiosperm The flowering plants, a group that includes most plants on Earth. These plants were named for the protective covering around the plant's embryo in the seed. Compare *gymnosperm*.

Animalia The kingdom that is made up of animals, from sponges to humans.

antenna complex An arrangement of chlorophyll molecules in the thylakoid membrane of a chloroplast that harvests energy from sunlight.

antibody A protein that is produced by a B lymphocyte in the human immune system and binds specifically to a particular antigen.

anticodon A sequence of three nucleotide bases on a transfer RNA molecule that can bind to a particular codon on a messenger RNA molecule.

antigen A characteristic protein in the plasma membrane of a cell or a chemical produced by an invading organism that is recognized by particular lymphocytes and the antibodies produced by those lymphocytes.

anus The opening at the end of the gut through which solid wastes leave an animal.

APC A tumor suppressor gene that is inactive in tumor cells associated with certain types of colon cancer.

apical meristem The region of rapidly dividing cells in the tips of plant branches and roots that gives rise to new stem and root tissues.

artery A blood vessel that carries blood from the heart. Compare *vein*.

arthropod Any of a group of animals that are characterized by a hard outer skeleton. Arthropods include millipedes, crustaceans, insects, and spiders.

artificial selection The process in which only individuals that possess certain characteristics are allowed to breed. Breeders use artificial selection to guide the evolution of crop and other species in ways that are advantageous for people.

asexual reproduction The production of genetically identical offspring without the exchange of genetic material with another individual. Compare *sexual reproduction*.

atherosclerosis A narrowing of the blood vessels in humans that results from the formation of fatty plaques.

atmospheric cycle A type of nutrient cycle in which the nutrient enters the atmosphere easily. Carbon, nitrogen, and sulfur are important nutrients that have atmospheric nutrient cycles. Compare *sedimentary cycle*.

atom The smallest unit of a chemical element that still has the properties of that element.

ATP Adenosine triphosphate, a molecule that is commonly used by cells to transfer energy from one chemical reaction to another.

atrium (pl. atria) The chamber in a chambered heart that receives blood from the body and pumps it into the ventricle.

autosome See *sex chromosome*.

axon The portion of a nerve cell that acts as a sort of biological wire to carry action potentials through the organism.

B lymphocyte   A lymphocyte that produces antibodies.

bacterial flagellum (pl. flagella)   A unique rotary motor that, in response to the flow of hydrogen ions across the bacterial plasma membrane, propels the bacterium. Compare *eukaryotic flagellum*.

base   (1) A chemical compound that can accept a hydrogen ion. Compare *acid* and *buffer*. (2) A nitrogen-containing molecule that is part of a nucleotide. See *nucleotide base*.

base pair   A process in which complementary bases (A and T, or C and G) form hydrogen bonds to each other, thereby causing two single strands of DNA to become double stranded.

behavior   A response to a stimulus, particularly a response that involves movement.

behavioral mutualism   A mutualism in which each of the two interacting species has altered its behavior to benefit the other species.

benign   Of or referring to a cancerous growth that is confined to a single tumor and does not spread to other tissues in the body. Compare *malignant*.

bile   A substance produced by the liver that helps break up large fat globules in the small intestine so that they move more easily in the food stream.

biodiversity   the variety of organisms on Earth or in a particular location, ranging from the genetic variety and behavioral diversity of individual organisms or species through the diversity of ecosystems.

biological evolution   (1) Change in the genetic characteristics of populations of organisms over time. (2) The history of the formation and extinction of species over time.

biome   A major terrestrial or aquatic life zone, defined either by its vegetation (terrestrial biomes) or by the physical characteristics of the environment (aquatic biomes). There are seven terrestrial biomes (tundra, boreal forest, temperate forest, chaparral, grassland, desert, and tropical forest) and eight aquatic biomes (lake, river, wetland, estuary, intertidal zone, coral reef, ocean, and benthic zone).

biosphere   All living organisms on Earth, together with the environments in which they live.

biosynthetic   Of or referring to chemical reactions that manufacture complex molecules in living cells.

bipedalism   The ability to walk on two legs.

bivalent   A pair of homologous chromosomes. Bivalents form during prophase I of meiosis.

blastula   The relatively undifferentiated stage early in animal development just prior to the formation of the germ lines.

blood clot   A mesh of protein that captures cells and seals a break in a blood vessel.

bone   A support structure found in most vertebrate animals that consists of connective cells that secrete a mineral matrix rich in calcium. The core of a bone serves as the site of red and white blood cell production. Compare *cartilage*.

BRCA1   One of the major and better known genes that causes breast cancer.

buffer   A chemical compound that can both release and accept hydrogen ions. Buffers can maintain the pH of water within specific limits. Compare *acid* and *base (1)*.

Cambrian explosion   The major increase in the diversity of life on Earth that occurred about 530 million years ago, during the Cambrian period. The Cambrian explosion lasted 5 to 10 million years; during this time larger and more complex forms of most living animal groups appeared suddenly in the fossil record.

canopy   The aerial habitat in the branches of forest trees. In the Tropics the canopy is very rich in species.

capillary   A tiny blood vessel in a closed circulatory system across the walls of which all exchange with the surrounding tissues takes place.

carbohydrate   Any of a class of organic compounds that includes monosaccharides and polysaccharides. See also *sugar*.

carbon fixation   The process by which carbon atoms from CO_2 gas are incorporated into sugars. Carbon fixation occurs in the chloroplasts of green plants.

carnivore   A consumer that relies on living animal tissues for nutrients. Compare *herbivore*.

carrier   An individual that carries a disease-causing allele but does not get the disease.

carrying capacity   The maximum population size that can be supported indefinitely by the environment in which the population is found.

cartilage   A support structure found in animals that consists of connective cells that secrete a matrix rich in the protein collagen. Compare *bone*.

catabolic   Of or referring to a chemical reaction that breaks down complex molecules to release energy for use by the cell.

catabolism   A general term for those chemical reactions which break down complex molecules to release energy for use by the cell.

catalysis   A process whereby the rate of a chemical reaction is accelerated.

catalyst   A molecule that speeds up a specific chemical reaction without being consumed in the process. Enzymes are protein catalysts.

cell   The smallest self-contained unit of life, enclosed by a membrane. Animals have many different cell types, including epithelial cells, muscle cells, and nerve cells.

cell communication   The process by which one cell can affect the activities of another via signaling molecules such as hormones and growth factors.

cell division cycle   A series of distinct stages in the life cycle of a cell that culminate in cell division.

cell specialization   The process by which a cell develops physical characteristics that are best suited for a specific function.

cell wall   A support layer that is secreted outside the cells of many single-celled organisms, fungi, and plants.

cellulose   A carbohydrate produced by plants and some other organisms that makes up much of their cell walls.

centromere   The physical constriction in a chromosome that holds sister chromatids together.

centrosome   A protein structure in the cytosol that helps organize the mitotic spindle and define the two poles of the dividing cell.

cerebral cortex   The highly folded outer portion of the human brain that is specialized to coordinate information from sensory cells with the activity of muscles.

cerebrum   The largest part of the human brain. It includes the cerebral cortex.

channel protein   A protein in the plasma membrane of a cell that forms an opening through which certain molecules can pass.

character   A feature of an organism, such as height, flower color, or the chemical structure of a protein.

character displacement   The process in which intense competition between species causes the forms of the competing species to evolve to become more different over time.

chemical compound   A combination of different types of atoms linked by covalent bonds.

chemical reaction   A process that rearranges atoms in chemical compounds.

chemoautotroph   Organisms that uses carbon dioxide as a carbon source and obtains energy not from the sun, but from chemical reactions that involve inorganic compounds. Only bacteria are known to be chemoautotrophs. Compare *chemoheterotroph, photoautotroph*, and *photoheterotroph*.

chemoheterotroph   Any organism, such as an animal, that obtains energy and carbon from organic compounds, typically by consuming other organisms. Compare *chemoautotroph, photoautotroph*, and *photoheterotroph*.

chitin   A carbohydrate that serves as an important support material in the cell walls of fungi and in the skeletons of animals.

chlorophyll   The green pigment that is used to capture energy from light in photosynthesis.

chloroplast   An organelle found in plants and algae that is the primary site of photosynthesis.

chromatid   Either of two identical, side-by-side copies of a chromosome that are linked at the centromere. The separation of chromatids during mitosis ensures equal sharing of DNA between daughter cells.

chromatin The combination of DNA and proteins that makes up chromosomes.

chromosome Any of several elongated structures found in the nucleus, each composed of DNA packaged with proteins. Chromosomes become visible in the microscope during mitosis and meiosis.

chromosome theory of inheritance A theory, supported by much experimental evidence, stating that genes are located on chromosomes.

cilium (pl. cilia) A hairlike structure found in some eukaryotes that moves in a paddle-like motion to propel the organism or to move fluid past the organism. Compare *eukaryotic flagellum*.

citric acid cycle A series of oxidation reactions that produce high-energy electrons stored in NADH, and CO_2 as waste. In eukaryotic cells, the citric acid cycle takes place in mitochondria.

cladistics The dominant school of systematics, in which shared derived features are used to determine evolutionary relationships. Compare *evolutionary taxonomy*.

climate Prevailing weather conditions experienced in an area over relatively long periods of time (30 years or more).

climax community For a given climate and soil type, a community whose species are not replaced by other species. A climax community is the end point of succession for a particular location; in many cases, however, ongoing disturbances such as fire or windstorms prevent the formation of a stable climax community.

clone (of a gene) See *DNA cloning*.

closed circulatory system An internal transport system found in animals in which blood vessels carry blood through all the organs. Compare *open circulatory system*.

clumped distribution A distribution of organisms in which the organisms tend to be clustered together throughout the environment. Compare *random distribution* and *regular distribution*.

cochlea A snail shell–shaped structure in the inner ear of mammals that allows them to distinguish between sounds of different frequencies.

codon A group of three nucleotide bases in a messenger RNA molecule. Each codon specifies either a particular amino acid or a signal to start or stop the construction of a protein. Compare *intron*.

collagen A protein produced by the connective cells of animals that resists tension.

colon The organ in which undigested food is prepared for release from the animal gut. Also called large intestine.

community An association of populations of different species that live in the same area.

companion cell A cell that helps actively pump sugars into phloem cells in the plant vascular system.

competition An interaction between two species in which each species has a negative impact on the other species.

complement A group of proteins that circulate in the blood and concentrate in inflamed areas. Complement stimulates macrophage and neutrophil production and activity.

compression A squeezing force that tends to push molecules together.

concentration gradient A change in the concentration of molecules from one area to another.

conductive heat Heat transferred by direct contact between two materials. Compare *radiant heat*.

connective cell A type of animal cell that often secretes a matrix of chemicals that play an important role in animal support systems.

consumer An organism that obtains its energy by eating other live organisms or the remains of dead organisms. Consumers include herbivores, carnivores, and decomposers. Compare *producer*.

consumer–victim interaction An interaction between two species in which one species benefits (the consumer) and the other species is harmed (the victim). Consumer–victim interactions include the killing of prey by predators, the eating of plants by herbivores, and the harming or killing of a host by a parasite or pathogen.

continental drift The movement of Earth's continents over time. The continents can be thought of as plates that "float" on a hot layer of semisolid rock. The continental plates move because hot plumes of liquid rock rise to the surface of Earth and push the continents apart and because one continental plate can sink below another, pulling the rest of the plate along with it.

convection The physical movement of heat in air or water.

convergent feature A feature shared by two groups of organisms not because it was inherited from a common ancestor, but because it arose independently in the two groups.

copulation A behavior in which a male animal places his sperm directly into the body of a female; the typical mode of sperm transfer in land animals.

covalent bond A strong chemical linkage between two atoms based on the sharing of electrons. Compare *hydrogen bond* and *ionic bond*.

crossing-over The physical exchange of genes between homologous chromosomes. As a result of crossing-over, part of the genetic material inherited from one parent is replaced with the corresponding genetic material inherited from the other parent.

cross-link A connection between two molecules that generally increases the resistance of the molecules to compressive or tensile forces, and that often increases rigidity.

cuticle The waxy covering on a plant's stem and leaves.

cytokinesis The final stage of mitosis, during which the cell physically divides into two daughter cells.

cytoplasm The contents of the cell enclosed by the plasma membrane, but in eukaryotes excluding the nucleus. Compare *cytosol*.

cytoskeleton A complex network of protein filaments found in the cytosol of eukaryotic cells. The cytoskeleton maintains cell shape and is necessary for the physical processes of cell division and movement.

cytosol The contents of the cell enclosed by the plasma membrane, but in eukaryotes excluding all organelles. Compare *cytoplasm*.

dark reactions A series of chemical reactions that directly use CO_2 to synthesize sugars. The dark reactions do not require light and take place in the stroma of chloroplasts. Compare *light reactions*.

decomposer An organism that breaks down dead tissues into simple chemical components, thereby returning nutrients to the physical environment. Compare *detritivore*.

deficiency anemia Low red blood cell production that results from inadequate iron in the diet.

deletion mutation A mutation in which one or more bases are removed from the DNA sequence of a gene. Compare *insertion mutation* and *substitution mutation*.

dendrite The portion of a nerve cell that interacts with other nerve cells by either producing or responding to neurotransmitters.

density-dependent Of or referring to a factor, such as food shortage, that limits the growth of a population more strongly as the density of the population increases. Compare *density-independent*.

density-independent Of or referring to a factor, such as weather, that can limit the size of a population but does not act more strongly as the density of the population increases. Compare *density-dependent*.

deoxyribonucleic acid See *DNA*.

dermal tissue system The tissue system in plants that forms their outer covering.

detritivore A consumer that relies on dead tissues for nutrients. Compare *decomposer*.

deuterostome Any of a group of animals, including sea stars and vertebrates like humans, birds, and fish, that are characterized by their early development, in which the second opening in the embryo becomes the mouth. Compare *protostome*.

development The process by which an organism grows from a single cell to its adult form.

differentiation The process by which cells develop specific physical characteristics and functions.

diffusion Passive movement of molecules from areas of high concentration of the molecule to areas of low concentration. Compare *osmosis*.

digestion The chemical breakdown of food.

diploid   Of or referring to a cell or organism that has two complete sets of homologous chromosomes (2*n*). Compare *haploid*.

directional selection   A type of natural selection in which individuals with one extreme of a heritable phenotypic characteristic have an advantage over other individuals in the population, as when large individuals produce more offspring than small and medium-sized individuals. Compare *disruptive selection* and *stabilizing selection*.

dispersal   An act in which individuals travel relatively long distances away from other members of their species. Dispersal may be active, as when individuals walk, swim, or fly, or it may be passive, as when organisms are transported by wind or water.

disruptive selection   A type of natural selection in which individuals with either extreme of a heritable phenotypic characteristic have an advantage over individuals with an intermediate phenotype, as when both small and large individuals survive at a higher rate than medium-sized individuals. Compare *directional selection* and *stabilizing selection*.

distribution   The geographic area over which a species is found. Also called range.

disturbance   Any event, such as a fire or windstorm, that kills or damages some organisms in an ecological community, thereby creating an opportunity for other organisms to become established.

DNA   Deoxyribonucleic acid, a polymer of nucleotides that stores the information needed to synthesize proteins in living organisms.

DNA cloning   A set of techniques in which a gene or other DNA sequence is isolated and then copied many times. DNA libraries and the polymerase chain reaction (PCR) are used to produce many copies of the gene or other DNA sequence of interest.

DNA fingerprinting   Genetic screening that creates a DNA profile that can be used to identify individuals and determine the relatedness of individuals.

DNA hybridization   Base pairing of DNA from two different sources.

DNA library   A collection of an organism's DNA fragments that are stored in a host organism, such as a bacterium.

DNA packing   The highly organized way in which large amounts of DNA are packed into the cells of eukaryotic organisms.

DNA polymerase   The key enzyme that cells use to copy their DNA. DNA polymerase is also used in DNA technology to make many copies of a gene or other DNA sequence in a test tube.

DNA probe   A short sequence of DNA (usually tens to hundreds of nucleotide bases long) that can base-pair with a particular gene or other specific region of DNA.

DNA repair   A three-step process in which damage to DNA is repaired. Damaged DNA is first recognized, then removed, and then replaced with newly synthesized DNA.

DNA replication   The duplication, or copying, of a DNA molecule. DNA replication begins when the hydrogen bonds connecting the two strands of DNA are broken, causing the strands to unwind and separate. Each strand is then used as a template for the construction of a new strand of DNA.

DNA segregation   The process by which the DNA of a dividing cell is divided equally between two daughter cells.

DNA technology   A powerful set of techniques that scientists use to manipulate DNA.

dominant   Of or referring to an allele that determines the phenotype of an organism when paired with a different (recessive) allele. Compare *recessive*.

donor   The bacterium individual that provides the genes incorporated into a recipient individual's genome. Compare *recipient*.

double helix   The structure of DNA. Two long strands of covalently bonded nucleotides are held together by hydrogen bonds and twisted into a spiral coil.

doubling time   The time it takes a population to double in size. Doubling time can be used as a measure of how fast a population is growing.

drag   A force that resists the motion of any moving object.

eardrum   A thin membrane separating the outer and middle ear of vertebrates that plays an important role in converting the rapid changes in air pressure that constitute sound into physical motion that the organism can detect.

ecology   The scientific study of interactions between organisms and their environment.

ecosystem   A community of organisms, together with the chemical and physical environment in which the organisms live. Global patterns of air and water circulation link all the world's organisms into one giant ecosystem, the biosphere.

ectoderm   The germ layer in animals that forms the exterior of the gastrula and that gives rise to the epidermis and nerve tissue. Compare *endoderm* and *mesoderm*.

ectotherm   An organism that relies on environmental heat for most of its heat input. Compare *endotherm*.

egg   The large, sedentary, haploid gamete that is produced by sexually reproducing female eukaryotes. Compare *sperm*.

El Niño   A seasonal, warm water current that flows south along the coast of Ecuador. Usually, El Niño has a temporary and local effect, but once every 2 to 10 years it triggers floods, fires, disease outbreaks, crop failures, and more throughout much of Earth.

electron   A negatively charged particle found in atoms. Each atom contains a characteristic number of electrons. Compare *proton*.

electron transport chain (ETC)   A group of membrane-associated proteins that can both accept and donate electrons. The transfer of electrons from one ETC protein to another releases energy that is used to manufacture ATP in both chloroplasts and mitochondria.

element   A substance made up of only one type of atom. The physical world is made up of 92 different elements.

embryo sac   The female gametophyte produced by flowering plants. The embryo sac remains within the ovaries of the flowers produced by a spore-producing individual. Compare *pollen grain*.

endocrine system   The collection of glands in animals that produce and release chemical messengers called hormones into the circulatory system.

endoderm   The germ layer in animals that forms the interior of the gastrula and that gives rise to the gut and associated organs. Compare *ectoderm* and *mesoderm*.

endoplasmic reticulum (ER)   An organelle composed of many interconnected membrane sacs and tubes. The ER is a major site of protein and lipid synthesis in eukaryotic cells.

endosperm   The nutritive tissue produced within a plant seed.

endotherm   An organism that relies on metabolic heat for most of its heat input. Compare *ectotherm*.

energy carrier   A molecule that can store and donate energy to another molecule or chemical reaction. ATP is the most commonly used energy carrier in living organisms.

enzyme   A protein that acts as a catalyst, speeding the progress of chemical reactions. All chemical reactions in living organisms are catalyzed by enzymes.

epithelial cell   A type of animal cell that forms the outer surface of the skin and lines the lungs and guts.

ER   See *endoplasmic reticulum*.

ErbB   A mutant form of the EGF (epidermal growth factor) receptor produced by an oncogene. ErbB no longer binds to growth factor and is always active.

esophagus (pl. esophagi)   The portion of the gut that is specialized to transport food from the mouth to the stomach. The esophagus is used as a site of food storage in some organisms.

essential amino acid   An amino acid that a consumer cannot synthesize and must therefore obtain from its food.

estrogen   Any of a group of steroid hormones that maintain female sexual and behavioral characteristics. Compare *androgen*.

ETC   See *electron transport chain*.

eukaryote   A single-celled or multicellular organism in which each cell has a distinct nucleus and cytoplasm. All organisms other than bacteria are eukaryotes. Compare *prokaryote*.

eukaryotic flagellum (pl. flagella)   A hair-like structure found in eukaryotes that propels organisms by means of waves passing from its base to its tip. Compare *bacterial flagellum* and *cilium*.

evaporative heat   Heat lost when liquid water is converted into water vapor.

evolution   Change over time in a lineage of organisms.

evolution of populations   Changes in the allele or genotype frequencies found in populations over generations.

evolutionary innovation   A key adaptation of a group that originated in that group.

evolutionary radiation   Rapid expansion of a group of organisms to form new species and higher taxonomic groups.

evolutionary taxonomy   The most traditional school of systematics, lacking a clearly defined method, in which scientists use expert knowledge of a group to determine relationships. Compare *cladistics*.

evolutionary tree   A diagrammatic representation showing the order in which different lineages arose, the lowest branches having arisen first.

experiment   A controlled manipulation of nature designed to test a hypothesis.

exploitation competition   A type of competition in which species compete indirectly for shared resources, each reducing the resource levels available to the other. Compare *interference competition*.

exponential growth   An important type of rapid population growth in which a population increases by a constant proportion from one generation to the next.

external skeleton   A skeleton that surrounds the soft tissues of the animal it supports. Compare *internal skeleton*.

F_1 generation   The first generation of offspring in a genetic cross. Compare *P generation*.

F_2 generation   The second generation of offspring in a genetic cross.

fast muscle fiber   A muscle fiber that contracts quickly, but that can sustain the contraction for only short amounts of time. Compare *slow muscle fiber*.

fat   A chemical compound that consists of glycerol linked to three fatty acids. Fats are solid at room temperature and can store energy in living organisms.

fatty acid   A chemical compound with a long hydrocarbon chain. Fatty acids are found in lipids and fats.

female   The individual in a sexually reproducing species that produces eggs. Compare *male*.

fermentation   A series of catabolic reactions that produce small amounts of ATP without oxygen. Fermentation is similar to glycolysis, with the exception that pyruvate is converted to other products, such as ethanol or lactic acid.

fertilization   The fusion of haploid gametes (egg and sperm) to produce a diploid zygote (the fertilized egg).

fetal biopsy   A procedure in which tissue is extracted directly from the body of a developing fetus while it is still in a woman's uterus. DNA from the tissue is used in genetic screening.

first law of thermodynamics   The law stating that energy can neither be created nor destroyed, only transformed or transferred from one molecule to another. Living organisms use energy by capturing it from the environment and transferring it from one molecule to another.

fixed behavior   A predictable response to a particular, often simple, stimulus. Compare *learned behavior*.

flower   A specialized structure for reproduction that is characteristic of the plant group known as the angiosperms, or flowering plants.

fluid mosaic model   A model that describes the cell membrane as a mobile phospholipid bilayer with imbedded proteins that can move laterally in the plane of the membrane.

food chain   A single sequence of which organism eats which organism in a community. Food chains represent a portion of a food web and are used to describe feeding relations in an ecological community. Compare *food web*.

food web   A summary of the movement of food through an ecological community. A food web is formed by connecting all of the different food chains in the community to each other. Compare *food chain*.

foreign species   See *nonnative species*.

fossil   Preserved remains from or an impression of a former living organism. Fossils document the history of life on Earth, showing that past organisms were unlike living forms, that many organisms have gone extinct, and that life has evolved through time.

frameshift   A change in how the information in a gene is translated by the cell. Frameshifts occur when a deletion or insertion mutation is not a multiple of three base pairs. For example, if a deletion mutation has the result of removing the G from the mRNA sequence GAACCUA, the first two codons change from GAA and CCU to AAC and CUA, thus changing the sequence of amino acids specified by the mRNA.

friction drag   Drag that results because of the interaction of a fluid with the surface of a moving object. Compare *pressure drag*.

fulcrum   The point around which a lever pivots.

functional group   A specific arrangement of atoms that helps define the properties of a chemical compound. Examples include phosphate and amino functional groups.

Fungi   The kingdom of mushroom-producing species, yeasts, and molds that usually live as decomposers.

G protein   Any of a class of proteins that participate in signal cascades, including those activated by hormones. G protein activity is regulated by the guanine nucleotides GDP (guanosine diphosphate) and GTP (guanosine triphosphate).

G_0 phase   The period of time during which the cell pauses in the cell division cycle between mitosis and S phase. No preparations for S phase are made during this period.

G_1 phase   The period of time following mitosis and before S phase of the cell division cycle. The cell makes preparations for DNA synthesis during the G_1 phase.

G_2 phase   The period of time following S phase and before mitosis of the cell division cycle.

gamete   A haploid sex cell, which fuses with another sex cell during fertilization. Eggs and sperm are gametes.

gametophyte   The egg- or sperm-producing individual in the life cycle of a vascular plant.

gastrula   An embryo forming the characteristic three cell layers (endoderm, mesoderm, and ectoderm) which will give rise to all of the major tissue systems of the adult animal.

gastrulation   The rearrangement of the germ layers (endoderm, mesoderm, and ectoderm) during animal development to the positions appropriate for the tissues to which these layers give rise.

gel electrophoresis   The process in which DNA fragments are placed in a gelatin-like substance (a gel) and subjected to an electrical charge, causing the DNA fragments to move through the gel. Small DNA fragments move farther than large DNA fragments, thus causing the fragments to separate by size.

gene   A sequence of DNA that contains information for the synthesis of one of several types of RNA molecules used to make protein. Genes are located on chromosomes.

gene cascade   The process in which the protein products of different genes interact with each other and with signals from the environment, thereby turning on other sets of genes in some cells but not in other cells. Organisms use gene cascades to control how genes are expressed during development.

gene expression   The synthesis of a gene's protein or RNA product. Gene expression is the means by which a gene influences the cell or organism in which it is found.

gene flow   The exchange of genes between populations.

gene pool   In principle, all alleles at all genetic loci in all individuals of a population. In practice, the alleles in a population for a restricted set of genetic loci.

gene therapy   A treatment approach that seeks to correct genetic disorders by fixing the genes that cause the disorders.

genetic bottleneck   A drop in the size of a population that causes low genetic variation and/or causes harmful alleles to reach a frequency of 100 percent in the population.

genetic code   A code in which each set of three nucleotide bases in messenger RNA specifies either an amino acid or a signal to start or stop the construction of a protein. The genetic code allows the cell to use the information in a gene to build the protein called for by that gene.

genetic cross   A controlled mating experiment, usually performed to examine the inheritance of a particular character.

genetic drift   The process in which alleles are sampled at random over time, as when chance events cause certain alleles to be favored from one generation to the next. In genetic drift, the genetic makeup of a population drifts at random over time rather than being shaped in a nonrandom way by natural selection.

genetic engineering   A three-step process in which a DNA sequence (often a gene) is isolated, modified, and inserted back into an individual of the same or a different species. Genetic engineering is commonly used to change the performance of the genetically modified organism, as when a crop plant is engineered to resist attack from a pest insect.

genetic linkage   The situation in which genes are located close to one another on the same chromosome and thus do not follow Mendel's law of independent assortment. Genes located far from one another on the same chromosome can follow Mendel's law of independent assortment.

genetic recombination   The combining of genetic material such that offspring have a different genotype from that of either parent. Genetic recombination is caused by crossing-over and independent assortment of chromosomes.

genetic screening   The examination of an individual's genes to assess current or future health risks and status.

genetic variation   Genetic differences among the individuals of a population.

genetics   The scientific study of genes.

genome   All the DNA of an organism, including its genes.

genotype   The genetic makeup of an organism. Compare *phenotype*.

genotype frequency   The proportion (percentage) at which a particular genotype is found in a population.

geographic isolation   The physical separation of populations from one another due to barriers such as a mountain chain or a river. Geographic isolation often causes the formation of new species, as when populations of a single species become physically separated from one another and then accumulate so many genetic differences that they become reproductively isolated from one another.

germ layer   Any of three layers of cells that differentiate early in animal development and give rise to predictable organs during subsequent development. The three layers are endoderm, mesoderm, and ectoderm.

germ line   The cells in animals that remain undifferentiated during development and give rise to the ovaries or testes at maturity.

giant convection cell   Any large region of the atmosphere that has consistent air circulation patterns in which moist, warm air rises in one area and cool dry air sinks in another area. Earth has four of these giant convection cells; these cells play a major role in shaping climate.

gill   The gas exchange surface of aquatic animals, formed by an outfolding of epidermal tissue. Compare *lung*.

gizzard   A portion of the animal gut specialized for grinding and mixing food with digestive enzymes.

gland   An organ specialized for producing and releasing chemicals, particularly hormones.

glial cell   Any of several cell types that support the function of nerve cells. Glial cells that form an insulating lining around the nerves of vertebrates are an important example of this type of cell.

global change   Worldwide change to the environment. There are many causes of global change, including climate change caused by the movement of continents and changes in land and water use caused by humans.

global warming   A global increase in temperature. Earth appears to be entering a period of global warming that is caused by humans, specifically by the release to the atmosphere of large quantities of greenhouse gases like carbon dioxide.

glyceraldehyde 3-phosphate   An important three-carbon sugar produced by carbon fixation in chloroplasts.

glycolysis   A series of catabolic reactions that split glucose to produce pyruvate, which is used in either oxidative phosphorylation or fermentation.

Golgi apparatus   An organelle composed of flattened membrane sacs. The Golgi apparatus sorts proteins and lipids to various parts of the eukaryotic cell.

gonadotropin   Any of a group of hormones produced by the pituitary glands that regulate the development and function of the reproductive organs.

greenhouse gas   Any of several gases in the atmosphere that function like the windows on a car: They let in sunlight but trap heat. The atmospheric concentration of greenhouse gases like carbon dioxide is increasing, a trend that is likely to contribute to an increase in global temperatures.

ground meristem   The tissue in the flowering plant embryo that gives rise to cells of the ground tissue system. Compare *procambium* and *protoderm*.

ground tissue system   The tissue system in plants that forms most of the photosynthetic tissue in leaves and that surrounds the vascular tissue system.

growth factor   Any of a class of signaling molecules that bind to cell surface receptors and promote cell division.

gymnosperm   Any of a group of plants that includes pine trees and other conifers, ginkgos, and cycads. Gymnosperms were the first plants to evolve seeds. Compare *angiosperm*.

habitat   A characteristic place or type of environment in which an organism lives.

haploid   Of or referring to a cell or organism that has only one complete set of homologous chromosomes (*n*). Compare *diploid*.

Hardy–Weinberg equation   An equation ($p^2 + 2pq + q^2 = 1$) that predicts the genotype frequencies in a population that does not evolve. In this equation, where p is the frequency of the *A* allele and q is the frequency of the *a* allele, p^2 is the predicted frequency of the *AA* genotype, $2pq$ is the predicted frequency of the *Aa* genotype, and q^2 is the predicted frequency of the *aa* genotype.

heart attack   The death of muscle tissue in the heart that results from the blockage of arteries supplying blood to the heart.

helper T lymphocyte   A lymphocyte that stimulates killer T lymphocytes and B lymphocytes.

heme   The iron complex in oxygen-binding pigments that provides the oxygen-binding sites.

hemoglobin   An oxygen-binding pigment common in animals, including humans, that changes from blue when not bound to oxygen to red when bound to oxygen.

herbivore   A consumer that relies on living plant tissues for nutrients. Compare *carnivore*.

hermaphrodite   An individual that can produce both eggs and sperm.

heterozygote   An individual that carries one copy of each of two different alleles (for example, an *Aa* individual). Compare *homozygote*.

heterozygote advantage   The evolutionary advantage that heterozygotes (*Aa* individuals) enjoy because they leave more offspring than homozygotes (*AA* or *aa* individuals).

homeotic gene   A master-switch gene that plays a key role in the control of gene expression during development. Each homeotic gene controls the expression of a series of other genes whose protein products direct the development of an organism.

hominid   Any of a group of primates that contains humans and our now extinct humanlike ancestors. Compare *hominoid*.

hominoid   Any of a group of primates whose living relatives include gibbons and the great apes (orangutans, gorillas, chimpanzees, and humans). Compare *hominid*.

homologous chromosomes   Two copies of a specific chromosome found in diploid cells, one received from the mother and the other from the father.

homozygote An individual that carries two copies of the same allele (for example, an *AA* or an *aa* individual). Compare *heterozygote*.

hormone A chemical messenger released into the circulatory system of an animal or the vascular system of a plant that, in small amounts, affects the functioning of other, target tissues.

host An organism in which a parasite or infectious disease organism lives. The invader often does not kill its host.

housekeeping gene A gene that has an essential role in the maintenance of cellular activities and that is expressed by most cells in the body.

Huntington's disease A deadly degenerative disease of the nervous system for which there is no cure. Huntington's Disease proceeds from a mild difficulty with walking to the complete loss of the ability to stand, walk, and speak, ending with dementia and death.

hybrid An offspring that results when two different species mate.

hybridize To cause hybrids to be produced, as when two different species mate and produce fertile offspring.

hydrocarbon A chemical compound that contains only hydrogen and carbon atoms.

hydrogen bond A chemical linkage between a hydrogen atom with a slight positive charge and another atom with a slight negative charge. Hydrogen bonds are weaker than covalent bonds. Compare *covalent bond* and *ionic bond*.

hydrophilic Of or referring to molecules or parts of molecules that interact freely with water. Hydrophilic molecules dissolve easily in water, but not in fats or oils. Compare *hydrophobic*.

hydrophobic Of or referring to molecules or parts of molecules that do not interact freely with water. Hydrophobic molecules dissolve easily in fats and oils, but not in water. Compare *hydrophilic*.

hydrostat A support structure that depends on the interaction between a fluid under pressure and an elastic membrane for rigidity.

hypersensitive response A response by a plant to invasion by disease organisms or parasites, in which the plant kills the infected cells to isolate the invading organisms from healthy tissues.

hypha (pl. hyphae) A threadlike, absorptive structure of a fungus. Mats of hyphae form mycelia, the main bodies of fungi.

hypothalamus (pl. hypothalami) A structure at the base of the vertebrate brain that controls the release of hormones by the pituitary gland. Along with the pituitary gland, the hypothalamus helps regulate interactions between the nervous and endocrine systems.

hypothesis (pl. hypotheses) A possible explanation of how a natural phenomenon works. A hypothesis must have logical consequences whose truth can be tested.

immune response The heightened response of the human immune system to the second exposure to a particular disease organism.

implantation A procedure in which eggs are fertilized outside a woman's body and implanted into her uterus (as with so-called test tube babies).

incomplete dominance The situation in which heterozygotes (*Aa* individuals) are intermediate in form between the two homozygotes (*AA* and *aa* individuals).

independent assortment of chromosomes The random distribution of maternal and paternal chromosomes into gametes during meiosis.

individual A single organism, usually physically separate and genetically distinct from other individuals.

infectious disease A disease that spreads by microorganisms that invade other organisms.

inflammation The series of responses to a wound, most of which isolate any invading organisms and recruit white blood cells and complement to the wound.

insect Any of the most species-rich group of animals on Earth. Insects are six-legged arthropods that include grasshoppers, beetles, ants, and butterflies.

insertion mutation A mutation in which one or more bases are inserted into the DNA sequence of a gene. Compare *deletion mutation* and *substitution mutation*.

insulin A hormone that promotes the uptake of glucose by vertebrate cells.

intercellular air space The spaces between the spongy parenchyma cells through which carbon dioxide enters and oxygen and water leave the leaf.

interference competition A type of competition in which one organism directly excludes another from the use of resources such as food or nesting sites. Compare *exploitation competition*.

interferon A protein that is released by vertebrate cells under attack by a virus.

intermembrane space The space between the inner and outer membranes of either the chloroplast or the mitochondrion.

internal skeleton A skeleton that lies within the soft tissues of the animal it supports. Compare *external skeleton*.

interphase The period of time between two successive mitotic divisions. Most of the preparations for cell division occur during interphase.

intron A sequence of bases in a gene that does not specify part of the gene's final protein or RNA product. Enzymes in the nucleus must remove introns from mRNA, tRNA, and rRNA molecules for these molecules to function properly.

ion An atom or group of atoms that has either gained or lost electrons, and therefore has a negative or positive charge.

ionic bond A chemical linkage between two atoms based on the electrical attraction between positive and negative charges. Compare *covalent bond* and *hydrogen bond*.

joint A flexible connection between the rigid elements that make up an animal skeleton.

karyotype The specific number and shapes of chromosomes found in the cells of a particular organism.

keystone species A species that, relative to its own abundance, has a large impact on the type and abundance of the other species in a community.

killer T lymphocyte A lymphocyte that binds to cells bearing a particular antigen and helps destroy those cells.

kinase An enzyme that catalyzes the addition of phosphate groups to proteins.

kinetochore A plaque of protein on a chromosome where spindle microtubules attach during mitosis and meiosis.

kingdom The largest taxonomic category in the Linnaean hierarchy. Generally five kingdoms are recognized: Animalia, Plantae, Fungi, Protista, and Monera.

land transformation Changes humans have made to the land surface of Earth that alter the physical or biological characteristics of the affected regions. Together with water transformation, human impacts on the world's lands are thought to be the main cause of the current high rate of extinction of species.

language A means of communicating complex and often abstract ideas to others. Language may be unique to humans.

large intestine See *colon*.

law of equal segregation Gregor Mendel's first law, which states that the two copies of a gene separate during meiosis and end up in different gametes.

law of independent assortment Gregor Mendel's second law, which states that when gametes form, the separation of alleles for one gene is independent of the separation of alleles for other genes. We now know that this law does not apply to genes that are linked.

learned behavior A behavior acquired by trial and error or by watching others. Compare *fixed behavior*.

lever A rigid structure that pivots about a fulcrum, and that can either increase the force generated at the expense of speed or increase the speed generated at the expense of force.

ligament   A collagen-rich connective structure that attaches bone to bone in vertebrate skeletons. Compare *tendon*.

ligase   A key enzyme of DNA technology, used to connect two DNA fragments to each other.

light reactions   A group of chemical reactions that harvest energy from sunlight and use it to produce energy-rich compounds like ATP and NADPH. The light reactions occur at the thylakoid membranes of chloroplasts and produce oxygen gas as a waste product. Compare *dark reactions*.

lignin   A complex molecule largely responsible for the rigidity of the cell walls of woody plants.

lineage   A group of closely related individuals, species, genera, or the like, depicted as a branch on an evolutionary tree.

linkage group   A group consisting of all genes on the same chromosome.

linked genes   See *genetic linkage*.

lipid   A hydrophobic molecule that contains fatty acids. Lipids are key components of cell membranes (see *phospholipid bilayer*).

load arm   The section of a lever that extends from the fulcrum to the point at which work is done. Compare *power arm*.

locus (pl. loci)   The physical location of a gene on a chromosome.

lumen   The space enclosed by the membrane of an organelle.

lung   The gas exchange structure of terrestrial animals, formed by an infolding of epidermal tissue. Compare *gill*.

lymph   The fluid that filters out of the capillaries and flows between the cells in vertebrates.

lymph node   A structure in the lymphatic system of mammals and birds that houses macrophages and lymphocytes that destroy invading cells circulating in the lymph.

lymphatic duct   A tube through which the fluid that collects between the cells of vertebrates drains back into the circulatory system. Lymphatic ducts play an important role in the movement and activity of many of the white blood cells of the vertebrate immune system.

lymphocyte   Any of various white blood cells that bind to specific antigens and then contribute in various ways to the destruction of the cells that bear the antigens.

lysosome   A specialized vesicle with an acidic lumen. Lysosomes contain enzymes that break down macromolecules.

M phase   See *mitosis*.

macroevolution   The rise and fall of major groups of organisms due to the evolutionary radiations that bring new groups to prominence and the mass extinctions that greatly alter the diversity of life on Earth. Macroevolution is the history of large-scale evolu-tionary changes over time. Compare *microevolution*.

macrophage   A white blood cell that phago-cytoses invading cells marked by complement or antibodies.

male   The individual in a sexually reproducing species that produces sperm. Compare *female*.

malignant   Of or referring to a cancerous growth that begins as a single tumor and then spreads to other tissues in the body. Compare *benign*.

mass extinction   A period of time during which great numbers of species become extinct throughout most of Earth.

matrix (pl. matrices)   A nonliving coating of chemicals, released by the cells of multicellular animals, that often holds the cells together.

mechanoreceptor   A sensory cell that detects motion or pressure.

meiosis   The specialized process of cell division in eukaryotes during which diploid cells divide to produce haploid cells. Meiosis has two division cycles and occurs exclusively in cells that produce gametes. Compare *mitosis*.

meiosis I   The first cycle of cell division in meiosis. Meiosis I produces haploid daughter cells, each with half the chromosome number of the diploid parent cell.

meiosis II   The second cycle of cell division in meiosis. Meiosis II is essentially mitosis in haploid cells.

membrane   A phospholipid bilayer with associated proteins that encloses both cells and organelles.

menstrual cycle   A series of hormonally controlled, cyclical changes that take place in the female reproductive system of humans. It includes the shedding of the lining of the uterus (menstruation) about every 28 days.

mesoderm   The germ layer present in most animals that lies between the endo- and ectoderm in the gastrula and that gives rise to muscle tissue, connective tissue, and the kidney. Compare *ectoderm* and *endoderm*.

messenger RNA (mRNA)   The type of RNA that specifies the order of amino acids in a protein.

metabolic heat   Heat generated as a product of respiration.

metabolism   All the chemical reactions that occur in the cell.

metamorphosis (pl. metamorphoses)   A dramatic transformation in many animals from a reproductively immature to a reproductively mature form. Metamorphosis involves great change in the form and function of the animal.

metaphase   The stage of mitosis during which chromosomes become aligned at the equator of the cell.

micelle   A globule formed by fat in the water-based food stream.

microevolution   Changes in allele or genotype frequencies in a population over time. Microevolution is the smallest scale at which evolution occurs. Compare *macroevolution*.

microtubule   A protein fiber composed of tubulin monomers. Microtubules are part of the cell's cytoskeleton.

milk-letdown reflex   The reflexive milk release by a human mother in response to the suckling behavior of an infant.

mismatch error   The insertion of an incorrect base during the DNA replication process that is not detected and corrected.

mitochondrion (pl. mitochondria)   An organelle with a double membrane that is the site of oxidative phosphorylation. Mitochondria break down simple sugars to produce most of the ATP needed by eukaryotic cells.

mitosis (M phase)   The process of cell division in eukaryotes that produces two daughter cells, each with the same chromosome number as the parent cell. Compare *meiosis*.

mitotic spindle   Basketlike arrangement of microtubules that guide the movement of chromosomes during mitosis.

module   A portion of a modular organism like a plant that is repeated during development.

molecule   An arrangement of two or more atoms linked by chemical bonds.

Monera   The kingdom made up of all the bacteria, or prokaryotes.

monomer   A molecule that can be linked with other related molecules to form a larger polymer.

monosaccharide   A simple sugar that can be linked to other sugars, forming a polysaccharide. Glucose is the most common monosaccharide.

morphogen   A chemical signal that influences the developmental fate of a cell.

most recent common ancestor   The ancestral organism from which a group of descendants arose.

motor protein   A protein that uses the energy of ATP to move organelles or other proteins along cytoskeletal filaments.

mouth   The portion of the animal gut where food enters the body.

mRNA   See *messenger RNA*.

multicellular   Made up of more than one cell.

multiregional model   A hypothesis stating that anatomically modern humans evolved from *Homo erectus* populations scattered throughout the world. According to this idea, worldwide gene flow caused different human populations to evolve modern characteristics simultaneously and to remain a single species. Compare *out-of-Africa model*.

muscle cell   A type of animal cell that can contract and plays a central role in the ability of animals to move.

muscle fiber The basic unit of muscle tissue, consisting of a collection of myofibrils.

mutation A change in the sequence of an organism's DNA. New alleles arise only by mutation, so mutations are the original source of all genetic variation.

mutualism An interaction between two species in which both species benefit.

mutualist An organism that lives in association with another organism to the mutual benefit of both.

mycelium (pl. mycelia) The main body of a fungus, composed of hyphae.

mycorrhiza (pl. mycorrhizae) A close association between a fungus and a plant, in which the fungus provides the plant with mineral nutrients while receiving organic nutrients from the plant.

myelin A fatty material produced by some glial cells in vertebrates. Myelin acts as an insulator around nerve cells and greatly speeds the rate at which action potentials move along the axon.

myofibril A basic unit of muscle structure consisting of many sarcomeres attached end-to-end.

myoglobin An oxygen-binding pigment in muscle tissue that helps store blood.

myosin A protein found in muscle tissue. The heads of myosin walk along actin molecules to cause contraction.

natural selection The process in which individuals in a population that possess particular, heritable characteristics survive and reproduce at a higher rate than other individuals in the population. Natural selection is the only evolutionary mechanism that consistently improves the survival and reproduction of the organism in its environment.

neocortex The portion of the cerebrum of the vertebrate brain that interprets and responds to information gathered by the sensory structures.

nerve cell see *neuron*.

net primary productivity (NPP) The amount of energy that organisms fix in photosynthesis, minus the amount lost as heat in cellular respiration. NPP is a measure of the amount of new biomass produced by photosynthetic organisms and other producers during a specified period of time. Compare *secondary productivity*.

neuron A type of animal cell that can rapidly conduct action potentials. Also called a nerve cell.

neurotransmitter Any of various chemical messengers that carry the effects of action potentials across the gaps, or synapses, that separate adjacent nerve cells. Neurotransmitters may either stimulate or inhibit action potentials in the downstream nerve cell.

neutrophil A white blood cell, found in abundance in the human body, that phagocytoses invading bacterial cells.

nitric oxide A gas that functions as a signaling molecule.

nitrogen fixation The process by which nitrogen gas (N_2), a form of nitrogen that is readily available in the atmosphere but that cannot be used by plants, is converted to ammonium (NH_4^+), a form of nitrogen that can be used by plants. Nitrogen fixation is accomplished naturally by bacteria and by lightning, and by humans in industrial processes such as the production of fertilizer.

node of Ranvier A gap in the myelin sheath around vertebrate nerve cells. Action potentials form in these nodes.

noncoding DNA A segment of DNA that does not code for protein or RNA. Introns and spacer DNA are two common types of noncoding DNA.

noncovalent bond Any chemical linkage between two atoms that does not involve the sharing of electrons. Hydrogen bonds and ionic bonds are examples of noncovalent bonds.

nonnative species A species that does not naturally live in an area but was brought there either accidentally or on purpose by humans. Also called foreign species.

nonpolar Of or referring to molecules or parts of molecules that have an equal distribution of electrical charge across all constituent atoms. Nonpolar molecules do not form hydrogen bonds and therefore tend not to dissolve in water. Compare *polar*.

nonrandom mating The situation in which individuals within a subset of a population are more likely to mate with each other than they are to mate with a randomly selected member of the population.

nonspecific response A defensive response of most animals that leads to the destruction of cells not recognized as belonging to the organism. Compare *specific response*.

NPP See *net primary productivity*.

nuclear envelope The double membrane that encloses the nucleus of a eukaryotic cell.

nuclear pore A channel in the nuclear envelope that allows selected molecules to move into and out of the nucleus.

nucleotide The chemical building block of nucleic acids such as DNA and RNA. A nucleotide has a sugar–phosphate backbone and one of four nitrogen-containing molecules called bases (see *nucleotide base*). Nucleotides are linked together to form a single strand of DNA or RNA.

nucleotide base Any of various nitrogen-rich compounds found in nucleotides. The four nucleotide bases found in DNA are adenine, cytosine, guanine, and thymine; in RNA, uracil replaces thymine.

nucleus (pl. nuclei) The single organelle in a eukaryotic cell that contains the genetic blueprint in the form of DNA.

nutrient In an ecosystem context, an essential element required by a producer. Such essential elements include carbon, hydrogen, oxygen, nitrogen, phosphorus, and many others.

nutrient cycle The cyclic movement of a nutrient between organisms and the physical environment. There are two main types of nutrient cycles: atmospheric and sedimentary.

oncogene A mutated gene that promotes excessive cell division, leading to cancer.

open circulatory system An internal transport system found in animal circulatory system in which blood flows through the body cavity that surrounds the organs. Compare *closed circulatory system*.

operator In prokaryotes, a sequence of DNA that functions to control the transcription of a gene or group of genes. An operator is an example of a type of regulatory DNA sequence.

opposable thumb A thumb positioned as in humans, opposite the other fingers, to allow an organism to manipulate an object with its hand.

organ A self-contained collection of tissues, usually of a characteristic size and shape, that is organized for a particular function, such as the heart or stomach.

organ system A group of organs that can work together for a particular function, such as the digestive or reproductive system.

organelle A distinct, membrane-enclosed structure in eukaryotic cells that has a specific function. The nucleus and endoplasmic reticulum are examples of organelles.

organic Of or referring to material of biological origin.

orientation behavior A sequence of behaviors that leads an organism toward the source of a stimulus.

osmosis Passive movement of water from areas of low solute concentration to areas of high solute concentration. Compare *diffusion*.

osmotic pressure The pressure generated when water diffuses down a concentration gradient.

out-of-Africa model A hypothesis stating that anatomically modern humans evolved in Africa within the past 200,000 years, then spread throughout the rest of the world. According to this idea, as they spread from Africa, modern humans completely replaced older forms of *Homo sapiens*, including advanced forms such as the Neanderthals. Compare *multiregional model*.

ovary (1) The organ in female animals that produces eggs. (2) The structure in spore-producing plant individuals that produces the female gametophyte.

oviduct The tube down which eggs produced by female animals pass after leaving the ovary.

oxidation The process by which a chemical compound loses electrons. Compare *reduction*.

oxidative phosphorylation The production of ATP in mitochondria that results from the shuttling of electrons down an electron transport chain.

oxygen-binding pigment Any of various complex molecules used by animals to increase the oxygen capacity of their body fluids.

P generation The parent generation of a genetic cross. Compare *F₁ generation and F₂ generation.*

p53 A tumor suppressor gene that is inactive in many different types of cancer cells.

palisade parenchyma The tissue in plant leaves that is specialized for photosynthesis.

parasite An organism that lives in or on another organism (the host), obtaining nutrients from the host. Parasites harm and may eventually kill their hosts.

passive carrier protein A protein in the plasma membrane of a cell that, without the input of energy, changes shape to transport a molecule across the membrane from the side of high concentration to the side of low concentration. Compare *active carrier protein.*

passive movement Movement from areas of low concentration to areas of high concentration without the expenditure of energy. Compare *active movement.*

pathogen A foreign organism or virus that infects a host and causes disease, harming and in some cases killing the host.

pathway A series of chemical reactions in which the product of one reaction becomes the starting material for the next.

PCR See *polymerase chain reaction.*

pedigree A chart that shows genetic relationships among family members over two or more generations of a family's history.

pelvic aperture The passage in the pelvic bone through which an infant must pass during birth.

penis The structure used by male animals to introduce sperm directly into the female.

pH A scale from 1 to 14 that indicates the concentration of hydrogen ions in water. A pH of 7 is neutral; values below 7 indicate acids, and values above 7 indicate bases.

phagocytosis The process by which one cell engulfs and digests another.

phenotype The observable physical characteristics of an organism. Compare *genotype.*

pheromone A chemical produced by one individual to signal its identity and location to another individual.

phloem tissue The tissue composed of living cells through which a plant transports the products of photosynthesis. Compare *xylem tissue.*

phospholipid A lipid molecule with an attached phosphate group. Phospholipids are the major components of all cell membranes.

phospholipid bilayer A double layer of phospholipid molecules arranged so that the hydrophobic "tails" lie sandwiched between the hydrophilic "heads." This bilayer forms the backbone of the plasma membrane.

photoautotroph Any organism, such as a plant, that uses light as an energy source and obtains its carbon from the gas carbon dioxide in the air. Compare *chemoautotroph, chemoheterotroph,* and *photoheterotroph.*

photoheterotroph Any bacterium that uses light as an energy source and obtains its carbon from organic compounds. Compare *chemoautotroph, chemoheterotroph,* and *photoautotroph.*

photosynthesis The process by which the chloroplasts of plants capture energy from sunlight and use it to synthesize sugars from carbon dioxide and water.

photosystem A large complex of proteins and chlorophyll that captures energy from sunlight. Two distinct photosystems (I and II) are present in the thylakoid membranes of chloroplasts.

pituitary gland A gland associated with the vertebrate brain that releases hormones that control the release of hormones by other glands. Along with the hypothalamus, the pituitary helps regulate interactions between the nervous and endocrine systems.

placenta The structure that in most mammals transfers nutrients and gases from the blood of the mother to the blood of the developing infant.

Plantae The kingdom made up of plants.

plaque A fatty growth that develops on the walls of arteries as a result of excess cholesterol in the diet.

plasma The fluid that flows through animal circulatory systems.

plasma membrane The phospholipid bilayer that surrounds the cell.

plasmid Any of the small circular segments of DNA that are found naturally in bacteria and that are used as vectors in the formation of a DNA library.

polar Of or referring to molecules or parts of molecules that have an uneven distribution of electrical charge. Polar molecules can easily interact with water molecules and are therefore soluble. Compare *nonpolar.*

pollen grain The mobile male gametophyte produced by plants. Compare *embryo sac.*

pollen tube A structure produced by pollen grains to enable the two sperm cells to reach an unfertilized embryo sac.

pollinator mutualism A mutualism in which an animal transfers male reproductive cells from one plant to the female reproductive organs of another plant of the same species. The animal usually receives food as reward for this service.

polymer A large molecule composed of many smaller monomers linked together.

polymerase chain reaction (PCR) A method that uses the DNA polymerase enzyme to make billions of copies of a targeted sequence of DNA in a test tube in just a few hours.

polyploidy The condition in which an organism has more than two entire sets of chromosomes. Polyploidy can cause new species to form rapidly without geographic isolation.

polysaccharide A polymer composed of many linked monosaccharides. Starch and cellulose are examples of polysaccharides.

population A group of interacting individuals of a single species located within a particular area.

population density The number of individuals in a population, divided by the area covered by the population.

population size The total number of individuals in a population.

power arm The section of a lever that extends from the fulcrum to the point at which force is applied. Compare *load arm.*

predator An organism that kills other organisms for food.

prenatal screening Genetic screening of developing fetuses still in the uterus which can be used to diagnose diseases or the likelihood an individual will develop a disease .

pressure drag Drag that results because of pressure differences that develop in front of and behind a moving object. Compare *friction drag.*

primary consumer An organism that eats a producer, such as a cow that eats grass. Compare *secondary consumer.*

primary succession Ecological succession that occurs in newly created habitat, as when an island rises from the sea or a glacier retreats, exposing newly available bare ground. Compare *secondary succession.*

primate An order of mammals whose living members include lemurs, tarsiers, monkeys, humans, and other apes. Primates share characteristics such as flexible shoulder and elbow joints, opposable thumbs or big toes, forward-facing eyes, and brains that are large relative to body size.

procambium The tissue in the developing plant embryo that gives rise to cells of the vascular tissue system. Compare *ground meristem* and *protoderm.*

producer An organism that uses energy from an external source like the sun to produce its own food without having to eat other organisms or their remains. Compare *consumer.*

productivity The mass of plant matter grown in a particular area or by a particular plant or plants.

prokaryote A single-celled organism that does not have a nucleus. All prokaryotes are bacteria. Compare *eukaryote.*

prometaphase The stage of mitosis during which chromosomes become attached to the mitotic spindle.

promoter The region of DNA in a gene to which the enzyme that performs transcription (RNA polymerase) can bind. Transcription begins when RNA polymerase binds to the promoter.

prophase The stage of mitosis during which chromosomes first become visible in the microscope.

protein A linear polymer of amino acids linked together in a specific sequence. Most proteins fold up into complex three-dimensional shapes.

protein phosphorylation The enzymatic process by which phosphate groups are attached to proteins.

Protista The oldest eukaryotic kingdom, consisting of a diverse collection of mostly single-celled but some multicellular organisms.

protoderm The tissue in the developing plant embryo that gives rise to cells of the dermal tissue system. Compare *ground meristem* and *procambium*.

proton A positively charged particle found in atoms. Each atom contains a characteristic number of protons. Compare *electron*.

proton gradient An imbalance in the concentration of protons across a membrane.

proto-oncogene A gene that promotes cell division in response to normal growth signals.

protostome Any of a group of animals, including insects, worms, and snails, that are characterized by their early development, in which the first opening in the embryo becomes the mouth. Compare *deuterostome*.

proximate cause An explanation that addresses how organisms respond to their immediate environment. Compare *ultimate cause*.

pseudopodium (pl. pseudopodia) A dynamic protrusion of the plasma membrane that enables some cells to crawl. The extension of pseudopodia depends on actin filaments inside the cell.

Punnett square A graphical construction in which the possible types of male and female gametes are listed on two sides of a square, providing a visual way to predict the genotypes of the offspring produced in a genetic cross.

radiant heat Heat transferred in the form of light energy. Compare *conductive heat*.

rain shadow A condition in which little rain or snow falls on the side of a mountain that faces away from moist prevailing winds.

rainforest A forest that receives high rainfall. Tropical rainforests harbor huge numbers of species.

random distribution A distribution of organisms in which the organisms tend to be positioned in an unpredictable manner throughout the environment. Compare *clumped distribution* and *regular distribution*.

range See *distribution*.

Ras An oncogene that is found in many different types of cancer cells.

Rb A tumor suppressor that is inactive in retinoblastoma cancer cells.

real group A complete group of descendants of a single common ancestor.

receptor A protein that facilitates the transmission of a signal after binding to a specific signaling molecule. Receptors are found either inside the cell or embedded in the plasma membrane.

recessive Of or referring an allele that does not have a phenotypic effect when paired with a dominant allele. Compare *dominant*.

recipient The bacterium that incorporates donor genes into its genome. Compare *donor*.

reduction The process by which a chemical compound gains electrons. Compare *oxidation*.

reflex arc A nervous connection between a sensory cell and a muscle cell that allows a simple stimulus to be translated rapidly into a motion.

regular distribution A distribution of organisms in which the organisms tend to be positioned in an even or uniform manner throughout the environment. Compare *clumped distribution* and *random distribution*.

regulatory DNA sequence A sequence of DNA that functions to turn the expression of a particular gene or group of genes on or off. Regulatory DNA sequences interact with regulatory proteins in the control of gene expression.

regulatory protein A protein that signals whether or not a particular gene or group of genes should be expressed. Regulatory proteins interact with regulatory DNA sequences in the control of gene expression.

repressor protein A protein that prevents the expression of a particular gene or group of genes.

reproduction The act of producing more individuals.

reproductive isolation The condition in which barriers prevent or strongly limit reproduction between species. Reproductive isolation can occur in many different ways, but it always has the same effect: No or few genes are exchanged between reproductively isolated species.

restriction enzyme A key tool of DNA technology, restriction enzymes cut DNA at highly specific sites. For example, the restriction enzyme *Alu*I cuts DNA wherever the sequence AGCT occurs, but not at any other location.

retina A field of light-sensing cells in the eye that allows organisms to form an image.

retinoblastoma A rare childhood cancer of the retina.

ribonucleic acid See *RNA*.

ribosomal RNA (rRNA) The type of RNA that is an important component of ribosomes.

ribosome A particle composed of proteins and RNA that synthesizes new proteins. Ribosomes can be either attached to the endoplasmic reticulum or free in the cytosol.

rivet hypothesis The hypothesis that likens an ecosystem losing species to an airplane slowly losing its parts and suggests that once enough species are lost, an ecosystem can suddenly collapse.

RNA Ribonucleic acid, a polymer of nucleotides that is necessary for the synthesis of proteins in living organisms.

root hair The absorptive structure formed by epidermal cells produced near the root tips.

root system A collection of highly branched growths that a plant uses to absorb water and nutrients from the soil.

Rooting reflex The reflexive search for a nipple by a newborn when its cheek or lips are touched lightly.

Rough ER A region of the endoplasmic reticulum that has attached ribosomes. Compare *smooth ER*.

Rous sarcoma virus A virus that causes cancer in chickens.

rRNA See *ribosomal RNA*.

rubisco The enzyme that catalyzes the first reaction of carbon fixation in chloroplasts.

S phase The stage of the cell division cycle during which the cell's DNA is duplicated.

sap The fluid that carries nutrients through the vascular system of plants.

sarcomere The collection of actin and myosin fibers that extend between two Z discs.

saturated Of or referring to a fatty acid that has no double bonds between its carbon atoms. Compare *unsaturated*.

schizophrenia A genetically based psychological disorder of humans that is characterized by hallucinations and personality changes.

science A method of inquiry that provides a rational way to discover truths about the natural world.

scientific method A series of steps in which a scientist develops a hypothesis, tests its predictions by performing experiments, and then changes or discards the hypothesis if the predictions of the hypothesis are not supported by the results of the experiments.

second law of thermodynamics The law stating that all closed systems, such as the cell or the universe, tend to become more disordered. The creation and maintenance of order in cells requires the transfer of disorder to the environment.

secondary chemical Any of various chemicals, often used in defense by plants and animals, that are not essential to the basic metabolism of an organism.

secondary consumer An organism that eats a primary consumer, such as a wolf that eats a deer. Compare *primary consumer.*

secondary productivity The rate of new biotmass production by consumers. Compare *net primary productivity.*

secondary succession Ecological succession in which communities recover from disturbance, as when a forest grows back when a field ceases to be used for agriculture. Compare *primary succession.*

sedimentary cycle A type of nutrient cycle in which the nutrient does not enter the atmosphere easily. Phosphorus is an important nutrient that has a sedimentary nutrient cycle. Compare *atmospheric cycle.*

seed A structure produced by a flower in which plant young (an embryo) is encased in a protective covering.

seed leaf Any of the first leaves to develop in a flowering plant. Seed leaves provide nutrients during the initial stages of germination.

semicircular canals Three small tubes associated with the vertebrate ear that allow the animal to detect changes in movement. The semicircular canals are positioned to detect front-to-back, up-and-down, and side-to-side motions.

senescence The aging process that takes place during the lifetime of an organism or during the lifetime of individual organs in modular organisms.

sensory receptor A cell or organ that converts a stimulus into a signal that can be transmitted through the organism.

sex chromosome Either of a pair of chromosomes that determine the gender of an individual. All other chromosomes are called autosomes.

sex-linked Of or referring to genes located on a sex chromosome. Genes located on the X chromosome are called X-linked; genes located on the Y chromosome are called Y-linked.

sexual reproduction Reproduction in which genes from two individuals are combined. Compare *asexual reproduction.*

shared, ancestral feature A feature shared by members of a group of organisms that was not unique to their most recent common ancestor but is shared by many other organisms as well. Compare *shared, derived feature.*

shared, derived feature A feature unique to a common ancestor that is passed down to all descendants, clearly defining them as a group. Compare *shared, ancestral feature.*

signal cascade A series of stepwise protein activations triggered by a signaling molecule. Signal cascades occur inside the cell and greatly amplify the effect of the original signal.

signal integration The process by which several different signaling molecules can affect components of one signal cascade.

signal transduction The triggering of a specific chemical reaction inside the cell in response to an external signal.

signaling molecule Any of various small molecules, produced and released by one cell, that affect the activities of another cell. Signaling molecules enable cells to communicate with each other.

sit-and-wait forager A forager that waits in one place for food to come to it.

slow muscle fiber A type of muscle fiber that is specialized for slow, sustained contraction. Compare *fast muscle fiber.*

small intestine The portion of the animal gut where digestion and most nutrient absorption takes place.

smooth ER A region of the endoplasmic reticulum that does not have attached ribosomes. Compare *rough ER.*

soluble Of or referring to a chemical compound that will dissolve in water.

spacer DNA A region of noncoding DNA that separates two genes. Spacer DNA is very common in eukaryotes but is not common in prokaryotes.

speciation The process in which one species splits to form two or more species that are reproductively isolated from one another.

species A group of interbreeding natural populations that that is reproductively isolated from other such groups.

specific response A response, so far documented only for the human immune system, in which an organism can distinguish between different invading species to which it has had previous exposure. Compare *nonspecific response.*

sperm The small, mobile, haploid gamete that is produced by sexually reproducing male eukaryotes. Compare *egg.*

spiracle Any of the openings to the system of tracheoles that supplies oxygen to insect tissues.

spongy parenchyma The leaf tissue specialized for gas exchange and the evaporation of water that drives the movement of xylem sap.

sporadic mutation A mutation that is not inherited from a parent but that arises first in an individual's genome. Most cancers are caused by sporadic mutations.

spore (1) The reproductive cell of a fungus that is typically encased in a protective coating that shields it from drying or rotting. (2) A mobile structure that germinates to produce a gametophyte in vascular plants such as ferns and mosses.

stabilizing selection A type of natural selection in which individuals with intermediate values of a heritable phenotypic characteristic have an advantage over other individuals in the population, as when medium-sized individuals survive at a higher rate than small or large individuals. Compare *directional selection* and *disruptive selection.*

stamen The structure on spore-producing individuals of flowering plants that produces the pollen grain.

stem cell A cell that can divide to produce one or more specialized cell types.

steroid Any of a class of hydrophobic signaling molecules that can pass through the plasma membrane of the target cell.

stigma The structure on a spore-producing flowering plant individual on which pollen grains land and germinate.

stimulus (pl. stimuli) A detectable change in the environment of an organism or in the organism itself.

stoma (pl. stomata) Any of the openings to the intercellular air spaces in plant leaves that open and close to regulate gas exchange and evaporative water loss.

stomach The portion of the animal gut that is specialized for mixing and digesting food.

stroma The space enclosed by the inner membrane of the chloroplast. The thylakoids are situated inside the stroma.

substitution mutation A mutation in which one base is replaced by another at a single position in the DNA sequence of a gene. Compare *deletion mutation* and *insertion mutation.*

substrate The specific molecule on which an enzyme acts. Only the substrate will bind to the active site of the enzyme.

succession The process in which species in an ecological community are replaced over time. For a given location, the order in which species are replaced over time is fairly predictable.

sugar An organic compound that has the general chemical formula $(CH_2O)_n$. All carbohydrates are sugars.

surface area–to–volume ratio The ratio between the surface area of an organism and the volume of cells that it must supply.

sustainable In ecology, of or referring to an action or process that can continue indefinitely without using up resources or causing serious damage to the environment.

synapse The narrow gap that separates adjacent nerve cells.

synovial fluid A fluid that forms a cushion in vertebrate joints.

systematics The study of evolutionary relationships between organisms and the building of evolutionary trees.

T cell Any of a specialized class of immune system cells. Some T cells promote the recognition of foreign pathogens while others destroy infected cells.

target cell The cell that receives and responds to a signaling molecule.

telophase The stage of mitosis during which chromosomes arrive at the opposite poles of the cell and new nuclear envelopes begin to form around each set of chromosomes.

tendon A collagen-rich connective structure that attaches muscle to bone in vertebrate skeletons. Compare *ligament*.

tensile strength The ability of a structure to resist breaking when under tension.

tension The stretching force that tends to pull molecules apart.

testis (pl. testes) The structure in male animals that produces sperm.

thalamus (pl. thalami) The portion of the vertebrate brain that relays sensory information to the cerebral cortex.

thrust The force that propels an organism.

thylakoid membrane The membrane that encloses the thylakoid space inside chloroplasts. The thylakoid membrane houses both photosystems and their associated electron transport chains.

thylakoid space The space enclosed by the thylakoid membrane inside chloroplasts. This is the innermost compartment of the chloroplast.

tissue A group of specialized cell types that together fulfill a particular function for the body. Animals have four basic tissue types, corresponding to the four basic cell types: connective tissue, epithelial tissue, muscle tissue, and nerve tissue.

tissue system Any of three basic collections of different kinds of plant cells that, in various arrangements, make up all parts of a plant. The three are the dermal, ground, and vascular tissue systems.

tracheole Any of the tubes that carry gases to and from insect cells.

transcription Synthesis of an RNA molecule from a DNA template. Transcription is the first major step in the process by which genes specify proteins; it is used to produce mRNA, tRNA, and rRNA molecules, all of which are essential in the production of proteins. Compare *translation*.

transfer RNA (tRNA) The type of RNA that during protein synthesis transfers the correct amino acid to the ribosome for each amino acid specified by a messenger RNA.

transformation A change in the genotype of a cell or organism as a result of the incorporation of donor genes.

translation The conversion of a sequence of bases in messenger RNA to a sequence of amino acids in a protein. Translation occurs at the ribosomes and is the second major step in the process by which genes specify proteins. Compare *transcription*.

transposon A DNA sequence that can move from one position on a chromosome to another, or from one chromosome to another. Known informally as a "jumping gene."

tRNA See *transfer RNA*.

trophic level A level or step in a food chain. Trophic levels begin with producers and end with predators such as lions that eat other organisms but are not fed upon by other predators.

true-breeding variety A subset of individuals within a species that are homozygous for a particular trait. When individuals of a true-breeding variety mate with other individuals of the same variety, all offspring resemble the parents for the trait that defines the variety.

tubulin The protein monomer that makes up microtubules.

tumor suppressor A gene that inhibits cell division under normal conditions.

ultimate cause An explanation that addresses the historical and evolutionary reasons why organisms respond as they do to their immediate environment. Compare *proximate cause*.

unsaturated Of or referring to a fatty acid that has one or more double bonds between its carbon atoms. Compare *saturated*.

uterus (pl. uteri) The organ in female mammals in which the fertilized egg implants to continue development.

vaccine A preparation of killed or weakened organisms that is used to stimulate the vertebrate immune system as a protection against future attack by those organisms.

vacuole A large water-filled vesicle found in plant cells. Vacuoles help maintain the shape of plant cells, and they can also be used to store food molecules.

vagina The structure in female animals into which the penis of the male deposits sperm.

vascular system A network of specialized cells that transports water and nutrients from the roots throughout a plant's body.

vascular tissue system The tissue system in plants that is devoted to internal transport.

vasodilation The expansion of blood vessels, leading to increased blood flow.

vector A piece of DNA that is used to transfer a gene or other DNA fragment from one organism to another.

vein A blood vessel that carries blood to the heart. Compare *artery*.

ventricle The chamber of the heart that receives blood from the atrium and contracts to propel blood away from the heart.

vertebrate Any of a group of animals that have backbones. Vertebrates include fish, amphibians, mammals, birds, and reptiles.

vesicle A small, membrane-enclosed sac found in the cytosol of eukaryotic cells. Vesicles are used to transport proteins and lipids to various destinations in the cell.

virus An infectious particle consisting of nucleic acids and protein. A virus cannot reproduce on its own, and must instead use the proteins of infected cells to reproduce.

viscous Of or referring to fluids in which adjacent molecules stick to each other, leading to a syrupy consistency.

vitamin An essential nutrient that is needed in small amounts by an organism. Most vitamins speed essential biochemical events.

warning coloration Bright colors or striking patterns on an organism that warn potential predators that the organism is well defended, usually by chemical means.

water transformation Changes humans have made to the waters of Earth that alter their physical or biological characteristics. Together with land transformation, human impacts on the world's waters are thought to be the main cause of the current high rate of extinction of species.

weather Temperature, precipitation, wind speed, humidity, cloud cover, and other physical conditions of the lower atmosphere at a specific place over a short period of time.

white blood cell Any of a variety of cells that are part of the vertebrate immune system and provide protection against invaders.

X-linked See *sex-linked*.

xylem tissue The tissue composed of dead cells through which a plant transports mineral nutrients and water from the soil to the leaves. Compare *phloem tissue*.

Z disc The disc that separates the sarcomeres in a myofibril.

zygote The diploid product formed by the fusion of two haploid gametes.

Index

David Chen
630-778-8256